Algebraic Topology

This geometrically flavored introduction to algebraic topology has the dual goals of serving as a textbook for a standard graduate-level course and as a background reference for many additional topics that do not usually fit into such a course. The broad coverage includes both the homological and homotopical sides of the subject. Care has been taken to present a readable, self-contained exposition, with many examples and exercises, aimed at the student or the researcher from another area of mathematics seeing the subject for the first time.

The four main chapters present the basic core material of algebraic topology: fundamental groups, homology, cohomology, and higher homotopy groups. Each chapter concludes with a generous selection of optional topics, accounting for nearly half the book altogether.

Allen Hatcher is Professor of Mathematics at Cornell University.

Algebraic Topology

ALLEN HATCHER
Cornell University

CAMBRIDGE
UNIVERSITY PRESS

CAMBRIDGE
UNIVERSITY PRESS

University Printing House, Cambridge CB2 8BS, United Kingdom

One Liberty Plaza, 20th Floor, New York, NY 10006, USA

477 Williamstown Road, Port Melbourne, VIC 3207, Australia

314-321, 3rd Floor, Plot 3, Splendor Forum, Jasola District Centre, New Delhi - 110025, India

79 Anson Road, #06-04/06, Singapore 079906

Cambridge University Press is part of the University of Cambridge.

It furthers the University's mission by disseminating knowledge in the pursuit of education, learning and research at the highest international levels of excellence.

www.cambridge.org
Information on this title: www.cambridge.org/9780521795401

First published 2001
21st printing 2019

A catalogue record for this publication is available from the British Library

Library of Congress Cataloging in Publication data
Hatcher, Allen.
Algebraic topology / Allen Hatcher.
  p. cm.
Includes bibliographical references and index.
ISBN 0-521-79160-X – ISBN 0-521-79540-0 (pbk.)
1. Algebraic topology. I. Title.
QA612 H42 2001
514´.2 – dc21   00-065166

ISBN 978-0-521-79160-1 Hardback
ISBN 978-0-521-79540-1 Paperback

Table of Contents

Preface

This book was written to be a readable introduction to algebraic topology with rather broad coverage of the subject. The viewpoint is quite classical in spirit, and stays well within the confines of pure algebraic topology. In a sense, the book could have been written thirty or forty years ago since virtually everything in it is at least that old. However, the passage of the intervening years has helped clarify what are the most important results and techniques. For example, CW complexes have proved over time to be the most natural class of spaces for algebraic topology, so they are emphasized here much more than in the books of an earlier generation. This emphasis also illustrates the book's general slant towards geometric, rather than algebraic, aspects of the subject. The geometry of algebraic topology is so pretty, it would seem a pity to slight it and to miss all the intuition it provides.

At the elementary level, algebraic topology separates naturally into the two broad channels of homology and homotopy. This material is here divided into four chapters, roughly according to increasing sophistication, with homotopy split between Chapters 1 and 4, and homology and its mirror variant cohomology in Chapters 2 and 3. These four chapters do not have to be read in this order, however. One could begin with homology and perhaps continue with cohomology before turning to homotopy. In the other direction, one could postpone homology and cohomology until after parts of Chapter 4. If this latter strategy is pushed to its natural limit, homology and cohomology can be developed just as branches of homotopy theory. Appealing as this approach is from a strictly logical point of view, it places more demands on the reader, and since readability is one of the first priorities of the book, this homotopic interpretation of homology and cohomology is described only after the latter theories have been developed independently of homotopy theory.

Preceding the four main chapters there is a preliminary Chapter 0 introducing some of the basic geometric concepts and constructions that play a central role in both the homological and homotopical sides of the subject. This can either be read before the other chapters or skipped and referred back to later for specific topics as they become needed in the subsequent chapters.

Each of the four main chapters concludes with a selection of additional topics that the reader can sample at will, independent of the basic core of the book contained in the earlier parts of the chapters. Many of these extra topics are in fact rather important in the overall scheme of algebraic topology, though they might not fit into the time constraints of a first course. Altogether, these additional topics amount to nearly half the book, and they are included here both to make the book more comprehensive and to give the reader who takes the time to delve into them a more substantial sample of the true richness and beauty of the subject.

There is also an Appendix dealing mainly with a number of matters of a point-set topological nature that arise in algebraic topology. Since this is a textbook on algebraic topology, details involving point-set topology are often treated lightly or skipped entirely in the body of the text.

Not included in this book is the important but somewhat more sophisticated topic of spectral sequences. It was very tempting to include something about this marvelous tool here, but spectral sequences are such a big topic that it seemed best to start with them afresh in a new volume. This is tentatively titled 'Spectral Sequences in Algebraic Topology' and is referred to herein as [SSAT]. There is also a third book in progress, on vector bundles, characteristic classes, and K–theory, which will be largely independent of [SSAT] and also of much of the present book. This is referred to as [VBKT], its provisional title being 'Vector Bundles and K–Theory.'

In terms of prerequisites, the present book assumes the reader has some familiarity with the content of the standard undergraduate courses in algebra and point-set topology. In particular, the reader should know about quotient spaces, or identification spaces as they are sometimes called, which are quite important for algebraic topology. Good sources for this concept are the textbooks [Armstrong 1983] and [Jänich 1984] listed in the Bibliography.

A book such as this one, whose aim is to present classical material from a rather classical viewpoint, is not the place to indulge in wild innovation. There is, however, one small novelty in the exposition that may be worth commenting upon, even though in the book as a whole it plays a relatively minor role. This is the use of what we call Δ-complexes, which are a mild generalization of the classical notion of a simplicial complex. The idea is to decompose a space into simplices allowing different faces of a simplex to coincide and dropping the requirement that simplices are uniquely determined by their vertices. For example, if one takes the standard picture of the torus as a square with opposite edges identified and divides the square into two triangles by cutting along a diagonal, then the result is a Δ-complex structure on the torus having 2 triangles, 3 edges, and 1 vertex. By contrast, a simplicial complex structure on the torus must have at least 14 triangles, 21 edges, and 7 vertices. So Δ-complexes provide a significant improvement in efficiency, which is nice from a pedagogical viewpoint since it simplifies calculations in examples. A more fundamental reason for considering Δ-complexes is that they seem to be very natural objects from the viewpoint of algebraic topology. They are the natural domain of definition for simplicial homology, and a number of standard constructions produce Δ-complexes rather than simplicial complexes. Historically, Δ-complexes were first introduced by

Eilenberg and Zilber in 1950 under the name of semisimplicial complexes. Soon after this, additional structure in the form of certain 'degeneracy maps' was introduced, leading to a very useful class of objects that came to be called simplicial sets. The semisimplicial complexes of Eilenberg and Zilber then became 'semisimplicial sets', but in this book we have chosen to use the shorter term '∆-complex'.

This book will remain available online in electronic form after it has been printed in the traditional fashion. The web address is

http://www.math.cornell.edu/~hatcher

One can also find here the parts of the other two books in the sequence that are currently available. Although the present book has gone through countless revisions, including the correction of many small errors both typographical and mathematical found by careful readers of earlier versions, it is inevitable that some errors remain, so the web page includes a list of corrections to the printed version. With the electronic version of the book it will be possible not only to incorporate corrections but also to make more substantial revisions and additions. Readers are encouraged to send comments and suggestions as well as corrections to the email address posted on the web page.

Note on the 2015 reprinting. A large number of corrections are included in this reprinting. In addition there are two places in the book where the material was re-arranged to an extent requiring renumbering of theorems, etc. In §3.2 starting on page 210 the renumbering is the following:

old	3.11	3.12	3.13	3.14	3.15	3.16	3.17	3.18	3.19	3.20	3.21
new	3.16	3.19	3.14	3.11	3.13	3.15	3.20	3.16	3.17	3.21	3.18

And in §4.1 the following renumbering occurs in pages 352–355:

old	4.13	4.14	4.15	4.16	4.17
new	4.17	4.13	4.14	4.15	4.16

Standard Notations

$\mathbb{Z}, \mathbb{Q}, \mathbb{R}, \mathbb{C}, \mathbb{H}, \mathbb{O}$: the integers, rationals, reals, complexes, quaternions, and octonions.

$\mathbb{Z}_n$: the integers $\bmod\, n$.

$\mathbb{R}^n$: n-dimensional Euclidean space.

$\mathbb{C}^n$: complex n-space.

In particular, $\mathbb{R}^0 = \{0\} = \mathbb{C}^0$, zero-dimensional vector spaces.

$I = [0,1]$: the unit interval.

S^n : the unit sphere in $\mathbb{R}^{n+1}$, all points of distance 1 from the origin.

D^n : the unit disk or ball in $\mathbb{R}^n$, all points of distance ≤ 1 from the origin.

$\partial D^n = S^{n-1}$: the boundary of the n-disk.

e^n : an n-cell, homeomorphic to the open n-disk $D^n - \partial D^n$.

In particular, D^0 and e^0 consist of a single point since $\mathbb{R}^0 = \{0\}$.

But S^0 consists of two points since it is ∂D^1.

$\mathbb{1}$: the identity function from a set to itself.

$\coprod$: disjoint union of sets or spaces.

$\times, \prod$: product of sets, groups, or spaces.

$\approx$: isomorphism.

$A \subset B$ or $B \supset A$: set-theoretic containment, not necessarily proper.

$A \hookrightarrow B$: the inclusion map $A \to B$ when $A \subset B$.

$A - B$: set-theoretic difference, all points in A that are not in B.

iff : if and only if.

There are also a few notations used in this book that are not completely standard. The union of a set X with a family of sets Y_i, with i ranging over some index set, is usually written simply as $X \cup_i Y_i$ rather than something more elaborate such as $X \cup (\bigcup_i Y_i)$. Intersections and other similar operations are treated in the same way.

Definitions of mathematical terms are generally given within paragraphs of text, rather than displayed separately like theorems, and these definitions are indicated by the use of **boldface type** for the term being defined. Some authors use italics for this purpose, but in this book italics usually denote simply emphasis, as in standard written prose. Each term defined using the boldface convention is listed in the Index, with the page number where the definition occurs.

Chapter 0

Some Underlying Geometric Notions

The aim of this short preliminary chapter is to introduce a few of the most common geometric concepts and constructions in algebraic topology. The exposition is somewhat informal, with no theorems or proofs until the last couple pages, and it should be read in this informal spirit, skipping bits here and there. In fact, this whole chapter could be skipped now, to be referred back to later for basic definitions.

To avoid overusing the word 'continuous' we adopt the convention that maps between spaces are always assumed to be continuous unless otherwise stated.

Homotopy and Homotopy Type

One of the main ideas of algebraic topology is to consider two spaces to be equivalent if they have 'the same shape' in a sense that is much broader than homeomorphism. To take an everyday example, the letters of the alphabet can be written either as unions of finitely many straight and curved line segments, or in thickened forms that are compact regions in the plane bounded by one or more simple closed curves. In each case the thin letter is a subspace of 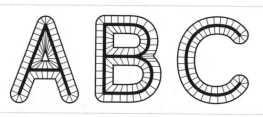 the thick letter, and we can continuously shrink the thick letter to the thin one. A nice way to do this is to decompose a thick letter, call it **X**, into line segments connecting each point on the outer boundary of **X** to a unique point of the thin subletter X, as indicated in the figure. Then we can shrink **X** to X by sliding each point of **X** − X into X along the line segment that contains it. Points that are already in X do not move.

We can think of this shrinking process as taking place during a time interval $0 \leq t \leq 1$, and then it defines a family of functions $f_t : \mathbf{X} \rightarrow \mathbf{X}$ parametrized by $t \in I = [0, 1]$, where $f_t(x)$ is the point to which a given point $x \in \mathbf{X}$ has moved at time t. Naturally we would like $f_t(x)$ to depend continuously on both t and x, and this will be true if we have each $x \in \mathbf{X} - X$ move along its line segment at constant speed so as to reach its image point in X at time $t = 1$, while points $x \in X$ are stationary, as remarked earlier.

Examples of this sort lead to the following general definition. A **deformation retraction** of a space X onto a subspace A is a family of maps $f_t : X \rightarrow X$, $t \in I$, such that $f_0 = \mathbb{1}$ (the identity map), $f_1(X) = A$, and $f_t | A = \mathbb{1}$ for all t. The family f_t should be continuous in the sense that the associated map $X \times I \rightarrow X$, $(x,t) \mapsto f_t(x)$, is continuous.

It is easy to produce many more examples similar to the letter examples, with the deformation retraction f_t obtained by sliding along line segments. The figure on the left below shows such a deformation retraction of a Möbius band onto its core circle.

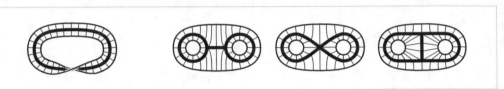

The three figures on the right show deformation retractions in which a disk with two smaller open subdisks removed shrinks to three different subspaces.

In all these examples the structure that gives rise to the deformation retraction can be described by means of the following definition. For a map $f : X \rightarrow Y$, the **mapping cylinder** M_f is the quotient space of the disjoint union $(X \times I) \amalg Y$ obtained by identifying each $(x,1) \in X \times I$ with $f(x) \in Y$. In the letter examples, the space X is the outer boundary of the thick letter, Y is the thin letter, and $f : X \rightarrow Y$ sends

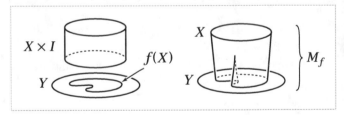

the outer endpoint of each line segment to its inner endpoint. A similar description applies to the other examples. Then it is a general fact that a mapping cylinder M_f deformation retracts to the subspace Y by sliding each point (x,t) along the segment $\{x\} \times I \subset M_f$ to the endpoint $f(x) \in Y$. Continuity of this deformation retraction is evident in the specific examples above, and for a general $f : X \rightarrow Y$ it can be verified using Proposition A.17 in the Appendix concerning the interplay between quotient spaces and product spaces.

Not all deformation retractions arise in this simple way from mapping cylinders. For example, the thick **X** deformation retracts to the thin X, which in turn deformation retracts to the point of intersection of its two crossbars. The net result is a deformation retraction of **X** onto a point, during which certain pairs of points follow paths that merge before reaching their final destination. Later in this section we will describe a considerably more complicated example, the so-called 'house with two rooms.'

A deformation retraction $f_t : X \to X$ is a special case of the general notion of a **homotopy**, which is simply any family of maps $f_t : X \to Y$, $t \in I$, such that the associated map $F : X \times I \to Y$ given by $F(x, t) = f_t(x)$ is continuous. One says that two maps $f_0, f_1 : X \to Y$ are **homotopic** if there exists a homotopy f_t connecting them, and one writes $f_0 \simeq f_1$.

In these terms, a deformation retraction of X onto a subspace A is a homotopy from the identity map of X to a **retraction** of X onto A, a map $r : X \to X$ such that $r(X) = A$ and $r | A = \mathbb{1}$. One could equally well regard a retraction as a map $X \to A$ restricting to the identity on the subspace $A \subset X$. From a more formal viewpoint a retraction is a map $r : X \to X$ with $r^2 = r$, since this equation says exactly that r is the identity on its image. Retractions are the topological analogs of projection operators in other parts of mathematics.

Not all retractions come from deformation retractions. For example, a space X always retracts onto any point $x_0 \in X$ via the constant map sending all of X to x_0, but a space that deformation retracts onto a point must be path-connected since a deformation retraction of X to x_0 gives a path joining each $x \in X$ to x_0. It is less trivial to show that there are path-connected spaces that do not deformation retract onto a point. One would expect this to be the case for the letters 'with holes,' A, B, D, O, P, Q, R. In Chapter 1 we will develop techniques to prove this.

A homotopy $f_t : X \to X$ that gives a deformation retraction of X onto a subspace A has the property that $f_t | A = \mathbb{1}$ for all t. In general, a homotopy $f_t : X \to Y$ whose restriction to a subspace $A \subset X$ is independent of t is called a **homotopy relative to** A, or more concisely, a homotopy rel A. Thus, a deformation retraction of X onto A is a homotopy rel A from the identity map of X to a retraction of X onto A.

If a space X deformation retracts onto a subspace A via $f_t : X \to X$, then if $r : X \to A$ denotes the resulting retraction and $i : A \to X$ the inclusion, we have $ri = \mathbb{1}$ and $ir \simeq \mathbb{1}$, the latter homotopy being given by f_t. Generalizing this situation, a map $f : X \to Y$ is called a **homotopy equivalence** if there is a map $g : Y \to X$ such that $fg \simeq \mathbb{1}$ and $gf \simeq \mathbb{1}$. The spaces X and Y are said to be **homotopy equivalent** or to have the same **homotopy type**. The notation is $X \simeq Y$. It is an easy exercise to check that this is an equivalence relation, in contrast with the nonsymmetric notion of deformation retraction. For example, the three graphs ○─○ ∞ ⊂⊃ are all homotopy equivalent since they are deformation retracts of the same space, as we saw earlier, but none of the three is a deformation retract of any other.

It is true in general that two spaces X and Y are homotopy equivalent if and only if there exists a third space Z containing both X and Y as deformation retracts. For the less trivial implication one can in fact take Z to be the mapping cylinder M_f of any homotopy equivalence $f : X \to Y$. We observed previously that M_f deformation retracts to Y, so what needs to be proved is that M_f also deformation retracts to its other end X if f is a homotopy equivalence. This is shown in Corollary 0.21.

A space having the homotopy type of a point is called **contractible**. This amounts to requiring that the identity map of the space be **nullhomotopic**, that is, homotopic to a constant map. In general, this is slightly weaker than saying the space deformation retracts to a point; see the exercises at the end of the chapter for an example distinguishing these two notions.

Let us describe now an example of a 2-dimensional subspace of $\mathbb{R}^3$, known as the *house with two rooms*, which is contractible but not in any obvious way. To build this

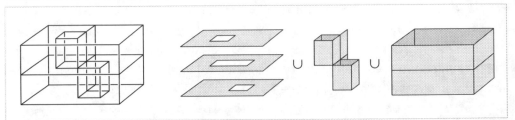

space, start with a box divided into two chambers by a horizontal rectangle, where by a 'rectangle' we mean not just the four edges of a rectangle but also its interior. Access to the two chambers from outside the box is provided by two vertical tunnels. The upper tunnel is made by punching out a square from the top of the box and another square directly below it from the middle horizontal rectangle, then inserting four vertical rectangles, the walls of the tunnel. This tunnel allows entry to the lower chamber from outside the box. The lower tunnel is formed in similar fashion, providing entry to the upper chamber. Finally, two vertical rectangles are inserted to form 'support walls' for the two tunnels. The resulting space X thus consists of three horizontal pieces homeomorphic to annuli plus all the vertical rectangles that form the walls of the two chambers.

To see that X is contractible, consider a closed ε-neighborhood $N(X)$ of X. This clearly deformation retracts onto X if ε is sufficiently small. In fact, $N(X)$ is the mapping cylinder of a map from the boundary surface of $N(X)$ to X. Less obvious is the fact that $N(X)$ is homeomorphic to D^3, the unit ball in $\mathbb{R}^3$. To see this, imagine forming $N(X)$ from a ball of clay by pushing a finger into the ball to create the upper tunnel, then gradually hollowing out the lower chamber, and similarly pushing a finger in to create the lower tunnel and hollowing out the upper chamber. Mathematically, this process gives a family of embeddings $h_t : D^3 \to \mathbb{R}^3$ starting with the usual inclusion $D^3 \hookrightarrow \mathbb{R}^3$ and ending with a homeomorphism onto $N(X)$.

Thus we have $X \simeq N(X) = D^3 \simeq point$, so X is contractible since homotopy equivalence is an equivalence relation. In fact, X deformation retracts to a point. For if f_t is a deformation retraction of the ball $N(X)$ to a point $x_0 \in X$ and if $r : N(X) \to X$ is a retraction, for example the end result of a deformation retraction of $N(X)$ to X, then the restriction of the composition rf_t to X is a deformation retraction of X to x_0. However, it is quite a challenging exercise to see exactly what this deformation retraction looks like.

Cell Complexes

A familiar way of constructing the torus $S^1 \times S^1$ is by identifying opposite sides of a square. More generally, an orientable surface M_g of genus g can be constructed from a polygon with $4g$ sides by identifying pairs of edges, as shown in the figure in the first three cases $g = 1, 2, 3$. The $4g$ edges of the polygon become a union of $2g$ circles in the surface, all intersecting in a single point. The interior of the polygon can be thought of as an open disk, or a **2-cell**, attached to the union of the $2g$ circles. One can also regard the union of the circles as being obtained from their common point of intersection, by attaching $2g$ open arcs, or **1-cells**. Thus the surface can be built up in stages: Start with a point, attach 1-cells to this point, then attach a 2-cell.

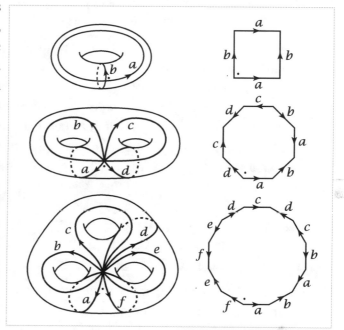

A natural generalization of this is to construct a space by the following procedure:

(1) Start with a discrete set X^0, whose points are regarded as 0-cells.

(2) Inductively, form the **n-skeleton** X^n from X^{n-1} by attaching n-cells e_α^n via maps $\varphi_\alpha : S^{n-1} \to X^{n-1}$. This means that X^n is the quotient space of the disjoint union $X^{n-1} \coprod_\alpha D_\alpha^n$ of X^{n-1} with a collection of n-disks D_α^n under the identifications $x \sim \varphi_\alpha(x)$ for $x \in \partial D_\alpha^n$. Thus as a set, $X^n = X^{n-1} \coprod_\alpha e_\alpha^n$ where each e_α^n is an open n-disk.

(3) One can either stop this inductive process at a finite stage, setting $X = X^n$ for some $n < \infty$, or one can continue indefinitely, setting $X = \bigcup_n X^n$. In the latter case X is given the weak topology: A set $A \subset X$ is open (or closed) iff $A \cap X^n$ is open (or closed) in X^n for each n.

A space X constructed in this way is called a **cell complex** or **CW complex**. The explanation of the letters 'CW' is given in the Appendix, where a number of basic topological properties of cell complexes are proved. The reader who wonders about various point-set topological questions lurking in the background of the following discussion should consult the Appendix for details.

If $X = X^n$ for some n, then X is said to be finite-dimensional, and the smallest such n is the **dimension** of X, the maximum dimension of cells of X.

Example 0.1. A 1-dimensional cell complex $X = X^1$ is what is called a **graph** in algebraic topology. It consists of vertices (the 0-cells) to which edges (the 1-cells) are attached. The two ends of an edge can be attached to the same vertex.

Example 0.2. The house with two rooms, pictured earlier, has a visually obvious 2-dimensional cell complex structure. The 0-cells are the vertices where three or more of the depicted edges meet, and the 1-cells are the interiors of the edges connecting these vertices. This gives the 1-skeleton X^1, and the 2-cells are the components of the remainder of the space, $X - X^1$. If one counts up, one finds there are 29 0-cells, 51 1-cells, and 23 2-cells, with the alternating sum $29 - 51 + 23$ equal to 1. This is the **Euler characteristic**, which for a cell complex with finitely many cells is defined to be the number of even-dimensional cells minus the number of odd-dimensional cells. As we shall show in Theorem 2.44, the Euler characteristic of a cell complex depends only on its homotopy type, so the fact that the house with two rooms has the homotopy type of a point implies that its Euler characteristic must be 1, no matter how it is represented as a cell complex.

Example 0.3. The sphere S^n has the structure of a cell complex with just two cells, e^0 and e^n, the n-cell being attached by the constant map $S^{n-1} \rightarrow e^0$. This is equivalent to regarding S^n as the quotient space $D^n/\partial D^n$.

Example 0.4. Real projective n-space $\mathbb{R}\mathrm{P}^n$ is defined to be the space of all lines through the origin in $\mathbb{R}^{n+1}$. Each such line is determined by a nonzero vector in $\mathbb{R}^{n+1}$, unique up to scalar multiplication, and $\mathbb{R}\mathrm{P}^n$ is topologized as the quotient space of $\mathbb{R}^{n+1} - \{0\}$ under the equivalence relation $v \sim \lambda v$ for scalars $\lambda \neq 0$. We can restrict to vectors of length 1, so $\mathbb{R}\mathrm{P}^n$ is also the quotient space $S^n/(v \sim -v)$, the sphere with antipodal points identified. This is equivalent to saying that $\mathbb{R}\mathrm{P}^n$ is the quotient space of a hemisphere D^n with antipodal points of ∂D^n identified. Since ∂D^n with antipodal points identified is just $\mathbb{R}\mathrm{P}^{n-1}$, we see that $\mathbb{R}\mathrm{P}^n$ is obtained from $\mathbb{R}\mathrm{P}^{n-1}$ by attaching an n-cell, with the quotient projection $S^{n-1} \rightarrow \mathbb{R}\mathrm{P}^{n-1}$ as the attaching map. It follows by induction on n that $\mathbb{R}\mathrm{P}^n$ has a cell complex structure $e^0 \cup e^1 \cup \cdots \cup e^n$ with one cell e^i in each dimension $i \leq n$.

Example 0.5. Since $\mathbb{R}\mathrm{P}^n$ is obtained from $\mathbb{R}\mathrm{P}^{n-1}$ by attaching an n-cell, the infinite union $\mathbb{R}\mathrm{P}^\infty = \bigcup_n \mathbb{R}\mathrm{P}^n$ becomes a cell complex with one cell in each dimension. We can view $\mathbb{R}\mathrm{P}^\infty$ as the space of lines through the origin in $\mathbb{R}^\infty = \bigcup_n \mathbb{R}^n$.

Example 0.6. Complex projective n-space $\mathbb{C}\mathrm{P}^n$ is the space of complex lines through the origin in $\mathbb{C}^{n+1}$, that is, 1-dimensional vector subspaces of $\mathbb{C}^{n+1}$. As in the case of $\mathbb{R}\mathrm{P}^n$, each line is determined by a nonzero vector in $\mathbb{C}^{n+1}$, unique up to scalar multiplication, and $\mathbb{C}\mathrm{P}^n$ is topologized as the quotient space of $\mathbb{C}^{n+1} - \{0\}$ under the equivalence relation $v \sim \lambda v$ for $\lambda \neq 0$. Equivalently, this is the quotient of the unit sphere $S^{2n+1} \subset \mathbb{C}^{n+1}$ with $v \sim \lambda v$ for $|\lambda| = 1$. It is also possible to obtain $\mathbb{C}\mathrm{P}^n$ as a quotient space of the disk D^{2n} under the identifications $v \sim \lambda v$ for $v \in \partial D^{2n}$, in the following way. The vectors in $S^{2n+1} \subset \mathbb{C}^{n+1}$ with last coordinate real and nonnegative are precisely the vectors of the form $(w, \sqrt{1 - |w|^2}) \in \mathbb{C}^n \times \mathbb{C}$ with $|w| \leq 1$. Such vectors form the graph of the function $w \mapsto \sqrt{1 - |w|^2}$. This is a disk D_+^{2n} bounded by the sphere $S^{2n-1} \subset S^{2n+1}$ consisting of vectors $(w, 0) \in \mathbb{C}^n \times \mathbb{C}$ with $|w| = 1$. Each vector in S^{2n+1} is equivalent under the identifications $v \sim \lambda v$ to a vector in D_+^{2n}, and the latter vector is unique if its last coordinate is nonzero. If the last coordinate is zero, we have just the identifications $v \sim \lambda v$ for $v \in S^{2n-1}$.

From this description of $\mathbb{C}\mathrm{P}^n$ as the quotient of D_+^{2n} under the identifications $v \sim \lambda v$ for $v \in S^{2n-1}$ it follows that $\mathbb{C}\mathrm{P}^n$ is obtained from $\mathbb{C}\mathrm{P}^{n-1}$ by attaching a cell e^{2n} via the quotient map $S^{2n-1} \to \mathbb{C}\mathrm{P}^{n-1}$. So by induction on n we obtain a cell structure $\mathbb{C}\mathrm{P}^n = e^0 \cup e^2 \cup \cdots \cup e^{2n}$ with cells only in even dimensions. Similarly, $\mathbb{C}\mathrm{P}^\infty$ has a cell structure with one cell in each even dimension.

After these examples we return now to general theory. Each cell e_α^n in a cell complex X has a **characteristic map** $\Phi_\alpha : D_\alpha^n \to X$ which extends the attaching map φ_α and is a homeomorphism from the interior of D_α^n onto e_α^n. Namely, we can take Φ_α to be the composition $D_\alpha^n \hookrightarrow X^{n-1} \coprod_\alpha D_\alpha^n \to X^n \hookrightarrow X$ where the middle map is the quotient map defining X^n. For example, in the canonical cell structure on S^n described in Example 0.3, a characteristic map for the n-cell is the quotient map $D^n \to S^n$ collapsing ∂D^n to a point. For $\mathbb{R}\mathrm{P}^n$ a characteristic map for the cell e^i is the quotient map $D^i \to \mathbb{R}\mathrm{P}^i \subset \mathbb{R}\mathrm{P}^n$ identifying antipodal points of ∂D^i, and similarly for $\mathbb{C}\mathrm{P}^n$.

A **subcomplex** of a cell complex X is a closed subspace $A \subset X$ that is a union of cells of X. Since A is closed, the characteristic map of each cell in A has image contained in A, and in particular the image of the attaching map of each cell in A is contained in A, so A is a cell complex in its own right. A pair (X, A) consisting of a cell complex X and a subcomplex A will be called a **CW pair**.

For example, each skeleton X^n of a cell complex X is a subcomplex. Particular cases of this are the subcomplexes $\mathbb{R}\mathrm{P}^k \subset \mathbb{R}\mathrm{P}^n$ and $\mathbb{C}\mathrm{P}^k \subset \mathbb{C}\mathrm{P}^n$ for $k \leq n$. These are in fact the only subcomplexes of $\mathbb{R}\mathrm{P}^n$ and $\mathbb{C}\mathrm{P}^n$.

There are natural inclusions $S^0 \subset S^1 \subset \cdots \subset S^n$, but these subspheres are not subcomplexes of S^n in its usual cell structure with just two cells. However, we can give S^n a different cell structure in which each of the subspheres S^k is a subcomplex, by regarding each S^k as being obtained inductively from the equatorial S^{k-1} by attaching two k-cells, the components of $S^k - S^{k-1}$. The infinite-dimensional sphere $S^\infty = \bigcup_n S^n$ then becomes a cell complex as well. Note that the two-to-one quotient map $S^\infty \to \mathbb{R}\mathrm{P}^\infty$ that identifies antipodal points of S^∞ identifies the two n-cells of S^∞ to the single n-cell of $\mathbb{R}\mathrm{P}^\infty$.

In the examples of cell complexes given so far, the closure of each cell is a sub-complex, and more generally the closure of any collection of cells is a subcomplex. Most naturally arising cell structures have this property, but it need not hold in general. For example, if we start with S^1 with its minimal cell structure and attach to this a 2-cell by a map $S^1 \to S^1$ whose image is a nontrivial subarc of S^1, then the closure of the 2-cell is not a subcomplex since it contains only a part of the 1-cell.

Operations on Spaces

Cell complexes have a very nice mixture of rigidity and flexibility, with enough rigidity to allow many arguments to proceed in a combinatorial cell-by-cell fashion and enough flexibility to allow many natural constructions to be performed on them. Here are some of those constructions.

Products. If X and Y are cell complexes, then $X \times Y$ has the structure of a cell complex with cells the products $e_\alpha^m \times e_\beta^n$ where e_α^m ranges over the cells of X and e_β^n ranges over the cells of Y. For example, the cell structure on the torus $S^1 \times S^1$ described at the beginning of this section is obtained in this way from the standard cell structure on S^1. For completely general CW complexes X and Y there is one small complication: The topology on $X \times Y$ as a cell complex is sometimes finer than the product topology, with more open sets than the product topology has, though the two topologies coincide if either X or Y has only finitely many cells, or if both X and Y have countably many cells. This is explained in the Appendix. In practice this subtle issue of point-set topology rarely causes problems, however.

Quotients. If (X, A) is a CW pair consisting of a cell complex X and a subcomplex A, then the quotient space X/A inherits a natural cell complex structure from X. The cells of X/A are the cells of $X - A$ plus one new 0-cell, the image of A in X/A. For a cell e_α^n of $X - A$ attached by $\varphi_\alpha : S^{n-1} \to X^{n-1}$, the attaching map for the corresponding cell in X/A is the composition $S^{n-1} \to X^{n-1} \to X^{n-1}/A^{n-1}$.

For example, if we give S^{n-1} any cell structure and build D^n from S^{n-1} by attaching an n-cell, then the quotient D^n/S^{n-1} is S^n with its usual cell structure. As another example, take X to be a closed orientable surface with the cell structure described at the beginning of this section, with a single 2-cell, and let A be the complement of this 2-cell, the 1-skeleton of X. Then X/A has a cell structure consisting of a 0-cell with a 2-cell attached, and there is only one way to attach a cell to a 0-cell, by the constant map, so X/A is S^2.

Suspension. For a space X, the **suspension** SX is the quotient of $X \times I$ obtained by collapsing $X \times \{0\}$ to one point and $X \times \{1\}$ to another point. The motivating example is $X = S^n$, when $SX = S^{n+1}$ with the two 'suspension points' at the north and south poles of S^{n+1}, the points $(0, \cdots, 0, \pm 1)$. One can regard SX as a double cone on X, the union of two copies of the **cone** $CX = (X \times I)/(X \times \{0\})$. If X is a CW complex, so are SX and CX as quotients of $X \times I$ with its product cell structure, I being given the standard cell structure of two 0-cells joined by a 1-cell.

Suspension becomes increasingly important the farther one goes into algebraic topology, though why this should be so is certainly not evident in advance. One especially useful property of suspension is that not only spaces but also maps can be suspended. Namely, a map $f : X \to Y$ suspends to $Sf : SX \to SY$, the quotient map of $f \times \mathbb{1} : X \times I \to Y \times I$.

Join. The cone CX is the union of all line segments joining points of X to an external vertex, and similarly the suspension SX is the union of all line segments joining points of X to two external vertices. More generally, given X and a second space Y, one can define the space of all line segments joining points in X to points in Y. This is the **join** $X * Y$, the quotient space of $X \times Y \times I$ under the identifications $(x, y_1, 0) \sim (x, y_2, 0)$ and $(x_1, y, 1) \sim (x_2, y, 1)$. Thus we are collapsing the subspace $X \times Y \times \{0\}$ to X and $X \times Y \times \{1\}$ to Y. For example, if X and Y are both closed intervals, then we are collapsing two opposite faces of a cube onto line segments so that the cube becomes a tetrahedron. In the general case, $X * Y$ contains copies of X and Y at its two ends,

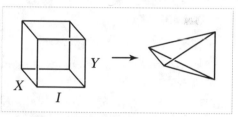

and every other point (x, y, t) in $X * Y$ is on a unique line segment joining the point $x \in X \subset X * Y$ to the point $y \in Y \subset X * Y$, the segment obtained by fixing x and y and letting the coordinate t in (x, y, t) vary.

A nice way to write points of $X * Y$ is as formal linear combinations $t_1 x + t_2 y$ with $0 \le t_i \le 1$ and $t_1 + t_2 = 1$, subject to the rules $0x + 1y = y$ and $1x + 0y = x$ that correspond exactly to the identifications defining $X * Y$. In much the same way, an iterated join $X_1 * \cdots * X_n$ can be constructed as the space of formal linear combinations $t_1 x_1 + \cdots + t_n x_n$ with $0 \le t_i \le 1$ and $t_1 + \cdots + t_n = 1$, with the convention that terms $0x_i$ can be omitted. A very special case that plays a central role in algebraic topology is when each X_i is just a point. For example, the join of two points is a line segment, the join of three points is a triangle, and the join of four points is a tetrahedron. In general, the join of n points is a convex polyhedron of dimension $n - 1$ called a **simplex**. Concretely, if the n points are the n standard basis vectors for $\mathbb{R}^n$, then their join is the $(n - 1)$-dimensional simplex

$$\Delta^{n-1} = \{ (t_1, \cdots, t_n) \in \mathbb{R}^n \mid t_1 + \cdots + t_n = 1 \text{ and } t_i \ge 0 \}$$

Another interesting example is when each X_i is S^0, two points. If we take the two points of X_i to be the two unit vectors along the ith coordinate axis in $\mathbb{R}^n$, then the join $X_1 * \cdots * X_n$ is the union of 2^n copies of the simplex Δ^{n-1}, and radial projection from the origin gives a homeomorphism between $X_1 * \cdots * X_n$ and S^{n-1}.

If X and Y are CW complexes, then there is a natural CW structure on $X * Y$ having the subspaces X and Y as subcomplexes, with the remaining cells being the product cells of $X \times Y \times (0, 1)$. As usual with products, the CW topology on $X * Y$ may be finer than the quotient of the product topology on $X \times Y \times I$.

Wedge Sum. This is a rather trivial but still quite useful operation. Given spaces X and Y with chosen points $x_0 \in X$ and $y_0 \in Y$, then the **wedge sum** $X \vee Y$ is the quotient of the disjoint union $X \amalg Y$ obtained by identifying x_0 and y_0 to a single point. For example, $S^1 \vee S^1$ is homeomorphic to the figure '8,' two circles touching at a point. More generally one could form the wedge sum $\bigvee_\alpha X_\alpha$ of an arbitrary collection of spaces X_α by starting with the disjoint union $\coprod_\alpha X_\alpha$ and identifying points $x_\alpha \in X_\alpha$ to a single point. In case the spaces X_α are cell complexes and the points x_α are 0-cells, then $\bigvee_\alpha X_\alpha$ is a cell complex since it is obtained from the cell complex $\coprod_\alpha X_\alpha$ by collapsing a subcomplex to a point.

For any cell complex X, the quotient X^n/X^{n-1} is a wedge sum of n-spheres $\bigvee_\alpha S_\alpha^n$, with one sphere for each n-cell of X.

Smash Product. Like suspension, this is another construction whose importance becomes evident only later. Inside a product space $X \times Y$ there are copies of X and Y, namely $X \times \{y_0\}$ and $\{x_0\} \times Y$ for points $x_0 \in X$ and $y_0 \in Y$. These two copies of X and Y in $X \times Y$ intersect only at the point (x_0, y_0), so their union can be identified with the wedge sum $X \vee Y$. The **smash product** $X \wedge Y$ is then defined to be the quotient $X \times Y / X \vee Y$. One can think of $X \wedge Y$ as a reduced version of $X \times Y$ obtained by collapsing away the parts that are not genuinely a product, the separate factors X and Y.

The smash product $X \wedge Y$ is a cell complex if X and Y are cell complexes with x_0 and y_0 0-cells, assuming that we give $X \times Y$ the cell-complex topology rather than the product topology in cases when these two topologies differ. For example, $S^m \wedge S^n$ has a cell structure with just two cells, of dimensions 0 and $m+n$, hence $S^m \wedge S^n = S^{m+n}$. In particular, when $m = n = 1$ we see that collapsing longitude and meridian circles of a torus to a point produces a 2-sphere.

Two Criteria for Homotopy Equivalence

Earlier in this chapter the main tool we used for constructing homotopy equivalences was the fact that a mapping cylinder deformation retracts onto its 'target' end. By repeated application of this fact one can often produce homotopy equivalences between rather different-looking spaces. However, this process can be a bit cumbersome in practice, so it is useful to have other techniques available as well. We will describe two commonly used methods here. The first involves collapsing certain subspaces to points, and the second involves varying the way in which the parts of a space are put together.

Collapsing Subspaces

The operation of collapsing a subspace to a point usually has a drastic effect on homotopy type, but one might hope that if the subspace being collapsed already has the homotopy type of a point, then collapsing it to a point might not change the homotopy type of the whole space. Here is a positive result in this direction:

> If (X, A) is a CW pair consisting of a CW complex X and a contractible subcomplex A, then the quotient map $X \to X/A$ is a homotopy equivalence.

A proof will be given later in Proposition 0.17, but for now let us look at some examples showing how this result can be applied.

Example 0.7: Graphs. The three graphs ⚬–⚬ ⚬⚬ ⚬⚬ are homotopy equivalent since each is a deformation retract of a disk with two holes, but we can also deduce this from the collapsing criterion above since collapsing the middle edge of the first and third graphs produces the second graph.

More generally, suppose X is any graph with finitely many vertices and edges. If the two endpoints of any edge of X are distinct, we can collapse this edge to a point, producing a homotopy equivalent graph with one fewer edge. This simplification can be repeated until all edges of X are loops, and then each component of X is either an isolated vertex or a wedge sum of circles.

This raises the question of whether two such graphs, having only one vertex in each component, can be homotopy equivalent if they are not in fact just isomorphic graphs. Exercise 12 at the end of the chapter reduces the question to the case of connected graphs. Then the task is to prove that a wedge sum $\bigvee_m S^1$ of m circles is not homotopy equivalent to $\bigvee_n S^1$ if $m \neq n$. This sort of thing is hard to do directly. What one would like is some sort of algebraic object associated to spaces, depending only on their homotopy type, and taking different values for $\bigvee_m S^1$ and $\bigvee_n S^1$ if $m \neq n$. In fact the Euler characteristic does this since $\bigvee_m S^1$ has Euler characteristic $1 - m$. But it is a rather nontrivial theorem that the Euler characteristic of a space depends only on its homotopy type. A different algebraic invariant that works equally well for graphs, and whose rigorous development requires less effort than the Euler characteristic, is the fundamental group of a space, the subject of Chapter 1.

Example 0.8. Consider the space X obtained from S^2 by attaching the two ends of an arc A to two distinct points on the sphere, say the north and south poles. Let B be an arc in S^2 joining the two points where A attaches. Then X can be given a CW complex structure with the two endpoints of A and B as 0-cells, the interiors of A and B as 1-cells, and the rest of S^2 as a 2-cell. Since A and B are contractible,

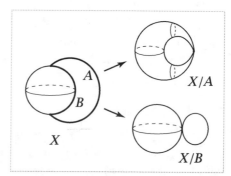

X/A and X/B are homotopy equivalent to X. The space X/A is the quotient S^2/S^0, the sphere with two points identified, and X/B is $S^1 \vee S^2$. Hence S^2/S^0 and $S^1 \vee S^2$ are homotopy equivalent, a fact which may not be entirely obvious at first glance.

Example 0.9. Let X be the union of a torus with n meridional disks. To obtain a CW structure on X, choose a longitudinal circle in the torus, intersecting each of the meridional disks in one point. These intersection points are then the 0-cells, the 1-cells are the rest of the longitudinal circle and the boundary circles of the meridional disks, and the 2-cells are the remaining regions of the torus and the interiors of the meridional disks. Collapsing each meridional disk to a point yields a homotopy

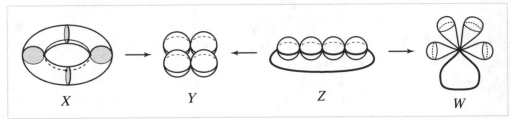

equivalent space Y consisting of n 2-spheres, each tangent to its two neighbors, a 'necklace with n beads.' The third space Z in the figure, a strand of n beads with a string joining its two ends, collapses to Y by collapsing the string to a point, so this collapse is a homotopy equivalence. Finally, by collapsing the arc in Z formed by the front halves of the equators of the n beads, we obtain the fourth space W, a wedge sum of S^1 with n 2-spheres. (One can see why a wedge sum is sometimes called a 'bouquet' in the older literature.)

Example 0.10: Reduced Suspension. Let X be a CW complex and $x_0 \in X$ a 0-cell. Inside the suspension SX we have the line segment $\{x_0\} \times I$, and collapsing this to a point yields a space ΣX homotopy equivalent to SX, called the **reduced suspension** of X. For example, if we take X to be $S^1 \vee S^1$ with x_0 the intersection point of the two circles, then the ordinary suspension SX is the union of two spheres intersecting along the arc $\{x_0\} \times I$, so the reduced suspension ΣX is $S^2 \vee S^2$, a slightly simpler space. More generally we have $\Sigma(X \vee Y) = \Sigma X \vee \Sigma Y$ for arbitrary CW complexes X and Y. Another way in which the reduced suspension ΣX is slightly simpler than SX is in its CW structure. In SX there are two 0-cells (the two suspension points) and an $(n+1)$-cell $e^n \times (0,1)$ for each n-cell e^n of X, whereas in ΣX there is a single 0-cell and an $(n+1)$-cell for each n-cell of X other than the 0-cell x_0.

The reduced suspension ΣX is actually the same as the smash product $X \wedge S^1$ since both spaces are the quotient of $X \times I$ with $X \times \partial I \cup \{x_0\} \times I$ collapsed to a point.

Attaching Spaces

Another common way to change a space without changing its homotopy type involves the idea of continuously varying how its parts are attached together. A general definition of 'attaching one space to another' that includes the case of attaching cells is the following. We start with a space X_0 and another space X_1 that we wish to attach to X_0 by identifying the points in a subspace $A \subset X_1$ with points of X_0. The data needed to do this is a map $f : A \rightarrow X_0$, for then we can form a quotient space of $X_0 \amalg X_1$ by identifying each point $a \in A$ with its image $f(a) \in X_0$. Let us denote this quotient space by $X_0 \sqcup_f X_1$, the space X_0 with X_1 **attached along** A **via** f. When $(X_1, A) = (D^n, S^{n-1})$ we have the case of attaching an n-cell to X_0 via a map $f : S^{n-1} \rightarrow X_0$.

Mapping cylinders are examples of this construction, since the mapping cylinder M_f of a map $f : X \rightarrow Y$ is the space obtained from Y by attaching $X \times I$ along $X \times \{1\}$ via f. Closely related to the mapping cylinder M_f is the **mapping cone** $C_f = Y \sqcup_f CX$ where CX is the cone $(X \times I)/(X \times \{0\})$ and we attach this to Y along $X \times \{1\}$ via the identifications $(x, 1) \sim f(x)$. For example, when X is a sphere S^{n-1} the mapping cone C_f is the space obtained from Y by attaching an n-cell via $f : S^{n-1} \rightarrow Y$. A mapping cone C_f can also be viewed as the quotient M_f/X of the mapping cylinder M_f with the subspace $X = X \times \{0\}$ collapsed to a point.

If one varies an attaching map f by a homotopy f_t, one gets a family of spaces whose shape is undergoing a continuous change, it would seem, and one might expect these spaces all to have the same homotopy type. This is often the case:

> If (X_1, A) is a CW pair and the two attaching maps $f, g : A \rightarrow X_0$ are homotopic, then $X_0 \sqcup_f X_1 \simeq X_0 \sqcup_g X_1$.

Again let us defer the proof and look at some examples.

Example 0.11. Let us rederive the result in Example 0.8 that a sphere with two points identified is homotopy equivalent to $S^1 \vee S^2$. The sphere with two points identified can be obtained by attaching S^2 to S^1 by a map that wraps a closed arc A in S^2 around S^1, as shown in the figure. Since A is contractible, this attaching map is homotopic to a constant map, and attaching S^2 to S^1 via a constant map of A yields $S^1 \vee S^2$. The result then follows since (S^2, A) is a CW pair, S^2 being obtained from A by attaching a 2-cell.

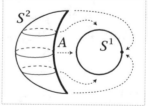

Example 0.12. In similar fashion we can see that the necklace in Example 0.9 is homotopy equivalent to the wedge sum of a circle with n 2-spheres. The necklace can be obtained from a circle by attaching n 2-spheres along arcs, so the necklace is homotopy equivalent to the space obtained by attaching n 2-spheres to a circle at points. Then we can slide these attaching points around the circle until they all coincide, producing the wedge sum.

Example 0.13. Here is an application of the earlier fact that collapsing a contractible subcomplex is a homotopy equivalence: If (X, A) is a CW pair, consisting of a cell complex X and a subcomplex A, then $X/A \simeq X \cup CA$, the mapping cone of the inclusion $A \hookrightarrow X$. For we have $X/A = (X \cup CA)/CA \simeq X \cup CA$ since CA is a contractible subcomplex of $X \cup CA$.

Example 0.14. If (X, A) is a CW pair and A is contractible in X, that is, the inclusion $A \hookrightarrow X$ is homotopic to a constant map, then $X/A \simeq X \vee SA$. Namely, by the previous example we have $X/A \simeq X \cup CA$, and then since A is contractible in X, the mapping cone $X \cup CA$ of the inclusion $A \hookrightarrow X$ is homotopy equivalent to the mapping cone of a constant map, which is $X \vee SA$. For example, $S^n/S^i \simeq S^n \vee S^{i+1}$ for $i < n$, since S^i is contractible in S^n if $i < n$. In particular this gives $S^2/S^0 \simeq S^2 \vee S^1$, which is Example 0.8 again.

The Homotopy Extension Property

In this final section of the chapter we will actually prove a few things, including the two criteria for homotopy equivalence described above. The proofs depend upon a technical property that arises in many other contexts as well. Consider the following problem. Suppose one is given a map $f_0 : X \to Y$, and on a subspace $A \subset X$ one is also given a homotopy $f_t : A \to Y$ of $f_0 | A$ that one would like to extend to a homotopy $f_t : X \to Y$ of the given f_0. If the pair (X, A) is such that this extension problem can always be solved, one says that (X, A) has the **homotopy extension property**. Thus (X, A) has the homotopy extension property if every pair of maps $X \times \{0\} \to Y$ and $A \times I \to Y$ that agree on $A \times \{0\}$ can be extended to a map $X \times I \to Y$.

‖ *A pair (X, A) has the homotopy extension property if and only if $X \times \{0\} \cup A \times I$ is a*
‖ *retract of $X \times I$.*

For one implication, the homotopy extension property for (X, A) implies that the identity map $X \times \{0\} \cup A \times I \to X \times \{0\} \cup A \times I$ extends to a map $X \times I \to X \times \{0\} \cup A \times I$, so $X \times \{0\} \cup A \times I$ is a retract of $X \times I$. The converse is equally easy when A is closed in X. Then any two maps $X \times \{0\} \to Y$ and $A \times I \to Y$ that agree on $A \times \{0\}$ combine to give a map $X \times \{0\} \cup A \times I \to Y$ which is continuous since it is continuous on the closed sets $X \times \{0\}$ and $A \times I$. By composing this map $X \times \{0\} \cup A \times I \to Y$ with a retraction $X \times I \to X \times \{0\} \cup A \times I$ we get an extension $X \times I \to Y$, so (X, A) has the homotopy extension property. The hypothesis that A is closed can be avoided by a more complicated argument given in the Appendix of the online version of the book. If $X \times \{0\} \cup A \times I$ is a retract of $X \times I$ and X is Hausdorff, then A must in fact be closed in X. For if $r : X \times I \to X \times I$ is a retraction onto $X \times \{0\} \cup A \times I$, then the image of r is the set of points $z \in X \times I$ with $r(z) = z$, a closed set if X is Hausdorff, so $X \times \{0\} \cup A \times I$ is closed in $X \times I$ and hence A is closed in X.

A simple example of a pair (X, A) with A closed for which the homotopy extension property fails is the pair (I, A) where $A = \{0, 1, 1/2, 1/3, 1/4, \cdots\}$. It is not hard to show that there is no continuous retraction $I \times I \to I \times \{0\} \cup A \times I$. The breakdown of homotopy extension here can be attributed to the bad structure of (X, A) near 0. With nicer local structure the homotopy extension property does hold, as the next example shows.

Example 0.15. A pair (X, A) has the homotopy extension property if A has a mapping cylinder neighborhood in X, by which we mean a closed neighborhood N containing a subspace B, thought of as the boundary of N, with $N - B$ an open neighborhood of A, such that there exists a map $f : B \to A$ and a homeomorphism $h : M_f \to N$ with $h \,|\, A \cup B = \mathbb{1}$. Mapping cylinder neighborhoods like this occur fairly often. For example, the thick letters discussed at the beginning of the chapter provide such

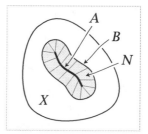

neighborhoods of the thin letters, regarded as subspaces of the plane. To verify the homotopy extension property, notice first that $I \times I$ retracts onto $I \times \{0\} \cup \partial I \times I$, hence $B \times I \times I$ retracts onto $B \times I \times \{0\} \cup B \times \partial I \times I$, and this retraction induces a retraction of $M_f \times I$ onto $M_f \times \{0\} \cup (A \cup B) \times I$. Thus $(M_f, A \cup B)$ has the homotopy extension property. Hence so does the homeomorphic pair $(N, A \cup B)$. Now given a map $X \to Y$ and a homotopy of its restriction to A, we can take the constant homotopy on $X - (N - B)$ and then extend over N by applying the homotopy extension property for $(N, A \cup B)$ to the given homotopy on A and the constant homotopy on B.

Proposition 0.16. *If (X, A) is a CW pair, then $X \times \{0\} \cup A \times I$ is a deformation retract of $X \times I$, hence (X, A) has the homotopy extension property.*

Proof: There is a retraction $r : D^n \times I \to D^n \times \{0\} \cup \partial D^n \times I$, for example the radial projection from the point $(0, 2) \in D^n \times \mathbb{R}$. Then setting $r_t = tr + (1 - t)\mathbb{1}$ gives a deformation retraction of $D^n \times I$ onto $D^n \times \{0\} \cup \partial D^n \times I$. This deformation retraction gives rise to a deformation retraction of $X^n \times I$ onto $X^n \times \{0\} \cup (X^{n-1} \cup A^n) \times I$

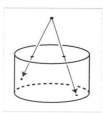

since $X^n \times I$ is obtained from $X^n \times \{0\} \cup (X^{n-1} \cup A^n) \times I$ by attaching copies of $D^n \times I$ along $D^n \times \{0\} \cup \partial D^n \times I$. If we perform the deformation retraction of $X^n \times I$ onto $X^n \times \{0\} \cup (X^{n-1} \cup A^n) \times I$ during the t-interval $[1/2^{n+1}, 1/2^n]$, this infinite concatenation of homotopies is a deformation retraction of $X \times I$ onto $X \times \{0\} \cup A \times I$. There is no problem with continuity of this deformation retraction at $t = 0$ since it is continuous on $X^n \times I$, being stationary there during the t-interval $[0, 1/2^{n+1}]$, and CW complexes have the weak topology with respect to their skeleta so a map is continuous iff its restriction to each skeleton is continuous. $\square$

Now we can prove a generalization of the earlier assertion that collapsing a contractible subcomplex is a homotopy equivalence.

Proposition 0.17. *If the pair (X, A) satisfies the homotopy extension property and A is contractible, then the quotient map $q : X \to X/A$ is a homotopy equivalence.*

Proof: Let $f_t : X \to X$ be a homotopy extending a contraction of A, with $f_0 = 1\!\!1$. Since $f_t(A) \subset A$ for all t, the composition $qf_t : X \to X/A$ sends A to a point and hence factors as a composition $X \xrightarrow{q} X/A \to X/A$. Denoting the latter map by $\overline{f}_t : X/A \to X/A$, we have $qf_t = \overline{f}_t q$ in the first of the two diagrams at the right. When $t = 1$ we have $f_1(A)$ equal to a point, the point to which A contracts, so f_1 induces a map $g : X/A \to X$ with $gq = f_1$, as in the second diagram. It

$$
\begin{array}{ccc}
X & \xrightarrow{\ f_t\ } & X \\
{\scriptstyle q}\downarrow & & \downarrow{\scriptstyle q} \\
X/A & \xrightarrow[\ \overline{f}_t\ ]{} & X/A
\end{array}
\qquad
\begin{array}{ccc}
X & \xrightarrow{\ f_1\ } & X \\
{\scriptstyle q}\downarrow & \nearrow{\scriptstyle g} & \downarrow{\scriptstyle q} \\
X/A & \xrightarrow[\ \overline{f}_1\ ]{} & X/A
\end{array}
$$

follows that $qg = \overline{f}_1$ since $qg(\overline{x}) = qgq(x) = qf_1(x) = \overline{f}_1 q(x) = \overline{f}_1(\overline{x})$. The maps g and q are inverse homotopy equivalences since $gq = f_1 \simeq f_0 = 1\!\!1$ via f_t and $qg = \overline{f}_1 \simeq \overline{f}_0 = 1\!\!1$ via $\overline{f}_t$. $\qquad\square$

Another application of the homotopy extension property, giving a slightly more refined version of one of our earlier criteria for homotopy equivalence, is the following:

Proposition 0.18. *If (X_1, A) is a CW pair and we have attaching maps $f, g : A \to X_0$ that are homotopic, then $X_0 \sqcup_f X_1 \simeq X_0 \sqcup_g X_1$ rel X_0.*

Here the definition of $W \simeq Z$ rel Y for pairs (W, Y) and (Z, Y) is that there are maps $\varphi : W \to Z$ and $\psi : Z \to W$ restricting to the identity on Y, such that $\psi\varphi \simeq 1\!\!1$ and $\varphi\psi \simeq 1\!\!1$ via homotopies that restrict to the identity on Y at all times.

Proof: If $F : A \times I \to X_0$ is a homotopy from f to g, consider the space $X_0 \sqcup_F (X_1 \times I)$. This contains both $X_0 \sqcup_f X_1$ and $X_0 \sqcup_g X_1$ as subspaces. A deformation retraction of $X_1 \times I$ onto $X_1 \times \{0\} \cup A \times I$ as in Proposition 0.16 induces a deformation retraction of $X_0 \sqcup_F (X_1 \times I)$ onto $X_0 \sqcup_f X_1$. Similarly $X_0 \sqcup_F (X_1 \times I)$ deformation retracts onto $X_0 \sqcup_g X_1$. Both these deformation retractions restrict to the identity on X_0, so together they give a homotopy equivalence $X_0 \sqcup_f X_1 \simeq X_0 \sqcup_g X_1$ rel X_0. $\qquad\square$

We finish this chapter with a technical result whose proof will involve several applications of the homotopy extension property:

Proposition 0.19. *Suppose (X, A) and (Y, A) satisfy the homotopy extension property, and $f : X \to Y$ is a homotopy equivalence with $f|A = 1\!\!1$. Then f is a homotopy equivalence rel A.*

Corollary 0.20. *If (X, A) satisfies the homotopy extension property and the inclusion $A \hookrightarrow X$ is a homotopy equivalence, then A is a deformation retract of X.*

Proof: Apply the proposition to the inclusion $A \hookrightarrow X$. $\qquad\square$

Corollary 0.21. *A map $f : X \to Y$ is a homotopy equivalence iff X is a deformation retract of the mapping cylinder M_f. Hence, two spaces X and Y are homotopy equivalent iff there is a third space containing both X and Y as deformation retracts.*

Proof: In the diagram at the right the maps i and j are the inclusions and r is the canonical retraction, so $f = ri$ and $i \simeq jf$. Since j and r are homotopy equivalences, it follows that f is a homotopy equivalence iff i is a homotopy equivalence, since the composition

of two homotopy equivalences is a homotopy equivalence and a map homotopic to a homotopy equivalence is a homotopy equivalence. Now apply the preceding corollary to the pair (M_f, X), which satisfies the homotopy extension property by Example 0.15 using the neighborhood $X \times [0, 1/2]$ of X in M_f. $\square$

Proof of 0.19: Let $g : Y \rightarrow X$ be a homotopy inverse for f. There will be three steps to the proof:

(1) Construct a homotopy from g to a map g_1 with $g_1 | A = \mathbb{1}$.

(2) Show $g_1 f \simeq \mathbb{1}$ rel A.

(3) Show $f g_1 \simeq \mathbb{1}$ rel A.

(1) Let $h_t : X \rightarrow X$ be a homotopy from $gf = h_0$ to $\mathbb{1} = h_1$. Since $f | A = \mathbb{1}$, we can view $h_t | A$ as a homotopy from $g | A$ to $\mathbb{1}$. Then since we assume (Y, A) has the homotopy extension property, we can extend this homotopy to a homotopy $g_t : Y \rightarrow X$ from $g = g_0$ to a map g_1 with $g_1 | A = \mathbb{1}$.

(2) A homotopy from $g_1 f$ to $\mathbb{1}$ is given by the formulas

$$k_t = \begin{cases} g_{1-2t} f, & 0 \le t \le 1/2 \\ h_{2t-1}, & 1/2 \le t \le 1 \end{cases}$$

Note that the two definitions agree when $t = 1/2$. Since $f | A = \mathbb{1}$ and $g_t = h_t$ on A, the homotopy $k_t | A$ starts and ends with the identity, and its second half simply retraces its first half, that is, $k_t = k_{1-t}$ on A. We will define a 'homotopy of homotopies' $k_{tu} : A \rightarrow X$ by means of the figure at the right showing the parameter domain $I \times I$ for the pairs (t, u), with the t-axis horizontal and the u-axis vertical. On the bottom edge of the square we define $k_{t0} = k_t | A$. Below the 'V' we define k_{tu} to be independent of u, and above the 'V' we define k_{tu} to be independent of t.

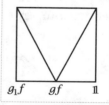

This is unambiguous since $k_t = k_{1-t}$ on A. Since $k_0 = \mathbb{1}$ on A, we have $k_{tu} = \mathbb{1}$ for (t, u) in the left, right, and top edges of the square. Next we extend k_{tu} over X, as follows. Since (X, A) has the homotopy extension property, so does $(X \times I, A \times I)$, as one can see from the equivalent retraction property. Viewing k_{tu} as a homotopy of $k_t | A$, we can therefore extend $k_{tu} : A \rightarrow X$ to $k_{tu} : X \rightarrow X$ with $k_{t0} = k_t$. If we restrict this k_{tu} to the left, top, and right edges of the (t, u)-square, we get a homotopy $g_1 f \simeq \mathbb{1}$ rel A.

(3) Since $g_1 \simeq g$, we have $f g_1 \simeq f g \simeq \mathbb{1}$, so $f g_1 \simeq \mathbb{1}$ and steps (1) and (2) can be repeated with the pair f, g replaced by g_1, f. The result is a map $f_1 : X \rightarrow Y$ with $f_1 | A = \mathbb{1}$ and $f_1 g_1 \simeq \mathbb{1}$ rel A. Hence $f_1 \simeq f_1(g_1 f) = (f_1 g_1) f \simeq f$ rel A. From this we deduce that $f g_1 \simeq f_1 g_1 \simeq \mathbb{1}$ rel A. $\square$

Exercises

1. Construct an explicit deformation retraction of the torus with one point deleted onto a graph consisting of two circles intersecting in a point, namely, longitude and meridian circles of the torus.

2. Construct an explicit deformation retraction of $\mathbb{R}^n - \{0\}$ onto S^{n-1}.

3. (a) Show that the composition of homotopy equivalences $X \to Y$ and $Y \to Z$ is a homotopy equivalence $X \to Z$. Deduce that homotopy equivalence is an equivalence relation.

(b) Show that the relation of homotopy among maps $X \to Y$ is an equivalence relation.

(c) Show that a map homotopic to a homotopy equivalence is a homotopy equivalence.

4. A **deformation retraction in the weak sense** of a space X to a subspace A is a homotopy $f_t : X \to X$ such that $f_0 = \mathbb{1}$, $f_1(X) \subset A$, and $f_t(A) \subset A$ for all t. Show that if X deformation retracts to A in this weak sense, then the inclusion $A \hookrightarrow X$ is a homotopy equivalence.

5. Show that if a space X deformation retracts to a point $x \in X$, then for each neighborhood U of x in X there exists a neighborhood $V \subset U$ of x such that the inclusion map $V \hookrightarrow U$ is nullhomotopic.

6. (a) Let X be the subspace of $\mathbb{R}^2$ consisting of the horizontal segment $[0,1] \times \{0\}$ together with all the vertical segments $\{r\} \times [0, 1-r]$ for r a rational number in $[0,1]$. Show that X deformation retracts to any point in the segment $[0,1] \times \{0\}$, but not to any other point. [See the preceding problem.]

(b) Let Y be the subspace of $\mathbb{R}^2$ that is the union of an infinite number of copies of X arranged as in the figure below. Show that Y is contractible but does not deformation retract onto any point.

(c) Let Z be the zigzag subspace of Y homeomorphic to $\mathbb{R}$ indicated by the heavier line. Show there is a deformation retraction in the weak sense (see Exercise 4) of Y onto Z, but no true deformation retraction.

7. Fill in the details in the following construction from [Edwards 1999] of a compact space $Y \subset \mathbb{R}^3$ with the same properties as the space Y in Exercise 6, that is, Y is contractible but does not deformation retract to any point. To begin, let X be the union of an infinite sequence of cones on the Cantor set arranged end-to-end, as in the figure. Next, form the one-point compactification of $X \times \mathbb{R}$. This embeds in $\mathbb{R}^3$ as a closed disk with curved 'fins' attached along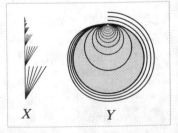

circular arcs, and with the one-point compactification of X as a cross-sectional slice. The desired space Y is then obtained from this subspace of $\mathbb{R}^3$ by wrapping one more cone on the Cantor set around the boundary of the disk.

8. For $n > 2$, construct an n-room analog of the house with two rooms.

9. Show that a retract of a contractible space is contractible.

10. Show that a space X is contractible iff every map $f:X \to Y$, for arbitrary Y, is nullhomotopic. Similarly, show X is contractible iff every map $f:Y \to X$ is nullhomotopic.

11. Show that $f:X \to Y$ is a homotopy equivalence if there exist maps $g, h:Y \to X$ such that $fg \simeq \mathbb{1}$ and $hf \simeq \mathbb{1}$. More generally, show that f is a homotopy equivalence if fg and hf are homotopy equivalences.

12. Show that a homotopy equivalence $f:X \to Y$ induces a bijection between the set of path-components of X and the set of path-components of Y, and that f restricts to a homotopy equivalence from each path-component of X to the corresponding path-component of Y. Prove also the corresponding statements with components instead of path-components. Deduce that if the components of a space X coincide with its path-components, then the same holds for any space Y homotopy equivalent to X.

13. Show that any two deformation retractions r_t^0 and r_t^1 of a space X onto a subspace A can be joined by a continuous family of deformation retractions r_t^s, $0 \leq s \leq 1$, of X onto A, where continuity means that the map $X \times I \times I \to X$ sending (x, s, t) to $r_t^s(x)$ is continuous.

14. Given positive integers v, e, and f satisfying $v - e + f = 2$, construct a cell structure on S^2 having v 0-cells, e 1-cells, and f 2-cells.

15. Enumerate all the subcomplexes of S^∞, with the cell structure on S^∞ that has S^n as its n-skeleton.

16. Show that S^∞ is contractible.

17. (a) Show that the mapping cylinder of every map $f:S^1 \to S^1$ is a CW complex.
(b) Construct a 2-dimensional CW complex that contains both an annulus $S^1 \times I$ and a Möbius band as deformation retracts.

18. Show that $S^1 * S^1 = S^3$, and more generally $S^m * S^n = S^{m+n+1}$.

19. Show that the space obtained from S^2 by attaching n 2-cells along any collection of n circles in S^2 is homotopy equivalent to the wedge sum of $n + 1$ 2-spheres.

20. Show that the subspace $X \subset \mathbb{R}^3$ formed by a Klein bottle intersecting itself in a circle, as shown in the figure, is homotopy equivalent to $S^1 \vee S^1 \vee S^2$.

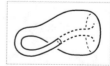

21. If X is a connected Hausdorff space that is a union of a finite number of 2-spheres, any two of which intersect in at most one point, show that X is homotopy equivalent to a wedge sum of S^1's and S^2's.

22. Let X be a finite graph lying in a half-plane $P \subset \mathbb{R}^3$ and intersecting the edge of P in a subset of the vertices of X. Describe the homotopy type of the 'surface of revolution' obtained by rotating X about the edge line of P.

23. Show that a CW complex is contractible if it is the union of two contractible subcomplexes whose intersection is also contractible.

24. Let X and Y be CW complexes with 0-cells x_0 and y_0. Show that the quotient spaces $X * Y/(X * \{y_0\} \cup \{x_0\} * Y)$ and $S(X \wedge Y)/S(\{x_0\} \wedge \{y_0\})$ are homeomorphic, and deduce that $X * Y \simeq S(X \wedge Y)$.

25. If X is a CW complex with components X_α, show that the suspension SX is homotopy equivalent to $Y \vee_\alpha SX_\alpha$ for some graph Y. In the case that X is a finite graph, show that SX is homotopy equivalent to a wedge sum of circles and 2-spheres.

26. Use Corollary 0.20 to show that if (X, A) has the homotopy extension property, then $X \times I$ deformation retracts to $X \times \{0\} \cup A \times I$. Deduce from this that Proposition 0.18 holds more generally for any pair (X_1, A) satisfying the homotopy extension property.

27. Given a pair (X, A) and a homotopy equivalence $f : A \to B$, show that the natural map $X \to B \sqcup_f X$ is a homotopy equivalence if (X, A) satisfies the homotopy extension property. [Hint: Consider $X \cup M_f$ and use the preceding problem.] An interesting case is when f is a quotient map, hence the map $X \to B \sqcup_f X$ is the quotient map identifying each set $f^{-1}(b)$ to a point. When B is a point this gives another proof of Proposition 0.17.

28. Show that if (X_1, A) satisfies the homotopy extension property, then so does every pair $(X_0 \sqcup_f X_1, X_0)$ obtained by attaching X_1 to a space X_0 via a map $f : A \to X_0$.

29. In case the CW complex X is obtained from a subcomplex A by attaching a single cell e^n, describe exactly what the extension of a homotopy $f_t : A \to Y$ to X given by the proof of Proposition 0.16 looks like. That is, for a point $x \in e^n$, describe the path $f_t(x)$ for the extended f_t.

Chapter 1

The Fundamental Group

Algebraic topology can be roughly defined as the study of techniques for forming algebraic images of topological spaces. Most often these algebraic images are groups, but more elaborate structures such as rings, modules, and algebras also arise. The mechanisms that create these images — the 'lanterns' of algebraic topology, one might say — are known formally as *functors* and have the characteristic feature that they form images not only of spaces but also of maps. Thus, continuous maps between spaces are projected onto homomorphisms between their algebraic images, so topologically related spaces have algebraically related images.

With suitably constructed lanterns one might hope to be able to form images with enough detail to reconstruct accurately the shapes of all spaces, or at least of large and interesting classes of spaces. This is one of the main goals of algebraic topology, and to a surprising extent this goal is achieved. Of course, the lanterns necessary to do this are somewhat complicated pieces of machinery. But this machinery also has a certain intrinsic beauty.

This first chapter introduces one of the simplest and most important functors of algebraic topology, the fundamental group, which creates an algebraic image of a space from the loops in the space, the paths in the space starting and ending at the same point.

The Idea of the Fundamental Group

To get a feeling for what the fundamental group is about, let us look at a few preliminary examples before giving the formal definitions.

Consider two linked circles A and B in $\mathbb{R}^3$, as shown
in the figure. Our experience with actual links and chains
tells us that since the two circles are linked, it is impossible to separate B from A by any continuous motion of B,
such as pushing, pulling, or twisting. We could even take

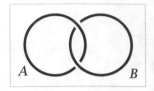

B to be made of rubber or stretchable string and allow completely general continuous deformations of B, staying in the complement of A at all times, and it would
still be impossible to pull B off A. At least that is what intuition suggests, and the
fundamental group will give a way of making this intuition mathematically rigorous.

Instead of having B link with A just once, we could
make it link with A two or more times, as in the figures to the
right. As a further variation, by assigning an orientation to B
we can speak of B linking A a positive or a negative number
of times, say positive when B comes forward through A and
negative for the reverse direction. Thus for each nonzero
integer n we have an oriented circle B_n linking A n times,
where by 'circle' we mean a curve homeomorphic to a circle.
To complete the scheme, we could let B_0 be a circle not
linked to A at all.

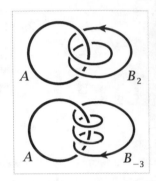

Now, integers not only measure quantity, but they form a group under addition.
Can the group operation be mimicked geometrically with some sort of addition operation on the oriented circles B linking A? An oriented circle B can be thought
of as a path traversed in time, starting and ending at the same point x_0, which we
can choose to be any point on the circle. Such a path starting and ending at the
same point is called a *loop*. Two different loops B and B' both starting and ending at the same point x_0 can be 'added' to form a new loop $B + B'$ that travels first
around B, then around B'. For example, if B_1 and B_1' are loops each linking A once in
the positive direction,
then their sum $B_1 + B_1'$
is deformable to B_2,
linking A twice. Simi-
larly, $B_1 + B_{-1}$ can be
deformed to the loop
B_0, unlinked from A.
More generally, we see
that $B_m + B_n$ can be
deformed to B_{m+n} for
arbitrary integers m and n.

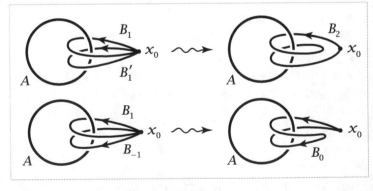

Note that in forming sums of loops we produce loops that pass through the basepoint more than once. This is one reason why loops are defined merely as continuous

paths, which are allowed to pass through the same point many times. So if one is thinking of a loop as something made of stretchable string, one has to give the string the magical power of being able to pass through itself unharmed. However, we must be sure not to allow our loops to intersect the fixed circle A at any time, otherwise we could always unlink them from A.

Next we consider a slightly more complicated sort of linking, involving three circles forming a configuration known as the Borromean rings, shown at the left in the figure below. The interesting feature here is that if any one of the three circles is removed, the other two are not linked. In the same spirit as before, let us regard one of the circles, say C, as a loop in the complement of the other two, A and

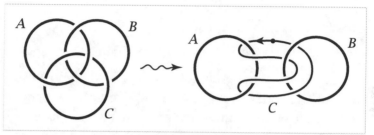

B, and we ask whether C can be continuously deformed to unlink it completely from A and B, always staying in the complement of A and B during the deformation. We can redraw the picture by pulling A and B apart, dragging C along, and then we see C winding back and forth between A and B as shown in the second figure above. In this new position, if we start at the point of C indicated by the dot and proceed in the direction given by the arrow, then we pass in sequence: (1) forward through A, (2) forward through B, (3) backward through A, and (4) backward through B. If we measure the linking of C with A and B by two integers, then the 'forwards' and 'backwards' cancel and both integers are zero. This reflects the fact that C is not linked with A or B individually.

To get a more accurate measure of how C links with A and B together, we regard the four parts (1)–(4) of C as an ordered sequence. Taking into account the directions in which these segments of C pass through A and B, we may deform C to the sum $a + b - a - b$ of four loops as in the figure. We write the third and fourth loops as the negatives of the first two since they can be deformed to the first two, but with the opposite orientations, and as we saw in the preceding example, the sum of two oppositely oriented loops is deformable to a trivial loop, not linked with anything. We would like to view the expression

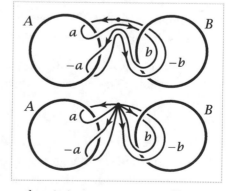

$a + b - a - b$ as lying in a nonabelian group, so that it is not automatically zero. Changing to the more usual multiplicative notation for nonabelian groups, it would be written $aba^{-1}b^{-1}$, the commutator of a and b.

To shed further light on this example, suppose we modify it slightly so that the circles A and B are now linked, as in the next figure. The circle C can then be deformed into the position shown at the right, where it again represents the composite loop $aba^{-1}b^{-1}$, where a and b are loops linking A and B. But from the picture on the left it is apparent that C can

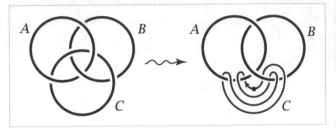

actually be unlinked completely from A and B. So in this case the product $aba^{-1}b^{-1}$ should be trivial.

The fundamental group of a space X will be defined so that its elements are loops in X starting and ending at a fixed basepoint $x_0 \in X$, but two such loops are regarded as determining the same element of the fundamental group if one loop can be continuously deformed to the other within the space X. (All loops that occur during deformations must also start and end at x_0.) In the first example above, X is the complement of the circle A, while in the other two examples X is the complement of the two circles A and B. In the second section in this chapter we will show:

- The fundamental group of the complement of the circle A in the first example is infinite cyclic with the loop B as a generator. This amounts to saying that every loop in the complement of A can be deformed to one of the loops B_n, and that B_n cannot be deformed to B_m if $n \neq m$.

- The fundamental group of the complement of the two unlinked circles A and B in the second example is the nonabelian free group on two generators, represented by the loops a and b linking A and B. In particular, the commutator $aba^{-1}b^{-1}$ is a nontrivial element of this group.

- The fundamental group of the complement of the two linked circles A and B in the third example is the free abelian group on two generators, represented by the loops a and b linking A and B.

As a result of these calculations, we have two ways to tell when a pair of circles A and B is linked. The direct approach is given by the first example, where one circle is regarded as an element of the fundamental group of the complement of the other circle. An alternative and somewhat more subtle method is given by the second and third examples, where one distinguishes a pair of linked circles from a pair of unlinked circles by the fundamental group of their complement, which is abelian in one case and nonabelian in the other. This method is much more general: One can often show that two spaces are not homeomorphic by showing that their fundamental groups are not isomorphic, since it will be an easy consequence of the definition of the fundamental group that homeomorphic spaces have isomorphic fundamental groups.

1.1 Basic Constructions

This first section begins with the basic definitions and constructions, and then proceeds quickly to an important calculation, the fundamental group of the circle, using notions developed more fully in §1.3. More systematic methods of calculation are given in §1.2. These are sufficient to show for example that every group is realized as the fundamental group of some space. This idea is exploited in the Additional Topics at the end of the chapter, which give some illustrations of how algebraic facts about groups can be derived topologically, such as the fact that every subgroup of a free group is free.

Paths and Homotopy

The fundamental group will be defined in terms of loops and deformations of loops. Sometimes it will be useful to consider more generally paths and their deformations, so we begin with this slight extra generality.

By a **path** in a space X we mean a continuous map $f : I \to X$ where I is the unit interval $[0, 1]$. The idea of continuously deforming a path, keeping its endpoints fixed, is made precise by the following definition. A **homotopy** of paths in X is a family $f_t : I \to X$, $0 \le t \le 1$, such that

(1) The endpoints $f_t(0) = x_0$ and $f_t(1) = x_1$ are independent of t.

(2) The associated map $F : I \times I \to X$ defined by $F(s, t) = f_t(s)$ is continuous.

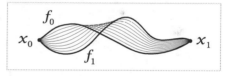

When two paths f_0 and f_1 are connected in this way by a homotopy f_t, they are said to be **homotopic**. The notation for this is $f_0 \simeq f_1$.

Example 1.1: Linear Homotopies. Any two paths f_0 and f_1 in $\mathbb{R}^n$ having the same endpoints x_0 and x_1 are homotopic via the homotopy $f_t(s) = (1 - t)f_0(s) + tf_1(s)$. During this homotopy each point $f_0(s)$ travels along the line segment to $f_1(s)$ at constant speed. This is because the line through $f_0(s)$ and $f_1(s)$ is linearly parametrized as $f_0(s) + t[f_1(s) - f_0(s)] = (1 - t)f_0(s) + tf_1(s)$, with the segment from $f_0(s)$ to $f_1(s)$ covered by t values in the interval from 0 to 1. If $f_1(s)$ happens to equal $f_0(s)$ then this segment degenerates to a point and $f_t(s) = f_0(s)$ for all t. This occurs in particular for $s = 0$ and $s = 1$, so each f_t is a path from x_0 to x_1. Continuity of the homotopy f_t as a map $I \times I \to \mathbb{R}^n$ follows from continuity of f_0 and f_1 since the algebraic operations of vector addition and scalar multiplication in the formula for f_t are continuous.

This construction shows more generally that for a convex subspace $X \subset \mathbb{R}^n$, all paths in X with given endpoints x_0 and x_1 are homotopic, since if f_0 and f_1 lie in X then so does the homotopy f_t.

Before proceeding further we need to verify a technical property:

Proposition 1.2. *The relation of homotopy on paths with fixed endpoints in any space is an equivalence relation.*

The equivalence class of a path f under the equivalence relation of homotopy will be denoted $[f]$ and called the **homotopy class** of f.

Proof: Reflexivity is evident since $f \simeq f$ by the constant homotopy $f_t = f$. Symmetry is also easy since if $f_0 \simeq f_1$ via f_t, then $f_1 \simeq f_0$ via the inverse homotopy f_{1-t}. For transitivity, if $f_0 \simeq f_1$ via f_t and if $f_1 = g_0$ with $g_0 \simeq g_1$ via g_t, then $f_0 \simeq g_1$ via the homotopy h_t that equals f_{2t} for $0 \le t \le 1/2$ and g_{2t-1} for $1/2 \le t \le 1$. These two definitions agree for $t = 1/2$ since we assume $f_1 = g_0$. Continuity of the associated map $H(s,t) = h_t(s)$ comes from the elementary 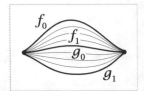 fact, which will be used frequently without explicit mention, that a function defined on the union of two closed sets is continuous if it is continuous when restricted to each of the closed sets separately. In the case at hand we have $H(s,t) = F(s, 2t)$ for $0 \le t \le 1/2$ and $H(s,t) = G(s, 2t - 1)$ for $1/2 \le t \le 1$ where F and G are the maps $I \times I \to X$ associated to the homotopies f_t and g_t. Since H is continuous on $I \times [0, 1/2]$ and on $I \times [1/2, 1]$, it is continuous on $I \times I$.   □

Given two paths $f, g : I \to X$ such that $f(1) = g(0)$, there is a **composition** or **product path** $f \cdot g$ that traverses first f and then g, defined by the formula

$$f \cdot g(s) = \begin{cases} f(2s), & 0 \le s \le 1/2 \\ g(2s - 1), & 1/2 \le s \le 1 \end{cases}$$

Thus f and g are traversed twice as fast in order for $f \cdot g$ to be traversed in unit time. This product operation respects homotopy classes since if $f_0 \simeq f_1$ and $g_0 \simeq g_1$ via homotopies f_t and g_t, and if $f_0(1) = g_0(0)$ so that $f_0 \cdot g_0$ is defined, then $f_t \cdot g_t$ is defined and provides a homotopy $f_0 \cdot g_0 \simeq f_1 \cdot g_1$.

In particular, suppose we restrict attention to paths $f : I \to X$ with the same starting and ending point $f(0) = f(1) = x_0 \in X$. Such paths are called **loops**, and the common starting and ending point x_0 is referred to as the **basepoint**. The set of all homotopy classes $[f]$ of loops $f : I \to X$ at the basepoint x_0 is denoted $\pi_1(X, x_0)$.

Proposition 1.3. $\pi_1(X, x_0)$ *is a group with respect to the product* $[f][g] = [f \cdot g]$.

This group is called the **fundamental group** of X at the basepoint x_0. We will see in Chapter 4 that $\pi_1(X, x_0)$ is the first in a sequence of groups $\pi_n(X, x_0)$, called homotopy groups, which are defined in an entirely analogous fashion using the n-dimensional cube I^n in place of I.

Proof: By restricting attention to loops with a fixed basepoint $x_0 \in X$ we guarantee that the product $f \cdot g$ of any two such loops is defined. We have already observed that the homotopy class of $f \cdot g$ depends only on the homotopy classes of f and g, so the product $[f][g] = [f \cdot g]$ is well-defined. It remains to verify the three axioms for a group.

As a preliminary step, define a **reparametrization** of a path f to be a composition $f\varphi$ where $\varphi : I \to I$ is any continuous map such that $\varphi(0) = 0$ and $\varphi(1) = 1$. Reparametrizing a path preserves its homotopy class since $f\varphi \simeq f$ via the homotopy $f\varphi_t$ where $\varphi_t(s) = (1-t)\varphi(s) + ts$ so that $\varphi_0 = \varphi$ and $\varphi_1(s) = s$. Note that $(1-t)\varphi(s) + ts$ lies between $\varphi(s)$ and s, hence is in I, so the composition $f\varphi_t$ is defined.

If we are given paths f, g, h with $f(1) = g(0)$ and $g(1) = h(0)$, then both products $(f \cdot g) \cdot h$ and $f \cdot (g \cdot h)$ are defined, and $f \cdot (g \cdot h)$ is a reparametrization of $(f \cdot g) \cdot h$ by the piecewise linear function φ whose graph is shown in the figure at the right. Hence $(f \cdot g) \cdot h \simeq f \cdot (g \cdot h)$. Restricting attention to loops at the basepoint x_0, this says the product in $\pi_1(X, x_0)$ is associative.

Given a path $f : I \to X$, let c be the constant path at $f(1)$, defined by $c(s) = f(1)$ for all $s \in I$. Then $f \cdot c$ is a reparametrization of f via the function φ whose graph is shown in the first figure at the right, so $f \cdot c \simeq f$. Similarly, $c \cdot f \simeq f$ where c is now the constant path at $f(0)$, using the reparametrization function in the second figure. Taking f to be a loop, we deduce that the homotopy class of the constant path at x_0 is a two-sided identity in $\pi_1(X, x_0)$.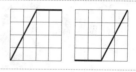

For a path f from x_0 to x_1, the **inverse path** $\overline{f}$ from x_1 back to x_0 is defined by $\overline{f}(s) = f(1-s)$. To see that $f \cdot \overline{f}$ is homotopic to a constant path we use the homotopy $h_t = f_t \cdot g_t$ where f_t is the path that equals f on the interval $[0, 1-t]$ and that is stationary at $f(1-t)$ on the interval $[1-t, 1]$, and g_t is the inverse path of f_t. We could also describe h_t in terms of the associated function $H : I \times I \to X$ using the decomposition of $I \times I$ shown in the figure. On the bottom edge of the square H is given by $f \cdot \overline{f}$ and below the 'V' we let $H(s, t)$ be independent of t, while above the 'V' we let $H(s, t)$ be independent of s. Going back to the first description of h_t, we see that since $f_0 = f$ and f_1 is the constant path c at x_0, h_t is a homotopy from $f \cdot \overline{f}$ to $c \cdot \overline{c} = c$. Replacing f by $\overline{f}$ gives $\overline{f} \cdot f \simeq c$ for c the constant path at x_1. Taking f to be a loop at the basepoint x_0, we deduce that $[\overline{f}]$ is a two-sided inverse for $[f]$ in $\pi_1(X, x_0)$. $\square$

Example 1.4. For a convex set X in $\mathbb{R}^n$ with basepoint $x_0 \in X$ we have $\pi_1(X, x_0) = 0$, the trivial group, since any two loops f_0 and f_1 based at x_0 are homotopic via the linear homotopy $f_t(s) = (1-t)f_0(s) + t f_1(s)$, as described in Example 1.1.

It is not so easy to show that a space has a nontrivial fundamental group since one must somehow demonstrate the nonexistence of homotopies between certain loops. We will tackle the simplest example shortly, computing the fundamental group of the circle.

It is natural to ask about the dependence of $\pi_1(X, x_0)$ on the choice of the basepoint x_0. Since $\pi_1(X, x_0)$ involves only the path-component of X containing x_0, it is clear that we can hope to find a relation between $\pi_1(X, x_0)$ and $\pi_1(X, x_1)$ for two basepoints x_0 and x_1 only if x_0 and x_1 lie in the same path-component of X. So let $h : I \to X$ be a path from x_0 to x_1, with the inverse path $\overline{h}(s) = h(1 - s)$ from x_1 back to x_0. We can then associate to each loop f based at x_1 the loop $h \cdot f \cdot \overline{h}$ based at x_0.

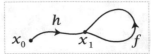

Strictly speaking, we should choose an order of forming the product $h \cdot f \cdot \overline{h}$, either $(h \cdot f) \cdot \overline{h}$ or $h \cdot (f \cdot \overline{h})$, but the two choices are homotopic and we are only interested in homotopy classes here. Alternatively, to avoid any ambiguity we could define a general n-fold product $f_1 \cdot \cdots \cdot f_n$ in which the path f_i is traversed in the time interval $[\frac{i-1}{n}, \frac{i}{n}]$. Either way, we define a **change-of-basepoint** map $\beta_h : \pi_1(X, x_1) \to \pi_1(X, x_0)$ by $\beta_h[f] = [h \cdot f \cdot \overline{h}]$. This is well-defined since if f_t is a homotopy of loops based at x_1 then $h \cdot f_t \cdot \overline{h}$ is a homotopy of loops based at x_0.

|| **Proposition 1.5.** *The map* $\beta_h : \pi_1(X, x_1) \to \pi_1(X, x_0)$ *is an isomorphism.*

Proof: We see first that β_h is a homomorphism since $\beta_h[f \cdot g] = [h \cdot f \cdot g \cdot \overline{h}] = [h \cdot f \cdot \overline{h} \cdot h \cdot g \cdot \overline{h}] = \beta_h[f]\beta_h[g]$. Further, β_h is an isomorphism with inverse $\beta_{\overline{h}}$ since $\beta_h \beta_{\overline{h}}[f] = \beta_h[\overline{h} \cdot f \cdot h] = [h \cdot \overline{h} \cdot f \cdot h \cdot \overline{h}] = [f]$, and similarly $\beta_{\overline{h}} \beta_h[f] = [f]$. $\square$

Thus if X is path-connected, the group $\pi_1(X, x_0)$ is, up to isomorphism, independent of the choice of basepoint x_0. In this case the notation $\pi_1(X, x_0)$ is often abbreviated to $\pi_1(X)$, or one could go further and write just $\pi_1 X$.

In general, a space is called **simply-connected** if it is path-connected and has trivial fundamental group. The following result explains the name.

|| **Proposition 1.6.** *A space X is simply-connected iff there is a unique homotopy class of paths connecting any two points in X.*

Proof: Path-connectedness is the existence of paths connecting every pair of points, so we need be concerned only with the uniqueness of connecting paths. Suppose $\pi_1(X) = 0$. If f and g are two paths from x_0 to x_1, then $f \simeq f \cdot \overline{g} \cdot g \simeq g$ since the loops $\overline{g} \cdot g$ and $f \cdot \overline{g}$ are each homotopic to constant loops, using the assumption $\pi_1(X, x_0) = 0$ in the latter case. Conversely, if there is only one homotopy class of paths connecting a basepoint x_0 to itself, then all loops at x_0 are homotopic to the constant loop and $\pi_1(X, x_0) = 0$. $\square$

The Fundamental Group of the Circle

Our first real theorem will be the calculation $\pi_1(S^1) \approx \mathbb{Z}$. Besides its intrinsic interest, this basic result will have several immediate applications of some substance, and it will be the starting point for many more calculations in the next section. It should be no surprise then that the proof will involve some genuine work.

Theorem 1.7. *$\pi_1(S^1)$ is an infinite cyclic group generated by the homotopy class of the loop $\omega(s) = (\cos 2\pi s, \sin 2\pi s)$ based at $(1,0)$.*

Note that $[\omega]^n = [\omega_n]$ where $\omega_n(s) = (\cos 2\pi n s, \sin 2\pi n s)$ for $n \in \mathbb{Z}$. The theorem is therefore equivalent to the statement that every loop in S^1 based at $(1,0)$ is homotopic to ω_n for a unique $n \in \mathbb{Z}$. To prove this the idea will be to compare paths in S^1 with paths in $\mathbb{R}$ via the map $p:\mathbb{R}\to S^1$ given by $p(s) = (\cos 2\pi s, \sin 2\pi s)$. This map can be visualized geometrically by embedding $\mathbb{R}$ in $\mathbb{R}^3$ as the helix parametrized by $s \mapsto (\cos 2\pi s, \sin 2\pi s, s)$, and then p is the restriction to the helix of the projection of $\mathbb{R}^3$ onto $\mathbb{R}^2$, $(x,y,z) \mapsto (x,y)$. Observe that the loop ω_n is the composition $p\widetilde{\omega}_n$ where $\widetilde{\omega}_n : I \to \mathbb{R}$ is the path $\widetilde{\omega}_n(s) = ns$, starting at 0 and ending at n, winding around the helix $|n|$ times, upward if $n > 0$ and downward if $n < 0$. The relation $\omega_n = p\widetilde{\omega}_n$ is expressed by saying that $\widetilde{\omega}_n$ is a **lift** of ω_n.

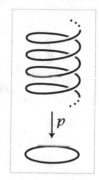

We will prove the theorem by studying how paths in S^1 lift to paths in $\mathbb{R}$. Most of the arguments will apply in much greater generality, and it is both more efficient and more enlightening to give them in the general context. The first step will be to define this context.

Given a space X, a **covering space** of X consists of a space $\widetilde{X}$ and a map $p:\widetilde{X}\to X$ satisfying the following condition:

(∗)　For each point $x \in X$ there is an open neighborhood U of x in X such that $p^{-1}(U)$ is a union of disjoint open sets each of which is mapped homeomorphically onto U by p.

Such a U will be called **evenly covered**. For example, for the previously defined map $p:\mathbb{R}\to S^1$ any open arc in S^1 is evenly covered.

To prove the theorem we will need just the following two facts about covering spaces $p:\widetilde{X}\to X$.

(a) For each path $f:I\to X$ starting at a point $x_0 \in X$ and each $\widetilde{x}_0 \in p^{-1}(x_0)$ there is a unique lift $\widetilde{f}:I\to\widetilde{X}$ starting at $\widetilde{x}_0$.

(b) For each homotopy $f_t:I\to X$ of paths starting at x_0 and each $\widetilde{x}_0 \in p^{-1}(x_0)$ there is a unique lifted homotopy $\widetilde{f}_t:I\to\widetilde{X}$ of paths starting at $\widetilde{x}_0$.

Before proving these facts, let us see how they imply the theorem.

Proof of Theorem 1.7: Let $f : I \to S^1$ be a loop at the basepoint $x_0 = (1,0)$, representing a given element of $\pi_1(S^1, x_0)$. By (a) there is a lift $\tilde{f}$ starting at 0. This path $\tilde{f}$ ends at some integer n since $p\tilde{f}(1) = f(1) = x_0$ and $p^{-1}(x_0) = \mathbb{Z} \subset \mathbb{R}$. Another path in $\mathbb{R}$ from 0 to n is $\tilde{\omega}_n$, and $\tilde{f} \simeq \tilde{\omega}_n$ via the linear homotopy $(1 - t)\tilde{f} + t\tilde{\omega}_n$. Composing this homotopy with p gives a homotopy $f \simeq \omega_n$ so $[f] = [\omega_n]$.

To show that n is uniquely determined by $[f]$, suppose that $f \simeq \omega_n$ and $f \simeq \omega_m$, so $\omega_m \simeq \omega_n$. Let f_t be a homotopy from $\omega_m = f_0$ to $\omega_n = f_1$. By (b) this homotopy lifts to a homotopy $\tilde{f}_t$ of paths starting at 0. The uniqueness part of (a) implies that $\tilde{f}_0 = \tilde{\omega}_m$ and $\tilde{f}_1 = \tilde{\omega}_n$. Since $\tilde{f}_t$ is a homotopy of paths, the endpoint $\tilde{f}_t(1)$ is independent of t. For $t = 0$ this endpoint is m and for $t = 1$ it is n, so $m = n$.

It remains to prove (a) and (b). Both statements can be deduced from a more general assertion about covering spaces $p : \tilde{X} \to X$:

  (c) Given a map $F : Y \times I \to X$ and a map $\tilde{F} : Y \times \{0\} \to \tilde{X}$ lifting $F|Y \times \{0\}$, then there is a unique map $\tilde{F} : Y \times I \to \tilde{X}$ lifting F and restricting to the given $\tilde{F}$ on $Y \times \{0\}$.

Statement (a) is the special case that Y is a point, and (b) is obtained by applying (c) with $Y = I$ in the following way. The homotopy f_t in (b) gives a map $F : I \times I \to X$ by setting $F(s,t) = f_t(s)$ as usual. A unique lift $\tilde{F} : I \times \{0\} \to \tilde{X}$ is obtained by an application of (a). Then (c) gives a unique lift $\tilde{F} : I \times I \to \tilde{X}$. The restrictions $\tilde{F}|\{0\} \times I$ and $\tilde{F}|\{1\} \times I$ are paths lifting constant paths, hence they must also be constant by the uniqueness part of (a). So $\tilde{f}_t(s) = \tilde{F}(s,t)$ is a homotopy of paths, and $\tilde{f}_t$ lifts f_t since $p\tilde{F} = F$.

To prove (c) we will first construct a lift $\tilde{F} : N \times I \to \tilde{X}$ for N some neighborhood in Y of a given point $y_0 \in Y$. Since F is continuous, every point $(y_0, t) \in Y \times I$ has a product neighborhood $N_t \times (a_t, b_t)$ such that $F(N_t \times (a_t, b_t))$ is contained in an evenly covered neighborhood of $F(y_0, t)$. By compactness of $\{y_0\} \times I$, finitely many such products $N_t \times (a_t, b_t)$ cover $\{y_0\} \times I$. This implies that we can choose a single neighborhood N of y_0 and a partition $0 = t_0 < t_1 < \cdots < t_m = 1$ of I so that for each i, $F(N \times [t_i, t_{i+1}])$ is contained in an evenly covered neighborhood U_i. Assume inductively that $\tilde{F}$ has been constructed on $N \times [0, t_i]$, starting with the given $\tilde{F}$ on $N \times \{0\}$. We have $F(N \times [t_i, t_{i+1}]) \subset U_i$, so since U_i is evenly covered there is an open set $\tilde{U}_i \subset \tilde{X}$ projecting homeomorphically onto U_i by p and containing the point $\tilde{F}(y_0, t_i)$. After replacing N by a smaller neighborhood of y_0 we may assume that $\tilde{F}(N \times \{t_i\})$ is contained in $\tilde{U}_i$, namely, replace $N \times \{t_i\}$ by its intersection with $(\tilde{F}|N \times \{t_i\})^{-1}(\tilde{U}_i)$. Now we can define $\tilde{F}$ on $N \times [t_i, t_{i+1}]$ to be the composition of F with the homeomorphism $p^{-1} : U_i \to \tilde{U}_i$. After a finite number of steps we eventually get a lift $\tilde{F} : N \times I \to \tilde{X}$ for some neighborhood N of y_0.

Next we show the uniqueness part of (c) in the special case that Y is a point. In this case we can omit Y from the notation. So suppose $\tilde{F}$ and $\tilde{F}'$ are two lifts of $F : I \to X$

such that $\tilde{F}(0) = \tilde{F}'(0)$. As before, choose a partition $0 = t_0 < t_1 < \cdots < t_m = 1$ of I so that for each i, $F([t_i, t_{i+1}])$ is contained in some evenly covered neighborhood U_i. Assume inductively that $\tilde{F} = \tilde{F}'$ on $[0, t_i]$. Since $[t_i, t_{i+1}]$ is connected, so is $\tilde{F}([t_i, t_{i+1}])$, which must therefore lie in a single one of the disjoint open sets $\tilde{U}_i$ projecting homeomorphically to U_i as in $(*)$. By the same token, $\tilde{F}'([t_i, t_{i+1}])$ lies in a single $\tilde{U}_i$, in fact in the same one that contains $\tilde{F}([t_i, t_{i+1}])$ since $\tilde{F}'(t_i) = \tilde{F}(t_i)$. Because p is injective on $\tilde{U}_i$ and $p\tilde{F} = p\tilde{F}'$, it follows that $\tilde{F} = \tilde{F}'$ on $[t_i, t_{i+1}]$, and the induction step is finished.

The last step in the proof of (c) is to observe that since the $\tilde{F}$'s constructed above on sets of the form $N \times I$ are unique when restricted to each segment $\{y\} \times I$, they must agree whenever two such sets $N \times I$ overlap. So we obtain a well-defined lift $\tilde{F}$ on all of $Y \times I$. This $\tilde{F}$ is continuous since it is continuous on each $N \times I$. And $\tilde{F}$ is unique since it is unique on each segment $\{y\} \times I$. $\square$

Now we turn to some applications of the calculation of $\pi_1(S^1)$, beginning with a proof of the Fundamental Theorem of Algebra.

Theorem 1.8. *Every nonconstant polynomial with coefficients in $\mathbb{C}$ has a root in $\mathbb{C}$.*

Proof: We may assume the polynomial is of the form $p(z) = z^n + a_1 z^{n-1} + \cdots + a_n$. If $p(z)$ has no roots in $\mathbb{C}$, then for each real number $r \geq 0$ the formula

$$f_r(s) = \frac{p(re^{2\pi i s})/p(r)}{|p(re^{2\pi i s})/p(r)|}$$

defines a loop in the unit circle $S^1 \subset \mathbb{C}$ based at 1. As r varies, f_r is a homotopy of loops based at 1. Since f_0 is the trivial loop, we deduce that the class $[f_r] \in \pi_1(S^1)$ is zero for all r. Now fix a large value of r, bigger than $|a_1| + \cdots + |a_n|$ and bigger than 1. Then for $|z| = r$ we have

$$|z^n| > (|a_1| + \cdots + |a_n|)|z^{n-1}| > |a_1 z^{n-1}| + \cdots + |a_n| \geq |a_1 z^{n-1} + \cdots + a_n|$$

From the inequality $|z^n| > |a_1 z^{n-1} + \cdots + a_n|$ it follows that the polynomial $p_t(z) = z^n + t(a_1 z^{n-1} + \cdots + a_n)$ has no roots on the circle $|z| = r$ when $0 \leq t \leq 1$. Replacing p by p_t in the formula for f_r above and letting t go from 1 to 0, we obtain a homotopy from the loop f_r to the loop $\omega_n(s) = e^{2\pi i n s}$. By Theorem 1.7, ω_n represents n times a generator of the infinite cyclic group $\pi_1(S^1)$. Since we have shown that $[\omega_n] = [f_r] = 0$, we conclude that $n = 0$. Thus the only polynomials without roots in $\mathbb{C}$ are constants. $\square$

Our next application is the Brouwer fixed point theorem in dimension 2.

Theorem 1.9. *Every continuous map $h : D^2 \to D^2$ has a fixed point, that is, a point $x \in D^2$ with $h(x) = x$.*

Here we are using the standard notation D^n for the closed unit disk in $\mathbb{R}^n$, all vectors x of length $|x| \leq 1$. Thus the boundary of D^n is the unit sphere S^{n-1}.

Proof: Suppose on the contrary that $h(x) \neq x$ for all $x \in D^2$. Then we can define a map $r : D^2 \to S^1$ by letting $r(x)$ be the point of S^1 where the ray in $\mathbb{R}^2$ starting at $h(x)$ and passing through x leaves D^2. Continuity of r is clear since small perturbations of x produce small perturbations of $h(x)$, hence also small perturbations of the ray through these two points.

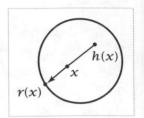

The crucial property of r, besides continuity, is that $r(x) = x$ if $x \in S^1$. Thus r is a retraction of D^2 onto S^1. We will show that no such retraction can exist.

Let f_0 be any loop in S^1. In D^2 there is a homotopy of f_0 to a constant loop, for example the linear homotopy $f_t(s) = (1 - t)f_0(s) + tx_0$ where x_0 is the basepoint of f_0. Since the retraction r is the identity on S^1, the composition rf_t is then a homotopy in S^1 from $rf_0 = f_0$ to the constant loop at x_0. But this contradicts the fact that $\pi_1(S^1)$ is nonzero. □

This theorem was first proved by Brouwer around 1910, quite early in the history of topology. Brouwer in fact proved the corresponding result for D^n, and we shall obtain this generalization in Corollary 2.15 using homology groups in place of π_1. One could also use the higher homotopy group π_n. Brouwer's original proof used neither homology nor homotopy groups, which had not been invented at the time. Instead it used the notion of degree for maps $S^n \to S^n$, which we shall define in §2.2 using homology but which Brouwer defined directly in more geometric terms.

These proofs are all arguments by contradiction, and so they show just the existence of fixed points without giving any clue as to how to find one in explicit cases. Our proof of the Fundamental Theorem of Algebra was similar in this regard. There exist other proofs of the Brouwer fixed point theorem that are somewhat more constructive, for example the elegant and quite elementary proof by Sperner in 1928, which is explained very nicely in [Aigner-Ziegler 1999].

The techniques used to calculate $\pi_1(S^1)$ can be applied to prove the Borsuk–Ulam theorem in dimension two:

Theorem 1.10. *For every continuous map $f : S^2 \to \mathbb{R}^2$ there exists a pair of antipodal points x and $-x$ in S^2 with $f(x) = f(-x)$.*

It may be that there is only one such pair of antipodal points x, $-x$, for example if f is simply orthogonal projection of the standard sphere $S^2 \subset \mathbb{R}^3$ onto a plane.

The Borsuk–Ulam theorem holds more generally for maps $S^n \to \mathbb{R}^n$, as we will show in Corollary 2B.7. The proof for $n = 1$ is easy since the difference $f(x) - f(-x)$ changes sign as x goes halfway around the circle, hence this difference must be zero for some x. For $n \geq 2$ the theorem is certainly less obvious. Is it apparent, for example, that at every instant there must be a pair of antipodal points on the surface of the earth having the same temperature and the same barometric pressure?

The theorem says in particular that there is no one-to-one continuous map from S^2 to $\mathbb{R}^2$, so S^2 is not homeomorphic to a subspace of $\mathbb{R}^2$, an intuitively obvious fact that is not easy to prove directly.

Proof: If the conclusion is false for $f:S^2 \to \mathbb{R}^2$, we can define a map $g:S^2 \to S^1$ by $g(x) = (f(x) - f(-x))/|f(x) - f(-x)|$. Define a loop η circling the equator of $S^2 \subset \mathbb{R}^3$ by $\eta(s) = (\cos 2\pi s, \sin 2\pi s, 0)$, and let $h:I \to S^1$ be the composed loop $g\eta$. Since $g(-x) = -g(x)$, we have the relation $h(s + 1/2) = -h(s)$ for all s in the interval $[0, 1/2]$. As we showed in the calculation of $\pi_1(S^1)$, the loop h can be lifted to a path $\tilde{h}:I \to \mathbb{R}$. The equation $h(s + 1/2) = -h(s)$ implies that $\tilde{h}(s + 1/2) = \tilde{h}(s) + q/2$ for some odd integer q that might conceivably depend on $s \in [0, 1/2]$. But in fact q is independent of s since by solving the equation $\tilde{h}(s + 1/2) = \tilde{h}(s) + q/2$ for q we see that q depends continuously on $s \in [0, 1/2]$, so q must be a constant since it is constrained to integer values. In particular, we have $\tilde{h}(1) = \tilde{h}(1/2) + q/2 = \tilde{h}(0) + q$. This means that h represents q times a generator of $\pi_1(S^1)$. Since q is odd, we conclude that h is not nullhomotopic. But h was the composition $g\eta:I \to S^2 \to S^1$, and η is obviously nullhomotopic in S^2, so $g\eta$ is nullhomotopic in S^1 by composing a nullhomotopy of η with g. Thus we have arrived at a contradiction. $\qquad\square$

Corollary 1.11. *Whenever S^2 is expressed as the union of three closed sets A_1, A_2, and A_3, then at least one of these sets must contain a pair of antipodal points $\{x, -x\}$.*

Proof: Let $d_i:S^2 \to \mathbb{R}$ measure distance to A_i, that is, $d_i(x) = \inf_{y \in A_i}|x - y|$. This is a continuous function, so we may apply the Borsuk–Ulam theorem to the map $S^2 \to \mathbb{R}^2$, $x \mapsto (d_1(x), d_2(x))$, obtaining a pair of antipodal points x and $-x$ with $d_1(x) = d_1(-x)$ and $d_2(x) = d_2(-x)$. If either of these two distances is zero, then x and $-x$ both lie in the same set A_1 or A_2 since these are closed sets. On the other hand, if the distances from x and $-x$ to A_1 and A_2 are both strictly positive, then x and $-x$ lie in neither A_1 nor A_2 so they must lie in A_3. $\qquad\square$

To see that the number 'three' in this result is best possible, consider a sphere inscribed in a tetrahedron. Projecting the four faces of the tetrahedron radially onto the sphere, we obtain a cover of S^2 by four closed sets, none of which contains a pair of antipodal points.

Assuming the higher-dimensional version of the Borsuk–Ulam theorem, the same arguments show that S^n cannot be covered by $n + 1$ closed sets without antipodal pairs of points, though it can be covered by $n + 2$ such sets, as the higher-dimensional analog of a tetrahedron shows. Even the case $n = 1$ is somewhat interesting: If the circle is covered by two closed sets, one of them must contain a pair of antipodal points. This is of course false for nonclosed sets since the circle is the union of two disjoint half-open semicircles.

The relation between the fundamental group of a product space and the fundamental groups of its factors is as simple as one could wish:

Proposition 1.12. $\pi_1(X \times Y)$ *is isomorphic to* $\pi_1(X) \times \pi_1(Y)$ *if* X *and* Y *are path-connected.*

Proof: A basic property of the product topology is that a map $f : Z \to X \times Y$ is continuous iff the maps $g : Z \to X$ and $h : Z \to Y$ defined by $f(z) = (g(z), h(z))$ are both continuous. Hence a loop f in $X \times Y$ based at (x_0, y_0) is equivalent to a pair of loops g in X and h in Y based at x_0 and y_0 respectively. Similarly, a homotopy f_t of a loop in $X \times Y$ is equivalent to a pair of homotopies g_t and h_t of the corresponding loops in X and Y. Thus we obtain a bijection $\pi_1(X \times Y, (x_0, y_0)) \approx \pi_1(X, x_0) \times \pi_1(Y, y_0)$, $[f] \mapsto ([g], [h])$. This is obviously a group homomorphism, and hence an isomorphism. $\qquad \square$

Example 1.13: The Torus. By the proposition we have an isomorphism $\pi_1(S^1 \times S^1) \approx \mathbb{Z} \times \mathbb{Z}$. Under this isomorphism a pair $(p, q) \in \mathbb{Z} \times \mathbb{Z}$ corresponds to a loop that winds p times around one S^1 factor of the torus and q times around the other S^1 factor, for example the loop $\omega_{pq}(s) = (\omega_p(s), \omega_q(s))$. Interestingly, this loop can be knotted, as the figure shows for the case $p = 3$, $q = 2$. The knots that arise in this fashion, the so-called *torus knots*, are studied in Example 1.24.

More generally, the n-dimensional torus, which is the product of n circles, has fundamental group isomorphic to the product of n copies of $\mathbb{Z}$. This follows by induction on n.

Induced Homomorphisms

Suppose $\varphi : X \to Y$ is a map taking the basepoint $x_0 \in X$ to the basepoint $y_0 \in Y$. For brevity we write $\varphi : (X, x_0) \to (Y, y_0)$ in this situation. Then φ induces a homomorphism $\varphi_* : \pi_1(X, x_0) \to \pi_1(Y, y_0)$, defined by composing loops $f : I \to X$ based at x_0 with φ, that is, $\varphi_*[f] = [\varphi f]$. This induced map φ_* is well-defined since a homotopy f_t of loops based at x_0 yields a composed homotopy φf_t of loops based at y_0, so $\varphi_*[f_0] = [\varphi f_0] = [\varphi f_1] = \varphi_*[f_1]$. Furthermore, φ_* is a homomorphism since $\varphi(f \cdot g) = (\varphi f) \cdot (\varphi g)$, both functions having the value $\varphi f(2s)$ for $0 \le s \le 1/2$ and the value $\varphi g(2s - 1)$ for $1/2 \le s \le 1$.

Two basic properties of induced homomorphisms are:

- $(\varphi \psi)_* = \varphi_* \psi_*$ for a composition $(X, x_0) \xrightarrow{\psi} (Y, y_0) \xrightarrow{\varphi} (Z, z_0)$.

- $\mathbb{1}_* = \mathbb{1}$, which is a concise way of saying that the identity map $\mathbb{1} : X \to X$ induces the identity map $\mathbb{1} : \pi_1(X, x_0) \to \pi_1(X, x_0)$.

The first of these follows from the fact that composition of maps is associative, so $(\varphi \psi) f = \varphi(\psi f)$, and the second is obvious. These two properties of induced homomorphisms are what makes the fundamental group a functor. The formal definition of a functor requires the introduction of certain other preliminary concepts, however, so we postpone this until it is needed in §2.3.

As an application we can deduce easily that if φ is a homeomorphism with inverse ψ then φ_* is an isomorphism with inverse ψ_* since $\varphi_* \psi_* = (\varphi \psi)_* = \mathbb{1}_* = \mathbb{1}$ and similarly $\psi_* \varphi_* = \mathbb{1}$. We will use this fact in the following calculation of the fundamental groups of higher-dimensional spheres:

Proposition 1.14. $\pi_1(S^n) = 0$ if $n \geq 2$.

The main step in the proof will be a general fact that will also play a key role in the next section:

Lemma 1.15. *If a space X is the union of a collection of path-connected open sets A_α each containing the basepoint $x_0 \in X$ and if each intersection $A_\alpha \cap A_\beta$ is path-connected, then every loop in X at x_0 is homotopic to a product of loops each of which is contained in a single A_α.*

Proof: Given a loop $f : I \to X$ at the basepoint x_0, we claim there is a partition $0 = s_0 < s_1 < \cdots < s_m = 1$ of I such that each subinterval $[s_{i-1}, s_i]$ is mapped by f to a single A_α. Namely, since f is continuous, each $s \in I$ has an open neighborhood V_s in I mapped by f to some A_α. We may in fact take V_s to be an interval whose closure is mapped to a single A_α. Compactness of I implies that a finite number of these intervals cover I. The endpoints of this finite set of intervals then define the desired partition of I.

Denote the A_α containing $f([s_{i-1}, s_i])$ by A_i, and let f_i be the path obtained by restricting f to $[s_{i-1}, s_i]$. Then f is the composition $f_1 \cdot \cdots \cdot f_m$ with f_i a path in A_i. Since we assume $A_i \cap A_{i+1}$ is path-connected, we may choose a path g_i in $A_i \cap A_{i+1}$ from x_0 to the point $f(s_i) \in A_i \cap A_{i+1}$. Consider the loop

$$(f_1 \cdot \overline{g}_1) \cdot (g_1 \cdot f_2 \cdot \overline{g}_2) \cdot (g_2 \cdot f_3 \cdot \overline{g}_3) \cdot \cdots \cdot (g_{m-1} \cdot f_m)$$

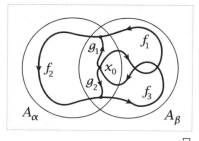

which is homotopic to f. This loop is a composition of loops each lying in a single A_i, the loops indicated by the parentheses. $\qquad \square$

Proof of Proposition 1.14: We can express S^n as the union of two open sets A_1 and A_2 each homeomorphic to $\mathbb{R}^n$ such that $A_1 \cap A_2$ is homeomorphic to $S^{n-1} \times \mathbb{R}$, for example by taking A_1 and A_2 to be the complements of two antipodal points in S^n. Choose a basepoint x_0 in $A_1 \cap A_2$. If $n \geq 2$ then $A_1 \cap A_2$ is path-connected. The lemma then applies to say that every loop in S^n based at x_0 is homotopic to a product of loops in A_1 or A_2. Both $\pi_1(A_1)$ and $\pi_1(A_2)$ are zero since A_1 and A_2 are homeomorphic to $\mathbb{R}^n$. Hence every loop in S^n is nullhomotopic. $\qquad \square$

‖ **Corollary 1.16.** $\mathbb{R}^2$ *is not homeomorphic to* $\mathbb{R}^n$ *for* $n \neq 2$.

Proof: Suppose $f : \mathbb{R}^2 \to \mathbb{R}^n$ is a homeomorphism. The case $n = 1$ is easily disposed of since $\mathbb{R}^2 - \{0\}$ is path-connected but the homeomorphic space $\mathbb{R}^n - \{f(0)\}$ is not path-connected when $n = 1$. When $n > 2$ we cannot distinguish $\mathbb{R}^2 - \{0\}$ from $\mathbb{R}^n - \{f(0)\}$ by the number of path-components, but we can distinguish them by their fundamental groups. Namely, for a point x in $\mathbb{R}^n$, the complement $\mathbb{R}^n - \{x\}$ is homeomorphic to $S^{n-1} \times \mathbb{R}$, so Proposition 1.12 implies that $\pi_1(\mathbb{R}^n - \{x\})$ is isomorphic to $\pi_1(S^{n-1}) \times \pi_1(\mathbb{R}) \approx \pi_1(S^{n-1})$. Hence $\pi_1(\mathbb{R}^n - \{x\})$ is $\mathbb{Z}$ for $n = 2$ and trivial for $n > 2$, using Proposition 1.14 in the latter case.   □

The more general statement that $\mathbb{R}^m$ is not homeomorphic to $\mathbb{R}^n$ if $m \neq n$ can be proved in the same way using either the higher homotopy groups or homology groups. In fact, nonempty open sets in $\mathbb{R}^m$ and $\mathbb{R}^n$ can be homeomorphic only if $m = n$, as we will show in Theorem 2.26 using homology.

Induced homomorphisms allow relations between spaces to be transformed into relations between their fundamental groups. Here is an illustration of this principle:

‖ **Proposition 1.17.** *If a space* X *retracts onto a subspace* A, *then the homomorphism* $i_* : \pi_1(A, x_0) \to \pi_1(X, x_0)$ *induced by the inclusion* $i : A \hookrightarrow X$ *is injective. If* A *is a deformation retract of* X, *then* i_* *is an isomorphism.*

Proof: If $r : X \to A$ is a retraction, then $ri = \mathbb{1}$, hence $r_* i_* = \mathbb{1}$, which implies that i_* is injective. If $r_t : X \to X$ is a deformation retraction of X onto A, so $r_0 = \mathbb{1}$, $r_t | A = \mathbb{1}$, and $r_1(X) \subset A$, then for any loop $f : I \to X$ based at $x_0 \in A$ the composition $r_t f$ gives a homotopy of f to a loop in A, so i_* is also surjective.   □

This gives another way of seeing that S^1 is not a retract of D^2, a fact we showed earlier in the proof of the Brouwer fixed point theorem, since the inclusion-induced map $\pi_1(S^1) \to \pi_1(D^2)$ is a homomorphism $\mathbb{Z} \to 0$ that cannot be injective.

The exact group-theoretic analog of a retraction is a homomorphism ρ of a group G onto a subgroup H such that ρ restricts to the identity on H. In the notation above, if we identify $\pi_1(A)$ with its image under i_*, then r_* is such a homomorphism from $\pi_1(X)$ onto the subgroup $\pi_1(A)$. The existence of a retracting homomorphism $\rho : G \to H$ is quite a strong condition on H. If H is a normal subgroup, it implies that G is the direct product of H and the kernel of ρ. If H is not normal, then G is what is called in group theory the semi-direct product of H and the kernel of ρ.

Recall from Chapter 0 the general definition of a homotopy as a family $\varphi_t : X \to Y$, $t \in I$, such that the associated map $\Phi : X \times I \to Y$, $\Phi(x, t) = \varphi_t(x)$, is continuous. If φ_t takes a subspace $A \subset X$ to a subspace $B \subset Y$ for all t, then we speak of a homotopy of maps of pairs, $\varphi_t : (X, A) \to (Y, B)$. In particular, a **basepoint-preserving homotopy**

$\varphi_t : (X, x_0) \to (Y, y_0)$ is the case that $\varphi_t(x_0) = y_0$ for all t. Another basic property of induced homomorphisms is their invariance under such homotopies:

- If $\varphi_t : (X, x_0) \to (Y, y_0)$ is a basepoint-preserving homotopy, then $\varphi_{0*} = \varphi_{1*}$.

This holds since $\varphi_{0*}[f] = [\varphi_0 f] = [\varphi_1 f] = \varphi_{1*}[f]$, the middle equality coming from the homotopy $\varphi_t f$.

There is a notion of homotopy equivalence for spaces with basepoints. One says $(X, x_0) \simeq (Y, y_0)$ if there are maps $\varphi : (X, x_0) \to (Y, y_0)$ and $\psi : (Y, y_0) \to (X, x_0)$ with homotopies $\varphi \psi \simeq \mathbb{1}$ and $\psi \varphi \simeq \mathbb{1}$ through maps fixing the basepoints. In this case the induced maps on π_1 satisfy $\varphi_* \psi_* = (\varphi \psi)_* = \mathbb{1}_* = \mathbb{1}$ and likewise $\psi_* \varphi_* = \mathbb{1}$, so φ_* and ψ_* are inverse isomorphisms $\pi_1(X, x_0) \approx \pi_1(Y, y_0)$. This somewhat formal argument gives another proof that a deformation retraction induces an isomorphism on fundamental groups, since if X deformation retracts onto A then $(X, x_0) \simeq (A, x_0)$ for any choice of basepoint $x_0 \in A$.

Having to pay so much attention to basepoints when dealing with the fundamental group is something of a nuisance. For homotopy equivalences one does not have to be quite so careful, as the conditions on basepoints can actually be dropped:

Proposition 1.18. *If $\psi : X \to Y$ is a homotopy equivalence, then the induced homomorphism $\varphi_* : \pi_1(X, x_0) \to \pi_1(Y, \varphi(x_0))$ is an isomorphism for all $x_0 \in X$.*

The proof will use a simple fact about homotopies that do not fix the basepoint:

Lemma 1.19. *If $\varphi_t : X \to Y$ is a homotopy and h is the path $\varphi_t(x_0)$ formed by the images of a basepoint $x_0 \in X$, then the three maps in the diagram at the right satisfy $\varphi_{0*} = \beta_h \varphi_{1*}$.*

$$\pi_1(X, x_0) \xrightarrow{\varphi_{1*}} \pi_1(Y, \varphi_1(x_0)) \xrightarrow{\beta_h} \pi_1(Y, \varphi_0(x_0)) \qquad \varphi_{0*}$$

Proof: Let h_t be the restriction of h to the interval $[0, t]$, with a reparametrization so that the domain of h_t is still $[0, 1]$. Explicitly, we can take $h_t(s) = h(ts)$. Then if f is a loop in X at the basepoint x_0, the product $h_t \cdot (\varphi_t f) \cdot \overline{h}_t$ gives a homotopy of loops at $\varphi_0(x_0)$. Restricting this homotopy to $t = 0$ and $t = 1$, we see that $\varphi_{0*}([f]) = \beta_h(\varphi_{1*}([f]))$. $\square$

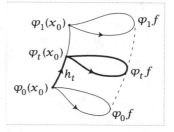

Proof of 1.18: Let $\psi : Y \to X$ be a homotopy-inverse for φ, so that $\varphi \psi \simeq \mathbb{1}$ and $\psi \varphi \simeq \mathbb{1}$. Consider the maps

$$\pi_1(X, x_0) \xrightarrow{\varphi_*} \pi_1(Y, \varphi(x_0)) \xrightarrow{\psi_*} \pi_1(X, \psi \varphi(x_0)) \xrightarrow{\varphi_*} \pi_1(Y, \varphi \psi \varphi(x_0))$$

The composition of the first two maps is an isomorphism since $\psi \varphi \simeq \mathbb{1}$ implies that $\psi_* \varphi_* = \beta_h$ for some h, by the lemma. In particular, since $\psi_* \varphi_*$ is an isomorphism,

φ_* is injective. The same reasoning with the second and third maps shows that ψ_* is injective. Thus the first two of the three maps are injections and their composition is an isomorphism, so the first map φ_* must be surjective as well as injective. $\square$

Exercises

1. Show that composition of paths satisfies the following cancellation property: If $f_0 \cdot g_0 \simeq f_1 \cdot g_1$ and $g_0 \simeq g_1$ then $f_0 \simeq f_1$.

2. Show that the change-of-basepoint homomorphism β_h depends only on the homotopy class of h.

3. For a path-connected space X, show that $\pi_1(X)$ is abelian iff all basepoint-change homomorphisms β_h depend only on the endpoints of the path h.

4. A subspace $X \subset \mathbb{R}^n$ is said to be *star-shaped* if there is a point $x_0 \in X$ such that, for each $x \in X$, the line segment from x_0 to x lies in X. Show that if a subspace $X \subset \mathbb{R}^n$ is locally star-shaped, in the sense that every point of X has a star-shaped neighborhood in X, then every path in X is homotopic in X to a piecewise linear path, that is, a path consisting of a finite number of straight line segments traversed at constant speed. Show this applies in particular when X is open or when X is a union of finitely many closed convex sets.

5. Show that for a space X, the following three conditions are equivalent:
 (a) Every map $S^1 \to X$ is homotopic to a constant map, with image a point.
 (b) Every map $S^1 \to X$ extends to a map $D^2 \to X$.
 (c) $\pi_1(X, x_0) = 0$ for all $x_0 \in X$.
Deduce that a space X is simply-connected iff all maps $S^1 \to X$ are homotopic. [In this problem, 'homotopic' means 'homotopic without regard to basepoints.']

6. We can regard $\pi_1(X, x_0)$ as the set of basepoint-preserving homotopy classes of maps $(S^1, s_0) \to (X, x_0)$. Let $[S^1, X]$ be the set of homotopy classes of maps $S^1 \to X$, with no conditions on basepoints. Thus there is a natural map $\Phi : \pi_1(X, x_0) \to [S^1, X]$ obtained by ignoring basepoints. Show that Φ is onto if X is path-connected, and that $\Phi([f]) = \Phi([g])$ iff $[f]$ and $[g]$ are conjugate in $\pi_1(X, x_0)$. Hence Φ induces a one-to-one correspondence between $[S^1, X]$ and the set of conjugacy classes in $\pi_1(X)$, when X is path-connected.

7. Define $f : S^1 \times I \to S^1 \times I$ by $f(\theta, s) = (\theta + 2\pi s, s)$, so f restricts to the identity on the two boundary circles of $S^1 \times I$. Show that f is homotopic to the identity by a homotopy f_t that is stationary on one of the boundary circles, but not by any homotopy f_t that is stationary on both boundary circles. [Consider what f does to the path $s \mapsto (\theta_0, s)$ for fixed $\theta_0 \in S^1$.]

8. Does the Borsuk–Ulam theorem hold for the torus? In other words, for every map $f : S^1 \times S^1 \to \mathbb{R}^2$ must there exist $(x, y) \in S^1 \times S^1$ such that $f(x, y) = f(-x, -y)$?

9. Let A_1, A_2, A_3 be compact sets in $\mathbb{R}^3$. Use the Borsuk–Ulam theorem to show that there is one plane $P \subset \mathbb{R}^3$ that simultaneously divides each A_i into two pieces of equal measure.

10. From the isomorphism $\pi_1(X \times Y, (x_0, y_0)) \approx \pi_1(X, x_0) \times \pi_1(Y, y_0)$ it follows that loops in $X \times \{y_0\}$ and $\{x_0\} \times Y$ represent commuting elements of $\pi_1(X \times Y, (x_0, y_0))$. Construct an explicit homotopy demonstrating this.

11. If X_0 is the path-component of a space X containing the basepoint x_0, show that the inclusion $X_0 \hookrightarrow X$ induces an isomorphism $\pi_1(X_0, x_0) \to \pi_1(X, x_0)$.

12. Show that every homomorphism $\pi_1(S^1) \to \pi_1(S^1)$ can be realized as the induced homomorphism φ_* of a map $\varphi : S^1 \to S^1$.

13. Given a space X and a path-connected subspace A containing the basepoint x_0, show that the map $\pi_1(A, x_0) \to \pi_1(X, x_0)$ induced by the inclusion $A \hookrightarrow X$ is surjective iff every path in X with endpoints in A is homotopic to a path in A.

14. Show that the isomorphism $\pi_1(X \times Y) \approx \pi_1(X) \times \pi_1(Y)$ in Proposition 1.12 is given by $[f] \mapsto (p_{1*}([f]), p_{2*}([f]))$ where p_1 and p_2 are the projections of $X \times Y$ onto its two factors.

15. Given a map $f : X \to Y$ and a path $h : I \to X$ from x_0 to x_1, show that $f_* \beta_h = \beta_{fh} f_*$ in the diagram at the right.

$$\pi_1(X, x_1) \xrightarrow{\ \beta_h\ } \pi_1(X, x_0)$$
$$f_* \downarrow \qquad\qquad \downarrow f_*$$
$$\pi_1(Y, f(x_1)) \xrightarrow{\ \beta_{fh}\ } \pi_1(Y, f(x_0))$$

16. Show that there are no retractions $r : X \to A$ in the following cases:
(a) $X = \mathbb{R}^3$ with A any subspace homeomorphic to S^1.
(b) $X = S^1 \times D^2$ with A its boundary torus $S^1 \times S^1$.
(c) $X = S^1 \times D^2$ and A the circle shown in the figure.
(d) $X = D^2 \vee D^2$ with A its boundary $S^1 \vee S^1$.

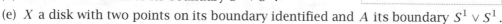

(e) X a disk with two points on its boundary identified and A its boundary $S^1 \vee S^1$.
(f) X the Möbius band and A its boundary circle.

17. Construct infinitely many nonhomotopic retractions $S^1 \vee S^1 \to S^1$.

18. Using Lemma 1.15, show that if a space X is obtained from a path-connected subspace A by attaching a cell e^n with $n \geq 2$, then the inclusion $A \hookrightarrow X$ induces a surjection on π_1. Apply this to show:
(a) The wedge sum $S^1 \vee S^2$ has fundamental group $\mathbb{Z}$.
(b) For a path-connected CW complex X the inclusion map $X^1 \hookrightarrow X$ of its 1-skeleton induces a surjection $\pi_1(X^1) \to \pi_1(X)$. [For the case that X has infinitely many cells, see Proposition A.1 in the Appendix.]

19. Show that if X is a path-connected 1-dimensional CW complex with basepoint x_0 a 0-cell, then every loop in X is homotopic to a loop consisting of a finite sequence of edges traversed monotonically. [See the proof of Lemma 1.15. This exercise gives an elementary proof that $\pi_1(S^1)$ is cyclic generated by the standard loop winding once around the circle. The more difficult part of the calculation of $\pi_1(S^1)$ is therefore the fact that no iterate of this loop is nullhomotopic.]

20. Suppose $f_t : X \to X$ is a homotopy such that f_0 and f_1 are each the identity map. Use Lemma 1.19 to show that for any $x_0 \in X$, the loop $f_t(x_0)$ represents an element of the center of $\pi_1(X, x_0)$. [One can interpret the result as saying that a loop represents an element of the center of $\pi_1(X)$ if it extends to a loop of maps $X \to X$.]

1.2 Van Kampen's Theorem

The van Kampen theorem gives a method for computing the fundamental groups of spaces that can be decomposed into simpler spaces whose fundamental groups are already known. By systematic use of this theorem one can compute the fundamental groups of a very large number of spaces. We shall see for example that for every group G there is a space X_G whose fundamental group is isomorphic to G.

To give some idea of how one might hope to compute fundamental groups by decomposing spaces into simpler pieces, let us look at an example. Consider the space X formed by two circles A and B intersecting in a single point, which we choose as the basepoint x_0. By our preceding calculations we know that $\pi_1(A)$ is infinite cyclic, generated by a loop a that goes once around A. Similarly, $\pi_1(B)$ is a copy of $\mathbb{Z}$ generated by a loop b going once around B. Each product of powers of a and b then gives  an element of $\pi_1(X)$. For example, the product $a^5 b^2 a^{-3} b a^2$ is the loop that goes five times around A, then twice around B, then three times around A in the opposite direction, then once around B, then twice around A. The set of all words like this consisting of powers of a alternating with powers of b forms a group usually denoted $\mathbb{Z} * \mathbb{Z}$. Multiplication in this group is defined just as one would expect, for example $(b^4 a^5 b^2 a^{-3})(a^4 b^{-1} a b^3) = b^4 a^5 b^2 a b^{-1} a b^3$. The identity element is the empty word, and inverses are what they have to be, for example $(a b^2 a^{-3} b^{-4})^{-1} = b^4 a^3 b^{-2} a^{-1}$. It would be very nice if such words in a and b corresponded exactly to elements of $\pi_1(X)$, so that $\pi_1(X)$ was isomorphic to the group $\mathbb{Z} * \mathbb{Z}$. The van Kampen theorem will imply that this is indeed the case.

Similarly, if X is the union of three circles touching at a single point, the van Kampen theorem will imply that $\pi_1(X)$ is $\mathbb{Z} * \mathbb{Z} * \mathbb{Z}$, the group consisting of words in powers of three letters a, b, c. The generalization to a union of any number of circles touching at one point will also follow.

The group $\mathbb{Z} * \mathbb{Z}$ is an example of a general construction called the *free product* of groups. The statement of van Kampen's theorem will be in terms of free products, so before stating the theorem we will make an algebraic digression to describe the construction of free products in some detail.

Free Products of Groups

Suppose one is given a collection of groups G_α and one wishes to construct a single group containing all these groups as subgroups. One way to do this would be to take the product group $\prod_\alpha G_\alpha$, whose elements can be regarded as the functions $\alpha \mapsto g_\alpha \in G_\alpha$. Or one could restrict to functions taking on nonidentity values at most finitely often, forming the direct sum $\bigoplus_\alpha G_\alpha$. Both these constructions produce groups containing all the G_α's as subgroups, but with the property that elements of different subgroups G_α commute with each other. In the realm of nonabelian groups this commutativity is unnatural, and so one would like a 'nonabelian' version of $\prod_\alpha G_\alpha$ or $\bigoplus_\alpha G_\alpha$. Since the sum $\bigoplus_\alpha G_\alpha$ is smaller and presumably simpler than $\prod_\alpha G_\alpha$, it should be easier to construct a nonabelian version of $\bigoplus_\alpha G_\alpha$, and this is what the free product $*_\alpha G_\alpha$ achieves.

Here is the precise definition. As a set, the free product $*_\alpha G_\alpha$ consists of all words $g_1 g_2 \cdots g_m$ of arbitrary finite length $m \geq 0$, where each letter g_i belongs to a group G_{α_i} and is not the identity element of G_{α_i}, and adjacent letters g_i and g_{i+1} belong to different groups G_α, that is, $\alpha_i \neq \alpha_{i+1}$. Words satisfying these conditions are called *reduced*, the idea being that unreduced words can always be simplified to reduced words by writing adjacent letters that lie in the same G_{α_i} as a single letter and by canceling trivial letters. The empty word is allowed, and will be the identity element of $*_\alpha G_\alpha$. The group operation in $*_\alpha G_\alpha$ is juxtaposition, $(g_1 \cdots g_m)(h_1 \cdots h_n) = g_1 \cdots g_m h_1 \cdots h_n$. This product may not be reduced, however: If g_m and h_1 belong to the same G_α, they should be combined into a single letter $(g_m h_1)$ according to the multiplication in G_α, and if this new letter $g_m h_1$ happens to be the identity of G_α, it should be canceled from the product. This may allow g_{m-1} and h_2 to be combined, and possibly canceled too. Repetition of this process eventually produces a reduced word. For example, in the product $(g_1 \cdots g_m)(g_m^{-1} \cdots g_1^{-1})$ everything cancels and we get the identity element of $*_\alpha G_\alpha$, the empty word.

Verifying directly that this multiplication is associative would be rather tedious, but there is an indirect approach that avoids most of the work. Let W be the set of reduced words $g_1 \cdots g_m$ as above, including the empty word. To each $g \in G_\alpha$ we associate the function $L_g : W \to W$ given by multiplication on the left, $L_g(g_1 \cdots g_m) = g g_1 \cdots g_m$ where we combine g with g_1 if $g_1 \in G_\alpha$ to make $g g_1 \cdots g_m$ a reduced word. A key property of the association $g \mapsto L_g$ is the formula $L_{gg'} = L_g L_{g'}$ for $g, g' \in G_\alpha$, that is, $g(g'(g_1 \cdots g_m)) = (gg')(g_1 \cdots g_m)$. This special case of associativity follows rather trivially from associativity in G_α. The formula $L_{gg'} = L_g L_{g'}$ implies that L_g is invertible with inverse $L_{g^{-1}}$. Therefore the association $g \mapsto L_g$ defines a homomorphism from G_α to the group $P(W)$ of all permutations of W. More generally, we can define $L : W \to P(W)$ by $L(g_1 \cdots g_m) = L_{g_1} \cdots L_{g_m}$ for each reduced word $g_1 \cdots g_m$. This function L is injective since the permutation $L(g_1 \cdots g_m)$ sends the empty word to $g_1 \cdots g_m$. The product operation in W corresponds under L to composition in $P(W)$, because of the relation $L_{gg'} = L_g L_{g'}$. Since composition in $P(W)$ is associative, we conclude that the product in W is associative.

In particular, we have the free product $\mathbb{Z} * \mathbb{Z}$ as described earlier. This is an example of a *free group*, the free product of any number of copies of $\mathbb{Z}$, finite or infinite. The elements of a free group are uniquely representable as reduced words in powers of generators for the various copies of $\mathbb{Z}$, with one generator for each $\mathbb{Z}$, just as in the case of $\mathbb{Z} * \mathbb{Z}$. These generators are called a *basis* for the free group, and the number of basis elements is the *rank* of the free group. The abelianization of a free group is a free abelian group with basis the same set of generators, so since the rank of a free abelian group is well-defined, independent of the choice of basis, the same is true for the rank of a free group.

An interesting example of a free product that is not a free group is $\mathbb{Z}_2 * \mathbb{Z}_2$. This is like $\mathbb{Z} * \mathbb{Z}$ but simpler since $a^2 = e = b^2$, so powers of a and b are not needed, and $\mathbb{Z}_2 * \mathbb{Z}_2$ consists of just the alternating words in a and b: $a, b, ab, ba, aba, bab,$ $abab, baba, ababa, \cdots$, together with the empty word. The structure of $\mathbb{Z}_2 * \mathbb{Z}_2$ can be elucidated by looking at the homomorphism $\varphi : \mathbb{Z}_2 * \mathbb{Z}_2 \to \mathbb{Z}_2$ associating to each word its length mod 2. Obviously φ is surjective, and its kernel consists of the words of even length. These form an infinite cyclic subgroup generated by ab since $ba = (ab)^{-1}$ in $\mathbb{Z}_2 * \mathbb{Z}_2$. In fact, $\mathbb{Z}_2 * \mathbb{Z}_2$ is the semi-direct product of the subgroups $\mathbb{Z}$ and $\mathbb{Z}_2$ generated by ab and a, with the conjugation relation $a(ab)a^{-1} = (ab)^{-1}$. This group is sometimes called the infinite dihedral group.

For a general free product $*_\alpha G_\alpha$, each group G_α is naturally identified with a subgroup of $*_\alpha G_\alpha$, the subgroup consisting of the empty word and the nonidentity one-letter words $g \in G_\alpha$. From this viewpoint the empty word is the common identity element of all the subgroups G_α, which are otherwise disjoint. A consequence of associativity is that any product $g_1 \cdots g_m$ of elements g_i in the groups G_α has a unique reduced form, the element of $*_\alpha G_\alpha$ obtained by performing the multiplications in any order. Any sequence of reduction operations on an unreduced product $g_1 \cdots g_m$, combining adjacent letters g_i and g_{i+1} that lie in the same G_α or canceling a g_i that is the identity, can be viewed as a way of inserting parentheses into $g_1 \cdots g_m$ and performing the resulting sequence of multiplications. Thus associativity implies that any two sequences of reduction operations performed on the same unreduced word always yield the same reduced word.

A basic property of the free product $*_\alpha G_\alpha$ is that any collection of homomorphisms $\varphi_\alpha : G_\alpha \to H$ extends uniquely to a homomorphism $\varphi : *_\alpha G_\alpha \to H$. Namely, the value of φ on a word $g_1 \cdots g_n$ with $g_i \in G_{\alpha_i}$ must be $\varphi_{\alpha_1}(g_1) \cdots \varphi_{\alpha_n}(g_n)$, and using this formula to define φ gives a well-defined homomorphism since the process of reducing an unreduced product in $*_\alpha G_\alpha$ does not affect its image under φ. For example, for a free product $G * H$ the inclusions $G \hookrightarrow G \times H$ and $H \hookrightarrow G \times H$ induce a surjective homomorphism $G * H \to G \times H$.

The van Kampen Theorem

Suppose a space X is decomposed as the union of a collection of path-connected open subsets A_α, each of which contains the basepoint $x_0 \in X$. By the remarks in the preceding paragraph, the homomorphisms $j_\alpha : \pi_1(A_\alpha) \to \pi_1(X)$ induced by the inclusions $A_\alpha \hookrightarrow X$ extend to a homomorphism $\Phi : *_\alpha \pi_1(A_\alpha) \to \pi_1(X)$. The van Kampen theorem will say that Φ is very often surjective, but we can expect Φ to have a nontrivial kernel in general. For if $i_{\alpha\beta} : \pi_1(A_\alpha \cap A_\beta) \to \pi_1(A_\alpha)$ is the homomorphism induced by the inclusion $A_\alpha \cap A_\beta \hookrightarrow A_\alpha$ then $j_\alpha i_{\alpha\beta} = j_\beta i_{\beta\alpha}$, both these compositions being induced by the inclusion $A_\alpha \cap A_\beta \hookrightarrow X$, so the kernel of Φ contains all the elements of the form $i_{\alpha\beta}(\omega) i_{\beta\alpha}(\omega)^{-1}$ for $\omega \in \pi_1(A_\alpha \cap A_\beta)$. Van Kampen's theorem asserts that under fairly broad hypotheses this gives a full description of Φ:

Theorem 1.20. *If X is the union of path-connected open sets A_α each containing the basepoint $x_0 \in X$ and if each intersection $A_\alpha \cap A_\beta$ is path-connected, then the homomorphism $\Phi : *_\alpha \pi_1(A_\alpha) \to \pi_1(X)$ is surjective. If in addition each intersection $A_\alpha \cap A_\beta \cap A_\gamma$ is path-connected, then the kernel of Φ is the normal subgroup N generated by all elements of the form $i_{\alpha\beta}(\omega) i_{\beta\alpha}(\omega)^{-1}$ for $\omega \in \pi_1(A_\alpha \cap A_\beta)$, and hence Φ induces an isomorphism $\pi_1(X) \approx *_\alpha \pi_1(A_\alpha)/N$.*

Example 1.21: Wedge Sums. In Chapter 0 we defined the wedge sum $\bigvee_\alpha X_\alpha$ of a collection of spaces X_α with basepoints $x_\alpha \in X_\alpha$ to be the quotient space of the disjoint union $\coprod_\alpha X_\alpha$ in which all the basepoints x_α are identified to a single point. If each x_α is a deformation retract of an open neighborhood U_α in X_α, then X_α is a deformation retract of its open neighborhood $A_\alpha = X_\alpha \bigvee_{\beta \neq \alpha} U_\beta$. The intersection of two or more distinct A_α's is $\bigvee_\alpha U_\alpha$, which deformation retracts to a point. Van Kampen's theorem then implies that $\Phi : *_\alpha \pi_1(X_\alpha) \to \pi_1(\bigvee_\alpha X_\alpha)$ is an isomorphism.

Thus for a wedge sum $\bigvee_\alpha S_\alpha^1$ of circles, $\pi_1(\bigvee_\alpha S_\alpha^1)$ is a free group, the free product of copies of $\mathbb{Z}$, one for each circle S_α^1. In particular, $\pi_1(S^1 \vee S^1)$ is the free group $\mathbb{Z} * \mathbb{Z}$, as in the example at the beginning of this section.

It is true more generally that the fundamental group of any connected graph is free, as we show in §1.A. Here is an example illustrating the general technique.

Example 1.22. Let X be the graph shown in the figure, consisting of the twelve edges of a cube. The seven heavily shaded edges form a maximal tree $T \subset X$, a contractible subgraph containing all the vertices of X. We claim that $\pi_1(X)$ is the free product of five copies of $\mathbb{Z}$, one for each edge not in T. To deduce this from van 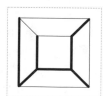 Kampen's theorem, choose for each edge e_α of $X - T$ an open neighborhood A_α of $T \cup e_\alpha$ in X that deformation retracts onto $T \cup e_\alpha$. The intersection of two or more A_α's deformation retracts onto T, hence is contractible. The A_α's form a cover of X satisfying the hypotheses of van Kampen's theorem, and since the intersection of any two of them is simply-connected we obtain an isomorphism $\pi_1(X) \approx *_\alpha \pi_1(A_\alpha)$. Each A_α deformation retracts onto a circle, so $\pi_1(X)$ is free on five generators, as claimed. As explicit generators we can choose for each edge e_α of $X - T$ a loop f_α that starts at a basepoint in T, travels in T to one end of e_α, then across e_α, then back to the basepoint along a path in T.

Van Kampen's theorem is often applied when there are just two sets A_α and A_β in the cover of X, so the condition on triple intersections $A_\alpha \cap A_\beta \cap A_y$ is superfluous and one obtains an isomorphism $\pi_1(X) \approx (\pi_1(A_\alpha) * \pi_1(A_\beta))/N$, under the assumption that $A_\alpha \cap A_\beta$ is path-connected. The proof in this special case is virtually identical with the proof in the general case, however.

One can see that the intersections $A_\alpha \cap A_\beta$ need to be path-connected by considering the example of S^1 decomposed as the union of two open arcs. In this case Φ is not surjective. For an example showing that triple intersections $A_\alpha \cap A_\beta \cap A_y$ need to be path-connected, let X be the suspension of three points a, b, c, and let A_α, A_β, and A_y be the complements of these three points. The theorem does apply to the covering $\{A_\alpha, A_\beta\}$, so there are isomorphisms $\pi_1(X) \approx \pi_1(A_\alpha) * \pi_1(A_\beta) \approx \mathbb{Z} * \mathbb{Z}$ since $A_\alpha \cap A_\beta$ is contractible. If we tried to use the covering $\{A_\alpha, A_\beta, A_y\}$, which has each of the 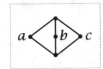 twofold intersections path-connected but not the triple intersection, then we would get $\pi_1(X) \approx \mathbb{Z} * \mathbb{Z} * \mathbb{Z}$, but this is not isomorphic to $\mathbb{Z} * \mathbb{Z}$ since it has a different abelianization.

Proof of van Kampen's theorem: We have already proved the first part of the theorem concerning surjectivity of Φ in Lemma 1.15. The harder part of the proof is to show that the kernel of Φ is N. It may clarify matters to introduce some terminology. By a *factorization* of an element $[f] \in \pi_1(X)$ we shall mean a formal product $[f_1] \cdots [f_k]$ where:

- Each f_i is a loop in some A_α at the basepoint x_0, and $[f_i] \in \pi_1(A_\alpha)$ is the homotopy class of f_i.
- The loop f is homotopic to $f_1 \cdot \cdots \cdot f_k$ in X.

A factorization of $[f]$ is thus a word in $*_\alpha \pi_1(A_\alpha)$, possibly unreduced, that is mapped to $[f]$ by Φ. Surjectivity of Φ is equivalent to saying that every $[f] \in \pi_1(X)$ has a factorization.

We will be concerned with the uniqueness of factorizations. Call two factorizations of $[f]$ *equivalent* if they are related by a sequence of the following two sorts of moves or their inverses:

- Combine adjacent terms $[f_i][f_{i+1}]$ into a single term $[f_i \cdot f_{i+1}]$ if $[f_i]$ and $[f_{i+1}]$ lie in the same group $\pi_1(A_\alpha)$.
- Regard the term $[f_i] \in \pi_1(A_\alpha)$ as lying in the group $\pi_1(A_\beta)$ rather than $\pi_1(A_\alpha)$ if f_i is a loop in $A_\alpha \cap A_\beta$.

The first move does not change the element of $*_\alpha \pi_1(A_\alpha)$ defined by the factorization. The second move does not change the image of this element in the quotient group $Q = *_\alpha \pi_1(A_\alpha)/N$, by the definition of N. So equivalent factorizations give the same element of Q.

If we can show that any two factorizations of $[f]$ are equivalent, this will say that the map $Q \to \pi_1(X)$ induced by Φ is injective, hence the kernel of Φ is exactly N, and the proof will be complete.

Let $[f_1] \cdots [f_k]$ and $[f_1'] \cdots [f_\ell']$ be two factorizations of $[f]$. The composed paths $f_1 \cdot \cdots \cdot f_k$ and $f_1' \cdot \cdots \cdot f_\ell'$ are then homotopic, so let $F : I \times I \to X$ be a homotopy from $f_1 \cdot \cdots \cdot f_k$ to $f_1' \cdot \cdots \cdot f_\ell'$. There exist partitions $0 = s_0 < s_1 < \cdots < s_m = 1$ and $0 = t_0 < t_1 < \cdots < t_n = 1$ such that each rectangle $R_{ij} = [s_{i-1}, s_i] \times [t_{j-1}, t_j]$ is mapped by F into a single A_α, which we label A_{ij}. These partitions may be obtained by covering $I \times I$ by finitely many rectangles $[a, b] \times [c, d]$ each mapping to a single A_α, using a compactness argument, then partitioning $I \times I$ by the union of all the horizontal and vertical lines containing edges of these rectangles. We may assume the s-partition subdivides the partitions giving the products

$f_1 \cdot \cdots \cdot f_k$ and $f_1' \cdot \cdots \cdot f_\ell'$. Since F maps a neighborhood of R_{ij} to A_{ij}, we may perturb the vertical sides of the rectangles R_{ij} so that each point of $I \times I$ lies in at most three R_{ij}'s. We may assume there are at least three rows of rectangles, so we can do this perturbation just on the rectangles in the intermediate rows, leaving the top and bottom rows unchanged. Let us relabel

9	10	11	12
5	6	7	8
1	2	3	4

the new rectangles $R_1, R_2, \cdots, R_{mn}$, ordering them as in the figure.

If γ is a path in $I \times I$ from the left edge to the right edge, then the restriction $F | \gamma$ is a loop at the basepoint x_0 since F maps both the left and right edges of $I \times I$ to x_0. Let γ_r be the path separating the first r rectangles $R_1, \cdots, R_r$ from the remaining rectangles. Thus γ_0 is the bottom edge of $I \times I$ and γ_{mn} is the top edge. We pass from γ_r to γ_{r+1} by pushing across the rectangle R_{r+1}.

Let us call the corners of the R_r's $vertices$. For each vertex v with $F(v) \neq x_0$ we can choose a path g_v from x_0 to $F(v)$ that lies in the intersection of the two or three A_{ij}'s corresponding to the R_r's containing v, since we assume the intersection of any two or three A_{ij}'s is path-connected. Then we obtain a factorization of $[F | \gamma_r]$ by inserting the appropriate paths $\overline{g}_v g_v$ into $F | \gamma_r$ at successive vertices, as in the proof of surjectivity of Φ in Lemma 1.15. This factorization depends on certain choices, since the loop corresponding to a segment between two successive vertices can lie in two different A_{ij}'s when there are two different rectangles R_{ij} containing this edge. Different choices of these A_{ij}'s change the factorization of $[F | \gamma_r]$ to an equivalent factorization, however. Furthermore, the factorizations associated to successive paths γ_r and γ_{r+1} are equivalent since pushing γ_r across R_{r+1} to γ_{r+1} changes $F | \gamma_r$ to $F | \gamma_{r+1}$ by a homotopy within the A_{ij} corresponding to R_{r+1}, and we can choose this A_{ij} for all the segments of γ_r and γ_{r+1} in R_{r+1}.

We can arrange that the factorization associated to γ_0 is equivalent to the factorization $[f_1] \cdots [f_k]$ by choosing the path g_v for each vertex v along the lower edge of $I \times I$ to lie not just in the two A_{ij}'s corresponding to the R_s's containing v, but also to lie in the A_α for the f_i containing v in its domain. In case v is the common endpoint of the domains of two consecutive f_i's we have $F(v) = x_0$, so there is no need to choose a g_v for such v's. In similar fashion we may assume that the factorization associated to the final γ_{mn} is equivalent to $[f_1'] \cdots [f_\ell']$. Since the factorizations associated to all the γ_r's are equivalent, we conclude that the factorizations $[f_1] \cdots [f_k]$ and $[f_1'] \cdots [f_\ell']$ are equivalent. $\qquad\qquad\square$

Example 1.23: Linking of Circles. We can apply van Kampen's theorem to calculate the fundamental groups of three spaces discussed in the introduction to this chapter, the complements in $\mathbb{R}^3$ of a single circle, two unlinked circles, and two linked circles.

The complement $\mathbb{R}^3 - A$ of a single circle A deformation retracts onto a wedge sum $S^1 \vee S^2$ embedded in $\mathbb{R}^3 - A$ as shown in the first of the two figures at the right. It may be easier to see that $\mathbb{R}^3 - A$ deformation retracts onto the union of S^2 with a diameter, as in the second figure, 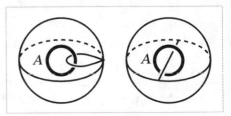 where points outside S^2 deformation retract onto S^2, and points inside S^2 and not in A can be pushed away from A toward S^2 or the diameter. Having this deformation retraction in mind, one can then see how it must be modified if the two endpoints of the diameter are gradually moved toward each other along the equator until they coincide, forming the S^1 summand of $S^1 \vee S^2$. Another way of seeing the deformation retraction of $\mathbb{R}^3 - A$ onto $S^1 \vee S^2$ is to note first that an open ε-neighborhood of $S^1 \vee S^2$ obviously deformation retracts onto $S^1 \vee S^2$ if ε is sufficiently small. Then observe that this neighborhood is homeomorphic to $\mathbb{R}^3 - A$ by a homeomorphism that is the identity on $S^1 \vee S^2$. In fact, the neighborhood can be gradually enlarged by homeomorphisms until it becomes all of $\mathbb{R}^3 - A$.

In any event, once we see that $\mathbb{R}^3 - A$ deformation retracts to $S^1 \vee S^2$, then we immediately obtain isomorphisms $\pi_1(\mathbb{R}^3 - A) \approx \pi_1(S^1 \vee S^2) \approx \mathbb{Z}$ since $\pi_1(S^2) = 0$.

In similar fashion, the complement $\mathbb{R}^3 - (A \cup B)$ of two unlinked circles A and B deformation retracts onto $S^1 \vee S^1 \vee S^2 \vee S^2$, as in the figure to the right. From 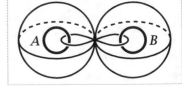 this we get $\pi_1(\mathbb{R}^3 - (A \cup B)) \approx \mathbb{Z} * \mathbb{Z}$. On the other hand, if A  and B are linked, then $\mathbb{R}^3 - (A \cup B)$ deformation retracts onto the wedge sum of S^2 and a torus $S^1 \times S^1$ separating A and B, as shown in the figure to the left, hence $\pi_1(\mathbb{R}^3 - (A \cup B)) \approx \pi_1(S^1 \times S^1) \approx \mathbb{Z} \times \mathbb{Z}$.

Example 1.24: Torus Knots. For relatively prime positive integers m and n, the **torus knot** $K = K_{m,n} \subset \mathbb{R}^3$ is the image of the embedding $f : S^1 \to S^1 \times S^1 \subset \mathbb{R}^3$, $f(z) = (z^m, z^n)$, where the torus $S^1 \times S^1$ is embedded in $\mathbb{R}^3$ in the standard way. The knot K winds around the torus a total of m times in the longitudinal direction and n times in the meridional direction, as shown in the figure for the cases $(m, n) = (2, 3)$ and $(3, 4)$. One needs to assume that m and n are relatively prime in order

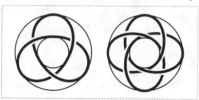

for the map f to be injective. Without this assumption f would be d-to-1 where d is the greatest common divisor of m and n, and the image of f would be the knot $K_{m/d, n/d}$. One could also allow negative values for m or n, but this would only change K to a mirror-image knot.

Let us compute $\pi_1(\mathbb{R}^3 - K)$. It is slightly easier to do the calculation with $\mathbb{R}^3$ replaced by its one-point compactification S^3. An application of van Kampen's theorem shows that this does not affect π_1. Namely, write $S^3 - K$ as the union of $\mathbb{R}^3 - K$ and an open ball B formed by the compactification point together with the complement of a large closed ball in $\mathbb{R}^3$ containing K. Both B and $B \cap (\mathbb{R}^3 - K)$ are simply-connected, the latter space being homeomorphic to $S^2 \times \mathbb{R}$. Hence van Kampen's theorem implies that the inclusion $\mathbb{R}^3 - K \hookrightarrow S^3 - K$ induces an isomorphism on π_1.

We compute $\pi_1(S^3 - K)$ by showing that it deformation retracts onto a 2-dimensional complex $X = X_{m,n}$ homeomorphic to the quotient space of a cylinder $S^1 \times I$ under the identifications $(z, 0) \sim (e^{2\pi i/m} z, 0)$ and $(z, 1) \sim (e^{2\pi i/n} z, 1)$. If we let X_m and X_n be the two halves of X formed by the quotients of $S^1 \times [0, 1/2]$ and $S^1 \times [1/2, 1]$, then X_m and X_n are the mapping cylinders of $z \mapsto z^m$ and $z \mapsto z^n$. The intersection $X_m \cap X_n$ is the circle $S^1 \times \{1/2\}$, the domain end of each mapping cylinder.

To obtain an embedding of X in $S^3 - K$ as a deformation retract we will use the standard decomposition of S^3 into two solid tori $S^1 \times D^2$ and $D^2 \times S^1$, the result of regarding S^3 as $\partial D^4 = \partial (D^2 \times D^2) = \partial D^2 \times D^2 \cup D^2 \times \partial D^2$. Geometrically, the first solid torus $S^1 \times D^2$ can be identified with the compact region in $\mathbb{R}^3$ bounded by the standard torus $S^1 \times S^1$ containing K, and the second solid torus $D^2 \times S^1$ is then the closure of the complement of the first solid torus, together with the compactification point at infinity. Notice that meridional circles in $S^1 \times S^1$ bound disks in the first solid torus, while it is longitudinal circles that bound disks in the second solid torus.

In the first solid torus, K intersects each of the meridian circles $\{x\} \times \partial D^2$ in m equally spaced points, as indicated in the figure at the right, which shows a meridian disk $\{x\} \times D^2$. These m points can be separated by a union of m radial line segments. Letting x vary, these radial segments then trace out a copy of the mapping cylinder X_m in the first solid torus. Sym-

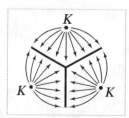

metrically, there is a copy of the other mapping cylinder X_n in the second solid torus.

The complement of K in the first solid torus deformation retracts onto X_m by flowing within each meridian disk as shown. In similar fashion the complement of K in the second solid torus deformation retracts onto X_n. These two deformation retractions do not agree on their common domain of definition $S^1 \times S^1 - K$, but this is easy to correct by distorting the flows in the two solid tori so that in $S^1 \times S^1 - K$ both flows are orthogonal to K. After this modification we now have a well-defined deformation retraction of $S^3 - K$ onto X. Another way of describing the situation would be to say that for an open ε-neighborhood N of K bounded by a torus T, the complement $S^3 - N$ is the mapping cylinder of a map $T \to X$.

To compute $\pi_1(X)$ we apply van Kampen's theorem to the decomposition of X as the union of X_m and X_n, or more properly, open neighborhoods of these two sets that deformation retract onto them. Both X_m and X_n are mapping cylinders that deformation retract onto circles, and $X_m \cap X_n$ is a circle, so all three of these spaces have fundamental group $\mathbb{Z}$. A loop in $X_m \cap X_n$ representing a generator of $\pi_1(X_m \cap X_n)$ is homotopic in X_m to a loop representing m times a generator, and in X_n to a loop representing n times a generator. Van Kampen's theorem then says that $\pi_1(X)$ is the quotient of the free group on generators a and b obtained by factoring out the normal subgroup generated by the element $a^m b^{-n}$.

Let us denote by $G_{m,n}$ this group $\pi_1(X_{m,n})$ defined by two generators a and b and one relation $a^m = b^n$. If m or n is 1, then $G_{m,n}$ is infinite cyclic since in these cases the relation just expresses one generator as a power of the other. To describe the structure of $G_{m,n}$ when $m, n > 1$ let us first compute the center of $G_{m,n}$, the subgroup consisting of elements that commute with all elements of $G_{m,n}$. The element $a^m = b^n$ commutes with a and b, so the cyclic subgroup C generated by this element lies in the center. In particular, C is a normal subgroup, so we can pass to the quotient group $G_{m,n}/C$, which is the free product $\mathbb{Z}_m * \mathbb{Z}_n$. According to Exercise 1 at the end of this section, a free product of nontrivial groups has trivial center. From this it follows that C is exactly the center of $G_{m,n}$. As we will see in Example 1.44, the elements a and b have infinite order in $G_{m,n}$, so C is infinite cyclic, but we will not need this fact here.

We will show now that the integers m and n are uniquely determined by the group $\mathbb{Z}_m * \mathbb{Z}_n$, hence also by $G_{m,n}$. The abelianization of $\mathbb{Z}_m * \mathbb{Z}_n$ is $\mathbb{Z}_m \times \mathbb{Z}_n$, of order mn, so the product mn is uniquely determined by $\mathbb{Z}_m * \mathbb{Z}_n$. To determine m and n individually, we use another assertion from Exercise 1 at the end of the section, that all torsion elements of $\mathbb{Z}_m * \mathbb{Z}_n$ are conjugate to elements of one of the subgroups $\mathbb{Z}_m$ and $\mathbb{Z}_n$, hence have order dividing m or n. Thus the maximum order of torsion elements of $\mathbb{Z}_m * \mathbb{Z}_n$ is the larger of m and n. The larger of these two numbers is therefore uniquely determined by the group $\mathbb{Z}_m * \mathbb{Z}_n$, hence also the smaller since the product is uniquely determined.

The preceding analysis of $\pi_1(X_{m,n})$ did not need the assumption that m and n

are relatively prime, which was used only to relate $X_{m,n}$ to torus knots. An interesting fact is that $X_{m,n}$ can be embedded in $\mathbb{R}^3$ only when m and n are relatively prime. This is shown in the remarks following Corollary 3.45. For example, $X_{2,2}$ is the Klein bottle since it is the union of two copies of the Möbius band X_2 with their boundary circles identified, so this nonembeddability statement generalizes the fact that the Klein bottle cannot be embedded in $\mathbb{R}^3$.

An algorithm for computing a presentation for $\pi_1(\mathbb{R}^3 - K)$ for an arbitrary smooth or piecewise linear knot K is described in the exercises, but the problem of determining when two of these fundamental groups are isomorphic is generally much more difficult than in the special case of torus knots.

Example 1.25: The Shrinking Wedge of Circles. Consider the subspace $X \subset \mathbb{R}^2$ that is the union of the circles C_n of radius $1/n$ and center $(1/n, 0)$ for $n = 1, 2, \cdots$. At first glance one might confuse X with the wedge sum of an infinite sequence of circles, but we will show that X has a much larger fundamental group than the wedge sum. Consider the retractions $r_n : X \to C_n$ collapsing all C_i's except C_n to the origin. Each r_n induces a surjection $\rho_n : \pi_1(X) \to \pi_1(C_n) \approx \mathbb{Z}$, where we take the origin as the basepoint. The product of the ρ_n's is a homomorphism $\rho : \pi_1(X) \to \prod_\infty \mathbb{Z}$ to the direct product (not the direct sum) of infinitely many copies of $\mathbb{Z}$, and ρ is surjective since for every sequence of integers k_n we can construct a loop $f : I \to X$ that wraps k_n times around C_n in the time interval $[1 - 1/n, 1 - 1/_{n+1}]$. This infinite composition of loops is certainly continuous at each time less than 1, and it is continuous at time 1 since every neighborhood of the basepoint in X contains all but finitely many of the circles C_n. Since $\pi_1(X)$ maps onto the uncountable group $\prod_\infty \mathbb{Z}$, it is uncountable. On the other hand, the fundamental group of a wedge sum of countably many circles is countably generated, hence countable.

The group $\pi_1(X)$ is actually far more complicated than $\prod_\infty \mathbb{Z}$. For one thing, it is nonabelian, since the retraction $X \to C_1 \cup \cdots \cup C_n$ that collapses all the circles smaller than C_n to the basepoint induces a surjection from $\pi_1(X)$ to a free group on n generators. For a complete description of $\pi_1(X)$ see [Cannon & Conner 2000].

It is a theorem of [Shelah 1988] that for a path-connected, locally path-connected compact metric space X, $\pi_1(X)$ is either finitely generated or uncountable.

Applications to Cell Complexes

For the remainder of this section we shall be interested in cell complexes, and in particular in how the fundamental group is affected by attaching 2-cells.

Suppose we attach a collection of 2-cells e_α^2 to a path-connected space X via maps $\varphi_\alpha : S^1 \to X$, producing a space Y. If s_0 is a basepoint of S^1 then φ_α determines a loop at $\varphi_\alpha(s_0)$ that we shall call φ_α, even though technically loops are maps $I \to X$ rather than $S^1 \to X$. For different α's the basepoints $\varphi_\alpha(s_0)$ of these loops φ_α may not all coincide. To remedy this, choose a basepoint $x_0 \in X$ and a path γ_α in X from x_0 to $\varphi_\alpha(s_0)$ for each α. Then $\gamma_\alpha\varphi_\alpha\overline{\gamma}_\alpha$ is a loop at x_0. This loop may not be nullhomotopic in X, but it will certainly be nullhomotopic after the cell e_α^2 is attached. Thus the normal subgroup $N \subset \pi_1(X, x_0)$ generated by all the loops $\gamma_\alpha\varphi_\alpha\overline{\gamma}_\alpha$ for varying α lies in the kernel of the map $\pi_1(X, x_0) \to \pi_1(Y, x_0)$ induced by the inclusion $X \hookrightarrow Y$.

Proposition 1.26. (a) *If Y is obtained from X by attaching 2-cells as described above, then the inclusion $X \hookrightarrow Y$ induces a surjection $\pi_1(X, x_0) \to \pi_1(Y, x_0)$ whose kernel is N. Thus $\pi_1(Y) \approx \pi_1(X)/N$.*

(b) *If Y is obtained from X by attaching n-cells for a fixed $n > 2$, then the inclusion $X \hookrightarrow Y$ induces an isomorphism $\pi_1(X, x_0) \approx \pi_1(Y, x_0)$.*

(c) *For a path-connected cell complex X the inclusion of the 2-skeleton $X^2 \hookrightarrow X$ induces an isomorphism $\pi_1(X^2, x_0) \approx \pi_1(X, x_0)$.*

It follows from (a) that N is independent of the choice of the paths γ_α, but this can also be seen directly: If we replace γ_α by another path η_α having the same endpoints, then $\gamma_\alpha\varphi_\alpha\overline{\gamma}_\alpha$ changes to $\eta_\alpha\varphi_\alpha\overline{\eta}_\alpha = (\eta_\alpha\overline{\gamma}_\alpha)\gamma_\alpha\varphi_\alpha\overline{\gamma}_\alpha(\gamma_\alpha\overline{\eta}_\alpha)$, so $\gamma_\alpha\varphi_\alpha\overline{\gamma}_\alpha$ and $\eta_\alpha\varphi_\alpha\overline{\eta}_\alpha$ define conjugate elements of $\pi_1(X, x_0)$.

Proof: (a) Let us expand Y to a slightly larger space Z that deformation retracts onto Y and is more convenient for applying van Kampen's theorem. The space Z is obtained from Y by attaching rectangular strips $S_\alpha = I \times I$, with the lower edge $I \times \{0\}$

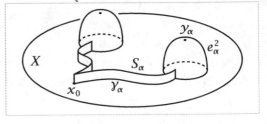

attached along γ_α, the right edge $\{1\} \times I$ attached along an arc in e_α^2, and all the left edges $\{0\} \times I$ of the different strips identified together. The top edges of the strips are not attached to anything, and this allows us to deformation retract Z onto Y.

In each cell e_α^2 choose a point y_α not in the arc along which S_α is attached. Let $A = Z - \bigcup_\alpha \{y_\alpha\}$ and let $B = Z - X$. Then A deformation retracts onto X, and B is contractible. Since $\pi_1(B) = 0$, van Kampen's theorem applied to the cover $\{A, B\}$ says that $\pi_1(Z)$ is isomorphic to the quotient of $\pi_1(A)$ by the normal subgroup generated by the image of the map $\pi_1(A \cap B) \to \pi_1(A)$. More specifically, choose a basepoint $z_0 \in A \cap B$ near x_0 on the segment where all the strips S_a intersect, and choose loops δ_α in $A \cap B$ based at z_0 representing the elements of $\pi_1(A, z_0)$ corresponding to $[\gamma_\alpha\varphi_\alpha\overline{\gamma}_\alpha] \in \pi_1(A, x_0)$ under the basepoint-change isomorphism β_h for h the line segment connecting z_0 to x_0 in the intersection of the S_α's. To finish the proof we just need to check that $\pi_1(A \cap B, z_0)$ is generated by the loops δ_α. This can be done by another application of van Kampen's theorem, this time to the cover of $A \cap B$ by the open sets $A_\alpha = A \cap B - \bigcup_{\beta \neq \alpha} e_\beta^2$. Since A_α deformation retracts onto a circle in $e_\alpha^2 - \{y_\alpha\}$, we have $\pi_1(A_\alpha, z_0) \approx \mathbb{Z}$ generated by δ_α, and we are done.

The proof of (b) follows the same plan with cells e_α^n instead of e_α^2. The only difference is that A_α deformation retracts onto a sphere S^{n-1} so $\pi_1(A_a) = 0$ if $n > 2$ by Proposition 1.14. Hence $\pi_1(A \cap B) = 0$ and the result follows.

Part (c) follows from (b) by induction when X is finite-dimensional, so $X = X^n$ for some n. When X is not finite-dimensional we argue as follows. Let $f : I \to X$ be a loop at the basepoint $x_0 \in X^2$. This has compact image, which must lie in X^n for some n by Proposition A.1 in the Appendix. Part (b) then implies that f is homotopic to a loop in X^2. Thus $\pi_1(X^2, x_0) \to \pi_1(X, x_0)$ is surjective. To see that it is also injective, suppose that f is a loop in X^2 which is nullhomotopic in X via a homotopy $F : I \times I \to X$. This has compact image lying in some X^n, and we can assume $n > 2$. Since $\pi_1(X^2, x_0) \to \pi_1(X^n, x_0)$ is injective by (b), we conclude that f is nullhomotopic in X^2. $\qquad\square$

As a first application we compute the fundamental group of the orientable surface M_g of genus g. This has a cell structure with one 0-cell, $2g$ 1-cells, and one 2-cell, as we saw in Chapter 0. The 1-skeleton is a wedge sum of $2g$ circles, with fundamental group free on $2g$ generators. The 2-cell is attached along the loop given by the product of the commutators of these generators, say $[a_1, b_1] \cdots [a_g, b_g]$. Therefore

$$\pi_1(M_g) \approx \langle\, a_1, b_1, \cdots, a_g, b_g \mid [a_1, b_1] \cdots [a_g, b_g] \,\rangle$$

where $\langle\, g_\alpha \mid r_\beta \,\rangle$ denotes the group with generators g_α and relators r_β, in other words, the free group on the generators g_α modulo the normal subgroup generated by the words r_β in these generators.

Corollary 1.27. *The surface M_g is not homeomorphic, or even homotopy equivalent, to M_h if $g \neq h$.*

Proof: The abelianization of $\pi_1(M_g)$ is the direct sum of $2g$ copies of $\mathbb{Z}$. So if $M_g \simeq M_h$ then $\pi_1(M_g) \approx \pi_1(M_h)$, hence the abelianizations of these groups are isomorphic, which implies $g = h$. $\qquad\square$

Nonorientable surfaces can be treated in the same way. If we attach a 2-cell to the wedge sum of g circles by the word $a_1^2 \cdots a_g^2$ we obtain a nonorientable surface N_g. For example, N_1 is the projective plane $\mathbb{R}P^2$, the quotient of D^2 with antipodal points of ∂D^2 identified, and N_2 is the Klein bottle, though the more usual representation of the Klein bottle is as a square with opposite sides identified via the word $aba^{-1}b$.

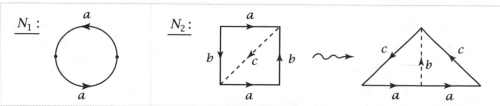

If one cuts the square along a diagonal and reassembles the resulting two triangles as shown in the figure, one obtains the other representation as a square with sides identified via the word $a^2 c^2$. By the proposition, $\pi_1(N_g) \approx \langle a_1, \cdots, a_g \mid a_1^2 \cdots a_g^2 \rangle$. This abelianizes to the direct sum of $\mathbb{Z}_2$ with $g-1$ copies of $\mathbb{Z}$ since in the abelianization we can rechoose the generators to be $a_1, \cdots, a_{g-1}$ and $a_1 + \cdots + a_g$, with $2(a_1 + \cdots + a_g) = 0$. Hence N_g is not homotopy equivalent to N_h if $g \neq h$, nor is N_g homotopy equivalent to any orientable surface M_h.

Here is another application of the preceding proposition:

Corollary 1.28. *For every group G there is a 2-dimensional cell complex X_G with $\pi_1(X_G) \approx G$.*

Proof: Choose a presentation $G = \langle g_\alpha \mid r_\beta \rangle$. This exists since every group is a quotient of a free group, so the g_α's can be taken to be the generators of this free group with the r_β's generators of the kernel of the map from the free group to G. Now construct X_G from $\bigvee_\alpha S_\alpha^1$ by attaching 2-cells e_β^2 by the loops specified by the words r_β. □

Example 1.29. If $G = \langle a \mid a^n \rangle = \mathbb{Z}_n$ then X_G is S^1 with a cell e^2 attached by the map $z \mapsto z^n$, thinking of S^1 as the unit circle in $\mathbb{C}$. When $n = 2$ we get $X_G = \mathbb{R}P^2$, but for $n > 2$ the space X_G is not a surface since there are n 'sheets' of e^2 attached at each point of the circle $S^1 \subset X_G$. For example, when $n = 3$ one can construct a neighborhood N of S^1 in X_G by taking the product of the graph Y with the interval I, and then identifying the two ends of this product via a one-third twist as shown in the figure. The boundary of N consists of a single circle, formed by the three endpoints of 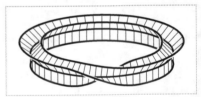 each Y cross section of N. To complete the construction of X_G from N one attaches a disk along the boundary circle of N. This cannot be done in $\mathbb{R}^3$, though it can in $\mathbb{R}^4$. For $n = 4$ one would use the graph X instead of Y, with a one-quarter twist instead of a one-third twist. For larger n one would use an n-pointed 'asterisk' and a $1/n$ twist.

Exercises

1. Show that the free product $G * H$ of nontrivial groups G and H has trivial center, and that the only elements of $G * H$ of finite order are the conjugates of finite-order elements of G and H.

2. Let $X \subset \mathbb{R}^m$ be the union of convex open sets $X_1, \cdots, X_n$ such that $X_i \cap X_j \cap X_k \neq \varnothing$ for all i, j, k. Show that X is simply-connected.

3. Show that the complement of a finite set of points in $\mathbb{R}^n$ is simply-connected if $n \geq 3$.

4. Let $X \subset \mathbb{R}^3$ be the union of n lines through the origin. Compute $\pi_1(\mathbb{R}^3 - X)$.

5. Let $X \subset \mathbb{R}^2$ be a connected graph that is the union of a finite number of straight line segments. Show that $\pi_1(X)$ is free with a basis consisting of loops formed by the boundaries of the bounded complementary regions of X, joined to a basepoint by suitably chosen paths in X. [Assume the Jordan curve theorem for polygonal simple closed curves, which is equivalent to the case that X is homeomorphic to S^1.]

6. Use Proposition 1.26 to show that the complement of a closed discrete subspace of $\mathbb{R}^n$ is simply-connected if $n \geq 3$.

7. Let X be the quotient space of S^2 obtained by identifying the north and south poles to a single point. Put a cell complex structure on X and use this to compute $\pi_1(X)$.

8. Compute the fundamental group of the space obtained from two tori $S^1 \times S^1$ by identifying a circle $S^1 \times \{x_0\}$ in one torus with the corresponding circle $S^1 \times \{x_0\}$ in the other torus.

9. In the surface M_g of genus g, let C be a circle that separates M_g into two compact subsurfaces M_h' and M_k' obtained from the closed surfaces M_h and M_k by deleting an open disk from

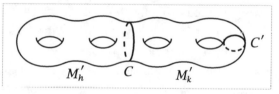

each. Show that M_h' does not retract onto its boundary circle C, and hence M_g does not retract onto C. [Hint: abelianize π_1.] But show that M_g does retract onto the nonseparating circle C' in the figure.

10. Consider two arcs α and β embedded in $D^2 \times I$ as shown in the figure. The loop y is obviously nullhomotopic in $D^2 \times I$, but show that there is no nullhomotopy of y in the complement of $\alpha \cup \beta$.

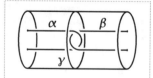

11. The **mapping torus** T_f of a map $f : X \to X$ is the quotient of $X \times I$ obtained by identifying each point $(x, 0)$ with $(f(x), 1)$. In the case $X = S^1 \vee S^1$ with f basepoint-preserving, compute a presentation for $\pi_1(T_f)$ in terms of the induced map $f_* : \pi_1(X) \to \pi_1(X)$. Do the same when $X = S^1 \times S^1$. [One way to do this is to regard T_f as built from $X \vee S^1$ by attaching cells.]

12. The Klein bottle is usually pictured as a subspace of $\mathbb{R}^3$ like the subspace $X \subset \mathbb{R}^3$ shown in the first figure at the right. If one wanted a model that could actually function as a bottle, one would delete the open disk bounded by the circle of self-

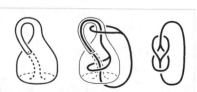

intersection of X, producing a subspace $Y \subset X$. Show that $\pi_1(X) \approx \mathbb{Z} * \mathbb{Z}$ and that

$\pi_1(Y)$ has the presentation $\langle\, a,b,c \mid aba^{-1}b^{-1}cb^{\varepsilon}c^{-1} \,\rangle$ for $\varepsilon = \pm 1$. (Changing the sign of ε gives an isomorphic group, as it happens.) Show also that $\pi_1(Y)$ is isomorphic to $\pi_1(\mathbb{R}^3 - Z)$ for Z the graph shown in the figure. The groups $\pi_1(X)$ and $\pi_1(Y)$ are not isomorphic, but this is not easy to prove; see the discussion in Example 1B.13.

13. The space Y in the preceding exercise can be obtained from a disk with two holes by identifying its three boundary circles. There are only two essentially different ways of identifying the three boundary circles. Show that the other way yields a space Z with $\pi_1(Z)$ not isomorphic to $\pi_1(Y)$. [Abelianize the fundamental groups to show they are not isomorphic.]

14. Consider the quotient space of a cube I^3 obtained by identifying each square face with the opposite square face via the right-handed screw motion consisting of a translation by one unit in the direction perpendicular to the face combined with a one-quarter twist of the face about its center point. Show this quotient space X is a cell complex with two 0-cells, four 1-cells, three 2-cells, and one 3-cell. Using this structure, show that $\pi_1(X)$ is the quaternion group $\{\pm 1, \pm i, \pm j, \pm k\}$, of order eight.

15. Given a space X with basepoint $x_0 \in X$, we may construct a CW complex $L(X)$ having a single 0-cell, a 1-cell e_{γ}^1 for each loop γ in X based at x_0, and a 2-cell e_{τ}^2 for each map τ of a standard triangle PQR into X taking the three vertices P, Q, and R of the triangle to x_0. The 2-cell e_{τ}^2 is attached to the three 1-cells that are the loops obtained by restricting τ to the three oriented edges PQ, PR, and QR. Show that the natural map $L(X) \to X$ induces an isomorphism $\pi_1(L(X)) \approx \pi_1(X, x_0)$.

16. Show that the fundamental group of the surface of infinite genus shown below is free on an infinite number of generators.

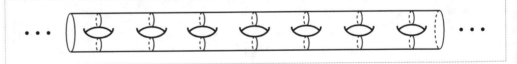

17. Show that $\pi_1(\mathbb{R}^2 - \mathbb{Q}^2)$ is uncountable.

18. In this problem we use the notions of suspension, reduced suspension, cone, and mapping cone defined in Chapter 0. Let X be the subspace of $\mathbb{R}$ consisting of the sequence $1, 1/2, 1/3, 1/4, \cdots$ together with its limit point 0.

(a) For the suspension SX, show that $\pi_1(SX)$ is free on a countably infinite set of generators, and deduce that $\pi_1(SX)$ is countable. In contrast to this, the reduced suspension ΣX, obtained from SX by collapsing the segment $\{0\} \times I$ to a point, is the shrinking wedge of circles in Example 1.25, with an uncountable fundamental group.

(b) Let C be the mapping cone of the quotient map $SX \to \Sigma X$. Show that $\pi_1(C)$ is uncountable by constructing a homomorphism from $\pi_1(C)$ onto $\prod_{\infty}\mathbb{Z}/\bigoplus_{\infty}\mathbb{Z}$. Note that C is the reduced suspension of the cone CX. Thus the reduced suspension of a contractible space need not be contractible, unlike the unreduced suspension.

19. Show that the subspace of $\mathbb{R}^3$ that is the union of the spheres of radius $1/n$ and center $(1/n, 0, 0)$ for $n = 1, 2, \cdots$ is simply-connected.

20. Let X be the subspace of $\mathbb{R}^2$ that is the union of the circles C_n of radius n and center $(n, 0)$ for $n = 1, 2, \cdots$. Show that $\pi_1(X)$ is the free group $*_n \pi_1(C_n)$, the same as for the infinite wedge sum $\bigvee_\infty S^1$. Show that X and $\bigvee_\infty S^1$ are in fact homotopy equivalent, but not homeomorphic.

21. Show that the join $X * Y$ of two nonempty spaces X and Y is simply-connected if X is path-connected.

22. In this exercise we describe an algorithm for computing a presentation of the fundamental group of the complement of a smooth or piecewise linear knot K in $\mathbb{R}^3$, called the *Wirtinger presentation*. To begin, we position the knot to lie almost flat on a table, so that K consists of finitely many disjoint arcs α_i where it intersects the table top together with finitely many disjoint arcs β_ℓ where K crosses over itself. The configuration at such a crossing is shown in the first figure below. We build a

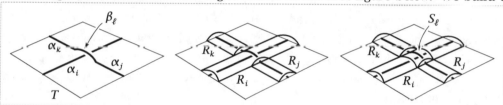

2-dimensional complex X that is a deformation retract of $\mathbb{R}^3 - K$ by the following three steps. First, start with the rectangle T formed by the table top. Next, just above each arc α_i place a long, thin rectangular strip R_i, curved to run parallel to α_i along the full length of α_i and arched so that the two long edges of R_i are identified with points of T, as in the second figure. Any arcs β_ℓ that cross over α_i are positioned to lie in R_i. Finally, over each arc β_ℓ put a square S_ℓ, bent downward along its four edges so that these edges are identified with points of three strips R_i, R_j, and R_k as in the third figure; namely, two opposite edges of S_ℓ are identified with short edges of R_j and R_k and the other two opposite edges of S_ℓ are identified with two arcs crossing the interior of R_i. The knot K is now a subspace of X, but after we lift K up slightly into the complement of X, it becomes evident that X is a deformation retract of $\mathbb{R}^3 - K$.

(a) Assuming this bit of geometry, show that $\pi_1(\mathbb{R}^3 - K)$ has a presentation with one generator x_i for each strip R_i and one relation of the form $x_i x_j x_i^{-1} = x_k$ for each square S_ℓ, where the indices are as in the figures above. [To get the correct signs it is helpful to use an orientation of K.]

(b) Use this presentation to show that the abelianization of $\pi_1(\mathbb{R}^3 - K)$ is $\mathbb{Z}$.

1.3 Covering Spaces

We come now to the second main topic of this chapter, covering spaces. We have already encountered these briefly in our calculation of $\pi_1(S^1)$ which used the example of the projection $\mathbb{R} \to S^1$ of a helix onto a circle. As we will see, covering spaces can be used to calculate fundamental groups of other spaces as well. But the connection between the fundamental group and covering spaces runs much deeper than this, and in many ways they can be regarded as two viewpoints toward the same thing. Algebraic aspects of the fundamental group can often be translated into the geometric language of covering spaces. This is exemplified in one of the main results in this section, an exact correspondence between connected covering spaces of a given space X and subgroups of $\pi_1(X)$. This is strikingly reminiscent of Galois theory, with its correspondence between field extensions and subgroups of the Galois group.

Let us recall the definition. A **covering space** of a space X is a space $\widetilde{X}$ together with a map $p : \widetilde{X} \to X$ satisfying the following condition: Each point $x \in X$ has an open neighborhood U in X such that $p^{-1}(U)$ is a union of disjoint open sets in $\widetilde{X}$, each of which is mapped homeomorphically onto U by p. Such a U is called **evenly covered** and the disjoint open sets in $\widetilde{X}$ that project homeomorphically to U by p are called **sheets** of $\widetilde{X}$ over U. If U is connected these sheets are the connected components of $p^{-1}(U)$ so in this case they are uniquely determined by U, but when U is not connected the decomposition of $p^{-1}(U)$ into sheets may not be unique. We allow $p^{-1}(U)$ to be empty, the union of an empty collection of sheets over U, so p need not be surjective. The number of sheets over U is the cardinality of $p^{-1}(x)$ for $x \in U$. As x varies over X this number is locally constant, so it is constant if X is connected.

An example related to the helix example is the helicoid surface $S \subset \mathbb{R}^3$ consisting of points of the form $(s \cos 2\pi t, s \sin 2\pi t, t)$ for $(s, t) \in (0, \infty) \times \mathbb{R}$. This projects onto $\mathbb{R}^2 - \{0\}$ via the map $(x, y, z) \mapsto (x, y)$, and this projection defines a covering space $p : S \to \mathbb{R}^2 - \{0\}$ since each point of $\mathbb{R}^2 - \{0\}$ is contained in an open disk U in $\mathbb{R}^2 - \{0\}$ with $p^{-1}(U)$ consisting of countably many disjoint open disks in S projecting homeomorphically onto U.

Another example is the map $p : S^1 \to S^1$, $p(z) = z^n$ where we view z as a complex number with $|z| = 1$ and n is any positive integer. The closest one can come to realizing this covering space as a linear projection in 3-space analogous to the projection of the helix is to draw a circle wrapping around a cylinder n times and intersecting itself in $n - 1$ points that one has to imagine are not really intersections. For an alternative picture without this defect, embed S^1 in the boundary torus of a solid torus $S^1 \times D^2$ so that it winds n times

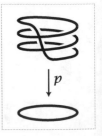

monotonically around the S^1 factor without self-intersections, then restrict the projection $S^1 \times D^2 \rightarrow S^1 \times \{0\}$ to this embedded circle. The figure for Example 1.29 in the preceding section illustrates the case $n = 3$.

These n-sheeted covering spaces $S^1 \rightarrow S^1$ for $n \geq 1$ together with the infinite-sheeted helix example exhaust all the connected coverings spaces of S^1, as our general theory will show. There are many other disconnected covering spaces of S^1, such as n disjoint circles each mapped homeomorphically onto S^1, but these disconnected covering spaces are just disjoint unions of connected ones. We will usually restrict our attention to connected covering spaces as these contain most of the interesting features of covering spaces.

The covering spaces of $S^1 \vee S^1$ form a remarkably rich family illustrating most of the general theory very concretely, so let us look at a few of these covering spaces to get an idea of what is going on. To abbreviate notation, set $X = S^1 \vee S^1$. We view this as a graph with one vertex and two edges. We label the edges a and b and we choose orientations for a and b. Now let $\tilde{X}$ be any other graph with four edges meeting at each vertex, and suppose the edges of $\tilde{X}$ have been assigned labels a and b and orientations in such a way that the local picture near each vertex is the same as in X, so there is an a-edge oriented toward the vertex, an a-edge oriented away from the vertex, a b-edge oriented toward the vertex, and a b-edge oriented away from the vertex. To give a name to this structure, let us call $\tilde{X}$ a 2-*oriented* graph.

The table on the next page shows just a small sample of the infinite variety of possible examples.

Given a 2-oriented graph $\tilde{X}$ we can construct a map $p: \tilde{X} \rightarrow X$ sending all vertices of $\tilde{X}$ to the vertex of X and sending each edge of $\tilde{X}$ to the edge of X with the same label by a map that is a homeomorphism on the interior of the edge and preserves orientation. It is clear that the covering space condition is satisfied for p. Conversely, every covering space of X is a graph that inherits a 2-orientation from X.

As the reader will discover by experimentation, it seems that every graph having four edges incident at each vertex can be 2-oriented. This can be proved for finite graphs as follows. A very classical and easily shown fact is that every finite connected graph with an even number of edges incident at each vertex has an Eulerian circuit, a loop traversing each edge exactly once. If there are four edges at each vertex, then labeling the edges of an Eulerian circuit alternately a and b produces a labeling with two a and two b edges at each vertex. The union of the a edges is then a collection of disjoint circles, as is the union of the b edges. Choosing orientations for all these circles gives a 2-orientation. It is a theorem in graph theory that infinite graphs with four edges incident at each vertex can also be 2-oriented; see Chapter 13 of [König 1990] for a proof. There is also a generalization to n-oriented graphs, which are covering spaces of the wedge sum of n circles.

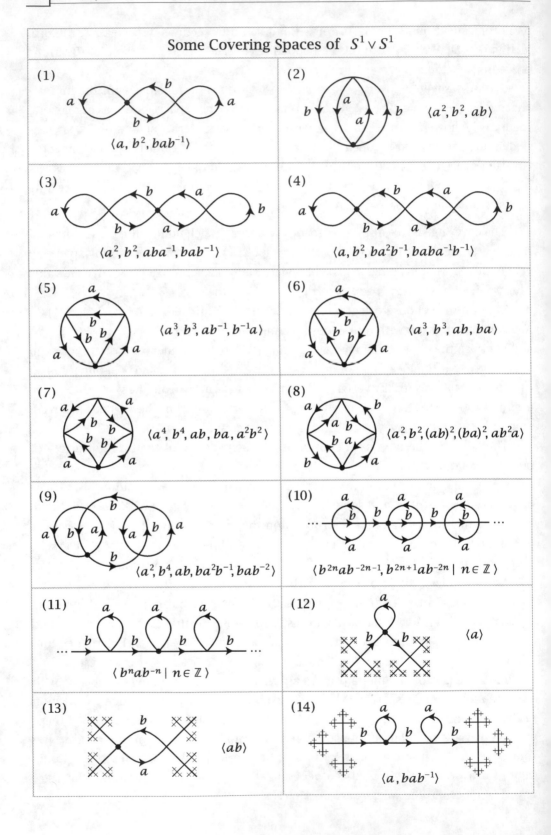

Some Covering Spaces of $S^1 \vee S^1$

(1) $\langle a, b^2, bab^{-1} \rangle$

(2) $\langle a^2, b^2, ab \rangle$

(3) $\langle a^2, b^2, aba^{-1}, bab^{-1} \rangle$

(4) $\langle a, b^2, ba^2b^{-1}, baba^{-1}b^{-1} \rangle$

(5) $\langle a^3, b^3, ab^{-1}, b^{-1}a \rangle$

(6) $\langle a^3, b^3, ab, ba \rangle$

(7) $\langle a^4, b^4, ab, ba, a^2b^2 \rangle$

(8) $\langle a^2, b^2, (ab)^2, (ba)^2, ab^2a \rangle$

(9) $\langle a^2, b^4, ab, ba^2b^{-1}, bab^{-2} \rangle$

(10) $\langle b^{2n}ab^{-2n-1}, b^{2n+1}ab^{-2n} \mid n \in \mathbb{Z} \rangle$

(11) $\langle b^nab^{-n} \mid n \in \mathbb{Z} \rangle$

(12) $\langle a \rangle$

(13) $\langle ab \rangle$

(14) $\langle a, bab^{-1} \rangle$

A simply-connected covering space of $X = S^1 \vee S^1$ can be constructed in the following way. Start with the open intervals $(-1, 1)$ in the coordinate axes of $\mathbb{R}^2$. Next, for a fixed number λ, $0 < \lambda < 1/2$, for example $\lambda = 1/3$, adjoin four open segments of length 2λ, at distance λ from the ends of the previous segments and perpendicular to them, the new shorter segments being bisected by the older ones. For the third stage, add perpendicular open segments of length $2\lambda^2$ at distance λ^2 from the endpoints of all the previous segments and bisected by them. The process 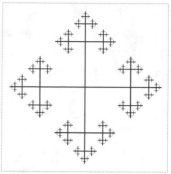 is now repeated indefinitely, at the nth stage adding open segments of length $2\lambda^{n-1}$ at distance λ^{n-1} from all the previous endpoints. The union of all these open segments is a graph, with vertices the intersection points of horizontal and vertical segments, and edges the subsegments between adjacent vertices. We label all the horizontal edges a, oriented to the right, and all the vertical edges b, oriented upward.

This covering space is called the *universal cover* of X because, as our general theory will show, it is a covering space of every other connected covering space of X.

The covering spaces (1)–(14) in the table are all nonsimply-connected. Their fundamental groups are free with bases represented by the loops specified by the listed words in a and b, starting at the basepoint $\tilde{x}_0$ indicated by the heavily shaded vertex. This can be proved in each case by applying van Kampen's theorem. One can also interpret the list of words as generators of the image subgroup $p_*(\pi_1(\tilde{X}, \tilde{x}_0))$ in $\pi_1(X, x_0) = \langle a, b \rangle$. A general fact we shall prove about covering spaces is that the induced map $p_* : \pi_1(\tilde{X}, \tilde{x}_0) \to \pi_1(X, x_0)$ is always injective. Thus we have the at-first-glance paradoxical fact that the free group on two generators can contain as a subgroup a free group on any finite number of generators, or even on a countably infinite set of generators as in examples (10) and (11).

Changing the basepoint vertex changes the subgroup $p_*(\pi_1(\tilde{X}, \tilde{x}_0))$ to a conjugate subgroup in $\pi_1(X, x_0)$. The conjugating element of $\pi_1(X, x_0)$ is represented by any loop that is the projection of a path in $\tilde{X}$ joining one basepoint to the other. For example, the covering spaces (3) and (4) differ only in the choice of basepoints, and the corresponding subgroups of $\pi_1(X, x_0)$ differ by conjugation by b.

The main classification theorem for covering spaces says that by associating the subgroup $p_*(\pi_1(\tilde{X}, \tilde{x}_0))$ to the covering space $p : \tilde{X} \to X$, we obtain a one-to-one correspondence between all the different connected covering spaces of X and the conjugacy classes of subgroups of $\pi_1(X, x_0)$. If one keeps track of the basepoint vertex $\tilde{x}_0 \in \tilde{X}$, then this is a one-to-one correspondence between covering spaces $p : (\tilde{X}, \tilde{x}_0) \to (X, x_0)$ and actual subgroups of $\pi_1(X, x_0)$, not just conjugacy classes. Of course, for these statements to make sense one has to have a precise notion of when two covering spaces are the same, or 'isomorphic.' In the case at hand, an isomorphism between covering spaces of X is just a graph isomorphism that preserves the labeling and orientations of edges. Thus the covering spaces in (3) and (4) are isomorphic, but not by an isomorphism preserving basepoints, so the two subgroups of $\pi_1(X, x_0)$ corresponding to these covering spaces are distinct but conjugate. On the other hand, the two covering spaces in (5) and (6) are not isomorphic, though the graphs are homeomorphic, so the corresponding subgroups of $\pi_1(X, x_0)$ are isomorphic but not conjugate.

Some of the covering spaces (1)–(14) are more symmetric than others, where by a 'symmetry' we mean an automorphism of the graph preserving the labeling and orientations. The most symmetric covering spaces are those having symmetries taking any one vertex onto any other. The examples (1), (2), (5)–(8), and (11) are the ones with this property. We shall see that a covering space of X has maximal symmetry exactly when the corresponding subgroup of $\pi_1(X, x_0)$ is a normal subgroup, and in this case the symmetries form a group isomorphic to the quotient group of $\pi_1(X, x_0)$ by the normal subgroup. Since every group generated by two elements is a quotient group of $\mathbb{Z} * \mathbb{Z}$, this implies that every two-generator group is the symmetry group of some covering space of X.

Lifting Properties

Covering spaces are defined in fairly geometric terms, as maps $p : \widetilde{X} \to X$ that are local homeomorphisms in a rather strong sense. But from the viewpoint of algebraic topology, the distinctive feature of covering spaces is their behavior with respect to lifting of maps. Recall the terminology from the proof of Theorem 1.7: A **lift** of a map $f : Y \to X$ is a map $\widetilde{f} : Y \to \widetilde{X}$ such that $p\widetilde{f} = f$. We will describe three special lifting properties of covering spaces and derive a few applications of these.

First we have the **homotopy lifting property**, also known as the **covering homotopy property**:

Proposition 1.30. *Given a covering space $p : \widetilde{X} \to X$, a homotopy $f_t : Y \to X$, and a map $\widetilde{f}_0 : Y \to \widetilde{X}$ lifting f_0, then there exists a unique homotopy $\widetilde{f}_t : Y \to \widetilde{X}$ of $\widetilde{f}_0$ that lifts f_t.*

Proof: This was proved as property (c) in the proof of Theorem 1.7. $\square$

Taking Y to be a point gives the **path lifting property** for a covering space $p : \widetilde{X} \to X$, which says that for each path $f : I \to X$ and each lift $\widetilde{x}_0$ of the starting point $f(0) = x_0$ there is a unique path $\widetilde{f} : I \to \widetilde{X}$ lifting f starting at $\widetilde{x}_0$. In particular, the uniqueness of lifts implies that every lift of a constant path is constant, but this could be deduced more simply from the fact that $p^{-1}(x_0)$ has the discrete topology, by the definition of a covering space.

Taking Y to be I, we see that every homotopy f_t of a path f_0 in X lifts to a homotopy $\tilde{f}_t$ of each lift $\tilde{f}_0$ of f_0. The lifted homotopy $\tilde{f}_t$ is a homotopy of paths, fixing the endpoints, since as t varies each endpoint of $\tilde{f}_t$ traces out a path lifting a constant path, which must therefore be constant.

Here is a simple application:

Proposition 1.31. *The map $p_* : \pi_1(\tilde{X}, \tilde{x}_0) \to \pi_1(X, x_0)$ induced by a covering space $p : (\tilde{X}, \tilde{x}_0) \to (X, x_0)$ is injective. The image subgroup $p_*(\pi_1(\tilde{X}, \tilde{x}_0))$ in $\pi_1(X, x_0)$ consists of the homotopy classes of loops in X based at x_0 whose lifts to $\tilde{X}$ starting at $\tilde{x}_0$ are loops.*

Proof: An element of the kernel of p_* is represented by a loop $\tilde{f}_0 : I \to \tilde{X}$ with a homotopy $f_t : I \to X$ of $f_0 = p\tilde{f}_0$ to the trivial loop f_1. By the remarks preceding the proposition, there is a lifted homotopy of loops $\tilde{f}_t$ starting with $\tilde{f}_0$ and ending with a constant loop. Hence $[\tilde{f}_0] = 0$ in $\pi_1(\tilde{X}, \tilde{x}_0)$ and p_* is injective.

For the second statement of the proposition, loops at x_0 lifting to loops at $\tilde{x}_0$ certainly represent elements of the image of $p_* : \pi_1(\tilde{X}, \tilde{x}_0) \to \pi_1(X, x_0)$. Conversely, a loop representing an element of the image of p_* is homotopic to a loop having such a lift, so by homotopy lifting, the loop itself must have such a lift. $\qquad\square$

Proposition 1.32. *The number of sheets of a covering space $p : (\tilde{X}, \tilde{x}_0) \to (X, x_0)$ with X and $\tilde{X}$ path-connected equals the index of $p_*(\pi_1(\tilde{X}, \tilde{x}_0))$ in $\pi_1(X, x_0)$.*

Proof: For a loop g in X based at x_0, let $\tilde{g}$ be its lift to $\tilde{X}$ starting at $\tilde{x}_0$. A product $h \cdot g$ with $[h] \in H = p_*(\pi_1(\tilde{X}, \tilde{x}_0))$ has the lift $\tilde{h} \cdot \tilde{g}$ ending at the same point as $\tilde{g}$ since $\tilde{h}$ is a loop. Thus we may define a function Φ from cosets $H[g]$ to $p^{-1}(x_0)$ by sending $H[g]$ to $\tilde{g}(1)$. The path-connectedness of $\tilde{X}$ implies that Φ is surjective since $\tilde{x}_0$ can be joined to any point in $p^{-1}(x_0)$ by a path $\tilde{g}$ projecting to a loop g at x_0. To see that Φ is injective, observe that $\Phi(H[g_1]) = \Phi(H[g_2])$ implies that $g_1 \cdot \bar{g}_2$ lifts to a loop in $\tilde{X}$ based at $\tilde{x}_0$, so $[g_1][g_2]^{-1} \in H$ and hence $H[g_1] = H[g_2]$. $\qquad\square$

It is important also to know about the existence and uniqueness of lifts of general maps, not just lifts of homotopies. For the existence question an answer is provided by the following **lifting criterion**:

Proposition 1.33. *Suppose given a covering space $p : (\tilde{X}, \tilde{x}_0) \to (X, x_0)$ and a map $f : (Y, y_0) \to (X, x_0)$ with Y path-connected and locally path-connected. Then a lift $\tilde{f} : (Y, y_0) \to (\tilde{X}, \tilde{x}_0)$ of f exists iff $f_*(\pi_1(Y, y_0)) \subset p_*(\pi_1(\tilde{X}, \tilde{x}_0))$.*

When we say a space has a certain property locally, such as being locally path-connected, we usually mean that each point has arbitrarily small open neighborhoods with this property. Thus for Y to be locally path-connected means that for each point $y \in Y$ and each neighborhood U of y there is an open neighborhood $V \subset U$ of

y that is path-connected. Some authors weaken the requirement that V be path-connected to the condition that any two points in V be joinable by a path in U. This broader definition would work just as well for our purposes, necessitating only small adjustments in the proofs, but for simplicity we shall use the more restrictive definition.

Proof: The 'only if' statement is obvious since $f_* = p_* \tilde{f}_*$. For the converse, let $y \in Y$ and let γ be a path in Y from y_0 to y. The path $f\gamma$ in X starting at x_0 has a unique lift $\widetilde{f\gamma}$ starting at $\tilde{x}_0$. Define $\tilde{f}(y) = \widetilde{f\gamma}(1)$. To show this is well-defined, independent of the choice of γ, let γ' be another path from y_0 to y. Then $(f\gamma') \cdot (\overline{f\gamma})$ is a loop h_0 at x_0 with $[h_0] \in f_*(\pi_1(Y, y_0)) \subset p_*(\pi_1(\tilde{X}, \tilde{x}_0))$. This means there is a homotopy h_t of h_0 to a loop h_1 that lifts to a loop $\tilde{h}_1$ in $\tilde{X}$ based at $\tilde{x}_0$. Apply the covering homotopy property to h_t to get a lifting $\tilde{h}_t$. Since $\tilde{h}_1$ is a loop at $\tilde{x}_0$, so is $\tilde{h}_0$. By the uniqueness of lifted paths, the first half of $\tilde{h}_0$ is $\widetilde{f\gamma'}$ and the second half is $\widetilde{f\gamma}$ traversed backwards, with the common midpoint $\widetilde{f\gamma}(1) = \widetilde{f\gamma'}(1)$. This shows that $\tilde{f}$ is well-defined.

To see that $\tilde{f}$ is continuous, let $U \subset X$ be an open neighborhood of $f(y)$ having a lift $\tilde{U} \subset \tilde{X}$ containing $\tilde{f}(y)$ such that $p : \tilde{U} \to U$ is a homeomorphism. Choose a path-connected open neighborhood V of y with $f(V) \subset U$. For paths from y_0 to points $y' \in V$ we can take a fixed path γ from y_0 to y followed by paths η in V from y to the points y'. Then the paths $(f\gamma) \cdot (f\eta)$ in X have lifts $(\widetilde{f\gamma}) \cdot (\widetilde{f\eta})$ where $\widetilde{f\eta} = p^{-1} f\eta$ and $p^{-1} : U \to \tilde{U}$ is the inverse of $p : \tilde{U} \to U$. Thus $\tilde{f}(V) \subset \tilde{U}$ and $\tilde{f}|V = p^{-1} f$, hence $\tilde{f}$ is continuous at y. $\square$

An example showing the necessity of the local path-connectedness assumption on Y is described in Exercise 7 at the end of this section.

Next we have the **unique lifting property**:

Proposition 1.34. *Given a covering space $p : \tilde{X} \to X$ and a map $f : Y \to X$, if two lifts $\tilde{f}_1, \tilde{f}_2 : Y \to \tilde{X}$ of f agree at one point of Y and Y is connected, then $\tilde{f}_1$ and $\tilde{f}_2$ agree on all of Y.*

Proof: For a point $y \in Y$, let U be an evenly covered open neighborhood of $f(y)$ in X, so $p^{-1}(U)$ is decomposed into disjoint sheets each mapped homeomorphically onto U by p. Let $\tilde{U}_1$ and $\tilde{U}_2$ be the sheets containing $\tilde{f}_1(y)$ and $\tilde{f}_2(y)$, respectively. By continuity of $\tilde{f}_1$ and $\tilde{f}_2$ there is a neighborhood N of y mapped into $\tilde{U}_1$ by $\tilde{f}_1$ and into $\tilde{U}_2$ by $\tilde{f}_2$. If $\tilde{f}_1(y) \neq \tilde{f}_2(y)$ then $\tilde{U}_1 \neq \tilde{U}_2$, hence $\tilde{U}_1$ and $\tilde{U}_2$ are disjoint and $\tilde{f}_1 \neq \tilde{f}_2$ throughout the neighborhood N. On the other hand, if $\tilde{f}_1(y) = \tilde{f}_2(y)$ then

$\widetilde{U}_1 = \widetilde{U}_2$ so $\widetilde{f}_1 = \widetilde{f}_2$ on N since $p\widetilde{f}_1 = p\widetilde{f}_2$ and p is injective on $\widetilde{U}_1 = \widetilde{U}_2$. Thus the set of points where $\widetilde{f}_1$ and $\widetilde{f}_2$ agree is both open and closed in Y. $\qquad\square$

The Classification of Covering Spaces

We consider next the problem of classifying all the different covering spaces of a fixed space X. Since the whole chapter is about paths, it should not be surprising that we will restrict attention to spaces X that are at least locally path-connected. Path-components of X are then the same as components, and for the purpose of classifying the covering spaces of X there is no loss in assuming that X is connected, or equivalently, path-connected. Local path-connectedness is inherited by covering spaces, so these too are connected iff they are path-connected. The main thrust of the classification will be the Galois correspondence between connected covering spaces of X and subgroups of $\pi_1(X)$, but when this is finished we will also describe a different method of classification that includes disconnected covering spaces as well.

The Galois correspondence arises from the function that assigns to each covering space $p:(\widetilde{X},\widetilde{x}_0)\to(X,x_0)$ the subgroup $p_*(\pi_1(\widetilde{X},\widetilde{x}_0))$ of $\pi_1(X,x_0)$. First we consider whether this function is surjective. That is, we ask whether every subgroup of $\pi_1(X,x_0)$ is realized as $p_*(\pi_1(\widetilde{X},\widetilde{x}_0))$ for some covering space $p:(\widetilde{X},\widetilde{x}_0)\to(X,x_0)$. In particular we can ask whether the trivial subgroup is realized. Since p_* is always injective, this amounts to asking whether X has a simply-connected covering space. Answering this will take some work.

A necessary condition for X to have a simply-connected covering space is the following: Each point $x \in X$ has a neighborhood U such that the inclusion-induced map $\pi_1(U,x)\to\pi_1(X,x)$ is trivial; one says X is **semilocally simply-connected** if this holds. To see the necessity of this condition, suppose $p:\widetilde{X}\to X$ is a covering space with $\widetilde{X}$ simply-connected. Every point $x \in X$ has a neighborhood U having a lift $\widetilde{U}\subset\widetilde{X}$ projecting homeomorphically to U by p. Each loop in U lifts to a loop in $\widetilde{U}$, and the lifted loop is nullhomotopic in $\widetilde{X}$ since $\pi_1(\widetilde{X})=0$. So, composing this nullhomotopy with p, the original loop in U is nullhomotopic in X.

A locally simply-connected space is certainly semilocally simply-connected. For example, CW complexes have the much stronger property of being locally contractible, as we show in the Appendix. An example of a space that is not semilocally simply-connected is the shrinking wedge of circles, the subspace $X\subset\mathbb{R}^2$ consisting of the circles of radius $1/n$ centered at the point $(1/n,0)$ for $n=1,2,\cdots$, introduced in Example 1.25. On the other hand, the cone $CX = (X\times I)/(X\times\{0\})$ is semilocally simply-connected since it is contractible, but it is not locally simply-connected.

We shall now show how to construct a simply-connected covering space of X if X is path-connected, locally path-connected, and semilocally simply-connected. To motivate the construction, suppose $p:(\widetilde{X},\widetilde{x}_0)\to(X,x_0)$ is a simply-connected covering space. Each point $\widetilde{x}\in\widetilde{X}$ can then be joined to $\widetilde{x}_0$ by a unique homotopy class of paths, by Proposition 1.6, so we can view points of $\widetilde{X}$ as homotopy classes of paths starting at $\widetilde{x}_0$. The advantage of this is that, by the homotopy lifting property, homotopy classes of paths in $\widetilde{X}$ starting at $\widetilde{x}_0$ are the same as homotopy classes of paths in X starting at x_0. This gives a way of describing $\widetilde{X}$ purely in terms of X.

Given a path-connected, locally path-connected, semilocally simply-connected space X with a basepoint $x_0 \in X$, we are therefore led to define

$$\widetilde{X} = \{\, [\gamma] \mid \gamma \text{ is a path in } X \text{ starting at } x_0 \,\}$$

where, as usual, $[\gamma]$ denotes the homotopy class of γ with respect to homotopies that fix the endpoints $\gamma(0)$ and $\gamma(1)$. The function $p : \widetilde{X} \rightarrow X$ sending $[\gamma]$ to $\gamma(1)$ is then well-defined. Since X is path-connected, the endpoint $\gamma(1)$ can be any point of X, so p is surjective.

Before we define a topology on $\widetilde{X}$ we make a few preliminary observations. Let $\mathcal{U}$ be the collection of path-connected open sets $U \subset X$ such that $\pi_1(U) \rightarrow \pi_1(X)$ is trivial. Note that if the map $\pi_1(U) \rightarrow \pi_1(X)$ is trivial for one choice of basepoint in U, it is trivial for all choices of basepoint since U is path-connected. A path-connected open subset $V \subset U \in \mathcal{U}$ is also in $\mathcal{U}$ since the composition $\pi_1(V) \rightarrow \pi_1(U) \rightarrow \pi_1(X)$ will also be trivial. It follows that $\mathcal{U}$ is a basis for the topology on X if X is locally path-connected and semilocally simply-connected.

Given a set $U \in \mathcal{U}$ and a path γ in X from x_0 to a point in U, let

$$U_{[\gamma]} = \{\, [\gamma \cdot \eta] \mid \eta \text{ is a path in } U \text{ with } \eta(0) = \gamma(1) \,\}$$

As the notation indicates, $U_{[\gamma]}$ depends only on the homotopy class $[\gamma]$. Observe that $p : U_{[\gamma]} \rightarrow U$ is surjective since U is path-connected and injective since different choices of η joining $\gamma(1)$ to a fixed $x \in U$ are all homotopic in X, the map $\pi_1(U) \rightarrow \pi_1(X)$ being trivial. Another property is

(∗) $U_{[\gamma]} = U_{[\gamma']}$ if $[\gamma'] \in U_{[\gamma]}$. For if $\gamma' = \gamma \cdot \eta$ then elements of $U_{[\gamma']}$ have the form $[\gamma \cdot \eta \cdot \mu]$ and hence lie in $U_{[\gamma]}$, while elements of $U_{[\gamma]}$ have the form $[\gamma \cdot \mu] = [\gamma \cdot \eta \cdot \overline{\eta} \cdot \mu] = [\gamma' \cdot \overline{\eta} \cdot \mu]$ and hence lie in $U_{[\gamma']}$.

This can be used to show that the sets $U_{[\gamma]}$ form a basis for a topology on $\widetilde{X}$. For if we are given two such sets $U_{[\gamma]}$, $V_{[\gamma']}$ and an element $[\gamma''] \in U_{[\gamma]} \cap V_{[\gamma']}$, we have $U_{[\gamma]} = U_{[\gamma'']}$ and $V_{[\gamma']} = V_{[\gamma'']}$ by (∗). So if $W \in \mathcal{U}$ is contained in $U \cap V$ and contains $\gamma''(1)$ then $W_{[\gamma'']} \subset U_{[\gamma'']} \cap V_{[\gamma'']}$ and $[\gamma''] \in W_{[\gamma'']}$.

The bijection $p : U_{[\gamma]} \rightarrow U$ is a homeomorphism since it gives a bijection between the subsets $V_{[\gamma']} \subset U_{[\gamma]}$ and the sets $V \in \mathcal{U}$ contained in U. Namely, in one direction we have $p(V_{[\gamma']}) = V$ and in the other direction we have $p^{-1}(V) \cap U_{[\gamma]} = V_{[\gamma']}$ for any $[\gamma'] \in U_{[\gamma]}$ with endpoint in V, since $V_{[\gamma']} \subset U_{[\gamma']} = U_{[\gamma]}$ and $V_{[\gamma']}$ maps onto V by the bijection p.

The preceding paragraph implies that $p : \widetilde{X} \rightarrow X$ is continuous. We can also deduce that this is a covering space since for fixed $U \in \mathcal{U}$, the sets $U_{[\gamma]}$ for varying $[\gamma]$ partition $p^{-1}(U)$ because if $[\gamma''] \in U_{[\gamma]} \cap U_{[\gamma']}$ then $U_{[\gamma]} = U_{[\gamma'']} = U_{[\gamma']}$ by (∗).

It remains only to show that $\tilde{X}$ is simply-connected. For a point $[y] \in \tilde{X}$ let y_t be the path in X that equals y on $[0, t]$ and is stationary at $y(t)$ on $[t, 1]$. Then the function $t \mapsto [y_t]$ is a path in $\tilde{X}$ lifting y that starts at $[x_0]$, the homotopy class of the constant path at x_0, and ends at $[y]$. Since $[y]$ was an arbitrary point in $\tilde{X}$, this shows that $\tilde{X}$ is path-connected. To show that $\pi_1(\tilde{X}, [x_0]) = 0$ it suffices to show that the image of this group under p_* is trivial since p_* is injective. Elements in the image of p_* are represented by loops y at x_0 that lift to loops in $\tilde{X}$ at $[x_0]$. We have observed that the path $t \mapsto [y_t]$ lifts y starting at $[x_0]$, and for this lifted path to be a loop means that $[y_1] = [x_0]$. Since $y_1 = y$, this says that $[y] = [x_0]$, so y is nullhomotopic and the image of p_* is trivial.

This completes the construction of a simply-connected covering space $\tilde{X} \to X$.

In concrete cases one usually constructs a simply-connected covering space by more direct methods. For example, suppose X is the union of subspaces A and B for which simply-connected covering spaces $\tilde{A} \to A$ and $\tilde{B} \to B$ are already known. Then one can attempt to build a simply-connected covering space $\tilde{X} \to X$ by assembling copies of $\tilde{A}$ and $\tilde{B}$. For example, for $X = S^1 \vee S^1$, if we take A and B to be the two circles, then $\tilde{A}$ and $\tilde{B}$ are each $\mathbb{R}$, and we can build the simply-connected cover $\tilde{X}$ described earlier in this section by glueing together infinitely many copies of $\tilde{A}$ and $\tilde{B}$, the horizontal and vertical lines in $\tilde{X}$. Here is another illustration of this method:

Example 1.35. For integers $m, n \geq 2$, let $X_{m,n}$ be the quotient space of a cylinder $S^1 \times I$ under the identifications $(z, 0) \sim (e^{2\pi i/m} z, 0)$ and $(z, 1) \sim (e^{2\pi i/n} z, 1)$. Let $A \subset X$ and $B \subset X$ be the quotients of $S^1 \times [0, \frac{1}{2}]$ and $S^1 \times [\frac{1}{2}, 1]$, so A and B are the mapping cylinders of $z \mapsto z^m$ and $z \mapsto z^n$, with $A \cap B = S^1$. The simplest case is $m = n = 2$, when A and B are Möbius bands and $X_{2,2}$ is the Klein bottle. We encountered the complexes $X_{m,n}$ previously in analyzing torus knot complements in Example 1.24.

The figure for Example 1.29 at the end of the preceding section shows what A looks like in the typical case $m = 3$. We have $\pi_1(A) \approx \mathbb{Z}$, and the universal cover $\tilde{A}$ is homeomorphic to a product $C_m \times \mathbb{R}$ where C_m is the graph that is a cone on m points, as shown in the figure to the right. The situation for B is similar, and $\tilde{B}$ is homeomorphic to $C_n \times \mathbb{R}$. Now we attempt to build the universal cover $\tilde{X}_{m,n}$ from copies of $\tilde{A}$ and $\tilde{B}$. Start with a copy of $\tilde{A}$. Its boundary, the outer edges of its fins, consists of m copies of $\mathbb{R}$. Along each of these m boundary 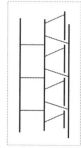 lines we attach a copy of $\tilde{B}$. Each of these copies of $\tilde{B}$ has one of its boundary lines attached to the initial copy of $\tilde{A}$, leaving $n - 1$ boundary lines free, and we attach a new copy of $\tilde{A}$ to each of these free boundary lines. Thus we now have $m(n - 1) + 1$ copies of $\tilde{A}$. Each of the newly attached copies of $\tilde{A}$ has $m - 1$ free boundary lines, and to each of these lines we attach a new copy of $\tilde{B}$. The process is now repeated ad infinitum in the evident way. Let $\tilde{X}_{m,n}$ be the resulting space.

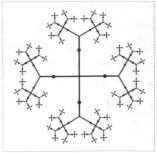

The product structures $\tilde{A} = C_m \times \mathbb{R}$ and $\tilde{B} = C_n \times \mathbb{R}$ give $\tilde{X}_{m,n}$ the structure of a product $T_{m,n} \times \mathbb{R}$ where $T_{m,n}$ is an infinite graph constructed by an inductive scheme just like the construction of $\tilde{X}_{m,n}$. Thus $T_{m,n}$ is the union of a sequence of finite subgraphs, each obtained from the preceding by attaching new copies of C_m or C_n. Each of these finite subgraphs deformation retracts onto the preceding one. The infinite concatenation of these deformation retractions, with the kth graph deformation retracting to the previous one during the time interval $[1/2^k, 1/2^{k-1}]$, gives a deformation retraction of $T_{m,n}$ onto the initial stage C_m. Since C_m is contractible, this means $T_{m,n}$ is contractible, hence also $\tilde{X}_{m,n}$, which is the product $T_{m,n} \times \mathbb{R}$. In particular, $\tilde{X}_{m,n}$ is simply-connected.

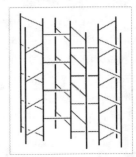

The map that projects each copy of $\tilde{A}$ in $\tilde{X}_{m,n}$ to A and each copy of $\tilde{B}$ to B is a covering space. To define this map precisely, choose a point $x_0 \in S^1$, and then the image of the line segment $\{x_0\} \times I$ in $X_{m,n}$ meets A in a line segment whose preimage in $\tilde{A}$ consists of an infinite number of line segments, appearing in the earlier figure as the horizontal segments spiraling around the central vertical axis. The picture in $\tilde{B}$ is similar, and when we glue together all the copies of $\tilde{A}$ and $\tilde{B}$ to form $\tilde{X}_{m,n}$, we do so in such a way that these horizontal segments always line up exactly. This decomposes $\tilde{X}_{m,n}$ into infinitely many rectangles, each formed from a rectangle in an $\tilde{A}$ and a rectangle in a $\tilde{B}$. The covering projection $\tilde{X}_{m,n} \to X_{m,n}$ is the quotient map that identifies all these rectangles.

Now we return to the general theory. The hypotheses for constructing a simply-connected covering space of X in fact suffice for constructing covering spaces realizing arbitrary subgroups of $\pi_1(X)$:

Proposition 1.36. *Suppose X is path-connected, locally path-connected, and semilocally simply-connected. Then for every subgroup $H \subset \pi_1(X, x_0)$ there is a covering space $p : X_H \to X$ such that $p_*(\pi_1(X_H, \tilde{x}_0)) = H$ for a suitably chosen basepoint $\tilde{x}_0 \in X_H$.*

Proof: For points $[\gamma]$, $[\gamma']$ in the simply-connected covering space $\tilde{X}$ constructed above, define $[\gamma] \sim [\gamma']$ to mean $\gamma(1) = \gamma'(1)$ and $[\gamma \cdot \overline{\gamma'}] \in H$. It is easy to see that this is an equivalence relation since H is a subgroup: it is reflexive since H contains the identity element, symmetric since H is closed under inverses, and transitive since H is closed under multiplication. Let X_H be the quotient space of $\tilde{X}$ obtained by identifying $[\gamma]$ with $[\gamma']$ if $[\gamma] \sim [\gamma']$. Note that if $\gamma(1) = \gamma'(1)$, then $[\gamma] \sim [\gamma']$ iff $[\gamma \cdot \eta] \sim [\gamma' \cdot \eta]$. This means that if any two points in basic neighborhoods $U_{[\gamma]}$

and $U_{[y']}$ are identified in X_H then the whole neighborhoods are identified. Hence the natural projection $X_H \to X$ induced by $[y] \mapsto y(1)$ is a covering space.

If we choose for the basepoint $\tilde{x}_0 \in X_H$ the equivalence class of the constant path c at x_0, then the image of $p_* : \pi_1(X_H, \tilde{x}_0) \to \pi_1(X, x_0)$ is exactly H. This is because for a loop y in X based at x_0, its lift to $\tilde{X}$ starting at $[c]$ ends at $[y]$, so the image of this lifted path in X_H is a loop iff $[y] \sim [c]$, or equivalently, $[y] \in H$. $\qquad \square$

Having taken care of the existence of covering spaces of X corresponding to all subgroups of $\pi_1(X)$, we turn now to the question of uniqueness. More specifically, we are interested in uniqueness up to isomorphism, where an **isomorphism** between covering spaces $p_1 : \tilde{X}_1 \to X$ and $p_2 : \tilde{X}_2 \to X$ is a homeomorphism $f : \tilde{X}_1 \to \tilde{X}_2$ such that $p_1 = p_2 f$. This condition means exactly that f preserves the covering space structures, taking $p_1^{-1}(x)$ to $p_2^{-1}(x)$ for each $x \in X$. The inverse f^{-1} is then also an isomorphism, and the composition of two isomorphisms is an isomorphism, so we have an equivalence relation.

Proposition 1.37. *If X is path-connected and locally path-connected, then two path-connected covering spaces $p_1 : \tilde{X}_1 \to X$ and $p_2 : \tilde{X}_2 \to X$ are isomorphic via an isomorphism $f : \tilde{X}_1 \to \tilde{X}_2$ taking a basepoint $\tilde{x}_1 \in p_1^{-1}(x_0)$ to a basepoint $\tilde{x}_2 \in p_2^{-1}(x_0)$ iff $p_{1*}(\pi_1(\tilde{X}_1, \tilde{x}_1)) = p_{2*}(\pi_1(\tilde{X}_2, \tilde{x}_2))$.*

Proof: If there is an isomorphism $f : (\tilde{X}_1, \tilde{x}_1) \to (\tilde{X}_2, \tilde{x}_2)$, then from the two relations $p_1 = p_2 f$ and $p_2 = p_1 f^{-1}$ it follows that $p_{1*}(\pi_1(\tilde{X}_1, \tilde{x}_1)) = p_{2*}(\pi_1(\tilde{X}_2, \tilde{x}_2))$. Conversely, suppose that $p_{1*}(\pi_1(\tilde{X}_1, \tilde{x}_1)) = p_{2*}(\pi_1(\tilde{X}_2, \tilde{x}_2))$. By the lifting criterion, we may lift p_1 to a map $\tilde{p}_1 : (\tilde{X}_1, \tilde{x}_1) \to (\tilde{X}_2, \tilde{x}_2)$ with $p_2 \tilde{p}_1 = p_1$. Symmetrically, we obtain $\tilde{p}_2 : (\tilde{X}_2, \tilde{x}_2) \to (\tilde{X}_1, \tilde{x}_1)$ with $p_1 \tilde{p}_2 = p_2$. Then by the unique lifting property, $\tilde{p}_1 \tilde{p}_2 = \mathbb{1}$ and $\tilde{p}_2 \tilde{p}_1 = \mathbb{1}$ since these composed lifts fix the basepoints. Thus $\tilde{p}_1$ and $\tilde{p}_2$ are inverse isomorphisms. $\qquad \square$

We have proved the first half of the following classification theorem:

Theorem 1.38. *Let X be path-connected, locally path-connected, and semilocally simply-connected. Then there is a bijection between the set of basepoint-preserving isomorphism classes of path-connected covering spaces $p : (\tilde{X}, \tilde{x}_0) \to (X, x_0)$ and the set of subgroups of $\pi_1(X, x_0)$, obtained by associating the subgroup $p_*(\pi_1(\tilde{X}, \tilde{x}_0))$ to the covering space $(\tilde{X}, \tilde{x}_0)$. If basepoints are ignored, this correspondence gives a bijection between isomorphism classes of path-connected covering spaces $p : \tilde{X} \to X$ and conjugacy classes of subgroups of $\pi_1(X, x_0)$.*

Proof: It remains only to prove the last statement. We show that for a covering space $p : (\tilde{X}, \tilde{x}_0) \to (X, x_0)$, changing the basepoint $\tilde{x}_0$ within $p^{-1}(x_0)$ corresponds exactly to changing $p_*(\pi_1(\tilde{X}, \tilde{x}_0))$ to a conjugate subgroup of $\pi_1(X, x_0)$. Suppose that $\tilde{x}_1$ is another basepoint in $p^{-1}(x_0)$, and let $\tilde{y}$ be a path from $\tilde{x}_0$ to $\tilde{x}_1$. Then $\tilde{y}$ projects to a loop y in X representing some element $g \in \pi_1(X, x_0)$. Set $H_i = p_*(\pi_1(\tilde{X}, \tilde{x}_i))$ for $i = 0, 1$. We have an inclusion $g^{-1} H_0 g \subset H_1$ since for $\tilde{f}$ a loop at $\tilde{x}_0$, $\bar{\tilde{y}} \cdot \tilde{f} \cdot \tilde{y}$ is a loop at $\tilde{x}_1$. Similarly we have $g H_1 g^{-1} \subset H_0$. Conjugating the latter relation by g^{-1} gives $H_1 \subset g^{-1} H_0 g$, so $g^{-1} H_0 g = H_1$. Thus, changing the basepoint from $\tilde{x}_0$ to $\tilde{x}_1$ changes H_0 to the conjugate subgroup $H_1 = g^{-1} H_0 g$.

Conversely, to change H_0 to a conjugate subgroup $H_1 = g^{-1} H_0 g$, choose a loop y representing g, lift this to a path $\tilde{y}$ starting at $\tilde{x}_0$, and let $\tilde{x}_1 = \tilde{y}(1)$. The preceding argument then shows that we have the desired relation $H_1 = g^{-1} H_0 g$. $\qquad\square$

A consequence of the lifting criterion is that a simply-connected covering space of a path-connected, locally path-connected space X is a covering space of every other path-connected covering space of X. A simply-connected covering space of X is therefore called a **universal cover**. It is unique up to isomorphism, so one is justified in calling it *the* universal cover.

More generally, there is a partial ordering on the various path-connected covering spaces of X, according to which ones cover which others. This corresponds to the partial ordering by inclusion of the corresponding subgroups of $\pi_1(X)$, or conjugacy classes of subgroups if basepoints are ignored.

Representing Covering Spaces by Permutations

We wish to describe now another way of classifying the different covering spaces of a connected, locally path-connected, semilocally simply-connected space X, without restricting just to connected covering spaces. To give the idea, consider the 3-sheeted covering spaces of S^1. There are three of these, $\tilde{X}_1$, $\tilde{X}_2$, and $\tilde{X}_3$, with the subscript indicating the number of components. For each of these covering spaces $p : \tilde{X}_i \to S^1$ the three different lifts of a loop in S^1 generating $\pi_1(S^1, x_0)$ determine a permutation of $p^{-1}(x_0)$ sending the starting point of the lift to the ending point of the lift. For $\tilde{X}_1$ this is a cyclic permutation, for $\tilde{X}_2$ it is a transposition of two points fixing the third point, and for $\tilde{X}_3$ it is the identity permutation. These permutations obviously determine the covering spaces uniquely, up to isomorphism. The same would be true for n-sheeted covering spaces of S^1 for arbitrary n, even for n infinite.

The covering spaces of $S^1 \vee S^1$ can be encoded using the same idea. Referring back to the large table of examples near the beginning of this section, we see in the covering space (1) that the loop a lifts to the identity permutation of the two vertices and b lifts to the permutation that transposes the two vertices. In (2), both a and b lift to transpositions of the two vertices. In (3) and (4), a and b lift to transpositions of different pairs of the three vertices, while in (5) and (6) they lift to cyclic permutations of the vertices. In (11) the vertices can be labeled by $\mathbb{Z}$, with a lifting to the identity permutation and b lifting to the shift $n \mapsto n + 1$. Indeed, one can see from these examples that a covering space of $S^1 \vee S^1$ is nothing more than an efficient graphical representation of a pair of permutations of a given set.

This idea of lifting loops to permutations generalizes to arbitrary covering spaces. For a covering space $p : \tilde{X} \rightarrow X$, a path y in X has a unique lift $\tilde{y}$ starting at a given point of $p^{-1}(y(0))$, so we obtain a well-defined map $L_y : p^{-1}(y(0)) \rightarrow p^{-1}(y(1))$ by sending the starting point $\tilde{y}(0)$ of each lift $\tilde{y}$ to its ending point $\tilde{y}(1)$. It is evident that L_y is a bijection since $L_{\bar{y}}$ is its inverse. For a composition of paths $y \cdot \eta$ we have $L_{y \cdot \eta} = L_\eta L_y$, rather than $L_y L_\eta$, since composition of paths is written from left to right while composition of functions is written from right to left. To compensate for this, let us modify the definition by replacing L_y by its inverse. Thus the new L_y is a bijection $p^{-1}(y(1)) \rightarrow p^{-1}(y(0))$, and $L_{y \cdot \eta} = L_y L_\eta$. Since L_y depends only on the homotopy class of y, this means that if we restrict attention to loops at a basepoint $x_0 \in X$, then the association $y \mapsto L_y$ gives a homomorphism from $\pi_1(X, x_0)$ to the group of permutations of $p^{-1}(x_0)$. This is called *the action of $\pi_1(X, x_0)$ on the fiber* $p^{-1}(x_0)$.

Let us see how the covering space $p : \tilde{X} \rightarrow X$ can be reconstructed from the associated action of $\pi_1(X, x_0)$ on the fiber $F = p^{-1}(x_0)$, assuming that X is path-connected, locally path-connected, and semilocally simply-connected, so it has a universal cover $\tilde{X}_0 \rightarrow X$. We can take the points of $\tilde{X}_0$ to be homotopy classes of paths in X starting at x_0, as in the general construction of a universal cover. Define a map $h : \tilde{X}_0 \times F \rightarrow \tilde{X}$ sending a pair $([y], \tilde{x}_0)$ to $\tilde{y}(1)$ where $\tilde{y}$ is the lift of y to $\tilde{X}$ starting at $\tilde{x}_0$. Then h is continuous, and in fact a local homeomorphism, since a neighborhood of $([y], \tilde{x}_0)$ in $\tilde{X}_0 \times F$ consists of the pairs $([y \cdot \eta], \tilde{x}_0)$ with η a path in a suitable neighborhood of $y(1)$. It is obvious that h is surjective since X is path-connected. If h were injective as well, it would be a homeomorphism, which is unlikely since $\tilde{X}$ is probably not homeomorphic to $\tilde{X}_0 \times F$. Even if h is not injective, it will induce a homeomorphism from some quotient space of $\tilde{X}_0 \times F$ onto $\tilde{X}$. To see what this quotient space is, suppose $h([y], \tilde{x}_0) = h([y'], \tilde{x}_0')$. Then y and y' are both paths from x_0 to the same endpoint, and from the figure we see that $\tilde{x}_0' = L_{y' \cdot \bar{y}}(\tilde{x}_0)$. Letting λ be the loop $y' \cdot \bar{y}$, this means that $h([y], \tilde{x}_0) = h([\lambda \cdot y], L_\lambda(\tilde{x}_0))$. Conversely, for any loop λ we have $h([y], \tilde{x}_0) = h([\lambda \cdot y], L_\lambda(\tilde{x}_0))$. Thus h induces a well-defined map to $\tilde{X}$ from the quotient space of $\tilde{X}_0 \times F$ obtained by identifying $([y], \tilde{x}_0)$ with $([\lambda \cdot y], L_\lambda(\tilde{x}_0))$

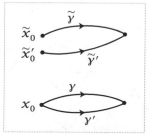

for each $[\lambda] \in \pi_1(X, x_0)$. Let this quotient space be denoted $\tilde{X}_\rho$ where ρ is the homomorphism from $\pi_1(X, x_0)$ to the permutation group of F specified by the action.

Notice that the definition of $\tilde{X}_\rho$ makes sense whenever we are given an action ρ of $\pi_1(X, x_0)$ on a set F. There is a natural projection $\tilde{X}_\rho \rightarrow X$ sending $([y], \tilde{x}_0)$ to $y(1)$, and this is a covering space since if $U \subset X$ is an open set over which the universal cover $\tilde{X}_0$ is a product $U \times \pi_1(X, x_0)$, then the identifications defining $\tilde{X}_\rho$

simply collapse $U \times \pi_1(X, x_0) \times F$ to $U \times F$.

Returning to our given covering space $\tilde{X} \to X$ with associated action ρ, the map $\tilde{X}_\rho \to \tilde{X}$ induced by h is a bijection and therefore a homeomorphism since h was a local homeomorphism. Since this homeomorphism $\tilde{X}_\rho \to \tilde{X}$ takes each fiber of $\tilde{X}_\rho$ to the corresponding fiber of $\tilde{X}$, it is an isomorphism of covering spaces.

If two covering spaces $p_1 : \tilde{X}_1 \to X$ and $p_2 : \tilde{X}_2 \to X$ are isomorphic, one may ask how the corresponding actions of $\pi_1(X, x_0)$ on the fibers F_1 and F_2 over x_0 are related. An isomorphism $h : \tilde{X}_1 \to \tilde{X}_2$ restricts to a bijection $F_1 \to F_2$, and evidently $L_y(h(\tilde{x}_0)) = h(L_y(\tilde{x}_0))$. Using the less cumbersome notation $y\tilde{x}_0$ for $L_y(\tilde{x}_0)$, this relation can be written more concisely as $yh(\tilde{x}_0) = h(y\tilde{x}_0)$. A bijection $F_1 \to F_2$ with this property is what one would naturally call an *isomorphism of sets with* $\pi_1(X, x_0)$ *action*. Thus isomorphic covering spaces have isomorphic actions on fibers. The converse is also true, and easy to prove. One just observes that for isomorphic actions ρ_1 and ρ_2, an isomorphism $h : F_1 \to F_2$ induces a map $\tilde{X}_{\rho_1} \to \tilde{X}_{\rho_2}$ and h^{-1} induces a similar map in the opposite direction, such that the compositions of these two maps, in either order, are the identity.

This shows that n-sheeted covering spaces of X are classified by equivalence classes of homomorphisms $\pi_1(X, x_0) \to \Sigma_n$, where Σ_n is the symmetric group on n symbols and the equivalence relation identifies a homomorphism ρ with each of its conjugates $h^{-1}\rho h$ by elements $h \in \Sigma_n$. The study of the various homomorphisms from a given group to Σ_n is a very classical topic in group theory, so we see that this algebraic question has a nice geometric interpretation.

Deck Transformations and Group Actions

For a covering space $p : \tilde{X} \to X$ the isomorphisms $\tilde{X} \to \tilde{X}$ are called **deck transformations** or **covering transformations**. These form a group $G(\tilde{X})$ under composition. For example, for the covering space $p : \mathbb{R} \to S^1$ projecting a vertical helix onto a circle, the deck transformations are the vertical translations taking the helix onto itself, so $G(\tilde{X}) \approx \mathbb{Z}$ in this case. For the n-sheeted covering space $S^1 \to S^1$, $z \mapsto z^n$, the deck transformations are the rotations of S^1 through angles that are multiples of $2\pi/n$, so $G(\tilde{X}) = \mathbb{Z}_n$.

By the unique lifting property, a deck transformation is completely determined by where it sends a single point, assuming $\tilde{X}$ is path-connected. In particular, only the identity deck transformation can fix a point of $\tilde{X}$.

A covering space $p : \tilde{X} \to X$ is called **normal** if for each $x \in X$ and each pair of lifts $\tilde{x}, \tilde{x}'$ of x there is a deck transformation taking $\tilde{x}$ to $\tilde{x}'$. For example, the covering space $\mathbb{R} \to S^1$ and the n-sheeted covering spaces $S^1 \to S^1$ are normal. Intuitively, a normal covering space is one with maximal symmetry. This can be seen in the covering spaces of $S^1 \vee S^1$ shown in the table earlier in this section, where the normal covering spaces are (1), (2), (5)–(8), and (11). Note that in (7) the group of deck transformations is $\mathbb{Z}_4$ while in (8) it is $\mathbb{Z}_2 \times \mathbb{Z}_2$.

Sometimes normal covering spaces are called regular covering spaces. The term 'normal' is motivated by the following result.

Proposition 1.39. *Let $p : (\tilde{X}, \tilde{x}_0) \to (X, x_0)$ be a path-connected covering space of the path-connected, locally path-connected space X, and let H be the subgroup $p_*(\pi_1(\tilde{X}, \tilde{x}_0)) \subset \pi_1(X, x_0)$. Then:*

(a) *This covering space is normal iff H is a normal subgroup of $\pi_1(X, x_0)$.*

(b) *$G(\tilde{X})$ is isomorphic to the quotient $N(H)/H$ where $N(H)$ is the normalizer of H in $\pi_1(X, x_0)$.*

In particular, $G(\tilde{X})$ is isomorphic to $\pi_1(X, x_0)/H$ if $\tilde{X}$ is a normal covering. Hence for the universal cover $\tilde{X} \to X$ we have $G(\tilde{X}) \approx \pi_1(X)$.

Proof: We observed earlier in the proof of the classification theorem that changing the basepoint $\tilde{x}_0 \in p^{-1}(x_0)$ to $\tilde{x}_1 \in p^{-1}(x_0)$ corresponds precisely to conjugating H by an element $[\gamma] \in \pi_1(X, x_0)$ where γ lifts to a path $\tilde{\gamma}$ from $\tilde{x}_0$ to $\tilde{x}_1$. Thus $[\gamma]$ is in the normalizer $N(H)$ iff $p_*(\pi_1(\tilde{X}, \tilde{x}_0)) = p_*(\pi_1(\tilde{X}, \tilde{x}_1))$, which by the lifting criterion is equivalent to the existence of a deck transformation taking $\tilde{x}_0$ to $\tilde{x}_1$. Hence the covering space is normal iff $N(H) = \pi_1(X, x_0)$, that is, iff H is a normal subgroup of $\pi_1(X, x_0)$.

Define $\varphi : N(H) \to G(\tilde{X})$ sending $[\gamma]$ to the deck transformation τ taking $\tilde{x}_0$ to $\tilde{x}_1$, in the notation above. Then φ is a homomorphism, for if γ' is another loop corresponding to the deck transformation τ' taking $\tilde{x}_0$ to $\tilde{x}_1'$ then $\gamma \cdot \gamma'$ lifts to $\tilde{\gamma} \cdot (\tau(\tilde{\gamma}'))$, a path from $\tilde{x}_0$ to $\tau(\tilde{x}_1') = \tau\tau'(\tilde{x}_0)$, so $\tau\tau'$ is the deck transformation corresponding to $[\gamma][\gamma']$. By the preceding paragraph φ is surjective. Its kernel consists of classes $[\gamma]$ lifting to loops in $\tilde{X}$. These are exactly the elements of $p_*(\pi_1(\tilde{X}, \tilde{x}_0)) = H$. $\square$

The group of deck transformations is a special case of the general notion of 'groups acting on spaces.' Given a group G and a space Y, then an **action** of G on Y is a homomorphism ρ from G to the group Homeo(Y) of all homeomorphisms from Y to itself. Thus to each $g \in G$ is associated a homeomorphism $\rho(g) : Y \to Y$, which for notational simplicity we write simply as $g : Y \to Y$. For ρ to be a homomorphism amounts to requiring that $g_1(g_2(y)) = (g_1 g_2)(y)$ for all $g_1, g_2 \in G$ and $y \in Y$. If ρ is injective then it identifies G with a subgroup of Homeo(Y), and in practice not much is lost in assuming ρ is an inclusion $G \hookrightarrow$ Homeo(Y) since in any case the subgroup $\rho(G) \subset$ Homeo(Y) contains all the topological information about the action.

We shall be interested in actions satisfying the following condition:

(∗)   Each $y \in Y$ has a neighborhood U such that all the images $g(U)$ for varying $g \in G$ are disjoint. In other words, $g_1(U) \cap g_2(U) \neq \varnothing$ implies $g_1 = g_2$.

The action of the deck transformation group $G(\tilde{X})$ on $\tilde{X}$ satisfies (∗). To see this, let $\tilde{U} \subset \tilde{X}$ project homeomorphically to $U \subset X$. If $g_1(\tilde{U}) \cap g_2(\tilde{U}) \neq \varnothing$ for some $g_1, g_2 \in G(\tilde{X})$, then $g_1(\tilde{x}_1) = g_2(\tilde{x}_2)$ for some $\tilde{x}_1, \tilde{x}_2 \in \tilde{U}$. Since $\tilde{x}_1$ and $\tilde{x}_2$ must lie in the same set $p^{-1}(x)$, which intersects $\tilde{U}$ in only one point, we must have $\tilde{x}_1 = \tilde{x}_2$. Then $g_1^{-1}g_2$ fixes this point, so $g_1^{-1}g_2 = \mathbb{1}$ and $g_1 = g_2$.

Note that in (∗) it suffices to take g_1 to be the identity since $g_1(U) \cap g_2(U) \neq \varnothing$ is equivalent to $U \cap g_1^{-1}g_2(U) \neq \varnothing$. Thus we have the equivalent condition that $U \cap g(U) \neq \varnothing$ only when g is the identity.

Given an action of a group G on a space Y, we can form a space Y/G, the quotient space of Y in which each point y is identified with all its images $g(y)$ as g ranges over G. The points of Y/G are thus the **orbits** $Gy = \{g(y) \mid g \in G\}$ in Y, and Y/G is called the **orbit space** of the action. For example, for a normal covering space $\tilde{X} \to X$, the orbit space $\tilde{X}/G(\tilde{X})$ is just X.

Proposition 1.40. *If an action of a group G on a space Y satisfies* (∗), *then:*
(a) *The quotient map $p : Y \to Y/G$, $p(y) = Gy$, is a normal covering space.*
(b) *G is the group of deck transformations of this covering space $Y \to Y/G$ if Y is path-connected.*
(c) *G is isomorphic to $\pi_1(Y/G)/p_*(\pi_1(Y))$ if Y is path-connected and locally path-connected.*

Proof: Given an open set $U \subset Y$ as in condition (∗), the quotient map p simply identifies all the disjoint homeomorphic sets $\{g(U) \mid g \in G\}$ to a single open set $p(U)$ in Y/G. By the definition of the quotient topology on Y/G, p restricts to a homeomorphism from $g(U)$ onto $p(U)$ for each $g \in G$ so we have a covering space. Each element of G acts as a deck transformation, and the covering space is normal since $g_2 g_1^{-1}$ takes $g_1(U)$ to $g_2(U)$. The deck transformation group contains G as a subgroup, and equals this subgroup if Y is path-connected, since if f is any deck transformation, then for an arbitrarily chosen point $y \in Y$, y and $f(y)$ are in the same orbit and there is a $g \in G$ with $g(y) = f(y)$, hence $f = g$ since deck transformations of a path-connected covering space are uniquely determined by where they send a point. The final statement of the proposition is immediate from part (b) of Proposition 1.39.   □

In view of the preceding proposition, we shall call an action satisfying (∗) a **covering space action**. This is not standard terminology, but there does not seem to be a universally accepted name for actions satisfying (∗). Sometimes these are called 'properly discontinuous' actions, but more often this rather unattractive term means something weaker: Every point $x \in X$ has a neighborhood U such that $U \cap g(U)$ is nonempty for only finitely many $g \in G$. Many symmetry groups have this proper discontinuity property without satisfying $(*)$, for example the group of symmetries of the familiar tiling of $\mathbb{R}^2$ by regular hexagons. The reason why the action of this group on $\mathbb{R}^2$ fails to satisfy $(*)$ is that there are **fixed points**: points y for which there is a nontrivial element $g \in G$ with $g(y) = y$. For example, the vertices of the hexagons are fixed by the 120 degree rotations about these points, and the midpoints of edges are fixed by 180 degree rotations. An action without fixed points is called a **free** action. Thus for a free action of G on Y, only the identity element of G fixes any point of Y. This is equivalent to requiring that all the images $g(y)$ of each $y \in Y$ are distinct, or in other words $g_1(y) = g_2(y)$ only when $g_1 = g_2$, since $g_1(y) = g_2(y)$ is equivalent to $g_1^{-1}g_2(y) = y$. Though condition $(*)$ implies freeness, the converse is not always true. An example is the action of $\mathbb{Z}$ on S^1 in which a generator of $\mathbb{Z}$ acts by rotation through an angle α that is an irrational multiple of 2π. In this case each orbit $\mathbb{Z}y$ is dense in S^1, so condition $(*)$ cannot hold since it implies that orbits are discrete subspaces. An exercise at the end of the section is to show that for actions on Hausdorff spaces, freeness plus proper discontinuity implies condition $(*)$. Note that proper discontinuity is automatic for actions by a finite group.

Example 1.41. Let Y be the closed orientable surface of genus 11, an '11-hole torus' as shown in the figure. This has a 5-fold rotational symme-
try, generated by a rotation of angle $2\pi/5$. Thus we have the cyclic group $\mathbb{Z}_5$ acting on Y, and the condition $(*)$ is obviously satisfied. The quotient space $Y/\mathbb{Z}_5$ is a surface of genus 3, obtained from one of the five subsurfaces of Y cut off by the circles $C_1, \cdots, C_5$ by identifying its two boundary circles C_i and C_{i+1} to form the circle C as shown. Thus we have a covering space $M_{11} \to M_3$ where M_g denotes the closed orientable surface of genus g. In particular, we see that $\pi_1(M_3)$ contains the 'larger' group $\pi_1(M_{11})$ as a normal subgroup of index 5, with quotient $\mathbb{Z}_5$. This example obviously generalizes by re-

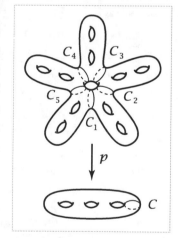

placing the two holes in each 'arm' of M_{11} by m holes and the 5-fold symmetry by n-fold symmetry. This gives a covering space $M_{mn+1} \to M_{m+1}$. An exercise in §2.2 is to show by an Euler characteristic argument that if there is a covering space $M_g \to M_h$ then $g = mn + 1$ and $h = m + 1$ for some m and n.

As a special case of the final statement of the preceding proposition we see that for a covering space action of a group G on a simply-connected locally path-connected space Y, the orbit space Y/G has fundamental group isomorphic to G. Under this isomorphism an element $g \in G$ corresponds to a loop in Y/G that is the projection of a path in Y from a chosen basepoint y_0 to $g(y_0)$. Any two such paths are homotopic since Y is simply-connected, so we get a well-defined element of $\pi_1(Y/G)$ associated to g.

This method for computing fundamental groups via group actions on simply-connected spaces is essentially how we computed $\pi_1(S^1)$ in §1.1, via the covering space $\mathbb{R} \to S^1$ arising from the action of $\mathbb{Z}$ on $\mathbb{R}$ by translations. This is a useful general technique for computing fundamental groups, in fact. Here are some examples illustrating this idea.

Example 1.42. Consider the grid in $\mathbb{R}^2$ formed by the horizontal and vertical lines through points in $\mathbb{Z}^2$. Let us decorate this grid with arrows in either of the two ways shown in the figure, the difference between the two cases being that in the second case the horizontal arrows in adjacent lines point in opposite directions. The group G consisting of all symmetries of the first decorated grid is isomorphic to $\mathbb{Z} \times \mathbb{Z}$ 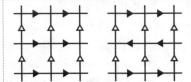 since it consists of all translations $(x, y) \mapsto (x + m, y + n)$ for $m, n \in \mathbb{Z}$. For the second grid the symmetry group G contains a subgroup of translations of the form $(x, y) \mapsto (x + m, y + 2n)$ for $m, n \in \mathbb{Z}$, but there are also glide-reflection symmetries consisting of vertical translation by an odd integer distance followed by reflection across a vertical line, either a vertical line of the grid or a vertical line halfway between two adjacent grid lines. For both decorated grids there are elements of G taking any square to any other, but only the identity element of G takes a square to itself. The minimum distance any point is moved by a nontrivial element of G is 1, which easily implies the covering space condition $(*)$. The orbit space $\mathbb{R}^2/G$ is the quotient space of a square in the grid with opposite edges identified according to the arrows. Thus we see that the fundamental groups of the torus and the Klein bottle are the symmetry groups G in the two cases. In the second case the subgroup of G formed by the translations has index two, and the orbit space for this subgroup is a torus forming a two-sheeted covering space of the Klein bottle.

Example 1.43: $\mathbb{RP}^n$. The antipodal map of S^n, $x \mapsto -x$, generates an action of $\mathbb{Z}_2$ on S^n with orbit space $\mathbb{RP}^n$, real projective n-space, as defined in Example 0.4. The action is a covering space action since each open hemisphere in S^n is disjoint from its antipodal image. As we saw in Proposition 1.14, S^n is simply-connected if $n \geq 2$, so from the covering space $S^n \to \mathbb{RP}^n$ we deduce that $\pi_1(\mathbb{RP}^n) \approx \mathbb{Z}_2$ for $n \geq 2$. A generator for $\pi_1(\mathbb{RP}^n)$ is any loop obtained by projecting a path in S^n connecting two antipodal points. One can see explicitly that such a loop y has order two in $\pi_1(\mathbb{RP}^n)$ if $n \geq 2$ since the composition $y \cdot y$ lifts to a loop in S^n, and this can be homotoped to the trivial loop since $\pi_1(S^n) = 0$, so the projection of this homotopy into $\mathbb{RP}^n$ gives a nullhomotopy of $y \cdot y$.

One may ask whether there are other finite groups that act freely on S^n, defining covering spaces $S^n \to S^n/G$. We will show in Proposition 2.29 that $\mathbb{Z}_2$ is the only possibility when n is even, but for odd n the question is much more difficult. It is easy to construct a free action of any cyclic group $\mathbb{Z}_m$ on S^{2k-1}, the action generated by the rotation $v \mapsto e^{2\pi i/m} v$ of the unit sphere S^{2k-1} in $\mathbb{C}^k = \mathbb{R}^{2k}$. This action is free since an equation $v = e^{2\pi i \ell/m} v$ with $0 < \ell < m$ implies $v = 0$, but 0 is not a point of S^{2k-1}. The orbit space $S^{2k-1}/\mathbb{Z}_m$ is one of a family of spaces called *lens spaces* defined in Example 2.43.

There are also noncyclic finite groups that act freely as rotations of S^n for odd $n > 1$. These actions are classified quite explicitly in [Wolf 1984]. Examples in the simplest case $n = 3$ can be produced as follows. View $\mathbb{R}^4$ as the quaternion algebra $\mathbb{H}$. Multiplication of quaternions satisfies $|ab| = |a||b|$ where $|a|$ denotes the usual Euclidean length of a vector $a \in \mathbb{R}^4$. Thus if a and b are unit vectors, so is ab, and hence quaternion multiplication defines a map $S^3 \times S^3 \to S^3$. This in fact makes S^3 into a group, though associativity is all we need now since associativity implies that any subgroup G of S^3 acts on S^3 by left-multiplication, $g(x) = gx$. This action is free since an equation $x = gx$ in the division algebra $\mathbb{H}$ implies $g = 1$ or $x = 0$. As a concrete example, G could be the familiar quaternion group $Q_8 = \{\pm 1, \pm i, \pm j, \pm k\}$ from group theory. More generally, for a positive integer m, let Q_{4m} be the subgroup of S^3 generated by the two quaternions $a = e^{\pi i/m}$ and $b = j$. Thus a has order $2m$ and b has order 4. The easily verified relations $a^m = b^2 = -1$ and $bab^{-1} = a^{-1}$ imply that the subgroup $\mathbb{Z}_{2m}$ generated by a is normal and of index 2 in Q_{4m}. Hence Q_{4m} is a group of order $4m$, called the *generalized quaternion group*. Another common name for this group is the *binary dihedral group* D_{4m}^* since its quotient by the subgroup $\{\pm 1\}$ is the ordinary dihedral group D_{2m} of order $2m$.

Besides the groups $Q_{4m} = D_{4m}^*$ there are just three other noncyclic finite subgroups of S^3: the binary tetrahedral, octahedral, and icosahedral groups T_{24}^*, O_{48}^*, and I_{120}^*, of orders indicated by the subscripts. These project two-to-one onto the groups of rotational symmetries of a regular tetrahedron, octahedron (or cube), and icosahedron (or dodecahedron). In fact, it is not hard to see that the homomorphism $S^3 \to SO(3)$ sending $u \in S^3 \subset \mathbb{H}$ to the isometry $v \to u^{-1}vu$ of $\mathbb{R}^3$, viewing $\mathbb{R}^3$ as the 'pure imaginary' quaternions $v = ai + bj + ck$, is surjective with kernel $\{\pm 1\}$. Then the groups D_{4m}^*, T_{24}^*, O_{48}^*, I_{120}^* are the preimages in S^3 of the groups of rotational symmetries of a regular polygon or polyhedron in $\mathbb{R}^3$.

There are two conditions that a finite group G acting freely on S^n must satisfy:

(a) Every abelian subgroup of G is cyclic. This is equivalent to saying that G contains no subgroup $\mathbb{Z}_p \times \mathbb{Z}_p$ with p prime.

(b) G contains at most one element of order 2.

A proof of (a) is sketched in an exercise for §4.2. For a proof of (b) the original source [Milnor 1957] is recommended reading. The groups satisfying (a) have been completely classified; see [Brown 1982], section VI.9, for details. An example of a group satisfying (a) but not (b) is the dihedral group D_{2m} for odd $m > 1$.

There is also a much more difficult converse: A finite group satisfying (a) and (b) acts freely on S^n for some n. References for this are [Madsen, Thomas, & Wall 1976] and [Davis & Milgram 1985]. There is also almost complete information about which n's are possible for a given group.

Example 1.44. In Example 1.35 we constructed a contractible 2-complex $\widetilde{X}_{m,n} = T_{m,n} \times \mathbb{R}$ as the universal cover of a finite 2-complex $X_{m,n}$ that was the union of the mapping cylinders of the two maps $S^1 \to S^1$, $z \mapsto z^m$ and $z \mapsto z^n$. The group of deck transformations of this covering space is therefore the fundamental group $\pi_1(X_{m,n})$. From van Kampen's theorem applied to the decomposition of $X_{m,n}$ into the two mapping cylinders we have the presentation $\langle a, b \mid a^m b^{-n} \rangle$ for this group $G_{m,n} = \pi_1(X_{m,n})$. It is interesting to look at the action of $G_{m,n}$ on $\widetilde{X}_{m,n}$ more closely. We described a decomposition of $\widetilde{X}_{m,n}$ into rectangles, with $X_{m,n}$ the quotient of one rectangle. These rectangles in fact define a cell structure on $\widetilde{X}_{m,n}$ lifting a cell structure on $X_{m,n}$ with two vertices, three edges, and one 2-cell. The group $G_{m,n}$ is thus a group of symmetries of this cell structure on $\widetilde{X}_{m,n}$. If we orient the three edges of $X_{m,n}$ and lift these orientations to the edges of $\widetilde{X}_{m,n}$, then $G_{m,n}$ is the group of all symmetries of $\widetilde{X}_{m,n}$ preserving the orientations of edges. For example, the element a acts as a 'screw motion' about an axis that is a vertical line $\{v_a\} \times \mathbb{R}$ with v_a a vertex of $T_{m,n}$, and b acts similarly for a vertex v_b.

Since the action of $G_{m,n}$ on $\widetilde{X}_{m,n}$ preserves the cell structure, it also preserves the product structure $T_{m,n} \times \mathbb{R}$. This means that there are actions of $G_{m,n}$ on $T_{m,n}$ and $\mathbb{R}$ such that the action on the product $X_{m,n} = T_{m,n} \times \mathbb{R}$ is the diagonal action $g(x, y) = (g(x), g(y))$ for $g \in G_{m,n}$. If we make the rectangles of unit height in the $\mathbb{R}$ coordinate, then the element $a^m = b^n$ acts on $\mathbb{R}$ as unit translation, while a acts by $1/m$ translation and b by $1/n$ translation. The translation actions of a and b on $\mathbb{R}$ generate a group of translations of $\mathbb{R}$ that is infinite cyclic, generated by translation by the reciprocal of the least common multiple of m and n.

The action of $G_{m,n}$ on $T_{m,n}$ has kernel consisting of the powers of the element $a^m = b^n$. This infinite cyclic subgroup is precisely the center of $G_{m,n}$, as we saw in Example 1.24. There is an induced action of the quotient group $\mathbb{Z}_m * \mathbb{Z}_n$ on $T_{m,n}$, but this is not a free action since the elements a and b and all their conjugates fix vertices of $T_{m,n}$. On the other hand, if we restrict the action of $G_{m,n}$ on $T_{m,n}$ to the kernel K of the map $G_{m,n} \to \mathbb{Z}$ given by the action of $G_{m,n}$ on the $\mathbb{R}$ factor of $X_{m,n}$, then we do obtain a free action of K on $T_{m,n}$. Since this action takes vertices to vertices and edges to edges, it is a covering space action, so K is a free group, the fundamental group of the graph $T_{m,n}/K$. An exercise at the end of the section is to determine $T_{m,n}/K$ explicitly and compute the number of generators of K.

Cayley Complexes

Covering spaces can be used to describe a very classical method for viewing groups geometrically as graphs. Recall from Corollary 1.28 how we associated to each group presentation $G = \langle g_\alpha \mid r_\beta \rangle$ a 2-dimensional cell complex X_G with $\pi_1(X_G) \approx G$ by taking a wedge-sum of circles, one for each generator g_α, and then attaching a 2-cell for each relator r_β. We can construct a cell complex $\tilde{X}_G$ with a covering space action of G such that $\tilde{X}_G/G = X_G$ in the following way. Let the vertices of $\tilde{X}_G$ be the elements of G themselves. Then, at each vertex $g \in G$, insert an edge joining g to gg_α for each of the chosen generators g_α. The resulting graph is known as the **Cayley graph** of G with respect to the generators g_α. This graph is connected since every element of G is a product of g_α's, so there is a path in the graph joining each vertex to the identity vertex e. Each relation r_β determines a loop in the graph, starting at any vertex g, and we attach a 2-cell for each such loop. The resulting cell complex $\tilde{X}_G$ is the **Cayley complex** of G. The group G acts on $\tilde{X}_G$ by multiplication on the left. Thus, an element $g \in G$ sends a vertex $g' \in G$ to the vertex gg', and the edge from g' to $g'g_\alpha$ is sent to the edge from gg' to $gg'g_\alpha$. The action extends to 2-cells in the obvious way. This is clearly a covering space action, and the orbit space is just X_G.

In fact $\tilde{X}_G$ is the universal cover of X_G since it is simply-connected. This can be seen by considering the homomorphism $\varphi : \pi_1(X_G) \to G$ defined in the proof of Proposition 1.39. For an edge e_α in X_G corresponding to a generator g_α of G, it is clear from the definition of φ that $\varphi([e_\alpha]) = g_\alpha$, so φ is an isomorphism. In particular the kernel of φ, $p_*(\pi_1(\tilde{X}_G))$, is zero, hence also $\pi_1(\tilde{X}_G)$ since p_* is injective.

Let us look at some examples of Cayley complexes.

Example 1.45. When G is the free group on two generators a and b, X_G is $S^1 \vee S^1$ and $\tilde{X}_G$ is the Cayley graph of $\mathbb{Z} * \mathbb{Z}$ pictured at the right. The action of a on this graph is a rightward shift along the central horizontal axis, while b acts by an upward shift along the central vertical axis. The composition ab of these two shifts then takes the vertex e to the vertex ab. Similarly, the action of any $w \in \mathbb{Z} * \mathbb{Z}$ takes e to the vertex w.

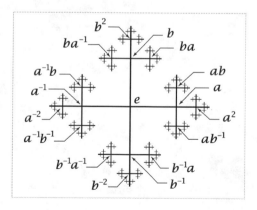

Example 1.46. The group $G = \mathbb{Z} \times \mathbb{Z}$ with presentation $\langle x, y \mid xyx^{-1}y^{-1} \rangle$ has X_G the torus $S^1 \times S^1$, and $\tilde{X}_G$ is $\mathbb{R}^2$ with vertices the integer lattice $\mathbb{Z}^2 \subset \mathbb{R}^2$ and edges the horizontal and vertical segments between these lattice points. The action of G is by translations $(x, y) \mapsto (x + m, y + n)$.

Example 1.47. For $G = \mathbb{Z}_2 = \langle x \mid x^2 \rangle$, X_G is $\mathbb{R}P^2$ and $\tilde{X}_G = S^2$. More generally, for $\mathbb{Z}_n = \langle x \mid x^n \rangle$, X_G is S^1 with a disk attached by the map $z \mapsto z^n$ and $\tilde{X}_G$ consists of n disks $D_1, \cdots, D_n$ with their boundary circles identified. A generator of $\mathbb{Z}_n$ acts on this union of disks by sending D_i to D_{i+1} via a $2\pi/n$ rotation, the subscript i being taken mod n. The common boundary circle of the disks is rotated by $2\pi/n$.

Example 1.48. If $G = \mathbb{Z}_2 * \mathbb{Z}_2 = \langle a, b \mid a^2, b^2 \rangle$ then the Cayley graph is a union of an infinite sequence of circles each tangent to its two neighbors.

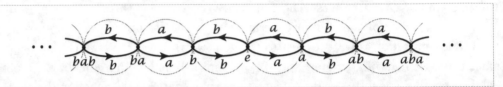

We obtain $\tilde{X}_G$ from this graph by making each circle the equator of a 2-sphere, yielding an infinite sequence of tangent 2-spheres. Elements of the index-two normal subgroup $\mathbb{Z} \subset \mathbb{Z}_2 * \mathbb{Z}_2$ generated by ab act on $\tilde{X}_G$ as translations by an even number of units, while each of the remaining elements of $\mathbb{Z}_2 * \mathbb{Z}_2$ acts as the antipodal map on one of the spheres and flips the whole chain of spheres end-for-end about this sphere. The orbit space X_G is $\mathbb{R}P^2 \vee \mathbb{R}P^2$.

It is not hard to see the generalization of this example to $\mathbb{Z}_m * \mathbb{Z}_n$ with the presentation $\langle a, b \mid a^m, b^n \rangle$, so that $\tilde{X}_G$ consists of an infinite union of copies of the Cayley complexes for $\mathbb{Z}_m$ and $\mathbb{Z}_n$ constructed in Example 1.47, arranged in a tree-like pattern. The case of $\mathbb{Z}_2 * \mathbb{Z}_3$ is pictured below.

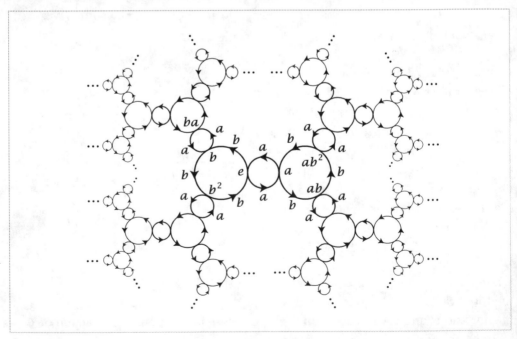

Exercises

1. For a covering space $p:\tilde{X}\to X$ and a subspace $A\subset X$, let $\tilde{A}=p^{-1}(A)$. Show that the restriction $p:\tilde{A}\to A$ is a covering space.

2. Show that if $p_1:\tilde{X}_1\to X_1$ and $p_2:\tilde{X}_2\to X_2$ are covering spaces, so is their product $p_1\times p_2:\tilde{X}_1\times\tilde{X}_2\to X_1\times X_2$.

3. Let $p:\tilde{X}\to X$ be a covering space with $p^{-1}(x)$ finite and nonempty for all $x\in X$. Show that $\tilde{X}$ is compact Hausdorff iff X is compact Hausdorff.

4. Construct a simply-connected covering space of the space $X\subset\mathbb{R}^3$ that is the union of a sphere and a diameter. Do the same when X is the union of a sphere and a circle intersecting it in two points.

5. Let X be the subspace of $\mathbb{R}^2$ consisting of the four sides of the square $[0,1]\times[0,1]$ together with the segments of the vertical lines $x=\frac{1}{2},\frac{1}{3},\frac{1}{4},\cdots$ inside the square. Show that for every covering space $\tilde{X}\to X$ there is some neighborhood of the left edge of X that lifts homeomorphically to $\tilde{X}$. Deduce that X has no simply-connected covering space.

6. Let X be the shrinking wedge of circles in Example 1.25, and let $\tilde{X}$ be its covering space shown in the figure below.

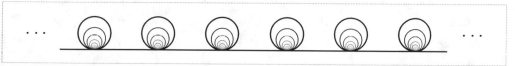

Construct a two-sheeted covering space $Y\to\tilde{X}$ such that the composition $Y\to\tilde{X}\to X$ of the two covering spaces is not a covering space. Note that a composition of two covering spaces does have the unique path lifting property, however.

7. Let Y be the *quasi-circle* shown in the figure, a closed subspace of $\mathbb{R}^2$ consisting of a portion of the graph of $y=\sin(1/x)$, the segment $[-1,1]$ in the y-axis, and an arc connecting these two pieces. Collapsing the segment of Y in the y-axis to a point gives a quotient map $f:Y\to S^1$. Show that f does not lift to the covering space $\mathbb{R}\to S^1$, even though $\pi_1(Y)=0$. Thus local 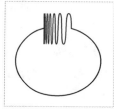 path-connectedness of Y is a necessary hypothesis in the lifting criterion.

8. Let $\tilde{X}$ and $\tilde{Y}$ be simply-connected covering spaces of the path-connected, locally path-connected spaces X and Y. Show that if $X\simeq Y$ then $\tilde{X}\simeq\tilde{Y}$. [Exercise 11 in Chapter 0 may be helpful.]

9. Show that if a path-connected, locally path-connected space X has $\pi_1(X)$ finite, then every map $X\to S^1$ is nullhomotopic. [Use the covering space $\mathbb{R}\to S^1$.]

10. Find all the connected 2-sheeted and 3-sheeted covering spaces of $S^1\vee S^1$, up to isomorphism of covering spaces without basepoints.

11. Construct finite graphs X_1 and X_2 having a common finite-sheeted covering space $\tilde{X}_1 = \tilde{X}_2$, but such that there is no space having both X_1 and X_2 as covering spaces.

12. Let a and b be the generators of $\pi_1(S^1 \vee S^1)$ corresponding to the two S^1 summands. Draw a picture of the covering space of $S^1 \vee S^1$ corresponding to the normal subgroup generated by a^2, b^2, and $(ab)^4$, and prove that this covering space is indeed the correct one.

13. Determine the covering space of $S^1 \vee S^1$ corresponding to the subgroup of $\pi_1(S^1 \vee S^1)$ generated by the cubes of all elements. The covering space is 27-sheeted and can be drawn on a torus so that the complementary regions are nine triangles with edges labeled aaa, nine triangles with edges labeled bbb, and nine hexagons with edges labeled $ababab$. [For the analogous problem with sixth powers instead of cubes, the resulting covering space would have $2^{28}3^{25}$ sheets! And for k^{th} powers with k sufficiently large, the covering space would have infinitely many sheets. The underlying group theory question here, whether the quotient of $\mathbb{Z} * \mathbb{Z}$ obtained by factoring out all k^{th} powers is finite, is known as Burnside's problem. It can also be asked for a free group on n generators.]

14. Find all the connected covering spaces of $\mathbb{RP}^2 \vee \mathbb{RP}^2$.

15. Let $p : \tilde{X} \to X$ be a simply-connected covering space of X and let $A \subset X$ be a path-connected, locally path-connected subspace, with $\tilde{A} \subset \tilde{X}$ a path-component of $p^{-1}(A)$. Show that $p : \tilde{A} \to A$ is the covering space corresponding to the kernel of the map $\pi_1(A) \to \pi_1(X)$.

16. Given maps $X \to Y \to Z$ such that both $Y \to Z$ and the composition $X \to Z$ are covering spaces, show that $X \to Y$ is a covering space if Z is locally path-connected, and show that this covering space is normal if $X \to Z$ is a normal covering space.

17. Given a group G and a normal subgroup N, show that there exists a normal covering space $\tilde{X} \to X$ with $\pi_1(X) \approx G$, $\pi_1(\tilde{X}) \approx N$, and deck transformation group $G(\tilde{X}) \approx G/N$.

18. For a path-connected, locally path-connected, and semilocally simply-connected space X, call a path-connected covering space $\tilde{X} \to X$ *abelian* if it is normal and has abelian deck transformation group. Show that X has an abelian covering space that is a covering space of every other abelian covering space of X, and that such a 'universal' abelian covering space is unique up to isomorphism. Describe this covering space explicitly for $X = S^1 \vee S^1$ and $X = S^1 \vee S^1 \vee S^1$.

19. Use the preceding problem to show that a closed orientable surface M_g of genus g has a connected normal covering space with deck transformation group isomorphic to $\mathbb{Z}^n$ (the product of n copies of $\mathbb{Z}$) iff $n \le 2g$. For $n = 3$ and $g \ge 3$, describe such a covering space explicitly as a subspace of $\mathbb{R}^3$ with translations of $\mathbb{R}^3$ as deck transformations. Show that such a covering space in $\mathbb{R}^3$ exists iff there is an embedding of M_g in the 3-torus $T^3 = S^1 \times S^1 \times S^1$ such that the induced map $\pi_1(M_g) \to \pi_1(T^3)$ is surjective.

20. Construct nonnormal covering spaces of the Klein bottle by a Klein bottle and by a torus.

21. Let X be the space obtained from a torus $S^1 \times S^1$ by attaching a Möbius band via a homeomorphism from the boundary circle of the Möbius band to the circle $S^1 \times \{x_0\}$ in the torus. Compute $\pi_1(X)$, describe the universal cover of X, and describe the action of $\pi_1(X)$ on the universal cover. Do the same for the space Y obtained by attaching a Möbius band to $\mathbb{RP}^2$ via a homeomorphism from its boundary circle to the circle in $\mathbb{RP}^2$ formed by the 1-skeleton of the usual CW structure on $\mathbb{RP}^2$.

22. Given covering space actions of groups G_1 on X_1 and G_2 on X_2, show that the action of $G_1 \times G_2$ on $X_1 \times X_2$ defined by $(g_1, g_2)(x_1, x_2) = (g_1(x_1), g_2(x_2))$ is a covering space action, and that $(X_1 \times X_2)/(G_1 \times G_2)$ is homeomorphic to $X_1/G_1 \times X_2/G_2$.

23. Show that if a group G acts freely and properly discontinuously on a Hausdorff space X, then the action is a covering space action. (Here 'properly discontinuously' means that each $x \in X$ has a neighborhood U such that $\{g \in G \mid U \cap g(U) \neq \varnothing\}$ is finite.) In particular, a free action of a finite group on a Hausdorff space is a covering space action.

24. Given a covering space action of a group G on a path-connected, locally path-connected space X, then each subgroup $H \subset G$ determines a composition of covering spaces $X \to X/H \to X/G$. Show:

(a) Every path-connected covering space between X and X/G is isomorphic to X/H for some subgroup $H \subset G$.

(b) Two such covering spaces X/H_1 and X/H_2 of X/G are isomorphic iff H_1 and H_2 are conjugate subgroups of G.

(c) The covering space $X/H \to X/G$ is normal iff H is a normal subgroup of G, in which case the group of deck transformations of this cover is G/H.

25. Let $\varphi : \mathbb{R}^2 \to \mathbb{R}^2$ be the linear transformation $\varphi(x, y) = (2x, y/2)$. This generates an action of $\mathbb{Z}$ on $X = \mathbb{R}^2 - \{0\}$. Show this action is a covering space action and compute $\pi_1(X/\mathbb{Z})$. Show the orbit space $X/\mathbb{Z}$ is non-Hausdorff, and describe how it is a union of four subspaces homeomorphic to $S^1 \times \mathbb{R}$, coming from the complementary components of the x-axis and the y-axis.

26. For a covering space $p : \widetilde{X} \to X$ with X connected, locally path-connected, and semilocally simply-connected, show:

(a) The components of $\widetilde{X}$ are in one-to-one correspondence with the orbits of the action of $\pi_1(X, x_0)$ on the fiber $p^{-1}(x_0)$.

(b) Under the Galois correspondence between connected covering spaces of X and subgroups of $\pi_1(X, x_0)$, the subgroup corresponding to the component of $\widetilde{X}$

containing a given lift $\tilde{x}_0$ of x_0 is the *stabilizer* of $\tilde{x}_0$, the subgroup consisting of elements whose action on the fiber leaves $\tilde{x}_0$ fixed.

27. For a universal cover $p : \tilde{X} \to X$ we have two actions of $\pi_1(X, x_0)$ on the fiber $p^{-1}(x_0)$, namely the action given by lifting loops at x_0 and the action given by restricting deck transformations to the fiber. Are these two actions the same when $X = S^1 \vee S^1$ or $X = S^1 \times S^1$? Do the actions always agree when $\pi_1(X, x_0)$ is abelian?

28. Show that for a covering space action of a group G on a simply-connected space Y, $\pi_1(Y/G)$ is isomorphic to G. [If Y is locally path-connected, this is a special case of part (b) of Proposition 1.40.]

29. Let Y be path-connected, locally path-connected, and simply-connected, and let G_1 and G_2 be subgroups of $\mathrm{Homeo}(Y)$ defining covering space actions on Y. Show that the orbit spaces Y/G_1 and Y/G_2 are homeomorphic iff G_1 and G_2 are conjugate subgroups of $\mathrm{Homeo}(Y)$.

30. Draw the Cayley graph of the group $\mathbb{Z} * \mathbb{Z}_2 = \langle a, b \mid b^2 \rangle$.

31. Show that the normal covering spaces of $S^1 \vee S^1$ are precisely the graphs that are Cayley graphs of groups with two generators. More generally, the normal covering spaces of the wedge sum of n circles are the Cayley graphs of groups with n generators.

32. Consider covering spaces $p : \tilde{X} \to X$ with $\tilde{X}$ and X connected CW complexes, the cells of $\tilde{X}$ projecting homeomorphically onto cells of X. Restricting p to the 1-skeleton then gives a covering space $\tilde{X}^1 \to X^1$ over the 1-skeleton of X. Show:
  (a) Two such covering spaces $\tilde{X}_1 \to X$ and $\tilde{X}_2 \to X$ are isomorphic iff the restrictions $\tilde{X}_1^1 \to X^1$ and $\tilde{X}_2^1 \to X^1$ are isomorphic.
  (b) $\tilde{X} \to X$ is a normal covering space iff $\tilde{X}^1 \to X^1$ is normal.
  (c) The groups of deck transformations of the coverings $\tilde{X} \to X$ and $\tilde{X}^1 \to X^1$ are isomorphic, via the restriction map.

33. In Example 1.44 let d be the greatest common divisor of m and n, and let $m' = m/d$ and $n' = n/d$. Show that the graph $T_{m,n}/K$ consists of m' vertices labeled a, n' vertices labeled b, together with d edges joining each a vertex to each b vertex. Deduce that the subgroup $K \subset G_{m,n}$ is free on $dm'n' - m' - n' + 1$ generators.

Additional Topics

1.A Graphs and Free Groups

Since all groups can be realized as fundamental groups of spaces, this opens the way for using topology to study algebraic properties of groups. The topics in this section and the next give some illustrations of this principle, mainly using covering space theory.

We remind the reader that the Additional Topics which form the remainder of this chapter are not to be regarded as an essential part of the basic core of the book. Readers who are eager to move on to new topics should feel free to skip ahead.

By definition, a **graph** is a 1-dimensional CW complex, in other words, a space X obtained from a discrete set X^0 by attaching a collection of 1-cells e_α. Thus X is obtained from the disjoint union of X^0 with closed intervals I_α by identifying the two endpoints of each I_α with points of X^0. The points of X^0 are the **vertices** and the 1-cells the **edges** of X. Note that with this definition an edge does not include its endpoints, so an edge is an open subset of X. The two endpoints of an edge can be the same vertex, so the closure $\bar{e}_\alpha$ of an edge e_α is homeomorphic either to I or S^1.

Since X has the quotient topology from the disjoint union $X^0 \coprod_\alpha I_\alpha$, a subset of X is open (or closed) iff it intersects the closure $\bar{e}_\alpha$ of each edge e_α in an open (or closed) set in $\bar{e}_\alpha$. One says that X has the **weak topology** with respect to the subspaces $\bar{e}_\alpha$. In this topology a sequence of points in the interiors of distinct edges forms a closed subset, hence never converges. This is true in particular if the edges containing the sequence all have a common vertex and one tries to choose the sequence so that it gets 'closer and closer' to the vertex. Thus if there is a vertex that is the endpoint of infinitely many edges, then the weak topology cannot be a metric topology. An exercise at the end of this section is to show the converse, that the weak topology is a metric topology if each vertex is an endpoint of only finitely many edges.

A basis for the topology of X consists of the open intervals in the edges together with the path-connected neighborhoods of the vertices. A neighborhood of the latter sort about a vertex v is the union of connected open neighborhoods U_α of v in $\bar{e}_\alpha$ for all $\bar{e}_\alpha$ containing v. In particular, we see that X is locally path-connected. Hence a graph is connected iff it is path-connected.

If X has only finitely many vertices and edges, then X is compact, being the continuous image of the compact space $X^0 \coprod_\alpha I_\alpha$. The converse is also true, and more generally, a compact subset C of a graph X can meet only finitely many vertices and edges of X. To see this, let the subspace $D \subset C$ consist of the vertices in C together with one point in each edge that C meets. Then D is a closed subset of X since it meets each $\overline{e}_\alpha$ in a closed set. For the same reason, any subset of D is closed, so D has the discrete topology. But D is compact, being a closed subset of the compact space C, so D must be finite. By the definition of D this means that C can meet only finitely many vertices and edges.

A **subgraph** of a graph X is a subspace $Y \subset X$ that is a union of vertices and edges of X, such that $e_\alpha \subset Y$ implies $\overline{e}_\alpha \subset Y$. The latter condition just says that Y is a closed subspace of X. A **tree** is a contractible graph. By a tree in a graph X we mean a subgraph that is a tree. We call a tree in X **maximal** if it contains all the vertices of X. This is equivalent to the more obvious meaning of maximality, as we will see below.

Proposition 1A.1. *Every connected graph contains a maximal tree, and in fact any tree in the graph is contained in a maximal tree.*

Proof: Let X be a connected graph. We will describe a construction that embeds an arbitrary subgraph $X_0 \subset X$ as a deformation retract of a subgraph $Y \subset X$ that contains all the vertices of X. By choosing X_0 to be any subtree of X, for example a single vertex, this will prove the proposition.

As a preliminary step, we construct a sequence of subgraphs $X_0 \subset X_1 \subset X_2 \subset \cdots$, letting X_{i+1} be obtained from X_i by adjoining the closures $\overline{e}_\alpha$ of all edges $e_\alpha \subset X - X_i$ having at least one endpoint in X_i. The union $\bigcup_i X_i$ is open in X since a neighborhood of a point in X_i is contained in X_{i+1}. Furthermore, $\bigcup_i X_i$ is closed since it is a union of closed edges and X has the weak topology. So $X = \bigcup_i X_i$ since X is connected.

Now to construct Y we begin by setting $Y_0 = X_0$. Then inductively, assuming that $Y_i \subset X_i$ has been constructed so as to contain all the vertices of X_i, let Y_{i+1} be obtained from Y_i by adjoining one edge connecting each vertex of $X_{i+1} - X_i$ to Y_i, and let $Y = \bigcup_i Y_i$. It is evident that Y_{i+1} deformation retracts to Y_i, and we may obtain a deformation retraction of Y to $Y_0 = X_0$ by performing the deformation retraction of Y_{i+1} to Y_i during the time interval $[1/2^{i+1}, 1/2^i]$. Thus a point $x \in Y_{i+1} - Y_i$ is stationary until this interval, when it moves into Y_i and thereafter continues moving until it reaches Y_0. The resulting homotopy $h_t : Y \to Y$ is continuous since it is continuous on the closure of each edge and Y has the weak topology. $\quad\square$

Given a maximal tree $T \subset X$ and a base vertex $x_0 \in T$, then each edge e_α of $X - T$ determines a loop f_α in X that goes first from x_0 to one endpoint of e_α by a path in T, then across e_α, then back to x_0 by a path in T. Strictly speaking, we should first orient the edge e_α in order to specify which direction to cross it. Note that the homotopy class of f_α is independent of the choice of the paths in T since T is simply-connected.

Proposition 1A.2. *For a connected graph X with maximal tree T, $\pi_1(X)$ is a free group with basis the classes $[f_\alpha]$ corresponding to the edges e_α of $X - T$.*

In particular this implies that a maximal tree is maximal in the sense of not being contained in any larger tree, since adjoining any edge to a maximal tree produces a graph with nontrivial fundamental group. Another consequence is that a graph is a tree iff it is simply-connected.

Proof: The quotient map $X \to X/T$ is a homotopy equivalence by Proposition 0.17. The quotient X/T is a graph with only one vertex, hence is a wedge sum of circles, whose fundamental group we showed in Example 1.21 to be free with basis the loops given by the edges of X/T, which are the images of the loops f_α in X.   □

Here is a very useful fact about graphs:

Lemma 1A.3. *Every covering space of a graph is also a graph, with vertices and edges the lifts of the vertices and edges in the base graph.*

Proof: Let $p : \tilde{X} \to X$ be the covering space. For the vertices of $\tilde{X}$ we take the discrete set $\tilde{X}^0 = p^{-1}(X^0)$. Writing X as a quotient space of $X^0 \coprod_\alpha I_\alpha$ as in the definition of a graph and applying the path lifting property to the resulting maps $I_\alpha \to X$, we get a unique lift $I_\alpha \to \tilde{X}$ passing through each point in $p^{-1}(x)$, for $x \in e_\alpha$. These lifts define the edges of a graph structure on $\tilde{X}$. The resulting topology on $\tilde{X}$ is the same as its original topology since both topologies have the same basic open sets, the covering projection $\tilde{X} \to X$ being a local homeomorphism.   □

We can now apply what we have proved about graphs and their fundamental groups to prove a basic fact of group theory:

Theorem 1A.4. *Every subgroup of a free group is free.*

Proof: Given a free group F, choose a graph X with $\pi_1(X) \approx F$, for example a wedge of circles corresponding to a basis for F. For each subgroup G of F there is by Proposition 1.36 a covering space $p : \tilde{X} \to X$ with $p_*(\pi_1(\tilde{X})) = G$, hence $\pi_1(\tilde{X}) \approx G$ since p_* is injective by Proposition 1.31. Since $\tilde{X}$ is a graph by the preceding lemma, the group $G \approx \pi_1(\tilde{X})$ is free by Proposition 1A.2.   □

The structure of trees can be elucidated by looking more closely at the constructions in the proof of Proposition 1A.1. If X is a tree and v_0 is any vertex of X, then the construction of a maximal tree $Y \subset X$ starting with $Y_0 = \{v_0\}$ yields an increasing sequence of subtrees $Y_n \subset X$ whose union is all of X since a tree has only one maximal subtree, namely itself. We can think of the vertices in $Y_n - Y_{n-1}$ as being at 'height' n, with the edges of $Y_n - Y_{n-1}$ connecting these vertices to vertices of height $n-1$. In this way we get a 'height function' $h : X \to \mathbb{R}$ assigning to each vertex its height, and monotone on edges.

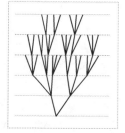

For each vertex v of X there is exactly one edge leading downward from v, so by following these downward edges we obtain a path from v to the base vertex v_0. This is an example of an **edgepath**, which is a composition of finitely many paths each consisting of a single edge traversed monotonically. For any edgepath joining v to v_0 other than the downward edgepath, the height function would not be monotone and hence would have local maxima, occurring when the edgepath backtracked, retracing some edge it had just crossed. Thus in a tree there is a unique nonbacktracking edgepath joining any two points. All the vertices and edges along this edgepath are distinct.

A tree can contain no subgraph homeomorphic to a circle, since two vertices in such a subgraph could be joined by more than one nonbacktracking edgepath. Conversely, if a connected graph X contains no circle subgraph, then it must be a tree. For if T is a maximal tree in X that is not equal to X, then the union of an edge of $X - T$ with the nonbacktracking edgepath in T joining the endpoints of this edge is a circle subgraph of X. So if there are no circle subgraphs of X, we must have $X = T$, a tree.

For an arbitrary connected graph X and a pair of vertices v_0 and v_1 in X there is a unique nonbacktracking edgepath in each homotopy class of paths from v_0 to v_1. This can be seen by lifting to the universal cover $\tilde{X}$, which is a tree since it is simply-connected. Choosing a lift $\tilde{v}_0$ of v_0, a homotopy class of paths from v_0 to v_1 lifts to a homotopy class of paths starting at $\tilde{v}_0$ and ending at a unique lift $\tilde{v}_1$ of v_1. Then the unique nonbacktracking edgepath in $\tilde{X}$ from $\tilde{v}_0$ to $\tilde{v}_1$ projects to the desired nonbacktracking edgepath in X.

Exercises

1. Let X be a graph in which each vertex is an endpoint of only finitely many edges. Show that the weak topology on X is a metric topology.

2. Show that a connected graph retracts onto any connected subgraph.

3. For a finite graph X define the Euler characteristic $\chi(X)$ to be the number of vertices minus the number of edges. Show that $\chi(X) = 1$ if X is a tree, and that the rank (number of elements in a basis) of $\pi_1(X)$ is $1 - \chi(X)$ if X is connected.

4. If X is a finite graph and Y is a subgraph homeomorphic to S^1 and containing the basepoint x_0, show that $\pi_1(X, x_0)$ has a basis in which one element is represented by the loop Y.

5. Construct a connected graph X and maps $f, g : X \to X$ such that $fg = \mathbb{1}$ but f and g do not induce isomorphisms on π_1. [Note that $f_* g_* = \mathbb{1}$ implies that f_* is surjective and g_* is injective.]

6. Let F be the free group on two generators and let F' be its commutator subgroup. Find a set of free generators for F' by considering the covering space of the graph $S^1 \vee S^1$ corresponding to F'.

7. If F is a finitely generated free group and N is a nontrivial normal subgroup of infinite index, show, using covering spaces, that N is not finitely generated.

8. Show that a finitely generated group has only a finite number of subgroups of a given finite index. [First do the case of free groups, using covering spaces of graphs. The general case then follows since every group is a quotient group of a free group.]

9. Using covering spaces, show that an index n subgroup H of a group G has at most n conjugate subgroups gHg^{-1} in G. Apply this to show that there exists a normal subgroup $K \subset G$ of finite index with $K \subset H$. [For the latter statement, consider the intersection of all the conjugate subgroups gHg^{-1}. This is the maximal normal subgroup of G contained in H.]

10. Let X be the wedge sum of n circles, with its natural graph structure, and let $\tilde{X} \to X$ be a covering space with $Y \subset \tilde{X}$ a finite connected subgraph. Show there is a finite graph $Z \supset Y$ having the same vertices as Y, such that the projection $Y \to X$ extends to a covering space $Z \to X$.

11. Apply the two preceding problems to show that if F is a finitely generated free group and $x \in F$ is not the identity element, then there is a normal subgroup $H \subset F$ of finite index such that $x \notin H$. Hence x has nontrivial image in a finite quotient group of F. In this situation one says F is *residually finite*.

12. Let F be a finitely generated free group, $H \subset F$ a finitely generated subgroup, and $x \in F - H$. Show there is a subgroup K of finite index in F such that $K \supset H$ and $x \notin K$. [Apply Exercise 10.]

13. Let x be a nontrivial element of a finitely generated free group F. Show there is a finite-index subgroup $H \subset F$ in which x is one element of a basis. [Exercises 4 and 10 may be helpful.]

14. Show that the existence of maximal trees is equivalent to the Axiom of Choice.

1.B K(G,1) Spaces and Graphs of Groups

In this section we introduce a class of spaces whose homotopy type depends only on their fundamental group. These spaces arise many places in topology, especially in its interactions with group theory.

A path-connected space whose fundamental group is isomorphic to a given group G and which has a contractible universal covering space is called a **K(G,1) space**. The '1' here refers to π_1. More general $K(G,n)$ spaces are studied in §4.2. All these spaces are called Eilenberg–MacLane spaces, though in the case $n = 1$ they were studied by

Hurewicz before Eilenberg and MacLane took up the general case. Here are some examples:

Example 1B.1. S^1 is a $K(\mathbb{Z}, 1)$. More generally, a connected graph is a $K(G, 1)$ with G a free group, since by the results of §1.A its universal cover is a tree, hence contractible.

Example 1B.2. Closed surfaces with infinite π_1, in other words, closed surfaces other than S^2 and $\mathbb{R}P^2$, are $K(G, 1)$'s. This will be shown in Example 1B.14 below. It also follows from the theorem in surface theory that the only simply-connected surfaces without boundary are S^2 and $\mathbb{R}^2$, so the universal cover of a closed surface with infinite fundamental group must be $\mathbb{R}^2$ since it is noncompact. Nonclosed surfaces deformation retract onto graphs, so such surfaces are $K(G, 1)$'s with G free.

Example 1B.3. The infinite-dimensional projective space $\mathbb{R}P^\infty$ is a $K(\mathbb{Z}_2, 1)$ since its universal cover is S^∞, which is contractible. To show the latter fact, a homotopy from the identity map of S^∞ to a constant map can be constructed in two stages as follows. First, define $f_t : \mathbb{R}^\infty \to \mathbb{R}^\infty$ by $f_t(x_1, x_2, \cdots) = (1 - t)(x_1, x_2, \cdots) + t(0, x_1, x_2, \cdots)$. This takes nonzero vectors to nonzero vectors for all $t \in [0, 1]$, so $f_t / |f_t|$ gives a homotopy from the identity map of S^∞ to the map $(x_1, x_2, \cdots) \mapsto (0, x_1, x_2, \cdots)$. Then a homotopy from this map to a constant map is given by $g_t / |g_t|$ where $g_t(x_1, x_2, \cdots) = (1 - t)(0, x_1, x_2, \cdots) + t(1, 0, 0, \cdots)$.

Example 1B.4. Generalizing the preceding example, we can construct a $K(\mathbb{Z}_m, 1)$ as an infinite-dimensional lens space $S^\infty / \mathbb{Z}_m$, where $\mathbb{Z}_m$ acts on S^∞, regarded as the unit sphere in $\mathbb{C}^\infty$, by scalar multiplication by m^{th} roots of unity, a generator of this action being the map $(z_1, z_2, \cdots) \mapsto e^{2\pi i / m}(z_1, z_2, \cdots)$. It is not hard to check that this is a covering space action.

Example 1B.5. A product $K(G, 1) \times K(H, 1)$ is a $K(G \times H, 1)$ since its universal cover is the product of the universal covers of $K(G, 1)$ and $K(H, 1)$. By taking products of circles and infinite-dimensional lens spaces we therefore get $K(G, 1)$'s for arbitrary finitely generated abelian groups G. For example the n-dimensional torus T^n, the product of n circles, is a $K(\mathbb{Z}^n, 1)$.

Example 1B.6. For a closed connected subspace K of S^3 that is nonempty, the complement $S^3 - K$ is a $K(G, 1)$. This is a theorem in 3-manifold theory, but in the special case that K is a torus knot the result follows from our study of torus knot complements in Examples 1.24 and 1.35. Namely, we showed that for K the torus knot $K_{m,n}$ there is a deformation retraction of $S^3 - K$ onto a certain 2-dimensional complex $X_{m,n}$ having contractible universal cover. The homotopy lifting property then implies that the universal cover of $S^3 - K$ is homotopy equivalent to the universal cover of $X_{m,n}$, hence is also contractible.

Example 1B.7. It is not hard to construct a $K(G,1)$ for an arbitrary group G, using the notion of a Δ-complex defined in §2.1. Let EG be the Δ-complex whose n-simplices are the ordered $(n+1)$-tuples $[g_0, \cdots, g_n]$ of elements of G. Such an n-simplex attaches to the $(n-1)$-simplices $[g_0, \cdots, \hat{g}_i, \cdots, g_n]$ in the obvious way, just as a standard simplex attaches to its faces. (The notation $\hat{g}_i$ means that this vertex is deleted.) The complex EG is contractible by the homotopy h_t that slides each point $x \in [g_0, \cdots, g_n]$ along the line segment in $[e, g_0, \cdots, g_n]$ from x to the vertex $[e]$, where e is the identity element of G. This is well-defined in EG since when we restrict to a face $[g_0, \cdots, \hat{g}_i, \cdots, g_n]$ we have the linear deformation to $[e]$ in $[e, g_0, \cdots, \hat{g}_i, \cdots, g_n]$. Note that h_t carries $[e]$ around the loop $[e, e]$, so h_t is not actually a deformation retraction of EG onto $[e]$.

The group G acts on EG by left multiplication, an element $g \in G$ taking the simplex $[g_0, \cdots, g_n]$ linearly onto the simplex $[gg_0, \cdots, gg_n]$. Only the identity e takes any simplex to itself, so by an exercise at the end of this section, the action of G on EG is a covering space action. Hence the quotient map $EG \to EG/G$ is the universal cover of the orbit space $BG = EG/G$, and BG is a $K(G,1)$.

Since G acts on EG by freely permuting simplices, BG inherits a Δ-complex structure from EG. The action of G on EG identifies all the vertices of EG, so BG has just one vertex. To describe the Δ-complex structure on BG explicitly, note first that every n-simplex of EG can be written uniquely in the form

$$[g_0, g_0g_1, g_0g_1g_2, \cdots, g_0g_1 \cdots g_n] = g_0[e, g_1, g_1g_2, \cdots, g_1 \cdots g_n]$$

The image of this simplex in BG may be denoted unambiguously by the symbol $[g_1|g_2|\cdots|g_n]$. In this 'bar' notation the g_i's and their ordered products can be used to label edges, viewing an edge label as the ratio between the two labels on the vertices at the endpoints of the edge, as indicated in the figure. With this notation, the boundary of a simplex $[g_1|\cdots|g_n]$ of BG

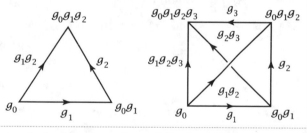

consists of the simplices $[g_2|\cdots|g_n]$, $[g_1|\cdots|g_{n-1}]$, and $[g_1|\cdots|g_ig_{i+1}|\cdots|g_n]$ for $i = 1, \cdots, n-1$.

This construction of a $K(G,1)$ produces a rather large space, since BG is always infinite-dimensional, and if G is infinite, BG has an infinite number of cells in each positive dimension. For example, $B\mathbb{Z}$ is much bigger than S^1, the most efficient $K(\mathbb{Z}, 1)$. On the other hand, BG has the virtue of being functorial: A homomorphism $f: G \to H$ induces a map $Bf: BG \to BH$ sending a simplex $[g_1|\cdots|g_n]$ to the simplex $[f(g_1)|\cdots|f(g_n)]$. A different construction of a $K(G,1)$ is given in §4.2. Here one starts with any 2-dimensional complex having fundamental group G, for example the complex X_G associated to a presentation of G, and then one attaches cells of dimension 3 and higher to make the universal cover contractible without affecting π_1. In general, it is hard to get any control on the number of higher-dimensional cells needed in this construction, so it too can be rather inefficient. Indeed, finding an efficient $K(G, 1)$ for a given group G is often a difficult problem.

It is a curious and almost paradoxical fact that if G contains any elements of finite order, then every $K(G, 1)$ CW complex must be infinite-dimensional. This is shown in Proposition 2.45. In particular the infinite-dimensional lens space $K(\mathbb{Z}_m, 1)$'s in Example 1B.4 cannot be replaced by any finite-dimensional complex.

In spite of the great latitude possible in the construction of $K(G, 1)$'s, there is a very nice homotopical uniqueness property that accounts for much of the interest in $K(G, 1)$'s:

Theorem 1B.8. *The homotopy type of a CW complex $K(G, 1)$ is uniquely determined by G.*

Having a unique homotopy type of $K(G, 1)$'s associated to each group G means that algebraic invariants of spaces that depend only on homotopy type, such as homology and cohomology groups, become invariants of groups. This has proved to be a quite fruitful idea, and has been much studied both from the algebraic and topological viewpoints. The discussion following Proposition 2.45 gives a few references.

The preceding theorem will follow easily from:

Proposition 1B.9. *Let X be a connected CW complex and let Y be a $K(G, 1)$. Then every homomorphism $\pi_1(X, x_0) \to \pi_1(Y, y_0)$ is induced by a map $(X, x_0) \to (Y, y_0)$ that is unique up to homotopy fixing x_0.*

To deduce the theorem from this, let X and Y be CW complex $K(G, 1)$'s with isomorphic fundamental groups. The proposition gives maps $f : (X, x_0) \to (Y, y_0)$ and $g : (Y, y_0) \to (X, x_0)$ inducing inverse isomorphisms $\pi_1(X, x_0) \approx \pi_1(Y, y_0)$. Then fg and gf induce the identity on π_1 and hence are homotopic to the identity maps.

Proof of 1B.9: Let us first consider the case that X has a single 0-cell, the basepoint x_0. Given a homomorphism $\varphi : \pi_1(X, x_0) \to \pi_1(Y, y_0)$, we begin the construction of a map $f : (X, x_0) \to (Y, y_0)$ with $f_* = \varphi$ by setting $f(x_0) = y_0$. Each 1-cell e_α^1 of X has closure a circle determining an element $[e_\alpha^1] \in \pi_1(X, x_0)$, and we let f on the closure of e_α^1 be a map representing $\varphi([e_\alpha^1])$. If $i : X^1 \hookrightarrow X$ denotes the inclusion, then $\varphi i_* = f_*$ since $\pi_1(X^1, x_0)$ is generated by the elements $[e_\alpha^1]$.

$$\begin{array}{ccc} \pi_1(X^1, x_0) & \xrightarrow{\ f_*\ } & \pi_1(Y, y_0) \\ {\scriptstyle i_*}\searrow & & \nearrow{\scriptstyle \varphi} \\ & \pi_1(X, x_0) & \end{array}$$

To extend f over a cell e_β^2 with attaching map $\psi_\beta : S^1 \to X^1$, all we need is for the composition $f\psi_\beta$ to be nullhomotopic. Choosing a basepoint $s_0 \in S^1$ and a path in X^1 from $\psi_\beta(s_0)$ to x_0, ψ_β determines an element $[\psi_\beta] \in \pi_1(X^1, x_0)$, and the existence of a nullhomotopy of $f\psi_\beta$ is equivalent to $f_*([\psi_\beta])$ being zero in $\pi_1(Y, y_0)$. We have $i_*([\psi_\beta]) = 0$ since the cell e_β^2 provides a nullhomotopy of ψ_β in X. Hence $f_*([\psi_\beta]) = \varphi i_*([\psi_\beta]) = 0$, and so f can be extended over e_β^2.

Extending f inductively over cells e_γ^n with $n > 2$ is possible since the attaching maps $\psi_\gamma : S^{n-1} \to X^{n-1}$ have nullhomotopic compositions $f\psi_\gamma : S^{n-1} \to Y$. This is because $f\psi_\gamma$ lifts to the universal cover of Y if $n > 2$, and this cover is contractible by hypothesis, so the lift of $f\psi_\gamma$ is nullhomotopic, hence also $f\psi_\gamma$ itself.

Turning to the uniqueness statement, if two maps $f_0, f_1 : (X, x_0) \to (Y, y_0)$ induce the same homomorphism on π_1, then we see immediately that their restrictions to X^1 are homotopic, fixing x_0. To extend the resulting map $X^1 \times I \cup X \times \partial I \to Y$ over the remaining cells $e^n \times (0, 1)$ of $X \times I$ we can proceed just as in the preceding paragraph since these cells have dimension $n + 1 > 2$. Thus we obtain a homotopy $f_t : (X, x_0) \to (Y, y_0)$, finishing the proof in the case that X has a single 0-cell.

The case that X has more than one 0-cell can be treated by a small elaboration on this argument. Choose a maximal tree $T \subset X$. To construct a map f realizing a given φ, begin by setting $f(T) = y_0$. Then each edge e_α^1 in $X - T$ determines an element $[e_\alpha^1] \in \pi_1(X, x_0)$, and we let f on the closure of e_α^1 be a map representing $\varphi([e_\alpha^1])$. Extending f over higher-dimensional cells then proceeds just as before. Constructing a homotopy f_t joining two given maps f_0 and f_1 with $f_{0*} = f_{1*}$ also has an extra step. Let $h_t : X^1 \to X^1$ be a homotopy starting with $h_0 = \mathbb{1}$ and restricting to a deformation retraction of T onto x_0. (It is easy to extend such a deformation retraction to a homotopy defined on all of X^1.) We can construct a homotopy from $f_0|X^1$ to $f_1|X^1$ by first deforming $f_0|X^1$ and $f_1|X^1$ to take T to y_0 by composing with h_t, then applying the earlier argument to obtain a homotopy between the modified $f_0|X^1$ and $f_1|X^1$. Having a homotopy $f_0|X^1 \simeq f_1|X^1$ we extend this over all of X in the same way as before. $\square$

The first part of the preceding proof also works for the 2-dimensional complexes X_G associated to presentations of groups. Thus every homomorphism $G \to H$ is realized as the induced homomorphism of some map $X_G \to X_H$. However, there is no uniqueness statement for this map, and it can easily happen that different presentations of a group G give X_G's that are not homotopy equivalent.

Graphs of Groups

As an illustration of how $K(G, 1)$ spaces can be useful in group theory, we shall describe a procedure for assembling a collection of $K(G, 1)$'s together into a $K(G, 1)$ for a larger group G. Group-theoretically, this gives a method for assembling smaller groups together to form a larger group, generalizing the notion of free products.

Let Γ be a graph that is connected and oriented, that is, its edges are viewed as arrows, each edge having a specified direction. Suppose that at each vertex v of Γ we place a group G_v and along each edge e of Γ we put a homomorphism φ_e from the group at the tail of the edge to the group at the head of the edge. We call this data a **graph of groups**. Now build a space $B\Gamma$ by putting the space BG_v from Example 1B.7 at each vertex v of Γ and then filling in a mapping cylinder of the map $B\varphi_e$ along each edge e of Γ, identifying the two ends of the mapping cylinder with the two BG_v's at the ends of e. The resulting space $B\Gamma$ is then a CW complex since the maps $B\varphi_e$ take n-cells homeomorphically onto n-cells. In fact, the cell structure on $B\Gamma$ can be canonically subdivided into a Δ-complex structure using the prism construction from the proof of Theorem 2.10, but we will not need to do this here.

More generally, instead of BG_v one could take any CW complex $K(G_v, 1)$ at the vertex v, and then along edges put mapping cylinders of maps realizing the homomorphisms φ_e. We leave it for the reader to check that the resulting space $K\Gamma$ is homotopy equivalent to the $B\Gamma$ constructed above.

Example 1B.10. Suppose Γ consists of one central vertex with a number of edges radiating out from it, and the group G_v at this central vertex is trivial, hence also all the edge homomorphisms. Then van Kampen's theorem implies that $\pi_1(K\Gamma)$ is the free product of the groups at all the outer vertices.

In view of this example, we shall call $\pi_1(K\Gamma)$ for a general graph of groups Γ the **graph product** of the vertex groups G_v with respect to the edge homomorphisms φ_e. The name for $\pi_1(K\Gamma)$ that is generally used in the literature is the rather awkward phrase, 'the fundamental group of the graph of groups.'

Here is the main result we shall prove about graphs of groups:

Theorem 1B.11. *If all the edge homomorphisms φ_e are injective, then $K\Gamma$ is a $K(G, 1)$ and the inclusions $K(G_v, 1) \hookrightarrow K\Gamma$ induce injective maps on π_1.*

Before giving the proof, let us look at some interesting special cases:

Example 1B.12: Free Products with Amalgamation. Suppose the graph of groups is $A \leftarrow C \rightarrow B$, with the two maps monomorphisms. One can regard this data as specifying embeddings of C as subgroups of A and B. Applying van Kampen's theorem to the decomposition of $K\Gamma$ into its two mapping cylinders, we see that $\pi_1(K\Gamma)$ is the quotient of $A * B$ obtained by identifying the subgroup $C \subset A$ with the subgroup $C \subset B$. The standard notation for this group is $A *_C B$, the free product of A and B **amalgamated** along the subgroup C. According to the theorem, $A *_C B$ contains both A and B as subgroups.

For example, a free product with amalgamation $\mathbb{Z} *_{\mathbb{Z}} \mathbb{Z}$ can be realized by mapping cylinders of the maps $S^1 \leftarrow S^1 \rightarrow S^1$ that are m-sheeted and n-sheeted covering spaces, respectively. We studied this case in Examples 1.24 and 1.35 where we showed that the complex $K\Gamma$ is a deformation retract of the complement of a torus knot in S^3 if m and n are relatively prime. It is a basic result in 3-manifold theory that the complement of every smooth knot in S^3 can be built up by iterated graph of groups constructions with injective edge homomorphisms, starting with free groups, so the theorem implies that these knot complements are $K(G, 1)$'s. Their universal covers are all $\mathbb{R}^3$, in fact.

Example 1B.13: HNN Extensions. Consider a graph of groups $C \underset{\psi}{\overset{\varphi}{\rightrightarrows}} A$ with φ and ψ both monomorphisms. This is analogous to the previous case $A \leftarrow C \rightarrow B$, but with the two groups A and B coalesced to a single group. The group $\pi_1(K\Gamma)$, which was denoted $A *_C B$ in the previous case, is now denoted $A*_C$. To see what this group looks like, let us regard $K\Gamma$ as being obtained from $K(A, 1)$ by attaching $K(C, 1) \times I$ along the two ends $K(C, 1) \times \partial I$ via maps realizing the monomorphisms φ and ψ. Using a $K(C, 1)$ with a single 0-cell, we see that $K\Gamma$ can be obtained from $K(A, 1) \vee S^1$ by attaching cells of dimension two and greater, so $\pi_1(K\Gamma)$ is a quotient of $A * \mathbb{Z}$, and it is not hard to figure out that the relations defining this quotient are of the form $t\varphi(c)t^{-1} = \psi(c)$ where t is a generator of the $\mathbb{Z}$ factor and c ranges over C, or a set of generators for C. We leave the verification of this for the Exercises.

As a very special case, taking $\varphi = \psi = \mathbb{1}$ gives $A*_A = A \times \mathbb{Z}$ since we can take $K\Gamma = K(A, 1) \times S^1$ in this case. More generally, taking $\varphi = \mathbb{1}$ with ψ an arbitrary automorphism of A, we realize any semidirect product of A and $\mathbb{Z}$ as $A*_A$. For example, the Klein bottle occurs this way, with φ realized by the identity map of S^1 and ψ by a reflection. In these cases when $\varphi = \mathbb{1}$ we could realize the same group $\pi_1(K\Gamma)$ using a slightly simpler graph of groups, with a single vertex, labeled A, and a single edge, labeled ψ.

Here is another special case. Suppose we take a torus, delete a small open disk, then identify the resulting boundary circle with a longitudinal circle of the torus. This produces a space X that happens to be homeomorphic to a subspace of the standard picture of a Klein bottle in $\mathbb{R}^3$; see Exercise 12 of §1.2. The fundamental group $\pi_1(X)$ has the form $(\mathbb{Z} * \mathbb{Z}) *_\mathbb{Z} \mathbb{Z}$ with the defining relation $tb^{\pm 1}t^{-1} = aba^{-1}b^{-1}$ where a is a meridional loop and b is a longitudinal loop on the torus. The sign of the exponent in the term $b^{\pm 1}$ is immaterial since the two ways of glueing the boundary circle to the longitude produce homeomorphic spaces. The group $\pi_1(X) = \langle a, b, t \mid tbt^{-1}aba^{-1}b^{-1} \rangle$ abelianizes to $\mathbb{Z} \times \mathbb{Z}$, but to show that $\pi_1(X)$ is not isomorphic to $\mathbb{Z} * \mathbb{Z}$ takes some work. There is a surjection $\pi_1(X) \rightarrow \mathbb{Z} * \mathbb{Z}$ obtained by setting $b = 1$. This has nontrivial kernel since b is nontrivial in $\pi_1(X)$ by the preceding theorem. If $\pi_1(X)$ were isomorphic to $\mathbb{Z} * \mathbb{Z}$ we would then have a surjective homomorphism $\mathbb{Z} * \mathbb{Z} \rightarrow \mathbb{Z} * \mathbb{Z}$ that was not an isomorphism. However, it is a theorem in group theory that a free group F is *hopfian* — every surjective homomorphism $F \rightarrow F$ must be injective. Hence $\pi_1(X)$ is not free.

Example 1B.14: Closed Surfaces. A closed orientable surface M of genus two or greater can be cut along a circle into two compact surfaces M_1 and M_2 such that the closed surfaces obtained from M_1 and M_2 by filling in their boundary circle with a disk have smaller genus than M. Each of M_1 and M_2 is the mapping cylinder of a map from S^1 to a finite graph. Namely, view M_i as obtained from a closed surface by deleting an open disk in the interior of the 2-cell in the standard CW structure described in Chapter 0, so that M_i becomes the mapping cylinder of the attaching map of the 2-cell. This attaching map is not nullhomotopic, so it induces an injection on π_1 since free groups are torsionfree. Thus we have realized the original surface M as $K\Gamma$ for Γ a graph of groups of the form $F_1 \longleftarrow \mathbb{Z} \longrightarrow F_2$ with F_1 and F_2 free and the two maps injective. The theorem then says that M is a $K(G,1)$.

A similar argument works for closed nonorientable surfaces other than $\mathbb{R}\mathrm{P}^2$. For example, the Klein bottle is obtained from two Möbius bands by identifying their boundary circles, and a Möbius band is the mapping cylinder of the 2-sheeted covering space $S^1 \rightarrow S^1$.

Proof of 1B.11: We shall construct a covering space $\widetilde{K} \rightarrow K\Gamma$ by gluing together copies of the universal covering spaces of the various mapping cylinders in $K\Gamma$ in such a way that $\widetilde{K}$ will be contractible. Hence $\widetilde{K}$ will be the universal cover of $K\Gamma$, which will therefore be a $K(G,1)$.

A preliminary observation: Given a universal covering space $p : \widetilde{X} \rightarrow X$ and a connected, locally path-connected subspace $A \subset X$ such that the inclusion $A \hookrightarrow X$ induces an injection on π_1, then each component $\widetilde{A}$ of $p^{-1}(A)$ is a universal cover of A. To see this, note that $p : \widetilde{A} \rightarrow A$ is a covering space, so we have injective maps $\pi_1(\widetilde{A}) \rightarrow \pi_1(A) \rightarrow \pi_1(X)$ whose composition factors through $\pi_1(\widetilde{X}) = 0$, hence $\pi_1(\widetilde{A}) = 0$. For example, if X is the torus $S^1 \times S^1$ and A is the circle $S^1 \times \{x_0\}$, then $p^{-1}(A)$ consists of infinitely many parallel lines in $\mathbb{R}^2$, each a universal cover of A.

For a map $f : A \rightarrow B$ between connected CW complexes, let $p : \widetilde{M}_f \rightarrow M_f$ be the universal cover of the mapping cylinder M_f. Then $\widetilde{M}_f$ is itself the mapping cylinder of a map $\widetilde{f} : p^{-1}(A) \rightarrow p^{-1}(B)$ since the line segments in the mapping cylinder structure on M_f lift to line segments in $\widetilde{M}_f$ defining a mapping cylinder structure. Since $\widetilde{M}_f$ is a mapping cylinder, it deformation retracts onto $p^{-1}(B)$, so $p^{-1}(B)$ is also simply-connected, hence is the universal cover of B. If f induces an injection on π_1, then the remarks in the preceding paragraph apply, and the components of $p^{-1}(A)$ are universal covers of A. If we assume further that A and B are $K(G,1)$'s, then $\widetilde{M}_f$ and the components of $p^{-1}(A)$ are contractible, and we claim that $\widetilde{M}_f$ deformation retracts onto each component $\widetilde{A}$ of $p^{-1}(A)$. Namely, the inclusion $\widetilde{A} \hookrightarrow \widetilde{M}_f$ is a homotopy equivalence since both spaces are contractible, and then Corollary 0.20 implies that $\widetilde{M}_f$ deformation retracts onto $\widetilde{A}$ since the pair $(\widetilde{M}_f, \widetilde{A})$ satisfies the homotopy extension property, as shown in Example 0.15.

Now we can describe the construction of the covering space $\widetilde{K}$ of $K\Gamma$. It will be the union of an increasing sequence of spaces $\widetilde{K}_1 \subset \widetilde{K}_2 \subset \cdots$. For the first stage, let $\widetilde{K}_1$ be the universal cover of one of the mapping cylinders M_f of $K\Gamma$. By the preceding remarks, this contains various disjoint copies of universal covers of the two $K(G_v, 1)$'s at the ends of M_f. We build $\widetilde{K}_2$ from $\widetilde{K}_1$ by attaching to each of these universal covers of $K(G_v, 1)$'s a copy of the universal cover of each mapping cylinder M_g of $K\Gamma$ meeting M_f at the end of M_f in question. Now repeat the process to construct $\widetilde{K}_3$ by attaching universal covers of mapping cylinders at all the universal covers of $K(G_v, 1)$'s created in the previous step. In the same way, we construct $\widetilde{K}_{n+1}$ from $\widetilde{K}_n$ for all n, and then we set $\widetilde{K} = \bigcup_n \widetilde{K}_n$.

Note that $\widetilde{K}_{n+1}$ deformation retracts onto $\widetilde{K}_n$ since it is formed by attaching pieces to $\widetilde{K}_n$ that deformation retract onto the subspaces along which they attach, by our earlier remarks. It follows that $\widetilde{K}$ is contractible since we can deformation retract $\widetilde{K}_{n+1}$ onto $\widetilde{K}_n$ during the time interval $[1/2^{n+1}, 1/2^n]$, and then finish with a contraction of $\widetilde{K}_1$ to a point during the time interval $[1/2, 1]$.

The natural projection $\widetilde{K} \to K\Gamma$ is clearly a covering space, so this finishes the proof that $K\Gamma$ is a $K(G, 1)$.

The remaining statement that each inclusion $K(G_v, 1) \hookrightarrow K\Gamma$ induces an injection on π_1 can easily be deduced from the preceding constructions. For suppose a loop $\gamma : S^1 \to K(G_v, 1)$ is nullhomotopic in $K\Gamma$. By the lifting criterion for covering spaces, there is a lift $\widetilde{\gamma} : S^1 \to \widetilde{K}$. This has image contained in one of the copies of the universal cover of $K(G_v, 1)$, so $\widetilde{\gamma}$ is nullhomotopic in this universal cover, and hence γ is nullhomotopic in $K(G_v, 1)$. $\qquad\qquad\square$

The various mapping cylinders that make up the universal cover of $K\Gamma$ are arranged in a treelike pattern. The tree in question, call it $T\Gamma$, has one vertex for each copy of a universal cover of a $K(G_v, 1)$ in $\widetilde{K}$, and two vertices are joined by an edge whenever the two universal covers of $K(G_v, 1)$'s corresponding to these vertices are connected by a line segment lifting a line segment in the mapping cylinder structure of a mapping cylinder of $K\Gamma$. The inductive construction of $\widetilde{K}$ is reflected in an inductive construction of $T\Gamma$ as a union of an increasing sequence of subtrees $T_1 \subset T_2 \subset \cdots$. Corresponding to $\widetilde{K}_1$ is a subtree $T_1 \subset T\Gamma$ consisting of a central vertex with a number of edges radiating out from it, an 'asterisk' with possibly an infinite number of edges. When we enlarge $\widetilde{K}_1$ to $\widetilde{K}_2$, T_1 is correspondingly enlarged to a tree T_2 by attaching a similar asterisk at the end of each outer vertex of T_1, and each subsequent enlargement is handled in the same way. The action of $\pi_1(K\Gamma)$ on $\widetilde{K}$ as deck transformations induces an action on $T\Gamma$, permuting its vertices and edges, and the orbit space of $T\Gamma$ under this action is just the original graph Γ. The action on $T\Gamma$ will not generally be a free action since the elements of a subgroup $G_v \subset \pi_1(K\Gamma)$ fix the vertex of $T\Gamma$ corresponding to one of the universal covers of $K(G_v, 1)$.

There is in fact an exact correspondence between graphs of groups and groups acting on trees. See [Scott & Wall 1979] for an exposition of this rather nice theory. From the viewpoint of groups acting on trees, the definition of a graph of groups is usually taken to be slightly more restrictive than the one we have given here, namely, one considers only oriented graphs obtained from an unoriented graph by subdividing each edge by adding a vertex at its midpoint, then orienting the two resulting edges outward, away from the new vertex.

Exercises

1. Suppose a group G acts simplicially on a Δ-complex X, where 'simplicially' means that each element of G takes each simplex of X onto another simplex by a linear homeomorphism. If the action is free, show it is a covering space action.

2. Let X be a connected CW complex and G a group such that every homomorphism $\pi_1(X) \to G$ is trivial. Show that every map $X \to K(G, 1)$ is nullhomotopic.

3. Show that every graph product of trivial groups is free.

4. Use van Kampen's theorem to compute $A*_C$ as a quotient of $A * \mathbb{Z}$, as stated in the text.

5. Consider the graph of groups Γ having one vertex, $\mathbb{Z}$, and one edge, the map $\mathbb{Z} \to \mathbb{Z}$ that is multiplication by 2, realized by the 2-sheeted covering space $S^1 \to S^1$. Show that $\pi_1(K\Gamma)$ has presentation $\langle a, b \mid bab^{-1}a^{-2} \rangle$ and describe the universal cover of $K\Gamma$ explicitly as a product $T \times \mathbb{R}$ with T a tree. [The group $\pi_1(K\Gamma)$ is the first in a family of groups called Baumslag-Solitar groups, having presentations of the form $\langle a, b \mid ba^m b^{-1} a^{-n} \rangle$. These are HNN extensions $\mathbb{Z}*_{\mathbb{Z}}$.]

6. Show that for a graph of groups all of whose edge homomorphisms are injective maps $\mathbb{Z} \to \mathbb{Z}$, we can choose $K\Gamma$ to have universal cover a product $T \times \mathbb{R}$ with T a tree. Work out in detail the case that the graph of groups is the infinite sequence $\mathbb{Z} \xrightarrow{2} \mathbb{Z} \xrightarrow{3} \mathbb{Z} \xrightarrow{4} \mathbb{Z} \to \cdots$ where the map $\mathbb{Z} \xrightarrow{n} \mathbb{Z}$ is multiplication by n. Show that $\pi_1(K\Gamma)$ is isomorphic to $\mathbb{Q}$ in this case. How would one modify this example to get $\pi_1(K\Gamma)$ isomorphic to the subgroup of $\mathbb{Q}$ consisting of rational numbers with denominator a power of 2?

7. Show that every graph product of groups can be realized by a graph whose vertices are partitioned into two subsets, with every oriented edge going from a vertex in the first subset to a vertex in the second subset.

8. Show that a finite graph product of finitely generated groups is finitely generated, and similarly for finitely presented groups.

9. If Γ is a finite graph of finite groups with injective edge homomorphisms, show that the graph product of the groups has a free subgroup of finite index by constructing a suitable finite-sheeted covering space of $K\Gamma$ from universal covers of the mapping cylinders in $K\Gamma$. [The converse is also true: A finitely generated group having a free subgroup of finite index is isomorphic to such a graph product. For a proof of this see [Scott & Wall 1979], Theorem 7.3.]

Chapter 2

Homology

The fundamental group $\pi_1(X)$ is especially useful when studying spaces of low dimension, as one would expect from its definition which involves only maps from low-dimensional spaces into X, namely loops $I \to X$ and homotopies of loops, maps $I \times I \to X$. The definition in terms of objects that are at most 2-dimensional manifests itself for example in the fact that when X is a CW complex, $\pi_1(X)$ depends only on the 2-skeleton of X. In view of the low-dimensional nature of the fundamental group, we should not expect it to be a very refined tool for dealing with high-dimensional spaces. Thus it cannot distinguish between spheres S^n with $n \geq 2$. This limitation to low dimensions can be removed by considering the natural higher-dimensional analogs of $\pi_1(X)$, the homotopy groups $\pi_n(X)$, which are defined in terms of maps of the n-dimensional cube I^n into X and homotopies $I^n \times I \to X$ of such maps. Not surprisingly, when X is a CW complex, $\pi_n(X)$ depends only on the $(n+1)$-skeleton of X. And as one might hope, homotopy groups do indeed distinguish spheres of all dimensions since $\pi_i(S^n)$ is 0 for $i < n$ and $\mathbb{Z}$ for $i = n$.

However, the higher-dimensional homotopy groups have the serious drawback that they are extremely difficult to compute in general. Even for simple spaces like spheres, the calculation of $\pi_i(S^n)$ for $i > n$ turns out to be a huge problem. Fortunately there is a more computable alternative to homotopy groups: the homology groups $H_n(X)$. Like $\pi_n(X)$, the homology group $H_n(X)$ for a CW complex X depends only on the $(n+1)$-skeleton. For spheres, the homology groups $H_i(S^n)$ are isomorphic to the homotopy groups $\pi_i(S^n)$ in the range $1 \leq i \leq n$, but homology groups have the advantage that $H_i(S^n) = 0$ for $i > n$.

The computability of homology groups does not come for free, unfortunately. The definition of homology groups is decidedly less transparent than the definition of homotopy groups, and once one gets beyond the definition there is a certain amount of technical machinery to be set up before any real calculations and applications can be given. In the exposition below we approach the definition of $H_n(X)$ by two preliminary stages, first giving a few motivating examples nonrigorously, then constructing a restricted model of homology theory called simplicial homology, before plunging into the general theory, known as singular homology. After the definition of singular homology has been assimilated, the real work of establishing its basic properties begins. This takes close to 20 pages, and there is no getting around the fact that it is a substantial effort. This takes up most of the first section of the chapter, with small digressions only for two applications to classical theorems of Brouwer: the fixed point theorem and 'invariance of dimension.'

The second section of the chapter gives more applications, including the homology definition of Euler characteristic and Brouwer's notion of degree for maps $S^n \to S^n$. However, the main thrust of this section is toward developing techniques for calculating homology groups efficiently. The maximally efficient method is known as cellular homology, whose power comes perhaps from the fact that it is 'homology squared' — homology defined in terms of homology. Another quite useful tool is Mayer–Vietoris sequences, the analog for homology of van Kampen's theorem for the fundamental group.

An interesting feature of homology that begins to emerge after one has worked with it for a while is that it is the basic properties of homology that are used most often, and not the actual definition itself. This suggests that an axiomatic approach to homology might be possible. This is indeed the case, and in the third section of the chapter we list axioms which completely characterize homology groups for CW complexes. One could take the viewpoint that these rather algebraic axioms are all that really matters about homology groups, that the geometry involved in the definition of homology is secondary, needed only to show that the axiomatic theory is not vacuous. The extent to which one adopts this viewpoint is a matter of taste, and the route taken here of postponing the axioms until the theory is well-established is just one of several possible approaches.

The chapter then concludes with three optional sections of Additional Topics. The first is rather brief, relating $H_1(X)$ to $\pi_1(X)$, while the other two contain a selection of classical applications of homology. These include the n-dimensional version of the Jordan curve theorem and the 'invariance of domain' theorem, both due to Brouwer, along with the Lefschetz fixed point theorem.

The Idea of Homology

The difficulty with the higher homotopy groups π_n is that they are not directly computable from a cell structure as π_1 is. For example, the 2-sphere has no cells in dimensions greater than 2, yet its n-dimensional homotopy group $\pi_n(S^2)$ is nonzero for infinitely many values of n. Homology groups, by contrast, are quite directly related to cell structures, and may indeed be regarded as simply an algebraization of the first layer of geometry in cell structures: how cells of dimension n attach to cells of dimension $n - 1$.

Let us look at some examples to see what the idea is. Consider the graph X_1 shown in the figure, consisting of two vertices joined by four edges. When studying the fundamental group of X_1 we consider loops formed by sequences of edges, starting and ending at a fixed basepoint. For example, at the basepoint x, the loop ab^{-1} travels forward along the edge a, then backward along b, as indicated by the exponent -1. A more compli- 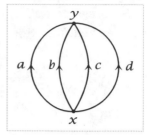 cated loop would be $ac^{-1}bd^{-1}ca^{-1}$. A salient feature of the fundamental group is that it is generally nonabelian, which both enriches and compli- cates the theory. Suppose we simplify matters by abelianizing. Thus for example the two loops ab^{-1} and $b^{-1}a$ are to be regarded as equal if we make a commute with b^{-1}. These two loops ab^{-1} and $b^{-1}a$ are really the same circle, just with a different choice of starting and ending point: x for ab^{-1} and y for $b^{-1}a$. The same thing happens for all loops: Rechoosing the basepoint in a loop just permutes its letters cyclically, so a byproduct of abelianizing is that we no longer have to pin all our loops down to a fixed basepoint. Thus loops become *cycles*, without a chosen basepoint.

Having abelianized, let us switch to additive notation, so cycles become linear combinations of edges with integer coefficients, such as $a - b + c - d$. Let us call these linear combinations *chains* of edges. Some chains can be decomposed into cycles in several different ways, for example $(a - c) + (b - d) = (a - d) + (b - c)$, and if we adopt an algebraic viewpoint then we do not want to distinguish between these different decompositions. Thus we broaden the meaning of the term 'cycle' to be simply any linear combination of edges for which at least one decomposition into cycles in the previous more geometric sense exists.

What is the condition for a chain to be a cycle in this more algebraic sense? A geometric cycle, thought of as a path traversed in time, is distinguished by the prop- erty that it enters each vertex the same number of times that it leaves the vertex. For an arbitrary chain $ka + \ell b + mc + nd$, the net number of times this chain enters y is $k + \ell + m + n$ since each of a, b, c, and d enters y once. Similarly, each of the four edges leaves x once, so the net number of times the chain $ka + \ell b + mc + nd$ enters x is $-k - \ell - m - n$. Thus the condition for $ka + \ell b + mc + nd$ to be a cycle is simply $k + \ell + m + n = 0$.

To describe this result in a way that would generalize to all graphs, let C_1 be the free abelian group with basis the edges a, b, c, d and let C_0 be the free abelian group with basis the vertices x, y. Elements of C_1 are chains of edges, or 1-dimensional chains, and elements of C_0 are linear combinations of vertices, or 0-dimensional chains. Define a homomorphism $\partial : C_1 \to C_0$ by sending each basis element a, b, c, d to $y - x$, the vertex at the head of the edge minus the vertex at the tail. Thus we have $\partial(ka + \ell b + mc + nd) = (k + \ell + m + n)y - (k + \ell + m + n)x$, and the cycles are precisely the kernel of ∂. It is a simple calculation to verify that $a - b$, $b - c$, and $c - d$

form a basis for this kernel. Thus every cycle in X_1 is a unique linear combination of these three most obvious cycles. By means of these three basic cycles we convey the geometric information that the graph X_1 has three visible 'holes,' the empty spaces between the four edges.

Let us now enlarge the preceding graph X_1 by attaching a 2-cell A along the cycle $a - b$, producing a 2-dimensional cell complex X_2. If we think of the 2-cell A as being oriented clockwise, then we can regard its boundary as the cycle $a - b$. This cycle is now homotopically trivial since we can contract it to a point by sliding over A. In other words, it no longer encloses a hole in X_2. This suggests that we form a quotient of the group of cycles in the preceding example by factoring out the subgroup generated by $a - b$. In this quotient the cycles $a - c$ and $b - c$, for example, become equivalent, consistent with the fact that they are homotopic in X_2.

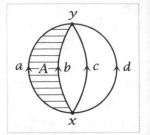

Algebraically, we can define now a pair of homomorphisms $C_2 \xrightarrow{\partial_2} C_1 \xrightarrow{\partial_1} C_0$ where C_2 is the infinite cyclic group generated by A and $\partial_2(A) = a - b$. The map ∂_1 is the boundary homomorphism in the previous example. The quotient group we are interested in is $\operatorname{Ker} \partial_1 / \operatorname{Im} \partial_2$, the kernel of ∂_1 modulo the image of ∂_2, or in other words, the 1-dimensional cycles modulo those that are boundaries, the multiples of $a - b$. This quotient group is the *homology group* $H_1(X_2)$. The previous example can be fit into this scheme too by taking C_2 to be zero since there are no 2-cells in X_1, so in this case $H_1(X_1) = \operatorname{Ker} \partial_1 / \operatorname{Im} \partial_2 = \operatorname{Ker} \partial_1$, which as we saw was free abelian on three generators. In the present example, $H_1(X_2)$ is free abelian on two generators, $b - c$ and $c - d$, expressing the geometric fact that by filling in the 2-cell A we have reduced the number of 'holes' in our space from three to two.

Suppose we enlarge X_2 to a space X_3 by attaching a second 2-cell B along the same cycle $a - b$. This gives a 2-dimensional chain group C_2 consisting of linear combinations of A and B, and the boundary homomorphism $\partial_2 : C_2 \rightarrow C_1$ sends both A and B to $a - b$. The homology group $H_1(X_3) = \operatorname{Ker} \partial_1 / \operatorname{Im} \partial_2$ is the same as for X_2, but now ∂_2 has a nontrivial kernel, the infinite cyclic group generated by $A - B$. We view $A - B$ as a 2-dimensional cycle, generating the homology group $H_2(X_3) = \operatorname{Ker} \partial_2 \approx \mathbb{Z}$.

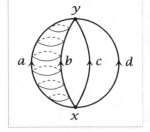

Topologically, the cycle $A - B$ is the sphere formed by the cells A and B together with their common boundary circle. This spherical cycle detects the presence of a 'hole' in X_3, the missing interior of the sphere. However, since this hole is enclosed by a sphere rather than a circle, it is of a different sort from the holes detected by $H_1(X_3) \approx \mathbb{Z} \times \mathbb{Z}$, which are detected by the cycles $b - c$ and $c - d$.

Let us continue one more step and construct a complex X_4 from X_3 by attaching a 3-cell C along the 2-sphere formed by A and B. This creates a chain group C_3

generated by this 3-cell C, and we define a boundary homomorphism $\partial_3 : C_3 \to C_2$ sending C to $A - B$ since the cycle $A - B$ should be viewed as the boundary of C in the same way that the 1-dimensional cycle $a - b$ is the boundary of A. Now we have a sequence of three boundary homomorphisms $C_3 \xrightarrow{\partial_3} C_2 \xrightarrow{\partial_2} C_1 \xrightarrow{\partial_1} C_0$ and the quotient $H_2(X_4) = \operatorname{Ker} \partial_2 / \operatorname{Im} \partial_3$ has become trivial. Also $H_3(X_4) = \operatorname{Ker} \partial_3 = 0$. The group $H_1(X_4)$ is the same as $H_1(X_3)$, namely $\mathbb{Z} \times \mathbb{Z}$, so this is the only nontrivial homology group of X_4.

It is clear what the general pattern of the examples is. For a cell complex X one has chain groups $C_n(X)$ which are free abelian groups with basis the n-cells of X, and there are boundary homomorphisms $\partial_n : C_n(X) \to C_{n-1}(X)$, in terms of which one defines the homology group $H_n(X) = \operatorname{Ker} \partial_n / \operatorname{Im} \partial_{n+1}$. The major difficulty is how to define ∂_n in general. For $n = 1$ this is easy: The boundary of an oriented edge is the vertex at its head minus the vertex at its tail. The next case $n = 2$ is also not hard, at least for cells attached along cycles that are simply loops of edges, for then the boundary of the cell is this cycle of edges, with the appropriate signs taking orientations into account. But for larger n, matters become more complicated. Even if one restricts attention to cell complexes formed from polyhedral cells with nice attaching maps, there is still the matter of orientations to sort out.

The best solution to this problem seems to be to adopt an indirect approach. Arbitrary polyhedra can always be subdivided into special polyhedra called simplices (the triangle and the tetrahedron are the 2-dimensional and 3-dimensional instances) so there is no loss of generality, though initially there is some loss of efficiency, in restricting attention entirely to simplices. For simplices there is no difficulty in defining boundary maps or in handling orientations. So one obtains a homology theory, called simplicial homology, for cell complexes built from simplices. Still, this is a rather restricted class of spaces, and the theory itself has a certain rigidity that makes it awkward to work with.

The way around these obstacles is to step back from the geometry of spaces decomposed into simplices and to consider instead something which at first glance seems wildly more complicated, the collection of all possible continuous maps of simplices into a given space X. These maps generate tremendously large chain groups $C_n(X)$, but the quotients $H_n(X) = \operatorname{Ker} \partial_n / \operatorname{Im} \partial_{n+1}$, called singular homology groups, turn out to be much smaller, at least for reasonably nice spaces X. In particular, for spaces like those in the four examples above, the singular homology groups coincide with the homology groups we computed from the cellular chains. And as we shall see later in this chapter, singular homology allows one to define these nice cellular homology groups for all cell complexes, and in particular to solve the problem of defining the boundary maps for cellular chains.

2.1 Simplicial and Singular Homology

The most important homology theory in algebraic topology, and the one we shall be studying almost exclusively, is called singular homology. Since the technical apparatus of singular homology is somewhat complicated, we will first introduce a more primitive version called simplicial homology in order to see how some of the apparatus works in a simpler setting before beginning the general theory.

The natural domain of definition for simplicial homology is a class of spaces we call Δ-complexes, which are a mild generalization of the more classical notion of a simplicial complex. Historically, the modern definition of singular homology was first given in [Eilenberg 1944], and Δ-complexes were introduced soon thereafter in [Eilenberg-Zilber 1950] where they were called semisimplicial complexes. Within a few years this term came to be applied to what Eilenberg and Zilber called complete semisimplicial complexes, and later there was yet another shift in terminology as the latter objects came to be called simplicial sets. In theory this frees up the term semisimplicial complex to have its original meaning, but to avoid potential confusion it seems best to introduce a new name, and the term Δ-complex has at least the virtue of brevity.

Δ-Complexes

The torus, the projective plane, and the Klein bottle can each be obtained from a square by identifying opposite edges in the way indicated by the arrows in the following figures:

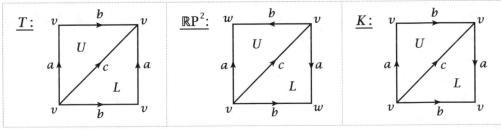

Cutting a square along a diagonal produces two triangles, so each of these surfaces can also be built from two triangles by identifying their edges in pairs. In similar fashion a polygon with any number of sides can be cut along diagonals into triangles, so in fact all closed surfaces can be constructed from triangles by identifying edges. Thus we have a single building block, the triangle, from which all surfaces can be constructed. Using only triangles we could also construct a large class of 2-dimensional spaces that are not surfaces in the strict sense, by allowing more than two edges to be identified together at a time.

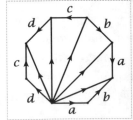

The idea of a Δ-complex is to generalize constructions like these to any number of dimensions. The n-dimensional analog of the triangle is the **n-simplex**. This is the smallest convex set in a Euclidean space $\mathbb{R}^m$ containing $n + 1$ points $v_0, \cdots, v_n$ that do not lie in a hyperplane of dimension less than n, where by a hyperplane we mean the set of solutions of a system of linear equations. An equivalent condition would be that the difference vectors

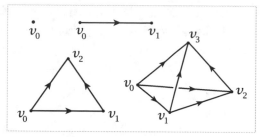

$v_1 - v_0, \cdots, v_n - v_0$ are linearly independent. The points v_i are the **vertices** of the simplex, and the simplex itself is denoted $[v_0, \cdots, v_n]$. For example, there is the standard n-simplex

$$\Delta^n = \{ (t_0, \cdots, t_n) \in \mathbb{R}^{n+1} \mid \textstyle\sum_i t_i = 1 \text{ and } t_i \geq 0 \text{ for all } i \}$$

whose vertices are the unit vectors along the coordinate axes.

For purposes of homology it will be important to keep track of the order of the vertices of a simplex, so 'n-simplex' will really mean 'n-simplex with an ordering of its vertices.' A by-product of ordering the vertices of a simplex $[v_0, \cdots, v_n]$ is that this determines orientations of the edges $[v_i, v_j]$ according to increasing subscripts, as shown in the two preceding figures. Specifying the ordering of the vertices also determines a canonical linear homeomorphism from the standard n-simplex Δ^n onto any other n-simplex $[v_0, \cdots, v_n]$, preserving the order of vertices, namely, $(t_0, \cdots, t_n) \mapsto \sum_i t_i v_i$. The coefficients t_i are the **barycentric coordinates** of the point $\sum_i t_i v_i$ in $[v_0, \cdots, v_n]$.

If we delete one of the $n + 1$ vertices of an n-simplex $[v_0, \cdots, v_n]$, then the remaining n vertices span an $(n - 1)$-simplex, called a **face** of $[v_0, \cdots, v_n]$. We adopt the following convention:

> The vertices of a face, or of any subsimplex spanned by a subset of the vertices, will always be ordered according to their order in the larger simplex.

The union of all the faces of Δ^n is the **boundary** of Δ^n, written $\partial\Delta^n$. The **open simplex** $\mathring{\Delta}^n$ is $\Delta^n - \partial\Delta^n$, the interior of Δ^n.

A **Δ-complex** structure on a space X is a collection of maps $\sigma_\alpha : \Delta^n \to X$, with n depending on the index α, such that:

(i) The restriction $\sigma_\alpha | \mathring{\Delta}^n$ is injective, and each point of X is in the image of exactly one such restriction $\sigma_\alpha | \mathring{\Delta}^n$.

(ii) Each restriction of σ_α to a face of Δ^n is one of the maps $\sigma_\beta : \Delta^{n-1} \to X$. Here we are identifying the face of Δ^n with Δ^{n-1} by the canonical linear homeomorphism between them that preserves the ordering of the vertices.

(iii) A set $A \subset X$ is open iff $\sigma_\alpha^{-1}(A)$ is open in Δ^n for each σ_α.

Among other things, this last condition rules out trivialities like regarding all the points of X as individual vertices. The earlier decompositions of the torus, projective plane, and Klein bottle into two triangles, three edges, and one or two vertices define Δ-complex structures with a total of six σ_α's for the torus and Klein bottle and seven for the projective plane. The orientations on the edges in the pictures are compatible with a unique ordering of the vertices of each simplex, and these orderings determine the maps σ_α.

A consequence of (iii) is that X can be built as a quotient space of a collection of disjoint simplices Δ_α^n, one for each $\sigma_\alpha : \Delta^n \to X$, the quotient space obtained by identifying each face of a Δ_α^n with the Δ_β^{n-1} corresponding to the restriction σ_β of σ_α to the face in question, as in condition (ii). One can think of building the quotient space inductively, starting with a discrete set of vertices, then attaching edges to these to produce a graph, then attaching 2-simplices to the graph, and so on. From this viewpoint we see that the data specifying a Δ-complex can be described purely combinatorially as collections of n-simplices Δ_α^n for each n together with functions associating to each face of each n-simplex Δ_α^n an $(n-1)$-simplex Δ_β^{n-1}.

More generally, Δ-complexes can be built from collections of disjoint simplices by identifying various subsimplices spanned by subsets of the vertices, where the identifications are performed using the canonical linear homeomorphisms that preserve the orderings of the vertices. The earlier Δ-complex structures on a torus, projective plane, or Klein bottle can be obtained in this way, by identifying pairs of edges of two 2-simplices. If one starts with a single 2-simplex and identifies all three edges to a single edge, preserving the orientations given by the ordering of the vertices, this produces a Δ-complex known as the 'dunce hat.' By contrast, if the three edges of a 2-simplex are identified preserving a cyclic orientation of the three edges, as in the first figure at the right, this does not produce a Δ-complex structure, although if the 2-simplex is subdivided into three smaller 2-simplices about a central vertex, then one does obtain a Δ-complex structure on the quotient space.

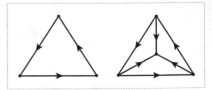

Thinking of a Δ-complex X as a quotient space of a collection of disjoint simplices, it is not hard to see that X must be a Hausdorff space. Condition (iii) then implies that each restriction $\sigma_\alpha | \mathring{\Delta}^n$ is a homeomorphism onto its image, which is thus an open simplex in X. It follows from Proposition A.2 in the Appendix that these open simplices $\sigma_\alpha(\mathring{\Delta}^n)$ are the cells e_α^n of a CW complex structure on X with the σ_α's as characteristic maps. We will not need this fact at present, however.

Simplicial Homology

Our goal now is to define the simplicial homology groups of a Δ-complex X. Let $\Delta_n(X)$ be the free abelian group with basis the open n-simplices e_α^n of X. Elements of $\Delta_n(X)$, called n-**chains**, can be written as finite formal sums $\sum_\alpha n_\alpha e_\alpha^n$ with coefficients $n_\alpha \in \mathbb{Z}$. Equivalently, we could write $\sum_\alpha n_\alpha \sigma_\alpha$ where $\sigma_\alpha : \Delta^n \to X$ is the characteristic map of e_α^n, with image the closure of e_α^n as described above. Such a sum $\sum_\alpha n_\alpha \sigma_\alpha$ can be thought of as a finite collection, or 'chain,' of n-simplices in X with integer multiplicities, the coefficients n_α.

As one can see in the next figure, the boundary of the n-simplex $[v_0, \cdots, v_n]$ consists of the various $(n-1)$-dimensional simplices $[v_0, \cdots, \hat{v}_i, \cdots, v_n]$, where the 'hat' symbol $\hat{\ }$ over v_i indicates that this vertex is deleted from the sequence $v_0, \cdots, v_n$. In terms of chains, we might then wish to say that the boundary of $[v_0, \cdots, v_n]$ is the $(n-1)$-chain formed by the sum of the faces $[v_0, \cdots, \hat{v}_i, \cdots, v_n]$. However, it turns out to be better to insert certain signs and instead let the boundary of $[v_0, \cdots, v_n]$ be $\sum_i (-1)^i [v_0, \cdots, \hat{v}_i, \cdots, v_n]$. Heuristically, the signs are inserted to take orientations into account, so that all the faces of a simplex are coherently oriented, as indicated in the following figure:

$$\partial[v_0, v_1] = [v_1] - [v_0]$$

$$\partial[v_0, v_1, v_2] = [v_1, v_2] - [v_0, v_2] + [v_0, v_1]$$

$$\partial[v_0, v_1, v_2, v_3] = [v_1, v_2, v_3] - [v_0, v_2, v_3]$$
$$+ [v_0, v_1, v_3] - [v_0, v_1, v_2]$$

In the last case, the orientations of the two hidden faces are also counterclockwise when viewed from outside the 3-simplex.

With this geometry in mind we define for a general Δ-complex X a **boundary homomorphism** $\partial_n : \Delta_n(X) \to \Delta_{n-1}(X)$ by specifying its values on basis elements:

$$\partial_n(\sigma_\alpha) = \sum_i (-1)^i \sigma_\alpha | [v_0, \cdots, \hat{v}_i, \cdots, v_n]$$

Note that the right side of this equation does indeed lie in $\Delta_{n-1}(X)$ since each restriction $\sigma_\alpha | [v_0, \cdots, \hat{v}_i, \cdots, v_n]$ is the characteristic map of an $(n-1)$-simplex of X.

‖ **Lemma 2.1.** *The composition* $\Delta_n(X) \xrightarrow{\partial_n} \Delta_{n-1}(X) \xrightarrow{\partial_{n-1}} \Delta_{n-2}(X)$ *is zero.*

Proof: We have $\partial_n(\sigma) = \sum_i (-1)^i \sigma | [v_0, \cdots, \hat{v}_i, \cdots, v_n]$, and hence

$$\partial_{n-1}\partial_n(\sigma) = \sum_{j<i} (-1)^i (-1)^j \sigma | [v_0, \cdots, \hat{v}_j, \cdots, \hat{v}_i, \cdots, v_n]$$

$$+ \sum_{j>i} (-1)^i (-1)^{j-1} \sigma | [v_0, \cdots, \hat{v}_i, \cdots, \hat{v}_j, \cdots, v_n]$$

The latter two summations cancel since after switching i and j in the second sum, it becomes the negative of the first. □

The algebraic situation we have now is a sequence of homomorphisms of abelian groups

$$\cdots \longrightarrow C_{n+1} \xrightarrow{\partial_{n+1}} C_n \xrightarrow{\partial_n} C_{n-1} \longrightarrow \cdots \longrightarrow C_1 \xrightarrow{\partial_1} C_0 \xrightarrow{\partial_0} 0$$

with $\partial_n \partial_{n+1} = 0$ for each n. Such a sequence is called a **chain complex**. Note that we have extended the sequence by a 0 at the right end, with $\partial_0 = 0$. The equation $\partial_n \partial_{n+1} = 0$ is equivalent to the inclusion $\operatorname{Im} \partial_{n+1} \subset \operatorname{Ker} \partial_n$, where Im and Ker denote image and kernel. So we can define the nth **homology group** of the chain complex to be the quotient group $H_n = \operatorname{Ker} \partial_n / \operatorname{Im} \partial_{n+1}$. Elements of $\operatorname{Ker} \partial_n$ are called **cycles** and elements of $\operatorname{Im} \partial_{n+1}$ are called **boundaries**. Elements of H_n are cosets of $\operatorname{Im} \partial_{n+1}$, called **homology classes**. Two cycles representing the same homology class are said to be **homologous**. This means their difference is a boundary.

Returning to the case that $C_n = \Delta_n(X)$, the homology group $\operatorname{Ker} \partial_n / \operatorname{Im} \partial_{n+1}$ will be denoted $H_n^{\Delta}(X)$ and called the nth **simplicial homology group** of X.

Example 2.2. $X = S^1$, with one vertex v and one edge e. Then $\Delta_0(S^1)$ and $\Delta_1(S^1)$ are both $\mathbb{Z}$ and the boundary map ∂_1 is zero since $\partial e = v - v$. The groups $\Delta_n(S^1)$ are 0 for $n \geq 2$ since there are no simplices in these dimensions. Hence

$$H_n^{\Delta}(S^1) \approx \begin{cases} \mathbb{Z} & \text{for } n = 0, 1 \\ 0 & \text{for } n \geq 2 \end{cases}$$

This is an illustration of the general fact that if the boundary maps in a chain complex are all zero, then the homology groups of the complex are isomorphic to the chain groups themselves.

Example 2.3. $X = T$, the torus with the Δ-complex structure pictured earlier, having one vertex, three edges a, b, and c, and two 2-simplices U and L. As in the previous example, $\partial_1 = 0$ so $H_0^{\Delta}(T) \approx \mathbb{Z}$. Since $\partial_2 U = a + b - c = \partial_2 L$ and $\{a, b, a + b - c\}$ is a basis for $\Delta_1(T)$, it follows that $H_1^{\Delta}(T) \approx \mathbb{Z} \oplus \mathbb{Z}$ with basis the homology classes $[a]$ and $[b]$. Since there are no 3-simplices, $H_2^{\Delta}(T)$ is equal to $\operatorname{Ker} \partial_2$, which is infinite cyclic generated by $U - L$ since $\partial(pU + qL) = (p + q)(a + b - c) = 0$ only if $p = -q$. Thus

$$H_n^{\Delta}(T) \approx \begin{cases} \mathbb{Z} \oplus \mathbb{Z} & \text{for } n = 1 \\ \mathbb{Z} & \text{for } n = 0, 2 \\ 0 & \text{for } n \geq 3 \end{cases}$$

Example 2.4. $X = \mathbb{RP}^2$, as pictured earlier, with two vertices v and w, three edges a, b, and c, and two 2-simplices U and L. Then $\operatorname{Im} \partial_1$ is generated by $w - v$, so $H_0^{\Delta}(X) \approx \mathbb{Z}$ with either vertex as a generator. Since $\partial_2 U = -a + b + c$ and $\partial_2 L = a - b + c$, we see that ∂_2 is injective, so $H_2^{\Delta}(X) = 0$. Further, $\operatorname{Ker} \partial_1 \approx \mathbb{Z} \oplus \mathbb{Z}$ with basis $a - b$ and c, and $\operatorname{Im} \partial_2$ is an index-two subgroup of $\operatorname{Ker} \partial_1$ since we can choose c and $a - b + c$

as a basis for $\mathrm{Ker}\,\partial_1$ and $a - b + c$ and $2c = (a - b + c) + (-a + b + c)$ as a basis for $\mathrm{Im}\,\partial_2$. Thus $H_1^\Delta(X) \approx \mathbb{Z}_2$.

Example 2.5. We can obtain a Δ-complex structure on S^n by taking two copies of Δ^n and identifying their boundaries via the identity map. Labeling these two n-simplices U and L, then it is obvious that $\mathrm{Ker}\,\partial_n$ is infinite cyclic generated by $U - L$. Thus $H_n^\Delta(S^n) \approx \mathbb{Z}$ for this Δ-complex structure on S^n. Computing the other homology groups would be more difficult.

Many similar examples could be worked out without much trouble, such as the other closed orientable and nonorientable surfaces. However, the calculations do tend to increase in complexity before long, particularly for higher-dimensional complexes.

Some obvious general questions arise: Are the groups $H_n^\Delta(X)$ independent of the choice of Δ-complex structure on X? In other words, if two Δ-complexes are homeomorphic, do they have isomorphic homology groups? More generally, do they have isomorphic homology groups if they are merely homotopy equivalent? To answer such questions and to develop a general theory it is best to leave the rather rigid simplicial realm and introduce the singular homology groups. These have the added advantage that they are defined for all spaces, not just Δ-complexes. At the end of this section, after some theory has been developed, we will show that simplicial and singular homology groups coincide for Δ-complexes.

Traditionally, simplicial homology is defined for **simplicial complexes,** which are the Δ-complexes whose simplices are uniquely determined by their vertices. This amounts to saying that each n-simplex has $n + 1$ distinct vertices, and that no other n-simplex has this same set of vertices. Thus a simplicial complex can be described combinatorially as a set X_0 of vertices together with sets X_n of n-simplices, which are $(n + 1)$-element subsets of X_0. The only requirement is that each $(k + 1)$-element subset of the vertices of an n-simplex in X_n is a k-simplex, in X_k. From this combinatorial data a Δ-complex X can be constructed, once we choose a partial ordering of the vertices X_0 that restricts to a linear ordering on the vertices of each simplex in X_n. For example, we could just choose a linear ordering of all the vertices. This might perhaps involve invoking the Axiom of Choice for large vertex sets.

An exercise at the end of this section is to show that every Δ-complex can be subdivided to be a simplicial complex. In particular, every Δ-complex is then homeomorphic to a simplicial complex.

Compared with simplicial complexes, Δ-complexes have the advantage of simpler computations since fewer simplices are required. For example, to put a simplicial complex structure on the torus one needs at least 14 triangles, 21 edges, and 7 vertices, and for $\mathbb{RP}^2$ one needs at least 10 triangles, 15 edges, and 6 vertices. This would slow down calculations considerably!

Singular Homology

A **singular n-simplex** in a space X is by definition just a map $\sigma : \Delta^n \to X$. The word 'singular' is used here to express the idea that σ need not be a nice embedding but can have 'singularities' where its image does not look at all like a simplex. All that is required is that σ be continuous. Let $C_n(X)$ be the free abelian group with basis the set of singular n-simplices in X. Elements of $C_n(X)$, called **n-chains**, or more precisely singular n-chains, are finite formal sums $\sum_i n_i \sigma_i$ for $n_i \in \mathbb{Z}$ and $\sigma_i : \Delta^n \to X$. A boundary map $\partial_n : C_n(X) \to C_{n-1}(X)$ is defined by the same formula as before:

$$\partial_n(\sigma) = \sum_i (-1)^i \sigma \,|\, [v_0, \cdots, \hat{v}_i, \cdots, v_n]$$

Implicit in this formula is the canonical identification of $[v_0, \cdots, \hat{v}_i, \cdots, v_n]$ with Δ^{n-1}, preserving the ordering of vertices, so that $\sigma \,|\, [v_0, \cdots, \hat{v}_i, \cdots, v_n]$ is regarded as a map $\Delta^{n-1} \to X$, that is, a singular $(n-1)$-simplex.

Often we write the boundary map ∂_n from $C_n(X)$ to $C_{n-1}(X)$ simply as ∂ when this does not lead to serious ambiguities. The proof of Lemma 2.1 applies equally well to singular simplices, showing that $\partial_n \partial_{n+1} = 0$ or more concisely $\partial^2 = 0$, so we can define the **singular homology group** $H_n(X) = \operatorname{Ker} \partial_n / \operatorname{Im} \partial_{n+1}$.

It is evident from the definition that homeomorphic spaces have isomorphic singular homology groups H_n, in contrast with the situation for H_n^Δ. On the other hand, since the groups $C_n(X)$ are so large, the number of singular n-simplices in X usually being uncountable, it is not at all clear that for a Δ-complex X with finitely many simplices, $H_n(X)$ should be finitely generated for all n, or that $H_n(X)$ should be zero for n larger than the dimension of X — two properties that are trivial for $H_n^\Delta(X)$.

Though singular homology looks so much more general than simplicial homology, it can actually be regarded as a special case of simplicial homology by means of the following construction. For an arbitrary space X, define the **singular complex** $S(X)$ to be the Δ-complex with one n-simplex Δ_σ^n for each singular n-simplex $\sigma : \Delta^n \to X$, with Δ_σ^n attached in the obvious way to the $(n-1)$-simplices of $S(X)$ that are the restrictions of σ to the various $(n-1)$-simplices in $\partial \Delta^n$. It is clear from the definitions that $H_n^\Delta(S(X))$ is identical with $H_n(X)$ for all n, and in this sense the singular homology group $H_n(X)$ is a special case of a simplicial homology group. One can regard $S(X)$ as a Δ-complex model for X, although it is usually an extremely large object compared to X.

Cycles in singular homology are defined algebraically, but they can be given a somewhat more geometric interpretation in terms of maps from finite Δ-complexes. To see this, note first that a singular n-chain ξ can always be written in the form $\sum_i \varepsilon_i \sigma_i$ with $\varepsilon_i = \pm 1$, allowing repetitions of the singular n-simplices σ_i. Given such an n-chain $\xi = \sum_i \varepsilon_i \sigma_i$, when we compute $\partial \xi$ as a sum of singular $(n-1)$-simplices with signs ± 1, there may be some *canceling pairs* consisting of two identical singular $(n-1)$-simplices with opposite signs. Choosing a maximal collection of such canceling pairs, construct an n-dimensional Δ-complex K_ξ from a disjoint union of n-simplices Δ_i^n, one for each σ_i, by identifying the pairs of $(n-1)$-dimensional faces corresponding to the chosen canceling pairs. The σ_i's then induce a map $K_\xi \to X$. If ξ is a cycle, all the $(n-1)$-simplices of K_ξ come from canceling pairs, hence are faces of exactly two n-simplices of K_ξ. Thus K_ξ is a manifold, locally homeomorphic to $\mathbb{R}^n$, except at a subcomplex of dimension at most $n-2$. All the n-simplices of K_ξ can be coherently oriented by taking the signs of the σ_i's into account, so K_ξ is actually an oriented manifold away from its nonmanifold points. A closer inspection shows that K_ξ is also a manifold near points in the interiors of $(n-2)$-simplices, so the nonmanifold points of K_ξ in fact have dimension at most $n-3$. However, near the interiors of $(n-3)$-simplices it can very well happen that K_ξ is not a manifold.

In particular, elements of $H_1(X)$ are represented by collections of oriented loops in X, and elements of $H_2(X)$ are represented by maps of closed oriented surfaces into X. With a bit more work it can be shown that an oriented 1-cycle $\coprod_\alpha S^1_\alpha \to X$ is zero in $H_1(X)$ iff it extends to a map of an oriented surface into X, and there is an analogous statement for 2-cycles. In the early days of homology theory it may have been believed, or at least hoped, that this close connection with manifolds continued in all higher dimensions, but this has turned out not to be the case. There is a sort of homology theory built from manifolds, called *bordism*, but it is quite a bit more complicated than the homology theory we are studying here.

After these preliminary remarks let us begin to see what can be proved about singular homology.

Proposition 2.6. *Corresponding to the decomposition of a space X into its path-components X_α there is an isomorphism of $H_n(X)$ with the direct sum $\bigoplus_\alpha H_n(X_\alpha)$.*

Proof: Since a singular simplex always has path-connected image, $C_n(X)$ splits as the direct sum of its subgroups $C_n(X_\alpha)$. The boundary maps ∂_n preserve this direct sum decomposition, taking $C_n(X_\alpha)$ to $C_{n-1}(X_\alpha)$, so Ker ∂_n and Im ∂_{n+1} split similarly as direct sums, hence the homology groups also split, $H_n(X) \approx \bigoplus_\alpha H_n(X_\alpha)$. $\qquad\square$

Proposition 2.7. *If X is nonempty and path-connected, then $H_0(X) \approx \mathbb{Z}$. Hence for any space X, $H_0(X)$ is a direct sum of $\mathbb{Z}$'s, one for each path-component of X.*

Proof: By definition, $H_0(X) = C_0(X)/\operatorname{Im}\partial_1$ since $\partial_0 = 0$. Define a homomorphism $\varepsilon : C_0(X) \to \mathbb{Z}$ by $\varepsilon(\sum_i n_i \sigma_i) = \sum_i n_i$. This is obviously surjective if X is nonempty. The claim is that Ker $\varepsilon = \operatorname{Im}\partial_1$ if X is path-connected, and hence ε induces an iso-morphism $H_0(X) \approx \mathbb{Z}$.

To verify the claim, observe first that $\operatorname{Im}\partial_1 \subset \operatorname{Ker}\varepsilon$ since for a singular 1-simplex $\sigma : \Delta^1 \to X$ we have $\varepsilon\partial_1(\sigma) = \varepsilon(\sigma|[v_1] - \sigma|[v_0]) = 1 - 1 = 0$. For the reverse inclusion Ker $\varepsilon \subset \operatorname{Im}\partial_1$, suppose $\varepsilon(\sum_i n_i \sigma_i) = 0$, so $\sum_i n_i = 0$. The σ_i's are singular 0-simplices, which are simply points of X. Choose a path $\tau_i : I \to X$ from a basepoint

x_0 to $\sigma_i(v_0)$ and let σ_0 be the singular 0-simplex with image x_0. We can view τ_i as a singular 1-simplex, a map $\tau_i : [v_0, v_1] \rightarrow X$, and then we have $\partial \tau_i = \sigma_i - \sigma_0$. Hence $\partial(\sum_i n_i \tau_i) = \sum_i n_i \sigma_i - \sum_i n_i \sigma_0 = \sum_i n_i \sigma_i$ since $\sum_i n_i = 0$. Thus $\sum_i n_i \sigma_i$ is a boundary, which shows that $\mathrm{Ker}\, \varepsilon \subset \mathrm{Im}\, \partial_1$. $\hfill\square$

‖ Proposition 2.8. *If X is a point, then $H_n(X) = 0$ for $n > 0$ and $H_0(X) \approx \mathbb{Z}$.*

Proof: In this case there is a unique singular n-simplex σ_n for each n, and $\partial(\sigma_n) = \sum_i (-1)^i \sigma_{n-1}$, a sum of $n + 1$ terms, which is therefore 0 for n odd and σ_{n-1} for n even, $n \neq 0$. Thus we have the chain complex

$$\cdots \longrightarrow \mathbb{Z} \xrightarrow{\approx} \mathbb{Z} \xrightarrow{0} \mathbb{Z} \xrightarrow{\approx} \mathbb{Z} \xrightarrow{0} \mathbb{Z} \longrightarrow 0$$

with boundary maps alternately isomorphisms and trivial maps, except at the last $\mathbb{Z}$. The homology groups of this complex are trivial except for $H_0 \approx \mathbb{Z}$. $\hfill\square$

It is often very convenient to have a slightly modified version of homology for which a point has trivial homology groups in all dimensions, including zero. This is done by defining the **reduced homology groups** $\tilde{H}_n(X)$ to be the homology groups of the augmented chain complex

$$\cdots \longrightarrow C_2(X) \xrightarrow{\partial_2} C_1(X) \xrightarrow{\partial_1} C_0(X) \xrightarrow{\varepsilon} \mathbb{Z} \longrightarrow 0$$

where $\varepsilon(\sum_i n_i \sigma_i) = \sum_i n_i$ as in the proof of Proposition 2.7. Here we had better require X to be nonempty, to avoid having a nontrivial homology group in dimension -1. Since $\varepsilon \partial_1 = 0$, ε vanishes on $\mathrm{Im}\, \partial_1$ and hence induces a map $H_0(X) \rightarrow \mathbb{Z}$ with kernel $\tilde{H}_0(X)$, so $H_0(X) \approx \tilde{H}_0(X) \oplus \mathbb{Z}$. Obviously $H_n(X) \approx \tilde{H}_n(X)$ for $n > 0$.

Formally, one can think of the extra $\mathbb{Z}$ in the augmented chain complex as generated by the unique map $[\varnothing] \rightarrow X$ where $[\varnothing]$ is the empty simplex, with no vertices. The augmentation map ε is then the usual boundary map since $\partial[v_0] = [\hat{v}_0] = [\varnothing]$.

Readers who know about the fundamental group $\pi_1(X)$ may wish to make a detour here to look at §2.A where it is shown that $H_1(X)$ is the abelianization of $\pi_1(X)$ whenever X is path-connected. This result will not be needed elsewhere in the chapter, however.

Homotopy Invariance

The first substantial result we will prove about singular homology is that homotopy equivalent spaces have isomorphic homology groups. This will be done by showing that a map $f : X \rightarrow Y$ induces a homomorphism $f_* : H_n(X) \rightarrow H_n(Y)$ for each n, and that f_* is an isomorphism if f is a homotopy equivalence.

For a map $f : X \rightarrow Y$, an induced homomorphism $f_\sharp : C_n(X) \rightarrow C_n(Y)$ is defined by composing each singular n-simplex $\sigma : \Delta^n \rightarrow X$ with f to get a singular n-simplex

$f_\sharp(\sigma) = f\sigma : \Delta^n \to Y$, then extending $f_\sharp$ linearly via $f_\sharp(\sum_i n_i \sigma_i) = \sum_i n_i f_\sharp(\sigma_i) = \sum_i n_i f \sigma_i$. The maps $f_\sharp : C_n(X) \to C_n(Y)$ satisfy $f_\sharp \partial = \partial f_\sharp$ since

$$f_\sharp \partial(\sigma) = f_\sharp(\sum_i (-1)^i \sigma | [v_0, \cdots, \hat{v}_i, \cdots, v_n])$$
$$= \sum_i (-1)^i f \sigma | [v_0, \cdots, \hat{v}_i, \cdots, v_n] = \partial f_\sharp(\sigma)$$

Thus we have a diagram

$$\cdots \longrightarrow C_{n+1}(X) \xrightarrow{\ \partial\ } C_n(X) \xrightarrow{\ \partial\ } C_{n-1}(X) \longrightarrow \cdots$$
$$\downarrow f_\sharp \qquad\qquad \downarrow f_\sharp \qquad\qquad \downarrow f_\sharp$$
$$\cdots \longrightarrow C_{n+1}(Y) \xrightarrow{\ \partial\ } C_n(Y) \xrightarrow{\ \partial\ } C_{n-1}(Y) \longrightarrow \cdots$$

such that in each square the composition $f_\sharp \partial$ equals the composition $\partial f_\sharp$. A diagram of maps with the property that any two compositions of maps starting at one point in the diagram and ending at another are equal is called a **commutative diagram**. In the present case commutativity of the diagram is equivalent to the commutativity relation $f_\sharp \partial = \partial f_\sharp$, but commutative diagrams can contain commutative triangles, pentagons, etc., as well as commutative squares.

The fact that the maps $f_\sharp : C_n(X) \to C_n(Y)$ satisfy $f_\sharp \partial = \partial f_\sharp$ is also expressed by saying that the $f_\sharp$'s define a **chain map** from the singular chain complex of X to that of Y. The relation $f_\sharp \partial = \partial f_\sharp$ implies that $f_\sharp$ takes cycles to cycles since $\partial \alpha = 0$ implies $\partial(f_\sharp \alpha) = f_\sharp(\partial \alpha) = 0$. Also, $f_\sharp$ takes boundaries to boundaries since $f_\sharp(\partial \beta) = \partial(f_\sharp \beta)$. Hence $f_\sharp$ induces a homomorphism $f_* : H_n(X) \to H_n(Y)$. An algebraic statement of what we have just proved is:

Proposition 2.9. *A chain map between chain complexes induces homomorphisms between the homology groups of the two complexes.* □

Two basic properties of induced homomorphisms which are important in spite of being rather trivial are:

(i) $(fg)_* = f_* g_*$ for a composed mapping $X \xrightarrow{g} Y \xrightarrow{f} Z$. This follows from associativity of compositions $\Delta^n \xrightarrow{\sigma} X \xrightarrow{g} Y \xrightarrow{f} Z$.

(ii) $\mathbb{1}_* = \mathbb{1}$ where $\mathbb{1}$ denotes the identity map of a space or a group.

Less trivially, we have:

Theorem 2.10. *If two maps $f, g : X \to Y$ are homotopic, then they induce the same homomorphism $f_* = g_* : H_n(X) \to H_n(Y)$.*

In view of the formal properties $(fg)_* = f_* g_*$ and $\mathbb{1}_* = \mathbb{1}$, this immediately implies:

Corollary 2.11. *The maps $f_* : H_n(X) \to H_n(Y)$ induced by a homotopy equivalence $f : X \to Y$ are isomorphisms for all n.* □

For example, if X is contractible then $\tilde{H}_n(X) = 0$ for all n.

Proof of 2.10: The essential ingredient is a procedure for subdividing $\Delta^n \times I$ into simplices. The figure shows the cases $n = 1, 2$. In $\Delta^n \times I$, let $\Delta^n \times \{0\} = [v_0, \cdots, v_n]$ and $\Delta^n \times \{1\} = [w_0, \cdots, w_n]$, where v_i and w_i have the same image under the projection $\Delta^n \times I \to \Delta^n$. We can pass from $[v_0, \cdots, v_n]$ to $[w_0, \cdots, w_n]$ by interpolating a sequence of n-simplices, each obtained from the preceding one by moving one vertex v_i up to w_i, starting with v_n and working backwards to v_0. Thus the first step is to move $[v_0, \cdots, v_n]$ up to $[v_0, \cdots, v_{n-1}, w_n]$, then the second step is to move this up to $[v_0, \cdots, v_{n-2}, w_{n-1}, w_n]$, and so on. In the typical step $[v_0, \cdots, v_i, w_{i+1}, \cdots, w_n]$

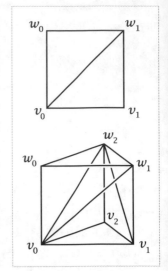

moves up to $[v_0, \cdots, v_{i-1}, w_i, \cdots, w_n]$. The region between these two n-simplices is exactly the $(n+1)$-simplex $[v_0, \cdots, v_i, w_i, \cdots, w_n]$ which has $[v_0, \cdots, v_i, w_{i+1}, \cdots, w_n]$ as its lower face and $[v_0, \cdots, v_{i-1}, w_i, \cdots, w_n]$ as its upper face. Altogether, $\Delta^n \times I$ is the union of the $(n+1)$-simplices $[v_0, \cdots, v_i, w_i, \cdots, w_n]$, each intersecting the next in an n-simplex face.

Given a homotopy $F : X \times I \to Y$ from f to g and a singular simplex $\sigma : \Delta^n \to X$, we can form the composition $F \circ (\sigma \times \mathbb{1}) : \Delta^n \times I \to X \times I \to Y$. Using this, we can define *prism operators* $P : C_n(X) \to C_{n+1}(Y)$ by the following formula:

$$P(\sigma) = \sum_i (-1)^i F \circ (\sigma \times \mathbb{1}) \,|\, [v_0, \cdots, v_i, w_i, \cdots, w_n]$$

We will show that these prism operators satisfy the basic relation

$$\partial P = g_\sharp - f_\sharp - P \partial$$

Geometrically, the left side of this equation represents the boundary of the prism, and the three terms on the right side represent the top $\Delta^n \times \{1\}$, the bottom $\Delta^n \times \{0\}$, and the sides $\partial \Delta^n \times I$ of the prism. To prove the relation we calculate

$$\partial P(\sigma) = \sum_{j \le i} (-1)^i (-1)^j F \circ (\sigma \times \mathbb{1}) \,|\, [v_0, \cdots, \hat{v}_j, \cdots, v_i, w_i, \cdots, w_n]$$
$$+ \sum_{j \ge i} (-1)^i (-1)^{j+1} F \circ (\sigma \times \mathbb{1}) \,|\, [v_0, \cdots, v_i, w_i, \cdots, \widehat{w}_j, \cdots, w_n]$$

The terms with $i = j$ in the two sums cancel except for $F \circ (\sigma \times \mathbb{1}) \,|\, [\hat{v}_0, w_0, \cdots, w_n]$, which is $g \circ \sigma = g_\sharp(\sigma)$, and $-F \circ (\sigma \times \mathbb{1}) \,|\, [v_0, \cdots, v_n, \widehat{w}_n]$, which is $-f \circ \sigma = -f_\sharp(\sigma)$. The terms with $i \ne j$ are exactly $-P \partial(\sigma)$ since

$$P \partial(\sigma) = \sum_{i < j} (-1)^i (-1)^j F \circ (\sigma \times \mathbb{1}) \,|\, [v_0, \cdots, v_i, w_i, \cdots, \widehat{w}_j, \cdots, w_n]$$
$$+ \sum_{i > j} (-1)^{i-1} (-1)^j F \circ (\sigma \times \mathbb{1}) \,|\, [v_0, \cdots, \hat{v}_j, \cdots, v_i, w_i, \cdots, w_n]$$

Now we can finish the proof of the theorem. If $\alpha \in C_n(X)$ is a cycle, then we have $g_\sharp(\alpha) - f_\sharp(\alpha) = \partial P(\alpha) + P\partial(\alpha) = \partial P(\alpha)$ since $\partial \alpha = 0$. Thus $g_\sharp(\alpha) - f_\sharp(\alpha)$ is a boundary, so $g_\sharp(\alpha)$ and $f_\sharp(\alpha)$ determine the same homology class, which means that g_* equals f_* on the homology class of α. □

The relationship $\partial P + P\partial = g_\sharp - f_\sharp$ is expressed by saying P is a **chain homotopy** between the chain maps $f_\sharp$ and $g_\sharp$. We have just shown:

Proposition 2.12. *Chain-homotopic chain maps induce the same homomorphism on homology.* □

There are also induced homomorphisms $f_* : \tilde{H}_n(X) \to \tilde{H}_n(Y)$ for reduced homology groups since $f_\sharp \varepsilon = \varepsilon f_\sharp$ where $f_\sharp$ is the identity map on the added groups $\mathbb{Z}$ in the augmented chain complexes. The properties of induced homomorphisms we proved above hold equally well in the setting of reduced homology, with the same proofs.

Exact Sequences and Excision

If there was always a simple relationship between the homology groups of a space X, a subspace A, and the quotient space X/A, then this could be a very useful tool in understanding the homology groups of spaces such as CW complexes that can be built inductively from successively more complicated subspaces. Perhaps the simplest possible relationship would be if $H_n(X)$ contained $H_n(A)$ as a subgroup and the quotient group $H_n(X)/H_n(A)$ was isomorphic to $H_n(X/A)$. While this does hold in some cases, if it held in general then homology theory would collapse totally since every space X can be embedded as a subspace of a space with trivial homology groups, namely the cone $CX = (X \times I)/(X \times \{0\})$, which is contractible.

It turns out that this overly simple model does not have to be modified too much to get a relationship that is valid in fair generality. The novel feature of the actual relationship is that it involves the groups $H_n(X)$, $H_n(A)$, and $H_n(X/A)$ for all values of n simultaneously. In practice this is not as bad as it might sound, and in addition it has the pleasant side effect of sometimes allowing higher-dimensional homology groups to be computed in terms of lower-dimensional groups which may already be known, for example by induction.

In order to formulate the relationship we are looking for, we need an algebraic definition which is central to algebraic topology. A sequence of homomorphisms

$$\cdots \longrightarrow A_{n+1} \xrightarrow{\alpha_{n+1}} A_n \xrightarrow{\alpha_n} A_{n-1} \longrightarrow \cdots$$

is said to be **exact** if $\operatorname{Ker} \alpha_n = \operatorname{Im} \alpha_{n+1}$ for each n. The inclusions $\operatorname{Im} \alpha_{n+1} \subset \operatorname{Ker} \alpha_n$ are equivalent to $\alpha_n \alpha_{n+1} = 0$, so the sequence is a chain complex, and the opposite inclusions $\operatorname{Ker} \alpha_n \subset \operatorname{Im} \alpha_{n+1}$ say that the homology groups of this chain complex are trivial.

A number of basic algebraic concepts can be expressed in terms of exact sequences, for example:

(i) $0 \longrightarrow A \xrightarrow{\alpha} B$ is exact iff $\operatorname{Ker} \alpha = 0$, i.e., α is injective.

(ii) $A \xrightarrow{\alpha} B \longrightarrow 0$ is exact iff $\operatorname{Im} \alpha = B$, i.e., α is surjective.

(iii) $0 \longrightarrow A \xrightarrow{\alpha} B \longrightarrow 0$ is exact iff α is an isomorphism, by (i) and (ii).

(iv) $0 \longrightarrow A \xrightarrow{\alpha} B \xrightarrow{\beta} C \longrightarrow 0$ is exact iff α is injective, β is surjective, and $\operatorname{Ker} \beta = \operatorname{Im} \alpha$, so β induces an isomorphism $C \approx B / \operatorname{Im} \alpha$. This can be written $C \approx B / A$ if we think of α as an inclusion of A as a subgroup of B.

An exact sequence $0 \rightarrow A \rightarrow B \rightarrow C \rightarrow 0$ as in (iv) is called a **short exact sequence**.

Exact sequences provide the right tool to relate the homology groups of a space, a subspace, and the associated quotient space:

Theorem 2.13. *If X is a space and A is a nonempty closed subspace that is a deformation retract of some neighborhood in X, then there is an exact sequence*

$$\cdots \longrightarrow \tilde{H}_n(A) \xrightarrow{i_*} \tilde{H}_n(X) \xrightarrow{j_*} \tilde{H}_n(X/A) \xrightarrow{\partial} \tilde{H}_{n-1}(A) \xrightarrow{i_*} \tilde{H}_{n-1}(X) \longrightarrow \cdots$$

$$\cdots \longrightarrow \tilde{H}_0(X/A) \longrightarrow 0$$

where i is the inclusion $A \hookrightarrow X$ and j is the quotient map $X \rightarrow X/A$.

The map ∂ will be constructed in the course of the proof. The idea is that an element $x \in \tilde{H}_n(X/A)$ can be represented by a chain α in X with $\partial \alpha$ a cycle in A whose homology class is $\partial x \in \tilde{H}_{n-1}(A)$.

Pairs of spaces (X, A) satisfying the hypothesis of the theorem will be called **good pairs**. For example, if X is a CW complex and A is a nonempty subcomplex, then (X, A) is a good pair by Proposition A.5 in the Appendix.

Corollary 2.14. $\tilde{H}_n(S^n) \approx \mathbb{Z}$ *and* $\tilde{H}_i(S^n) = 0$ *for* $i \ne n$.

Proof: For $n > 0$ take $(X, A) = (D^n, S^{n-1})$ so $X/A = S^n$. The terms $\tilde{H}_i(D^n)$ in the long exact sequence for this pair are zero since D^n is contractible. Exactness of the sequence then implies that the maps $\tilde{H}_i(S^n) \xrightarrow{\partial} \tilde{H}_{i-1}(S^{n-1})$ are isomorphisms for $i > 0$ and that $\tilde{H}_0(S^n) = 0$. The result now follows by induction on n, starting with the case of S^0 where the result holds by Propositions 2.6 and 2.8. $\square$

As an application of this calculation we have the following classical theorem of Brouwer, the 2-dimensional case of which was proved in §1.1.

Corollary 2.15. ∂D^n *is not a retract of D^n. Hence every map $f : D^n \rightarrow D^n$ has a fixed point.*

Proof: If $r : D^n \rightarrow \partial D^n$ is a retraction, then $ri = \mathbb{1}$ for $i : \partial D^n \rightarrow D^n$ the inclusion map. The composition $\tilde{H}_{n-1}(\partial D^n) \xrightarrow{i_*} \tilde{H}_{n-1}(D^n) \xrightarrow{r_*} \tilde{H}_{n-1}(\partial D^n)$ is then the identity map on $\tilde{H}_{n-1}(\partial D^n) \approx \mathbb{Z}$. But i_* and r_* are both 0 since $\tilde{H}_{n-1}(D^n) = 0$, and we have a contradiction. The statement about fixed points follows as in Theorem 1.9. $\square$

The derivation of the exact sequence of homology groups for a good pair (X, A) will be rather a long story. We will in fact derive a more general exact sequence which holds for arbitrary pairs (X, A), but with the homology groups of the quotient space X/A replaced by *relative homology groups*, denoted $H_n(X, A)$. These turn out to be quite useful for many other purposes as well.

Relative Homology Groups

It sometimes happens that by ignoring a certain amount of data or structure one obtains a simpler, more flexible theory which, almost paradoxically, can give results not readily obtainable in the original setting. A familiar instance of this is arithmetic mod n, where one ignores multiples of n. Relative homology is another example. In this case what one ignores is all singular chains in a subspace of the given space.

Relative homology groups are defined in the following way. Given a space X and a subspace $A \subset X$, let $C_n(X, A)$ be the quotient group $C_n(X)/C_n(A)$. Thus chains in A are trivial in $C_n(X, A)$. Since the boundary map $\partial : C_n(X) \to C_{n-1}(X)$ takes $C_n(A)$ to $C_{n-1}(A)$, it induces a quotient boundary map $\partial : C_n(X, A) \to C_{n-1}(X, A)$. Letting n vary, we have a sequence of boundary maps

$$\cdots \longrightarrow C_n(X, A) \xrightarrow{\ \partial\ } C_{n-1}(X, A) \longrightarrow \cdots$$

The relation $\partial^2 = 0$ holds for these boundary maps since it holds before passing to quotient groups. So we have a chain complex, and the homology groups $\operatorname{Ker} \partial / \operatorname{Im} \partial$ of this chain complex are by definition the **relative homology groups** $H_n(X, A)$. By considering the definition of the relative boundary map we see:

- Elements of $H_n(X, A)$ are represented by **relative cycles**: n-chains $\alpha \in C_n(X)$ such that $\partial \alpha \in C_{n-1}(A)$.
- A relative cycle α is trivial in $H_n(X, A)$ iff it is a **relative boundary**: $\alpha = \partial \beta + \gamma$ for some $\beta \in C_{n+1}(X)$ and $\gamma \in C_n(A)$.

These properties make precise the intuitive idea that $H_n(X, A)$ is 'homology of X modulo A.'

The quotient $C_n(X)/C_n(A)$ could also be viewed as a subgroup of $C_n(X)$, the subgroup with basis the singular n-simplices $\sigma : \Delta^n \to X$ whose image is not contained in A. However, the boundary map does not take this subgroup of $C_n(X)$ to the corresponding subgroup of $C_{n-1}(X)$, so it is usually better to regard $C_n(X, A)$ as a quotient rather than a subgroup of $C_n(X)$.

Our goal now is to show that the relative homology groups $H_n(X, A)$ for any pair (X, A) fit into a long exact sequence

$$\cdots \longrightarrow H_n(A) \longrightarrow H_n(X) \longrightarrow H_n(X, A) \longrightarrow H_{n-1}(A) \longrightarrow H_{n-1}(X) \longrightarrow \cdots$$

$$\cdots \longrightarrow H_0(X, A) \longrightarrow 0$$

This will be entirely a matter of algebra. To start the process, consider the diagram

$$0 \longrightarrow C_n(A) \xrightarrow{\ i\ } C_n(X) \xrightarrow{\ j\ } C_n(X,A) \longrightarrow 0$$
$$\Big\downarrow\partial \qquad\qquad \Big\downarrow\partial \qquad\qquad \Big\downarrow\partial$$
$$0 \longrightarrow C_{n-1}(A) \xrightarrow{\ i\ } C_{n-1}(X) \xrightarrow{\ j\ } C_{n-1}(X,A) \longrightarrow 0$$

where i is inclusion and j is the quotient map. The diagram is commutative by the definition of the boundary maps. Letting n vary, and drawing these short exact sequences vertically rather than horizontally, we have a large commutative diagram of the form shown at the right, where the columns are exact and the rows are chain complexes which we denote A, B, and C. Such a diagram is called a **short exact sequence of chain complexes**. We will show that when we pass to homology groups, this short exact sequence of chain complexes stretches out into a long exact sequence of homology groups

$$\cdots \longrightarrow H_n(A) \xrightarrow{\ i_*\ } H_n(B) \xrightarrow{\ j_*\ } H_n(C) \xrightarrow{\ \partial\ } H_{n-1}(A) \xrightarrow{\ i_*\ } H_{n-1}(B) \longrightarrow \cdots$$

where $H_n(A)$ denotes the homology group $\mathrm{Ker}\,\partial/\,\mathrm{Im}\,\partial$ at A_n in the chain complex A, and $H_n(B)$ and $H_n(C)$ are defined similarly.

The commutativity of the squares in the short exact sequence of chain complexes means that i and j are chain maps. These therefore induce maps i_* and j_* on homology. To define the boundary map $\partial : H_n(C) \to H_{n-1}(A)$, let $c \in C_n$ be a cycle. Since j is onto, $c = j(b)$ for some $b \in B_n$. The element $\partial b \in B_{n-1}$ is in $\mathrm{Ker}\,j$ since $j(\partial b) = \partial j(b) = \partial c = 0$. So $\partial b = i(a)$ for some $a \in A_{n-1}$ since $\mathrm{Ker}\,j = \mathrm{Im}\,i$. Note that $\partial a = 0$ since $i(\partial a) = \partial i(a) = \partial\partial b = 0$ and i is injective. We define $\partial : H_n(C) \to H_{n-1}(A)$ by sending the homology class of c to the homology class of a, $\partial[c] = [a]$. This is well-defined since:

- The element a is uniquely determined by ∂b since i is injective.
- A different choice b' for b would have $j(b') = j(b)$, so $b' - b$ is in $\mathrm{Ker}\,j = \mathrm{Im}\,i$. Thus $b' - b = i(a')$ for some a', hence $b' = b + i(a')$. The effect of replacing b by $b + i(a')$ is to change a to the homologous element $a + \partial a'$ since $i(a + \partial a') = i(a) + i(\partial a') = \partial b + \partial i(a') = \partial(b + i(a'))$.
- A different choice of c within its homology class would have the form $c + \partial c'$. Since $c' = j(b')$ for some b', we then have $c + \partial c' = c + \partial j(b') = c + j(\partial b') = j(b + \partial b')$, so b is replaced by $b + \partial b'$, which leaves ∂b and therefore also a unchanged.

The map $\partial : H_n(C) \rightarrow H_{n-1}(A)$ is a homomorphism since if $\partial[c_1] = [a_1]$ and $\partial[c_2] = [a_2]$ via elements b_1 and b_2 as above, then $j(b_1 + b_2) = j(b_1) + j(b_2) = c_1 + c_2$ and $i(a_1 + a_2) = i(a_1) + i(a_2) = \partial b_1 + \partial b_2 = \partial(b_1 + b_2)$, so $\partial([c_1] + [c_2]) = [a_1] + [a_2]$.

Theorem 2.16. *The sequence of homology groups*

$$\cdots \longrightarrow H_n(A) \xrightarrow{\ i_*\ } H_n(B) \xrightarrow{\ j_*\ } H_n(C) \xrightarrow{\ \partial\ } H_{n-1}(A) \xrightarrow{\ i_*\ } H_{n-1}(B) \longrightarrow \cdots$$

is exact.

Proof: There are six things to verify:

Im $i_* \subset$ Ker j_*. This is immediate since $ji = 0$ implies $j_* i_* = 0$.

Im $j_* \subset$ Ker ∂. We have $\partial j_* = 0$ since in this case $\partial b = 0$ in the definition of ∂.

Im $\partial \subset$ Ker i_*. Here $i_* \partial = 0$ since $i_* \partial$ takes $[c]$ to $[\partial b] = 0$.

Ker $j_* \subset$ Im i_*. A homology class in Ker j_* is represented by a cycle $b \in B_n$ with $j(b)$ a boundary, so $j(b) = \partial c'$ for some $c' \in C_{n+1}$. Since j is surjective, $c' = j(b')$ for some $b' \in B_{n+1}$. We have $j(b - \partial b') = j(b) - j(\partial b') = j(b) - \partial j(b') = 0$ since $\partial j(b') = \partial c' = j(b)$. So $b - \partial b' = i(a)$ for some $a \in A_n$. This a is a cycle since $i(\partial a) = \partial i(a) = \partial(b - \partial b') = \partial b = 0$ and i is injective. Thus $i_*[a] = [b - \partial b'] = [b]$, showing that i_* maps onto Ker j_*.

Ker $\partial \subset$ Im j_*. In the notation used in the definition of ∂, if c represents a homology class in Ker ∂, then $a = \partial a'$ for some $a' \in A_n$. The element $b - i(a')$ is a cycle since $\partial(b - i(a')) = \partial b - \partial i(a') = \partial b - i(\partial a') = \partial b - i(a) = 0$. And $j(b - i(a')) = j(b) - ji(a') = j(b) = c$, so j_* maps $[b - i(a')]$ to $[c]$.

Ker $i_* \subset$ Im ∂. Given a cycle $a \in A_{n-1}$ such that $i(a) = \partial b$ for some $b \in B_n$, then $j(b)$ is a cycle since $\partial j(b) = j(\partial b) = ji(a) = 0$, and ∂ takes $[j(b)]$ to $[a]$.   □

This theorem represents the beginnings of the subject of homological algebra. The method of proof is sometimes called *diagram chasing*.

Returning to topology, the preceding algebraic theorem yields a long exact sequence of homology groups:

$$\cdots \longrightarrow H_n(A) \xrightarrow{\ i_*\ } H_n(X) \xrightarrow{\ j_*\ } H_n(X, A) \xrightarrow{\ \partial\ } H_{n-1}(A) \xrightarrow{\ i_*\ } H_{n-1}(X) \longrightarrow \cdots$$

$$\cdots \longrightarrow H_0(X, A) \longrightarrow 0$$

The boundary map $\partial : H_n(X, A) \rightarrow H_{n-1}(A)$ has a very simple description: If a class $[\alpha] \in H_n(X, A)$ is represented by a relative cycle α, then $\partial[\alpha]$ is the class of the cycle $\partial \alpha$ in $H_{n-1}(A)$. This is immediate from the algebraic definition of the boundary homomorphism in the long exact sequence of homology groups associated to a short exact sequence of chain complexes.

This long exact sequence makes precise the idea that the groups $H_n(X, A)$ measure the difference between the groups $H_n(X)$ and $H_n(A)$. In particular, exactness implies that if $H_n(X, A) = 0$ for all n, then the inclusion $A \hookrightarrow X$ induces isomorphisms $H_n(A) \approx H_n(X)$ for all n, by the remark (iii) following the definition of exactness. The converse is also true according to an exercise at the end of this section.

There is a completely analogous long exact sequence of reduced homology groups for a pair (X, A) with $A \neq \varnothing$. This comes from applying the preceding algebraic machinery to the short exact sequence of chain complexes formed by the short exact sequences $0 \rightarrow C_n(A) \rightarrow C_n(X) \rightarrow C_n(X, A) \rightarrow 0$ in nonnegative dimensions, augmented by the short exact sequence $0 \longrightarrow \mathbb{Z} \xrightarrow{\;1\!1\;} \mathbb{Z} \longrightarrow 0 \longrightarrow 0$ in dimension -1. In particular this means that $\tilde{H}_n(X, A)$ is the same as $H_n(X, A)$ for all n, when $A \neq \varnothing$.

Example 2.17. In the long exact sequence of reduced homology groups for the pair $(D^n, \partial D^n)$, the maps $H_i(D^n, \partial D^n) \xrightarrow{\;\partial\;} \tilde{H}_{i-1}(S^{n-1})$ are isomorphisms for all $i > 0$ since the remaining terms $\tilde{H}_i(D^n)$ are zero for all i. Thus we obtain the calculation

$$H_i(D^n, \partial D^n) \approx \begin{cases} \mathbb{Z} & \text{for } i = n \\ 0 & \text{otherwise} \end{cases}$$

Example 2.18. Applying the long exact sequence of reduced homology groups to a pair (X, x_0) with $x_0 \in X$ yields isomorphisms $H_n(X, x_0) \approx \tilde{H}_n(X)$ for all n since $\tilde{H}_n(x_0) = 0$ for all n.

There are induced homomorphisms for relative homology just as there are in the nonrelative, or 'absolute,' case. A map $f : X \rightarrow Y$ with $f(A) \subset B$, or more concisely $f : (X, A) \rightarrow (Y, B)$, induces homomorphisms $f_\sharp : C_n(X, A) \rightarrow C_n(Y, B)$ since the chain map $f_\sharp : C_n(X) \rightarrow C_n(Y)$ takes $C_n(A)$ to $C_n(B)$, so we get a well-defined map on quotients, $f_\sharp : C_n(X, A) \rightarrow C_n(Y, B)$. The relation $f_\sharp \partial = \partial f_\sharp$ holds for relative chains since it holds for absolute chains. By Proposition 2.9 we then have induced homomorphisms $f_* : H_n(X, A) \rightarrow H_n(Y, B)$.

Proposition 2.19. *If two maps* $f, g : (X, A) \rightarrow (Y, B)$ *are homotopic through maps of pairs* $(X, A) \rightarrow (Y, B)$, *then* $f_* = g_* : H_n(X, A) \rightarrow H_n(Y, B)$.

Proof: The prism operator P from the proof of Theorem 2.10 takes $C_n(A)$ to $C_{n+1}(B)$, hence induces a relative prism operator $P : C_n(X, A) \rightarrow C_{n+1}(Y, B)$. Since we are just passing to quotient groups, the formula $\partial P + P \partial = g_\sharp - f_\sharp$ remains valid. Thus the maps $f_\sharp$ and $g_\sharp$ on relative chain groups are chain homotopic, and hence they induce the same homomorphism on relative homology groups. $\qquad\square$

An easy generalization of the long exact sequence of a pair (X, A) is the long exact sequence of a triple (X, A, B), where $B \subset A \subset X$:

$$\cdots \longrightarrow H_n(A, B) \longrightarrow H_n(X, B) \longrightarrow H_n(X, A) \longrightarrow H_{n-1}(A, B) \longrightarrow \cdots$$

This is the long exact sequence of homology groups associated to the short exact sequence of chain complexes formed by the short exact sequences

$$0 \longrightarrow C_n(A, B) \longrightarrow C_n(X, B) \longrightarrow C_n(X, A) \longrightarrow 0$$

For example, taking B to be a point, the long exact sequence of the triple (X, A, B) becomes the long exact sequence of reduced homology for the pair (X, A).

Excision

A fundamental property of relative homology groups is given by the following **Excision Theorem**, describing when the relative groups $H_n(X, A)$ are unaffected by deleting, or excising, a subset $Z \subset A$.

Theorem 2.20. *Given subspaces $Z \subset A \subset X$ such that the closure of Z is contained in the interior of A, then the inclusion $(X - Z, A - Z) \hookrightarrow (X, A)$ induces isomorphisms $H_n(X - Z, A - Z) \rightarrow H_n(X, A)$ for all n. Equivalently, for subspaces $A, B \subset X$ whose interiors cover X, the inclusion $(B, A \cap B) \hookrightarrow (X, A)$ induces isomorphisms $H_n(B, A \cap B) \rightarrow H_n(X, A)$ for all n.*

The translation between the two versions is obtained by setting $B = X - Z$ and $Z = X - B$. Then $A \cap B = A - Z$ and the condition $\operatorname{cl} Z \subset \operatorname{int} A$ is equivalent to $X = \operatorname{int} A \cup \operatorname{int} B$ since $X - \operatorname{int} B = \operatorname{cl} Z$.

The proof of the excision theorem will involve a rather lengthy technical detour involving a construction known as barycentric subdivision, which allows homology groups to be computed using small singular simplices. In a metric space 'smallness' can be defined in terms of diameters, but for general spaces it will be defined in terms of covers.

For a space X, let $\mathcal{U} = \{U_j\}$ be a collection of subspaces of X whose interiors form an open cover of X, and let $C_n^{\mathcal{U}}(X)$ be the subgroup of $C_n(X)$ consisting of chains $\sum_i n_i \sigma_i$ such that each σ_i has image contained in some set in the cover $\mathcal{U}$. The boundary map $\partial : C_n(X) \rightarrow C_{n-1}(X)$ takes $C_n^{\mathcal{U}}(X)$ to $C_{n-1}^{\mathcal{U}}(X)$, so the groups $C_n^{\mathcal{U}}(X)$ form a chain complex. We denote the homology groups of this chain complex by $H_n^{\mathcal{U}}(X)$.

Proposition 2.21. *The inclusion $\iota : C_n^{\mathcal{U}}(X) \hookrightarrow C_n(X)$ is a chain homotopy equivalence, that is, there is a chain map $\rho : C_n(X) \rightarrow C_n^{\mathcal{U}}(X)$ such that $\iota \rho$ and $\rho \iota$ are chain homotopic to the identity. Hence ι induces isomorphisms $H_n^{\mathcal{U}}(X) \approx H_n(X)$ for all n.*

Proof: The barycentric subdivision process will be performed at four levels, beginning with the most geometric and becoming increasingly algebraic.

(1) *Barycentric Subdivision of Simplices.* The points of a simplex $[v_0, \cdots, v_n]$ are the linear combinations $\sum_i t_i v_i$ with $\sum_i t_i = 1$ and $t_i \geq 0$ for each i. The **barycenter** or 'center of gravity' of the simplex $[v_0, \cdots, v_n]$ is the point $b = \sum_i t_i v_i$ whose barycentric coordinates t_i are all equal, namely $t_i = 1/(n+1)$ for each i. The **barycentric subdivision** of $[v_0, \cdots, v_n]$ is the decomposition of $[v_0, \cdots, v_n]$ into the n-simplices $[b, w_0, \cdots, w_{n-1}]$ where, inductively, $[w_0, \cdots, w_{n-1}]$ is an $(n-1)$-simplex in the barycentric subdivision of a face $[v_0, \cdots, \hat{v}_i, \cdots, v_n]$. The induction starts with the case $n = 0$ when the barycentric subdivision of $[v_0]$ is defined to be just $[v_0]$ itself. The next two cases $n = 1, 2$ and part of the case $n = 3$ are shown in the figure. It follows from the inductive definition that the vertices of simplices in the barycentric subdivision of $[v_0, \cdots, v_n]$ are exactly the barycenters of all

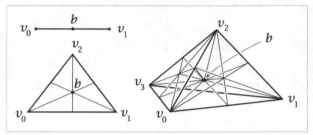

the k-dimensional faces $[v_{i_0}, \cdots, v_{i_k}]$ of $[v_0, \cdots, v_n]$ for $0 \le k \le n$. When $k = 0$ this gives the original vertices v_i since the barycenter of a 0-simplex is itself. The barycenter of $[v_{i_0}, \cdots, v_{i_k}]$ has barycentric coordinates $t_i = 1/(k + 1)$ for $i = i_0, \cdots, i_k$ and $t_i = 0$ otherwise.

The n-simplices of the barycentric subdivision of Δ^n, together with all their faces, do in fact form a Δ-complex structure on Δ^n, indeed a simplicial complex structure, though we shall not need to know this in what follows.

A fact we will need is that the diameter of each simplex of the barycentric subdivision of $[v_0, \cdots, v_n]$ is at most $n/(n+1)$ times the diameter of $[v_0, \cdots, v_n]$. Here the diameter of a simplex is by definition the maximum distance between any two of its points, and we are using the metric from the ambient Euclidean space $\mathbb{R}^m$ containing $[v_0, \cdots, v_n]$. The diameter of a simplex equals the maximum distance between any of its vertices because the distance between two points v and $\sum_i t_i v_i$ of $[v_0, \cdots, v_n]$ satisfies the inequality

$$\left| v - \textstyle\sum_i t_i v_i \right| = \left| \textstyle\sum_i t_i (v - v_i) \right| \le \textstyle\sum_i t_i |v - v_i| \le \textstyle\sum_i t_i \max |v - v_i| = \max |v - v_i|$$

To obtain the bound $n/(n + 1)$ on the ratio of diameters, we therefore need to verify that the distance between any two vertices w_j and w_k of a simplex $[w_0, \cdots, w_n]$ of the barycentric subdivision of $[v_0, \cdots, v_n]$ is at most $n/(n+1)$ times the diameter of $[v_0, \cdots, v_n]$. If neither w_j nor w_k is the barycenter b of $[v_0, \cdots, v_n]$, then these two points lie in a proper face of $[v_0, \cdots, v_n]$ and we are done by induction on n. So we may suppose w_j, say, is the barycenter b, and then by the previous displayed inequality we may take w_k to be a vertex v_i. Let b_i be the barycenter of $[v_0, \cdots, \hat{v}_i, \cdots, v_n]$, with all barycentric coordinates equal to $1/n$ except for $t_i = 0$. Then we have $b = \frac{1}{n+1} v_i + \frac{n}{n+1} b_i$. The sum of the two coefficients is 1, so b lies on the line segment $[v_i, b_i]$ from v_i to b_i, and the distance from

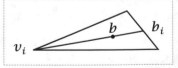

b to v_i is $n/(n + 1)$ times the length of $[v_i, b_i]$. Hence the distance from b to v_i is bounded by $n/(n + 1)$ times the diameter of $[v_0, \cdots, v_n]$.

The significance of the factor $n/(n+1)$ is that by repeated barycentric subdivision we can produce simplices of arbitrarily small diameter since $(n/(n+1))^r$ approaches

0 as r goes to infinity. It is important that the bound $n/(n+1)$ does not depend on the shape of the simplex since repeated barycentric subdivision produces simplices of many different shapes.

(2) *Barycentric Subdivision of Linear Chains.* The main part of the proof will be to construct a subdivision operator $S: C_n(X) \to C_n(X)$ and show this is chain homotopic to the identity map. First we will construct S and the chain homotopy in a more restricted linear setting.

For a convex set Y in some Euclidean space, the linear maps $\Delta^n \to Y$ generate a subgroup of $C_n(Y)$ that we denote $LC_n(Y)$, the *linear chains*. The boundary map $\partial: C_n(Y) \to C_{n-1}(Y)$ takes $LC_n(Y)$ to $LC_{n-1}(Y)$, so the linear chains form a subcomplex of the singular chain complex of Y. We can uniquely designate a linear map $\lambda: \Delta^n \to Y$ by $[w_0, \cdots, w_n]$ where w_i is the image under λ of the ith vertex of Δ^n. To avoid having to make exceptions for 0-simplices it will be convenient to augment the complex $LC(Y)$ by setting $LC_{-1}(Y) = \mathbb{Z}$ generated by the empty simplex $[\varnothing]$, with $\partial[w_0] = [\varnothing]$ for all 0-simplices $[w_0]$.

Each point $b \in Y$ determines a homomorphism $b: LC_n(Y) \to LC_{n+1}(Y)$ defined on basis elements by $b([w_0, \cdots, w_n]) = [b, w_0, \cdots, w_n]$. Geometrically, the homomorphism b can be regarded as a cone operator, sending a linear chain to the cone having the linear chain as the base of the cone and the point b as the tip of the cone. Applying the usual formula for ∂, we obtain the relation $\partial b([w_0, \cdots, w_n]) = [w_0, \cdots, w_n] - b(\partial[w_0, \cdots, w_n])$. By linearity it follows that $\partial b(\alpha) = \alpha - b(\partial\alpha)$ for all $\alpha \in LC_n(Y)$. This expresses algebraically the geometric fact that the boundary of a cone consists of its base together with the cone on the boundary of its base. The relation $\partial b(\alpha) = \alpha - b(\partial\alpha)$ can be rewritten as $\partial b + b\partial = \mathbb{1}$, so b is a chain homotopy between the identity map and the zero map on the augmented chain complex $LC(Y)$.

Now we define a subdivision homomorphism $S: LC_n(Y) \to LC_n(Y)$ by induction on n. Let $\lambda: \Delta^n \to Y$ be a generator of $LC_n(Y)$ and let b_λ be the image of the barycenter of Δ^n under λ. Then the inductive formula for S is $S(\lambda) = b_\lambda(S\partial\lambda)$ where $b_\lambda: LC_{n-1}(Y) \to LC_n(Y)$ is the cone operator defined in the preceding paragraph. The induction starts with $S([\varnothing]) = [\varnothing]$, so S is the identity on $LC_{-1}(Y)$. It is also the identity on $LC_0(Y)$, since when $n = 0$ the formula for S becomes $S([w_0]) = w_0(S\partial[w_0]) = w_0(S([\varnothing])) = w_0([\varnothing]) = [w_0]$. When λ is an embedding, with image a genuine n-simplex $[w_0, \cdots, w_n]$, then $S(\lambda)$ is the sum of the n-simplices in the barycentric subdivision of $[w_0, \cdots, w_n]$, with certain signs that could be computed explicitly. This is apparent by comparing the inductive definition of S with the inductive definition of the barycentric subdivision of a simplex.

Let us check that the maps S satisfy $\partial S = S\partial$, and hence give a chain map from the chain complex $LC(Y)$ to itself. Since $S = \mathbb{1}$ on $LC_0(Y)$ and $LC_{-1}(Y)$, we certainly have $\partial S = S\partial$ on $LC_0(Y)$. The result for larger n is given by the following calculation, in which we omit some parentheses to unclutter the formulas:

$$\partial S\lambda = \partial b_\lambda(S\partial\lambda)$$

$$= S\partial\lambda - b_\lambda\partial(S\partial\lambda) \qquad \text{since } \partial b_\lambda = 1\!\!1 - b_\lambda\partial$$

$$= S\partial\lambda - b_\lambda S(\partial\partial\lambda) \qquad \text{since } \partial S(\partial\lambda) = S\partial(\partial\lambda) \text{ by induction on } n$$

$$= S\partial\lambda \qquad \text{since } \partial\partial = 0$$

We next build a chain homotopy $T:LC_n(Y)\to LC_{n+1}(Y)$ between S and the identity, fitting into a diagram

$$\cdots \longrightarrow LC_2(Y) \longrightarrow LC_1(Y) \longrightarrow LC_0(Y) \longrightarrow LC_{-1}(Y) \longrightarrow 0$$
$$\Big\downarrow S \quad {}^T\!\!\diagup \quad \Big\downarrow S \quad {}^T\!\!\diagup \quad S\Big\downarrow 1\!\!1 \quad {}^T\!\!\diagup \quad S\Big\downarrow 1\!\!1$$
$$\cdots \longrightarrow LC_2(Y) \longrightarrow LC_1(Y) \longrightarrow LC_0(Y) \longrightarrow LC_{-1}(Y) \longrightarrow 0$$

We define T on $LC_n(Y)$ inductively by setting $T = 0$ for $n = -1$ and letting $T\lambda = b_\lambda(\lambda - T\partial\lambda)$ for $n \geq 0$. The geometric motivation for this formula is an inductively defined subdivision of $\Delta^n\times I$ obtained by joining all simplices in $\Delta^n\times\{0\} \cup \partial\Delta^n\times I$ to the barycenter of $\Delta^n\times\{1\}$, as indicated in the figure in the case $n = 2$. What T actually does is take the image of this subdivision under the projection $\Delta^n\times I\to\Delta^n$.

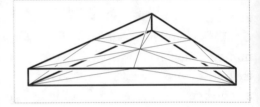

The chain homotopy formula $\partial T + T\partial = 1\!\!1 - S$ is trivial on $LC_{-1}(Y)$ where $T = 0$ and $S = 1\!\!1$. Verifying the formula on $LC_n(Y)$ with $n \geq 0$ is done by the calculation

$$\partial T\lambda = \partial b_\lambda(\lambda - T\partial\lambda)$$

$$= \lambda - T\partial\lambda - b_\lambda\partial(\lambda - T\partial\lambda) \qquad \text{since } \partial b_\lambda = 1\!\!1 - b_\lambda\partial$$

$$= \lambda - T\partial\lambda - b_\lambda[\partial\lambda - \partial T(\partial\lambda)]$$

$$= \lambda - T\partial\lambda - b_\lambda[S(\partial\lambda) + T\partial(\partial\lambda)] \qquad \text{by induction on } n$$

$$= \lambda - T\partial\lambda - S\lambda \qquad \text{since } \partial\partial = 0 \text{ and } S\lambda = b_\lambda(S\partial\lambda)$$

Now we can discard the group $LC_{-1}(Y)$ and the relation $\partial T + T\partial = 1\!\!1 - S$ still holds since T was zero on $LC_{-1}(Y)$.

(3) *Barycentric Subdivision of General Chains.* Define $S:C_n(X)\to C_n(X)$ by setting $S\sigma = \sigma_\sharp S\Delta^n$ for a singular n-simplex $\sigma:\Delta^n\to X$. Since $S\Delta^n$ is the sum of the n-simplices in the barycentric subdivision of Δ^n, with certain signs, $S\sigma$ is the corresponding signed sum of the restrictions of σ to the n-simplices of the barycentric subdivision of Δ^n. The operator S is a chain map since

$$\partial S\sigma = \partial\sigma_\sharp S\Delta^n = \sigma_\sharp\partial S\Delta^n = \sigma_\sharp S\partial\Delta^n$$

$$= \sigma_\sharp S(\textstyle\sum_i(-1)^i\Delta_i^n) \qquad \text{where } \Delta_i^n \text{ is the } i^{th} \text{ face of } \Delta^n$$

$$= \textstyle\sum_i(-1)^i\sigma_\sharp S\Delta_i^n$$

$$= \textstyle\sum_i(-1)^i S(\sigma|\Delta_i^n)$$

$$= S(\textstyle\sum_i(-1)^i\sigma|\Delta_i^n) = S(\partial\sigma)$$

In similar fashion we define $T : C_n(X) \to C_{n+1}(X)$ by $T\sigma = \sigma_\sharp T\Delta^n$, and this gives a chain homotopy between S and the identity, since the formula $\partial T + T\partial = 1\!\!1 - S$ holds by the calculation

$$\partial T\sigma = \partial\sigma_\sharp T\Delta^n = \sigma_\sharp \partial T\Delta^n = \sigma_\sharp(\Delta^n - S\Delta^n - T\partial\Delta^n) = \sigma - S\sigma - \sigma_\sharp T\partial\Delta^n$$
$$= \sigma - S\sigma - T(\partial\sigma)$$

where the last equality follows just as in the previous displayed calculation, with S replaced by T.

(4) *Iterated Barycentric Subdivision.* A chain homotopy between $1\!\!1$ and the iterate S^m is given by the operator $D_m = \sum_{0 \le i < m} TS^i$ since

$$\partial D_m + D_m\partial = \sum_{0 \le i < m} (\partial TS^i + TS^i\partial) = \sum_{0 \le i < m} (\partial TS^i + T\partial S^i) =$$
$$\sum_{0 \le i < m} (\partial T + T\partial)S^i = \sum_{0 \le i < m} (1\!\!1 - S)S^i = \sum_{0 \le i < m} (S^i - S^{i+1}) = 1\!\!1 - S^m$$

For each singular n-simplex $\sigma : \Delta^n \to X$ there exists an m such that $S^m(\sigma)$ lies in $C_n^{\mathcal{U}}(X)$ since the diameter of the simplices of $S^m(\Delta^n)$ will be less than a Lebesgue number of the cover of Δ^n by the open sets $\sigma^{-1}(\text{int } U_j)$ if m is large enough. (Recall that a Lebesgue number for an open cover of a compact metric space is a number $\varepsilon > 0$ such that every set of diameter less than ε lies in some set of the cover; such a number exists by an elementary compactness argument.) We cannot expect the same number m to work for all σ's, so let us define $m(\sigma)$ to be the smallest m such that $S^m\sigma$ is in $C_n^{\mathcal{U}}(X)$.

We now define $D : C_n(X) \to C_{n+1}(X)$ by setting $D\sigma = D_{m(\sigma)}\sigma$ for each singular n-simplex $\sigma : \Delta^n \to X$. For this D we would like to find a chain map $\rho : C_n(X) \to C_n(X)$ with image in $C_n^{\mathcal{U}}(X)$ satisfying the chain homotopy equation

$$(*) \qquad\qquad\qquad \partial D + D\partial = 1\!\!1 - \rho$$

A quick way to do this is simply to regard this equation as defining ρ, so we let $\rho = 1\!\!1 - \partial D - D\partial$. It follows easily that ρ is a chain map since

$$\partial\rho(\sigma) = \partial\sigma - \partial^2 D\sigma - \partial D\partial\sigma = \partial\sigma - \partial D\partial\sigma$$
$$\text{and} \qquad \rho(\partial\sigma) = \partial\sigma - \partial D\partial\sigma - D\partial^2\sigma = \partial\sigma - \partial D\partial\sigma$$

To check that ρ takes $C_n(X)$ to $C_n^{\mathcal{U}}(X)$ we compute $\rho(\sigma)$ more explicitly:

$$\rho(\sigma) = \sigma - \partial D\sigma - D(\partial\sigma)$$
$$= \sigma - \partial D_{m(\sigma)}\sigma - D(\partial\sigma)$$
$$= S^{m(\sigma)}\sigma + D_{m(\sigma)}(\partial\sigma) - D(\partial\sigma) \qquad \text{since} \quad \partial D_m + D_m\partial = 1\!\!1 - S^m$$

The term $S^{m(\sigma)}\sigma$ lies in $C_n^{\mathcal{U}}(X)$ by the definition of $m(\sigma)$. The remaining terms $D_{m(\sigma)}(\partial\sigma) - D(\partial\sigma)$ are linear combinations of terms $D_{m(\sigma)}(\sigma_j) - D_{m(\sigma_j)}(\sigma_j)$ for σ_j the restriction of σ to a face of Δ^n, so $m(\sigma_j) \le m(\sigma)$ and hence the difference

$D_{m(\sigma)}(\sigma_j) - D_{m(\sigma_j)}(\sigma_j)$ consists of terms $TS^i(\sigma_j)$ with $i \geq m(\sigma_j)$, and these terms lie in $C_n^{\mathcal{U}}(X)$ since T takes $C_{n-1}^{\mathcal{U}}(X)$ to $C_n^{\mathcal{U}}(X)$.

Viewing ρ as a chain map $C_n(X) \to C_n^{\mathcal{U}}(X)$, the equation $(*)$ says that $\partial D + D\partial = \mathbb{1} - \iota\rho$ for $\iota : C_n^{\mathcal{U}}(X) \hookrightarrow C_n(X)$ the inclusion. Furthermore, $\rho\iota = \mathbb{1}$ since D is identically zero on $C_n^{\mathcal{U}}(X)$, as $m(\sigma) = 0$ if σ is in $C_n^{\mathcal{U}}(X)$, hence the summation defining $D\sigma$ is empty. Thus we have shown that ρ is a chain homotopy inverse for ι. $\qquad \square$

Proof of the Excision Theorem: We prove the second version, involving a decomposition $X = A \cup B$. For the cover $\mathcal{U} = \{A, B\}$ we introduce the suggestive notation $C_n(A + B)$ for $C_n^{\mathcal{U}}(X)$, the sums of chains in A and chains in B. At the end of the preceding proof we had formulas $\partial D + D\partial = \mathbb{1} - \iota\rho$ and $\rho\iota = \mathbb{1}$. All the maps appearing in these formulas take chains in A to chains in A, so they induce quotient maps when we factor out chains in A. These quotient maps automatically satisfy the same two formulas, so the inclusion $C_n(A + B)/C_n(A) \hookrightarrow C_n(X)/C_n(A)$ induces an isomorphism on homology. The map $C_n(B)/C_n(A \cap B) \to C_n(A + B)/C_n(A)$ induced by inclusion is obviously an isomorphism since both quotient groups are free with basis the singular n-simplices in B that do not lie in A. Hence we obtain the desired isomorphism $H_n(B, A \cap B) \approx H_n(X, A)$ induced by inclusion. $\qquad \square$

All that remains in the proof of Theorem 2.13 is to replace relative homology groups with absolute homology groups. This is achieved by the following result.

Proposition 2.22. *For good pairs (X, A), the quotient map $q : (X, A) \to (X/A, A/A)$ induces isomorphisms $q_* : H_n(X, A) \to H_n(X/A, A/A) \approx \tilde{H}_n(X/A)$ for all n.*

Proof: Let V be a neighborhood of A in X that deformation retracts onto A. We have a commutative diagram

$$
\begin{array}{ccccc}
H_n(X,A) & \longrightarrow & H_n(X,V) & \longleftarrow & H_n(X-A, V-A) \\
\downarrow q_* & & \downarrow q_* & & \downarrow q_* \\
H_n(X/A, A/A) & \longrightarrow & H_n(X/A, V/A) & \longleftarrow & H_n(X/A - A/A, V/A - A/A)
\end{array}
$$

The upper left horizontal map is an isomorphism since in the long exact sequence of the triple (X, V, A) the groups $H_n(V, A)$ are zero for all n, because a deformation retraction of V onto A gives a homotopy equivalence of pairs $(V, A) \simeq (A, A)$, and $H_n(A, A) = 0$. The deformation retraction of V onto A induces a deformation retraction of V/A onto A/A, so the same argument shows that the lower left horizontal map is an isomorphism as well. The other two horizontal maps are isomorphisms directly from excision. The right-hand vertical map q_* is an isomorphism since q restricts to a homeomorphism on the complement of A. From the commutativity of the diagram it follows that the left-hand q_* is an isomorphism. $\qquad \square$

This proposition shows that relative homology can be expressed as reduced absolute homology in the case of good pairs (X, A), but in fact there is a way of doing this for arbitrary pairs. Consider the space $X \cup CA$ where CA is the cone $(A \times I)/(A \times \{0\})$ whose base $A \times \{1\}$ we identify with $A \subset X$. Using terminology introduced in Chapter 0, $X \cup CA$ can also be described as the mapping cone of the inclusion $A \hookrightarrow X$. The assertion is that $H_n(X, A)$ is isomorphic to $\tilde{H}_n(X \cup CA)$ for all n via the sequence of isomorphisms

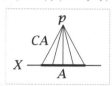

$$\tilde{H}_n(X \cup CA) \approx H_n(X \cup CA, CA) \approx H_n(X \cup CA - \{p\}, CA - \{p\}) \approx H_n(X, A)$$

where $p \in CA$ is the tip of the cone. The first isomorphism comes from the exact sequence of the pair, using the fact that CA is contractible. The second isomorphism is excision, and the third isomorphism comes from the deformation retraction of $CA - \{p\}$ onto A.

Here is an application of the preceding proposition:

Example 2.23. Let us find explicit cycles representing generators of the infinite cyclic groups $H_n(D^n, \partial D^n)$ and $\tilde{H}_n(S^n)$. Replacing $(D^n, \partial D^n)$ by the equivalent pair $(\Delta^n, \partial \Delta^n)$, we will show by induction on n that the identity map $i_n : \Delta^n \to \Delta^n$, viewed as a singular n-simplex, is a cycle generating $H_n(\Delta^n, \partial \Delta^n)$. That it is a cycle is clear since we are considering relative homology. When $n = 0$ it certainly represents a generator. For the induction step, let $\Lambda \subset \Delta^n$ be the union of all but one of the $(n-1)$-dimensional faces of Δ^n. Then we claim there are isomorphisms

$$H_n(\Delta^n, \partial \Delta^n) \xrightarrow{\approx} H_{n-1}(\partial \Delta^n, \Lambda) \xleftarrow{\approx} H_{n-1}(\Delta^{n-1}, \partial \Delta^{n-1})$$

The first isomorphism is a boundary map in the long exact sequence of the triple $(\Delta^n, \partial \Delta^n, \Lambda)$, whose third terms $H_i(\Delta^n, \Lambda)$ are zero since Δ^n deformation retracts onto Λ, hence $(\Delta^n, \Lambda) \simeq (\Lambda, \Lambda)$. The second isomorphism comes from the preceding proposition since we are dealing with good pairs and the inclusion $\Delta^{n-1} \hookrightarrow \partial \Delta^n$ as the face not contained in Λ induces a homeomorphism of quotients $\Delta^{n-1}/\partial \Delta^{n-1} \approx \partial \Delta^n/\Lambda$. The induction step then follows since the cycle i_n is sent under the first isomorphism to the cycle ∂i_n which equals $\pm i_{n-1}$ in $C_{n-1}(\partial \Delta^n, \Lambda)$.

To find a cycle generating $\tilde{H}_n(S^n)$ let us regard S^n as two n-simplices Δ_1^n and Δ_2^n with their boundaries identified in the obvious way, preserving the ordering of vertices. The difference $\Delta_1^n - \Delta_2^n$, viewed as a singular n-chain, is then a cycle, and we claim it represents a generator of $\tilde{H}_n(S^n)$. To see this, consider the isomorphisms

$$\tilde{H}_n(S^n) \xrightarrow{\approx} H_n(S^n, \Delta_2^n) \xleftarrow{\approx} H_n(\Delta_1^n, \partial \Delta_1^n)$$

where the first isomorphism comes from the long exact sequence of the pair (S^n, Δ_2^n) and the second isomorphism is justified by passing to quotients as before. Under these isomorphisms the cycle $\Delta_1^n - \Delta_2^n$ in the first group corresponds to the cycle Δ_1^n in the third group, which represents a generator of this group as we have seen, so $\Delta_1^n - \Delta_2^n$ represents a generator of $\tilde{H}_n(S^n)$.

The preceding proposition implies that the excision property holds also for sub-complexes of CW complexes:

Corollary 2.24. *If the CW complex X is the union of subcomplexes A and B, then the inclusion $(B, A \cap B) \hookrightarrow (X, A)$ induces isomorphisms $H_n(B, A \cap B) \to H_n(X, A)$ for all n.*

Proof: Since CW pairs are good, Proposition 2.22 allows us to pass to the quotient spaces $B/(A \cap B)$ and X/A which are homeomorphic, assuming we are not in the trivial case $A \cap B = \varnothing$. $\square$

Here is another application of the preceding proposition:

Corollary 2.25. *For a wedge sum $\bigvee_\alpha X_\alpha$, the inclusions $i_\alpha : X_\alpha \hookrightarrow \bigvee_\alpha X_\alpha$ induce an isomorphism $\bigoplus_\alpha i_{\alpha*} : \bigoplus_\alpha \tilde{H}_n(X_\alpha) \to \tilde{H}_n(\bigvee_\alpha X_\alpha)$, provided that the wedge sum is formed at basepoints $x_\alpha \in X_\alpha$ such that the pairs (X_α, x_α) are good.*

Proof: Since reduced homology is the same as homology relative to a basepoint, this follows from the proposition by taking $(X, A) = (\coprod_\alpha X_\alpha, \coprod_\alpha \{x_\alpha\})$. $\square$

Here is an application of the machinery we have developed, a classical result of Brouwer from around 1910 known as 'invariance of dimension,' which says in particular that $\mathbb{R}^m$ is not homeomorphic to $\mathbb{R}^n$ if $m \neq n$.

Theorem 2.26. *If nonempty open sets $U \subset \mathbb{R}^m$ and $V \subset \mathbb{R}^n$ are homeomorphic, then $m = n$.*

Proof: For $x \in U$ we have $H_k(U, U - \{x\}) \approx H_k(\mathbb{R}^m, \mathbb{R}^m - \{x\})$ by excision. From the long exact sequence for the pair $(\mathbb{R}^m, \mathbb{R}^m - \{x\})$ we get $H_k(\mathbb{R}^m, \mathbb{R}^m - \{x\}) \approx \tilde{H}_{k-1}(\mathbb{R}^m - \{x\})$. Since $\mathbb{R}^m - \{x\}$ deformation retracts onto a sphere S^{m-1}, we conclude that $H_k(U, U - \{x\})$ is $\mathbb{Z}$ for $k = m$ and 0 otherwise. By the same reasoning, $H_k(V, V - \{y\})$ is $\mathbb{Z}$ for $k = n$ and 0 otherwise. Since a homeomorphism $h : U \to V$ induces isomorphisms $H_k(U, U - \{x\}) \to H_k(V, V - \{h(x)\})$ for all k, we must have $m = n$. $\square$

Generalizing the idea of this proof, the **local homology groups** of a space X at a point $x \in X$ are defined to be the groups $H_n(X, X - \{x\})$. For any open neighborhood U of x, excision gives isomorphisms $H_n(X, X - \{x\}) \approx H_n(U, U - \{x\})$ assuming points are closed in X, and thus the groups $H_n(X, X - \{x\})$ depend only on the local topology of X near x. A homeomorphism $f : X \to Y$ must induce isomorphisms $H_n(X, X - \{x\}) \approx H_n(Y, Y - \{f(x)\})$ for all x and n, so the local homology groups can be used to tell when spaces are not locally homeomorphic at certain points, as in the preceding proof. The exercises give some further examples of this.

Naturality

The exact sequences we have been constructing have an extra property that will become important later at key points in many arguments, though at first glance this property may seem just an idle technicality, not very interesting. We shall discuss the property now rather than interrupting later arguments to check it when it is needed, but the reader may prefer to postpone a careful reading of this discussion.

The property is called **naturality**. For example, to say that the long exact sequence of a pair is natural means that for a map $f : (X, A) \to (Y, B)$, the diagram

$$\begin{array}{ccccccccc}
\cdots & \longrightarrow & H_n(A) & \xrightarrow{\ i_*\ } & H_n(X) & \xrightarrow{\ j_*\ } & H_n(X, A) & \xrightarrow{\ \partial\ } & H_{n-1}(A) & \longrightarrow & \cdots \\
 & & \downarrow{\scriptstyle f_*} & & \downarrow{\scriptstyle f_*} & & \downarrow{\scriptstyle f_*} & & \downarrow{\scriptstyle f_*} & & \\
\cdots & \longrightarrow & H_n(B) & \xrightarrow{\ i_*\ } & H_n(Y) & \xrightarrow{\ j_*\ } & H_n(Y, B) & \xrightarrow{\ \partial\ } & H_{n-1}(B) & \longrightarrow & \cdots
\end{array}$$

is commutative. Commutativity of the squares involving i_* and j_* follows from the obvious commutativity of the corresponding squares of chain groups, with C_n in place of H_n. For the other square, when we defined induced homomorphisms we saw that $f_\sharp \partial = \partial f_\sharp$ at the chain level. Then for a class $[\alpha] \in H_n(X, A)$ represented by a relative cycle α, we have $f_* \partial [\alpha] = f_* [\partial \alpha] = [f_\sharp \partial \alpha] = [\partial f_\sharp \alpha] = \partial [f_\sharp \alpha] = \partial f_* [\alpha]$.

Alternatively, we could appeal to the general algebraic fact that the long exact sequence of homology groups associated to a short exact sequence of chain complexes is natural: For a commutative diagram of short exact sequences of chain complexes

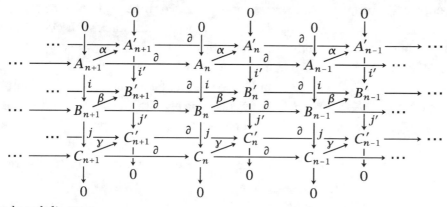

the induced diagram

$$\begin{array}{ccccccccc}
\cdots & \longrightarrow & H_n(A) & \xrightarrow{\ i_*\ } & H_n(B) & \xrightarrow{\ j_*\ } & H_n(C) & \xrightarrow{\ \partial\ } & H_{n-1}(A) & \longrightarrow & \cdots \\
 & & \downarrow{\scriptstyle \alpha_*} & & \downarrow{\scriptstyle \beta_*} & & \downarrow{\scriptstyle \gamma_*} & & \downarrow{\scriptstyle \alpha_*} & & \\
\cdots & \longrightarrow & H_n(A') & \xrightarrow{\ i'_*\ } & H_n(B') & \xrightarrow{\ j'_*\ } & H_n(C') & \xrightarrow{\ \partial\ } & H_{n-1}(A') & \longrightarrow & \cdots
\end{array}$$

is commutative. Commutativity of the first two squares is obvious since $\beta i = i' \alpha$ implies $\beta_* i_* = i'_* \alpha_*$ and $\gamma j = j' \beta$ implies $\gamma_* j_* = j'_* \beta_*$. For the third square, recall that the map $\partial : H_n(C) \to H_{n-1}(A)$ was defined by $\partial [c] = [a]$ where $c = j(b)$ and $i(a) = \partial b$. Then $\partial [\gamma(c)] = [\alpha(a)]$ since $\gamma(c) = \gamma j(b) = j'(\beta(b))$ and $i'(\alpha(a)) = \beta i(a) = \beta \partial(b) = \partial \beta(b)$. Hence $\partial \gamma_* [c] = \alpha_* [a] = \alpha_* \partial [c]$.

This algebraic fact also implies naturality of the long exact sequence of a triple and the long exact sequence of reduced homology of a pair.

Finally, there is the naturality of the long exact sequence in Theorem 2.13, that is, commutativity of the diagram

$$
\begin{array}{ccccccccc}
\cdots \longrightarrow & \widetilde{H}_n(A) & \xrightarrow{i_*} & \widetilde{H}_n(X) & \xrightarrow{q_*} & \widetilde{H}_n(X/A) & \xrightarrow{\partial} & \widetilde{H}_{n-1}(A) & \longrightarrow \cdots \\
& \downarrow{f_*} & & \downarrow{f_*} & & \downarrow{\overline{f}_*} & & \downarrow{f_*} & \\
\cdots \longrightarrow & \widetilde{H}_n(B) & \xrightarrow{i_*} & \widetilde{H}_n(Y) & \xrightarrow{q_*} & \widetilde{H}_n(Y/B) & \xrightarrow{\partial} & \widetilde{H}_{n-1}(B) & \longrightarrow \cdots
\end{array}
$$

where i and q denote inclusions and quotient maps, and $\overline{f} : X/A \to Y/B$ is induced by f. The first two squares commute since $fi = if$ and $\overline{f}q = qf$. The third square expands into

$$
\begin{array}{ccccccc}
\widetilde{H}_n(X/A) & \xrightarrow[\approx]{j_*} & H_n(X/A, A/A) & \xleftarrow[\approx]{q_*} & H_n(X,A) & \xrightarrow{\partial} & \widetilde{H}_{n-1}(A) \\
\downarrow{\overline{f}_*} & & \downarrow{\overline{f}_*} & & \downarrow{f_*} & & \downarrow{f_*} \\
\widetilde{H}_n(Y/B) & \xrightarrow[\approx]{j_*} & H_n(Y/B, B/B) & \xleftarrow[\approx]{q_*} & H_n(Y,B) & \xrightarrow{\partial} & \widetilde{H}_{n-1}(B)
\end{array}
$$

We have already shown commutativity of the first and third squares, and the second square commutes since $\overline{f}q = qf$.

The Equivalence of Simplicial and Singular Homology

We can use the preceding results to show that the simplicial and singular homology groups of Δ-complexes are always isomorphic. For the proof it will be convenient to consider the relative case as well, so let X be a Δ-complex with $A \subset X$ a subcomplex. Thus A is the Δ-complex formed by any union of simplices of X. Relative groups $H_n^\Delta(X, A)$ can be defined in the same way as for singular homology, via relative chains $\Delta_n(X, A) = \Delta_n(X)/\Delta_n(A)$, and this yields a long exact sequence of simplicial homology groups for the pair (X, A) by the same algebraic argument as for singular homology. There is a canonical homomorphism $H_n^\Delta(X, A) \to H_n(X, A)$ induced by the chain map $\Delta_n(X, A) \to C_n(X, A)$ sending each n-simplex of X to its characteristic map $\sigma : \Delta^n \to X$. The possibility $A = \varnothing$ is not excluded, in which case the relative groups reduce to absolute groups.

Theorem 2.27. *The homomorphisms $H_n^\Delta(X, A) \to H_n(X, A)$ are isomorphisms for all n and all Δ-complex pairs (X, A).*

Proof: First we do the case that X is finite-dimensional and A is empty. For X^k the k-skeleton of X, consisting of all simplices of dimension k or less, we have a commutative diagram of exact sequences:

$$
\begin{array}{ccccccccc}
H_{n+1}^\Delta(X^k, X^{k-1}) & \longrightarrow & H_n^\Delta(X^{k-1}) & \longrightarrow & H_n^\Delta(X^k) & \longrightarrow & H_n^\Delta(X^k, X^{k-1}) & \longrightarrow & H_{n-1}^\Delta(X^{k-1}) \\
\downarrow & & \downarrow & & \downarrow & & \downarrow & & \downarrow \\
H_{n+1}(X^k, X^{k-1}) & \longrightarrow & H_n(X^{k-1}) & \longrightarrow & H_n(X^k) & \longrightarrow & H_n(X^k, X^{k-1}) & \longrightarrow & H_{n-1}(X^{k-1})
\end{array}
$$

Let us first show that the first and fourth vertical maps are isomorphisms for all n. The simplicial chain group $\Delta_n(X^k, X^{k-1})$ is zero for $n \neq k$, and is free abelian with basis the k-simplices of X when $n = k$. Hence $H_n^\Delta(X^k, X^{k-1})$ has exactly the same description. The corresponding singular homology groups $H_n(X^k, X^{k-1})$ can be computed by considering the map $\Phi : \coprod_\alpha(\Delta_\alpha^k, \partial\Delta_\alpha^k) \to (X^k, X^{k-1})$ formed by the characteristic maps $\Delta^k \to X$ for all the k-simplices of X. Since Φ induces a homeomorphism of quotient spaces $\coprod_\alpha \Delta_\alpha^k / \coprod_\alpha \partial\Delta_\alpha^k \approx X^k / X^{k-1}$, it induces isomorphisms on all singular homology groups. Thus $H_n(X^k, X^{k-1})$ is zero for $n \neq k$, while for $n = k$ this group is free abelian with basis represented by the relative cycles given by the characteristic maps of all the k-simplices of X, in view of the fact that $H_k(\Delta^k, \partial\Delta^k)$ is generated by the identity map $\Delta^k \to \Delta^k$, as we showed in Example 2.23. Therefore the map $H_k^\Delta(X^k, X^{k-1}) \to H_k(X^k, X^{k-1})$ is an isomorphism.

By induction on k we may assume the second and fifth vertical maps in the preceding diagram are isomorphisms as well. The following frequently quoted basic algebraic lemma will then imply that the middle vertical map is an isomorphism, finishing the proof when X is finite-dimensional and $A = \varnothing$.

The Five-Lemma. *In a commutative diagram of abelian groups as at the right, if the two rows are exact and α, β, δ, and ε are isomorphisms, then γ is an isomorphism also.*

$$
\begin{array}{ccccccccc}
A & \xrightarrow{i} & B & \xrightarrow{j} & C & \xrightarrow{k} & D & \xrightarrow{\ell} & E \\
\downarrow{\alpha} & & \downarrow{\beta} & & \downarrow{\gamma} & & \downarrow{\delta} & & \downarrow{\varepsilon} \\
A' & \xrightarrow{i'} & B' & \xrightarrow{j'} & C' & \xrightarrow{k'} & D' & \xrightarrow{\ell'} & E'
\end{array}
$$

Proof: It suffices to show:

(a) γ is surjective if β and δ are surjective and ε is injective.

(b) γ is injective if β and δ are injective and α is surjective.

The proofs of these two statements are straightforward diagram chasing. There is really no choice about how the argument can proceed, and it would be a good exercise for the reader to close the book now and reconstruct the proofs without looking.

To prove (a), start with an element $c' \in C'$. Then $k'(c') = \delta(d)$ for some $d \in D$ since δ is surjective. Since ε is injective and $\varepsilon\ell(d) = \ell'\delta(d) = \ell'k'(c') = 0$, we deduce that $\ell(d) = 0$, hence $d = k(c)$ for some $c \in C$ by exactness of the upper row. The difference $c' - \gamma(c)$ maps to 0 under k' since $k'(c') - k'\gamma(c) = k'(c') - \delta k(c) = k'(c') - \delta(d) = 0$. Therefore $c' - \gamma(c) = j'(b')$ for some $b' \in B'$ by exactness. Since β is surjective, $b' = \beta(b)$ for some $b \in B$, and then $\gamma(c + j(b)) = \gamma(c) + \gamma j(b) = \gamma(c) + j'\beta(b) = \gamma(c) + j'(b') = c'$, showing that γ is surjective.

To prove (b), suppose that $\gamma(c) = 0$. Since δ is injective, $\delta k(c) = k'\gamma(c) = 0$ implies $k(c) = 0$, so $c = j(b)$ for some $b \in B$. The element $\beta(b)$ satisfies $j'\beta(b) = \gamma j(b) = \gamma(c) = 0$, so $\beta(b) = i'(a')$ for some $a' \in A'$. Since α is surjective, $a' = \alpha(a)$ for some $a \in A$. Since β is injective, $\beta(i(a) - b) = \beta i(a) - \beta(b) = i'\alpha(a) - \beta(b) = i'(a') - \beta(b) = 0$ implies $i(a) - b = 0$. Thus $b = i(a)$, and hence $c = j(b) = ji(a) = 0$ since $ji = 0$. This shows γ has trivial kernel. $\square$

Returning to the proof of the theorem, we next consider the case that X is infinite-dimensional, where we will use the following fact: A compact set in X can meet only finitely many open simplices of X, that is, simplices with their proper faces deleted. This is a general fact about CW complexes proved in the Appendix, but here is a direct proof for Δ-complexes. If a compact set C intersected infinitely many open simplices, it would contain an infinite sequence of points x_i each lying in a different open simplex. Then the sets $U_i = X - \bigcup_{j \neq i} \{x_j\}$, which are open since their preimages under the characteristic maps of all the simplices are clearly open, form an open cover of C with no finite subcover.

This can be applied to show the map $H_n^\Delta(X) \to H_n(X)$ is surjective. Represent a given element of $H_n(X)$ by a singular n-cycle z. This is a linear combination of finitely many singular simplices with compact images, meeting only finitely many open simplices of X, hence contained in X^k for some k. We have shown that $H_n^\Delta(X^k) \to H_n(X^k)$ is an isomorphism, in particular surjective, so z is homologous in X^k (hence in X) to a simplicial cycle. This gives surjectivity. Injectivity is similar: If a simplicial n-cycle z is the boundary of a singular chain in X, this chain has compact image and hence must lie in some X^k, so z represents an element of the kernel of $H_n^\Delta(X^k) \to H_n(X^k)$. But we know this map is injective, so z is a simplicial boundary in X^k, and therefore in X.

It remains to do the case of arbitrary X with $A \neq \varnothing$, but this follows from the absolute case by applying the five-lemma to the canonical map from the long exact sequence of simplicial homology groups for the pair (X, A) to the corresponding long exact sequence of singular homology groups. $\square$

We can deduce from this theorem that $H_n(X)$ is finitely generated whenever X is a Δ-complex with finitely many n-simplices, since in this case the simplicial chain group $\Delta_n(X)$ is finitely generated, hence also its subgroup of cycles and therefore also the latter group's quotient $H_n^\Delta(X)$. If we write $H_n(X)$ as the direct sum of cyclic groups, then the number of $\mathbb{Z}$ summands is known traditionally as the nth **Betti number** of X, and integers specifying the orders of the finite cyclic summands are called **torsion coefficients**.

It is a curious historical fact that homology was not thought of originally as a sequence of groups, but rather as Betti numbers and torsion coefficients. One can after all compute Betti numbers and torsion coefficients from the simplicial boundary maps without actually mentioning homology groups. This computational viewpoint, with homology being numbers rather than groups, prevailed from when Poincaré first started serious work on homology around 1900, up until the 1920s when the more abstract viewpoint of groups entered the picture. During this period 'homology' meant primarily 'simplicial homology,' and it was another 20 years before the shift to singular homology was complete, with the final definition of singular homology emerging only in a 1944 paper of Eilenberg, after contributions from quite a few others, particularly Alexander and Lefschetz. Within the next few years the rest of the basic structure of homology theory as we have presented it fell into place, and the first definitive treatment appeared in the classic book [Eilenberg & Steenrod 1952].

Exercises

1. What familiar space is the quotient Δ-complex of a 2-simplex $[v_0, v_1, v_2]$ obtained by identifying the edges $[v_0, v_1]$ and $[v_1, v_2]$, preserving the ordering of vertices?

2. Show that the Δ-complex obtained from Δ^3 by performing the edge identifications $[v_0, v_1] \sim [v_1, v_3]$ and $[v_0, v_2] \sim [v_2, v_3]$ deformation retracts onto a Klein bottle. Find other pairs of identifications of edges that produce Δ-complexes deformation retracting onto a torus, a 2-sphere, and $\mathbb{RP}^2$.

3. Construct a Δ-complex structure on $\mathbb{RP}^n$ as a quotient of a Δ-complex structure on S^n having vertices the two vectors of length 1 along each coordinate axis in $\mathbb{R}^{n+1}$.

4. Compute the simplicial homology groups of the triangular parachute obtained from Δ^2 by identifying its three vertices to a single point.

5. Compute the simplicial homology groups of the Klein bottle using the Δ-complex structure described at the beginning of this section.

6. Compute the simplicial homology groups of the Δ-complex obtained from $n + 1$ 2-simplices $\Delta_0^2, \cdots, \Delta_n^2$ by identifying all three edges of Δ_0^2 to a single edge, and for $i > 0$ identifying the edges $[v_0, v_1]$ and $[v_1, v_2]$ of Δ_i^2 to a single edge and the edge $[v_0, v_2]$ to the edge $[v_0, v_1]$ of Δ_{i-1}^2.

7. Find a way of identifying pairs of faces of Δ^3 to produce a Δ-complex structure on S^3 having a single 3-simplex, and compute the simplicial homology groups of this Δ-complex.

8. Construct a 3-dimensional Δ-complex X from n tetrahedra $T_1, \cdots, T_n$ by the following two steps. First arrange the tetrahedra in a cyclic pattern as in the figure, so that each T_i shares a common vertical face with its two neighbors T_{i-1} and T_{i+1}, subscripts being taken mod n. Then identify the 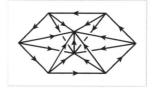 bottom face of T_i with the top face of T_{i+1} for each i. Show the simplicial homology groups of X in dimensions 0, 1, 2, 3 are $\mathbb{Z}$, $\mathbb{Z}_n$, 0, $\mathbb{Z}$, respectively. [The space X is an example of a *lens space*; see Example 2.43 for the general case.]

9. Compute the homology groups of the Δ-complex X obtained from Δ^n by identifying all faces of the same dimension. Thus X has a single k-simplex for each $k \le n$.

10. (a) Show the quotient space of a finite collection of disjoint 2-simplices obtained by identifying pairs of edges is always a surface, locally homeomorphic to $\mathbb{R}^2$.
(b) Show the edges can always be oriented so as to define a Δ-complex structure on the quotient surface. [This is more difficult.]

11. Show that if A is a retract of X then the map $H_n(A) \to H_n(X)$ induced by the inclusion $A \subset X$ is injective.

12. Show that chain homotopy of chain maps is an equivalence relation.

13. Verify that $f \simeq g$ implies $f_* = g_*$ for induced homomorphisms of reduced homology groups.

14. Determine whether there exists a short exact sequence $0 \to \mathbb{Z}_4 \to \mathbb{Z}_8 \oplus \mathbb{Z}_2 \to \mathbb{Z}_4 \to 0$. More generally, determine which abelian groups A fit into a short exact sequence $0 \to \mathbb{Z}_{p^m} \to A \to \mathbb{Z}_{p^n} \to 0$ with p prime. What about the case of short exact sequences $0 \to \mathbb{Z} \to A \to \mathbb{Z}_n \to 0$?

15. For an exact sequence $A \to B \to C \to D \to E$ show that $C = 0$ iff the map $A \to B$ is surjective and $D \to E$ is injective. Hence for a pair of spaces (X, A), the inclusion $A \hookrightarrow X$ induces isomorphisms on all homology groups iff $H_n(X, A) = 0$ for all n.

16. (a) Show that $H_0(X, A) = 0$ iff A meets each path-component of X.

(b) Show that $H_1(X, A) = 0$ iff $H_1(A) \to H_1(X)$ is surjective and each path-component of X contains at most one path-component of A.

17. (a) Compute the homology groups $H_n(X, A)$ when X is S^2 or $S^1 \times S^1$ and A is a finite set of points in X.

(b) Compute the groups $H_n(X, A)$ and $H_n(X, B)$ for X a closed orientable surface of genus two with A and B the circles shown. [What are X/A and X/B?]

18. Show that for the subspace $\mathbb{Q} \subset \mathbb{R}$, the relative homology group $H_1(\mathbb{R}, \mathbb{Q})$ is free abelian and find a basis.

19. Compute the homology groups of the subspace of $I \times I$ consisting of the four boundary edges plus all points in the interior whose first coordinate is rational.

20. Show that $\tilde{H}_n(X) \approx \tilde{H}_{n+1}(SX)$ for all n, where SX is the suspension of X. More generally, thinking of SX as the union of two cones CX with their bases identified, compute the reduced homology groups of the union of any finite number of cones CX with their bases identified.

21. Making the preceding problem more concrete, construct explicit chain maps $s : C_n(X) \to C_{n+1}(SX)$ inducing isomorphisms $\tilde{H}_n(X) \to \tilde{H}_{n+1}(SX)$.

22. Prove by induction on dimension the following facts about the homology of a finite-dimensional CW complex X, using the observation that X^n / X^{n-1} is a wedge sum of n-spheres:

(a) If X has dimension n then $H_i(X) = 0$ for $i > n$ and $H_n(X)$ is free.

(b) $H_n(X)$ is free with basis in bijective correspondence with the n-cells if there are no cells of dimension $n - 1$ or $n + 1$.

(c) If X has k n-cells, then $H_n(X)$ is generated by at most k elements.

23. Show that the second barycentric subdivision of a Δ-complex is a simplicial complex. Namely, show that the first barycentric subdivision produces a Δ-complex with the property that each simplex has all its vertices distinct, then show that for a Δ-complex with this property, barycentric subdivision produces a simplicial complex.

24. Show that each n-simplex in the barycentric subdivision of Δ^n is defined by n inequalities $t_{i_0} \le t_{i_1} \le \cdots \le t_{i_n}$ in its barycentric coordinates, where $(i_0, \cdots, i_n)$ is a permutation of $(0, \cdots, n)$.

25. Find an explicit, noninductive formula for the barycentric subdivision operator $S : C_n(X) \to C_n(X)$.

26. Show that $H_1(X, A)$ is not isomorphic to $\tilde{H}_1(X/A)$ if $X = [0, 1]$ and A is the sequence $1, \frac{1}{2}, \frac{1}{3}, \cdots$ together with its limit 0. [See Example 1.25.]

27. Let $f : (X, A) \to (Y, B)$ be a map such that both $f : X \to Y$ and the restriction $f : A \to B$ are homotopy equivalences.
(a) Show that $f_* : H_n(X, A) \to H_n(Y, B)$ is an isomorphism for all n.
(b) For the case of the inclusion $f : (D^n, S^{n-1}) \hookrightarrow (D^n, D^n - \{0\})$, show that f is not a homotopy equivalence of pairs — there is no $g : (D^n, D^n - \{0\}) \to (D^n, S^{n-1})$ such that fg and gf are homotopic to the identity through maps of pairs. [Observe that a homotopy equivalence of pairs $(X, A) \to (Y, B)$ is also a homotopy equivalence for the pairs obtained by replacing A and B by their closures.]

28. Let X be the cone on the 1-skeleton of Δ^3, the union of all line segments joining points in the six edges of Δ^3 to the barycenter of Δ^3. Compute the local homology groups $H_n(X, X - \{x\})$ for all $x \in X$. Define ∂X to be the subspace of points x such that $H_n(X, X - \{x\}) = 0$ for all n, and compute the local homology groups $H_n(\partial X, \partial X - \{x\})$. Use these calculations to determine which subsets $A \subset X$ have the property that $f(A) \subset A$ for all homeomorphisms $f : X \to X$.

29. Show that $S^1 \times S^1$ and $S^1 \vee S^1 \vee S^2$ have isomorphic homology groups in all dimensions, but their universal covering spaces do not.

30. In each of the following commutative diagrams assume that all maps but one are isomorphisms. Show that the remaining map must be an isomorphism as well.

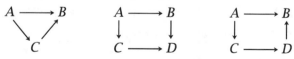

31. Using the notation of the five-lemma, give an example where the maps α, β, δ, and ε are zero but γ is nonzero. This can be done with short exact sequences in which all the groups are either $\mathbb{Z}$ or 0.

2.2 Computations and Applications

Now that the basic properties of homology have been established, we can begin to move a little more freely. Our first topic, exploiting the calculation of $H_n(S^n)$, is Brouwer's notion of degree for maps $S^n \to S^n$. Historically, Brouwer's introduction of this concept in the years 1910–12 preceded the rigorous development of homology, so his definition was rather different, using the technique of simplicial approximation which we explain in §2.C. The later definition in terms of homology is certainly more elegant, though perhaps with some loss of geometric intuition. More in the spirit of Brouwer's definition is a third approach using differential topology, presented very lucidly in [Milnor 1965].

Degree

For a map $f : S^n \to S^n$ with $n > 0$, the induced map $f_* : H_n(S^n) \to H_n(S^n)$ is a homomorphism from an infinite cyclic group to itself and so must be of the form $f_*(\alpha) = d\alpha$ for some integer d depending only on f. This integer is called the **degree** of f, with the notation $\deg f$. Here are some basic properties of degree:

(a) $\deg \mathbb{1} = 1$, since $\mathbb{1}_* = \mathbb{1}$.

(b) $\deg f = 0$ if f is not surjective. For if we choose a point $x_0 \in S^n - f(S^n)$ then f can be factored as a composition $S^n \to S^n - \{x_0\} \hookrightarrow S^n$ and $H_n(S^n - \{x_0\}) = 0$ since $S^n - \{x_0\}$ is contractible. Hence $f_* = 0$.

(c) If $f \simeq g$ then $\deg f = \deg g$ since $f_* = g_*$. The converse statement, that $f \simeq g$ if $\deg f = \deg g$, is a fundamental theorem of Hopf from around 1925 which we prove in Corollary 4.25.

(d) $\deg fg = \deg f \deg g$, since $(fg)_* = f_* g_*$. As a consequence, $\deg f = \pm 1$ if f is a homotopy equivalence since $fg \simeq \mathbb{1}$ implies $\deg f \deg g = \deg \mathbb{1} = 1$.

(e) $\deg f = -1$ if f is a reflection of S^n, fixing the points in a subsphere S^{n-1} and interchanging the two complementary hemispheres. For we can give S^n a Δ-complex structure with these two hemispheres as its two n-simplices Δ_1^n and Δ_2^n, and the n-chain $\Delta_1^n - \Delta_2^n$ represents a generator of $H_n(S^n)$ as we saw in Example 2.23, so the reflection interchanging Δ_1^n and Δ_2^n sends this generator to its negative.

(f) The antipodal map $-\mathbb{1} : S^n \to S^n$, $x \mapsto -x$, has degree $(-1)^{n+1}$ since it is the composition of $n + 1$ reflections, each changing the sign of one coordinate in $\mathbb{R}^{n+1}$.

(g) If $f : S^n \to S^n$ has no fixed points then $\deg f = (-1)^{n+1}$. For if $f(x) \neq x$ then the line segment from $f(x)$ to $-x$, defined by $t \mapsto (1 - t)f(x) - tx$ for $0 \leq t \leq 1$, does not pass through the origin. Hence if f has no fixed points, the formula $f_t(x) = [(1 - t)f(x) - tx]/|(1 - t)f(x) - tx|$ defines a homotopy from f to the antipodal map. Note that the antipodal map has no fixed points, so the fact that maps without fixed points are homotopic to the antipodal map is a sort of converse statement.

Here is an interesting application of degree:

|| **Theorem 2.28.** S^n *has a continuous field of nonzero tangent vectors iff* n *is odd.*

Proof: Suppose $x \mapsto v(x)$ is a tangent vector field on S^n, assigning to a vector $x \in S^n$ the vector $v(x)$ tangent to S^n at x. Regarding $v(x)$ as a vector at the origin instead of at x, tangency just means that x and $v(x)$ are orthogonal in $\mathbb{R}^{n+1}$. If $v(x) \neq 0$ for all x, we may normalize so that $|v(x)| = 1$ for all x by replacing $v(x)$ by $v(x)/|v(x)|$. Assuming this has been done, the vectors $(\cos t)x + (\sin t)v(x)$ lie in the unit circle in the plane spanned by x and $v(x)$. Letting t go from 0 to π, we obtain a homotopy $f_t(x) = (\cos t)x + (\sin t)v(x)$ from the identity map of S^n to the antipodal map $-\mathbb{1}$. This implies that $\deg(-\mathbb{1}) = \deg \mathbb{1}$, hence $(-1)^{n+1} = 1$ and n must be odd.

Conversely, if n is odd, say $n = 2k - 1$, we can define $v(x_1, x_2, \cdots, x_{2k-1}, x_{2k}) = (-x_2, x_1, \cdots, -x_{2k}, x_{2k-1})$. Then $v(x)$ is orthogonal to x, so v is a tangent vector field on S^n, and $|v(x)| = 1$ for all $x \in S^n$. $\qquad\square$

For the much more difficult problem of finding the maximum number of tangent vector fields on S^n that are linearly independent at each point, see [VBKT] or [Husemoller 1966].

Another nice application of degree, giving a partial answer to a question raised in Example 1.43, is the following result:

|| **Proposition 2.29.** $\mathbb{Z}_2$ *is the only nontrivial group that can act freely on* S^n *if* n *is even.*

Recall that an action of a group G on a space X is a homomorphism from G to the group $\text{Homeo}(X)$ of homeomorphisms $X \to X$, and the action is free if the homeomorphism corresponding to each nontrivial element of G has no fixed points. In the case of S^n, the antipodal map $x \mapsto -x$ generates a free action of $\mathbb{Z}_2$.

Proof: Since the degree of a homeomorphism must be ± 1, an action of a group G on S^n determines a degree function $d : G \to \{\pm 1\}$. This is a homomorphism since $\deg fg = \deg f \deg g$. If the action is free, then d sends every nontrivial element of G to $(-1)^{n+1}$ by property (g) above. Thus when n is even, d has trivial kernel, so $G \subset \mathbb{Z}_2$. $\qquad\square$

Next we describe a technique for computing degrees which can be applied to most maps that arise in practice. Suppose $f : S^n \to S^n$, $n > 0$, has the property that for some point $y \in S^n$, the preimage $f^{-1}(y)$ consists of only finitely many points, say $x_1, \cdots, x_m$. Let $U_1, \cdots, U_m$ be disjoint neighborhoods of these points, mapped by f into a neighborhood V of y. Then $f(U_i - x_i) \subset V - y$ for each i, and we have a diagram

$$
\begin{array}{ccc}
& H_n(U_i, U_i - x_i) & \xrightarrow{\;f_*\;} & H_n(V, V - y) \\
\approx \nearrow & \downarrow k_i & & \downarrow \approx \\
H_n(S^n, S^n - x_i) \xleftarrow{\;p_i\;} & H_n(S^n, S^n - f^{-1}(y)) & \xrightarrow{\;f_*\;} & H_n(S^n, S^n - y) \\
\nwarrow & \uparrow j & & \uparrow \approx \\
\approx & H_n(S^n) & \xrightarrow{\;f_*\;} & H_n(S^n)
\end{array}
$$

where all the maps are the obvious ones, and in particular k_i and p_i are induced by inclusions, so the triangles and squares commute. The two isomorphisms in the upper half of the diagram come from excision, while the lower two isomorphisms come from exact sequences of pairs. Via these four isomorphisms, the top two groups in the diagram can be identified with $H_n(S^n) \approx \mathbb{Z}$, and the top homomorphism f_* becomes multiplication by an integer called the **local degree** of f at x_i, written $\deg f | x_i$.

For example, if f is a homeomorphism, then y can be any point and there is only one corresponding x_i, so all the maps in the diagram are isomorphisms and $\deg f | x_i = \deg f = \pm 1$. More generally, if f maps each U_i homeomorphically onto V, then $\deg f | x_i = \pm 1$ for each i. This situation occurs quite often in applications, and it is usually not hard to determine the correct signs.

Here is the formula that reduces degree calculations to computing local degrees:

∥ Proposition 2.30. $\deg f = \sum_i \deg f | x_i$.

Proof: By excision, the central term $H_n(S^n, S^n - f^{-1}(y))$ in the preceding diagram is the direct sum of the groups $H_n(U_i, U_i - x_i) \approx \mathbb{Z}$, with k_i the inclusion of the ith summand and p_i the projection onto the ith summand. Identifying the outer groups in the diagram with $\mathbb{Z}$ as before, commutativity of the lower triangle says that $p_i j(1) = 1$, hence $j(1) = (1, \cdots, 1) = \sum_i k_i(1)$. Commutativity of the upper square says that the middle f_* takes $k_i(1)$ to $\deg f | x_i$, hence the sum $\sum_i k_i(1) = j(1)$ is taken to $\sum_i \deg f | x_i$. Commutativity of the lower square then gives the formula $\deg f = \sum_i \deg f | x_i$. $\qquad\square$

Example 2.31. We can use this result to construct a map $S^n \to S^n$ of any given degree, for each $n \geq 1$. Let $q: S^n \to \bigvee_k S^n$ be the quotient map obtained by collapsing the complement of k disjoint open balls B_i in S^n to a point, and let $p: \bigvee_k S^n \to S^n$ identify all the summands to a single sphere. Consider the composition $f = pq$. For almost all $y \in S^n$ we have $f^{-1}(y)$ consisting of one point x_i in each B_i. The local degree of f at x_i is ± 1 since f is a homeomorphism near x_i. By precomposing p with reflections of the summands of $\bigvee_k S^n$ if necessary, we can make each local degree either $+1$ or -1, whichever we wish. Thus we can produce a map $S^n \to S^n$ of degree $\pm k$.

Example 2.32. In the case of S^1, the map $f(z) = z^k$, where we view S^1 as the unit circle in $\mathbb{C}$, has degree k. This is evident in the case $k = 0$ since f is then constant. The case $k < 0$ reduces to the case $k > 0$ by composing with $z \mapsto z^{-1}$, which is a reflection, of degree -1. To compute the degree when $k > 0$, observe first that for any $y \in S^1$, $f^{-1}(y)$ consists of k points $x_1, \cdots, x_k$ near each of which f is a local homeomorphism, stretching a circular arc by a factor of k. This local stretching can be eliminated by a deformation of f near x_i that does not change local degree, so the local degree at x_i is the same as for a rotation of S^1. A rotation is a homeomorphism so its local degree at any point equals its global degree, which is $+1$ since a rotation is homotopic to the identity. Hence $\deg f \,|\, x_i = 1$ and $\deg f = k$.

Another way of obtaining a map $S^n \to S^n$ of degree k is to take a repeated suspension of the map $z \mapsto z^k$ in Example 2.32, since suspension preserves degree:

Proposition 2.33. $\deg Sf = \deg f$, where $Sf : S^{n+1} \to S^{n+1}$ is the suspension of the map $f : S^n \to S^n$.

Proof: Let CS^n denote the cone $(S^n \times I)/(S^n \times 1)$ with base $S^n = S^n \times 0 \subset CS^n$, so CS^n / S^n is the suspension of S^n. The map f induces $Cf : (CS^n, S^n) \to (CS^n, S^n)$ with quotient Sf. The naturality of the boundary maps in the long exact sequence of the pair (CS^n, S^n) then gives commutativity of the diagram at the right. Hence if f_* is multiplication by d, so is Sf_*. $\qquad\square$

$$\begin{array}{ccc} \widetilde{H}_{n+1}(S^{n+1}) & \xrightarrow[\approx]{\partial} & \widetilde{H}_n(S^n) \\ \downarrow Sf_* & & \downarrow f_* \\ \widetilde{H}_{n+1}(S^{n+1}) & \xrightarrow[\approx]{\partial} & \widetilde{H}_n(S^n) \end{array}$$

Note that for $f : S^n \to S^n$, the suspension Sf maps only one point to each of the two 'poles' of S^{n+1}. This implies that the local degree of Sf at each pole must equal the global degree of Sf. Thus the local degree of a map $S^n \to S^n$ can be any integer if $n \geq 2$, just as the degree itself can be any integer when $n \geq 1$.

Cellular Homology

Cellular homology is a very efficient tool for computing the homology groups of CW complexes, based on degree calculations. Before giving the definition of cellular homology, we first establish a few preliminary facts:

Lemma 2.34. If X is a CW complex, then:

(a) $H_k(X^n, X^{n-1})$ is zero for $k \neq n$ and is free abelian for $k = n$, with a basis in one-to-one correspondence with the n-cells of X.

(b) $H_k(X^n) = 0$ for $k > n$. In particular, if X is finite-dimensional then $H_k(X) = 0$ for $k > \dim X$.

(c) The inclusion $i : X^n \hookrightarrow X$ induces an isomorphism $i_* : H_k(X^n) \to H_k(X)$ if $k < n$.

Proof: Statement (a) follows immediately from the observation that (X^n, X^{n-1}) is a good pair and X^n / X^{n-1} is a wedge sum of n-spheres, one for each n-cell of X. Here we are using Proposition 2.22 and Corollary 2.25.

To prove (b), consider the long exact sequence of the pair (X^n, X^{n-1}), which contains the segments

$$H_{k+1}(X^n, X^{n-1}) \longrightarrow H_k(X^{n-1}) \longrightarrow H_k(X^n) \longrightarrow H_k(X^n, X^{n-1})$$

If k is not equal to n or $n-1$ then the outer two groups are zero by part (a), so we have isomorphisms $H_k(X^{n-1}) \approx H_k(X^n)$ for $k \neq n$, $n-1$. Thus if $k > n$ we have $H_k(X^n) \approx H_k(X^{n-1}) \approx H_k(X^{n-2}) \approx \cdots \approx H_k(X^0) = 0$, proving (b). Further, if $k < n$ then $H_k(X^n) \approx H_k(X^{n+1}) \approx \cdots \approx H_k(X^{n+m})$ for all $m \geq 0$, proving (c) if X is finite-dimensional.

The proof of (c) when X is infinite-dimensional requires more work, and this can be done in two different ways. The more direct approach is to descend to the chain level and use the fact that a singular chain in X has compact image, hence meets only finitely many cells of X by Proposition A.1 in the Appendix. Thus each chain lies in a finite skeleton X^m. So a k-cycle in X is a cycle in some X^m, and then by the finite-dimensional case of (c), the cycle is homologous to a cycle in X^n if $n > k$, so $i_* : H_k(X^n) \rightarrow H_k(X)$ is surjective. Similarly for injectivity, if a k-cycle in X^n bounds a chain in X, this chain lies in some X^m with $m \geq n$, so by the finite-dimensional case the cycle bounds a chain in X^n if $n > k$.

The other approach is more general. From the long exact sequence of the pair (X, X^n) it suffices to show $H_k(X, X^n) = 0$ for $k \leq n$. Since $H_k(X, X^n) \approx \tilde{H}_k(X/X^n)$, this reduces the problem to showing:

$(*)$ $\tilde{H}_k(X) = 0$ for $k \leq n$ if the n-skeleton of X is a point.

When X is finite-dimensional, $(*)$ is immediate from the finite-dimensional case of (c) which we have already shown. It will suffice therefore to reduce the infinite-dimensional case to the finite-dimensional case. This reduction will be achieved by stretching X out to a complex that is at least locally finite-dimensional, using a special case of the 'mapping telescope' construction described in greater generality in §3.F.

Consider $X \times [0, \infty)$ with its product cell structure, where we give $[0, \infty)$ the cell structure with the integer points as 0-cells. Let $T = \bigcup_i X^i \times [i, \infty)$, a subcomplex 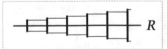 of $X \times [0, \infty)$. The figure shows a schematic picture of T with $[0, \infty)$ in the horizontal direction and the subcomplexes $X^i \times [i, i+1]$ as rectangles whose size increases with i since $X^i \subset X^{i+1}$. The line labeled R can be ignored for now. We claim that $T \simeq X$, hence $H_k(X) \approx H_k(T)$ for all k. Since X is a deformation retract of $X \times [0, \infty)$, it suffices to show that $X \times [0, \infty)$ also deformation retracts onto T. Let $Y_i = T \cup (X \times [i, \infty))$. Then Y_i deformation retracts onto Y_{i+1} since $X \times [i, i+1]$ deformation retracts onto $X^i \times [i, i+1] \cup X \times \{i+1\}$ by Proposition 0.16. If we perform the deformation retraction of Y_i onto Y_{i+1} during the t-interval $[1 - 1/2^i, 1 - 1/2^{i+1}]$, then this gives a deformation retraction f_t of $X \times [0, \infty)$ onto T, with points in $X^i \times [0, \infty)$ stationary under f_t for $t \geq 1 - 1/2^{i+1}$. Continuity follows from the fact that CW complexes have the weak topology with respect to their skeleta, so a map is continuous if its restriction to each skeleton is continuous.

Recalling that X^0 is a point, let $R \subset T$ be the ray $X^0 \times [0, \infty)$ and let $Z \subset T$ be the union of this ray with all the subcomplexes $X^i \times \{i\}$. Then Z/R is homeomorphic to $\bigvee_i X^i$, a wedge sum of finite-dimensional complexes with n-skeleton a point, so the finite-dimensional case of $(*)$ together with Corollary 2.25 describing the homology of wedge sums implies that $\tilde{H}_k(Z/R) = 0$ for $k \le n$. The same is therefore true for Z, from the long exact sequence of the pair (Z, R), since R is contractible. Similarly, T/Z is a wedge sum of finite-dimensional complexes with $(n+1)$-skeleton a point, since if we first collapse each subcomplex $X^i \times \{i\}$ of T to a point, we obtain the infinite sequence of suspensions SX^i 'skewered' along the ray R, and then if we collapse R to a point we obtain $\bigvee_i \Sigma X^i$ where ΣX^i is the reduced suspension of X^i, obtained from SX^i by collapsing the line segment $X^0 \times [i, i+1]$ to a point, so ΣX^i has $(n+1)$-skeleton a point. Thus $\tilde{H}_k(T/Z) = 0$ for $k \le n+1$, and then the long exact sequence of the pair (T, Z) implies that $\tilde{H}_k(T) = 0$ for $k \le n$, and we have proved $(*)$.   □

Let X be a CW complex. Using Lemma 2.34, portions of the long exact sequences for the pairs (X^{n+1}, X^n), (X^n, X^{n-1}), and (X^{n-1}, X^{n-2}) fit into a diagram

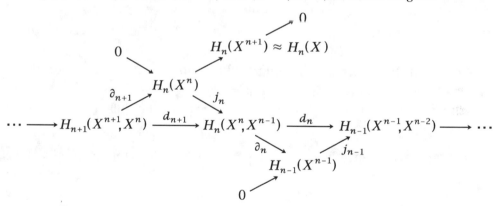

where d_{n+1} and d_n are defined as the compositions $j_n \partial_{n+1}$ and $j_{n-1} \partial_n$, which are just 'relativizations' of the boundary maps ∂_{n+1} and ∂_n. The composition $d_n d_{n+1}$ includes two successive maps in one of the exact sequences, hence is zero. Thus the horizontal row in the diagram is a chain complex, called the **cellular chain complex** of X since $H_n(X^n, X^{n-1})$ is free with basis in one-to-one correspondence with the n-cells of X, so one can think of elements of $H_n(X^n, X^{n-1})$ as linear combinations of n-cells of X. The homology groups of this cellular chain complex are called the **cellular homology groups** of X. Temporarily we denote them $H_n^{CW}(X)$.

‖ **Theorem 2.35.** $H_n^{CW}(X) \approx H_n(X)$.

Proof: From the diagram above, $H_n(X)$ can be identified with $H_n(X^n)/\operatorname{Im} \partial_{n+1}$. Since j_n is injective, it maps $\operatorname{Im} \partial_{n+1}$ isomorphically onto $\operatorname{Im}(j_n \partial_{n+1}) = \operatorname{Im} d_{n+1}$

and $H_n(X^n)$ isomorphically onto $\operatorname{Im} j_n = \operatorname{Ker} \partial_n$. Since j_{n-1} is injective, $\operatorname{Ker} \partial_n = \operatorname{Ker} d_n$. Thus j_n induces an isomorphism of the quotient $H_n(X^n)/\operatorname{Im} \partial_{n+1}$ onto $\operatorname{Ker} d_n/\operatorname{Im} d_{n+1}$. $\square$

Here are a few immediate applications:

(i) $H_n(X) = 0$ if X is a CW complex with no n-cells.

(ii) More generally, if X is a CW complex with k n-cells, then $H_n(X)$ is generated by at most k elements. For since $H_n(X^n, X^{n-1})$ is free abelian on k generators, the subgroup $\operatorname{Ker} d_n$ must be generated by at most k elements, hence also the quotient $\operatorname{Ker} d_n/\operatorname{Im} d_{n+1}$.

(iii) If X is a CW complex having no two of its cells in adjacent dimensions, then $H_n(X)$ is free abelian with basis in one-to-one correspondence with the n-cells of X. This is because the cellular boundary maps d_n are automatically zero in this case.

This last observation applies for example to $\mathbb{C}\mathrm{P}^n$, which has a CW structure with one cell of each even dimension $2k \leq 2n$ as we saw in Example 0.6. Thus

$$H_i(\mathbb{C}\mathrm{P}^n) \approx \begin{cases} \mathbb{Z} & \text{for } i = 0, 2, 4, \cdots, 2n \\ 0 & \text{otherwise} \end{cases}$$

Another simple example is $S^n \times S^n$ with $n > 1$, using the product CW structure consisting of a 0-cell, two n-cells, and a $2n$-cell.

It is possible to prove the statements (i)–(iii) for finite-dimensional CW complexes by induction on the dimension, without using cellular homology but only the basic results from the previous section. However, the viewpoint of cellular homology makes (i)–(iii) quite transparent.

Next we describe how the cellular boundary maps d_n can be computed. When $n = 1$ this is easy since the boundary map $d_1 : H_1(X^1, X^0) \to H_0(X^0)$ is the same as the simplicial boundary map $\Delta_1(X) \to \Delta_0(X)$. In case X is connected and has only one 0-cell, then d_1 must be 0, otherwise $H_0(X)$ would not be $\mathbb{Z}$. When $n > 1$ we will show that d_n can be computed in terms of degrees:

Cellular Boundary Formula. $d_n(e_\alpha^n) = \sum_\beta d_{\alpha\beta} e_\beta^{n-1}$ *where* $d_{\alpha\beta}$ *is the degree of the map* $S_\alpha^{n-1} \to X^{n-1} \to S_\beta^{n-1}$ *that is the composition of the attaching map of* e_α^n *with the quotient map collapsing* $X^{n-1} - e_\beta^{n-1}$ *to a point.*

Here we are identifying the cells e_α^n and e_β^{n-1} with generators of the corresponding summands of the cellular chain groups. The summation in the formula contains only finitely many terms since the attaching map of e_α^n has compact image, so this image meets only finitely many cells e_β^{n-1}.

To derive the cellular boundary formula, consider the commutative diagram

$$
\begin{array}{ccccc}
H_n(D_\alpha^n, \partial D_\alpha^n) & \xrightarrow[\approx]{\partial} & \widetilde{H}_{n-1}(\partial D_\alpha^n) & \xrightarrow{\Delta_{\alpha\beta*}} & \widetilde{H}_{n-1}(S_\beta^{n-1}) \\
\downarrow{\Phi_{\alpha*}} & & \downarrow{\varphi_{\alpha*}} & & \uparrow{q_{\beta*}} \\
H_n(X^n, X^{n-1}) & \xrightarrow{\partial_n} & \widetilde{H}_{n-1}(X^{n-1}) & \xrightarrow{q_*} & \widetilde{H}_{n-1}(X^{n-1}/X^{n-2})
\end{array}
$$

$$
\begin{array}{c}
\searrow{d_n} \\
 H_{n-1}(X^{n-1}, X^{n-2}) \xrightarrow{\approx} H_{n-1}(X^{n-1}/X^{n-2}, X^{n-2}/X^{n-2})
\end{array}
$$

where:

- Φ_α is the characteristic map of the cell e_α^n and φ_α is its attaching map.
- $q : X^{n-1} \to X^{n-1}/X^{n-2}$ is the quotient map.
- $q_\beta : X^{n-1}/X^{n-2} \to S_\beta^{n-1}$ collapses the complement of the cell e_β^{n-1} to a point, the resulting quotient sphere being identified with $S_\beta^{n-1} = D_\beta^{n-1}/\partial D_\beta^{n-1}$ via the characteristic map Φ_β.
- $\Delta_{\alpha\beta} : \partial D_\alpha^n \to S_\beta^{n-1}$ is the composition $q_\beta q \varphi_\alpha$, in other words, the attaching map of e_α^n followed by the quotient map $X^{n-1} \to S_\beta^{n-1}$ collapsing the complement of e_β^{n-1} in X^{n-1} to a point.

The map $\Phi_{\alpha*}$ takes a chosen generator $[D_\alpha^n] \in H_n(D_\alpha^n, \partial D_\alpha^n)$ to a generator of the $\mathbb{Z}$ summand of $H_n(X^n, X^{n-1})$ corresponding to e_α^n. Letting e_α^n denote this generator, commutativity of the left half of the diagram then gives $d_n(e_\alpha^n) = j_{n-1}\varphi_{\alpha*}\partial[D_\alpha^n]$. In terms of the basis for $H_{n-1}(X^{n-1}, X^{n-2})$ corresponding to the cells e_β^{n-1}, the map $q_{\beta*}$ is the projection of $\widetilde{H}_{n-1}(X^{n-1}/X^{n-2})$ onto its $\mathbb{Z}$ summand corresponding to e_β^{n-1}. Commutativity of the diagram then yields the formula for d_n given above.

Example 2.36. Let M_g be the closed orientable surface of genus g with its usual CW structure consisting of one 0-cell, $2g$ 1-cells, and one 2-cell attached by the product of commutators $[a_1, b_1] \cdots [a_g, b_g]$. The associated cellular chain complex is

$$
0 \longrightarrow \mathbb{Z} \xrightarrow{d_2} \mathbb{Z}^{2g} \xrightarrow{d_1} \mathbb{Z} \longrightarrow 0
$$

As observed above, d_1 must be 0 since there is only one 0-cell. Also, d_2 is 0 because each a_i or b_i appears with its inverse in $[a_1, b_1] \cdots [a_g, b_g]$, so the maps $\Delta_{\alpha\beta}$ are homotopic to constant maps. Since d_1 and d_2 are both zero, the homology groups of M_g are the same as the cellular chain groups, namely, $\mathbb{Z}$ in dimensions 0 and 2, and $\mathbb{Z}^{2g}$ in dimension 1.

Example 2.37. The closed nonorientable surface N_g of genus g has a cell structure with one 0-cell, g 1-cells, and one 2-cell attached by the word $a_1^2 a_2^2 \cdots a_g^2$. Again $d_1 = 0$, and $d_2 : \mathbb{Z} \to \mathbb{Z}^g$ is specified by the equation $d_2(1) = (2, \cdots, 2)$ since each a_i appears in the attaching word of the 2-cell with total exponent 2, which means that each $\Delta_{\alpha\beta}$ is homotopic to the map $z \mapsto z^2$, of degree 2. Since $d_2(1) = (2, \cdots, 2)$, we have d_2 injective and hence $H_2(N_g) = 0$. If we change the basis for $\mathbb{Z}^g$ by replacing the last standard basis element $(0, \cdots, 0, 1)$ by $(1, \cdots, 1)$, we see that $H_1(N_g) \approx \mathbb{Z}^{g-1} \oplus \mathbb{Z}_2$.

These two examples illustrate the general fact that the orientability of a closed connected manifold M of dimension n is detected by $H_n(M)$, which is $\mathbb{Z}$ if M is orientable and 0 otherwise. This is shown in Theorem 3.26.

Example 2.38: An Acyclic Space. Let X be obtained from $S^1 \vee S^1$ by attaching two 2-cells by the words $a^5 b^{-3}$ and $b^3(ab)^{-2}$. Then $d_2 : \mathbb{Z}^2 \to \mathbb{Z}^2$ has matrix $\left(\begin{smallmatrix} 5 & -2 \\ -3 & 1 \end{smallmatrix} \right)$, with the two columns coming from abelianizing $a^5 b^{-3}$ and $b^3(ab)^{-2}$ to $5a - 3b$ and $-2a + b$, in additive notation. The matrix has determinant -1, so d_2 is an isomorphism and $\tilde{H}_i(X) = 0$ for all i. Such a space X is called **acyclic**.

We can see that this acyclic space is not contractible by considering $\pi_1(X)$, which has the presentation $\langle a, b \mid a^5 b^{-3}, b^3(ab)^{-2} \rangle$. There is a nontrivial homomorphism from this group to the group G of rotational symmetries of a regular dodecahedron, sending a to the rotation ρ_a through angle $2\pi/5$ about the axis through the center of a pentagonal face, and b to the rotation ρ_b through angle $2\pi/3$ about the axis through a vertex of this face. The composition $\rho_a \rho_b$ is a rotation through angle π about the axis through the midpoint of an edge abutting this vertex. Thus the relations $a^5 = b^3 = (ab)^2$ defining $\pi_1(X)$ become $\rho_a^5 = \rho_b^3 = (\rho_a \rho_b)^2 = 1$ in G, which means there is a well-defined homomorphism $\rho : \pi_1(X) \to G$ sending a to ρ_a and b to ρ_b. It is not hard to see that G is generated by ρ_a and ρ_b, so ρ is surjective. With more work one can compute that the kernel of ρ is $\mathbb{Z}_2$, generated by the element $a^5 = b^3 = (ab)^2$, and this $\mathbb{Z}_2$ is in fact the center of $\pi_1(X)$. In particular, $\pi_1(X)$ has order 120 since G has order 60.

After these 2-dimensional examples, let us now move up to three dimensions, where we have the additional task of computing the cellular boundary map d_3.

Example 2.39. A 3-dimensional torus $T^3 = S^1 \times S^1 \times S^1$ can be constructed from a cube by identifying each pair of opposite square faces as in the first of the two figures. The second figure shows a slightly different pattern of 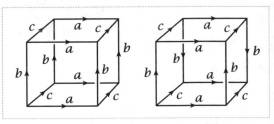 identifications of opposite faces, with the front and back faces now identified via a rotation of the cube around a horizontal left-right axis. The space produced by these identifications is the product $K \times S^1$ of a Klein bottle and a circle. For both T^3 and $K \times S^1$ we have a CW structure with one 3-cell, three 2-cells, three 1-cells, and one 0-cell. The cellular chain complexes thus have the form

$$0 \longrightarrow \mathbb{Z} \xrightarrow{\ d_3\ } \mathbb{Z}^3 \xrightarrow{\ d_2\ } \mathbb{Z}^3 \xrightarrow{\ 0\ } \mathbb{Z} \longrightarrow 0$$

In the case of the 3-torus T^3 the cellular boundary map d_2 is zero by the same calculation as for the 2-dimensional torus. We claim that d_3 is zero as well. This amounts to saying that the three maps $\Delta_{\alpha\beta} : S^2 \to S^2$ corresponding to the three 2-cells have degree zero. Each $\Delta_{\alpha\beta}$ maps the interiors of two opposite faces of the cube homeomorphically onto the complement of a point in the target S^2 and sends the remaining four faces to this point. Computing local degrees at the center points of the two opposite faces, we see that the local degree is $+1$ at one of these points and -1 at the other, since the restrictions of $\Delta_{\alpha\beta}$ to these two faces differ by a reflection of the boundary of the cube across the plane midway between them, and a reflection has degree -1. Since the cellular boundary maps are all zero, we deduce that $H_i(T^3)$ is $\mathbb{Z}$ for $i = 0, 3$, $\mathbb{Z}^3$ for $i = 1, 2$, and 0 for $i > 3$.

For $K \times S^1$, when we compute local degrees for the front and back faces we find that the degrees now have the same rather than opposite signs since the map $\Delta_{\alpha\beta}$ on these two faces differs not by a reflection but by a rotation of the boundary of the cube. The local degrees for the other faces are the same as before. Using the letters A, B, C to denote the 2-cells given by the faces orthogonal to the edges a, b, c, respectively, we have the boundary formulas $d_3 e^3 = 2C$, $d_2 A = 2b$, $d_2 B = 0$, and $d_2 C = 0$. It follows that $H_3(K \times S^1) = 0$, $H_2(K \times S^1) = \mathbb{Z} \oplus \mathbb{Z}_2$, and $H_1(K \times S^1) = \mathbb{Z} \oplus \mathbb{Z} \oplus \mathbb{Z}_2$.

Many more examples of a similar nature, quotients of a cube or other polyhedron with faces identified in some pattern, could be worked out in similar fashion. But let us instead turn to some higher-dimensional examples.

Example 2.40: Moore Spaces. Given an abelian group G and an integer $n \geq 1$, we will construct a CW complex X such that $H_n(X) \approx G$ and $\tilde{H}_i(X) = 0$ for $i \neq n$. Such a space is called a **Moore space**, commonly written $M(G, n)$ to indicate the dependence on G and n. It is probably best for the definition of a Moore space to include the condition that $M(G, n)$ be simply-connected if $n > 1$. The spaces we construct will have this property.

As an easy special case, when $G = \mathbb{Z}_m$ we can take X to be S^n with a cell e^{n+1} attached by a map $S^n \to S^n$ of degree m. More generally, any finitely generated G can be realized by taking wedge sums of examples of this type for finite cyclic summands of G, together with copies of S^n for infinite cyclic summands of G.

In the general nonfinitely generated case let $F \to G$ be a homomorphism of a free abelian group F onto G, sending a basis for F onto some set of generators of G. The kernel K of this homomorphism is a subgroup of a free abelian group, hence is itself free abelian. Choose bases $\{x_\alpha\}$ for F and $\{y_\beta\}$ for K, and write $y_\beta = \sum_\alpha d_{\beta\alpha} x_\alpha$. Let $X^n = \bigvee_\alpha S_\alpha^n$, so $H_n(X^n) \approx F$ via Corollary 2.25. We will construct X from X^n by attaching cells e_β^{n+1} via maps $f_\beta : S^n \to X^n$ such that the composition of f_β with the projection onto the summand S_α^n has degree $d_{\beta\alpha}$. Then the cellular boundary map d_{n+1} will be the inclusion $K \hookrightarrow F$, hence X will have the desired homology groups.

The construction of f_β generalizes the construction in Example 2.31 of a map $S^n \to S^n$ of given degree. Namely, we can let f_β map the complement of $\sum_\alpha |d_{\beta\alpha}|$

disjoint balls in S^n to the 0-cell of X^n while sending $|d_{\beta\alpha}|$ of the balls onto the summand S^n_α by maps of degree $+1$ if $d_{\beta\alpha} > 0$, or degree -1 if $d_{\beta\alpha} < 0$.

Example 2.41. By taking a wedge sum of the Moore spaces constructed in the preceding example for varying n we obtain a connected CW complex with any prescribed sequence of homology groups in dimensions $1, 2, 3, \cdots$.

Example 2.42: Real Projective Space $\mathbb{R}\mathrm{P}^n$. As we saw in Example 0.4, $\mathbb{R}\mathrm{P}^n$ has a CW structure with one cell e^k in each dimension $k \le n$, and the attaching map for e^k is the 2-sheeted covering projection $\varphi : S^{k-1} \to \mathbb{R}\mathrm{P}^{k-1}$. To compute the boundary map d_k we compute the degree of the composition $S^{k-1} \xrightarrow{\varphi} \mathbb{R}\mathrm{P}^{k-1} \xrightarrow{q} \mathbb{R}\mathrm{P}^{k-1}/\mathbb{R}\mathrm{P}^{k-2} = S^{k-1}$, with q the quotient map. The map $q\varphi$ is a homeomorphism when restricted to each component of $S^{k-1} - S^{k-2}$, and these two homeomorphisms are obtained from each other by precomposing with the antipodal map of S^{k-1}, which has degree $(-1)^k$. Hence $\deg q\varphi = \deg \mathbb{1} + \deg(-\mathbb{1}) = 1 + (-1)^k$, and so d_k is either 0 or multiplication by 2 according to whether k is odd or even. Thus the cellular chain complex for $\mathbb{R}\mathrm{P}^n$ is

$$0 \longrightarrow \mathbb{Z} \xrightarrow{2} \mathbb{Z} \xrightarrow{0} \cdots \xrightarrow{2} \mathbb{Z} \xrightarrow{0} \mathbb{Z} \xrightarrow{2} \mathbb{Z} \xrightarrow{0} \mathbb{Z} \longrightarrow 0 \qquad \text{if } n \text{ is even}$$

$$0 \longrightarrow \mathbb{Z} \xrightarrow{0} \mathbb{Z} \xrightarrow{2} \cdots \xrightarrow{2} \mathbb{Z} \xrightarrow{0} \mathbb{Z} \xrightarrow{2} \mathbb{Z} \xrightarrow{0} \mathbb{Z} \longrightarrow 0 \qquad \text{if } n \text{ is odd}$$

From this it follows that

$$H_k(\mathbb{R}\mathrm{P}^n) = \begin{cases} \mathbb{Z} & \text{for } k = 0 \text{ and for } k = n \text{ odd} \\ \mathbb{Z}_2 & \text{for } k \text{ odd}, 0 < k < n \\ 0 & \text{otherwise} \end{cases}$$

Example 2.43: Lens Spaces. This example is somewhat more complicated. Given an integer $m > 1$ and integers $\ell_1, \cdots, \ell_n$ relatively prime to m, define the **lens space** $L = L_m(\ell_1, \cdots, \ell_n)$ to be the orbit space $S^{2n-1}/\mathbb{Z}_m$ of the unit sphere $S^{2n-1} \subset \mathbb{C}^n$ with the action of $\mathbb{Z}_m$ generated by the rotation $\rho(z_1, \cdots, z_n) = (e^{2\pi i \ell_1/m} z_1, \cdots, e^{2\pi i \ell_n/m} z_n)$, rotating the jth $\mathbb{C}$ factor of $\mathbb{C}^n$ by the angle $2\pi \ell_j/m$. In particular, when $m = 2$, ρ is the antipodal map, so $L = \mathbb{R}\mathrm{P}^{2n-1}$ in this case. In the general case, the projection $S^{2n-1} \to L$ is a covering space since the action of $\mathbb{Z}_m$ on S^{2n-1} is free: Only the identity element fixes any point of S^{2n-1} since each point of S^{2n-1} has some coordinate z_j nonzero and then $e^{2\pi i k \ell_j/m} z_j \ne z_j$ for $0 < k < m$, as a result of the assumption that ℓ_j is relatively prime to m.

We shall construct a CW structure on L with one cell e^k for each $k \le 2n - 1$ and show that the resulting cellular chain complex is

$$0 \longrightarrow \mathbb{Z} \xrightarrow{0} \mathbb{Z} \xrightarrow{m} \mathbb{Z} \xrightarrow{0} \cdots \xrightarrow{0} \mathbb{Z} \xrightarrow{m} \mathbb{Z} \xrightarrow{0} \mathbb{Z} \longrightarrow 0$$

with boundary maps alternately 0 and multiplication by m. Hence

$$H_k(L_m(\ell_1, \cdots, \ell_n)) = \begin{cases} \mathbb{Z} & \text{for } k = 0, 2n - 1 \\ \mathbb{Z}_m & \text{for } k \text{ odd}, 0 < k < 2n - 1 \\ 0 & \text{otherwise} \end{cases}$$

To obtain the CW structure, first subdivide the unit circle C in the nth $\mathbb{C}$ factor of $\mathbb{C}^n$ by taking the points $e^{2\pi i j/m} \in C$ as vertices, $j = 1, \cdots, m$. Joining the jth vertex of C to the unit sphere $S^{2n-3} \subset \mathbb{C}^{n-1}$ by arcs of great circles in S^{2n-1} yields a $(2n-2)$-dimensional ball B_j^{2n-2} bounded by S^{2n-3}. Specifically, B_j^{2n-2} consists of the points $\cos \theta \, (0, \cdots, 0, e^{2\pi i j/m}) + \sin \theta \, (z_1, \cdots, z_{n-1}, 0)$ for $0 \le \theta \le \pi/2$. Similarly, joining the jth edge of C to S^{2n-3} gives a ball B_j^{2n-1} bounded by B_j^{2n-2} and B_{j+1}^{2n-2}, subscripts being taken mod m. The rotation ρ carries S^{2n-3} to itself and rotates C by the angle $2\pi\ell_n/m$, hence ρ permutes the B_j^{2n-2}'s and the B_j^{2n-1}'s. A suitable power of ρ, namely ρ^r where $r\ell_n \equiv 1 \mod m$, takes each B_j^{2n-2} and B_j^{2n-1} to the next one. Since ρ^r has order m, it is also a generator of the rotation group $\mathbb{Z}_m$, and hence we may obtain L as the quotient of one B_j^{2n-1} by identifying its two faces B_j^{2n-2} and B_{j+1}^{2n-2} together via ρ^r.

In particular, when $n = 2$, B_j^{2n-1} is a lens-shaped 3-ball and L is obtained from this ball by identifying its two curved disk faces via ρ^r, which may be described as the composition of the reflection across the plane containing the rim of the lens, taking one face of the lens to the other, followed by a rotation of this face through the angle $2\pi\ell/m$ where $\ell - r\ell_1$. The figure illustrates the 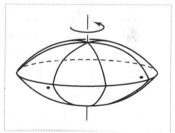 case $(m, \ell) = (7, 2)$, with the two dots indicating a typical pair of identified points in the upper and lower faces of the lens. Since the lens space L is determined by the rotation angle $2\pi\ell/m$, it is conveniently written $L_{\ell/m}$. Clearly only the mod m value of ℓ matters. It is a classical theorem of Reidemeister from the 1930s that $L_{\ell/m}$ is homeomorphic to $L_{\ell'/m'}$ iff $m' = m$ and $\ell' \equiv \pm\ell^{\pm 1} \mod m$. For example, when $m = 7$ there are only two distinct lens spaces $L_{1/7}$ and $L_{2/7}$. The 'if' part of this theorem is easy: Reflecting the lens through a mirror shows that $L_{\ell/m} \approx L_{-\ell/m}$, and by interchanging the roles of the two $\mathbb{C}$ factors of $\mathbb{C}^2$ one obtains $L_{\ell/m} \approx L_{\ell^{-1}/m}$. In the converse direction, $L_{\ell/m} \approx L_{\ell'/m'}$ clearly implies $m = m'$ since $\pi_1(L_{\ell/m}) \approx \mathbb{Z}_m$. The rest of the theorem takes considerably more work, involving either special 3-dimensional techniques or more algebraic methods that generalize to classify the higher-dimensional lens spaces as well. The latter approach is explained in [Cohen 1973].

Returning to the construction of a CW structure on $L_m(\ell_1, \cdots, \ell_n)$, observe that the $(2n-3)$-dimensional lens space $L_m(\ell_1, \cdots, \ell_{n-1})$ sits in $L_m(\ell_1, \cdots, \ell_n)$ as the quotient of S^{2n-3}, and $L_m(\ell_1, \cdots, \ell_n)$ is obtained from this subspace by attaching two cells, of dimensions $2n-2$ and $2n-1$, coming from the interiors of B_j^{2n-1} and its two identified faces B_j^{2n-2} and B_{j+1}^{2n-2}. Inductively this gives a CW structure on $L_m(\ell_1, \cdots, \ell_n)$ with one cell e^k in each dimension $k \le 2n-1$.

The boundary maps in the associated cellular chain complex are computed as follows. The first one, d_{2n-1}, is zero since the identification of the two faces of B_j^{2n-1} is via a reflection (degree -1) across B_j^{2n-1} fixing S^{2n-3}, followed by a rotation (degree $+1$), so $d_{2n-1}(e^{2n-1}) = e^{2n-2} - e^{2n-2} = 0$. The next boundary map d_{2n-2} takes e^{2n-2} to me^{2n-3} since the attaching map for e^{2n-2} is the quotient map $S^{2n-3} \to L_m(\ell_1, \cdots, \ell_{n-1})$ and the balls B_j^{2n-3} in S^{2n-3} which project down onto e^{2n-3} are permuted cyclically by the rotation ρ of degree $+1$. Inductively, the subsequent boundary maps d_k then alternate between 0 and multiplication by m.

Also of interest are the infinite-dimensional lens spaces $L_m(\ell_1, \ell_2, \cdots) = S^\infty/\mathbb{Z}_m$ defined in the same way as in the finite-dimensional case, starting from a sequence of integers $\ell_1, \ell_2, \cdots$ relatively prime to m. The space $L_m(\ell_1, \ell_2, \cdots)$ is the union of the increasing sequence of finite-dimensional lens spaces $L_m(\ell_1, \cdots, \ell_n)$ for $n = 1, 2, \cdots$, each of which is a subcomplex of the next in the cell structure we have just constructed, so $L_m(\ell_1, \ell_2, \cdots)$ is also a CW complex. Its cellular chain complex consists of a $\mathbb{Z}$ in each dimension with boundary maps alternately 0 and m, so its reduced homology consists of a $\mathbb{Z}_m$ in each odd dimension.

In the terminology of §1.B, the infinite-dimensional lens space $L_m(\ell_1, \ell_2, \cdots)$ is an Eilenberg–MacLane space $K(\mathbb{Z}_m, 1)$ since its universal cover S^∞ is contractible, as we showed there. By Theorem 1B.8 the homotopy type of $L_m(\ell_1, \ell_2, \cdots)$ depends only on m, and not on the ℓ_i's. This is not true in the finite-dimensional case, when two lens spaces $L_m(\ell_1, \cdots, \ell_n)$ and $L_m(\ell_1', \cdots, \ell_n')$ have the same homotopy type iff $\ell_1 \cdots \ell_n \equiv \pm k^n \ell_1' \cdots \ell_n'$ mod m for some integer k. A proof of this is outlined in Exercise 2 in §3.E and Exercise 29 in §4.2. For example, the 3-dimensional lens spaces $L_{1/5}$ and $L_{2/5}$ are not homotopy equivalent, though they have the same fundamental group and the same homology groups. On the other hand, $L_{1/7}$ and $L_{2/7}$ are homotopy equivalent but not homeomorphic.

Euler Characteristic

For a finite CW complex X, the **Euler characteristic** $\chi(X)$ is defined to be the alternating sum $\sum_n (-1)^n c_n$ where c_n is the number of n-cells of X, generalizing the familiar formula *vertices − edges + faces* for 2-dimensional complexes. The following result shows that $\chi(X)$ can be defined purely in terms of homology, and hence depends only on the homotopy type of X. In particular, $\chi(X)$ is independent of the choice of CW structure on X.

‖ **Theorem 2.44.** $\chi(X) = \sum_n (-1)^n \operatorname{rank} H_n(X)$.

Here the **rank** of a finitely generated abelian group is the number of $\mathbb{Z}$ summands when the group is expressed as a direct sum of cyclic groups. We shall need the following fact, whose proof we leave as an exercise: If $0 \to A \to B \to C \to 0$ is a short exact sequence of finitely generated abelian groups, then $\operatorname{rank} B = \operatorname{rank} A + \operatorname{rank} C$.

Proof of 2.44: This is purely algebraic. Let

$$0 \longrightarrow C_k \xrightarrow{d_k} C_{k-1} \longrightarrow \cdots \longrightarrow C_1 \xrightarrow{d_1} C_0 \longrightarrow 0$$

be a chain complex of finitely generated abelian groups, with cycles $Z_n = \operatorname{Ker} d_n$, boundaries $B_n = \operatorname{Im} d_{n+1}$, and homology $H_n = Z_n/B_n$. Thus we have short exact sequences $0 \to Z_n \to C_n \to B_{n-1} \to 0$ and $0 \to B_n \to Z_n \to H_n \to 0$, hence

$$\operatorname{rank} C_n = \operatorname{rank} Z_n + \operatorname{rank} B_{n-1}$$

$$\operatorname{rank} Z_n = \operatorname{rank} B_n + \operatorname{rank} H_n$$

Now substitute the second equation into the first, multiply the resulting equation by $(-1)^n$, and sum over n to get $\sum_n (-1)^n \operatorname{rank} C_n = \sum_n (-1)^n \operatorname{rank} H_n$. Applying this with $C_n = H_n(X^n, X^{n-1})$ then gives the theorem. $\qquad \square$

For example, the surfaces M_g and N_g have Euler characteristics $\chi(M_g) = 2 - 2g$ and $\chi(N_g) = 2 - g$. Thus all the orientable surfaces M_g are distinguished from each other by their Euler characteristics, as are the nonorientable surfaces N_g, and there are only the relations $\chi(M_g) = \chi(N_{2g})$.

Split Exact Sequences

Suppose one has a retraction $r : X \to A$, so $ri = 1\!\!1$ where $i : A \to X$ is the inclusion. The induced map $i_* : H_n(A) \to H_n(X)$ is then injective since $r_* i_* = 1\!\!1$. From this it follows that the boundary maps in the long exact sequence for (X, A) are zero, so the long exact sequence breaks up into short exact sequences

$$0 \longrightarrow H_n(A) \xrightarrow{\; i_* \;} H_n(X) \xrightarrow{\; j_* \;} H_n(X, A) \longrightarrow 0$$

The relation $r_* i_* = 1\!\!1$ actually gives more information than this, by the following piece of elementary algebra:

> **Splitting Lemma.** *For a short exact sequence* $0 \longrightarrow A \xrightarrow{\; i \;} B \xrightarrow{\; j \;} C \longrightarrow 0$ *of abelian groups the following statements are equivalent:*
>
> (a) *There is a homomorphism* $p : B \to A$ *such that* $pi = 1\!\!1 : A \to A$.
> (b) *There is a homomorphism* $s : C \to B$ *such that* $js = 1\!\!1 : C \to C$.
> (c) *There is an isomorphism* $B \approx A \oplus C$ *making a commutative diagram as at the right, where the maps in the lower row are the obvious ones,* $a \mapsto (a, 0)$ *and* $(a, c) \mapsto c$.

$$0 \longrightarrow A \underset{A \oplus C}{\overset{B}{\langle}} C \longrightarrow 0$$

If these conditions are satisfied, the exact sequence is said to **split**. Note that (c) is symmetric: There is no essential difference between the roles of A and C.

Sketch of Proof: For the implication (a) $\Rightarrow$ (c) one checks that the map $B \to A \oplus C$, $b \mapsto (p(b), j(b))$, is an isomorphism with the desired properties. For (b) $\Rightarrow$ (c) one uses instead the map $A \oplus C \to B$, $(a, c) \mapsto i(a) + s(c)$. The opposite implications (c) $\Rightarrow$ (a) and (c) $\Rightarrow$ (b) are fairly obvious. If one wants to show (b) $\Rightarrow$ (a) directly, one can define $p(b) = i^{-1}(b - sj(b))$. Further details are left to the reader. $\qquad \square$

Except for the implications (b) $\Rightarrow$ (a) and (b) $\Rightarrow$ (c), the proof works equally well for nonabelian groups. In the nonabelian case, (b) is definitely weaker than (a) and (c), and short exact sequences satisfying (b) only determine B as a semidirect product of A and C. The difficulty is that $s(C)$ might not be a normal subgroup of B. In the nonabelian case one defines 'splitting' to mean that (b) is satisfied.

In both the abelian and nonabelian contexts, if C is free then every exact sequence $0 \to A \xrightarrow{i} B \xrightarrow{j} C \to 0$ splits, since one can define $s : C \to B$ by choosing a basis $\{c_\alpha\}$ for C and letting $s(c_\alpha)$ be any element $b_\alpha \in B$ such that $j(b_\alpha) = c_\alpha$. The converse is also true: If every short exact sequence ending in C splits, then C is free. This is because for every C there is a short exact sequence $0 \to A \to B \to C \to 0$ with B free — choose generators for C and let B have a basis in one-to-one correspondence with these generators, then let $B \to C$ send each basis element to the corresponding generator — so if this sequence $0 \to A \to B \to C \to 0$ splits, C is isomorphic to a subgroup of a free group, hence is free.

From the Splitting Lemma and the remarks preceding it we deduce that a retraction $r : X \to A$ gives a splitting $H_n(X) \approx H_n(A) \oplus H_n(X, A)$. This can be used to show the nonexistence of such a retraction in some cases, for example in the situation of the Brouwer fixed point theorem, where a retraction $D^n \to S^{n-1}$ would give an impossible splitting $H_{n-1}(D^n) \approx H_{n-1}(S^{n-1}) \oplus H_{n-1}(D^n, S^{n-1})$. For a somewhat more subtle example, consider the mapping cylinder M_f of a degree m map $f : S^n \to S^n$ with $m > 1$. If M_f retracted onto the $S^n \subset M_f$ corresponding to the domain of f, we would have a split short exact sequence

$$
\begin{array}{ccccccccc}
0 & \longrightarrow & H_n(S^n) & \longrightarrow & H_n(M_f) & \longrightarrow & H_n(M_f, S^n) & \longrightarrow & 0 \\
 & & \| & & \| & & \| & & \\
0 & \longrightarrow & \mathbb{Z} & \xrightarrow{m} & \mathbb{Z} & \longrightarrow & \mathbb{Z}_m & \longrightarrow & 0
\end{array}
$$

But this sequence does not split since $\mathbb{Z}$ is not isomorphic to $\mathbb{Z} \oplus \mathbb{Z}_m$ if $m > 1$, so the retraction cannot exist. In the simplest case of the degree 2 map $S^1 \to S^1$, $z \mapsto z^2$, this says that the Möbius band does not retract onto its boundary circle.

Homology of Groups

In §1.B we constructed for each group G a CW complex $K(G, 1)$ having a contractible universal cover, and we showed that the homotopy type of such a space $K(G, 1)$ is uniquely determined by G. The homology groups $H_n(K(G, 1))$ therefore depend only on G, and are usually denoted simply $H_n(G)$. The calculations for lens spaces in Example 2.43 show that $H_n(\mathbb{Z}_m)$ is $\mathbb{Z}_m$ for odd n and 0 for even $n > 0$. Since S^1 is a $K(\mathbb{Z}, 1)$ and the torus is a $K(\mathbb{Z} \times \mathbb{Z}, 1)$, we also know the homology of these two groups. More generally, the homology of finitely generated abelian groups can be computed from these examples using the Künneth formula in §3.B and the fact that a product $K(G, 1) \times K(H, 1)$ is a $K(G \times H, 1)$.

Here is an application of the calculation of $H_n(\mathbb{Z}_m)$:

Proposition 2.45. *If a finite-dimensional CW complex X is a $K(G, 1)$, then the group $G = \pi_1(X)$ must be torsionfree.*

This applies to quite a few manifolds, for example closed surfaces other than S^2 and $\mathbb{RP}^2$, and also many 3-dimensional manifolds such as complements of knots in S^3.

Proof: If G had torsion, it would have a finite cyclic subgroup $\mathbb{Z}_m$ for some $m > 1$, and the covering space of X corresponding to this subgroup of $G = \pi_1(X)$ would be a $K(\mathbb{Z}_m, 1)$. Since X is a finite-dimensional CW complex, the same would be true of its covering space $K(\mathbb{Z}_m, 1)$, and hence the homology of the $K(\mathbb{Z}_m, 1)$ would be nonzero in only finitely many dimensions. But this contradicts the fact that $H_n(\mathbb{Z}_m)$ is nonzero for infinitely many values of n. □

Reflecting the richness of group theory, the homology of groups has been studied quite extensively. A good starting place for those wishing to learn more is the textbook [Brown 1982]. At a more advanced level the books [Adem & Milgram 1994] and [Benson 1992] treat the subject from a mostly topological viewpoint.

Mayer–Vietoris Sequences

In addition to the long exact sequence of homology groups for a pair (X, A), there is another sort of long exact sequence, known as a **Mayer–Vietoris sequence**, which is equally powerful but is sometimes more convenient to use. For a pair of subspaces A, $B \subset X$ such that X is the union of the interiors of A and B, this exact sequence has the form

$$\cdots \longrightarrow H_n(A \cap B) \overset{\Phi}{\longrightarrow} H_n(A) \oplus H_n(B) \overset{\Psi}{\longrightarrow} H_n(X) \overset{\partial}{\longrightarrow} H_{n-1}(A \cap B) \longrightarrow \cdots$$

$$\cdots \longrightarrow H_0(X) \longrightarrow 0$$

In addition to its usefulness for calculations, the Mayer–Vietoris sequence is also applied frequently in induction arguments, where one might know that a certain statement is true for A, B, and $A \cap B$ by induction and then deduce that it is true for $A \cup B$ by the exact sequence.

The Mayer–Vietoris sequence is easy to derive from the machinery of §2.1. Let $C_n(A + B)$ be the subgroup of $C_n(X)$ consisting of chains that are sums of chains in A and chains in B. The usual boundary map $\partial : C_n(X) \to C_{n-1}(X)$ takes $C_n(A + B)$ to $C_{n-1}(A+B)$, so the $C_n(A+B)$'s form a chain complex. According to Proposition 2.21, the inclusions $C_n(A + B) \hookrightarrow C_n(X)$ induce isomorphisms on homology groups. The Mayer–Vietoris sequence is then the long exact sequence of homology groups associated to the short exact sequence of chain complexes formed by the short exact sequences

$$0 \longrightarrow C_n(A \cap B) \overset{\varphi}{\longrightarrow} C_n(A) \oplus C_n(B) \overset{\psi}{\longrightarrow} C_n(A + B) \longrightarrow 0$$

where $\varphi(x) = (x, -x)$ and $\psi(x, y) = x + y$. The exactness of this short exact sequence can be checked as follows. First, $\text{Ker}\,\varphi = 0$ since a chain in $A \cap B$ that is zero as a chain in A (or in B) must be the zero chain. Next, $\text{Im}\,\varphi \subset \text{Ker}\,\psi$ since $\psi\varphi = 0$. Also, $\text{Ker}\,\psi \subset \text{Im}\,\varphi$ since for a pair $(x, y) \in C_n(A) \oplus C_n(B)$ the condition $x + y = 0$ implies $x = -y$, so x is a chain in both A and B, that is, $x \in C_n(A \cap B)$, and $(x, y) = (x, -x) \in \text{Im}\,\varphi$. Finally, exactness at $C_n(A + B)$ is immediate from the definition of $C_n(A + B)$.

The boundary map $\partial : H_n(X) \to H_{n-1}(A \cap B)$ can easily be made explicit. A class $\alpha \in H_n(X)$ is represented by a cycle z, and by barycentric subdivision or some other method we can choose z to be a sum $x + y$ of chains in A and B, respectively. It need not be true that x and y are cycles individually, but $\partial x = -\partial y$ since $\partial(x + y) = 0$, and the element $\partial\alpha \in H_{n-1}(A \cap B)$ is represented by the cycle $\partial x = -\partial y$, as is clear from the definition of the boundary map in the long exact sequence of homology groups associated to a short exact sequence of chain complexes.

There is also a formally identical Mayer–Vietoris sequence for reduced homology groups, obtained by augmenting the previous short exact sequence of chain complexes in the obvious way:

$$
\begin{array}{ccccccccc}
0 & \longrightarrow & C_0(A \cap B) & \overset{\varphi}{\longrightarrow} & C_0(A) \oplus C_0(B) & \overset{\psi}{\longrightarrow} & C_0(A + B) & \longrightarrow & 0 \\
& & \downarrow{\scriptstyle \varepsilon} & & \downarrow{\scriptstyle \varepsilon \oplus \varepsilon} & & \downarrow{\scriptstyle \varepsilon} & & \\
0 & \longrightarrow & \mathbb{Z} & \overset{\varphi}{\longrightarrow} & \mathbb{Z} \oplus \mathbb{Z} & \overset{\psi}{\longrightarrow} & \mathbb{Z} & \longrightarrow & 0
\end{array}
$$

Mayer–Vietoris sequences can be viewed as analogs of the van Kampen theorem since if $A \cap B$ is path-connected, the H_1 terms of the reduced Mayer–Vietoris sequence yield an isomorphism $H_1(X) \approx (H_1(A) \oplus H_1(B))/\text{Im}\,\Phi$. This is exactly the abelianized statement of the van Kampen theorem, and H_1 is the abelianization of π_1 for path-connected spaces, as we show in §2.A.

There are also Mayer–Vietoris sequences for decompositions $X = A \cup B$ such that A and B are deformation retracts of neighborhoods U and V with $U \cap V$ deformation retracting onto $A \cap B$. Under these assumptions the five-lemma implies that the maps $C_n(A + B) \to C_n(U + V)$ induce isomorphisms on homology, and hence so do the maps $C_n(A + B) \to C_n(X)$, which was all that we needed to obtain a Mayer–Vietoris sequence. For example, if X is a CW complex and A and B are subcomplexes, then we can choose for U and V neighborhoods of the form $N_\varepsilon(A)$ and $N_\varepsilon(B)$ constructed in the Appendix, which have the property that $N_\varepsilon(A) \cap N_\varepsilon(B) = N_\varepsilon(A \cap B)$.

Example 2.46. Take $X = S^n$ with A and B the northern and southern hemispheres, so that $A \cap B = S^{n-1}$. Then in the reduced Mayer–Vietoris sequence the terms $\tilde{H}_i(A) \oplus \tilde{H}_i(B)$ are zero, so we obtain isomorphisms $\tilde{H}_i(S^n) \approx \tilde{H}_{i-1}(S^{n-1})$. This gives another way of calculating the homology groups of S^n by induction.

Example 2.47. We can decompose the Klein bottle K as the union of two Möbius bands A and B glued together by a homeomorphism between their boundary circles.

Then A, B, and $A \cap B$ are homotopy equivalent to circles, so the interesting part of the reduced Mayer-Vietoris sequence for the decomposition $K = A \cup B$ is the segment

$$0 \longrightarrow H_2(K) \longrightarrow H_1(A \cap B) \xrightarrow{\Phi} H_1(A) \oplus H_1(B) \longrightarrow H_1(K) \longrightarrow 0$$

The map Φ is $\mathbb{Z} \to \mathbb{Z} \oplus \mathbb{Z}$, $1 \mapsto (2, -2)$, since the boundary circle of a Möbius band wraps twice around the core circle. Since Φ is injective we obtain $H_2(K) = 0$. Furthermore, we have $H_1(K) \approx \mathbb{Z} \oplus \mathbb{Z}_2$ since we can choose $(1, 0)$ and $(1, -1)$ as a basis for $\mathbb{Z} \oplus \mathbb{Z}$. All the higher homology groups of K are zero from the earlier part of the Mayer-Vietoris sequence.

Example 2.48. Let us describe an exact sequence which is somewhat similar to the Mayer-Vietoris sequence and which in some cases generalizes it. If we are given two maps $f, g : X \to Y$ then we can form a quotient space Z of the disjoint union of $X \times I$ and Y via the identifications $(x, 0) \sim f(x)$ and $(x, 1) \sim g(x)$, thus attaching one end of $X \times I$ to Y by f and the other end by g. For example, if f and g are each the identity map $X \to X$ then $Z = X \times S^1$. If only one of f and g, say f, is the identity map, then Z is homeomorphic to what is called the mapping torus of g, the quotient space of $X \times I$ under the identifications $(x, 0) \sim (g(x), 1)$. The Klein bottle is an example, with g a reflection $S^1 \to S^1$.

The exact sequence we want has the form

$$(*) \quad \cdots \longrightarrow H_n(X) \xrightarrow{f_* - g_*} H_n(Y) \xrightarrow{i_*} H_n(Z) \longrightarrow H_{n-1}(X) \xrightarrow{f_* - g_*} H_{n-1}(Y) \longrightarrow \cdots$$

where i is the evident inclusion $Y \hookrightarrow Z$. To derive this exact sequence, consider the map $q : (X \times I, X \times \partial I) \to (Z, Y)$ that is the restriction to $X \times I$ of the quotient map $X \times I \amalg Y \to Z$. The map q induces a map of long exact sequences:

$$\cdots \xrightarrow{0} H_{n+1}(X \times I, X \times \partial I) \xrightarrow{\partial} H_n(X \times \partial I) \xrightarrow{i_*} H_n(X \times I) \xrightarrow{0} \cdots$$
$$\Big\downarrow q_* \qquad\qquad \Big\downarrow q_* \qquad\qquad \Big\downarrow q_*$$
$$\cdots \longrightarrow H_{n+1}(Z, Y) \xrightarrow{\partial} H_n(Y) \xrightarrow{i_*} H_n(Z) \longrightarrow \cdots$$

In the upper row the middle term is the direct sum of two copies of $H_n(X)$, and the map i_* is surjective since $X \times I$ deformation retracts onto $X \times \{0\}$ and $X \times \{1\}$. Surjectivity of the maps i_* in the upper row implies that the next maps are 0, which in turn implies that the maps ∂ are injective. Thus the map ∂ in the upper row gives an isomorphism of $H_{n+1}(X \times I, X \times \partial I)$ onto the kernel of i_*, which consists of the pairs $(\alpha, -\alpha)$ for $\alpha \in H_n(X)$. This kernel is a copy of $H_n(X)$, and the middle vertical map q_* takes $(\alpha, -\alpha)$ to $f_*(\alpha) - g_*(\alpha)$. The left-hand q_* is an isomorphism since these are good pairs and q induces a homeomorphism of quotient spaces $(X \times I)/(X \times \partial I) \to Z/Y$. Hence if we replace $H_{n+1}(Z, Y)$ in the lower exact sequence by the isomorphic group $H_n(X) \approx \mathrm{Ker}\, i_*$ we obtain the long exact sequence we want.

In the case of the mapping torus of a reflection $g : S^1 \to S^1$, with Z a Klein bottle, the interesting portion of the exact sequence $(*)$ is

$$0 \longrightarrow H_2(Z) \longrightarrow H_1(S^1) \xrightarrow{\;1-g_*\;} H_1(S^1) \longrightarrow H_1(Z) \longrightarrow H_0(S^1) \xrightarrow{\;1-g_*\;} H_0(S^1)$$

$$\| \qquad\qquad\qquad \| \qquad\qquad\qquad\qquad\qquad\qquad \| \qquad\qquad\qquad \|$$

$$\mathbb{Z} \xrightarrow{\qquad 2 \qquad} \mathbb{Z} \qquad\qquad\qquad\qquad\qquad \mathbb{Z} \xrightarrow{\qquad 0 \qquad} \mathbb{Z}$$

Thus $H_2(Z) = 0$ and we have a short exact sequence $0 \to \mathbb{Z}_2 \to H_1(Z) \to \mathbb{Z} \to 0$. This splits since $\mathbb{Z}$ is free, so $H_1(Z) \approx \mathbb{Z}_2 \oplus \mathbb{Z}$. Other examples are given in the Exercises.

If Y is the disjoint union of spaces Y_1 and Y_2, with $f : X \to Y_1$ and $g : X \to Y_2$, then Z consists of the mapping cylinders of these two maps with their domain ends identified. For example, suppose we have a CW complex decomposed as the union of two subcomplexes A and B and we take f and g to be the inclusions $A \cap B \hookrightarrow A$ and $A \cap B \hookrightarrow B$. Then the double mapping cylinder Z is homotopy equivalent to $A \cup B$ since we can view Z as $(A \cap B) \times I$ with A and B attached at the two ends, and then slide the attaching of A down to the B end to produce $A \cup B$ with $(A \cap B) \times I$ attached at one of its ends. By Proposition 0.18 the sliding operation preserves homotopy type, so we obtain a homotopy equivalence $Z \simeq A \cup B$. The exact sequence $(*)$ in this case is the Mayer–Vietoris sequence.

A relative form of the Mayer–Vietoris sequence is sometimes useful. If one has a pair of spaces $(X, Y) = (A \cup B, C \cup D)$ with $C \subset A$ and $D \subset B$, such that X is the union of the interiors of A and B, and Y is the union of the interiors of C and D, then there is a relative Mayer–Vietoris sequence

$$\cdots \longrightarrow H_n(A \cap B, C \cap D) \xrightarrow{\;\Phi\;} H_n(A, C) \oplus H_n(B, D) \xrightarrow{\;\Psi\;} H_n(X, Y) \xrightarrow{\;\partial\;} \cdots$$

To derive this, consider the commutative diagram

$$
\begin{array}{ccccccccc}
& & 0 & & 0 & & 0 & & \\
& & \downarrow & & \downarrow & & \downarrow & & \\
0 & \longrightarrow & C_n(C \cap D) & \xrightarrow{\varphi} & C_n(C) \oplus C_n(D) & \xrightarrow{\psi} & C_n(C + D) & \longrightarrow & 0 \\
& & \downarrow & & \downarrow & & \downarrow & & \\
0 & \longrightarrow & C_n(A \cap B) & \xrightarrow{\varphi} & C_n(A) \oplus C_n(B) & \xrightarrow{\psi} & C_n(A + B) & \longrightarrow & 0 \\
& & \downarrow & & \downarrow & & \downarrow & & \\
0 & \longrightarrow & C_n(A \cap B, C \cap D) & \xrightarrow{\varphi} & C_n(A, C) \oplus C_n(B, D) & \xrightarrow{\psi} & C_n(A + B, C + D) & \longrightarrow & 0 \\
& & \downarrow & & \downarrow & & \downarrow & & \\
& & 0 & & 0 & & 0 & &
\end{array}
$$

where $C_n(A + B, C + D)$ is the quotient of the subgroup $C_n(A + B) \subset C_n(X)$ by its subgroup $C_n(C + D) \subset C_n(Y)$. Thus the three columns of the diagram are exact. We have seen that the first two rows are exact, and we claim that the third row is exact also, with the maps φ and ψ induced from the φ and ψ in the second row. Since $\psi\varphi = 0$ in the second row, this holds also in the third row, so the third row is at least a chain complex. Viewing the three rows as chain complexes, the diagram then represents a short exact sequence of chain complexes. The associated long exact sequence of homology groups has two out of every three terms zero since the first two rows of the diagram are exact. Hence the remaining homology groups are zero and the third row is exact.

The third column maps to $0 \to C_n(Y) \to C_n(X) \to C_n(X, Y) \to 0$, inducing maps of homology groups that are isomorphisms for the X and Y terms as we have seen above. So by the five-lemma the maps $C_n(A+B, C+D) \to C_n(X, Y)$ also induce isomorphisms on homology. The relative Mayer-Vietoris sequence is then the long exact sequence of homology groups associated to the short exact sequence of chain complexes given by the third row of the diagram.

Homology with Coefficients

There is an easy generalization of the homology theory we have considered so far that behaves in a very similar fashion and sometimes offers technical advantages. The generalization consists of using chains of the form $\sum_i n_i \sigma_i$ where each σ_i is a singular n-simplex in X as before, but now the coefficients n_i are taken to lie in a fixed abelian group G rather than $\mathbb{Z}$. Such n-chains form an abelian group $C_n(X; G)$, and there is the expected relative version $C_n(X, A; G) = C_n(X; G)/C_n(A; G)$. The old formula for the boundary maps ∂ can still be used for arbitrary G, namely $\partial(\sum_i n_i \sigma_i) = \sum_{i,j}(-1)^j n_i \sigma_i | [v_0, \cdots, \hat{v}_j, \cdots, v_n]$. Just as before, a calculation shows that $\partial^2 = 0$, so the groups $C_n(X; G)$ and $C_n(X, A; G)$ form chain complexes. The resulting homology groups $H_n(X; G)$ and $H_n(X, A; G)$ are called **homology groups with coefficients in G**. Reduced groups $\tilde{H}_n(X; G)$ are defined via the augmented chain complex $\cdots \to C_0(X; G) \xrightarrow{\varepsilon} G \to 0$ with ε again defined by summing coefficients.

The case $G = \mathbb{Z}_2$ is particularly simple since one is just considering sums of singular simplices with coefficients 0 or 1, so by discarding terms with coefficient 0 one can think of chains as just finite 'unions' of singular simplices. The boundary formulas also simplify since one no longer has to worry about signs. Since signs are an algebraic representation of orientation considerations, one can also ignore orientations. This means that homology with $\mathbb{Z}_2$ coefficients is often the most natural tool in the absence of orientability.

All the theory we developed in §2.1 for $\mathbb{Z}$ coefficients carries over directly to general coefficient groups G with no change in the proofs. The same is true for Mayer–Vietoris sequences. Differences between $H_n(X; G)$ and $H_n(X)$ begin to appear only when one starts making calculations. When X is a point, the method used to compute $H_n(X)$ shows that $H_n(X; G)$ is G for $n = 0$ and 0 for $n > 0$. From this it follows just as for $G = \mathbb{Z}$ that $\tilde{H}_n(S^k; G)$ is G for $n = k$ and 0 otherwise.

Cellular homology also generalizes to homology with coefficients, with the cellular chain group $H_n(X^n, X^{n-1})$ replaced by $H_n(X^n, X^{n-1}; G)$, which is a direct sum of G's, one for each n-cell. The proof that the cellular homology groups $H_n^{CW}(X)$ agree with singular homology $H_n(X)$ extends immediately to give $H_n^{CW}(X; G) \approx H_n(X; G)$. The cellular boundary maps are given by the same formula as for $\mathbb{Z}$ coefficients, $d_n(\sum_\alpha n_\alpha e_\alpha^n) = \sum_{\alpha,\beta} d_{\alpha\beta} n_\alpha e_\beta^{n-1}$. The old proof applies, but the following result is needed to know that the coefficients $d_{\alpha\beta}$ are the same as before:

‖ **Lemma 2.49.** *If $f : S^k \to S^k$ has degree m, then $f_* : H_k(S^k; G) \to H_k(S^k; G)$ is multiplication by m.*

Proof: As a preliminary observation, note that a homomorphism $\varphi : G_1 \to G_2$ induces maps $\varphi_\sharp : C_n(X, A; G_1) \to C_n(X, A; G_2)$ commuting with boundary maps, so there are induced homomorphisms $\varphi_* : H_n(X, A; G_1) \to H_n(X, A; G_2)$. These have various naturality properties. For example, they give a commutative diagram mapping the long exact sequence of homology for the pair (X, A) with G_1 coefficients to the corresponding sequence with G_2 coefficients. Also, the maps φ_* commute with homomorphisms f_* induced by maps $f : (X, A) \to (Y, B)$.

Now let $f : S^k \to S^k$ have degree m and let $\varphi : \mathbb{Z} \to G$ take 1 to a given element $g \in G$. Then we have a commutative diagram as at the right, where commutativity of the outer two squares comes from the inductive calculation of these

$$
\begin{array}{ccc}
\mathbb{Z} \approx \widetilde{H}_k(S^k; \mathbb{Z}) & \xrightarrow{\;f_*\;} & \widetilde{H}_k(S^k; \mathbb{Z}) \approx \mathbb{Z} \\
\downarrow{\varphi} \quad \downarrow{\varphi_*} & & \downarrow{\varphi_*} \quad \downarrow{\varphi} \\
G \approx \widetilde{H}_k(S^k; G) & \xrightarrow{\;f_*\;} & \widetilde{H}_k(S^k; G) \approx G
\end{array}
$$

homology groups, reducing to the case $k = 0$ when the commutativity is obvious.

Since the diagram commutes, the assumption that the map across the top takes 1 to m implies that the map across the bottom takes g to mg. ∎

Example 2.50. It is instructive to see what happens to the homology of $\mathbb{R}\mathrm{P}^n$ when the coefficient group G is chosen to be a field F. The cellular chain complex is

$$
\cdots \xrightarrow{\;0\;} F \xrightarrow{\;2\;} F \xrightarrow{\;0\;} F \xrightarrow{\;2\;} F \xrightarrow{\;0\;} F \to 0
$$

Hence if F has characteristic 2, for example if $F = \mathbb{Z}_2$, then $H_k(\mathbb{R}\mathrm{P}^n; F) \approx F$ for $0 \le k \le n$, a more uniform answer than with $\mathbb{Z}$ coefficients. On the other hand, if F has characteristic different from 2 then the boundary maps $F \xrightarrow{\;2\;} F$ are isomorphisms, hence $H_k(\mathbb{R}\mathrm{P}^n; F)$ is F for $k = 0$ and for $k = n$ odd, and is zero otherwise.

In §3.A we will see that there is a general algebraic formula expressing homology with arbitrary coefficients in terms of homology with $\mathbb{Z}$ coefficients. Some easy special cases that give much of the flavor of the general result are included in the Exercises.

In spite of the fact that homology with $\mathbb{Z}$ coefficients determines homology with other coefficient groups, there are many situations where homology with a suitably chosen coefficient group can provide more information than homology with $\mathbb{Z}$ coefficients. A good example of this is the proof of the Borsuk–Ulam theorem using $\mathbb{Z}_2$ coefficients in §2.B.

As another illustration, we will now give an example of a map $f : X \to Y$ with the property that the induced maps f_* are trivial for homology with $\mathbb{Z}$ coefficients but not for homology with $\mathbb{Z}_m$ coefficients for suitably chosen m. Thus homology with $\mathbb{Z}_m$ coefficients tells us that f is not homotopic to a constant map, which we would not know using only $\mathbb{Z}$ coefficients.

Example 2.51. Let X be a Moore space $M(\mathbb{Z}_m, n)$ obtained from S^n by attaching a cell e^{n+1} by a map of degree m. The quotient map $f : X \to X/S^n = S^{n+1}$ induces trivial homomorphisms on reduced homology with $\mathbb{Z}$ coefficients since the nonzero reduced homology groups of X and S^{n+1} occur in different dimensions. But with $\mathbb{Z}_m$ coefficients the story is different, as we can see by considering the long exact sequence of the pair (X, S^n), which contains the segment

$$0 = \tilde{H}_{n+1}(S^n; \mathbb{Z}_m) \longrightarrow \tilde{H}_{n+1}(X; \mathbb{Z}_m) \xrightarrow{f_*} \tilde{H}_{n+1}(X/S^n; \mathbb{Z}_m)$$

Exactness says that f_* is injective, hence nonzero since $\tilde{H}_{n+1}(X; \mathbb{Z}_m)$ is $\mathbb{Z}_m$, the cellular boundary map $H_{n+1}(X^{n+1}, X^n; \mathbb{Z}_m) \to H_n(X^n, X^{n-1}; \mathbb{Z}_m)$ being $\mathbb{Z}_m \xrightarrow{m} \mathbb{Z}_m$.

Exercises

1. Prove the Brouwer fixed point theorem for maps $f : D^n \to D^n$ by applying degree theory to the map $S^n \to S^n$ that sends both the northern and southern hemispheres of S^n to the southern hemisphere via f. [This was Brouwer's original proof.]

2. Given a map $f : S^{2n} \to S^{2n}$, show that there is some point $x \in S^{2n}$ with either $f(x) = x$ or $f(x) = -x$. Deduce that every map $\mathbb{RP}^{2n} \to \mathbb{RP}^{2n}$ has a fixed point. Construct maps $\mathbb{RP}^{2n-1} \to \mathbb{RP}^{2n-1}$ without fixed points from linear transformations $\mathbb{R}^{2n} \to \mathbb{R}^{2n}$ without eigenvectors.

3. Let $f : S^n \to S^n$ be a map of degree zero. Show that there exist points $x, y \in S^n$ with $f(x) = x$ and $f(y) = -y$. Use this to show that if F is a continuous vector field defined on the unit ball D^n in $\mathbb{R}^n$ such that $F(x) \neq 0$ for all x, then there exists a point on ∂D^n where F points radially outward and another point on ∂D^n where F points radially inward.

4. Construct a surjective map $S^n \to S^n$ of degree zero, for each $n \geq 1$.

5. Show that any two reflections of S^n across different n-dimensional hyperplanes are homotopic, in fact homotopic through reflections. [The linear algebra formula for a reflection in terms of inner products may be helpful.]

6. Show that every map $S^n \to S^n$ can be homotoped to have a fixed point if $n > 0$.

7. For an invertible linear transformation $f : \mathbb{R}^n \to \mathbb{R}^n$ show that the induced map on $H_n(\mathbb{R}^n, \mathbb{R}^n - \{0\}) \approx \tilde{H}_{n-1}(\mathbb{R}^n - \{0\}) \approx \mathbb{Z}$ is $\mathbb{1}$ or $-\mathbb{1}$ according to whether the determinant of f is positive or negative. [Use Gaussian elimination to show that the matrix of f can be joined by a path of invertible matrices to a diagonal matrix with ± 1's on the diagonal.]

8. A polynomial $f(z)$ with complex coefficients, viewed as a map $\mathbb{C} \to \mathbb{C}$, can always be extended to a continuous map of one-point compactifications $\hat{f} : S^2 \to S^2$. Show that the degree of $\hat{f}$ equals the degree of f as a polynomial. Show also that the local degree of $\hat{f}$ at a root of f is the multiplicity of the root.

9. Compute the homology groups of the following 2-complexes:

(a) The quotient of S^2 obtained by identifying north and south poles to a point.

(b) $S^1 \times (S^1 \vee S^1)$.

(c) The space obtained from D^2 by first deleting the interiors of two disjoint subdisks in the interior of D^2 and then identifying all three resulting boundary circles together via homeomorphisms preserving clockwise orientations of these circles.

(d) The quotient space of $S^1 \times S^1$ obtained by identifying points in the circle $S^1 \times \{x_0\}$ that differ by $2\pi/m$ rotation and identifying points in the circle $\{x_0\} \times S^1$ that differ by $2\pi/n$ rotation.

10. Let X be the quotient space of S^2 under the identifications $x \sim -x$ for x in the equator S^1. Compute the homology groups $H_i(X)$. Do the same for S^3 with antipodal points of the equatorial $S^2 \subset S^3$ identified.

11. In an exercise for §1.2 we described a 3-dimensional CW complex obtained from the cube I^3 by identifying opposite faces via a one-quarter twist. Compute the homology groups of this complex.

12. Show that the quotient map $S^1 \times S^1 \to S^2$ collapsing the subspace $S^1 \vee S^1$ to a point is not nullhomotopic by showing that it induces an isomorphism on H_2. On the other hand, show via covering spaces that any map $S^2 \to S^1 \times S^1$ is nullhomotopic.

13. Let X be the 2-complex obtained from S^1 with its usual cell structure by attaching two 2-cells by maps of degrees 2 and 3, respectively.

(a) Compute the homology groups of all the subcomplexes $A \subset X$ and the corresponding quotient complexes X/A.

(b) Show that $X \simeq S^2$ and that the only subcomplex $A \subset X$ for which the quotient map $X \to X/A$ is a homotopy equivalence is the trivial subcomplex, the 0-cell.

14. A map $f : S^n \to S^n$ satisfying $f(x) = f(-x)$ for all x is called an *even map*. Show that an even map $S^n \to S^n$ must have even degree, and that the degree must in fact be zero when n is even. When n is odd, show there exist even maps of any given even degree. [Hints: If f is even, it factors as a composition $S^n \to \mathbb{RP}^n \to S^n$. Using the calculation of $H_n(\mathbb{RP}^n)$ in the text, show that the induced map $H_n(S^n) \to H_n(\mathbb{RP}^n)$ sends a generator to twice a generator when n is odd. It may be helpful to show that the quotient map $\mathbb{RP}^n \to \mathbb{RP}^n/\mathbb{RP}^{n-1}$ induces an isomorphism on H_n when n is odd.]

15. Show that if X is a CW complex then $H_n(X^n)$ is free by identifying it with the kernel of the cellular boundary map $H_n(X^n, X^{n-1}) \to H_{n-1}(X^{n-1}, X^{n-2})$.

16. Let $\Delta^n = [v_0, \cdots, v_n]$ have its natural Δ-complex structure with k-simplices $[v_{i_0}, \cdots, v_{i_k}]$ for $i_0 < \cdots < i_k$. Compute the ranks of the simplicial (or cellular) chain groups $\Delta_i(\Delta^n)$ and the subgroups of cycles and boundaries. [Hint: Pascal's triangle.] Apply this to show that the k-skeleton of Δ^n has homology groups $\tilde{H}_i((\Delta^n)^k)$ equal to 0 for $i < k$, and free of rank $\binom{n}{k+1}$ for $i = k$.

17. Show the isomorphism between cellular and singular homology is natural in the following sense: A map $f: X \to Y$ that is *cellular* — satisfying $f(X^n) \subset Y^n$ for all n — induces a chain map f_* between the cellular chain complexes of X and Y, and the map $f_*: H_n^{CW}(X) \to H_n^{CW}(Y)$ induced by this chain map corresponds to $f_*: H_n(X) \to H_n(Y)$ under the isomorphism $H_n^{CW} \approx H_n$.

18. For a CW pair (X, A) show there is a relative cellular chain complex formed by the groups $H_i(X^i, X^{i-1} \cup A^i)$, having homology groups isomorphic to $H_n(X, A)$.

19. Compute $H_i(\mathbb{RP}^n / \mathbb{RP}^m)$ for $m < n$ by cellular homology, using the standard CW structure on $\mathbb{RP}^n$ with $\mathbb{RP}^m$ as its m-skeleton.

20. For finite CW complexes X and Y, show that $\chi(X \times Y) = \chi(X)\chi(Y)$.

21. If a finite CW complex X is the union of subcomplexes A and B, show that $\chi(X) = \chi(A) + \chi(B) - \chi(A \cap B)$.

22. For X a finite CW complex and $p: \tilde{X} \to X$ an n-sheeted covering space, show that $\chi(\tilde{X}) = n\chi(X)$.

23. Show that if the closed orientable surface M_g of genus g is a covering space of M_h, then $g = n(h - 1) + 1$ for some n, namely, n is the number of sheets in the covering. [Conversely, if $g = n(h - 1) + 1$ then there is an n-sheeted covering $M_g \to M_h$, as we saw in Example 1.41.]

24. Suppose we build S^2 from a finite collection of polygons by identifying edges in pairs. Show that in the resulting CW structure on S^2 the 1-skeleton cannot be either of the two graphs shown, with five and six vertices. [This is one step in a proof that neither of these graphs embeds in $\mathbb{R}^2$.]

25. Show that for each $n \in \mathbb{Z}$ there is a unique function φ assigning an integer to each finite CW complex, such that (a) $\varphi(X) = \varphi(Y)$ if X and Y are homeomorphic, (b) $\varphi(X) = \varphi(A) + \varphi(X/A)$ if A is a subcomplex of X, and (c) $\varphi(S^0) = n$. For such a function φ, show that $\varphi(X) = \varphi(Y)$ if $X \simeq Y$.

26. For a pair (X, A), let $X \cup CA$ be X with a cone on A attached.
 (a) Show that X is a retract of $X \cup CA$ iff A is *contractible in X*: There is a homotopy $f_t: A \to X$ with f_0 the inclusion $A \hookrightarrow X$ and f_1 a constant map.
 (b) Show that if A is contractible in X then $H_n(X, A) \approx \tilde{H}_n(X) \oplus \tilde{H}_{n-1}(A)$, using the fact that $(X \cup CA)/X$ is the suspension SA of A.

27. The short exact sequences $0 \to C_n(A) \to C_n(X) \to C_n(X, A) \to 0$ always split, but why does this not always yield splittings $H_n(X) \approx H_n(A) \oplus H_n(X, A)$?

28. (a) Use the Mayer–Vietoris sequence to compute the homology groups of the space obtained from a torus $S^1 \times S^1$ by attaching a Möbius band via a homeomorphism from the boundary circle of the Möbius band to the circle $S^1 \times \{x_0\}$ in the torus.
 (b) Do the same for the space obtained by attaching a Möbius band to $\mathbb{RP}^2$ via a homeomorphism of its boundary circle to the standard $\mathbb{RP}^1 \subset \mathbb{RP}^2$.

29. The surface M_g of genus g, embedded in $\mathbb{R}^3$ in the standard way, bounds a compact region R. Two copies of R, glued together by the identity map between their boundary surfaces M_g, form a closed 3-manifold X. Compute the homology groups of X via the Mayer–Vietoris sequence for this decomposition of X into two copies of R. Also compute the relative groups $H_i(R, M_g)$.

30. For the mapping torus T_f of a map $f : X \to X$, we constructed in Example 2.48 a long exact sequence $\cdots \to H_n(X) \xrightarrow{\mathbb{1} - f_*} H_n(X) \to H_n(T_f) \to H_{n-1}(X) \to \cdots$. Use this to compute the homology of the mapping tori of the following maps:
(a) A reflection $S^2 \to S^2$.
(b) A map $S^2 \to S^2$ of degree 2.
(c) The map $S^1 \times S^1 \to S^1 \times S^1$ that is the identity on one factor and a reflection on the other.
(d) The map $S^1 \times S^1 \to S^1 \times S^1$ that is a reflection on each factor.
(e) The map $S^1 \times S^1 \to S^1 \times S^1$ that interchanges the two factors and then reflects one of the factors.

31. Use the Mayer–Vietoris sequence to show there are isomorphisms $\tilde{H}_n(X \vee Y) \approx \tilde{H}_n(X) \oplus \tilde{H}_n(Y)$ if the basepoints of X and Y that are identified in $X \vee Y$ are deformation retracts of neighborhoods $U \subset X$ and $V \subset Y$.

32. For SX the suspension of X, show by a Mayer–Vietoris sequence that there are isomorphisms $\tilde{H}_n(SX) \approx \tilde{H}_{n-1}(X)$ for all n.

33. Suppose the space X is the union of open sets $A_1, \cdots, A_n$ such that each intersection $A_{i_1} \cap \cdots \cap A_{i_k}$ is either empty or has trivial reduced homology groups. Show that $\tilde{H}_i(X) = 0$ for $i \geq n - 1$, and give an example showing this inequality is best possible, for each n.

34. [Deleted — see the errata for comments.]

35. Use the Mayer–Vietoris sequence to show that a nonorientable closed surface, or more generally a finite simplicial complex X for which $H_1(X)$ contains torsion, cannot be embedded as a subspace of $\mathbb{R}^3$ in such a way as to have a neighborhood homeomorphic to the mapping cylinder of some map from a closed orientable surface to X. [This assumption on a neighborhood is in fact not needed if one deduces the result from Alexander duality in §3.3.]

36. Show that $H_i(X \times S^n) \approx H_i(X) \oplus H_{i-n}(X)$ for all i and n, where $H_i = 0$ for $i < 0$ by definition. Namely, show $H_i(X \times S^n) \approx H_i(X) \oplus H_i(X \times S^n, X \times \{x_0\})$ and $H_i(X \times S^n, X \times \{x_0\}) \approx H_{i-1}(X \times S^{n-1}, X \times \{x_0\})$. [For the latter isomorphism the relative Mayer–Vietoris sequence yields an easy proof.]

37. Give an elementary derivation for the Mayer–Vietoris sequence in simplicial homology for a Δ-complex X decomposed as the union of subcomplexes A and B.

38. Show that a commutative diagram

$$\cdots \longrightarrow C_{n+1} \searrow \qquad \nearrow B_n \longrightarrow C_n \searrow \qquad \nearrow B_{n-1} \longrightarrow \cdots$$

with the two sequences across the top and bottom exact, gives rise to an exact sequence $\cdots \longrightarrow E_{n+1} \longrightarrow B_n \longrightarrow C_n \oplus D_n \longrightarrow E_n \longrightarrow B_{n-1} \longrightarrow \cdots$ where the maps are obtained from those in the previous diagram in the obvious way, except that $B_n \to C_n \oplus D_n$ has a minus sign in one coordinate.

39. Use the preceding exercise to derive relative Mayer–Vietoris sequences for CW pairs $(X, Y) = (A \cup B, C \cup D)$ with $A = B$ or $C = D$.

40. From the long exact sequence of homology groups associated to the short exact sequence of chain complexes $0 \longrightarrow C_i(X) \xrightarrow{\ n\ } C_i(X) \longrightarrow C_i(X; \mathbb{Z}_n) \longrightarrow 0$ deduce immediately that there are short exact sequences

$$0 \longrightarrow H_i(X)/nH_i(X) \longrightarrow H_i(X; \mathbb{Z}_n) \longrightarrow n\text{-}Torsion(H_{i-1}(X)) \longrightarrow 0$$

where $n\text{-}Torsion(G)$ is the kernel of the map $G \xrightarrow{\ n\ } G$, $g \mapsto ng$. Use this to show that $\tilde{H}_i(X; \mathbb{Z}_p) = 0$ for all i and all primes p iff $\tilde{H}_i(X)$ is a vector space over $\mathbb{Q}$ for all i.

41. For X a finite CW complex and F a field, show that the Euler characteristic $\chi(X)$ can also be computed by the formula $\chi(X) = \sum_n (-1)^n \dim H_n(X; F)$, the alternating sum of the dimensions of the vector spaces $H_n(X; F)$.

42. Let X be a finite connected graph having no vertex that is the endpoint of just one edge, and suppose that $H_1(X; \mathbb{Z})$ is free abelian of rank $n > 1$, so the group of automorphisms of $H_1(X; \mathbb{Z})$ is $GL_n(\mathbb{Z})$, the group of invertible $n \times n$ matrices with integer entries whose inverse matrix also has integer entries. Show that if G is a finite group of homeomorphisms of X, then the homomorphism $G \to GL_n(\mathbb{Z})$ assigning to $g : X \to X$ the induced homomorphism $g_* : H_1(X; \mathbb{Z}) \to H_1(X; \mathbb{Z})$ is injective. Show the same result holds if the coefficient group $\mathbb{Z}$ is replaced by $\mathbb{Z}_m$ with $m > 2$. What goes wrong when $m = 2$?

43. (a) Show that a chain complex of free abelian groups C_n splits as a direct sum of subcomplexes $0 \to L_{n+1} \to K_n \to 0$ with at most two nonzero terms. [Show the short exact sequence $0 \to \operatorname{Ker} \partial \to C_n \to \operatorname{Im} \partial \to 0$ splits and take $K_n = \operatorname{Ker} \partial$.]
(b) In case the groups C_n are finitely generated, show there is a further splitting into summands $0 \to \mathbb{Z} \to 0$ and $0 \to \mathbb{Z} \xrightarrow{\ m\ } \mathbb{Z} \to 0$. [Reduce the matrix of the boundary map $L_{n+1} \to K_n$ to echelon form by elementary row and column operations.]
(c) Deduce that if X is a CW complex with finitely many cells in each dimension, then $H_n(X; G)$ is the direct sum of the following groups:
  - a copy of G for each $\mathbb{Z}$ summand of $H_n(X)$
  - a copy of G/mG for each $\mathbb{Z}_m$ summand of $H_n(X)$
  - a copy of the kernel of $G \xrightarrow{\ m\ } G$ for each $\mathbb{Z}_m$ summand of $H_{n-1}(X)$

2.3 The Formal Viewpoint

Sometimes it is good to step back from the forest of details and look for general patterns. In this rather brief section we will first describe the general pattern of homology by axioms, then we will look at some common formal features shared by many of the constructions we have made, using the language of categories and functors which has become common in much of modern mathematics.

Axioms for Homology

For simplicity let us restrict attention to CW complexes and focus on reduced homology to avoid mentioning relative homology. A (reduced) **homology theory** assigns to each nonempty CW complex X a sequence of abelian groups $\tilde{h}_n(X)$ and to each map $f:X \to Y$ between CW complexes a sequence of homomorphisms $f_* : \tilde{h}_n(X) \to \tilde{h}_n(Y)$ such that $(fg)_* = f_* g_*$ and $\mathbb{1}_* = \mathbb{1}$, and so that the following three axioms are satisfied.

(1) If $f \simeq g : X \to Y$, then $f_* = g_* : \tilde{h}_n(X) \to \tilde{h}_n(Y)$.

(2) There are boundary homomorphisms $\partial : \tilde{h}_n(X/A) \to \tilde{h}_{n-1}(A)$ defined for each CW pair (X, A), fitting into an exact sequence

$$\cdots \xrightarrow{\partial} \tilde{h}_n(A) \xrightarrow{i_*} \tilde{h}_n(X) \xrightarrow{q_*} \tilde{h}_n(X/A) \xrightarrow{\partial} \tilde{h}_{n-1}(A) \xrightarrow{i_*} \cdots$$

where i is the inclusion and q is the quotient map. Furthermore the boundary maps are natural: For $f:(X,A) \to (Y,B)$ inducing a quotient map $\bar{f}:X/A \to Y/B$, there are commutative diagrams

$$
\begin{array}{ccc}
\tilde{h}_n(X/A) & \xrightarrow{\ \partial\ } & \tilde{h}_{n-1}(A) \\
\downarrow{\bar{f}_*} & & \downarrow{f_*} \\
\tilde{h}_n(Y/B) & \xrightarrow{\ \partial\ } & \tilde{h}_{n-1}(B)
\end{array}
$$

(3) For a wedge sum $X = \bigvee_\alpha X_\alpha$ with inclusions $i_\alpha : X_\alpha \hookrightarrow X$, the direct sum map $\bigoplus_\alpha i_{\alpha*} : \bigoplus_\alpha \tilde{h}_n(X_\alpha) \to \tilde{h}_n(X)$ is an isomorphism for each n.

Negative values for the subscripts n are permitted. Ordinary singular homology is zero in negative dimensions by definition, but interesting homology theories with nontrivial groups in negative dimensions do exist.

The third axiom may seem less substantial than the first two, and indeed for finite wedge sums it can be deduced from the first two axioms, though not in general for infinite wedge sums, as an example in the Exercises shows.

It is also possible, and not much more difficult, to give axioms for unreduced homology theories. One supposes one has relative groups $h_n(X, A)$ defined, specializing to absolute groups by setting $h_n(X) = h_n(X, \varnothing)$. Axiom (1) is replaced by its obvious relative form, and axiom (2) is broken into two parts, the first hypothesizing a long exact sequence involving these relative groups, with natural boundary maps, the second stating some version of excision, for example $h_n(X, A) \approx h_n(X/A, A/A)$ if one is dealing with CW pairs. In axiom (3) the wedge sum is replaced by disjoint union.

These axioms for unreduced homology are essentially the same as those originally laid out in the highly influential book [Eilenberg & Steenrod 1952], except that axiom (3) was omitted since the focus there was on finite complexes, and there was another axiom specifying that the groups $h_n(point)$ are zero for $n \neq 0$, as is true for singular homology. This axiom was called the 'dimension axiom,' presumably because it specifies that a point has nontrivial homology only in dimension zero. It can be regarded as a normalization axiom, since one can trivially define a homology theory where it fails by setting $h_n(X, A) = H_{n+k}(X, A)$ for a fixed nonzero integer k. At the time there were no interesting homology theories known for which the dimension axiom did not hold, but soon thereafter topologists began studying a homology theory called 'bordism' having the property that the bordism groups of a point are nonzero in infinitely many dimensions. Axiom (3) seems to have appeared first in [Milnor 1962].

Reduced and unreduced homology theories are essentially equivalent. From an unreduced theory h one gets a reduced theory $\tilde{h}$ by setting $\tilde{h}_n(X)$ equal to the kernel of the canonical map $h_n(X) \to h_n(point)$. In the other direction, one sets $h_n(X) = \tilde{h}_n(X_+)$ where X_+ is the disjoint union of X with a point. We leave it as an exercise to show that these two transformations between reduced and unreduced homology are inverses of each other. Just as with ordinary homology, one has $h_n(X) \approx \tilde{h}_n(X) \oplus h_n(x_0)$ for any point $x_0 \in X$, since the long exact sequence of the pair (X, x_0) splits via the retraction of X onto x_0. Note that $\tilde{h}_n(x_0) = 0$ for all n, as can be seen by looking at the long exact sequence of reduced homology groups of the pair (x_0, x_0).

The groups $h_n(x_0) \approx \tilde{h}_n(S^0)$ are called the **coefficients** of the homology theories h and $\tilde{h}$, by analogy with the case of singular homology with coefficients. One can trivially realize any sequence of abelian groups G_i as the coefficient groups of a homology theory by setting $h_n(X, A) = \bigoplus_i H_{n-i}(X, A; G_i)$.

In general, homology theories are not uniquely determined by their coefficient groups, but this is true for singular homology: If h is a homology theory defined for CW pairs, whose coefficient groups $h_n(x_0)$ are zero for $n \neq 0$, then there are natural isomorphisms $h_n(X, A) \approx H_n(X, A; G)$ for all CW pairs (X, A) and all n, where $G = h_0(x_0)$. This will be proved in Theorem 4.59.

We have seen how Mayer–Vietoris sequences can be quite useful for singular homology, and in fact every homology theory has Mayer–Vietoris sequences, at least for CW complexes. These can be obtained directly from the axioms in the following way. For a CW complex $X = A \cup B$ with A and B subcomplexes, the inclusion $(B, A \cap B) \hookrightarrow (X, A)$ induces a commutative diagram of exact sequences

$$
\begin{array}{ccccccccc}
\cdots \longrightarrow & h_{n+1}(B, A \cap B) & \longrightarrow & h_n(A \cap B) & \longrightarrow & h_n(B) & \longrightarrow & h_n(B, A \cap B) & \longrightarrow \cdots \\
& \downarrow \approx & & \downarrow & & \downarrow & & \downarrow \approx & \\
\cdots \longrightarrow & h_{n+1}(X, A) & \longrightarrow & h_n(A) & \longrightarrow & h_n(X) & \longrightarrow & h_n(X, A) & \longrightarrow \cdots
\end{array}
$$

The vertical maps between relative groups are isomorphisms since $B/(A \cap B) = X/A$. Then it is a purely algebraic fact, whose proof is Exercise 38 at the end of the previous section, that a diagram such as this with every third vertical map an isomorphism gives rise to a long exact sequence involving the remaining nonisomorphic terms. In the present case this takes the form of a Mayer-Vietoris sequence

$$
\cdots \longrightarrow h_n(A \cap B) \xrightarrow{\varphi} h_n(A) \oplus h_n(B) \xrightarrow{\psi} h_n(X) \xrightarrow{\partial} h_{n-1}(A \cap B) \longrightarrow \cdots
$$

Categories and Functors

Formally, singular homology can be regarded as a sequence of functions H_n that assign to each space X an abelian group $H_n(X)$ and to each map $f : X \to Y$ a homomorphism $H_n(f) = f_* : H_n(X) \to H_n(Y)$, and similarly for relative homology groups. This sort of situation arises quite often, and not just in algebraic topology, so it is useful to introduce some general terminology for it. Roughly speaking, 'functions' like H_n are called 'functors,' and the domains and ranges of these functors are called 'categories.' Thus for H_n the domain category consists of topological spaces and continuous maps, or in the relative case, pairs of spaces and continuous maps of pairs, and the range category consists of abelian groups and homomorphisms. A key point is that one is interested not only in the objects in the category, for example spaces or groups, but also in the maps, or 'morphisms,' between these objects.

Now for the precise definitions. A **category** $\mathcal{C}$ consists of three things:

(1) A collection $\mathrm{Ob}(\mathcal{C})$ of **objects**.
(2) Sets $\mathrm{Mor}(X, Y)$ of **morphisms** for each pair $X, Y \in \mathrm{Ob}(\mathcal{C})$, including a distinguished 'identity' morphism $\mathbb{1} = \mathbb{1}_X \in \mathrm{Mor}(X, X)$ for each X.
(3) A 'composition of morphisms' function $\circ : \mathrm{Mor}(X, Y) \times \mathrm{Mor}(Y, Z) \to \mathrm{Mor}(X, Z)$ for each triple $X, Y, Z \in \mathrm{Ob}(\mathcal{C})$, satisfying $f \circ \mathbb{1} = f$, $\mathbb{1} \circ f = f$, and $(f \circ g) \circ h = f \circ (g \circ h)$.

There are plenty of obvious examples, such as:

- The category of topological spaces, with continuous maps as the morphisms. Or we could restrict to special classes of spaces such as CW complexes, keeping continuous maps as the morphisms. We could also restrict the morphisms, for example to homeomorphisms.
- The category of groups, with homomorphisms as morphisms. Or the subcategory of abelian groups, again with homomorphisms as the morphisms. Generalizing

this is the category of modules over a fixed ring, with morphisms the module homomorphisms.

- The category of sets, with arbitrary functions as the morphisms. Or the morphisms could be restricted to injections, surjections, or bijections.

There are also many categories where the morphisms are not simply functions, for example:

- Any group G can be viewed as a category with only one object and with G as the morphisms of this object, so that condition (3) reduces to two of the three axioms for a group. If we require only these two axioms, associativity and a left and right identity, we have a 'group without inverses,' usually called a *monoid* since it is the same thing as a category with one object.

- A partially ordered set $(X, \leq)$ can be considered a category where the objects are the elements of X and there is a unique morphism from x to y whenever $x \leq y$. The relation $x \leq x$ gives the morphism $\mathbb{1}$ and transitivity gives the composition $\mathrm{Mor}(x, y) \times \mathrm{Mor}(y, z) \to \mathrm{Mor}(x, z)$. The condition that $x \leq y$ and $y \leq x$ implies $x = y$ says that there is at most one morphism between any two objects.

- There is a 'homotopy category' whose objects are topological spaces and whose morphisms are homotopy classes of maps, rather than actual maps. This uses the fact that composition is well-defined on homotopy classes: $f_0 g_0 \simeq f_1 g_1$ if $f_0 \simeq f_1$ and $g_0 \simeq g_1$.

- Chain complexes are the objects of a category, with chain maps as morphisms. This category has various interesting subcategories, obtained by restricting the objects. For example, we could take chain complexes whose groups are zero in negative dimensions, or zero outside a finite range. Or we could restrict to exact sequences, or short exact sequences. In each case we take morphisms to be chain maps, which are commutative diagrams. Going a step further, there is a category whose objects are short exact sequences of chain complexes and whose morphisms are commutative diagrams of maps between such short exact sequences.

A **functor** F from a category $\mathcal{C}$ to a category $\mathcal{D}$ assigns to each object X in $\mathcal{C}$ an object $F(X)$ in $\mathcal{D}$ and to each morphism $f \in \mathrm{Mor}(X, Y)$ in $\mathcal{C}$ a morphism $F(f) \in \mathrm{Mor}(F(X), F(Y))$ in $\mathcal{D}$, such that $F(\mathbb{1}) = \mathbb{1}$ and $F(f \circ g) = F(f) \circ F(g)$. In the case of the singular homology functor H_n, the latter two conditions are the familiar properties $\mathbb{1}_* = \mathbb{1}$ and $(fg)_* = f_* g_*$ of induced maps. Strictly speaking, what we have just defined is a **covariant** functor. A **contravariant** functor would differ from this by assigning to $f \in \mathrm{Mor}(X, Y)$ a 'backwards' morphism $F(f) \in \mathrm{Mor}(F(Y), F(X))$ with $F(\mathbb{1}) = \mathbb{1}$ and $F(f \circ g) = F(g) \circ F(f)$. A classical example of this is the dual vector space functor, which assigns to a vector space V over a fixed scalar field K the dual vector space $F(V) = V^*$ of linear maps $V \to K$, and to each linear transformation

$f : V \to W$ the dual map $F(f) = f^* : W^* \to V^*$, going in the reverse direction. In the next chapter we will study the contravariant version of homology, called cohomology.

A number of the constructions we have studied in this chapter are functors:

- The singular chain complex functor assigns to a space X the chain complex of singular chains in X and to a map $f : X \to Y$ the induced chain map. This is a functor from the category of spaces and continuous maps to the category of chain complexes and chain maps.

- The algebraic homology functor assigns to a chain complex its sequence of homology groups and to a chain map the induced homomorphisms on homology. This is a functor from the category of chain complexes and chain maps to the category whose objects are sequences of abelian groups and whose morphisms are sequences of homomorphisms.

- The composition of the two preceding functors is the functor assigning to a space its singular homology groups.

- The first example above, the singular chain complex functor, can itself be regarded as the composition of two functors. The first functor assigns to a space X its singular complex $S(X)$, a Δ-complex, and the second functor assigns to a Δ-complex its simplicial chain complex. This is what the two functors do on objects, and what they do on morphisms can be described in the following way. A map of spaces $f : X \to Y$ induces a map $f_* : S(X) \to S(Y)$ by composing singular simplices $\Delta^n \to X$ with f. The map f_* is a map between Δ-complexes taking the distinguished characteristic maps in the domain Δ-complex to the distinguished characteristic maps in the target Δ-complex. Call such maps Δ-**maps** and let them be the morphisms in the category of Δ-complexes. Note that a Δ-map induces a chain map between simplicial chain complexes, taking basis elements to basis elements, so we have a simplicial chain complex functor taking the category of Δ-complexes and Δ-maps to the category of chain complexes and chain maps.

- There is a functor assigning to a pair of spaces (X, A) the associated long exact sequence of homology groups. Morphisms in the domain category are maps of pairs, and in the target category morphisms are maps between exact sequences forming commutative diagrams. This functor is the composition of two functors, the first assigning to (X, A) a short exact sequence of chain complexes, the second assigning to such a short exact sequence the associated long exact sequence of homology groups. Morphisms in the intermediate category are the evident commutative diagrams.

Another sort of process we have encountered is the transformation of one functor into another, for example:

- Boundary maps $H_n(X, A) \to H_{n-1}(A)$ in singular homology, or indeed in any homology theory.

- Change-of-coefficient homomorphisms $H_n(X; G_1) \rightarrow H_n(X; G_2)$ induced by a homomorphism $G_1 \rightarrow G_2$, as in the proof of Lemma 2.49.

In general, if one has two functors $F, G : \mathcal{C} \rightarrow \mathcal{D}$ then a **natural transformation** T from F to G assigns a morphism $T_X : F(X) \rightarrow G(X)$ to each object $X \in \mathcal{C}$, in such a way that for each morphism $f : X \rightarrow Y$ in $\mathcal{C}$ the square at the right commutes. The case that F and G are contravariant rather than covariant is similar.

$$\begin{array}{ccc} F(X) & \xrightarrow{F(f)} & F(Y) \\ \downarrow{T_X} & & \downarrow{T_Y} \\ G(X) & \xrightarrow{G(f)} & G(Y) \end{array}$$

We have been describing the passage from topology to the abstract world of categories and functors, but there is also a nice path in the opposite direction:

- To each category $\mathcal{C}$ there is associated a Δ-complex $B\mathcal{C}$ called the **classifying space** of $\mathcal{C}$, whose n-simplices are the strings $X_0 \rightarrow X_1 \rightarrow \cdots \rightarrow X_n$ of morphisms in $\mathcal{C}$. The faces of this simplex are obtained by deleting an X_i, and then composing the two adjacent morphisms if $i \neq 0, n$. Thus when $n = 2$ the three faces of $X_0 \rightarrow X_1 \rightarrow X_2$ are $X_0 \rightarrow X_1$, $X_1 \rightarrow X_2$, and the composed morphism $X_0 \rightarrow X_2$. In case $\mathcal{C}$ has a single object and the morphisms of $\mathcal{C}$ form a group G, then $B\mathcal{C}$ is the same as the Δ-complex BG constructed in Example 1B.7, a $K(G, 1)$. In general, the space $B\mathcal{C}$ need not be a $K(G, 1)$, however. For example, if we start with a Δ-complex X and regard its set of simplices as a partially ordered set $\mathcal{C}(X)$ under the relation of inclusion of faces, then $B\mathcal{C}(X)$ is the barycentric subdivision of X.

- A functor $F : \mathcal{C} \rightarrow \mathcal{D}$ induces a map $B\mathcal{C} \rightarrow B\mathcal{D}$. This is the Δ-map that sends an n-simplex $X_0 \rightarrow X_1 \rightarrow \cdots \rightarrow X_n$ to the n-simplex $F(X_0) \rightarrow F(X_1) \rightarrow \cdots \rightarrow F(X_n)$.

- A natural transformation from a functor F to a functor G induces a homotopy between the induced maps of classifying spaces. We leave this for the reader to make explicit, using the subdivision of $\Delta^n \times I$ into $(n + 1)$-simplices described earlier in the chapter.

Exercises

1. If $T_n(X, A)$ denotes the torsion subgroup of $H_n(X, A; \mathbb{Z})$, show that the functors $(X, A) \mapsto T_n(X, A)$, with the obvious induced homomorphisms $T_n(X, A) \rightarrow T_n(Y, B)$ and boundary maps $T_n(X, A) \rightarrow T_{n-1}(A)$, do not define a homology theory. Do the same for the 'mod torsion' functor $MT_n(X, A) = H_n(X, A; \mathbb{Z})/T_n(X, A)$.

2. Define a candidate for a reduced homology theory on CW complexes by $\tilde{h}_n(X) = \prod_i \tilde{H}_i(X) / \bigoplus_i \tilde{H}_i(X)$. Thus $\tilde{h}_n(X)$ is independent of n and is zero if X is finite-dimensional, but is not identically zero, for example for $X = \bigvee_i S^i$. Show that the axioms for a homology theory are satisfied except that the wedge axiom fails.

3. Show that if $\tilde{h}$ is a reduced homology theory, then $\tilde{h}_n(point) = 0$ for all n. Deduce that there are suspension isomorphisms $\tilde{h}_n(X) \approx \tilde{h}_{n+1}(SX)$ for all n.

4. Show that the wedge axiom for homology theories follows from the other axioms in the case of finite wedge sums.

Additional Topics

2.A Homology and Fundamental Group

There is a close connection between $H_1(X)$ and $\pi_1(X)$, arising from the fact that a map $f:I \to X$ can be viewed as either a path or a singular 1-simplex. If f is a loop, with $f(0) = f(1)$, this singular 1-simplex is a cycle since $\partial f = f(1) - f(0)$.

Theorem 2A.1. *By regarding loops as singular 1-cycles, we obtain a homomorphism $h:\pi_1(X,x_0) \to H_1(X)$. If X is path-connected, then h is surjective and has kernel the commutator subgroup of $\pi_1(X)$, so h induces an isomorphism from the abelianization of $\pi_1(X)$ onto $H_1(X)$.*

Proof: Recall the notation $f \simeq g$ for the relation of homotopy, fixing endpoints, between paths f and g. Regarding f and g as chains, the notation $f \sim g$ will mean that f is homologous to g, that is, $f - g$ is the boundary of some 2-chain. Here are some facts about this relation.

(i) If f is a constant path, then $f \sim 0$. Namely, f is a cycle since it is a loop, and since $H_1(point) = 0$, f must then be a boundary. Explicitly, f is the boundary of the constant singular 2-simplex σ having the same image as f since

$$\partial \sigma = \sigma \,|\, [v_1, v_2] - \sigma \,|\, [v_0, v_2] + \sigma \,|\, [v_0, v_1] = f - f + f = f$$

(ii) If $f \simeq g$ then $f \sim g$. To see this, consider a homotopy $F:I \times I \to X$ from f to g. This yields a pair of singular 2-simplices σ_1 and σ_2 in X by subdividing the square $I \times I$ into two triangles $[v_0, v_1, v_3]$ and $[v_0, v_2, v_3]$ as shown in the figure. When one computes $\partial(\sigma_1 - \sigma_2)$, the two restrictions of F to the diagonal of the square cancel, leaving $f - g$ together with two constant singular 1-simplices from the left and right edges of the square. By (i) these are boundaries, so $f - g$ is also a boundary.

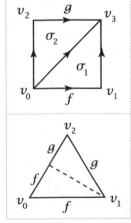

(iii) $f \cdot g \sim f + g$, where $f \cdot g$ denotes the product of the paths f and g. For if $\sigma:\Delta^2 \to X$ is the composition of orthogonal projection of $\Delta^2 = [v_0, v_1, v_2]$ onto the edge $[v_0, v_2]$ followed by $f \cdot g:[v_0, v_2] \to X$, then $\partial \sigma = g - f \cdot g + f$.

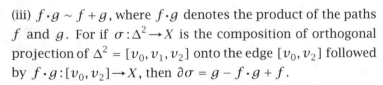

(iv) $\overline{f} \sim -f$, where $\overline{f}$ is the inverse path of f. This follows from the preceding three observations, which give $f + \overline{f} \sim f \cdot \overline{f} \sim 0$.

Applying (ii) and (iii) to loops, it follows that we have a well-defined homomorphism $h:\pi_1(X,x_0) \to H_1(X)$ sending the homotopy class of a loop f to the homology class of the 1-cycle f.

To show h is surjective when X is path-connected, let $\sum_i n_i \sigma_i$ be a 1-cycle representing a given element of $H_1(X)$. After relabeling the σ_i's we may assume each n_i is ± 1. By (iv) we may in fact take each n_i to be $+1$, so our 1-cycle is $\sum_i \sigma_i$. If some σ_i is not a loop, then the fact that $\partial(\sum_i \sigma_i) = 0$ means there must be another σ_j such that the composed path $\sigma_i \cdot \sigma_j$ is defined. By (iii) we may then combine the terms σ_i and σ_j into a single term $\sigma_i \cdot \sigma_j$. Iterating this, we reduce to the case that each σ_i is a loop. Since X is path-connected, we may choose a path γ_i from x_0 to the basepoint of σ_i. We have $\gamma_i \cdot \sigma_i \cdot \overline{\gamma}_i \sim \sigma_i$ by (iii) and (iv), so we may assume all σ_i's are loops at x_0. Then we can combine all the σ_i's into a single σ by (iii). This says the given element of $H_1(X)$ is in the image of h.

The commutator subgroup of $\pi_1(X)$ is contained in the kernel of h since $H_1(X)$ is abelian. To obtain the reverse inclusion we will show that every class $[f]$ in the kernel of h is trivial in the abelianization $\pi_1(X)_{ab}$ of $\pi_1(X)$.

If an element $[f] \in \pi_1(X)$ is in the kernel of h, then f, as a 1-cycle, is the boundary of a 2-chain $\sum_i n_i \sigma_i$. Again we may assume each n_i is ± 1. As in the discussion preceding Proposition 2.6, we can associate to the chain $\sum_i n_i \sigma_i$ a 2-dimensional Δ-complex K by taking a 2-simplex Δ_i^2 for each σ_i and identifying certain pairs of edges of these 2-simplices. Namely, if we apply the usual boundary formula to write $\partial \sigma_i = \tau_{i0} - \tau_{i1} + \tau_{i2}$ for singular 1-simplices τ_{ij}, then the formula

$$f = \partial\left(\sum_i n_i \sigma_i\right) = \sum_i n_i \partial \sigma_i = \sum_{i,j} (-1)^j n_i \tau_{ij}$$

implies that we can group all but one of the τ_{ij}'s into pairs for which the two coefficients $(-1)^j n_i$ in each pair are $+1$ and -1. The one remaining τ_{ij} is equal to f. We then identify edges of the Δ_j^2's corresponding to the paired τ_{ij}'s, preserving orientations of these edges so that we obtain a Δ-complex K.

The maps σ_i fit together to give a map $\sigma : K \to X$. We can deform σ, staying fixed on the edge corresponding to f, so that each vertex maps to the basepoint x_0, in the following way. Paths from the images of these vertices to x_0 define such a homotopy on the union of the 0-skeleton of K with the edge corresponding to f, and then we can appeal to the homotopy extension property in Proposition 0.16 to extend this homotopy to all of K. Alternatively, it is not hard to construct such an extension by hand. Restricting the new σ to the simplices Δ_i^2, we obtain a new chain $\sum_i n_i \sigma_i$ with boundary equal to f and with all τ_{ij}'s loops at x_0.

Using additive notation in the abelian group $\pi_1(X)_{ab}$, we have the formula $[f] = \sum_{i,j} (-1)^j n_i [\tau_{ij}]$ because of the canceling pairs of τ_{ij}'s. We can rewrite the summation $\sum_{i,j} (-1)^j n_i [\tau_{ij}]$ as $\sum_i n_i [\partial \sigma_i]$ where $[\partial \sigma_i] = [\tau_{i0}] - [\tau_{i1}] + [\tau_{i2}]$. Since σ_i gives a nullhomotopy of the composed loop $\tau_{i0} - \tau_{i1} + \tau_{i2}$, we conclude that $[f] = 0$ in $\pi_1(X)_{ab}$. $\qquad\square$

The end of this proof can be illuminated by looking more closely at the geometry. The complex K is in fact a compact surface with boundary consisting of a single circle formed by the edge corresponding to f. This is because any pattern of identifications of pairs of edges of a finite collection of disjoint 2-simplices produces a compact surface with boundary. We leave it as an exercise for the reader to check that the algebraic formula $f = \partial(\sum_i n_i \sigma_i)$ with each $n_i = \pm 1$ implies that K is an orientable surface. The component of K containing the boundary circle is a standard closed orientable surface of some genus g with an open disk removed, by the basic structure theorem for compact orientable surfaces. Giving this surface the cell structure indicated in the figure, it then becomes obvious that f is homotopic to a product of g commutators in $\pi_1(X)$.

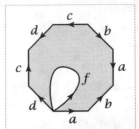

The map $h: \pi_1(X, x_0) \to H_1(X)$ can also be defined by $h([f]) = f_*(\alpha)$ where $f: S^1 \to X$ represents a given element of $\pi_1(X, x_0)$, f_* is the induced map on H_1, and α is the generator of $H_1(S^1) \approx \mathbb{Z}$ represented by the standard map $\sigma: I \to S^1$, $\sigma(s) = e^{2\pi i s}$. This is because both $[f] \in \pi_1(X, x_0)$ and $f_*(\alpha) \in H_1(X)$ are represented by the loop $f\sigma: I \to X$. A consequence of this definition is that $h([f]) = h([g])$ if f and g are homotopic maps $S^1 \to X$, since $f_* = g_*$ by Theorem 2.10.

Example 2A.2. For the closed orientable surface M of genus g, the abelianization of $\pi_1(M)$ is $\mathbb{Z}^{2g}$, the product of $2g$ copies of $\mathbb{Z}$, and a basis for $H_1(M)$ consists of the 1-cycles represented by the 1-cells of M in its standard CW structure. We can also represent a basis by the loops α_i and β_i shown in the figure below since these

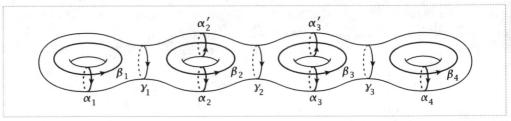

loops are homotopic to the loops represented by the 1-cells, as one can see in the picture of the cell structure in Chapter 0. The loops γ_i, on the other hand, are trivial in homology since the portion of M on one side of γ_i is a compact surface bounded by γ_i, so γ_i is homotopic to a loop that is a product of commutators, as we saw a couple paragraphs earlier. The loop α_i' represents the same homology class as α_i since the region between γ_i and $\alpha_i \cup \alpha_i'$ provides a homotopy between γ_i and a product of two loops homotopic to α_i and the inverse of α_i', so $\alpha_i - \alpha_i' \sim \gamma_i \sim 0$, hence $\alpha_i \sim \alpha_i'$.

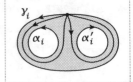

2.B Classical Applications

In this section we use homology theory to prove several interesting results in topology and algebra whose statements give no hint that algebraic topology might be involved.

To begin, we calculate the homology of complements of embedded spheres and disks in a sphere. Recall that an embedding is a map that is a homeomorphism onto its image.

Proposition 2B.1. **(a)** *For an embedding* $h : D^k \to S^n$, $\tilde{H}_i(S^n - h(D^k)) = 0$ *for all* i. **(b)** *For an embedding* $h : S^k \to S^n$ *with* $k < n$, $\tilde{H}_i(S^n - h(S^k))$ *is* $\mathbb{Z}$ *for* $i = n - k - 1$ *and* 0 *otherwise*.

As a special case of (b) we have the Jordan curve theorem: A subspace of S^2 homeomorphic to S^1 separates S^2 into two complementary components, or equivalently, path-components since open subsets of S^n are locally path-connected. One could just as well use $\mathbb{R}^2$ in place of S^2 here since deleting a point from an open set in S^2 does not affect its connectedness. More generally, (b) says that a subspace of S^n homeomorphic to S^{n-1} separates it into two components, and these components have the same homology groups as a point. Somewhat surprisingly, there are embeddings where these complementary components are not simply-connected as they are for the standard embedding. An example is the Alexander horned sphere in S^3 which we describe in detail following the proof of the proposition. These complications involving embedded S^{n-1}'s in S^n are all local in nature since it is known that any locally nicely embedded S^{n-1} in S^n is equivalent to the standard $S^{n-1} \subset S^n$, equivalent in the sense that there is a homeomorphism of S^n taking the given embedded S^{n-1} onto the standard S^{n-1}. In particular, both complementary regions are homeomorphic to open balls. See [Brown 1960] for a precise statement and proof. When $n = 2$ it is a classical theorem of Schoenflies that all embeddings $S^1 \hookrightarrow S^2$ are equivalent.

By contrast, when we come to embeddings of S^{n-2} in S^n, even locally nice embeddings need not be equivalent to the standard one. This is the subject of knot theory, including the classical case of knotted embeddings of S^1 in S^3 or $\mathbb{R}^3$. For embeddings of S^{n-2} in S^n the complement always has the same homology as S^1, according to the theorem, but the fundamental group can be quite different. In spite of the fact that the homology of a knot complement does not detect knottedness, it is still possible to use homology to distinguish different knots by looking at the homology of covering spaces of their complements.

Proof: We prove (a) by induction on k. When $k = 0$, $S^n - h(D^0)$ is homeomorphic to $\mathbb{R}^n$, so this case is trivial. For the induction step it will be convenient to replace the domain disk D^k of h by the cube I^k. Let $A = S^n - h(I^{k-1} \times [0, 1/2])$ and let

$B = S^n - h(I^{k-1} \times [1/2, 1])$, so $A \cap B = S^n - h(I^k)$ and $A \cup B = S^n - h(I^{k-1} \times \{1/2\})$. By induction $\widetilde{H}_i(A \cup B) = 0$ for all i, so the Mayer–Vietoris sequence gives isomorphisms $\Phi : \widetilde{H}_i(S^n - h(I^k)) \to \widetilde{H}_i(A) \oplus \widetilde{H}_i(B)$ for all i. Modulo signs, the two components of Φ are induced by the inclusions $S^n - h(I^k) \hookrightarrow A$ and $S^n - h(I^k) \hookrightarrow B$, so if there exists an i-dimensional cycle α in $S^n - h(I^k)$ that is not a boundary in $S^n - h(I^k)$, then α is also not a boundary in at least one of A and B. (When $i = 0$ the word 'cycle' here is to be interpreted in the sense of augmented chain complexes since we are dealing with reduced homology.) By iteration we can then produce a nested sequence of closed intervals $I_1 \supset I_2 \supset \cdots$ in the last coordinate of I^k shrinking down to a point $p \in I$, such that α is not a boundary in $S^n - h(I^{k-1} \times I_m)$ for any m. On the other hand, by induction on k we know that α is the boundary of a chain β in $S^n - h(I^{k-1} \times \{p\})$. This β is a finite linear combination of singular simplices with compact image in $S^n - h(I^{k-1} \times \{p\})$. The union of these images is covered by the nested sequence of open sets $S^n - h(I^{k-1} \times I_m)$, so by compactness β must actually be a chain in $S^n - h(I^{k-1} \times I_m)$ for some m. This contradiction shows that α must be a boundary in $S^n - h(I^k)$, finishing the induction step.

Part (b) is also proved by induction on k, starting with the trivial case $k = 0$ when $S^n - h(S^0)$ is homeomorphic to $S^{n-1} \times \mathbb{R}$. For the induction step, write S^k as the union of hemispheres D_+^k and D_-^k intersecting in S^{k-1}. The Mayer–Vietoris sequence for $A = S^n - h(D_+^k)$ and $B = S^n - h(D_-^k)$, both of which have trivial reduced homology by part (a), then gives isomorphisms $\widetilde{H}_i(S^n - h(S^k)) \approx \widetilde{H}_{i+1}(S^n - h(S^{k-1}))$. $\qquad\qquad$ $\square$

If we apply the last part of this proof to an embedding $h : S^n \to S^n$, the Mayer-Vietoris sequence ends with the terms $\widetilde{H}_0(A) \oplus \widetilde{H}_0(B) \to \widetilde{H}_0(S^n - h(S^{n-1})) \to 0$. Both $\widetilde{H}_0(A)$ and $\widetilde{H}_0(B)$ are zero, so exactness would imply that $\widetilde{H}_0(S^n - h(S^{n-1})) = 0$ which appears to contradict the fact that $S^n - h(S^{n-1})$ has two path-components. The only way out of this dilemma is for h to be surjective, so that $A \cap B$ is empty and the 0 at the end of the Mayer-Vietoris sequence is $\widetilde{H}_{-1}(\varnothing)$ which is $\mathbb{Z}$ rather than 0.

In particular, this shows that S^n cannot be embedded in $\mathbb{R}^n$ since this would yield a nonsurjective embedding in S^n. A consequence is that there is no embedding $\mathbb{R}^m \hookrightarrow \mathbb{R}^n$ for $m > n$ since this would restrict to an embedding of $S^n \subset \mathbb{R}^m$ into $\mathbb{R}^n$. More generally there is no continuous injection $\mathbb{R}^m \to \mathbb{R}^n$ for $m > n$ since this too would give an embedding $S^n \hookrightarrow \mathbb{R}^n$.

Example 2B.2: The Alexander Horned Sphere. This is a subspace $S \subset \mathbb{R}^3$ homeomorphic to S^2 such that the unbounded component of $\mathbb{R}^3 - S$ is not simply-connected as it is for the standard $S^2 \subset \mathbb{R}^3$. We will construct S by defining a sequence of compact subspaces $X_0 \supset X_1 \supset \cdots$ of $\mathbb{R}^3$ whose intersection is homeomorphic to a ball, and then S will be the boundary sphere of this ball.

We begin with X_0 a solid torus $S^1 \times D^2$ obtained from a ball B_0 by attaching a handle $I \times D^2$ along $\partial I \times D^2$. In the figure this handle is shown as the union of

two 'horns' attached to the ball, together with
a shorter handle drawn as dashed lines. To
form the space $X_1 \subset X_0$ we delete part of the
short handle, so that what remains is a pair of
linked handles attached to the ball B_1 that is
the union of B_0 with the two horns. To form
X_2 the process is repeated: Decompose each of
the second stage handles as a pair of horns and
a short handle, then delete a part of the short
handle. In the same way X_n is constructed inductively from X_{n-1}. Thus X_n is a ball B_n with
2^n handles attached, and B_n is obtained from
B_{n-1} by attaching 2^n horns. There are homeomorphisms $h_n : B_{n-1} \to B_n$ that are the identity

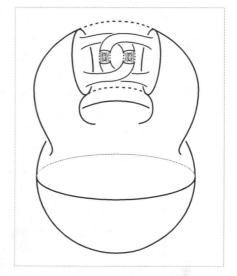

outside a small neighborhood of $B_n - B_{n-1}$. As n goes to infinity, the composition
$h_n \cdots h_1$ approaches a map $f : B_0 \to \mathbb{R}^3$ which is continuous since the convergence is
uniform. The set of points in B_0 where f is not equal to $h_n \cdots h_1$ for large n is a
Cantor set, whose image under f is the intersection of all the handles. It is not hard
to see that f is one-to-one. By compactness it follows that f is a homeomorphism
onto its image, a ball $B \subset \mathbb{R}^3$ whose boundary sphere $f(\partial B_0)$ is S, the Alexander
horned sphere.

Now we compute $\pi_1(\mathbb{R}^3 - B)$. Note that B is the intersection of the X_n's, so $\mathbb{R}^3 - B$
is the union of the complements Y_n of the X_n's, which form an increasing sequence
$Y_0 \subset Y_1 \subset \cdots$. We will show that the groups $\pi_1(Y_n)$ also form an increasing sequence
of successively larger groups, whose union is $\pi_1(\mathbb{R}^3 - B)$. To begin we have $\pi_1(Y_0) \approx \mathbb{Z}$
since X_0 is a solid torus embedded in $\mathbb{R}^3$ in a standard way. To compute $\pi_1(Y_1)$, let
$\overline{Y}_0$ be the closure of Y_0 in Y_1, so $\overline{Y}_0 - Y_0$ is an open annulus A and $\pi_1(\overline{Y}_0)$ is also $\mathbb{Z}$.
We obtain Y_1 from $\overline{Y}_0$ by attaching the space $Z = Y_1 - Y_0$ along A. The group $\pi_1(Z)$
is the free group F_2 on two generators α_1 and α_2 represented by loops linking the
two handles, since $Z - A$ is homeomorphic to an open ball with two straight tubes
deleted. A loop α generating $\pi_1(A)$ represents the commutator $[\alpha_1, \alpha_2]$, as one
can see by noting that the closure of Z is obtained from Z by adjoining two disjoint
surfaces, each homeomorphic to a torus with an open disk removed; the boundary
of this disk is homotopic to α and is also homotopic to the commutator of meridian
and longitude circles in the torus, which correspond to α_1 and α_2. Van Kampen's
theorem now implies that the inclusion $Y_0 \hookrightarrow Y_1$ induces an injection of $\pi_1(Y_0)$ into
$\pi_1(Y_1)$ as the infinite cyclic subgroup generated by $[\alpha_1, \alpha_2]$.

In a similar way we can regard Y_{n+1} as being obtained from Y_n by adjoining 2^n
copies of Z. Assuming inductively that $\pi_1(Y_n)$ is the free group F_{2^n} with generators
represented by loops linking the 2^n smallest handles of X_n, then each copy of Z ad-

joined to Y_n changes $\pi_1(Y_n)$ by making one of the generators into the commutator of two new generators. Note that adjoining a copy of Z induces an injection on π_1 since the induced homomorphism is the free product of the injection $\pi_1(A) \to \pi_1(Z)$ with the identity map on the complementary free factor. Thus the map $\pi_1(Y_n) \to \pi_1(Y_{n+1})$ is an injection $F_{2^n} \to F_{2^{n+1}}$. The group $\pi_1(\mathbb{R}^3 - B)$ is isomorphic to the union of this increasing sequence of groups by a compactness argument: Each loop in $\mathbb{R}^3 - B$ has compact image and hence must lie in some Y_n, and similarly for homotopies of loops.

In particular we see explicitly why $\pi_1(\mathbb{R}^3 - B)$ has trivial abelianization, because each of its generators is exactly equal to the commutator of two other generators. This inductive construction in which each generator of a free group is decreed to be the commutator of two new generators is perhaps the simplest way of building a nontrivial group with trivial abelianization, and for the construction to have such a nice geometric interpretation is something to marvel at. From a naive viewpoint it may seem a little odd that a highly nonfree group can be built as a union of an increasing sequence of free groups, but this can also easily happen for abelian groups, as $\mathbb{Q}$ for example is the union of an increasing sequence of infinite cyclic subgroups.

The next theorem says that for subspaces of $\mathbb{R}^n$, the property of being open is a topological invariant. This result is known classically as Invariance of Domain, the word 'domain' being an older designation for an open set in $\mathbb{R}^n$.

Theorem 2B.3. *If U is an open set in $\mathbb{R}^n$ and $h : U \to \mathbb{R}^n$ is an embedding, or more generally just a continuous injection, then the image $h(U)$ is an open set in $\mathbb{R}^n$.*

Proof: Viewing S^n as the one-point compactification of $\mathbb{R}^n$, an equivalent statement is that $h(U)$ is open in S^n, and this is what we will prove. Each $x \in U$ is the center point of a disk $D^n \subset U$. It will suffice to prove that $h(D^n - \partial D^n)$ is open in S^n. The hypothesis on h implies that its restrictions to D^n and ∂D^n are embeddings. By the previous proposition $S^n - h(\partial D^n)$ has two path-components. These path-components are $h(D^n - \partial D^n)$ and $S^n - h(D^n)$ since these two subspaces are disjoint and path-connected, the first since it is homeomorphic to $D^n - \partial D^n$ and the second by the proposition. Since $S^n - h(\partial D^n)$ is open in S^n, its path-components are the same as its components. The components of a space with finitely many components are open, so $h(D^n - \partial D^n)$ is open in $S^n - h(\partial D^n)$ and hence also in S^n. $\qquad\square$

Here is an application involving the notion of an n-manifold, which is a Hausdorff space locally homeomorphic to $\mathbb{R}^n$:

Corollary 2B.4. *If M is a compact n-manifold and N is a connected n-manifold, then an embedding $h : M \to N$ must be surjective, hence a homeomorphism.*

Proof: $h(M)$ is closed in N since it is compact and N is Hausdorff. Since N is connected it suffices to show $h(M)$ is also open in N, and this is immediate from the theorem. $\qquad\square$

The Invariance of Domain and the n-dimensional generalization of the Jordan curve theorem were first proved by Brouwer around 1910, at a very early stage in the development of algebraic topology.

Division Algebras

Here is an algebraic application of homology theory due to H. Hopf:

Theorem 2B.5. $\mathbb{R}$ *and* $\mathbb{C}$ *are the only finite-dimensional division algebras over* $\mathbb{R}$ *which are commutative and have an identity.*

By definition, an **algebra** structure on $\mathbb{R}^n$ is simply a bilinear multiplication map $\mathbb{R}^n \times \mathbb{R}^n \to \mathbb{R}^n$, $(a, b) \mapsto ab$. Thus the product satisfies left and right distributivity, $a(b+c) = ab + ac$ and $(a+b)c = ac + bc$, and scalar associativity, $\alpha(ab) = (\alpha a)b = a(\alpha b)$ for $\alpha \in \mathbb{R}$. Commutativity, full associativity, and an identity element are not assumed. An algebra is a **division algebra** if the equations $ax = b$ and $xa = b$ are always solvable whenever $a \neq 0$. In other words, the linear transformations $x \mapsto ax$ and $x \mapsto xa$ are surjective when $a \neq 0$. These are linear maps $\mathbb{R}^n \to \mathbb{R}^n$, so surjectivity is equivalent to having trivial kernel, which means there are no zero-divisors.

The four classical examples are $\mathbb{R}$, $\mathbb{C}$, the quaternion algebra $\mathbb{H}$, and the octonion algebra $\mathbb{O}$. Frobenius proved in 1877 that $\mathbb{R}$, $\mathbb{C}$, and $\mathbb{H}$ are the only finite-dimensional associative division algebras over $\mathbb{R}$ with an identity element. If the product satisfies $|ab| = |a||b|$ as in the classical examples, then Hurwitz showed in 1898 that the dimension of the algebra must be 1, 2, 4, or 8, and others subsequently showed that the only examples with an identity element are the classical ones. A full discussion of all this, including some examples showing the necessity of the hypothesis of an identity element, can be found in [Ebbinghaus 1991]. As one would expect, the proofs of these results are algebraic, but if one drops the condition that $|ab| = |a||b|$ it seems that more topological proofs are required. We will show in Theorem 3.21 that a finite-dimensional division algebra over $\mathbb{R}$ must have dimension a power of 2. The fact that the dimension can be at most 8 is a famous theorem of [Bott & Milnor 1958] and [Kervaire 1958]. See §4.B for a few more comments on this.

Proof: Suppose first that $\mathbb{R}^n$ has a commutative division algebra structure. Define a map $f : S^{n-1} \to S^{n-1}$ by $f(x) = x^2/|x^2|$. This is well-defined since $x \neq 0$ implies $x^2 \neq 0$ in a division algebra. The map f is continuous since the multiplication map $\mathbb{R}^n \times \mathbb{R}^n \to \mathbb{R}^n$ is bilinear, hence continuous. Since $f(-x) = f(x)$ for all x, f induces a quotient map $\overline{f} : \mathbb{R}\mathrm{P}^{n-1} \to S^{n-1}$. The following argument shows that $\overline{f}$ is injective. An equality $f(x) = f(y)$ implies $x^2 = \alpha^2 y^2$ for $\alpha = (|x^2|/|y^2|)^{1/2} > 0$. Thus we have $x^2 - \alpha^2 y^2 = 0$, which factors as $(x + \alpha y)(x - \alpha y) = 0$ using commutativity and the fact that α is a real scalar. Since there are no divisors of zero, we deduce that $x = \pm \alpha y$. Since x and y are unit vectors and α is real, this yields $x = \pm y$, so x and y determine the same point of $\mathbb{R}\mathrm{P}^{n-1}$, which means that $\overline{f}$ is injective.

Since $\overline{f}$ is an injective map of compact Hausdorff spaces, it must be a homeomorphism onto its image. By Corollary 2B.4, $\overline{f}$ must in fact be surjective if we are not in the trivial case $n = 1$. Thus we have a homeomorphism $\mathbb{RP}^{n-1} \approx S^{n-1}$. This implies $n = 2$ since if $n > 2$ the spaces $\mathbb{RP}^{n-1}$ and S^{n-1} have different homology groups (or different fundamental groups).

It remains to show that a 2-dimensional commutative division algebra A with identity is isomorphic to $\mathbb{C}$. This is elementary algebra: If $j \in A$ is not a real scalar multiple of the identity element $1 \in A$ and we write $j^2 = a + bj$ for $a, b \in \mathbb{R}$, then $(j - b/2)^2 = a + b^2/4$ so by rechoosing j we may assume that $j^2 = a \in \mathbb{R}$. If $a \geq 0$, say $a = c^2$, then $j^2 = c^2$ implies $(j + c)(j - c) = 0$, so $j = \pm c$, but this contradicts the choice of j. So $j^2 = -c^2$ and by rescaling j we may assume $j^2 = -1$, hence A is isomorphic to $\mathbb{C}$. $\qquad\qquad\qquad\qquad\qquad\qquad\qquad\qquad\square$

Leaving out the last paragraph, the proof shows that a finite-dimensional commutative division algebra, not necessarily with an identity, must have dimension at most 2. Oddly enough, there do exist 2-dimensional commutative division algebras without identity elements, for example $\mathbb{C}$ with the modified multiplication $z \cdot w = \overline{z}\overline{w}$, the bar denoting complex conjugation.

The Borsuk–Ulam Theorem

In Theorem 1.10 we proved the 2-dimensional case of the Borsuk–Ulam theorem, and now we will give a proof for all dimensions, using the following theorem of Borsuk:

Proposition 2B.6. *An odd map $f : S^n \rightarrow S^n$, satisfying $f(-x) = -f(x)$ for all x, must have odd degree.*

The corresponding result that even maps have even degree is easier, and was an exercise for §2.2.

The proof will show that using homology with a coefficient group other than $\mathbb{Z}$ can sometimes be a distinct advantage. The main ingredient will be a certain exact sequence associated to a two-sheeted covering space $p : \widetilde{X} \rightarrow X$,

$$\cdots \longrightarrow H_n(X; \mathbb{Z}_2) \xrightarrow{\tau_*} H_n(\widetilde{X}; \mathbb{Z}_2) \xrightarrow{p_*} H_n(X; \mathbb{Z}_2) \longrightarrow H_{n-1}(X; \mathbb{Z}_2) \longrightarrow \cdots$$

This is the long exact sequence of homology groups associated to a short exact sequence of chain complexes consisting of short exact sequences of chain groups

$$0 \longrightarrow C_n(X; \mathbb{Z}_2) \xrightarrow{\tau} C_n(\widetilde{X}; \mathbb{Z}_2) \xrightarrow{p_\sharp} C_n(X; \mathbb{Z}_2) \longrightarrow 0$$

The map $p_\sharp$ is surjective since singular simplices $\sigma : \Delta^n \rightarrow X$ always lift to $\widetilde{X}$, as Δ^n is simply-connected. Each σ has in fact precisely two lifts $\widetilde{\sigma}_1$ and $\widetilde{\sigma}_2$. Because we are using $\mathbb{Z}_2$ coefficients, the kernel of $p_\sharp$ is generated by the sums $\widetilde{\sigma}_1 + \widetilde{\sigma}_2$. So if we define τ to send each $\sigma : \Delta^n \rightarrow X$ to the sum of its two lifts to $\widetilde{\Delta}^n$, then the image of τ is the kernel of $p_\sharp$. Obviously τ is injective, so we have the short exact sequence indicated. Since τ and $p_\sharp$ commute with boundary maps, we have a short exact sequence of chain complexes, yielding the long exact sequence of homology groups.

The map τ_* is a special case of more general *transfer homomorphisms* considered in §3.G, so we will refer to the long exact sequence involving the maps τ_* as the *transfer sequence*. This sequence can also be viewed as a special case of the Gysin sequences discussed in §4.D. There is a generalization of the transfer sequence to homology with other coefficients, but this uses a more elaborate form of homology called homology with local coefficients, as we show in §3.H.

Proof of 2B.6: The proof will involve the transfer sequence for the covering space $p : S^n \to \mathbb{R}P^n$. This has the following form, where to simplify notation we abbreviate $\mathbb{R}P^n$ to P^n and we let the coefficient group $\mathbb{Z}_2$ be implicit:

$$0 \longrightarrow H_n(P^n) \xrightarrow[\approx]{\tau_*} H_n(S^n) \xrightarrow[0]{p_*} H_n(P^n) \xrightarrow{\approx} H_{n-1}(P^n) \longrightarrow 0 \longrightarrow \cdots$$

$$\cdots \longrightarrow 0 \longrightarrow H_i(P^n) \xrightarrow{\approx} H_{i-1}(P^n) \longrightarrow 0 \longrightarrow \cdots$$

$$\cdots \longrightarrow 0 \longrightarrow H_1(P^n) \xrightarrow{\approx} H_0(P^n) \xrightarrow{0} H_0(S^n) \xrightarrow[\approx]{p_*} H_0(P^n) \longrightarrow 0$$

The initial 0 is $H_{n+1}(P^n; \mathbb{Z}_2)$, which vanishes since P^n is an n-dimensional CW complex. The other terms that are zero are $H_i(S^n)$ for $0 < i < n$. We assume $n > 1$, leaving the minor modifications needed for the case $n = 1$ to the reader. All the terms that are not zero are $\mathbb{Z}_2$, by cellular homology. Alternatively, this exact sequence can be used to compute the homology groups $H_i(\mathbb{R}P^n; \mathbb{Z}_2)$ if one does not already know them. Since all the nonzero groups in the sequence are $\mathbb{Z}_2$, exactness forces the maps to be isomorphisms or zero as indicated.

An odd map $f : S^n \to S^n$ induces a quotient map $\overline{f} : \mathbb{R}P^n \to \mathbb{R}P^n$. These two maps induce a map from the transfer sequence to itself, and we will need to know that the squares in the resulting diagram commute. This follows from the naturality of the long exact sequence of homology associated to a short exact sequence of chain complexes, once we verify commutativity of the diagram

$$
\begin{array}{ccccccccc}
0 & \longrightarrow & C_i(P^n) & \xrightarrow{\tau} & C_i(S^n) & \xrightarrow{p_\sharp} & C_i(P^n) & \longrightarrow & 0 \\
 & & \downarrow{\overline{f}_\sharp} & & \downarrow{f_\sharp} & & \downarrow{\overline{f}_\sharp} & & \\
0 & \longrightarrow & C_i(P^n) & \xrightarrow{\tau} & C_i(S^n) & \xrightarrow{p_\sharp} & C_i(P^n) & \longrightarrow & 0
\end{array}
$$

Here the right-hand square commutes since $pf = \overline{f}p$. The left-hand square commutes since for a singular i-simplex $\sigma : \Delta^i \to P^n$ with lifts $\widetilde{\sigma}_1$ and $\widetilde{\sigma}_2$, the two lifts of $\overline{f}\sigma$ are $f\widetilde{\sigma}_1$ and $f\widetilde{\sigma}_2$ since f takes antipodal points to antipodal points.

Now we can see that all the maps f_* and $\overline{f}_*$ in the commutative diagram of transfer sequences are isomorphisms by induction on dimension, using the evident fact that if three maps in a commutative square are isomorphisms, so is the fourth. The induction starts with the trivial fact that f_* and $\overline{f}_*$ are isomorphisms in dimension zero.

In particular we deduce that the map $f_*:H_n(S^n;\mathbb{Z}_2) \to H_n(S^n;\mathbb{Z}_2)$ is an isomorphism. By Lemma 2.49 this map is multiplication by the degree of f mod 2, so the degree of f must be odd. $\qquad\square$

The fact that odd maps have odd degree easily implies the Borsuk–Ulam theorem:

Corollary 2B.7. *For every map $g:S^n \to \mathbb{R}^n$ there exists a point $x \in S^n$ with $g(x) = g(-x)$.*

Proof: Let $f(x) = g(x) - g(-x)$, so f is odd. We need to show that $f(x) = 0$ for some x. If this is not the case, we can replace $f(x)$ by $f(x)/|f(x)|$ to get a new map $f:S^n \to S^{n-1}$ which is still odd. The restriction of this f to the equator S^{n-1} then has odd degree by the proposition. But this restriction is nullhomotopic via the restriction of f to one of the hemispheres bounded by S^{n-1}. $\qquad\square$

Exercises

1. Compute $H_i(S^n - X)$ when X is a subspace of S^n homeomorphic to $S^k \vee S^\ell$ or to $S^k \amalg S^\ell$.

2. Show that $\tilde{H}_i(S^n - X) \approx \tilde{H}_{n-i-1}(X)$ when X is homeomorphic to a finite connected graph. [First do the case that the graph is a tree.]

3. Let $(D,S) \subset (D^n, S^{n-1})$ be a pair of subspaces homeomorphic to (D^k, S^{k-1}), with $D \cap S^{n-1} = S$. Show the inclusion $S^{n-1} - S \hookrightarrow D^n - D$ induces an isomorphism on homology. [Glue two copies of (D^n, D) to the two ends of $(S^{n-1} \times I, S \times I)$ to produce a k-sphere in S^n and look at a Mayer–Vietoris sequence for the complement of this k-sphere.]

4. In the unit sphere $S^{p+q-1} \subset \mathbb{R}^{p+q}$ let S^{p-1} and S^{q-1} be the subspheres consisting of points whose last q and first p coordinates are zero, respectively.
(a) Show that $S^{p+q-1} - S^{p-1}$ deformation retracts onto S^{q-1}, and is in fact homeomorphic to $S^{q-1} \times \mathbb{R}^p$.
(b) Show that S^{p-1} and S^{q-1} are not the boundaries of any pair of disjointly embedded disks D^p and D^q in D^{p+q}. [The preceding exercise may be useful.]

5. Let S be an embedded k-sphere in S^n for which there exists a disk $D^n \subset S^n$ intersecting S in the disk $D^k \subset D^n$ defined by the first k coordinates of D^n. Let $D^{n-k} \subset D^n$ be the disk defined by the last $n - k$ coordinates, with boundary sphere S^{n-k-1}. Show that the inclusion $S^{n-k-1} \hookrightarrow S^n - S$ induces an isomorphism on homology groups.

6. Modify the construction of the Alexander horned sphere to produce an embedding $S^2 \hookrightarrow \mathbb{R}^3$ for which neither component of $\mathbb{R}^3 - S^2$ is simply-connected.

7. Analyze what happens when the number of handles in the basic building block for the Alexander horned sphere is doubled, as in the figure at the right.

8. Show that $\mathbb{R}^{2n+1}$ is not a division algebra over $\mathbb{R}$ if $n > 0$ by considering how the determinant of the linear map $x \mapsto ax$ given by the multiplication in a division algebra structure would vary as a moves along a path in $\mathbb{R}^{2n+1} - \{0\}$ joining two antipodal points.

9. Make the transfer sequence explicit in the case of a trivial covering $\widetilde{X} \to X$, where $\widetilde{X} = X \times S^0$.

10. Use the transfer sequence for the covering $S^\infty \to \mathbb{R}\mathrm{P}^\infty$ to compute $H_n(\mathbb{R}\mathrm{P}^\infty; \mathbb{Z}_2)$.

11. Use the transfer sequence for the covering $X \times S^\infty \to X \times \mathbb{R}\mathrm{P}^\infty$ to produce isomorphisms $H_n(X \times \mathbb{R}\mathrm{P}^\infty; \mathbb{Z}_2) \approx \bigoplus_{i \leq n} H_i(X; \mathbb{Z}_2)$ for all n.

2.C Simplicial Approximation

Many spaces of interest in algebraic topology can be given the structure of simplicial complexes, and early in the history of the subject this structure was exploited as one of the main technical tools. Later, CW complexes largely superseded simplicial complexes in this role, but there are still some occasions when the extra structure of simplicial complexes can be quite useful. This will be illustrated nicely by the proof of the classical Lefschetz fixed point theorem in this section.

One of the good features of simplicial complexes is that arbitrary continuous maps between them can always be deformed to maps that are linear on the simplices of some subdivision of the domain complex. This is the idea of 'simplicial approximation,' developed by Brouwer and Alexander before 1920. Here is the relevant definition: If K and L are simplicial complexes, then a map $f : K \to L$ is **simplicial** if it sends each simplex of K to a simplex of L by a linear map taking vertices to vertices. In barycentric coordinates, a linear map of a simplex $[v_0, \cdots, v_n]$ has the form $\sum_i t_i v_i \mapsto \sum_i t_i f(v_i)$. Since a linear map from a simplex to a simplex is uniquely determined by its values on vertices, this means that a simplicial map is uniquely determined by its values on vertices. It is easy to see that a map from the vertices of K to the vertices of L extends to a simplicial map iff it sends the vertices of each simplex of K to the vertices of some simplex of L.

Here is the most basic form of the **Simplicial Approximation Theorem**:

Theorem 2C.1. *If K is a finite simplicial complex and L is an arbitrary simplicial complex, then any map $f : K \to L$ is homotopic to a map that is simplicial with respect to some iterated barycentric subdivision of K.*

To see that subdivision of K is essential, consider the case of maps $S^n \to S^n$. With fixed simplicial structures on the domain and range spheres there are only finitely many simplicial maps since there are only finitely many ways to map vertices to vertices. Hence only finitely many degrees are realized by maps that are simplicial with respect to fixed simplicial structures in both the domain and range spheres. This remains true even if the simplicial structure on the range sphere is allowed to vary, since if the range sphere has more vertices than the domain sphere then the map cannot be surjective, hence must have degree zero.

Before proving the simplicial approximation theorem we need some terminology and a lemma. The **star** St σ of a simplex σ in a simplicial complex X is defined to be the subcomplex consisting of all the simplices of X that contain σ. Closely related to this is the **open star** st σ, which is the union of the interiors of all simplices containing σ, where the interior of a simplex τ is by definition $\tau - \partial\tau$. Thus st σ is an open set in X whose closure is St σ.

Lemma 2C.2. *For vertices $v_1, \cdots, v_n$ of a simplicial complex X, the intersection* st $v_1 \cap \cdots \cap$ st v_n *is empty unless $v_1, \cdots, v_n$ are the vertices of a simplex σ of X, in which case* st $v_1 \cap \cdots \cap$ st $v_n =$ st σ.

Proof: The intersection st $v_1 \cap \cdots \cap$ st v_n consists of the interiors of all simplices τ whose vertex set contains $\{v_1, \cdots, v_n\}$. If st $v_1 \cap \cdots \cap$ st v_n is nonempty, such a τ exists and contains the simplex $\sigma = [v_1, \cdots, v_n] \subset X$. The simplices τ containing $\{v_1, \cdots, v_n\}$ are just the simplices containing σ, so st $v_1 \cap \cdots \cap$ st $v_n =$ st σ. $\square$

Proof of 2C.1: Choose a metric on K that restricts to the standard Euclidean metric on each simplex of K. For example, K can be viewed as a subcomplex of a simplex Δ^N whose vertices are all the vertices of K, and we can restrict a standard metric on Δ^N to give a metric on K. Let ε be a Lebesgue number for the open cover $\{ f^{-1}(\text{st } w) \mid w \text{ is a vertex of } L \}$ of K. After iterated barycentric subdivision of K we may assume that each simplex has diameter less than $\varepsilon/2$. The closed star of each vertex v of K then has diameter less than ε, hence this closed star maps by f to the open star of some vertex $g(v)$ of L. The resulting map $g: K^0 \to L^0$ thus satisfies $f(\text{St } v) \subset \text{st } g(v)$ for all vertices v of K.

To see that g extends to a simplicial map $g: K \to L$, consider the problem of extending g over a simplex $[v_1, \cdots, v_n]$ of K. An interior point x of this simplex lies in st v_i for each i, so $f(x)$ lies in st $g(v_i)$ for each i, since $f(\text{st } v_i) \subset \text{st } g(v_i)$ by the definition of $g(v_i)$. Thus st $g(v_1) \cap \cdots \cap$ st $g(v_n) \neq \varnothing$, so $[g(v_1), \cdots, g(v_n)]$ is a simplex of L by the lemma, and we can extend g linearly over $[v_1, \cdots, v_n]$. Both $f(x)$ and $g(x)$ lie in a single simplex of L since $g(x)$ lies in $[g(v_1), \cdots, g(v_n)]$ and $f(x)$ lies in the star of this simplex. So taking the linear path $(1-t)f(x) + tg(x)$, $0 \leq t \leq 1$, in the simplex containing $f(x)$ and $g(x)$ defines a homotopy from f to g. To check continuity of this homotopy it suffices to restrict to the simplex $[v_1, \cdots, v_n]$, where continuity is clear since $f(x)$ varies continuously in the star of $[g(v_1), \cdots, g(v_n)]$ and $g(x)$ varies continuously in $[g(v_1), \cdots, g(v_n)]$. $\square$

Notice that if f already sends some vertices of K to vertices of L then we may choose g to equal to f on these vertices, and hence the homotopy from f to g will be stationary on these vertices. This is convenient if one is in a situation where one wants maps and homotopies to preserve basepoints.

The proof makes it clear that the simplicial approximation g can be chosen not just homotopic to f but also close to f if we allow subdivisions of L as well as K.

The Lefschetz Fixed Point Theorem

This very classical application of homology is a considerable generalization of the Brouwer fixed point theorem. It is also related to the Euler characteristic formula.

For a homomorphism $\varphi : \mathbb{Z}^n \to \mathbb{Z}^n$ with matrix $[a_{ij}]$, the trace $\operatorname{tr} \varphi$ is defined to be $\sum_i a_{ii}$, the sum of the diagonal elements of $[a_{ij}]$. Since $\operatorname{tr}([a_{ij}][b_{ij}]) = \operatorname{tr}([b_{ij}][a_{ij}])$, conjugate matrices have the same trace, and it follows that $\operatorname{tr} \varphi$ is independent of the choice of basis for $\mathbb{Z}^n$. For a homomorphism $\varphi : A \to A$ of a finitely generated abelian group A we can then define $\operatorname{tr} \varphi$ to be the trace of the induced homomorphism $\overline{\varphi} : A/Torsion \to A/Torsion$.

For a map $f : X \to X$ of a finite CW complex X, or more generally any space whose homology groups are finitely generated and vanish in high dimensions, the **Lefschetz number** $\tau(f)$ is defined to be $\sum_n (-1)^n \operatorname{tr}(f_* : H_n(X) \to H_n(X))$. In particular, if f is the identity, or is homotopic to the identity, then $\tau(f)$ is the Euler characteristic $\chi(X)$ since the trace of the $n \times n$ identity matrix is n.

Here is the Lefschetz fixed point theorem:

Theorem 2C.3. *If X is a finite simplicial complex, or more generally a retract of a finite simplicial complex, and $f : X \to X$ is a map with $\tau(f) \neq 0$, then f has a fixed point.*

As we show in Theorem A.7 in the Appendix, every compact, locally contractible space that can be embedded in $\mathbb{R}^n$ for some n is a retract of a finite simplicial complex. This includes compact manifolds and finite CW complexes, for example. The compactness hypothesis is essential, since a translation of $\mathbb{R}$ has $\tau = 1$ but no fixed points. For an example showing that local properties are also significant, let X be the compact subspace of $\mathbb{R}^2$ consisting of two concentric circles together with a copy of $\mathbb{R}$ between them whose two ends spiral in to the two circles, wrapping around them infinitely often, and let $f : X \to X$ be a homeomorphism translating the copy of $\mathbb{R}$ along itself and rotating the circles, with no fixed points. Since f is homotopic to the identity, we have $\tau(f) = \chi(X)$, which equals 1 since the three path components of X are two circles and a line.

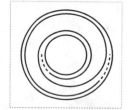

If X has the same homology groups as a point, at least modulo torsion, then the theorem says that every map $X \to X$ has a fixed point. This holds for example for $\mathbb{R}P^n$ if n is even. The case of projective spaces is interesting because of its connection with linear algebra. An invertible linear transformation $f : \mathbb{R}^n \to \mathbb{R}^n$ takes lines through 0 to lines through 0, hence induces a map $\bar{f} : \mathbb{R}P^{n-1} \to \mathbb{R}P^{n-1}$. Fixed points of $\bar{f}$ are equivalent to eigenvectors of f. The characteristic polynomial of f has odd degree if n is odd, hence has a real root, so an eigenvector exists in this case. This is in agreement with the observation above that every map $\mathbb{R}P^{2k} \to \mathbb{R}P^{2k}$ has a fixed point. On the other hand the rotation of $\mathbb{R}^{2k}$ defined by $f(x_1, \cdots, x_{2k}) = (x_2, -x_1, x_4, -x_3, \cdots, x_{2k}, -x_{2k-1})$ has no eigenvectors and its projectivization $\bar{f} : \mathbb{R}P^{2k-1} \to \mathbb{R}P^{2k-1}$ has no fixed points.

Similarly, in the complex case an invertible linear transformation $f : \mathbb{C}^n \to \mathbb{C}^n$ induces $\bar{f} : \mathbb{C}P^{n-1} \to \mathbb{C}P^{n-1}$, and this always has a fixed point since the characteristic polynomial always has a complex root. Nevertheless, as in the real case there is a map $\mathbb{C}P^{2k-1} \to \mathbb{C}P^{2k-1}$ without fixed points. Namely, consider $f : \mathbb{C}^{2k} \to \mathbb{C}^{2k}$ defined by $f(z_1, \cdots, z_{2k}) = (\bar{z}_2, -\bar{z}_1, \bar{z}_4, -\bar{z}_3, \cdots, \bar{z}_{2k}, -\bar{z}_{2k-1})$. This map is only 'conjugate-linear' over $\mathbb{C}$, but this is still good enough to imply that f induces a well-defined map $\bar{f}$ on $\mathbb{C}P^{2k-1}$, and it is easy to check that $\bar{f}$ has no fixed points. The similarity between the real and complex cases persists in the fact that every map $\mathbb{C}P^{2k} \to \mathbb{C}P^{2k}$ has a fixed point, though to deduce this from the Lefschetz fixed point theorem requires more structure than homology has, so this will be left as an exercise for §3.2, using cup products in cohomology.

One could go further and consider the quaternionic case. The antipodal map of $S^4 = \mathbb{H}P^1$ has no fixed points, but every map $\mathbb{H}P^n \to \mathbb{H}P^n$ with $n > 1$ does have a fixed point. This is shown in Example 4L.4 using considerably heavier machinery.

Proof of 2C.3: The general case easily reduces to the case of finite simplicial complexes, for suppose $r : K \to X$ is a retraction of the finite simplicial complex K onto X. For a map $f : X \to X$, the composition $fr : K \to X \subset K$ then has exactly the same fixed points as f. Since $r_* : H_n(K) \to H_n(X)$ is projection onto a direct summand, we clearly have $\mathrm{tr}(f_* r_*) = \mathrm{tr} f_*$, so $\tau(f_* r_*) = \tau(f_*)$.

For X a finite simplicial complex, suppose that $f : X \to X$ has no fixed points. We claim there is a subdivision L of X, a further subdivision K of L, and a simplicial map $g : K \to L$ homotopic to f such that $g(\sigma) \cap \sigma = \varnothing$ for each simplex σ of K. To see this, first choose a metric d on X as in the proof of the simplicial approximation theorem. Since f has no fixed points, $d(x, f(x)) > 0$ for all $x \in X$, so by the compactness of X there is an $\varepsilon > 0$ such that $d(x, f(x)) > \varepsilon$ for all x. Choose a subdivision L of X so that the stars of all simplices have diameter less than $\varepsilon/2$. Applying the simplicial approximation theorem, there is a subdivision K of L and a simplicial map $g : K \to L$ homotopic to f. By construction, g has the property that for each simplex σ of K, $f(\sigma)$ is contained in the star of the simplex $g(\sigma)$. Then $g(\sigma) \cap \sigma = \varnothing$

for each simplex σ of K since for any choice of $x \in \sigma$ we have $d(x, f(x)) > \varepsilon$, while $g(\sigma)$ lies within distance $\varepsilon/2$ of $f(x)$ and σ lies within distance $\varepsilon/2$ of x, as a consequence of the fact that σ is contained in a simplex of L, K being a subdivision of L.

The Lefschetz numbers $\tau(f)$ and $\tau(g)$ are equal since f and g are homotopic. Since g is simplicial, it takes the n-skeleton K^n of K to the n-skeleton L^n of L, for each n. Since K is a subdivision of L, L^n is contained in K^n, and hence $g(K^n) \subset K^n$ for all n. Thus g induces a chain map of the cellular chain complex $\{H_n(K^n, K^{n-1})\}$ to itself. This can be used to compute $\tau(g)$ according to the formula

$$\tau(g) = \sum_n (-1)^n \operatorname{tr}(g_* : H_n(K^n, K^{n-1}) \to H_n(K^n, K^{n-1}))$$

This is the analog of Theorem 2.44 for trace instead of rank, and is proved in precisely the same way, based on the elementary algebraic fact that trace is additive for endomorphisms of short exact sequences: Given a commutative diagram as at the right with exact rows, then $\operatorname{tr} \beta = \operatorname{tr} \alpha + \operatorname{tr} \gamma$. This algebraic fact can be proved by reducing to the easy case that A, B, and

$$\begin{array}{ccccccccc} 0 & \longrightarrow & A & \longrightarrow & B & \longrightarrow & C & \longrightarrow & 0 \\ & & \downarrow{\alpha} & & \downarrow{\beta} & & \downarrow{\gamma} & & \\ 0 & \longrightarrow & A & \longrightarrow & B & \longrightarrow & C & \longrightarrow & 0 \end{array}$$

C are free by first factoring out the torsion in B, hence also the torsion in A, then eliminating any remaining torsion in C by replacing A by a larger subgroup $A' \subset B$, with A having finite index in A'. The details of this argument are left to the reader.

Finally, note that $g_* : H_n(K^n, K^{n-1}) \to H_n(K^n, K^{n-1})$ has trace 0 since the matrix for g_* has zeros down the diagonal, in view of the fact that $g(\sigma) \cap \sigma = \varnothing$ for each n-simplex σ. So $\tau(f) = \tau(g) = 0$. $\qquad\square$

Example 2C.4. Let us verify the theorem in an example. Let X be the closed orientable surface of genus 3 as shown in the figure below, with $f : X \to X$ the 180 degree rotation about a vertical axis passing through the central hole of X. Since f has no fixed points, we should have $\tau(f) = 0$. The induced map $f_* : H_0(X) \to H_0(X)$ is the iden-

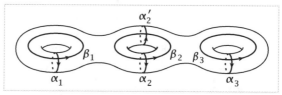

tity, as always for a path-connected space, so this contributes 1 to $\tau(f)$. For $H_1(X)$ we saw in Example 2A.2 that the six loops α_i and β_i represent a basis. The map f_* interchanges the homology classes of α_1 and α_3, and likewise for β_1 and β_3, while β_2 is sent to itself and α_2 is sent to α_2' which is homologous to α_2 as we saw in Example 2A.2. So $f_* : H_1(X) \to H_1(X)$ contributes -2 to $\tau(f)$. It remains to check that $f_* : H_2(X) \to H_2(X)$ is the identity, which we do by the commutative diagram at the right, where x is a point of X in the central torus and $y = f(x)$. We can see that the

$$\begin{array}{ccc} H_2(X) & \xrightarrow{\ f_*\ } & H_2(X) \\ \downarrow{\approx} & & \downarrow{\approx} \\ H_2(X, X - \{x\}) & \xrightarrow{\ f_*\ } & H_2(X, X - \{y\}) \end{array}$$

left-hand vertical map is an isomorphism by considering the long exact sequence of the triple $(X, X - \{x\}, X^1)$ where X^1 is the 1-skeleton of X in its usual CW structure and x is chosen in $X - X^1$, so that $X - \{x\}$ deformation retracts onto X^1 and $H_n(X - \{x\}, X^1) = 0$ for all n. The same reasoning shows the right-hand vertical map is an isomorphism. There is a similar commutative diagram with f replaced by a homeomorphism g that is homotopic to the identity and equals f in a neighborhood of x, with g the identity outside a disk in X containing x and y. Since g is homotopic to the identity, it induces the identity across the top row of the diagram, and since g equals f near x, it induces the same map as f in the bottom row of the diagram, by excision. It follows that the map f_* in the upper row is the identity.

This example generalizes to surfaces of any odd genus by adding symmetric pairs of tori at the left and right. Examples for even genus are described in one of the exercises.

Fixed point theory is a well-developed side branch of algebraic topology, but we touch upon it only occasionally in this book. For a nice introduction see [Brown 1971].

Simplicial Approximations to CW Complexes

The simplicial approximation theorem allows arbitrary continuous maps to be replaced by homotopic simplicial maps in many situations, and one might wonder about the analogous question for spaces: Which spaces are homotopy equivalent to simplicial complexes? We will show this is true for the most common class of spaces in algebraic topology, CW complexes. In the Appendix the question is answered for a few other classes of spaces as well.

Theorem 2C.5. *Every CW complex X is homotopy equivalent to a simplicial complex, which can be chosen to be of the same dimension as X, finite if X is finite, and countable if X is countable.*

We will build a simplicial complex $Y \simeq X$ inductively as an increasing union of subcomplexes Y_n homotopy equivalent to the skeleta X^n. For the inductive step, assuming we have already constructed $Y_n \simeq X^n$, let e^{n+1} be an $(n + 1)$-cell of X attached by a map $\varphi : S^n \to X^n$. The map $S^n \to Y_n$ corresponding to φ under the homotopy equivalence $Y_n \simeq X^n$ is homotopic to a simplicial map $f : S^n \to Y_n$ by the simplicial approximation theorem, and it is not hard to see that the spaces $X^n \cup_\varphi e^{n+1}$ and $Y_n \cup_f e^{n+1}$ are homotopy equivalent, where the subscripts denote attaching e^{n+1} via φ and f, respectively; see Proposition 0.18 for a proof. We can view $Y_n \cup_f e^{n+1}$ as the mapping cone C_f, obtained from the mapping cylinder of f by collapsing the domain end to a point. If we knew that the mapping cone of a simplicial map was a simplicial complex, then by performing the same construction for all the $(n+1)$-cells of X we would have completed the induction step. Unfortunately, and somewhat surprisingly, mapping cones and mapping cylinders are rather awkward objects in the simplicial category. To avoid this awkwardness we will instead construct simplicial analogs of mapping cones and cylinders that have all the essential features of actual mapping cones and cylinders.

Let us first construct the simplicial analog of a mapping cylinder. For a simplicial map $f:K\to L$ this will be a simplicial complex $M(f)$ containing both L and the barycentric subdivision K' of K as subcomplexes, and such that there is a deformation retraction r_t of $M(f)$ onto L with $r_1|K' = f$. The figure shows the case that f is a simplicial surjection $\Delta^2\to\Delta^1$. The construction proceeds one simplex of K at a time, by induction on dimension. To begin, the ordinary mapping cylinder of $f:K^0\to L$ suffices for $M(f|K^0)$. Assume inductively that we have already constructed $M(f|K^{n-1})$. Let σ be an n-simplex of

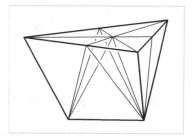

K and let $\tau = f(\sigma)$, a simplex of L of dimension n or less. By the inductive hypothesis we have already constructed $M(f:\partial\sigma\to\tau)$ with the desired properties, and we let $M(f:\sigma\to\tau)$ be the cone on $M(f:\partial\sigma\to\tau)$, as shown in the figure. The space $M(f:\partial\sigma\to\tau)$ is contractible since by induction it deformation retracts onto τ which is contractible. The cone $M(f:\sigma\to\tau)$ is of course contractible, so the inclusion of $M(f:\partial\sigma\to\tau)$ into $M(f:\sigma\to\tau)$ is a homotopy equivalence. This implies that $M(f:\sigma\to\tau)$ deformation retracts onto $M(f:\partial\sigma\to\tau)$ by Corollary 0.20, or one can give a direct argument using the fact that $M(f:\partial\sigma\to\tau)$ is contractible. By attaching $M(f:\sigma\to\tau)$ to $M(f|K^{n-1})$ along $M(f:\partial\sigma\to\tau)\subset M(f|K^{n-1})$ for all n-simplices σ of K we obtain $M(f|K^n)$ with a deformation retraction onto $M(f|K^{n-1})$. Taking the union over all n yields $M(f)$ with a deformation retraction r_t onto L, the infinite concatenation of the previous deformation retractions, with the deformation retraction of $M(f|K^n)$ onto $M(f|K^{n-1})$ performed in the t-interval $[1/2^{n+1},1/2^n]$. The map $r_1|K$ may not equal f, but it is homotopic to f via the linear homotopy $tf + (1-t)r_1$, which is defined since $r_1(\sigma)\subset f(\sigma)$ for all simplices σ of K. By applying the homotopy extension property to the homotopy of r_1 that equals $tf + (1-t)r_1$ on K and the identity map on L, we can improve our deformation retraction of $M(f)$ onto L so that its restriction to K at time 1 is f.

From the simplicial analog $M(f)$ of a mapping cylinder we construct the simplicial 'mapping cone' $C(f)$ by attaching the ordinary cone on K' to the subcomplex $K'\subset M(f)$.

Proof of 2C.5: We will construct for each n a CW complex Z_n containing X^n as a deformation retract and also containing as a deformation retract a subcomplex Y_n that is a simplicial complex. Beginning with $Y_0 = Z_0 = X^0$, suppose inductively that we have already constructed Y_n and Z_n. Let the cells e_α^{n+1} of X be attached by maps $\varphi_\alpha:S^n\to X^n$. Using the simplicial approximation theorem, there is a homotopy from φ_α to a simplicial map $f_\alpha:S^n\to Y_n$. The CW complex $W_n = Z_n\cup_\alpha M(f_\alpha)$ contains a simplicial subcomplex S_α^n homeomorphic to S^n at one end of $M(f_\alpha)$, and the homeomorphism $S^n \approx S_\alpha^n$ is homotopic in W_n to the map f_α, hence also to φ_α. Let Z_{n+1} be obtained from Z_n by attaching $D_\alpha^{n+1} \times I$'s via these homotopies between the φ_α's and the inclusions $S_\alpha^n \hookrightarrow W_n$. Thus Z_{n+1} contains X^{n+1} at one end, and at the other end we have a simplicial complex $Y_{n+1} = Y_n \bigcup_\alpha C(f_\alpha)$, where $C(f_\alpha)$ is obtained from $M(f_\alpha)$ by attaching a cone on the subcomplex S_α^n. Since $D^{n+1} \times I$ deformation retracts onto $\partial D^{n+1} \times I \cup D^{n+1} \times \{1\}$, we see that Z_{n+1} deformation retracts onto $Z_n \cup Y_{n+1}$, which in turn deformation retracts onto $Y_n \cup Y_{n+1} = Y_{n+1}$ by induction. Likewise, Z_{n+1} deformation retracts onto $X^{n+1} \cup W_n$ which deformation retracts onto $X^{n+1} \cup Z_n$ and hence onto $X^{n+1} \cup X^n = X^{n+1}$ by induction.

Let $Y = \bigcup_n Y_n$ and $Z = \bigcup_n Z_n$. The deformation retractions of Z_n onto X^n give deformation retractions of $X \cup Z_n$ onto X, and the infinite concatenation of the latter deformation retractions is a deformation retraction of Z onto X. Similarly, Z deformation retracts onto Y. $\square$

Exercises

1. What is the minimum number of edges in simplicial complex structures K and L on S^1 such that there is a simplicial map $K \to L$ of degree n?

2. Use the Lefschetz fixed point theorem to show that a map $S^n \to S^n$ has a fixed point unless its degree is equal to the degree of the antipodal map $x \mapsto -x$.

3. Verify that the formula $f(z_1, \cdots, z_{2k}) = (\overline{z}_2, -\overline{z}_1, \overline{z}_4, -\overline{z}_3, \cdots, \overline{z}_{2k}, -\overline{z}_{2k-1})$ defines a map $f : \mathbb{C}^{2k} \to \mathbb{C}^{2k}$ inducing a quotient map $\mathbb{C}\mathrm{P}^{2k-1} \to \mathbb{C}\mathrm{P}^{2k-1}$ without fixed points.

4. If X is a finite simplicial complex and $f : X \to X$ is a simplicial homeomorphism, show that the Lefschetz number $\tau(f)$ equals the Euler characteristic of the set of fixed points of f. In particular, $\tau(f)$ is the number of fixed points if the fixed points are isolated. [Hint: Barycentrically subdivide X to make the fixed point set a subcomplex.]

5. Let M be a closed orientable surface embedded in $\mathbb{R}^3$ in such a way that reflection across a plane P defines a homeomorphism $r : M \to M$ fixing $M \cap P$, a collection of circles. Is it possible to homotope r to have no fixed points?

6. Do an even-genus analog of Example 2C.4 by replacing the central torus by a sphere letting f be a homeomorphism that restricts to the antipodal map on this sphere.

7. Verify that the Lefschetz fixed point theorem holds also when $\tau(f)$ is defined using homology with coefficients in a field F.

8. Let X be homotopy equivalent to a finite simplicial complex and let Y be homotopy equivalent to a finite or countably infinite simplicial complex. Using the simplicial approximation theorem, show that there are at most countably many homotopy classes of maps $X \to Y$.

9. Show that there are only countably many homotopy types of finite CW complexes.

Chapter 3

Cohomology

Cohomology is an algebraic variant of homology, the result of a simple dualization in the definition. Not surprisingly, the cohomology groups $H^i(X)$ satisfy axioms much like the axioms for homology, except that induced homomorphisms go in the opposite direction as a result of the dualization. The basic distinction between homology and cohomology is thus that cohomology groups are contravariant functors while homology groups are covariant. In terms of intrinsic information, however, there is not a big difference between homology groups and cohomology groups. The homology groups of a space determine its cohomology groups, and the converse holds at least when the homology groups are finitely generated.

What is a little surprising is that contravariance leads to extra structure in cohomology. This first appears in a natural product, called *cup product*, which makes the cohomology groups of a space into a ring. This is an extremely useful piece of additional structure, and much of this chapter is devoted to studying cup products, which are considerably more subtle than the additive structure of cohomology.

How does contravariance lead to a product in cohomology that is not present in homology? Actually there is a natural product in homology, but it takes the somewhat different form of a map $H_i(X) \times H_j(Y) \longrightarrow H_{i+j}(X \times Y)$ called the *cross product*. If both X and Y are CW complexes, this cross product in homology is induced from a map of cellular chains sending a pair (e^i, e^j) consisting of a cell of X and a cell of Y to the product cell $e^i \times e^j$ in $X \times Y$. The details of the construction are described in §3.B. Taking $X = Y$, we thus have the first half of a hypothetical product

$$H_i(X) \times H_j(X) \longrightarrow H_{i+j}(X \times X) \longrightarrow H_{i+j}(X)$$

The difficulty is in defining the second map. The natural thing would be for this to be induced by a map $X \times X \to X$. The multiplication map in a topological group, or more generally an H–space, is such a map, and the resulting *Pontryagin product* can be quite useful when studying these spaces, as we show in §3.C. But for general X, the only natural maps $X \times X \to X$ are the projections onto one of the factors, and since these projections collapse the other factor to a point, the resulting product in homology is rather trivial.

With cohomology, however, the situation is better. One still has a cross product $H^i(X) \times H^j(Y) \to H^{i+j}(X \times Y)$ constructed in much the same way as in homology, so one can again take $X = Y$ and get the first half of a product

$$H^i(X) \times H^j(X) \to H^{i+j}(X \times X) \to H^{i+j}(X)$$

But now by contravariance the second map would be induced by a map $X \to X \times X$, and there is an obvious candidate for this map, the diagonal map $\Delta(x) = (x, x)$. This turns out to work very nicely, giving a well-behaved product in cohomology, the cup product.

Another sort of extra structure in cohomology whose existence is traceable to contravariance is provided by cohomology operations. These make the cohomology groups of a space into a module over a certain rather complicated ring. Cohomology operations lie at a depth somewhat greater than the cup product structure, so we defer their study to §4.L.

The extra layer of algebra in cohomology arising from the dualization in its definition may seem at first to be separating it further from topology, but there are many topological situations where cohomology arises quite naturally. One of these is Poincaré duality, the topic of the third section of this chapter. Another is obstruction theory, covered in §4.3. Characteristic classes in vector bundle theory (see [Milnor & Stasheff 1974] or [VBKT]) provide a further instance.

From the viewpoint of homotopy theory, cohomology is in some ways more basic than homology. As we shall see in §4.3, cohomology has a description in terms of homotopy classes of maps that is very similar to, and in a certain sense dual to, the definition of homotopy groups. There is an analog of this for homology, described in §4.F, but the construction is more complicated.

The Idea of Cohomology

Let us look at a few low-dimensional examples to get an idea of how one might be led naturally to consider cohomology groups, and to see what properties of a space they might be measuring. For the sake of simplicity we consider simplicial cohomology of Δ-complexes, rather than singular cohomology of more general spaces.

Taking the simplest case first, let X be a 1-dimensional Δ-complex, or in other words an oriented graph. For a fixed abelian group G, the set of all functions from vertices of X to G also forms an abelian group, which we denote by $\Delta^0(X; G)$. Similarly the set of all functions assigning an element of G to each edge of X forms an abelian group $\Delta^1(X; G)$. We will be interested in the homomorphism $\delta : \Delta^0(X; G) \to \Delta^1(X; G)$ sending $\varphi \in \Delta^0(X; G)$ to the function $\delta\varphi \in \Delta^1(X; G)$ whose value on an oriented edge $[v_0, v_1]$ is the difference $\varphi(v_1) - \varphi(v_0)$. For example, X might be the graph formed by a system of trails on a mountain, with vertices at the junctions between trails. The function φ could then assign to each junction its elevation above sea level, in which case $\delta\varphi$ would measure the net change in elevation along the trail from one junction to the next. Or X might represent a simple electrical circuit with φ measuring voltages at the connection points, the vertices, and $\delta\varphi$ measuring changes in voltage across the components of the circuit, represented by edges.

Regarding the map $\delta : \Delta^0(X; G) \to \Delta^1(X; G)$ as a chain complex with 0's before and after these two terms, the homology groups of this chain complex are by definition the simplicial cohomology groups of X, namely $H^0(X; G) = \operatorname{Ker} \delta \subset \Delta^0(X; G)$ and $H^1(X; G) = \Delta^1(X; G)/\operatorname{Im} \delta$. For simplicity we are using here the same notation as will be used for singular cohomology later in the chapter, in anticipation of the theorem that the two theories coincide for Δ-complexes, as we show in §3.1.

The group $H^0(X; G)$ is easy to describe explicitly. A function $\varphi \in \Delta^0(X; G)$ has $\delta\varphi = 0$ iff φ takes the same value at both ends of each edge of X. This is equivalent to saying that φ is constant on each component of X. So $H^0(X; G)$ is the group of all functions from the set of components of X to G. This is a direct product of copies of G, one for each component of X.

The cohomology group $H^1(X, G) = \Delta^1(X; G)/\operatorname{Im} \delta$ will be trivial iff the equation $\delta\varphi = \psi$ has a solution $\varphi \in \Delta^0(X; G)$ for each $\psi \in \Delta^1(X; G)$. Solving this equation means deciding whether specifying the change in φ across each edge of X determines an actual function $\varphi \in \Delta^0(X; G)$. This is rather like the calculus problem of finding a function having a specified derivative, with the difference operator δ playing the role of differentiation. As in calculus, if a solution of $\delta\varphi = \psi$ exists, it will be unique up to adding an element of the kernel of δ, that is, a function that is constant on each component of X.

The equation $\delta\varphi = \psi$ is always solvable if X is a tree since if we choose arbitrarily a value for φ at a basepoint vertex v_0, then if the change in φ across each edge of X is specified, this uniquely determines the value of φ at every other vertex v by induction along the unique path from v_0 to v in the tree. When X is not a tree, we first choose a maximal tree in each component of X. Then, since every vertex lies in one of these maximal trees, the values of ψ on the edges of the maximal trees determine φ uniquely up to a constant on each component of X. But in order for the equation $\delta\varphi = \psi$ to hold, the value of ψ on each edge not in any of the maximal trees must equal the difference in the already-determined values of φ at the two ends of the edge. This condition need not be satisfied since ψ can have arbitrary values on these edges. Thus we see that the cohomology group $H^1(X; G)$ is a direct product of copies of the group G, one copy for each edge of X not in one of the chosen maximal trees. This can be compared with the homology group $H_1(X; G)$ which consists of a direct *sum* of copies of G, one for each edge of X not in one of the maximal trees.

Note that the relation between $H^1(X;G)$ and $H_1(X;G)$ is the same as the relation between $H^0(X;G)$ and $H_0(X;G)$, with $H^0(X;G)$ being a direct product of copies of G and $H_0(X;G)$ a direct sum, with one copy for each component of X in either case.

Now let us move up a dimension, taking X to be a 2-dimensional Δ-complex. Define $\Delta^0(X;G)$ and $\Delta^1(X;G)$ as before, as functions from vertices and edges of X to the abelian group G, and define $\Delta^2(X;G)$ to be the functions from 2-simplices of X to G. A homomorphism $\delta:\Delta^1(X;G) \to \Delta^2(X;G)$ is defined by $\delta\psi([v_0,v_1,v_2]) = \psi([v_0,v_1]) + \psi([v_1,v_2]) - \psi([v_0,v_2])$, a signed sum of the values of ψ on the three edges in the boundary of $[v_0,v_1,v_2]$, just as $\delta\varphi([v_0,v_1])$ for $\varphi \in \Delta^0(X;G)$ was a signed sum of the values of φ on the boundary of $[v_0,v_1]$. The two homomorphisms $\Delta^0(X;G) \xrightarrow{\delta} \Delta^1(X;G) \xrightarrow{\delta} \Delta^2(X;G)$ form a chain complex since for $\varphi \in \Delta^0(X;G)$ we have $\delta\delta\varphi = (\varphi(v_1)-\varphi(v_0))+(\varphi(v_2)-\varphi(v_1))-(\varphi(v_2)-\varphi(v_0)) = 0$. Extending this chain complex by 0's on each end, the resulting homology groups are by definition the cohomology groups $H^i(X;G)$.

The formula for the map $\delta:\Delta^1(X;G) \to \Delta^2(X;G)$ can be looked at from several different viewpoints. Perhaps the simplest is the observation that $\delta\psi = 0$ iff ψ satisfies the additivity property $\psi([v_0,v_2]) = \psi([v_0,v_1]) + \psi([v_1,v_2])$, where we think of the edge $[v_0,v_2]$ as the sum of the edges $[v_0,v_1]$ and $[v_1,v_2]$. Thus $\delta\psi$ measures the deviation of ψ from being additive.

From another point of view, $\delta\psi$ can be regarded as an obstruction to finding $\varphi \in \Delta^0(X;G)$ with $\psi = \delta\varphi$, for if $\psi = \delta\varphi$ then $\delta\psi = 0$ since $\delta\delta\varphi = 0$ as we saw above. We can think of $\delta\psi$ as a local obstruction to solving $\psi = \delta\varphi$ since it depends only on the values of ψ within individual 2-simplices of X. If this local obstruction vanishes, then ψ defines an element of $H^1(X;G)$ which is zero iff $\psi = \delta\varphi$ has an actual solution. This class in $H^1(X;G)$ is thus the global obstruction to solving $\psi = \delta\varphi$. This situation is similar to the calculus problem of determining whether a given vector field is the gradient vector field of some function. The local obstruction here is the vanishing of the curl of the vector field, and the global obstruction is the vanishing of all line integrals around closed loops in the domain of the vector field.

The condition $\delta\psi = 0$ has an interpretation of a more geometric nature when X is a surface and the group G is $\mathbb{Z}$ or $\mathbb{Z}_2$. Consider first the simpler case $G = \mathbb{Z}_2$. The condition $\delta\psi = 0$ means that the number of times that ψ takes the value 1 on the edges of each 2-simplex is even, either 0 or 2. This means we can associate to ψ a collection C_ψ of disjoint curves in X crossing the 1-skeleton transversely, such that the number of intersections of C_ψ with each edge is equal to the value of ψ on that edge. If $\psi = \delta\varphi$ for some φ, then the curves of C_ψ divide X into two regions X_0 and X_1 where the subscript indicates the value of φ on all vertices in the region.

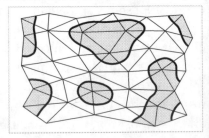

When $G = \mathbb{Z}$ we can refine this construction by building C_ψ from a number of arcs in each 2-simplex, each arc having a transverse orientation, the orientation which agrees or disagrees with the orientation of each edge according to the sign of the value of ψ on the edge, as in the figure at the right. The resulting collection C_ψ of disjoint curves in X can be thought of as something like level curves for a function φ with $\delta\varphi = \psi$, if such a function exists. The value of φ changes by 1 each time a curve of C_ψ is crossed. For example, if X is a disk then we will show that $H^1(X;\mathbb{Z}) = 0$, so $\delta\psi = 0$ implies $\psi = \delta\varphi$ for some φ, hence every

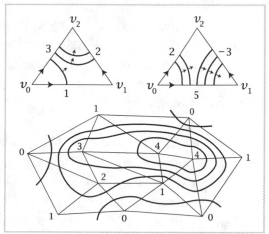

transverse curve system C_ψ forms the level curves of a function φ. On the other hand, if X is an annulus then this need no longer be true, as illustrated in the example shown in the figure at the left, where the equation $\psi = \delta\varphi$ obviously has no solution even though $\delta\psi = 0$. By identifying the inner and outer boundary circles of this annulus we obtain a similar example on the torus. Even with $G = \mathbb{Z}_2$ the equation $\psi = \delta\varphi$ has no solution since the curve C_ψ does not separate X into two regions X_0 and X_1.

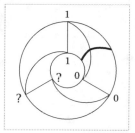

The key to relating cohomology groups to homology groups is the observation that a function from i-simplices of X to G is equivalent to a homomorphism from the simplicial chain group $\Delta_i(X)$ to G. This is because $\Delta_i(X)$ is free abelian with basis the i-simplices of X, and a homomorphism with domain a free abelian group is uniquely determined by its values on basis elements, which can be assigned arbitrarily. Thus we have an identification of $\Delta^i(X;G)$ with the group $\text{Hom}(\Delta_i(X), G)$ of homomorphisms $\Delta_i(X) \to G$, which is called the *dual group* of $\Delta_i(X)$. There is also a simple relationship of duality between the homomorphism $\delta : \Delta^i(X;G) \to \Delta^{i+1}(X;G)$ and the boundary homomorphism $\partial : \Delta_{i+1}(X) \to \Delta_i(X)$. The general formula for δ is

$$\delta\varphi([v_0, \cdots, v_{i+1}]) = \sum_j (-1)^j \varphi([v_0, \cdots, \hat{v}_j, \cdots, v_{i+1}])$$

and the latter sum is just $\varphi(\partial[v_0, \cdots, v_{i+1}])$. Thus we have $\delta\varphi = \varphi\partial$. In other words, δ sends each $\varphi \in \text{Hom}(\Delta_i(X), G)$ to the composition $\Delta_{i+1}(X) \xrightarrow{\partial} \Delta_i(X) \xrightarrow{\varphi} G$, which in the language of linear algebra means that δ is the dual map of ∂.

Thus we have the algebraic problem of understanding the relationship between the homology groups of a chain complex and the homology groups of the dual complex obtained by applying the functor $C \mapsto \text{Hom}(C, G)$. This is the first topic of the chapter.

3.1 Cohomology Groups

Homology groups $H_n(X)$ are the result of a two-stage process: First one forms a chain complex $\cdots \longrightarrow C_n \overset{\partial}{\longrightarrow} C_{n-1} \longrightarrow \cdots$ of singular, simplicial, or cellular chains, then one takes the homology groups of this chain complex, $\operatorname{Ker}\partial/\operatorname{Im}\partial$. To obtain the cohomology groups $H^n(X;G)$ we interpolate an intermediate step, replacing the chain groups C_n by the dual groups $\operatorname{Hom}(C_n, G)$ and the boundary maps ∂ by their dual maps δ, before forming the cohomology groups $\operatorname{Ker}\delta/\operatorname{Im}\delta$. The plan for this section is first to sort out the algebra of this dualization process and show that the cohomology groups are determined algebraically by the homology groups, though in a somewhat subtle way. Then after this algebraic excursion we will define the cohomology groups of spaces and show that these satisfy basic properties very much like those for homology. The payoff for all this formal work will begin to be apparent in subsequent sections.

The Universal Coefficient Theorem

Let us begin with a simple example. Consider the chain complex

$$0 \longrightarrow \underset{\substack{\| \\ C_3}}{\mathbb{Z}} \overset{0}{\longrightarrow} \underset{\substack{\| \\ C_2}}{\mathbb{Z}} \overset{2}{\longrightarrow} \underset{\substack{\| \\ C_1}}{\mathbb{Z}} \overset{0}{\longrightarrow} \underset{\substack{\| \\ C_0}}{\mathbb{Z}} \longrightarrow 0$$

where $\mathbb{Z} \overset{2}{\longrightarrow} \mathbb{Z}$ is the map $x \mapsto 2x$. If we dualize by taking $\operatorname{Hom}(-, G)$ with $G = \mathbb{Z}$, we obtain the cochain complex

$$0 \longleftarrow \underset{\substack{\| \\ C_3^*}}{\mathbb{Z}} \overset{0}{\longleftarrow} \underset{\substack{\| \\ C_2^*}}{\mathbb{Z}} \overset{2}{\longleftarrow} \underset{\substack{\| \\ C_1^*}}{\mathbb{Z}} \overset{0}{\longleftarrow} \underset{\substack{\| \\ C_0^*}}{\mathbb{Z}} \longleftarrow 0$$

In the original chain complex the homology groups are $\mathbb{Z}$'s in dimensions 0 and 3, together with a $\mathbb{Z}_2$ in dimension 1. The homology groups of the dual cochain complex, which are called cohomology groups to emphasize the dualization, are again $\mathbb{Z}$'s in dimensions 0 and 3, but the $\mathbb{Z}_2$ in the 1-dimensional homology of the original complex has shifted up a dimension to become a $\mathbb{Z}_2$ in 2-dimensional cohomology.

More generally, consider any chain complex of finitely generated free abelian groups. Such a chain complex always splits as the direct sum of elementary complexes of the forms $0 \to \mathbb{Z} \to 0$ and $0 \to \mathbb{Z} \overset{m}{\longrightarrow} \mathbb{Z} \to 0$, according to Exercise 43 in §2.2. Applying $\operatorname{Hom}(-, \mathbb{Z})$ to this direct sum of elementary complexes, we obtain the direct sum of the corresponding dual complexes $0 \leftarrow \mathbb{Z} \leftarrow 0$ and $0 \leftarrow \mathbb{Z} \overset{m}{\longleftarrow} \mathbb{Z} \leftarrow 0$. Thus the cohomology groups are the same as the homology groups except that torsion is shifted up one dimension. We will see later in this section that the same relation between homology and cohomology holds whenever the homology groups are finitely generated, even when the chain groups are not finitely generated. It would also be quite easy to see in this example what happens if $\text{Hom}(-,\mathbb{Z})$ is replaced by $\text{Hom}(-,G)$, since the dual elementary cochain complexes would then be $0 \leftarrow G \leftarrow 0$ and $0 \leftarrow G \xleftarrow{m} G \leftarrow 0$.

Consider now a completely general chain complex C of free abelian groups

$$\cdots \longrightarrow C_{n+1} \xrightarrow{\ \partial\ } C_n \xrightarrow{\ \partial\ } C_{n-1} \longrightarrow \cdots$$

To dualize this complex we replace each chain group C_n by its dual **cochain group** $C_n^* = \text{Hom}(C_n, G)$, the group of homomorphisms $C_n \to G$, and we replace each boundary map $\partial : C_n \to C_{n-1}$ by its dual **coboundary map** $\delta = \partial^* : C_{n-1}^* \to C_n^*$. The reason why δ goes in the opposite direction from ∂, increasing rather than decreasing dimension, is purely formal: For a homomorphism $\alpha : A \to B$, the dual homomorphism $\alpha^* : \text{Hom}(B,G) \to \text{Hom}(A,G)$ is defined by $\alpha^*(\varphi) = \varphi\alpha$, so α^* sends $B \xrightarrow{\varphi} G$ to the composition $A \xrightarrow{\alpha} B \xrightarrow{\varphi} G$. Dual homomorphisms obviously satisfy $(\alpha\beta)^* = \beta^*\alpha^*$, $\mathbb{1}^* = \mathbb{1}$, and $0^* = 0$. In particular, since $\partial\partial = 0$ it follows that $\delta\delta = 0$, and the **cohomology group** $H^n(C;G)$ can be defined as the 'homology group' $\text{Ker}\,\delta / \text{Im}\,\delta$ at C_n^* in the cochain complex

$$\cdots \longleftarrow C_{n+1}^* \xleftarrow{\ \delta\ } C_n^* \xleftarrow{\ \delta\ } C_{n-1}^* \longleftarrow \cdots$$

Our goal is to show that the cohomology groups $H^n(C;G)$ are determined solely by G and the homology groups $H_n(C) = \text{Ker}\,\partial / \text{Im}\,\partial$. A first guess might be that $H^n(C;G)$ is isomorphic to $\text{Hom}(H_n(C),G)$, but this is overly optimistic, as shown by the example above where H_2 was zero while H^2 was nonzero. Nevertheless, there is a natural map $h : H^n(C;G) \to \text{Hom}(H_n(C),G)$, defined as follows. Denote the cycles and boundaries by $Z_n = \text{Ker}\,\partial \subset C_n$ and $B_n = \text{Im}\,\partial \subset C_n$. A class in $H^n(C;G)$ is represented by a homomorphism $\varphi : C_n \to G$ such that $\delta\varphi = 0$, that is, $\varphi\partial = 0$, or in other words, φ vanishes on B_n. The restriction $\varphi_0 = \varphi | Z_n$ then induces a quotient homomorphism $\overline{\varphi}_0 : Z_n / B_n \to G$, an element of $\text{Hom}(H_n(C),G)$. If φ is in $\text{Im}\,\delta$, say $\varphi = \delta\psi = \psi\partial$, then φ is zero on Z_n, so $\varphi_0 = 0$ and hence also $\overline{\varphi}_0 = 0$. Thus there is a well-defined quotient map $h : H^n(C;G) \to \text{Hom}(H_n(C),G)$ sending the cohomology class of φ to $\overline{\varphi}_0$. Obviously h is a homomorphism.

It is not hard to see that h is surjective. The short exact sequence

$$0 \longrightarrow Z_n \longrightarrow C_n \xrightarrow{\ \partial\ } B_{n-1} \longrightarrow 0$$

splits since B_{n-1} is free, being a subgroup of the free abelian group C_{n-1}. Thus there is a projection homomorphism $p : C_n \to Z_n$ that restricts to the identity on Z_n. Composing with p gives a way of extending homomorphisms $\varphi_0 : Z_n \to G$ to homomorphisms $\varphi = \varphi_0 p : C_n \to G$. In particular, this extends homomorphisms $Z_n \to G$ that vanish on B_n to homomorphisms $C_n \to G$ that still vanish on B_n, or in other words, it extends homomorphisms $H_n(C) \to G$ to elements of $\text{Ker}\,\delta$. Thus we have a homomorphism $\text{Hom}(H_n(C),G) \to \text{Ker}\,\delta$. Composing this with the quotient map $\text{Ker}\,\delta \to H^n(C;G)$ gives a homomorphism from $\text{Hom}(H_n(C),G)$ to $H^n(C;G)$. If we follow this map by h we get the identity map on $\mathrm{Hom}(H_n(C), G)$ since the effect of composing with h is simply to undo the effect of extending homomorphisms via p. This shows that h is surjective. In fact it shows that we have a split short exact sequence

$$0 \longrightarrow \mathrm{Ker}\, h \longrightarrow H^n(C;G) \xrightarrow{\;h\;} \mathrm{Hom}(H_n(C), G) \longrightarrow 0$$

The remaining task is to analyze $\mathrm{Ker}\, h$. A convenient way to start the process is to consider not just the chain complex C, but also its subcomplexes consisting of the cycles and the boundaries. Thus we consider the commutative diagram of short exact sequences

(i)
$$\begin{array}{ccccccccc}
0 & \longrightarrow & Z_{n+1} & \longrightarrow & C_{n+1} & \xrightarrow{\;\partial\;} & B_n & \longrightarrow & 0 \\
 & & \downarrow{\scriptstyle 0} & & \downarrow{\scriptstyle \partial} & & \downarrow{\scriptstyle 0} & & \\
0 & \longrightarrow & Z_n & \longrightarrow & C_n & \xrightarrow{\;\partial\;} & B_{n-1} & \longrightarrow & 0
\end{array}$$

where the vertical boundary maps on Z_{n+1} and B_n are the restrictions of the boundary map in the complex C, hence are zero. Dualizing (i) gives a commutative diagram

(ii)
$$\begin{array}{ccccccccc}
0 & \longleftarrow & Z_{n+1}^* & \longleftarrow & C_{n+1}^* & \longleftarrow & B_n^* & \longleftarrow & 0 \\
 & & \uparrow{\scriptstyle 0} & & \uparrow{\scriptstyle \delta} & & \uparrow{\scriptstyle 0} & & \\
0 & \longleftarrow & Z_n^* & \longleftarrow & C_n^* & \longleftarrow & B_{n-1}^* & \longleftarrow & 0
\end{array}$$

The rows here are exact since, as we have already remarked, the rows of (i) split, and the dual of a split short exact sequence is a split short exact sequence because of the natural isomorphism $\mathrm{Hom}(A \oplus B, G) \approx \mathrm{Hom}(A, G) \oplus \mathrm{Hom}(B, G)$.

We may view (ii), like (i), as part of a short exact sequence of chain complexes. Since the coboundary maps in the Z_n^* and B_n^* complexes are zero, the associated long exact sequence of homology groups has the form

(iii)
$$\cdots \longleftarrow B_n^* \longleftarrow Z_n^* \longleftarrow H^n(C;G) \longleftarrow B_{n-1}^* \longleftarrow Z_{n-1}^* \longleftarrow \cdots$$

The 'boundary maps' $Z_n^* \to B_n^*$ in this long exact sequence are in fact the dual maps i_n^* of the inclusions $i_n : B_n \to Z_n$, as one sees by recalling how these boundary maps are defined: In (ii) one takes an element of Z_n^*, pulls this back to C_n^*, applies δ to get an element of C_{n+1}^*, then pulls this back to B_n^*. The first of these steps extends a homomorphism $\varphi_0 : Z_n \to G$ to $\varphi : C_n \to G$, the second step composes this φ with ∂, and the third step undoes this composition and restricts φ to B_n. The net effect is just to restrict φ_0 from Z_n to B_n.

A long exact sequence can always be broken up into short exact sequences, and doing this for the sequence (iii) yields short exact sequences

(iv)
$$0 \longleftarrow \mathrm{Ker}\, i_n^* \longleftarrow H^n(C;G) \longleftarrow \mathrm{Coker}\, i_{n-1}^* \longleftarrow 0$$

The group $\mathrm{Ker}\, i_n^*$ can be identified naturally with $\mathrm{Hom}(H_n(C), G)$ since elements of $\mathrm{Ker}\, i_n^*$ are homomorphisms $Z_n \to G$ that vanish on the subgroup B_n, and such homomorphisms are the same as homomorphisms $Z_n/B_n \to G$. Under this identification of

Ker i_n^* with $\text{Hom}(H_n(C), G)$, the map $H^n(C; G) \to \text{Ker } i_n^*$ in (iv) becomes the map h considered earlier. Thus we can rewrite (iv) as a split short exact sequence

(v) $$0 \longrightarrow \text{Coker } i_{n-1}^* \longrightarrow H^n(C; G) \xrightarrow{\ h\ } \text{Hom}(H_n(C), G) \longrightarrow 0$$

Our objective now is to show that the more mysterious term $\text{Coker } i_{n-1}^*$ depends only on $H_{n-1}(C)$ and G, in a natural, functorial way. First let us observe that $\text{Coker } i_{n-1}^*$ would be zero if it were always true that the dual of a short exact sequence was exact, since the dual of the short exact sequence

(vi) $$0 \longrightarrow B_{n-1} \xrightarrow{\ i_{n-1}\ } Z_{n-1} \longrightarrow H_{n-1}(C) \longrightarrow 0$$

is the sequence

(vii) $$0 \longleftarrow B_{n-1}^* \xleftarrow{\ i_{n-1}^*\ } Z_{n-1}^* \longleftarrow H_{n-1}(C)^* \longleftarrow 0$$

and if this were exact at B_{n-1}^*, then i_{n-1}^* would be surjective, hence $\text{Coker } i_{n-1}^*$ would be zero. This argument does apply if $H_{n-1}(C)$ happens to be free, since (vi) splits in this case, which implies that (vii) is also split exact. So in this case the map h in (v) is an isomorphism. However, in the general case it is easy to find short exact sequences whose duals are not exact. For example, if we dualize $0 \to \mathbb{Z} \xrightarrow{n} \mathbb{Z} \to \mathbb{Z}_n \to 0$ by applying $\text{Hom}(-, \mathbb{Z})$ we get $0 \leftarrow \mathbb{Z} \xleftarrow{n} \mathbb{Z} \leftarrow 0 \leftarrow 0$ which fails to be exact at the left-hand $\mathbb{Z}$, precisely the place we are interested in for $\text{Coker } i_{n-1}^*$.

We might mention in passing that the loss of exactness at the left end of a short exact sequence after dualization is in fact all that goes wrong, in view of the following:

Exercise. If $A \to B \to C \to 0$ is exact, then dualizing by applying $\text{Hom}(-, G)$ yields an exact sequence $A^* \leftarrow B^* \leftarrow C^* \leftarrow 0$.

However, we will not need this fact in what follows.

The exact sequence (vi) has the special feature that both B_{n-1} and Z_{n-1} are free, so (vi) can be regarded as a free resolution of $H_{n-1}(C)$, where a **free resolution** of an abelian group H is an exact sequence

$$\cdots \longrightarrow F_2 \xrightarrow{\ f_2\ } F_1 \xrightarrow{\ f_1\ } F_0 \xrightarrow{\ f_0\ } H \longrightarrow 0$$

with each F_n free. If we dualize this free resolution by applying $\text{Hom}(-, G)$, we may lose exactness, but at least we get a chain complex — or perhaps we should say 'cochain complex,' but algebraically there is no difference. This dual complex has the form

$$\cdots \longleftarrow F_2^* \xleftarrow{\ f_2^*\ } F_1^* \xleftarrow{\ f_1^*\ } F_0^* \xleftarrow{\ f_0^*\ } H^* \longleftarrow 0$$

Let us use the temporary notation $H^n(F; G)$ for the homology group $\text{Ker } f_{n+1}^* / \text{Im } f_n^*$ of this dual complex. Note that the group $\text{Coker } i_{n-1}^*$ that we are interested in is $H^1(F; G)$ where F is the free resolution in (vi). Part (b) of the following lemma therefore shows that $\text{Coker } i_{n-1}^*$ depends only on $H_{n-1}(C)$ and G.

Lemma 3.1. (a) *Given free resolutions F and F' of abelian groups H and H', then every homomorphism $\alpha : H \to H'$ can be extended to a chain map from F to F':*

$$\begin{array}{ccccccccc}
\cdots & \longrightarrow & F_2 & \xrightarrow{f_2} & F_1 & \xrightarrow{f_1} & F_0 & \xrightarrow{f_0} & H & \longrightarrow & 0 \\
& & \downarrow{\alpha_2} & & \downarrow{\alpha_1} & & \downarrow{\alpha_0} & & \downarrow{\alpha} & & \\
\cdots & \longrightarrow & F_2' & \xrightarrow{f_2'} & F_1' & \xrightarrow{f_1'} & F_0' & \xrightarrow{f_0'} & H' & \longrightarrow & 0
\end{array}$$

Furthermore, any two such chain maps extending α are chain homotopic.

(b) *For any two free resolutions F and F' of H, there are canonical isomorphisms $H^n(F; G) \approx H^n(F'; G)$ for all n.*

Proof: The α_i's will be constructed inductively. Since the F_i's are free, it suffices to define each α_i on a basis for F_i. To define α_0, observe that surjectivity of f_0' implies that for each basis element x of F_0 there exists $x' \in F_0'$ such that $f_0'(x') = \alpha f_0(x)$, so we define $\alpha_0(x) = x'$. We would like to define α_1 in the same way, sending a basis element $x \in F_1$ to an element $x' \in F_1'$ such that $f_1'(x') = \alpha_0 f_1(x)$. Such an x' will exist if $\alpha_0 f_1(x)$ lies in $\operatorname{Im} f_1' = \operatorname{Ker} f_0'$, which it does since $f_0' \alpha_0 f_1 = \alpha f_0 f_1 = 0$. The same procedure defines all the subsequent α_i's.

If we have another chain map extending α given by maps $\alpha_i' : F_i \to F_i'$, then the differences $\beta_i = \alpha_i - \alpha_i'$ define a chain map extending the zero map $\beta : H \to H'$. It will suffice to construct maps $\lambda_i : F_i \to F_{i+1}'$ defining a chain homotopy from β_i to 0, that is, with $\beta_i = f_{i+1}' \lambda_i + \lambda_{i-1} f_i$. The λ_i's are constructed inductively by a procedure much like the construction of the α_i's. When $i = 0$ we let $\lambda_{-1} : H \to F_0'$ be zero, and then the desired relation becomes $\beta_0 = f_1' \lambda_0$. We can achieve this by letting λ_0 send a basis element x to an element $x' \in F_1'$ such that $f_1'(x') = \beta_0(x)$. Such an x' exists since $\operatorname{Im} f_1' = \operatorname{Ker} f_0'$ and $f_0' \beta_0(x) = \beta f_0(x) = 0$. For the inductive step we wish to define λ_i to take a basis element $x \in F_i$ to an element $x' \in F_{i+1}'$ such that $f_{i+1}'(x') = \beta_i(x) - \lambda_{i-1} f_i(x)$. This will be possible if $\beta_i(x) - \lambda_{i-1} f_i(x)$ lies in $\operatorname{Im} f_{i+1}' = \operatorname{Ker} f_i'$, which will hold if $f_i'(\beta_i - \lambda_{i-1} f_i) = 0$. Using the relation $f_i' \beta_i = \beta_{i-1} f_i$ and the relation $\beta_{i-1} = f_i' \lambda_{i-1} + \lambda_{i-2} f_{i-1}$ which holds by induction, we have

$$\begin{aligned}
f_i'(\beta_i - \lambda_{i-1} f_i) &= f_i' \beta_i - f_i' \lambda_{i-1} f_i \\
&= \beta_{i-1} f_i - f_i' \lambda_{i-1} f_i = (\beta_{i-1} - f_i' \lambda_{i-1}) f_i = \lambda_{i-2} f_{i-1} f_i = 0
\end{aligned}$$

as desired. This finishes the proof of (a).

The maps α_n constructed in (a) dualize to maps $\alpha_n^* : F_n'^* \to F_n^*$ forming a chain map between the dual complexes F'^* and F^*. Therefore we have induced homomorphisms on cohomology $\alpha^* : H^n(F'; G) \to H^n(F; G)$. These do not depend on the choice of α_n's since any other choices α_n' are chain homotopic, say via chain homotopies λ_n, and then α_n^* and $\alpha_n'^*$ are chain homotopic via the dual maps λ_n^* since the dual of the relation $\alpha_i - \alpha_i' = f_{i+1}' \lambda_i + \lambda_{i-1} f_i$ is $\alpha_i^* - \alpha_i'^* = \lambda_i^* f_{i+1}'^* + f_i^* \lambda_{i-1}^*$.

The induced homomorphisms $\alpha^* : H^n(F'; G) \to H^n(F; G)$ satisfy $(\beta\alpha)^* = \alpha^* \beta^*$ for a composition $H \xrightarrow{\alpha} H' \xrightarrow{\beta} H''$ with a free resolution F'' of H'' also given, since one can choose the compositions $\beta_n \alpha_n$ of extensions α_n of α and β_n of β as an extension of $\beta\alpha$. In particular, if we take α to be an isomorphism and β to be its inverse, with $F'' = F$, then $\alpha^* \beta^* = (\beta\alpha)^* = \mathbb{1}$, the latter equality coming from the obvious extension of $\mathbb{1} : H \to H$ by the identity map of F. The same reasoning shows $\beta^* \alpha^* = \mathbb{1}$, so α^* is an isomorphism. Finally, if we specialize further, taking α to be the identity but with two different free resolutions F and F', we get a canonical isomorphism $\mathbb{1}^* : H^n(F'; G) \to H^n(F; G)$. $\square$

Every abelian group H has a free resolution of the form $0 \to F_1 \to F_0 \to H \to 0$, with $F_i = 0$ for $i > 1$, obtainable in the following way. Choose a set of generators for H and let F_0 be a free abelian group with basis in one-to-one correspondence with these generators. Then we have a surjective homomorphism $f_0 : F_0 \to H$ sending the basis elements to the chosen generators. The kernel of f_0 is free, being a subgroup of a free abelian group, so we can let F_1 be this kernel with $f_1 : F_1 \to F_0$ the inclusion, and we can then take $F_i = 0$ for $i > 1$. For this free resolution we obviously have $H^n(F; G) = 0$ for $n > 1$, so this must also be true for all free resolutions. Thus the only interesting group $H^n(F; G)$ is $H^1(F; G)$. As we have seen, this group depends only on H and G, and the standard notation for it is $\mathrm{Ext}(H, G)$. This notation arises from the fact that $\mathrm{Ext}(H, G)$ has an interpretation as the set of isomorphism classes of extensions of G by H, that is, short exact sequences $0 \to G \to J \to H \to 0$, with a natural definition of isomorphism between such exact sequences. This is explained in books on homological algebra, for example [Brown 1982], [Hilton & Stammbach 1970], or [MacLane 1963]. However, this interpretation of $\mathrm{Ext}(H, G)$ is rarely needed in algebraic topology.

Summarizing, we have established the following algebraic result:

Theorem 3.2. *If a chain complex C of free abelian groups has homology groups $H_n(C)$, then the cohomology groups $H^n(C; G)$ of the cochain complex $\mathrm{Hom}(C_n, G)$ are determined by split exact sequences*

$$0 \longrightarrow \mathrm{Ext}(H_{n-1}(C), G) \longrightarrow H^n(C; G) \overset{h}{\longrightarrow} \mathrm{Hom}(H_n(C), G) \longrightarrow 0 \qquad \square$$

This is known as the **universal coefficient theorem for cohomology** because it is formally analogous to the universal coefficient theorem for homology in §3.A which expresses homology with arbitrary coefficients in terms of homology with $\mathbb{Z}$ coefficients.

Computing $\mathrm{Ext}(H, G)$ for finitely generated H is not difficult using the following three properties:

- $\mathrm{Ext}(H \oplus H', G) \approx \mathrm{Ext}(H, G) \oplus \mathrm{Ext}(H', G)$.
- $\mathrm{Ext}(H, G) = 0$ if H is free.
- $\mathrm{Ext}(\mathbb{Z}_n, G) \approx G/nG$.

The first of these can be obtained by using the direct sum of free resolutions of H and H' as a free resolution for $H \oplus H'$. If H is free, the free resolution $0 \to H \to H \to 0$

yields the second property, while the third comes from dualizing the free resolution $0 \to \mathbb{Z} \xrightarrow{n} \mathbb{Z} \to \mathbb{Z}_n \to 0$ to produce an exact sequence

$$0 \longleftarrow \mathrm{Ext}(\mathbb{Z}_n, G) \longleftarrow \mathrm{Hom}(\mathbb{Z}, G) \xleftarrow{n} \mathrm{Hom}(\mathbb{Z}, G) \longleftarrow \mathrm{Hom}(\mathbb{Z}_n, G) \longleftarrow 0$$

$$\| \qquad\qquad\qquad \| \qquad\qquad\qquad \|$$

$$G/nG \longleftarrow \qquad\qquad G \xleftarrow{\quad n \quad} G$$

In particular, these three properties imply that $\mathrm{Ext}(H, \mathbb{Z})$ is isomorphic to the torsion subgroup of H if H is finitely generated. Since $\mathrm{Hom}(H, \mathbb{Z})$ is isomorphic to the free part of H if H is finitely generated, we have:

Corollary 3.3. *If the homology groups H_n and H_{n-1} of a chain complex C of free abelian groups are finitely generated, with torsion subgroups $T_n \subset H_n$ and $T_{n-1} \subset H_{n-1}$, then $H^n(C; \mathbb{Z}) \approx (H_n/T_n) \oplus T_{n-1}$.* $\qquad\square$

It is useful in many situations to know that the short exact sequences in the universal coefficient theorem are natural, meaning that a chain map α between chain complexes C and C' of free abelian groups induces a commutative diagram

$$
\begin{array}{ccccccccc}
0 & \longrightarrow & \mathrm{Ext}(H_{n-1}(C), G) & \longrightarrow & H^n(C; G) & \xrightarrow{\ h\ } & \mathrm{Hom}(H_n(C), G) & \longrightarrow & 0 \\
& & \uparrow{\scriptstyle (\alpha_*)^*} & & \uparrow{\scriptstyle \alpha^*} & & \uparrow{\scriptstyle (\alpha_*)^*} & & \\
0 & \longrightarrow & \mathrm{Ext}(H_{n-1}(C'), G) & \longrightarrow & H^n(C'; G) & \xrightarrow{\ h\ } & \mathrm{Hom}(H_n(C'), G) & \longrightarrow & 0
\end{array}
$$

This is apparent if one just thinks about the construction; one obviously obtains a map between the short exact sequences (iv) containing $\mathrm{Ker}\, i_n^*$ and $\mathrm{Coker}\, i_{n-1}^*$, the identification $\mathrm{Ker}\, i_n^* = \mathrm{Hom}(H_n(C), G)$ is certainly natural, and the proof of Lemma 3.1 shows that $\mathrm{Ext}(H, G)$ depends naturally on H.

However, the splitting in the universal coefficient theorem is not natural since it depends on the choice of the projections $p : C_n \to Z_n$. An exercise at the end of the section gives a topological example showing that the splitting in fact cannot be natural.

The naturality property together with the five-lemma proves:

Corollary 3.4. *If a chain map between chain complexes of free abelian groups induces an isomorphism on homology groups, then it induces an isomorphism on cohomology groups with any coefficient group G.* $\qquad\square$

One could attempt to generalize the algebraic machinery of the universal coefficient theorem by replacing abelian groups by modules over a chosen ring R and Hom by Hom_R, the R-module homomorphisms. The key fact about abelian groups that was needed was that subgroups of free abelian groups are free. Submodules of free R-modules are free if R is a principal ideal domain, so in this case the generalization is automatic. One obtains natural split short exact sequences

$$0 \longrightarrow \mathrm{Ext}_R(H_{n-1}(C), G) \longrightarrow H^n(C; G) \xrightarrow{\ h\ } \mathrm{Hom}_R(H_n(C), G) \longrightarrow 0$$

where C is a chain complex of free R-modules with boundary maps R-module homomorphisms, and the coefficient group G is also an R-module. If R is a field, for example, then R-modules are always free and so the Ext_R term is always zero since we may choose free resolutions of the form $0 \to F_0 \to H \to 0$.

It is interesting to note that the proof of Lemma 3.1 on the uniqueness of free resolutions is valid for modules over an arbitrary ring R. Moreover, every R-module H has a free resolution, which can be constructed in the following way. Choose a set of generators for H as an R-module, and let F_0 be a free R-module with basis in one-to-one correspondence with these generators. Thus we have a surjective homomorphism $f_0 : F_0 \to H$ sending the basis elements to the chosen generators. Now repeat the process with $\mathrm{Ker}\, f_0$ in place of H, constructing a homomorphism $f_1 : F_1 \to F_0$ sending a basis for a free R-module F_1 onto generators for $\mathrm{Ker}\, f_0$. And inductively, construct $f_n : F_n \to F_{n-1}$ with image equal to $\mathrm{Ker}\, f_{n-1}$ by the same procedure.

By Lemma 3.1 the groups $H^n(F; G)$ depend only on H and G, not on the free resolution F. The standard notation for $H^n(F; G)$ is $\mathrm{Ext}_R^n(H, G)$. For sufficiently complicated rings R the groups $\mathrm{Ext}_R^n(H, G)$ can be nonzero for $n > 1$. In certain more advanced topics in algebraic topology these Ext_R^n groups play an essential role.

A final remark about the definition of $\mathrm{Ext}_R^n(H, G)$: By the Exercise stated earlier, exactness of $F_1 \to F_0 \to H \to 0$ implies exactness of $F_1^* \leftarrow F_0^* \leftarrow H^* \leftarrow 0$. This means that $H^0(F; G)$ as defined above is zero. Rather than having $\mathrm{Ext}_R^0(H, G)$ be automatically zero, it is better to define $H^n(F; G)$ as the nth homology group of the complex $\cdots \leftarrow F_1^* \leftarrow F_0^* \leftarrow 0$ with the term H^* omitted. This can be viewed as defining the groups $H^n(F; G)$ to be unreduced cohomology groups. With this slightly modified definition we have $\mathrm{Ext}_R^0(H, G) = H^0(F; G) = H^* = \mathrm{Hom}_R(H, G)$ by the exactness of $F_1^* \leftarrow F_0^* \leftarrow H^* \leftarrow 0$. The real reason why unreduced Ext groups are better than reduced groups is perhaps to be found in certain exact sequences involving Ext and Hom derived in §3.F, which would not work with the Hom terms replaced by zeros.

Cohomology of Spaces

Now we return to topology. Given a space X and an abelian group G, we define the group $C^n(X; G)$ of **singular n-cochains with coefficients in G** to be the dual group $\mathrm{Hom}(C_n(X), G)$ of the singular chain group $C_n(X)$. Thus an n-cochain $\varphi \in C^n(X; G)$ assigns to each singular n-simplex $\sigma : \Delta^n \to X$ a value $\varphi(\sigma) \in G$. Since the singular n-simplices form a basis for $C_n(X)$, these values can be chosen arbitrarily, hence n-cochains are exactly equivalent to functions from singular n-simplices to G.

The **coboundary map** $\delta : C^n(X; G) \to C^{n+1}(X; G)$ is the dual ∂^*, so for a cochain $\varphi \in C^n(X; G)$, its coboundary $\delta\varphi$ is the composition $C_{n+1}(X) \xrightarrow{\partial} C_n(X) \xrightarrow{\varphi} G$. This means that for a singular $(n+1)$-simplex $\sigma : \Delta^{n+1} \to X$ we have

$$\delta\varphi(\sigma) = \sum_i (-1)^i \varphi(\sigma \,|\, [v_0, \cdots, \hat{v}_i, \cdots, v_{n+1}])$$

It is automatic that $\delta^2 = 0$ since δ^2 is the dual of $\partial^2 = 0$. Therefore we can define the **cohomology group $H^n(X; G)$ with coefficients in G** to be the quotient $\operatorname{Ker} \delta / \operatorname{Im} \delta$ at $C^n(X; G)$ in the cochain complex

$$\cdots \longleftarrow C^{n+1}(X; G) \xleftarrow{\ \delta\ } C^n(X; G) \xleftarrow{\ \delta\ } C^{n-1}(X; G) \longleftarrow \cdots \longleftarrow C^0(X; G) \longleftarrow 0$$

Elements of $\operatorname{Ker} \delta$ are **cocycles,** and elements of $\operatorname{Im} \delta$ are **coboundaries.** For a cochain φ to be a cocycle means that $\delta\varphi = \varphi\partial = 0$, or in other words, φ vanishes on boundaries.

Since the chain groups $C_n(X)$ are free, the algebraic universal coefficient theorem takes on the topological guise of split short exact sequences

$$0 \longrightarrow \operatorname{Ext}(H_{n-1}(X), G) \longrightarrow H^n(X; G) \longrightarrow \operatorname{Hom}(H_n(X), G) \longrightarrow 0$$

which describe how cohomology groups with arbitrary coefficients are determined purely algebraically by homology groups with $\mathbb{Z}$ coefficients. For example, if the homology groups of X are finitely generated then Corollary 3.3 tells how to compute the cohomology groups $H^n(X; \mathbb{Z})$ from the homology groups.

When $n = 0$ there is no Ext term, and the universal coefficient theorem reduces to an isomorphism $H^0(X; G) \approx \operatorname{Hom}(H_0(X), G)$. This can also be seen directly from the definitions. Since singular 0-simplices are just points of X, a cochain in $C^0(X; G)$ is an arbitrary function $\varphi : X \to G$, not necessarily continuous. For this to be a cocycle means that for each singular 1-simplex $\sigma : [v_0, v_1] \to X$ we have $\delta\varphi(\sigma) = \varphi(\partial\sigma) = \varphi(\sigma(v_1)) - \varphi(\sigma(v_0)) = 0$. This is equivalent to saying that φ is constant on path-components of X. Thus $H^0(X; G)$ is all the functions from path-components of X to G. This is the same as $\operatorname{Hom}(H_0(X), G)$.

Likewise in the case of $H^1(X; G)$ the universal coefficient theorem gives an isomorphism $H^1(X; G) \approx \operatorname{Hom}(H_1(X), G)$ since $\operatorname{Ext}(H_0(X), G) = 0$, the group $H_0(X)$ being free. If X is path-connected, $H_1(X)$ is the abelianization of $\pi_1(X)$ and we can identify $\operatorname{Hom}(H_1(X), G)$ with $\operatorname{Hom}(\pi_1(X), G)$ since G is abelian.

The universal coefficient theorem has a simpler form if we take coefficients in a field F for both homology and cohomology. In §2.2 we defined the homology groups $H_n(X; F)$ as the homology groups of the chain complex of free F-modules $C_n(X; F)$, where $C_n(X; F)$ has basis the singular n-simplices in X. The dual complex $\operatorname{Hom}_F(C_n(X; F), F)$ of F-module homomorphisms is the same as $\operatorname{Hom}(C_n(X), F)$ since both can be identified with the functions from singular n-simplices to F. Hence the homology groups of the dual complex $\operatorname{Hom}_F(C_n(X; F), F)$ are the cohomology groups $H^n(X; F)$. In the generalization of the universal coefficient theorem to the case of modules over a principal ideal domain, the Ext_F terms vanish since F is a field, so we obtain isomorphisms

$$H^n(X; F) \approx \operatorname{Hom}_F(H_n(X; F), F)$$

Thus, with field coefficients, cohomology is the exact dual of homology. Note that when $F = \mathbb{Z}_p$ or $\mathbb{Q}$ we have $\operatorname{Hom}_F(H, G) = \operatorname{Hom}(H, G)$, the group homomorphisms, for arbitrary F-modules G and H.

For the remainder of this section we will go through the main features of singular homology and check that they extend without much difficulty to cohomology.

Reduced Groups. Reduced cohomology groups $\tilde{H}^n(X; G)$ can be defined by dualizing the augmented chain complex $\cdots \to C_0(X) \xrightarrow{\varepsilon} \mathbb{Z} \to 0$, then taking Ker / Im. As with homology, this gives $\tilde{H}^n(X; G) = H^n(X; G)$ for $n > 0$, and the universal coefficient theorem identifies $\tilde{H}^0(X; G)$ with $\operatorname{Hom}(\tilde{H}_0(X), G)$. We can describe the difference between $\tilde{H}^0(X; G)$ and $H^0(X; G)$ more explicitly by using the interpretation of $H^0(X; G)$ as functions $X \to G$ that are constant on path-components. Recall that the augmentation map $\varepsilon : C_0(X) \to \mathbb{Z}$ sends each singular 0-simplex σ to 1, so the dual map ε^* sends a homomorphism $\varphi : \mathbb{Z} \to G$ to the composition $C_0(X) \xrightarrow{\varepsilon} \mathbb{Z} \xrightarrow{\varphi} G$, which is the function $\sigma \mapsto \varphi(1)$. This is a constant function $X \to G$, and since $\varphi(1)$ can be any element of G, the image of ε^* consists of precisely the constant functions. Thus $\tilde{H}^0(X; G)$ is all functions $X \to G$ that are constant on path-components modulo the functions that are constant on all of X.

Relative Groups and the Long Exact Sequence of a Pair. To define relative groups $H^n(X, A; G)$ for a pair (X, A) we first dualize the short exact sequence

$$0 \longrightarrow C_n(A) \xrightarrow{i} C_n(X) \xrightarrow{j} C_n(X, A) \longrightarrow 0$$

by applying $\operatorname{Hom}(-, G)$ to get

$$0 \longleftarrow C^n(A; G) \xleftarrow{i^*} C^n(X; G) \xleftarrow{j^*} C^n(X, A; G) \longleftarrow 0$$

where by definition $C^n(X, A; G) = \operatorname{Hom}(C_n(X, A), G)$. This sequence is exact by the following direct argument. The map i^* restricts a cochain on X to a cochain on A. Thus for a function from singular n-simplices in X to G, the image of this function under i^* is obtained by restricting the domain of the function to singular n-simplices in A. Every function from singular n-simplices in A to G can be extended to be defined on all singular n-simplices in X, for example by assigning the value 0 to all singular n-simplices not in A, so i^* is surjective. The kernel of i^* consists of cochains taking the value 0 on singular n-simplices in A. Such cochains are the same as homomorphisms $C_n(X, A) = C_n(X)/C_n(A) \to G$, so the kernel of i^* is exactly $C^n(X, A; G) = \operatorname{Hom}(C_n(X, A), G)$, giving the desired exactness. Notice that we can view $C^n(X, A; G)$ as the functions from singular n-simplices in X to G that vanish on simplices in A, since the basis for $C_n(X)$ consisting of singular n-simplices in X is the disjoint union of the simplices with image contained in A and the simplices with image not contained in A.

Relative coboundary maps $\delta : C^n(X, A; G) \to C^{n+1}(X, A; G)$ are obtained as restrictions of the absolute δ's, so relative cohomology groups $H^n(X, A; G)$ are defined. The fact that the relative cochain group is a subgroup of the absolute cochains, namely the cochains vanishing on chains in A, means that relative cohomology is conceptually a little simpler than relative homology.

The maps i^* and j^* commute with δ since i and j commute with ∂, so the preceding displayed short exact sequence of cochain groups is part of a short exact sequence of cochain complexes, giving rise to an associated long exact sequence of cohomology groups

$$\cdots \longrightarrow H^n(X,A;G) \xrightarrow{j^*} H^n(X;G) \xrightarrow{i^*} H^n(A;G) \xrightarrow{\delta} H^{n+1}(X,A;G) \longrightarrow \cdots$$

By similar reasoning one obtains a long exact sequence of reduced cohomology groups for a pair (X,A) with A nonempty, where $\tilde{H}^n(X,A;G) = H^n(X,A;G)$ for all n, as in homology. Taking A to be a point x_0, this exact sequence gives an identification of $\tilde{H}^n(X;G)$ with $H^n(X,x_0;G)$.

More generally there is a long exact sequence for a triple (X,A,B) coming from the short exact sequences

$$0 \longleftarrow C^n(A,B;G) \xleftarrow{i^*} C^n(X,B;G) \xleftarrow{j^*} C^n(X,A;G) \longleftarrow 0$$

The long exact sequence of reduced cohomology can be regarded as the special case that B is a point.

As one would expect, there is a duality relationship between the connecting homomorphisms $\delta : H^n(A;G) \to H^{n+1}(X,A;G)$ and $\partial : H_{n+1}(X,A) \to H_n(A)$. This takes the form of the commutative diagram shown at the right. To verify commutativity, recall how the two connecting homomorphisms are defined, via the diagrams

$$\begin{array}{ccc} H^n(A;G) & \xrightarrow{\;\;\delta\;\;} & H^{n+1}(X,A;G) \\ \downarrow{h} & & \downarrow{h} \\ \mathrm{Hom}\,(H_n(A),G) & \xrightarrow{\;\partial^*\;} & \mathrm{Hom}\,(H_{n+1}(X,A),G) \end{array}$$

$$\begin{array}{cc} C^{n+1}(X;G) \longleftarrow C^{n+1}(X,A;G) & \qquad C_{n+1}(X) \longrightarrow C_{n+1}(X,A) \\ C^n(A;G) \longleftarrow C^n(X;G) & \qquad C_n(A) \longrightarrow C_n(X) \end{array}$$

The connecting homomorphisms are represented by the dashed arrows, which are well-defined only when the chain and cochain groups are replaced by homology and cohomology groups. To show that $h\delta = \partial^* h$, start with an element $\alpha \in H^n(A;G)$ represented by a cocycle $\varphi \in C^n(A;G)$. To compute $\delta(\alpha)$ we first extend φ to a cochain $\overline{\varphi} \in C^n(X;G)$, say by letting it take the value 0 on singular simplices not in A. Then we compose $\overline{\varphi}$ with $\partial : C_{n+1}(X) \to C_n(X)$ to get a cochain $\overline{\varphi}\partial \in C^{n+1}(X;G)$, which actually lies in $C^{n+1}(X,A;G)$ since the original φ was a cocycle in A. This cochain $\overline{\varphi}\partial \in C^{n+1}(X,A;G)$ represents $\delta(\alpha)$ in $H^{n+1}(X,A;G)$. Now we apply the map h, which simply restricts the domain of $\overline{\varphi}\partial$ to relative cycles in $C_{n+1}(X,A)$, that is, $(n+1)$-chains in X whose boundary lies in A. On such chains we have $\overline{\varphi}\partial = \varphi\partial$ since the extension of φ to $\overline{\varphi}$ is irrelevant. The net result of all this is that $h\delta(\alpha)$

is represented by $\varphi\partial$. Let us compare this with $\partial^*h(\alpha)$. Applying h to φ restricts its domain to cycles in A. Then applying ∂^* composes with the map which sends a relative $(n+1)$-cycle in X to its boundary in A. Thus $\partial^*h(\alpha)$ is represented by $\varphi\partial$ just as $h\delta(\alpha)$ was, and so the square commutes.

Induced Homomorphisms. Dual to the chain maps $f_\sharp : C_n(X) \to C_n(Y)$ induced by $f : X \to Y$ are the cochain maps $f^\sharp : C^n(Y;G) \to C^n(X;G)$. The relation $f_\sharp\partial = \partial f_\sharp$ dualizes to $\delta f^\sharp = f^\sharp\delta$, so $f^\sharp$ induces homomorphisms $f^* : H^n(Y;G) \to H^n(X;G)$. In the relative case a map $f : (X,A) \to (Y,B)$ induces $f^* : H^n(Y,B;G) \to H^n(X,A;G)$ by the same reasoning, and in fact f induces a map between short exact sequences of cochain complexes, hence a map between long exact sequences of cohomology groups, with commuting squares. The properties $(fg)^\sharp = g^\sharp f^\sharp$ and $\mathbb{1}^\sharp = \mathbb{1}$ imply $(fg)^* = g^*f^*$ and $\mathbb{1}^* = \mathbb{1}$, so $X \mapsto H^n(X;G)$ and $(X,A) \mapsto H^n(X,A;G)$ are contravariant functors, the 'contra' indicating that induced maps go in the reverse direction.

The algebraic universal coefficient theorem applies also to relative cohomology since the relative chain groups $C_n(X,A)$ are free, and there is a naturality statement: A map $f : (X,A) \to (Y,B)$ induces a commutative diagram

$$
\begin{array}{ccccccccc}
0 & \longrightarrow & \mathrm{Ext}(H_{n-1}(X,A),G) & \longrightarrow & H^n(X,A;G) & \overset{h}{\longrightarrow} & \mathrm{Hom}(H_n(X,A),G) & \longrightarrow & 0 \\
 & & \big\uparrow {\scriptstyle (f_*)^*} & & \big\uparrow {\scriptstyle f^*} & & \big\uparrow {\scriptstyle (f_*)^*} & & \\
0 & \longrightarrow & \mathrm{Ext}(H_{n-1}(Y,B),G) & \longrightarrow & H^n(Y,B;G) & \overset{h}{\longrightarrow} & \mathrm{Hom}(H_n(Y,B),G) & \longrightarrow & 0
\end{array}
$$

This follows from the naturality of the algebraic universal coefficient sequences since the vertical maps are induced by the chain maps $f_\sharp : C_n(X,A) \to C_n(Y,B)$. When the subspaces A and B are empty we obtain the absolute forms of these results.

Homotopy Invariance. The statement is that if $f \simeq g : (X,A) \to (Y,B)$, then $f^* = g^* : H^n(Y,B) \to H^n(X,A)$. This is proved by direct dualization of the proof for homology. From the proof of Theorem 2.10 we have a chain homotopy P satisfying $g_\sharp - f_\sharp = \partial P + P\partial$. This relation dualizes to $g^\sharp - f^\sharp = P^*\delta + \delta P^*$, so P^* is a chain homotopy between the maps $f^\sharp, g^\sharp : C^n(Y;G) \to C^n(X;G)$. This restricts also to a chain homotopy between $f^\sharp$ and $g^\sharp$ on relative cochains, the cochains vanishing on singular simplices in the subspaces B and A. Since $f^\sharp$ and $g^\sharp$ are chain homotopic, they induce the same homomorphism $f^* = g^*$ on cohomology.

Excision. For cohomology this says that for subspaces $Z \subset A \subset X$ with the closure of Z contained in the interior of A, the inclusion $i : (X-Z, A-Z) \hookrightarrow (X,A)$ induces isomorphisms $i^* : H^n(X,A;G) \to H^n(X-Z, A-Z;G)$ for all n. This follows from the corresponding result for homology by the naturality of the universal coefficient theorem and the five-lemma. Alternatively, if one wishes to avoid appealing to the universal coefficient theorem, the proof of excision for homology dualizes easily to cohomology by the following argument. In the proof for homology there were chain maps $\iota : C_n(A+B) \to C_n(X)$ and $\rho : C_n(X) \to C_n(A+B)$ such that $\rho\iota = \mathbb{1}$ and $\mathbb{1} - \iota\rho = \partial D + D\partial$ for a chain homotopy D. Dualizing by taking $\mathrm{Hom}(-,G)$, we have maps

ρ^* and ι^* between $C^n(A + B; G)$ and $C^n(X; G)$, and these induce isomorphisms on cohomology since $\iota^*\rho^* = 1\!\!1$ and $1\!\!1 - \rho^*\iota^* = D^*\delta + \delta D^*$. By the five-lemma, the maps $C^n(X, A; G) \to C^n(A + B, A; G)$ also induce isomorphisms on cohomology. There is an obvious identification of $C^n(A+B, A; G)$ with $C^n(B, A \cap B; G)$, so we get isomorphisms $H^n(X, A; G)) \approx H^n(B, A \cap B; G)$ induced by the inclusion $(B, A \cap B) \hookrightarrow (X, A)$.

Axioms for Cohomology. These are exactly dual to the axioms for homology. Restricting attention to CW complexes again, a (reduced) **cohomology theory** is a sequence of contravariant functors $\tilde{h}^n$ from CW complexes to abelian groups, together with natural coboundary homomorphisms $\delta : \tilde{h}^n(A) \to \tilde{h}^{n+1}(X/A)$ for CW pairs (X, A), satisfying the following axioms:

(1) If $f \simeq g : X \to Y$, then $f^* = g^* : \tilde{h}^n(Y) \to \tilde{h}^n(X)$.

(2) For each CW pair (X, A) there is a long exact sequence

$$\cdots \xrightarrow{\delta} \tilde{h}^n(X/A) \xrightarrow{q^*} \tilde{h}^n(X) \xrightarrow{i^*} \tilde{h}^n(A) \xrightarrow{\delta} \tilde{h}^{n+1}(X/A) \xrightarrow{q^*} \cdots$$

where i is the inclusion and q is the quotient map.

(3) For a wedge sum $X = \bigvee_\alpha X_\alpha$ with inclusions $i_\alpha : X_\alpha \hookrightarrow X$, the product map $\prod_\alpha i_\alpha^* : \tilde{h}^n(X) \to \prod_\alpha \tilde{h}^n(X_\alpha)$ is an isomorphism for each n.

We have already seen that the first axiom holds for singular cohomology. The second axiom follows from excision in the same way as for homology, via isomorphisms $\tilde{H}^n(X/A; G) \approx H^n(X, A; G)$. Note that the third axiom involves direct product, rather than the direct sum appearing in the homology version. This is because of the natural isomorphism $\text{Hom}(\bigoplus_\alpha A_\alpha, G) \approx \prod_\alpha \text{Hom}(A_\alpha, G)$, which implies that the cochain complex of a disjoint union $\coprod_\alpha X_\alpha$ is the direct product of the cochain complexes of the individual X_α's, and this direct product splitting passes through to cohomology groups. The same argument applies in the relative case, so we get isomorphisms $H^n(\coprod_\alpha X_\alpha, \coprod_\alpha A_\alpha; G) \approx \prod_\alpha H^n(X_\alpha, A_\alpha; G)$. The third axiom is obtained by taking the A_α's to be basepoints x_α and passing to the quotient $\coprod_\alpha X_\alpha / \coprod_\alpha x_\alpha = \bigvee_\alpha X_\alpha$.

The relation between reduced and unreduced cohomology theories is the same as for homology, as described in §2.3.

Simplicial Cohomology. If X is a Δ-complex and $A \subset X$ is a subcomplex, then the simplicial chain groups $\Delta_n(X, A)$ dualize to simplicial cochain groups $\Delta^n(X, A; G) = \text{Hom}(\Delta_n(X, A), G)$, and the resulting cohomology groups are by definition the simplicial cohomology groups $H_\Delta^n(X, A; G)$. Since the inclusions $\Delta_n(X, A) \subset C_n(X, A)$ induce isomorphisms $H_n^\Delta(X, A) \approx H_n(X, A)$, Corollary 3.4 implies that the dual maps $C^n(X, A; G) \to \Delta^n(X, A; G)$ also induce isomorphisms $H^n(X, A; G) \approx H_\Delta^n(X, A; G)$.

Cellular Cohomology. For a CW complex X this is defined via the cellular cochain complex formed by the horizontal sequence in the following diagram, where coefficients in a given group G are understood, and the cellular coboundary maps d_n are the compositions $\delta_n j_n$, making the triangles commute. Note that $d_n d_{n-1} = 0$ since $j_n \delta_{n-1} = 0$.

$$
\begin{array}{c}
\nearrow 0 \\
H^{n-1}(X^{n-1}) \\
j_{n-1} \nearrow \qquad \searrow \delta_{n-1} \\
\cdots \longrightarrow H^{n-1}(X^{n-1}, X^{n-2}) \xrightarrow{d_{n-1}} H^n(X^n, X^{n-1}) \xrightarrow{d_n} H^{n+1}(X^{n+1}, X^n) \longrightarrow \cdots \\
j_n \searrow \qquad \nearrow \delta_n \\
H^n(X^n) \\
\nearrow \qquad \searrow \\
H^n(X) \approx H^n(X^{n+1}) \qquad \qquad 0 \\
\nearrow \\
0
\end{array}
$$

Theorem 3.5. $H^n(X; G) \approx \operatorname{Ker} d_n / \operatorname{Im} d_{n-1}$. *Furthermore, the cellular cochain complex $\{H^n(X^n, X^{n-1}; G), d_n\}$ is isomorphic to the dual of the cellular chain complex, obtained by applying* $\operatorname{Hom}(-, G)$.

Proof: The universal coefficient theorem implies that $H^k(X^n, X^{n-1}; G) = 0$ for $k \neq n$. The long exact sequence of the pair (X^n, X^{n-1}) then gives isomorphisms $H^k(X^n; G) \approx H^k(X^{n-1}; G)$ for $k \neq n$, $n - 1$. Hence by induction on n we obtain $H^k(X^n; G) = 0$ if $k > n$. Thus the diagonal sequences in the preceding diagram are exact. The universal coefficient theorem also gives $H^k(X, X^{n+1}; G) = 0$ for $k \leq n + 1$, so $H^n(X; G) \approx H^n(X^{n+1}; G)$. The diagram then yields isomorphisms

$$H^n(X; G) \approx H^n(X^{n+1}; G) \approx \operatorname{Ker} \delta_n \approx \operatorname{Ker} d_n / \operatorname{Im} \delta_{n-1} \approx \operatorname{Ker} d_n / \operatorname{Im} d_{n-1}$$

For the second statement in the theorem we have the diagram

$$
\begin{array}{ccccc}
H^k(X^k, X^{k-1}; G) & \longrightarrow & H^k(X^k; G) & \xrightarrow{\delta} & H^{k+1}(X^{k+1}, X^k; G) \\
\downarrow h & & \downarrow h & & \downarrow h \\
\operatorname{Hom}(H_k(X^k, X^{k-1}), G) & \longrightarrow & \operatorname{Hom}(H_k(X^k), G) & \xrightarrow{\partial^*} & \operatorname{Hom}(H_{k+1}(X^{k+1}, X^k), G)
\end{array}
$$

The cellular coboundary map is the composition across the top, and we want to see that this is the same as the composition across the bottom. The first and third vertical maps are isomorphisms by the universal coefficient theorem, so it suffices to show the diagram commutes. The first square commutes by naturality of h, and commutativity of the second square was shown in the discussion of the long exact sequence of cohomology groups of a pair (X, A). $\qquad \square$

Mayer–Vietoris Sequences. In the absolute case these take the form

$$\cdots \longrightarrow H^n(X; G) \xrightarrow{\Psi} H^n(A; G) \oplus H^n(B; G) \xrightarrow{\Phi} H^n(A \cap B; G) \longrightarrow H^{n+1}(X; G) \longrightarrow \cdots$$

where X is the union of the interiors of A and B. This is the long exact sequence associated to the short exact sequence of cochain complexes

$$0 \longrightarrow C^n(A + B; G) \xrightarrow{\psi} C^n(A; G) \oplus C^n(B; G) \xrightarrow{\varphi} C^n(A \cap B; G) \longrightarrow 0$$

Here $C^n(A + B; G)$ is the dual of the subgroup $C_n(A + B) \subset C_n(X)$ consisting of sums of singular n-simplices lying in A or in B. The inclusion $C_n(A + B) \subset C_n(X)$ is a chain homotopy equivalence by Proposition 2.21, so the dual restriction map $C^n(X; G) \to C^n(A + B; G)$ is also a chain homotopy equivalence, hence induces an isomorphism on cohomology as shown in the discussion of excision a couple pages back. The map ψ has coordinates the two restrictions to A and B, and φ takes the difference of the restrictions to $A \cap B$, so it is obvious that φ is onto with kernel the image of ψ.

There is a relative Mayer–Vietoris sequence

$$\cdots \to H^n(X, Y; G) \to H^n(A, C; G) \oplus H^n(B, D; G) \to H^n(A \cap B, C \cap D; G) \to \cdots$$

for a pair $(X, Y) = (A \cup B, C \cup D)$ with $C \subset A$ and $D \subset B$ such that X is the union of the interiors of A and B while Y is the union of the interiors of C and D. To derive this, consider first the map of short exact sequences of cochain complexes

$$
\begin{array}{ccccccccc}
0 & \longrightarrow & C^n(X, Y; G) & \longrightarrow & C^n(X; G) & \longrightarrow & C^n(Y; G) & \longrightarrow & 0 \\
& & \downarrow & & \downarrow & & \downarrow & & \\
0 & \longrightarrow & C^n(A+B, C+D; G) & \longrightarrow & C^n(A+B; G) & \longrightarrow & C^n(C+D; G) & \longrightarrow & 0
\end{array}
$$

Here $C^n(A + B, C + D; G)$ is defined as the kernel of $C^n(A + B; G) \to C^n(C + D; G)$, the restriction map, so the second sequence is exact. The vertical maps are restrictions. The second and third of these induce isomorphisms on cohomology, as we have seen, so by the five-lemma the first vertical map also induces isomorphisms on cohomology. The relative Mayer–Vietoris sequence is then the long exact sequence associated to the short exact sequence of cochain complexes

$$0 \to C^n(A + B, C + D; G) \xrightarrow{\psi} C^n(A, C; G) \oplus C^n(B, D; G) \xrightarrow{\varphi} C^n(A \cap B, C \cap D; G) \to 0$$

This is exact since it is the dual of the short exact sequence

$$0 \to C_n(A \cap B, C \cap D) \to C_n(A, C) \oplus C_n(B, D) \to C_n(A + B, C + D) \to 0$$

constructed in §2.2, which splits since $C_n(A + B, C + D)$ is free with basis the singular n-simplices in A or in B that do not lie in C or in D.

Exercises

1. Show that $\mathrm{Ext}(H, G)$ is a contravariant functor of H for fixed G, and a covariant functor of G for fixed H.

2. Show that the maps $G \xrightarrow{n} G$ and $H \xrightarrow{n} H$ multiplying each element by the integer n induce multiplication by n in $\mathrm{Ext}(H, G)$.

3. Regarding $\mathbb{Z}_2$ as a module over the ring $\mathbb{Z}_4$, construct a resolution of $\mathbb{Z}_2$ by free modules over $\mathbb{Z}_4$ and use this to show that $\mathrm{Ext}^n_{\mathbb{Z}_4}(\mathbb{Z}_2, \mathbb{Z}_2)$ is nonzero for all n.

4. What happens if one defines homology groups $h_n(X;G)$ as the homology groups of the chain complex $\cdots \to \text{Hom}(G, C_n(X)) \to \text{Hom}(G, C_{n-1}(X)) \to \cdots$? More specifically, what are the groups $h_n(X;G)$ when $G = \mathbb{Z}$, $\mathbb{Z}_m$, and $\mathbb{Q}$?

5. Regarding a cochain $\varphi \in C^1(X;G)$ as a function from paths in X to G, show that if φ is a cocycle, then

(a) $\varphi(f \cdot g) = \varphi(f) + \varphi(g)$,

(b) φ takes the value 0 on constant paths,

(c) $\varphi(f) = \varphi(g)$ if $f \simeq g$,

(d) φ is a coboundary iff $\varphi(f)$ depends only on the endpoints of f, for all f.

[In particular, (a) and (c) give a map $H^1(X;G) \to \text{Hom}(\pi_1(X), G)$, which the universal coefficient theorem says is an isomorphism if X is path-connected.]

6. (a) Directly from the definitions, compute the simplicial cohomology groups of $S^1 \times S^1$ with $\mathbb{Z}$ and $\mathbb{Z}_2$ coefficients, using the Δ-complex structure given in §2.1.

(b) Do the same for $\mathbb{R}P^2$ and the Klein bottle.

7. Show that the functors $h^n(X) = \text{Hom}(H_n(X), \mathbb{Z})$ do not define a cohomology theory on the category of CW complexes.

8. Many basic homology arguments work just as well for cohomology even though maps go in the opposite direction. Verify this in the following cases:

(a) Compute $H^i(S^n;G)$ by induction on n in two ways: using the long exact sequence of a pair, and using the Mayer–Vietoris sequence.

(b) Show that if A is a closed subspace of X that is a deformation retract of some neighborhood, then the quotient map $X \to X/A$ induces isomorphisms $H^n(X, A; G) \approx \tilde{H}^n(X/A; G)$ for all n.

(c) Show that if A is a retract of X then $H^n(X;G) \approx H^n(A;G) \oplus H^n(X, A; G)$.

9. Show that if $f : S^n \to S^n$ has degree d then $f^* : H^n(S^n;G) \to H^n(S^n;G)$ is multiplication by d.

10. For the lens space $L_m(\ell_1, \cdots, \ell_n)$ defined in Example 2.43, compute the cohomology groups using the cellular cochain complex and taking coefficients in $\mathbb{Z}$, $\mathbb{Q}$, $\mathbb{Z}_m$, and $\mathbb{Z}_p$ for p prime. Verify that the answers agree with those given by the universal coefficient theorem.

11. Let X be a Moore space $M(\mathbb{Z}_m, n)$ obtained from S^n by attaching a cell e^{n+1} by a map of degree m.

(a) Show that the quotient map $X \to X/S^n = S^{n+1}$ induces the trivial map on $\tilde{H}_i(-;\mathbb{Z})$ for all i, but not on $H^{n+1}(-;\mathbb{Z})$. Deduce that the splitting in the universal coefficient theorem for cohomology cannot be natural.

(b) Show that the inclusion $S^n \hookrightarrow X$ induces the trivial map on $\tilde{H}^i(-;\mathbb{Z})$ for all i, but not on $H_n(-;\mathbb{Z})$.

12. Show $H^k(X, X^n; G) = 0$ if X is a CW complex and $k \le n$, by using the cohomology version of the second proof of the corresponding result for homology in Lemma 2.34.

13. Let $\langle X, Y \rangle$ denote the set of basepoint-preserving homotopy classes of basepoint-preserving maps $X \to Y$. Using Proposition 1B.9, show that if X is a connected CW complex and G is an abelian group, then the map $\langle X, K(G,1) \rangle \to H^1(X; G)$ sending a map $f: X \to K(G,1)$ to the induced homomorphism $f_*: H_1(X) \to H_1(K(G,1)) \approx G$ is a bijection, where we identify $H^1(X; G)$ with $\mathrm{Hom}(H_1(X), G)$ via the universal coefficient theorem.

3.2 Cup Product

In the introduction to this chapter we sketched a definition of cup product in terms of another product called cross product. However, to define the cross product from scratch takes some work, so we will proceed in the opposite order, first giving an elementary definition of cup product by an explicit formula with simplices, then afterwards defining cross product in terms of cup product. The other approach of defining cup product via cross product is explained at the end of §3.B.

To define the cup product we consider cohomology with coefficients in a ring R, the most common choices being $\mathbb{Z}$, $\mathbb{Z}_n$, and $\mathbb{Q}$. For cochains $\varphi \in C^k(X; R)$ and $\psi \in C^\ell(X; R)$, the **cup product** $\varphi \smile \psi \in C^{k+\ell}(X; R)$ is the cochain whose value on a singular simplex $\sigma : \Delta^{k+\ell} \to X$ is given by the formula

$$(\varphi \smile \psi)(\sigma) = \varphi(\sigma | [v_0, \cdots, v_k]) \psi(\sigma | [v_k, \cdots, v_{k+\ell}])$$

where the right-hand side is the product in R. To see that this cup product of cochains induces a cup product of cohomology classes we need a formula relating it to the coboundary map:

‖ **Lemma** 3.6. $\delta(\varphi \smile \psi) = \delta\varphi \smile \psi + (-1)^k \varphi \smile \delta\psi$ for $\varphi \in C^k(X; R)$ and $\psi \in C^\ell(X; R)$.

Proof: For $\sigma : \Delta^{k+\ell+1} \to X$ we have

$$(\delta\varphi \smile \psi)(\sigma) = \sum_{i=0}^{k+1} (-1)^i \varphi(\sigma | [v_0, \cdots, \hat{v}_i, \cdots, v_{k+1}]) \psi(\sigma | [v_{k+1}, \cdots, v_{k+\ell+1}])$$

$$(-1)^k (\varphi \smile \delta\psi)(\sigma) = \sum_{i=k}^{k+\ell+1} (-1)^i \varphi(\sigma | [v_0, \cdots, v_k]) \psi(\sigma | [v_k, \cdots, \hat{v}_i, \cdots, v_{k+\ell+1}])$$

When we add these two expressions, the last term of the first sum cancels the first term of the second sum, and the remaining terms are exactly $\delta(\varphi \smile \psi)(\sigma) = (\varphi \smile \psi)(\partial \sigma)$ since $\partial \sigma = \sum_{i=0}^{k+\ell+1} (-1)^i \sigma | [v_0, \cdots, \hat{v}_i, \cdots, v_{k+\ell+1}]$.   □

From the formula $\delta(\varphi \smile \psi) = \delta\varphi \smile \psi \pm \varphi \smile \delta\psi$ it is apparent that the cup product of two cocycles is again a cocycle. Also, the cup product of a cocycle and a coboundary, in either order, is a coboundary since $\varphi \smile \delta\psi = \pm\delta(\varphi \smile \psi)$ if $\delta\varphi = 0$, and $\delta\varphi \smile \psi = \delta(\varphi \smile \psi)$ if $\delta\psi = 0$. It follows that there is an induced cup product

$$H^k(X;R) \times H^\ell(X;R) \xrightarrow{\smile} H^{k+\ell}(X;R)$$

This is associative and distributive since at the level of cochains the cup product obviously has these properties. If R has an identity element, then there is an identity element for cup product, the class $1 \in H^0(X;R)$ defined by the 0-cocycle taking the value 1 on each singular 0-simplex.

A cup product for simplicial cohomology can be defined by the same formula as for singular cohomology, so the canonical isomorphism between simplicial and singular cohomology respects cup products. Here are three examples of direct calculations of cup products using simplicial cohomology.

Example 3.7. Let M be the closed orientable surface of genus $g \geq 1$ with the Δ-complex structure shown in the figure for the case $g = 2$. The cup product of interest is $H^1(M) \times H^1(M) \rightarrow H^2(M)$. Taking $\mathbb{Z}$ coefficients, a basis for $H_1(M)$ is formed by the edges a_i and b_i, as we showed in Example 2.36 when we computed the homology of M using cellular homology. We have $H^1(M) \approx \operatorname{Hom}(H_1(M), \mathbb{Z})$ by cellular cohomology or the universal coefficient theorem. A basis for $H_1(M)$ determines a dual basis for $\operatorname{Hom}(H_1(M), \mathbb{Z})$, so dual to a_i is the cohomology class α_i assigning the value 1 to a_i and 0 to the other basis elements, and similarly we have cohomology classes β_i dual to b_i.

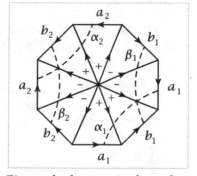

To represent α_i by a simplicial cocycle φ_i we need to choose values for φ_i on the edges radiating out from the central vertex in such a way that $\delta\varphi_i = 0$. This is the 'cocycle condition' discussed in the introduction to this chapter, where we saw that it has a geometric interpretation in terms of curves transverse to the edges of M. With this interpretation in mind, consider the arc labeled α_i in the figure, which represents a loop in M meeting a_i in one point and disjoint from all the other basis elements a_j and b_j. We define φ_i to have the value 1 on edges meeting the arc α_i and the value 0 on all other edges. Thus φ_i counts the number of intersections of each edge with the arc α_i. In similar fashion we obtain a cocycle ψ_i counting intersections with the arc β_i, and ψ_i represents the cohomology class β_i dual to b_i.

Now we can compute cup products by applying the definition. Keeping in mind that the ordering of the vertices of each 2-simplex is compatible with the indicated orientations of its edges, we see for example that $\varphi_1 \smile \psi_1$ takes the value 0 on all 2-simplices except the one with outer edge b_1 in the lower right part of the figure, where it takes the value 1. Thus $\varphi_1 \smile \psi_1$ takes the value 1 on the 2-chain c formed by the sum of all the 2-simplices with the signs indicated in the center of the figure. It is an easy calculation that $\partial c = 0$. Since there are no 3-simplices, c is not a boundary, so it represents a nonzero element of $H_2(M)$. The fact that $(\varphi_1 \smile \psi_1)(c)$ is a generator of $\mathbb{Z}$ implies both that c represents a generator of $H_2(M) \approx \mathbb{Z}$ and that $\varphi_1 \smile \psi_1$ represents the dual generator γ of $H^2(M) \approx \operatorname{Hom}(H_2(M), \mathbb{Z}) \approx \mathbb{Z}$. Thus $\alpha_1 \smile \beta_1 = \gamma$. In similar fashion one computes:

$$\alpha_i \smile \beta_j = \begin{cases} \gamma, & i = j \\ 0, & i \ne j \end{cases} = -(\beta_i \smile \alpha_j), \qquad \alpha_i \smile \alpha_j = 0, \qquad \beta_i \smile \beta_j = 0$$

These relations determine the cup product $H^1(M) \times H^1(M) \to H^2(M)$ completely since cup product is distributive. Notice that cup product is not commutative in this example since $\alpha_i \smile \beta_i = -(\beta_i \smile \alpha_i)$. We will show in Theorem 3.11 below that this is the worst that can happen: Cup product is commutative up to a sign depending only on dimension, assuming that the coefficient ring itself is commutative.

One can see in this example that nonzero cup products of distinct classes α_i or β_j occur precisely when the corresponding loops α_i or β_j intersect. This is also true for the cup product of α_i or β_i with itself if we allow ourselves to take two copies of the corresponding loop and deform one of them to be disjoint from the other.

Example 3.8. The closed nonorientable surface N of genus g can be treated in similar fashion if we use $\mathbb{Z}_2$ coefficients. Using the Δ-complex structure shown, the edges a_i give a basis for $H_1(N; \mathbb{Z}_2)$, and the dual basis elements $\alpha_i \in H^1(N; \mathbb{Z}_2)$ can be represented by cocycles with values given by counting intersections with the arcs labeled α_i in the figure. Then one computes that $\alpha_i \smile \alpha_i$ is the nonzero element of $H^2(N; \mathbb{Z}_2) \approx \mathbb{Z}_2$ and $\alpha_i \smile \alpha_j = 0$ for $i \ne j$. In particu-

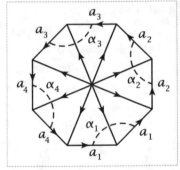

lar, when $g = 1$ we have $N = \mathbb{R}\mathrm{P}^2$, and the cup product of a generator of $H^1(\mathbb{R}\mathrm{P}^2; \mathbb{Z}_2)$ with itself is a generator of $H^2(\mathbb{R}\mathrm{P}^2; \mathbb{Z}_2)$.

The remarks in the paragraph preceding this example apply here also, but with the following difference: When one tries to deform a second copy of the loop α_i in the present example to be disjoint from the original copy, the best one can do is make it intersect the original in one point. This reflects the fact that $\alpha_i \smile \alpha_i$ is now nonzero.

Example 3.9. Let X be the 2-dimensional CW complex obtained by attaching a 2-cell to S^1 by the degree m map $S^1 \to S^1$, $z \mapsto z^m$. Using cellular cohomology, or cellular homology and the universal coefficient theorem, we see that $H^n(X; \mathbb{Z})$ consists of a $\mathbb{Z}$ for $n = 0$ and a $\mathbb{Z}_m$ for $n = 2$, so the cup product structure with $\mathbb{Z}$ coefficients is uninteresting. However, with $\mathbb{Z}_m$ coefficients we have $H^i(X; \mathbb{Z}_m) \approx \mathbb{Z}_m$ for $i = 0, 1, 2$, so there is the possibility that the cup product of two 1-dimensional classes can be nontrivial.

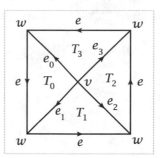

To obtain a Δ-complex structure on X, take a regular m-gon subdivided into m triangles T_i around a central vertex v, as shown in the figure for the case $m = 4$, then identify all the outer edges by rotations of the m-gon. This gives X a Δ-complex structure with 2 vertices, $m+1$ edges, and m 2-simplices. A generator α of $H^1(X;\mathbb{Z}_m)$ is represented by a cocycle φ assigning the value 1 to the edge e, which generates $H_1(X)$. The condition that φ be a cocycle means that $\varphi(e_i) + \varphi(e) = \varphi(e_{i+1})$ for all i, subscripts being taken mod m. So we may take $\varphi(e_i) = i \in \mathbb{Z}_m$. Hence $(\varphi \smile \varphi)(T_i) = \varphi(e_i)\varphi(e) = i$. The map $h : H^2(X;\mathbb{Z}_m) \to \operatorname{Hom}(H_2(X;\mathbb{Z}_m), \mathbb{Z}_m)$ is an isomorphism since $\sum_i T_i$ is a generator of $H_2(X;\mathbb{Z}_m)$ and there are 2-cocycles taking the value 1 on $\sum_i T_i$, for example the cocycle taking the value 1 on one T_i and 0 on all the others. The cocycle $\varphi \smile \varphi$ takes the value $0 + 1 + \cdots + (m - 1)$ on $\sum_i T_i$, hence represents $0 + 1 + \cdots + (m - 1)$ times a generator β of $H^2(X;\mathbb{Z}_m)$. In $\mathbb{Z}_m$ the sum $0 + 1 + \cdots + (m - 1)$ is 0 if m is odd and k if $m = 2k$ since the terms 1 and $m - 1$ cancel, 2 and $m - 2$ cancel, and so on. Thus, writing α^2 for $\alpha \smile \alpha$, we have $\alpha^2 = 0$ if m is odd and $\alpha^2 = k\beta$ if $m = 2k$.

In particular, if $m = 2$, X is $\mathbb{RP}^2$ and $\alpha^2 = \beta$ in $H^2(\mathbb{RP}^2;\mathbb{Z}_2)$, as we showed already in Example 3.8.

The cup product formula $(\varphi \smile \psi)(\sigma) = \varphi(\sigma \,|\, [v_0, \cdots, v_k])\psi(\sigma \,|\, [v_k, \cdots, v_{k+\ell}])$ also gives relative cup products

$$H^k(X;R) \times H^\ell(X,A;R) \xrightarrow{\ \smile\ } H^{k+\ell}(X,A;R)$$

$$H^k(X,A;R) \times H^\ell(X;R) \xrightarrow{\ \smile\ } H^{k+\ell}(X,A;R)$$

$$H^k(X,A;R) \times H^\ell(X,A;R) \xrightarrow{\ \smile\ } H^{k+\ell}(X,A;R)$$

since if φ or ψ vanishes on chains in A then so does $\varphi \smile \psi$. There is a more general relative cup product

$$H^k(X,A;R) \times H^\ell(X,B;R) \xrightarrow{\ \smile\ } H^{k+\ell}(X,A \cup B;R)$$

when A and B are open subsets of X or subcomplexes of the CW complex X. This is obtained in the following way. The absolute cup product restricts to a cup product $C^k(X,A;R) \times C^\ell(X,B;R) \to C^{k+\ell}(X,A + B;R)$ where $C^n(X,A + B;R)$ is the subgroup of $C^n(X;R)$ consisting of cochains vanishing on sums of chains in A and chains in B. If A and B are open in X, the inclusions $C^n(X,A \cup B;R) \hookrightarrow C^n(X,A + B;R)$ induce isomorphisms on cohomology, via the five-lemma and the fact that the restriction maps $C^n(A \cup B;R) \to C^n(A + B;R)$ induce isomorphisms on cohomology as we saw in the discussion of excision in the previous section. Therefore the cup product $C^k(X,A;R) \times C^\ell(X,B;R) \to C^{k+\ell}(X,A + B;R)$ induces the desired relative cup product

$H^k(X, A; R) \times H^\ell(X, B; R) \to H^{k+\ell}(X, A \cup B; R)$. This holds also if X is a CW complex with A and B subcomplexes since here again the maps $C^n(A \cup B; R) \to C^n(A + B; R)$ induce isomorphisms on cohomology, as we saw for homology in §2.2.   ·

Proposition 3.10. *For a map $f : X \to Y$, the induced maps $f^* : H^n(Y; R) \to H^n(X; R)$ satisfy $f^*(\alpha \smile \beta) = f^*(\alpha) \smile f^*(\beta)$, and similarly in the relative case.*

Proof: This comes from the cochain formula $f^\#(\varphi) \smile f^\#(\psi) = f^\#(\varphi \smile \psi)$:

$$
\begin{aligned}
(f^\#\varphi \smile f^\#\psi)(\sigma) &= f^\#\varphi(\sigma|[v_0, \cdots, v_k]) f^\#\psi(\sigma|[v_k, \cdots, v_{k+\ell}]) \\
&= \varphi(f\sigma|[v_0, \cdots, v_k])\psi(f\sigma|[v_k, \cdots, v_{k+\ell}]) \\
&= (\varphi \smile \psi)(f\sigma) = f^\#(\varphi \smile \psi)(\sigma) \qquad \square
\end{aligned}
$$

The natural question of whether the cup product is commutative is answered by the following:

Theorem 3.11. *The identity $\alpha \smile \beta = (-1)^{k\ell}\beta \smile \alpha$ holds for all $\alpha \in H^k(X, A; R)$ and $\beta \in H^\ell(X, A; R)$, when R is commutative.*

Taking $\alpha = \beta$, this implies in particular that if α is an element of $H^k(X, A; R)$ with k odd, then $2(\alpha \smile \alpha) = 0$ in $H^{2k}(X, A; R)$, or more concisely, $2\alpha^2 = 0$. Hence if $H^{2k}(X, A; R)$ has no elements of order two, then $\alpha^2 = 0$. For example, if X is the 2-complex obtained by attaching a disk to S^1 by a map of degree m as in Example 3.9 above, then we can deduce that the square of a generator of $H^1(X; \mathbb{Z}_m)$ is zero if m is odd, and is either zero or the unique element of $H^2(X; \mathbb{Z}_m) \approx \mathbb{Z}_m$ of order two if m is even. As we showed, the square is in fact nonzero when m is even.

Proof: Consider first the case $A = \varnothing$. For cochains $\varphi \in C^k(X; R)$ and $\psi \in C^\ell(X; R)$ one can see from the definition that the cup products $\varphi \smile \psi$ and $\psi \smile \varphi$ differ only by a permutation of the vertices of $\Delta^{k+\ell}$. The idea of the proof is to study a particularly nice permutation of vertices, namely the one that totally reverses their order. This has the convenient feature of also reversing the ordering of vertices in any face.

For a singular n-simplex $\sigma : [v_0, \cdots, v_n] \to X$, let $\overline{\sigma}$ be the singular n-simplex obtained by preceding σ by the linear homeomorphism of $[v_0, \cdots, v_n]$ reversing the order of the vertices. Thus $\overline{\sigma}(v_i) = \sigma(v_{n-i})$. This reversal of vertices is the product of $n + (n-1) + \cdots + 1 = n(n+1)/2$ transpositions of adjacent vertices, each of which reverses orientation of the n-simplex since it is a reflection across an $(n-1)$-dimensional hyperplane. So to take orientations into account we would expect that a sign $\varepsilon_n = (-1)^{n(n+1)/2}$ ought to be inserted. Hence we define a homomorphism $\rho : C_n(X) \to C_n(X)$ by $\rho(\sigma) = \varepsilon_n\overline{\sigma}$.

We will show that ρ is a chain map, chain homotopic to the identity, so it induces the identity on cohomology. From this the theorem quickly follows. Namely, the formulas

$$(\rho^*\varphi \smile \rho^*\psi)(\sigma) = \varphi(\varepsilon_k\sigma|[v_k, \cdots, v_0])\psi(\varepsilon_\ell\sigma|[v_{k+\ell}, \cdots, v_k])$$

$$\rho^*(\psi \smile \varphi)(\sigma) = \varepsilon_{k+\ell}\psi(\sigma|[v_{k+\ell}, \cdots, v_k])\varphi(\sigma|[v_k, \cdots, v_0])$$

show that $\varepsilon_k\varepsilon_\ell(\rho^*\varphi \smile \rho^*\psi) = \varepsilon_{k+\ell}\rho^*(\psi \smile \varphi)$, since we assume R is commutative. A trivial calculation gives $\varepsilon_{k+\ell} = (-1)^{k\ell}\varepsilon_k\varepsilon_\ell$, hence $\rho^*\varphi \smile \rho^*\psi = (-1)^{k\ell}\rho^*(\psi \smile \varphi)$. Since ρ is chain homotopic to the identity, the ρ^*'s disappear when we pass to cohomology classes, and so we obtain the desired formula $\alpha \smile \beta = (-1)^{k\ell}\beta \smile \alpha$.

The chain map property $\partial\rho = \rho\partial$ can be verified by calculating, for a singular n-simplex σ,

$$\partial\rho(\sigma) = \varepsilon_n \sum_i (-1)^i \sigma|[v_n, \cdots, \hat{v}_{n-i}, \cdots, v_0]$$

$$\rho\partial(\sigma) = \rho\left(\sum_i (-1)^i \sigma|[v_0, \cdots, \hat{v}_i, \cdots, v_n]\right)$$

$$= \varepsilon_{n-1} \sum_i (-1)^{n-i} \sigma|[v_n, \cdots, \hat{v}_{n-i}, \cdots, v_0]$$

so the result follows from the easily checked identity $\varepsilon_n = (-1)^n \varepsilon_{n-1}$.

To define a chain homotopy between ρ and the identity we are motivated by the construction of the prism operator P in the proof that homotopic maps induce the same homomorphism on homology, in Theorem 2.10. The main ingredient in the construction of P was a subdivision of $\Delta^n \times I$ into $(n+1)$-simplices with vertices v_i in $\Delta^n \times \{0\}$ and w_i in $\Delta^n \times \{1\}$, the vertex w_i lying directly above v_i. Using the same subdivision, and letting $\pi : \Delta^n \times I \to \Delta^n$ be the projection, we now define $P : C_n(X) \to C_{n+1}(X)$ by

$$P(\sigma) = \sum_i (-1)^i \varepsilon_{n-i}(\sigma\pi)|[v_0, \cdots, v_i, w_n, \cdots, w_i]$$

Thus the w-vertices are written in reverse order, and there is a compensating sign ε_{n-i}. One can view this formula as arising from the Δ-complex structure on $\Delta^n \times I$ in which the vertices are ordered $v_0, \cdots, v_n, w_n, \cdots, w_0$ rather than the more natural ordering $v_0, \cdots, v_n, w_0, \cdots, w_n$.

To show $\partial P + P\partial = \rho - \mathbb{1}$ we first calculate ∂P, leaving out σ's and $\sigma\pi$'s for notational simplicity:

$$\partial P = \sum_{j \le i} (-1)^i (-1)^j \varepsilon_{n-i}[v_0, \cdots, \hat{v}_j, \cdots, v_i, w_n, \cdots, w_i]$$

$$+ \sum_{j \ge i} (-1)^i (-1)^{i+1+n-j} \varepsilon_{n-i}[v_0, \cdots, v_i, w_n, \cdots, \widehat{w}_j, \cdots, w_i]$$

The $j = i$ terms in these two sums give

$$\varepsilon_n[w_n, \cdots, w_0] + \sum_{i>0} \varepsilon_{n-i}[v_0, \cdots, v_{i-1}, w_n, \cdots, w_i]$$

$$+ \sum_{i<n} (-1)^{n+i+1} \varepsilon_{n-i}[v_0, \cdots, v_i, w_n, \cdots, w_{i+1}] - [v_0, \cdots, v_n]$$

In this expression the two summation terms cancel since replacing i by $i - 1$ in the second sum produces a new sign $(-1)^{n+i}\varepsilon_{n-i+1} = -\varepsilon_{n-i}$. The remaining two terms $\varepsilon_n[w_n, \cdots, w_0]$ and $-[v_0, \cdots, v_n]$ represent $\rho(\sigma) - \sigma$. So in order to show that $\partial P + P\partial = \rho - \mathbb{1}$, it remains to check that in the formula for ∂P above, the terms with $j \neq i$ give $-P\partial$. Calculating $P\partial$ from the definitions, we have

$$P\partial = \sum_{i<j}(-1)^i(-1)^j\varepsilon_{n-i-1}[v_0, \cdots, v_i, w_n, \cdots, \widehat{w}_j, \cdots, w_i]$$

$$+ \sum_{i>j}(-1)^{i-1}(-1)^j\varepsilon_{n-i}[v_0, \cdots, \widehat{v}_j, \cdots, v_i, w_n, \cdots, w_i]$$

Since $\varepsilon_{n-i} = (-1)^{n-i}\varepsilon_{n-i-1}$, this finishes the verification that $\partial P + P\partial = \rho - \mathbb{1}$, and so the theorem is proved when $A = \varnothing$. The proof also applies when $A \neq \varnothing$ since the maps ρ and P take chains in A to chains in A, so the dual homomorphisms ρ^* and P^* act on relative cochains. $\qquad\square$

The Cohomology Ring

Since cup product is associative and distributive, it is natural to try to make it the multiplication in a ring structure on the cohomology groups of a space X. This is easy to do if we simply define $H^*(X;R)$ to be the direct sum of the groups $H^n(X;R)$. Elements of $H^*(X;R)$ are finite sums $\sum_i \alpha_i$ with $\alpha_i \in H^i(X;R)$, and the product of two such sums is defined to be $(\sum_i \alpha_i)(\sum_j \beta_j) = \sum_{i,j} \alpha_i\beta_j$. It is routine to check that this makes $H^*(X;R)$ into a ring, with identity if R has an identity. Similarly, $H^*(X,A;R)$ is a ring via the relative cup product. Taking scalar multiplication by elements of R into account, these rings can also be regarded as R-algebras.

For example, the calculations in Example 3.8 or 3.9 above show that $H^*(\mathbb{RP}^2;\mathbb{Z}_2)$ consists of the polynomials $a_0 + a_1\alpha + a_2\alpha^2$ with coefficients $a_i \in \mathbb{Z}_2$, so $H^*(\mathbb{RP}^2;\mathbb{Z}_2)$ is the quotient $\mathbb{Z}_2[\alpha]/(\alpha^3)$ of the polynomial ring $\mathbb{Z}_2[\alpha]$ by the ideal generated by α^3.

This example illustrates how $H^*(X;R)$ often has a more compact description than the sequence of individual groups $H^n(X;R)$, so there is a certain economy in the change of scale that comes from regarding all the groups $H^n(X;R)$ as part of a single object $H^*(X;R)$.

Adding cohomology classes of different dimensions to form $H^*(X;R)$ is a convenient formal device, but it has little topological significance. One always regards the cohomology ring as a **graded ring**: a ring A with a decomposition as a sum $\bigoplus_{k\geq0}A_k$ of additive subgroups A_k such that the multiplication takes $A_k \times A_\ell$ to $A_{k+\ell}$. To indicate that an element $a \in A$ lies in A_k we write $|a| = k$. This applies in particular to elements of $H^k(X;R)$. Some authors call $|a|$ the 'degree' of a, but we will use the term 'dimension' which is more geometric and avoids potential confusion with the degree of a polynomial.

A graded ring satisfying the commutativity property of Theorem 3.11, $ab = (-1)^{|a||b|}ba$, is usually called simply **commutative** in the context of algebraic topology, in spite of the potential for misunderstanding. In the older literature one finds less ambiguous terms such as *graded commutative, anticommutative,* or *skew commutative.*

Example 3.12: Polynomial Rings. Among the simplest graded rings are polynomial rings $R[\alpha]$ and their truncated versions $R[\alpha]/(\alpha^n)$, consisting of polynomials of degree less than n. The example we have seen is $H^*(\mathbb{R}P^2; \mathbb{Z}_2) \approx \mathbb{Z}_2[\alpha]/(\alpha^3)$. More generally we will show in Theorem 3.19 that $H^*(\mathbb{R}P^n; \mathbb{Z}_2) \approx \mathbb{Z}_2[\alpha]/(\alpha^{n+1})$ and $H^*(\mathbb{R}P^\infty; \mathbb{Z}_2) \approx \mathbb{Z}_2[\alpha]$. In these cases $|\alpha| = 1$. We will also show that $H^*(\mathbb{C}P^n; \mathbb{Z}) \approx \mathbb{Z}[\alpha]/(\alpha^{n+1})$ and $H^*(\mathbb{C}P^\infty; \mathbb{Z}) \approx \mathbb{Z}[\alpha]$ with $|\alpha| = 2$. The analogous results for quaternionic projective spaces are also valid, with $|\alpha| = 4$. The coefficient ring $\mathbb{Z}$ in the complex and quaternionic cases could be replaced by any commutative ring R, but not for $\mathbb{R}P^n$ and $\mathbb{R}P^\infty$ since a polynomial ring $R[\alpha]$ is strictly commutative, so for this to be a commutative ring in the graded sense we must have either $|\alpha|$ even or $2 = 0$ in R.

Polynomial rings in several variables also have graded ring structures, and these graded rings can sometimes be realized as cohomology rings of spaces. For example, $\mathbb{Z}_2[\alpha_1, \cdots, \alpha_n]$ is $H^*(X; \mathbb{Z}_2)$ for X the product of n copies of $\mathbb{R}P^\infty$, with $|\alpha_i| - 1$ for each i, as we will see in Example 3.20.

Example 3.13: Exterior Algebras. Another nice example of a commutative graded ring is the **exterior algebra** $\Lambda_R[\alpha_1, \cdots, \alpha_n]$ over a commutative ring R with identity. This is the free R-module with basis the finite products $\alpha_{i_1} \cdots \alpha_{i_k}$, $i_1 < \cdots < i_k$, with associative, distributive multiplication defined by the rules $\alpha_i \alpha_j = -\alpha_j \alpha_i$ for $i \neq j$ and $\alpha_i^2 = 0$. The empty product of α_i's is allowed, and provides an identity element 1 in $\Lambda_R[\alpha_1, \cdots, \alpha_n]$. The exterior algebra becomes a commutative graded ring by specifying odd dimensions for the generators α_i.

The example we have seen is the torus $T^2 = S^1 \times S^1$, where $H^*(T^2; \mathbb{Z}) \approx \Lambda_\mathbb{Z}[\alpha, \beta]$ with $|\alpha| = |\beta| = 1$ by the calculations in Example 3.7. More generally, for the n-torus T^n, $H^*(T^n; R)$ is the exterior algebra $\Lambda_R[\alpha_1, \cdots, \alpha_n]$ as we will see in Example 3.16. The same is true for any product of odd-dimensional spheres, where $|\alpha_i|$ is the dimension of the ith sphere.

Induced homomorphisms are ring homomorphisms by Proposition 3.10. Here is an example illustrating this fact.

Example 3.14: Product Rings. The isomorphism $H^*(\coprod_\alpha X_\alpha; R) \xrightarrow{\approx} \prod_\alpha H^*(X_\alpha; R)$ whose coordinates are induced by the inclusions $i_\alpha : X_\alpha \hookrightarrow \coprod_\alpha X_\alpha$ is a ring isomorphism with respect to the usual coordinatewise multiplication in a product ring, because each coordinate function i_α^* is a ring homomorphism. Similarly for a wedge sum the isomorphism $\tilde{H}^*(\bigvee_\alpha X_\alpha; R) \approx \prod_\alpha \tilde{H}^*(X_\alpha; R)$ is a ring isomorphism. Here we take reduced cohomology to be cohomology relative to a basepoint, and we use relative cup products. We should assume the basepoints $x_\alpha \in X_\alpha$ are deformation retracts of neighborhoods, to be sure that the claimed isomorphism does indeed hold.

This product ring structure for wedge sums can sometimes be used to rule out splittings of a space as a wedge sum up to homotopy equivalence. For example, consider $\mathbb{CP}^2$, which is S^2 with a cell e^4 attached by a certain map $f : S^3 \to S^2$. Using homology or just the additive structure of cohomology it is impossible to conclude that $\mathbb{CP}^2$ is not homotopy equivalent to $S^2 \vee S^4$, and hence that f is not homotopic to a constant map. However, with cup products we can distinguish these two spaces since the square of each element of $H^2(S^2 \vee S^4; \mathbb{Z})$ is zero in view of the ring isomorphism $\tilde{H}^*(S^2 \vee S^4; \mathbb{Z}) \approx \tilde{H}^*(S^2; \mathbb{Z}) \oplus \tilde{H}^*(S^4; \mathbb{Z})$, but the square of a generator of $H^2(\mathbb{CP}^2; \mathbb{Z})$ is nonzero since $H^*(\mathbb{CP}^2; \mathbb{Z}) \approx \mathbb{Z}[\alpha]/(\alpha^3)$.

More generally, cup products can be used to distinguish infinitely many different homotopy classes of maps $S^{4n-1} \to S^{2n}$ for all $n \geq 1$. This is systematized in the notion of the *Hopf invariant*, which is studied in §4.B.

Here is the evident general question raised by the preceding examples:

The Realization Problem. Which graded commutative R-algebras occur as cup product algebras $H^*(X; R)$ of spaces X?

This is a difficult problem, with the degree of difficulty depending strongly on the coefficient ring R. The most accessible case is $R = \mathbb{Q}$, where essentially every graded commutative $\mathbb{Q}$-algebra is realizable, as shown in [Quillen 1969]. Next in order of difficulty is $R = \mathbb{Z}_p$ with p prime. This is much harder than the case of $\mathbb{Q}$, and only partial results, obtained with much labor, are known. Finally there is $R = \mathbb{Z}$, about which very little is known beyond what is implied by the $\mathbb{Z}_p$ cases.

A Künneth Formula

One might guess that there should be some connection between cup product and product spaces, and indeed this is the case, as we will show in this subsection.

To begin, we define the **cross product**, or **external cup product** as it is sometimes called. This is the map

$$H^*(X; R) \times H^*(Y; R) \xrightarrow{\ \times\ } H^*(X \times Y; R)$$

given by $a \times b = p_1^*(a) \smile p_2^*(b)$ where p_1 and p_2 are the projections of $X \times Y$ onto X and Y. Since cup product is distributive, the cross product is bilinear, that is, linear in each variable separately. We might hope that the cross product map would be an isomorphism in many cases, thereby giving a nice description of the cohomology rings of these product spaces. However, a bilinear map is rarely a homomorphism, so it could hardly be an isomorphism. Fortunately there is a nice algebraic solution to this problem, and that is to replace the direct product $H^*(X;R) \times H^*(Y;R)$ by the tensor product $H^*(X;R) \otimes_R H^*(Y;R)$.

Let us review the definition and basic properties of tensor products. For abelian groups A and B the tensor product $A \otimes B$ is defined to be the abelian group with generators $a \otimes b$ for $a \in A$, $b \in B$, and relations $(a + a') \otimes b = a \otimes b + a' \otimes b$ and $a \otimes (b + b') = a \otimes b + a \otimes b'$. So the zero element of $A \otimes B$ is $0 \otimes 0 = 0 \otimes b = a \otimes 0$, and $-(a \otimes b) = -a \otimes b = a \otimes (-b)$. Some readily verified elementary properties are:

(1) $A \otimes B \approx B \otimes A$.

(2) $(\bigoplus_i A_i) \otimes B \approx \bigoplus_i (A_i \otimes B)$.

(3) $(A \otimes B) \otimes C \approx A \otimes (B \otimes C)$.

(4) $\mathbb{Z} \otimes A \approx A$.

(5) $\mathbb{Z}_n \otimes A \approx A/nA$.

(6) A pair of homomorphisms $f : A \to A'$ and $g : B \to B'$ induces a homomorphism $f \otimes g : A \otimes B \to A' \otimes B'$ via $(f \otimes g)(a \otimes b) = f(a) \otimes g(b)$.

(7) A bilinear map $\varphi : A \times B \to C$ induces a homomorphism $A \otimes B \to C$ sending $a \otimes b$ to $\varphi(a, b)$.

In (1)–(5) the isomorphisms are the obvious ones, for example $a \otimes b \mapsto b \otimes a$ in (1) and $n \otimes a \mapsto na$ in (4). Properties (1), (2), (4), and (5) allow the calculation of tensor products of finitely generated abelian groups.

The generalization to tensor products of modules over a commutative ring R is easy. One defines $A \otimes_R B$ for R-modules A and B to be the quotient of $A \otimes B$ obtained by imposing the further relations $ra \otimes b = a \otimes rb$ for $r \in R$, $a \in A$, and $b \in B$. This relation guarantees that $A \otimes_R B$ is again an R-module. In case R is not commutative, one assumes A is a right R-module and B is a left R-module, and the relation is written instead $ar \otimes b = a \otimes rb$, but now $A \otimes_R B$ is only an abelian group, not an R-module. However, we will restrict attention to the case that R is commutative in what follows.

It is an easy algebra exercise to see that $A \otimes_R B = A \otimes B$ when R is $\mathbb{Z}_m$ or $\mathbb{Q}$. But in general $A \otimes_R B$ is not the same as $A \otimes B$. For example, if $R = \mathbb{Q}(\sqrt{2})$, which is a 2-dimensional vector space over $\mathbb{Q}$, then $R \otimes_R R = R$ but $R \otimes R$ is a 4-dimensional vector space over $\mathbb{Q}$.

The statements (1)–(3), (6), and (7) remain valid for tensor products of R-modules. The generalization of (4) is the canonical isomorphism $R \otimes_R A \approx A$, $r \otimes a \mapsto ra$.

Property (7) of tensor products guarantees that the cross product as defined above gives rise to a homomorphism of R-modules

$$H^*(X;R) \otimes_R H^*(Y;R) \xrightarrow{\ \times\ } H^*(X \times Y;R), \qquad a \otimes b \mapsto a \times b$$

which we shall also call cross product. This map becomes a ring homomorphism if we define the multiplication in a tensor product of graded rings by $(a \otimes b)(c \otimes d) =$

$(-1)^{|b||c|} ac \otimes bd$ where $|x|$ denotes the dimension of x. Namely, if we denote the cross product map by μ and we define $(a \otimes b)(c \otimes d) = (-1)^{|b||c|} ac \otimes bd$, then

$$
\begin{aligned}
\mu((a \otimes b)(c \otimes d)) &= (-1)^{|b||c|} \mu(ac \otimes bd) \\
&= (-1)^{|b||c|}(a \smile c) \times (b \smile d) \\
&= (-1)^{|b||c|} p_1^*(a \smile c) \smile p_2^*(b \smile d) \\
&= (-1)^{|b||c|} p_1^*(a) \smile p_1^*(c) \smile p_2^*(b) \smile p_2^*(d) \\
&= p_1^*(a) \smile p_2^*(b) \smile p_1^*(c) \smile p_2^*(d) \\
&= (a \times b)(c \times d) = \mu(a \otimes b)\mu(c \otimes d)
\end{aligned}
$$

Theorem 3.15. *The cross product* $H^*(X; R) \otimes_R H^*(Y; R) \to H^*(X \times Y; R)$ *is an isomorphism of rings if X and Y are CW complexes and $H^k(Y; R)$ is a finitely generated free R-module for all k.*

Results of this type, computing homology or cohomology of a product space, are known as **Künneth formulas**. The hypothesis that X and Y are CW complexes will be shown to be unnecessary in §4.1 when we consider CW approximations to arbitrary spaces. On the other hand, the freeness hypothesis cannot always be dispensed with, as we shall see in §3.B when we obtain a completely general Künneth formula for the homology of a product space.

When the conclusion of the theorem holds, the ring structure in $H^*(X \times Y; R)$ is determined by the ring structures in $H^*(X; R)$ and $H^*(Y; R)$. Example 3E.6 shows that some hypotheses are necessary in order for this to be true.

Example 3.16. The exterior algebra $\Lambda_R[\alpha_1, \cdots, \alpha_n]$ is the graded tensor product over R of the one-variable exterior algebras $\Lambda_R[\alpha_i]$ where the α_i's have odd dimension. The Künneth formula then gives an isomorphism $H^*(S^{k_1} \times \cdots \times S^{k_n}; \mathbb{Z}) \approx \Lambda_{\mathbb{Z}}[\alpha_1, \cdots, \alpha_n]$ if the dimensions k_i are all odd. With some k_i's even, one would have the tensor product of an exterior algebra for the odd-dimensional spheres and truncated polynomial rings $\mathbb{Z}[\alpha]/(\alpha^2)$ for the even-dimensional spheres. Of course, $\Lambda_{\mathbb{Z}}[\alpha]$ and $\mathbb{Z}[\alpha]/(\alpha^2)$ are isomorphic as rings, but when one takes tensor products in the graded sense it becomes important to distinguish them as graded rings, with α odd-dimensional in $\Lambda_{\mathbb{Z}}[\alpha]$ and even-dimensional in $\mathbb{Z}[\alpha]/(\alpha^2)$. These remarks apply more generally with any coefficient ring R in place of $\mathbb{Z}$, though when $R = \mathbb{Z}_2$ there is no need to distinguish between the odd-dimensional and even-dimensional cases since signs become irrelevant.

The idea of the proof of the theorem will be to consider, for a fixed CW complex Y, the functors

$$
h^n(X, A) = \bigoplus_i (H^i(X, A; R) \otimes_R H^{n-i}(Y; R))
$$

$$
k^n(X, A) = H^n(X \times Y, A \times Y; R)
$$

The cross product, or a relative version of it, defines a map $\mu : h^n(X, A) \to k^n(X, A)$ which we would like to show is an isomorphism when X is a CW complex and $A = \varnothing$. We will show:

(1) h^* and k^* are cohomology theories on the category of CW pairs.

(2) μ is a natural transformation: It commutes with induced homomorphisms and with coboundary homomorphisms in long exact sequences of pairs.

It is obvious that $\mu : h^n(X) \to k^n(X)$ is an isomorphism when X is a point since it is just the scalar multiplication map $R \otimes_R H^n(Y; R) \to H^n(Y; R)$. The following general fact will then imply the theorem.

Proposition 3.17. *If a natural transformation between unreduced cohomology theories on the category of CW pairs is an isomorphism when the CW pair is (point, $\varnothing$), then it is an isomorphism for all CW pairs.*

Proof: Let $\mu : h^*(X, A) \to k^*(X, A)$ be the natural transformation. By the five-lemma it will suffice to show that μ is an isomorphism when $A = \varnothing$.

First we do the case of finite-dimensional X by induction on dimension. The induction starts with the case that X is 0-dimensional, where the result holds by hypothesis and by the axiom for disjoint unions. For the induction step, μ gives a map between the two long exact sequences for the pair (X^n, X^{n-1}), with commuting squares since μ is a natural transformation. The five-lemma reduces the inductive step to showing that μ is an isomorphism for $(X, A) = (X^n, X^{n-1})$. Let $\Phi : \coprod_\alpha (D^n_\alpha, \partial D^n_\alpha) \to (X^n, X^{n-1})$ be a collection of characteristic maps for all the n-cells of X. By excision, Φ^* is an isomorphism for h^* and k^*, so by naturality it suffices to show that μ is an isomorphism for $(X, A) = \coprod_\alpha (D^n_\alpha, \partial D^n_\alpha)$. The axiom for disjoint unions gives a further reduction to the case of the pair $(D^n, \partial D^n)$. Finally, this case follows by applying the five-lemma to the long exact sequences of this pair, since D^n is contractible and hence is covered by the 0-dimensional case, and ∂D^n is $(n-1)$-dimensional.

The case that X is infinite-dimensional reduces to the finite-dimensional case by a telescope argument as in the proof of Lemma 2.34. We leave this for the reader since the finite-dimensional case suffices for the special h^* and k^* we are considering, as the maps $h^i(X) \to h^i(X^n)$ and $k^i(X) \to k^i(X^n)$ induced by the inclusion $X^n \hookrightarrow X$ are isomorphisms when n is sufficiently large with respect to i. $\square$

Proof of 3.15: It remains to check that h^* and k^* are cohomology theories, and that μ is a natural transformation. Since we are dealing with unreduced cohomology theories there are four axioms to verify.

(1) Homotopy invariance: $f \simeq g$ implies $f^* = g^*$. This is obvious for both h^* and k^*.

(2) Excision: $h^*(X,A) \approx h^*(B, A\cap B)$ for A and B subcomplexes of the CW complex $X = A \cup B$. This is obvious, and so is the corresponding statement for k^* since $(A\times Y) \cup (B\times Y) = (A \cup B)\times Y$ and $(A\times Y) \cap (B\times Y) = (A \cap B)\times Y$.

(3) The long exact sequence of a pair. This is a triviality for k^*, but a few words of explanation are needed for h^*, where the desired exact sequence is obtained in two steps. For the first step, tensor the long exact sequence of ordinary cohomology groups for a pair (X,A) with the free R-module $H^n(Y;R)$, for a fixed n. This yields another exact sequence because $H^n(Y;R)$ is a direct sum of copies of R, so the result of tensoring an exact sequence with this direct sum is simply to produce a direct sum of copies of the exact sequence, which is again an exact sequence. The second step is to let n vary, taking a direct sum of the previously constructed exact sequences for each n, with the nth exact sequence shifted up by n dimensions.

(4) Disjoint unions. Again this axiom obviously holds for k^*, but some justification is required for h^*. What is needed is the algebraic fact that there is a canonical isomorphism $(\prod_\alpha M_\alpha) \otimes_R N \approx \prod_\alpha (M_\alpha \otimes_R N)$ for R-modules M_α and a finitely generated free R-module N. Since N is a direct product of finitely many copies R_β of R, $M_\alpha \otimes_R N$ is a direct product of corresponding copies $M_{\alpha\beta} = M_\alpha \otimes_R R_\beta$ of M_α and the desired relation becomes $\prod_\beta \prod_\alpha M_{\alpha\beta} \approx \prod_\alpha \prod_\beta M_{\alpha\beta}$, which is obviously true.

Finally there is naturality of μ to consider. Naturality with respect to maps between spaces is immediate from the naturality of cup products. Naturality with respect to coboundary maps in long exact sequences is commutativity of the following square:

$$\begin{array}{ccc} H^k(A;R) \times H^\ell(Y;R) & \xrightarrow{\delta\times\mathbb{1}} & H^{k+1}(X,A;R) \times H^\ell(Y;R) \\ \downarrow{\times} & & \downarrow{\times} \\ H^{k+\ell}(A\times Y;R) & \xrightarrow{\delta} & H^{k+\ell+1}(X\times Y, A\times Y;R) \end{array}$$

To check this, start with an element of the upper left product, represented by cocycles $\varphi \in C^k(A;R)$ and $\psi \in C^\ell(Y;R)$. Extend φ to a cochain $\overline{\varphi} \in C^k(X;R)$. Then the pair (φ,ψ) maps rightward to $(\delta\overline{\varphi}, \psi)$ and then downward to $p_1^\sharp(\delta\overline{\varphi}) \smile p_2^\sharp(\psi)$. Going the other way around the square, (φ,ψ) maps downward to $p_1^\sharp(\varphi) \smile p_2^\sharp(\psi)$ and then rightward to $\delta(p_1^\sharp(\overline{\varphi}) \smile p_2^\sharp(\psi))$ since $p_1^\sharp(\overline{\varphi}) \smile p_2^\sharp(\psi)$ extends $p_1^\sharp(\varphi) \smile p_2^\sharp(\psi)$ over $X\times Y$. Finally, $\delta(p_1^\sharp(\overline{\varphi}) \smile p_2^\sharp(\psi)) = p_1^\sharp(\delta\overline{\varphi}) \smile p_2^\sharp(\psi)$ since $\delta\psi = 0$. $\square$

It is sometimes important to have a relative version of the Künneth formula in Theorem 3.15. The relative cross product is

$$H^*(X,A;R) \otimes_R H^*(Y,B;R) \xrightarrow{\times} H^*(X\times Y, A\times Y \cup X\times B;R)$$

for CW pairs (X,A) and (Y,B), defined just as in the absolute case by $a\times b = p_1^*(a) \smile p_2^*(b)$ where $p_1^*(a) \in H^*(X\times Y, A\times Y;R)$ and $p_2^*(b) \in H^*(X\times Y, X\times B;R)$.

Theorem 3.18. *For CW pairs (X, A) and (Y, B) the cross product homomorphism $H^*(X, A; R) \otimes_R H^*(Y, B; R) \to H^*(X \times Y, A \times Y \cup X \times B; R)$ is an isomorphism of rings if $H^k(Y, B; R)$ is a finitely generated free R-module for each k.*

Proof: The case $B = \varnothing$ was covered in the course of the proof of the absolute case, so it suffices to deduce the case $B \neq \varnothing$ from the case $B = \varnothing$.

The following commutative diagram shows that collapsing B to a point reduces the proof to the case that B is a point:

$$
\begin{array}{ccc}
H^*(X, A) \otimes_R H^*(Y, B) & \xleftarrow{\;\approx\;} & H^*(X, A) \otimes_R H^*(Y/B, B/B) \\
\Big\downarrow{\scriptstyle \times} & & \Big\downarrow{\scriptstyle \times} \\
H^*(X \times Y, A \times Y \cup X \times B) & \xleftarrow{\;\approx\;} & H^*(X \times (Y/B), A \times (Y/B) \cup X \times (B/B))
\end{array}
$$

The lower map is an isomorphism since the quotient spaces $(X \times Y)/(A \times Y \cup X \times B)$ and $(X \times (Y/B))/(A \times (Y/B) \cup X \times (B/B))$ are the same.

In the case that B is a point $y_0 \in Y$, consider the commutative diagram

$$
\begin{array}{ccccc}
H^*(X, A) \otimes_R H^*(Y, y_0) & \longrightarrow & H^*(X, A) \otimes_R H^*(Y) & \longrightarrow & H^*(X, A) \otimes_R H^*(y_0) \\
\Big\downarrow{\scriptstyle \times} & & \Big\downarrow{\scriptstyle \times} & & \Big\downarrow{\scriptstyle \times} \\
 & & & & H^*(X \times y_0, A \times y_0) \\
 & & & & \Big\uparrow{\scriptstyle \approx} \\
H^*(X \times Y, X \times y_0 \cup A \times Y) & \longrightarrow & H^*(X \times Y, A \times Y) & \cdot & H^*(X \times y_0 \cup A \times Y, A \times Y)
\end{array}
$$

Since y_0 is a retract of Y, the upper row of this diagram is a split short exact sequence. The lower row is the long exact sequence of a triple, and it too is a split short exact sequence since $(X \times y_0, A \times y_0)$ is a retract of $(X \times Y, A \times Y)$. The middle and right cross product maps are isomorphisms by the case $B = \varnothing$ since $H^k(Y; R)$ is a finitely generated free R-module if $H^k(Y, y_0; R)$ is. The five-lemma then implies that the left-hand cross product map is an isomorphism as well. $\qquad\square$

The relative cross product for pairs (X, x_0) and (Y, y_0) gives a reduced cross product

$$
\widetilde{H}^*(X; R) \otimes_R \widetilde{H}^*(Y; R) \xrightarrow{\;\times\;} \widetilde{H}^*(X \wedge Y; R)
$$

where $X \wedge Y$ is the smash product $X \times Y / (X \times \{y_0\} \cup \{x_0\} \times Y)$. The preceding theorem implies that this reduced cross product is an isomorphism if $\widetilde{H}^*(X; R)$ or $\widetilde{H}^*(Y; R)$ is free and finitely generated in each dimension. For example, we have isomorphisms $\widetilde{H}^n(X; R) \approx \widetilde{H}^{n+k}(X \wedge S^k; R)$ via cross product with a generator of $H^k(S^k; R) \approx R$. The space $X \wedge S^k$ is the k-fold reduced suspension $\Sigma^k X$ of X, so we see that the suspension isomorphisms $\widetilde{H}^n(X; R) \approx \widetilde{H}^{n+k}(\Sigma^k X; R)$ derivable by elementary exact sequence arguments can also be obtained via cross product with a generator of $\widetilde{H}^*(S^k; R)$.

Spaces with Polynomial Cohomology

Earlier in this section we mentioned that projective spaces provide examples of spaces whose cohomology rings are polynomial rings. Here is the precise statement:

Theorem 3.19. $H^*(\mathbb{R}P^n; \mathbb{Z}_2) \approx \mathbb{Z}_2[\alpha]/(\alpha^{n+1})$ *and* $H^*(\mathbb{R}P^\infty; \mathbb{Z}_2) \approx \mathbb{Z}_2[\alpha]$, *where* $|\alpha| = 1$. *In the complex case,* $H^*(\mathbb{C}P^n; \mathbb{Z}) \approx \mathbb{Z}[\alpha]/(\alpha^{n+1})$ *and* $H^*(\mathbb{C}P^\infty; \mathbb{Z}) \approx \mathbb{Z}[\alpha]$ *where* $|\alpha| = 2$.

This turns out to be a quite important result, and it can be proved in a number of different ways. The proof we give here uses the geometry of projective spaces to reduce the result to a very special case of the Künneth formula. Another proof using Poincaré duality will be given in Example 3.40. A third proof is contained in Example 4D.5 as an application of the Gysin sequence.

Proof: Let us do the case of $\mathbb{R}P^n$ first. To simplify notation we abbreviate $\mathbb{R}P^n$ to P^n and we let the coefficient group $\mathbb{Z}_2$ be implicit. Since the inclusion $P^{n-1} \hookrightarrow P^n$ induces an isomorphism on H^i for $i \le n-1$, it suffices by induction on n to show that the cup product of a generator of $H^{n-1}(P^n)$ with a generator of $H^1(P^n)$ is a generator of $H^n(P^n)$. It will be no more work to show more generally that the cup product of a generator of $H^i(P^n)$ with a generator of $H^{n-i}(P^n)$ is a generator of $H^n(P^n)$. As a further notational aid, we let $j = n - i$, so $i + j = n$.

The proof uses some of the geometric structure of P^n. Recall that P^n consists of nonzero vectors $(x_0, \cdots, x_n) \in \mathbb{R}^{n+1}$ modulo multiplication by nonzero scalars. Inside P^n is a copy of P^i represented by vectors whose last j coordinates $x_{i+1}, \cdots, x_n$ are zero. We also have a copy of P^j represented by points whose first i coordinates $x_0, \cdots, x_{i-1}$ are zero. The intersection $P^i \cap P^j$ is a single point p, represented by vectors whose only nonzero coordinate is x_i. Let U be the subspace of P^n represented by vectors with nonzero coordinate x_i. Each point in U may be represented by a unique vector with $x_i = 1$ and the other n coordinates arbitrary, so U is homeomorphic to $\mathbb{R}^n$, with p corresponding to 0 under this homeomorphism.

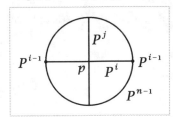

We can write this $\mathbb{R}^n$ as $\mathbb{R}^i \times \mathbb{R}^j$, with $\mathbb{R}^i$ as the coordinates $x_0, \cdots, x_{i-1}$ and $\mathbb{R}^j$ as the coordinates $x_{i+1}, \cdots, x_n$. In the figure P^n is represented as a disk with antipodal points of its boundary sphere identified to form a $P^{n-1} \subset P^n$ with $U = P^n - P^{n-1}$ the interior of the disk.

Consider the diagram

$$\text{(i)}\qquad
\begin{array}{ccc}
H^i(P^n) \times H^j(P^n) & \xrightarrow{\;\;\smile\;\;} & H^n(P^n) \\
\uparrow & & \uparrow \\
H^i(P^n, P^n - P^j) \times H^j(P^n, P^n - P^i) & \xrightarrow{\;\;\smile\;\;} & H^n(P^n, P^n - \{p\}) \\
\downarrow & & \downarrow \\
H^i(\mathbb{R}^n, \mathbb{R}^n - \mathbb{R}^j) \times H^j(\mathbb{R}^n, \mathbb{R}^n - \mathbb{R}^i) & \xrightarrow{\;\;\smile\;\;} & H^n(\mathbb{R}^n, \mathbb{R}^n - \{0\})
\end{array}$$

which commutes by naturality of cup product. We will show that the four vertical maps are isomorphisms and that the lower cup product map takes generator cross generator to generator. Commutativity of the diagram will then imply that the upper cup product map also takes generator cross generator to generator.

The lower map in the right column is an isomorphism by excision. For the upper map in this column, the fact that $P^n - \{p\}$ deformation retracts to a P^{n-1} gives an isomorphism $H^n(P^n, P^n - \{p\}) \approx H^n(P^n, P^{n-1})$ via the five-lemma applied to the long exact sequences for these pairs. And $H^n(P^n, P^{n-1}) \approx H^n(P^n)$ by cellular cohomology.

To see that the vertical maps in the left column of (i) are isomorphisms we will use the following commutative diagram:

(ii)
$$\begin{array}{ccccccc} H^i(P^n) & \longleftarrow & H^i(P^n, P^{i-1}) & \longleftarrow & H^i(P^n, P^n - P^j) & \longrightarrow & H^i(\mathbb{R}^n, \mathbb{R}^n - \mathbb{R}^j) \\ \downarrow & & \downarrow & & \downarrow & & \downarrow \\ H^i(P^i) & \longleftarrow & H^i(P^i, P^{i-1}) & \longleftarrow & H^i(P^i, P^i - \{p\}) & \longrightarrow & H^i(\mathbb{R}^i, \mathbb{R}^i - \{0\}) \end{array}$$

If we can show all these maps are isomorphisms, then the same argument will apply with i and j interchanged, and the vertical maps in the left column of (i) will be isomorphisms.

The left-hand square in (ii) consists of isomorphisms by cellular cohomology. The right-hand vertical map is obviously an isomorphism. The lower right horizontal map is an isomorphism by excision, and the map to the left of this is an isomorphism since $P^i - \{p\}$ deformation retracts onto P^{i-1}. The remaining maps will be isomorphisms if the middle map in the upper row is an isomorphism. And this map is in fact an isomorphism because $P^n - P^j$ deformation retracts onto P^{i-1} by the following argument. The subspace $P^n - P^j \subset P^n$ consists of points represented by vectors $v = (x_0, \cdots, x_n)$ with at least one of the coordinates $x_0, \cdots, x_{i-1}$ nonzero. The formula $f_t(v) = (x_0, \cdots, x_{i-1}, tx_i, \cdots, tx_n)$ for t decreasing from 1 to 0 gives a well-defined deformation retraction of $P^n - P^j$ onto P^{i-1} since $f_t(\lambda v) = \lambda f_t(v)$ for scalars $\lambda \in \mathbb{R}$.

The cup product map in the bottom row of (i) is equivalent to the cross product $H^i(I^i, \partial I^i) \times H^j(I^j, \partial I^j) \to H^n(I^n, \partial I^n)$, where the cross product of generators is a generator by the relative form of the Künneth formula in Theorem 3.18. Alternatively, if one wishes to use only the absolute Künneth formula, the cross product for cubes is equivalent to the cross product $H^i(S^i) \times H^j(S^j) \to H^n(S^i \times S^j)$ by means of the quotient maps $I^i \to S^i$ and $I^j \to S^j$ collapsing the boundaries of the cubes to points.

This finishes the proof for $\mathbb{R}P^n$. The case of $\mathbb{R}P^\infty$ follows from this since the inclusion $\mathbb{R}P^n \hookrightarrow \mathbb{R}P^\infty$ induces isomorphisms on $H^i(-; \mathbb{Z}_2)$ for $i \leq n$ by cellular cohomology.

Complex projective spaces are handled in precisely the same way, using $\mathbb{Z}$ coefficients and replacing each H^k by H^{2k} and $\mathbb{R}$ by $\mathbb{C}$. $\square$

There are also quaternionic projective spaces $\mathbb{H}P^n$ and $\mathbb{H}P^\infty$, defined exactly as in the complex case, with CW structures of the form $e^0 \cup e^4 \cup e^8 \cup \cdots$. Associativity of quaternion multiplication is needed for the identification $v \sim \lambda v$ to be an equivalence relation, so the definition does not extend to octonionic projective spaces, though there is an octonionic projective plane $\mathbb{O}P^2$ defined in Example 4.47. The cup product structure in quaternionic projective spaces is just like that in complex projective spaces, except that the generator is 4-dimensional:

$$H^*(\mathbb{H}P^\infty; \mathbb{Z}) \approx \mathbb{Z}[\alpha] \quad \text{and} \quad H^*(\mathbb{H}P^n; \mathbb{Z}) \approx \mathbb{Z}[\alpha]/(\alpha^{n+1}), \quad \text{with } |\alpha| = 4$$

The same proof as in the real and complex cases works here as well.

The cup product structure for $\mathbb{R}P^\infty$ with $\mathbb{Z}$ coefficients can easily be deduced from the cup product structure with $\mathbb{Z}_2$ coefficients, as follows. In general, a ring homomorphism $R \to S$ induces a ring homomorphism $H^*(X, A; R) \to H^*(X, A; S)$. In the case of the projection $\mathbb{Z} \to \mathbb{Z}_2$ we get for $\mathbb{R}P^\infty$ an induced chain map of cellular cochain complexes with $\mathbb{Z}$ and $\mathbb{Z}_2$ coefficients:

$$\cdots \longleftarrow \mathbb{Z} \xleftarrow{2} \mathbb{Z} \xleftarrow{0} \mathbb{Z} \xleftarrow{2} \mathbb{Z} \xleftarrow{0} \mathbb{Z} \longleftarrow 0$$
$$\Big\downarrow \qquad \Big\downarrow \qquad \Big\downarrow \qquad \Big\downarrow \qquad \Big\downarrow$$
$$\cdots \longleftarrow \mathbb{Z}_2 \xleftarrow{0} \mathbb{Z}_2 \xleftarrow{0} \mathbb{Z}_2 \xleftarrow{0} \mathbb{Z}_2 \xleftarrow{0} \mathbb{Z}_2 \longleftarrow 0$$

From this we see that the ring homomorphism $H^*(\mathbb{R}P^\infty; \mathbb{Z}) \to H^*(\mathbb{R}P^\infty; \mathbb{Z}_2)$ is injective in positive dimensions, with image the even-dimensional part of $H^*(\mathbb{R}P^\infty; \mathbb{Z}_2)$. Alternatively, this could be deduced from the universal coefficient theorem. Hence we have $H^*(\mathbb{R}P^\infty; \mathbb{Z}) \approx \mathbb{Z}[\alpha]/(2\alpha)$ with $|\alpha| = 2$.

The cup product structure in $H^*(\mathbb{R}P^n; \mathbb{Z})$ can be computed in a similar fashion, though the description is a little cumbersome:

$$H^*(\mathbb{R}P^{2k}; \mathbb{Z}) \approx \mathbb{Z}[\alpha]/(2\alpha, \alpha^{k+1}), \quad |\alpha| = 2$$
$$H^*(\mathbb{R}P^{2k+1}; \mathbb{Z}) \approx \mathbb{Z}[\alpha, \beta]/(2\alpha, \alpha^{k+1}, \beta^2, \alpha\beta), \quad |\alpha| = 2, \ |\beta| = 2k+1$$

Here β is a generator of $H^{2k+1}(\mathbb{R}P^{2k+1}; \mathbb{Z}) \approx \mathbb{Z}$. From this calculation we see that the rings $H^*(\mathbb{R}P^{2k+1}; \mathbb{Z})$ and $H^*(\mathbb{R}P^{2k} \vee S^{2k+1}; \mathbb{Z})$ are isomorphic, though with $\mathbb{Z}_2$ coefficients this is no longer true, as the generator $\alpha \in H^1(\mathbb{R}P^{2k+1}; \mathbb{Z}_2)$ has $\alpha^{2k+1} \neq 0$, while $\alpha^{2k+1} = 0$ for the generator $\alpha \in H^1(\mathbb{R}P^{2k} \vee S^{2k+1}; \mathbb{Z}_2)$.

Example 3.20. Combining the calculation $H^*(\mathbb{R}P^\infty; \mathbb{Z}_2) \approx \mathbb{Z}_2[\alpha]$ with the Künneth formula, we see that $H^*(\mathbb{R}P^\infty \times \mathbb{R}P^\infty; \mathbb{Z}_2)$ is isomorphic to $\mathbb{Z}_2[\alpha_1] \otimes \mathbb{Z}_2[\alpha_2]$, which is just the polynomial ring $\mathbb{Z}_2[\alpha_1, \alpha_2]$. More generally it follows by induction that for a product of n copies of $\mathbb{R}P^\infty$, the $\mathbb{Z}_2$-cohomology is a polynomial ring in n variables. Similar remarks apply to $\mathbb{C}P^\infty$ and $\mathbb{H}P^\infty$ with coefficients in $\mathbb{Z}$ or any commutative ring.

The following theorem of Hopf is a nice algebraic application of the cup product structure in $H^*(\mathbb{R}P^n \times \mathbb{R}P^n; \mathbb{Z}_2)$.

Theorem 3.21. *If $\mathbb{R}^n$ has the structure of a division algebra over the scalar field $\mathbb{R}$, then n must be a power of 2.*

Proof: For a division algebra structure on $\mathbb{R}^n$ the multiplication maps $x \mapsto ax$ and $x \mapsto xa$ are linear isomorphisms for each nonzero a, so the multiplication map $\mathbb{R}^n \times \mathbb{R}^n \to \mathbb{R}^n$ induces a map $h : \mathbb{R}P^{n-1} \times \mathbb{R}P^{n-1} \to \mathbb{R}P^{n-1}$ which is a homeomorphism when restricted to each subspace $\mathbb{R}P^{n-1} \times \{y\}$ and $\{x\} \times \mathbb{R}P^{n-1}$. The map h is continuous since it is a quotient of the multiplication map which is bilinear and hence continuous. The induced homomorphism h^* on $\mathbb{Z}_2$-cohomology is a ring homomorphism $\mathbb{Z}_2[\alpha]/(\alpha^n) \to \mathbb{Z}_2[\alpha_1, \alpha_2]/(\alpha_1^n, \alpha_2^n)$ determined by the element $h^*(\alpha) = k_1\alpha_1 + k_2\alpha_2$. The inclusion $\mathbb{R}P^{n-1} \hookrightarrow \mathbb{R}P^{n-1} \times \mathbb{R}P^{n-1}$ onto the first factor sends α_1 to α and α_2 to 0, as one sees by composing with the projections of $\mathbb{R}P^{n-1} \times \mathbb{R}P^{n-1}$ onto its two factors. The fact that h restricts to a homeomorphism on the first factor then implies that k_1 is nonzero. Similarly k_2 is nonzero, so since these coefficients lie in $\mathbb{Z}_2$ we have $h^*(\alpha) = \alpha_1 + \alpha_2$.

Since $\alpha^n = 0$ we must have $h^*(\alpha^n) = 0$, so $(\alpha_1 + \alpha_2)^n = \sum_k \binom{n}{k} \alpha_1^k \alpha_2^{n-k} = 0$. This is an equation in the ring $\mathbb{Z}_2[\alpha_1, \alpha_2]/(\alpha_1^n, \alpha_2^n)$, so the coefficient $\binom{n}{k}$ must be zero in $\mathbb{Z}_2$ for all k in the range $0 < k < n$. It is a rather easy number theory fact that this happens only when n is a power of 2. Namely, an obviously equivalent statement is that in the polynomial ring $\mathbb{Z}_2[x]$, the equality $(1+x)^n = 1+x^n$ holds only when n is a power of 2. To prove the latter statement, write n as a sum of powers of 2, $n = n_1 + \cdots + n_k$ with $n_1 < \cdots < n_k$. Then $(1+x)^n = (1+x)^{n_1} \cdots (1+x)^{n_k} = (1+x^{n_1}) \cdots (1+x^{n_k})$ since squaring is an additive homomorphism with $\mathbb{Z}_2$ coefficients. If one multiplies the product $(1+x^{n_1}) \cdots (1+x^{n_k})$ out, no terms combine or cancel since $n_i \geq 2n_{i-1}$ for each i, and so the resulting polynomial has 2^k terms. Thus if this polynomial equals $1 + x^n$ we must have $k = 1$, which means that n is a power of 2. $\square$

The same argument can be applied with $\mathbb{C}$ in place of $\mathbb{R}$, to show that if $\mathbb{C}^n$ is a division algebra over $\mathbb{C}$ then $\binom{n}{k} = 0$ for all k in the range $0 < k < n$, but now we can use $\mathbb{Z}$ rather than $\mathbb{Z}_2$ coefficients, so we deduce that $n = 1$. Thus there are no higher-dimensional division algebras over $\mathbb{C}$. This is assuming we are talking about finite-dimensional division algebras. For infinite dimensions there is for example the field of rational functions $\mathbb{C}(x)$.

We saw in Theorem 3.19 that $\mathbb{R}P^\infty$, $\mathbb{C}P^\infty$, and $\mathbb{H}P^\infty$ have cohomology rings that are polynomial algebras. We will describe now a construction for enlarging S^{2n} to a space $J(S^{2n})$ whose cohomology ring $H^*(J(S^{2n}); \mathbb{Z})$ is almost the polynomial ring $\mathbb{Z}[x]$ on a generator x of dimension $2n$. And if we change from $\mathbb{Z}$ to $\mathbb{Q}$ coefficients, then $H^*(J(S^{2n}); \mathbb{Q})$ is exactly the polynomial ring $\mathbb{Q}[x]$. This construction, known as the **James reduced product**, is also of interest because of its connections with loopspaces described in §4.J.

For a space X, let X^k be the product of k copies of X. From the disjoint union $\coprod_{k \geq 1} X^k$, let us form a quotient space $J(X)$ by identifying $(x_1, \cdots, x_i, \cdots, x_k)$ with $(x_1, \cdots, \hat{x}_i, \cdots, x_k)$ if $x_i = e$, a chosen basepoint of X. Points of $J(X)$ can thus be thought of as k-tuples $(x_1, \cdots, x_k)$, $k \geq 0$, with no $x_i = e$. Inside $J(X)$ is the subspace $J_m(X)$ consisting of the points $(x_1, \cdots, x_k)$ with $k \leq m$. This can be viewed as a quotient space of X^m under the identifications $(x_1, \cdots, x_i, e, \cdots, x_m) \sim (x_1, \cdots, e, x_i, \cdots, x_m)$. For example, $J_1(X) = X$ and $J_2(X) = X \times X/(x, e) \sim (e, x)$. If X is a CW complex with e a 0-cell, the quotient map $X^m \to J_m(X)$ glues together the m subcomplexes of the product complex X^m where one coordinate is e. These glueings are by homeomorphisms taking cells onto cells, so $J_m(X)$ inherits a CW structure from X^m. There are natural inclusions $J_m(X) \subset J_{m+1}(X)$ as subcomplexes, and $J(X)$ is the union of these subcomplexes, hence is also a CW complex.

Proposition 3.22. *For $n > 0$, $H^*(J(S^n); \mathbb{Z})$ consists of a $\mathbb{Z}$ in each dimension a multiple of n. If n is even, the ith power of a generator of $H^n(J(S^n); \mathbb{Z})$ is $i!$ times a generator of $H^{in}(J(S^n); \mathbb{Z})$, for each $i \geq 1$. When n is odd, $H^*(J(S^n); \mathbb{Z})$ is isomorphic as a graded ring to $H^*(S^n; \mathbb{Z}) \otimes H^*(J(S^{2n}); \mathbb{Z})$.*

It follows that for n even, $H^*(J(S^n); \mathbb{Z})$ can be identified with the subring of the polynomial ring $\mathbb{Q}[x]$ additively generated by the monomials $x^i/i!$. This subring is called a **divided polynomial algebra** and is denoted $\Gamma_{\mathbb{Z}}[x]$. Thus $H^*(J(S^n); \mathbb{Z})$ is isomorphic to $\Gamma_{\mathbb{Z}}[x]$ when n is even and to $\Lambda_{\mathbb{Z}}[x] \otimes \Gamma_{\mathbb{Z}}[y]$ when n is odd.

Proof: Giving S^n its usual CW structure, the resulting CW structure on $J(S^n)$ consists of exactly one cell in each dimension a multiple of n. If $n > 1$ we deduce immediately from cellular cohomology that $H^*(J(S^n); \mathbb{Z})$ consists exactly of $\mathbb{Z}$'s in dimensions a multiple of n. For an alternative argument that works also when $n = 1$, consider the quotient map $q : (S^n)^m \to J_m(S^n)$. This carries each cell of $(S^n)^m$ homeomorphically onto a cell of $J_m(S^n)$. In particular q is a cellular map, taking k-skeleton to k-skeleton for each k, so q induces a chain map of cellular chain complexes. This chain map is surjective since each cell of $J_m(S^n)$ is the homeomorphic image of a cell of $(S^n)^m$. Hence the cellular boundary maps for $J_m(S^n)$ will be trivial if they are trivial for $(S^n)^m$, as indeed they are since $H^*((S^n)^m; \mathbb{Z})$ is free with basis in one-to-one correspondence with the cells, by Theorem 3.16.

We can compute cup products in $H^*(J_m(S^n); \mathbb{Z})$ by computing their images under q^*. Let x_k denote the generator of $H^{kn}(J_m(S^n); \mathbb{Z})$ dual to the kn-cell, represented by the cellular cocycle assigning the value 1 to the kn-cell. Since q identifies all the n-cells of $(S^n)^m$ to form the n-cell of $J_m(S^n)$, we see from cellular cohomology that $q^*(x_1)$ is the sum $\alpha_1 + \cdots + \alpha_m$ of the generators of $H^n((S^n)^m; \mathbb{Z})$ dual to the n-cells of $(S^n)^m$. By the same reasoning we have $q^*(x_k) = \sum_{i_1 < \cdots < i_k} \alpha_{i_1} \cdots \alpha_{i_k}$.

If n is even, the cup product structure in $H^*((S^n)^m;\mathbb{Z})$ is strictly commutative and $H^*((S^n)^m;\mathbb{Z}) \approx \mathbb{Z}[\alpha_1, \cdots, \alpha_m]/(\alpha_1^2, \cdots, \alpha_m^2)$. Then we have

$$q^*(x_1^m) = (\alpha_1 + \cdots + \alpha_m)^m = m!\alpha_1 \cdots \alpha_m = m!q^*(x_m)$$

Since q^* is an isomorphism on H^{mn} this implies $x_1^m = m!x_m$ in $H^{mn}(J_m(S^n);\mathbb{Z})$. The inclusion $J_m(S^n) \hookrightarrow J(S^n)$ induces isomorphisms on H^i for $i \le mn$ so we have $x_1^m = m!x_m$ in $H^*(J(S^n);\mathbb{Z})$ as well, where x_1 and x_m are interpreted now as elements of $H^*(J(S^n);\mathbb{Z})$.

When n is odd we have $x_1^2 = 0$ by commutativity, and it will suffice to prove the following two formulas:

(a) $x_1 x_{2m} = x_{2m+1}$ in $H^*(J_{2m+1}(S^n);\mathbb{Z})$.

(b) $x_2 x_{2m-2} = m x_{2m}$ in $H^*(J_{2m}(S^n);\mathbb{Z})$.

For (a) we apply q^* and compute in the exterior algebra $\Lambda_{\mathbb{Z}}[\alpha_1, \cdots, \alpha_{2m+1}]$:

$$q^*(x_1 x_{2m}) = \left(\sum_i \alpha_i\right)\left(\sum_i \alpha_1 \cdots \hat{\alpha}_i \cdots \alpha_{2m+1}\right)$$
$$= \sum_i \alpha_i \alpha_1 \cdots \hat{\alpha}_i \cdots \alpha_{2m+1} = \sum_i (-1)^{i-1} \alpha_1 \cdots \alpha_{2m+1}$$

The coefficients in this last summation are $+1, -1, \cdots, +1$, so their sum is $+1$ and (a) follows. For (b) we have

$$q^*(x_2 x_{2m-2}) = \left(\sum_{i_1 < i_2} \alpha_{i_1} \alpha_{i_2}\right)\left(\sum_{i_1 < i_2} \alpha_1 \cdots \hat{\alpha}_{i_1} \cdots \hat{\alpha}_{i_2} \cdots \alpha_{2m}\right)$$
$$= \sum_{i_1 < i_2} \alpha_{i_1} \alpha_{i_2} \alpha_1 \cdots \hat{\alpha}_{i_1} \cdots \hat{\alpha}_{i_2} \cdots \alpha_{2m} = \sum_{i_1 < i_2} (-1)^{i_1-1}(-1)^{i_2-2} \alpha_1 \cdots \alpha_{2m}$$

The terms in the coefficient $\sum_{i_1 < i_2}(-1)^{i_1-1}(-1)^{i_2-2}$ for a fixed i_1 have i_2 varying from $i_1 + 1$ to $2m$. These terms are $+1, -1, \cdots$ and there are $2m - i_1$ of them, so their sum is 0 if i_1 is even and 1 if i_1 is odd. Now letting i_1 vary, it takes on the odd values $1, 3, \cdots, 2m - 1$, so the whole summation reduces to m 1's and we have the desired relation $x_2 x_{2m-2} = m x_{2m}$. $\square$

In $\Gamma_{\mathbb{Z}}[x] \subset \mathbb{Q}[x]$, if we let $x_i = x^i/i!$ then the multiplicative structure is given by $x_i x_j = \binom{i+j}{i} x_{i+j}$. More generally, for a commutative ring R we could define $\Gamma_R[x]$ to be the free R-module with basis $x_0 = 1, x_1, x_2, \cdots$ and multiplication defined by $x_i x_j = \binom{i+j}{i} x_{i+j}$. The preceding proposition implies that $H^*(J(S^{2n});R) \approx \Gamma_R[x]$. When $R = \mathbb{Q}$ it is clear that $\Gamma_{\mathbb{Q}}[x]$ is just $\mathbb{Q}[x]$. However, for $R = \mathbb{Z}_p$ with p prime something quite different happens: There is an isomorphism

$$\Gamma_{\mathbb{Z}_p}[x] \approx \mathbb{Z}_p[x_1, x_p, x_{p^2}, \cdots]/(x_1^p, x_p^p, x_{p^2}^p, \cdots) = \bigotimes_{i \ge 0} \mathbb{Z}_p[x_{p^i}]/(x_{p^i}^p)$$

as we show in §3.C, where we will also see that divided polynomial algebras are in a certain sense dual to polynomial algebras.

The examples of projective spaces lead naturally to the following question: Given a coefficient ring R and an integer $d > 0$, is there a space X having $H^*(X; R) \approx R[\alpha]$ with $|\alpha| = d$? Historically, it took major advances in the theory to answer this simple-looking question. Here is a table giving all the possible values of d for some of the most obvious and important choices of R, namely $\mathbb{Z}$, $\mathbb{Q}$, $\mathbb{Z}_2$, and $\mathbb{Z}_p$ with p an odd prime. As we have seen, projective

R	d
$\mathbb{Z}$	2, 4
$\mathbb{Q}$	any even number
$\mathbb{Z}_2$	1, 2, 4
$\mathbb{Z}_p$	any even divisor of $2(p-1)$

spaces give the examples for $\mathbb{Z}$ and $\mathbb{Z}_2$. Examples for $\mathbb{Q}$ are the spaces $J(S^d)$, and examples for $\mathbb{Z}_p$ are constructed in §3.G. Showing that no other d's are possible takes considerably more work. The fact that d must be even when $R \neq \mathbb{Z}_2$ is a consequence of the commutativity property of cup product. In Theorem 4L.9 and Corollary 4L.10 we will settle the case $R = \mathbb{Z}$ and show that d must be a power of 2 for $R = \mathbb{Z}_2$ and a power of p times an even divisor of $2(p-1)$ for $R = \mathbb{Z}_p$, p odd. Ruling out the remaining cases is best done using K–theory, as in [VBKT] or the classical reference [Adams & Atiyah 1966]. However there is one slightly anomalous case, $R = \mathbb{Z}_2$, $d = 8$, which must be treated by special arguments; see [Toda 1963].

It is an interesting fact that for each even d there exists a CW complex X_d which is simultaneously an example for all the admissible choices of coefficients R in the table. Moreover, X_d can be chosen to have the simplest CW structure consistent with its cohomology, namely a single cell in each dimension a multiple of d. For example, we may take $X_2 = \mathbb{C}P^\infty$ and $X_4 = \mathbb{H}P^\infty$. The next space X_6 would have $H^*(X_6; \mathbb{Z}_p) \approx \mathbb{Z}_p[\alpha]$ for $p = 7, 13, 19, 31, \cdots$, primes of the form $3s + 1$, the condition $6 | 2(p-1)$ being equivalent to $p = 3s + 1$. (By a famous theorem of Dirichlet there are infinitely many primes in any such arithmetic progression.) Note that, in terms of $\mathbb{Z}$ coefficients, X_d must have the property that for a generator α of $H^d(X_d; \mathbb{Z})$, each power α^i is an integer a_i times a generator of $H^{di}(X_d; \mathbb{Z})$, with $a_i \neq 0$ if $H^*(X_d; \mathbb{Q}) \approx \mathbb{Q}[\alpha]$ and a_i relatively prime to p if $H^*(X_d; \mathbb{Z}_p) \approx \mathbb{Z}_p[\alpha]$. A construction of X_d is given in [SSAT], or in the original source [Hoffman & Porter 1973].

One might also ask about realizing the truncated polynomial ring $R[\alpha]/(\alpha^{n+1})$, in view of the examples provided by $\mathbb{R}P^n$, $\mathbb{C}P^n$, and $\mathbb{H}P^n$, leaving aside the trivial case $n = 1$ where spheres provide examples. The analysis for polynomial rings also settles which truncated polynomial rings are realizable; there are just a few more than for the full polynomial rings.

There is also the question of realizing polynomial rings $R[\alpha_1, \cdots, \alpha_n]$ with generators α_i in specified dimensions d_i. Since $R[\alpha_1, \cdots, \alpha_m] \otimes_R R[\beta_1, \cdots, \beta_n]$ is equal to $R[\alpha_1, \cdots, \alpha_m, \beta_1, \cdots, \beta_n]$, the product of two spaces with polynomial cohomology is again a space with polynomial cohomology, assuming the number of polynomial generators is finite in each dimension. For example, the n-fold product $(\mathbb{C}P^\infty)^n$ has $H^*((\mathbb{C}P^\infty)^n; \mathbb{Z}) \approx \mathbb{Z}[\alpha_1, \cdots, \alpha_n]$ with each α_i 2-dimensional. Similarly, products of the spaces $J(S^{d_i})$ realize all choices of even d_i's with $\mathbb{Q}$ coefficients.

However, with $\mathbb{Z}$ and $\mathbb{Z}_p$ coefficients, products of one-variable examples do not exhaust all the possibilities. As we show in §4.D, there are three other basic examples with $\mathbb{Z}$ coefficients:

1. Generalizing the space $\mathbb{C}P^\infty$ of complex lines through the origin in $\mathbb{C}^\infty$, there is the **Grassmann manifold** $G_n(\mathbb{C}^\infty)$ of n-dimensional vector subspaces of $\mathbb{C}^\infty$, and this has $H^*(G_n(\mathbb{C}^\infty);\mathbb{Z}) \approx \mathbb{Z}[\alpha_1,\cdots,\alpha_n]$ with $|\alpha_i| = 2i$. This space is also known as $BU(n)$, the 'classifying space' of the unitary group $U(n)$. It is central to the study of vector bundles and K–theory.

2. Replacing $\mathbb{C}$ by $\mathbb{H}$, there is the quaternionic Grassmann manifold $G_n(\mathbb{H}^\infty)$, also known as $BSp(n)$, the classifying space for the symplectic group $Sp(n)$, with $H^*(G_n(\mathbb{H}^\infty);\mathbb{Z}) \approx \mathbb{Z}[\alpha_1,\cdots,\alpha_n]$ with $|\alpha_i| = 4i$.

3. There is a classifying space $BSU(n)$ for the special unitary group $SU(n)$, whose cohomology is the same as for $BU(n)$ but with the first generator α_1 omitted, so $H^*(BSU(n);\mathbb{Z}) \approx \mathbb{Z}[\alpha_2,\cdots,\alpha_n]$ with $|\alpha_i| = 2i$.

These examples and their products account for all the realizable polynomial cup product rings with $\mathbb{Z}$ coefficients, according to a theorem in [Andersen & Grodal 2008]. The situation for $\mathbb{Z}_p$ coefficients is more complicated and will be discussed in §3.G.

Polynomial algebras are examples of free graded commutative algebras, where 'free' means loosely 'having no unnecessary relations.' In general, a free graded commutative algebra is a tensor product of single-generator free graded commutative algebras. The latter are either polynomial algebras $R[\alpha]$ on even-dimension generators α or quotients $R[\alpha]/(2\alpha^2)$ with α odd-dimensional. Note that if R is a field then $R[\alpha]/(2\alpha^2)$ is either the exterior algebra $\Lambda_R[\alpha]$ if the characteristic of R is not 2, or the polynomial algebra $R[\alpha]$ otherwise. Every graded commutative algebra is a quotient of a free one, clearly.

Example 3.23: Subcomplexes of the n-Torus. To give just a small hint of the endless variety of nonfree cup product algebras that can be realized, consider subcomplexes of the n-torus T^n, the product of n copies of S^1. Here we give S^1 its standard minimal cell structure and T^n the resulting product cell structure. We know that $H^*(T^n;\mathbb{Z})$ is the exterior algebra $\Lambda_{\mathbb{Z}}[\alpha_1,\cdots,\alpha_n]$, with the monomial $\alpha_{i_1}\cdots\alpha_{i_k}$ corresponding via cellular cohomology to the k-cell $e_{i_1}^1\times\cdots\times e_{i_k}^1$. So if we pass to a subcomplex $X \subset T^n$ by omitting certain cells, then $H^*(X;\mathbb{Z})$ is the quotient of $\Lambda_{\mathbb{Z}}[\alpha_1,\cdots,\alpha_n]$ obtained by setting the monomials corresponding to the omitted cells equal to zero. Since we are dealing with rings, we are factoring out by an ideal in $\Lambda_{\mathbb{Z}}[\alpha_1,\cdots,\alpha_n]$, the ideal generated by the monomials corresponding to the 'minimal' omitted cells, those whose boundary is entirely contained in X. For example, if we take X to be the subcomplex of T^3 obtained by deleting the cells $e_1^1\times e_2^1\times e_3^1$ and $e_2^1\times e_3^1$, then $H^*(X;\mathbb{Z}) \approx \Lambda_{\mathbb{Z}}[\alpha_1,\alpha_2,\alpha_3]/(\alpha_2\alpha_3)$.

How many different subcomplexes of T^n are there? To each subcomplex $X \subset T^n$ we can associate a finite simplicial complex C_X by the following procedure. View T^n as the quotient of the n-cube $I^n = [0,1]^n \subset \mathbb{R}^n$ obtained by identifying opposite faces. If we intersect I^n with the hyperplane $x_1 + \cdots + x_n = \varepsilon$ for small $\varepsilon > 0$, we get a simplex Δ^{n-1}. Then for $q : I^n \to T^n$ the quotient map, we take C_X to be $\Delta^{n-1} \cap q^{-1}(X)$. This is a subcomplex of Δ^{n-1} whose k-simplices correspond exactly to the $(k+1)$-cells of X. In this way we get a one-to-one correspondence between subcomplexes $X \subset T^n$ and subcomplexes $C_X \subset \Delta^{n-1}$. Every simplicial complex with n vertices is a subcomplex of Δ^{n-1}, so we see that T^n has quite a large number of subcomplexes if n is not too small. The cohomology rings $H^*(X; \mathbb{Z})$ are of a type that was completely classified in [Gubeladze 1998], Theorem 3.1, and from this classification it follows that the ring $H^*(X; \mathbb{Z})$ (or even $H^*(X; \mathbb{Z}_2)$) determines the subcomplex X uniquely, up to permutation of the n circle factors of T^n.

More elaborate examples could be produced by looking at subcomplexes of the product of n copies of $\mathbb{CP}^\infty$. In this case the cohomology rings are isomorphic to polynomial rings modulo ideals generated by monomials, and it is again true that the cohomology ring determines the subcomplex up to permutation of factors. However, these cohomology rings are still a whole lot less complicated than the general case, where one takes free algebras modulo ideals generated by arbitrary polynomials having all their terms of the same dimension.

Let us conclude this section with an example of a cohomology ring that is not too far removed from a polynomial ring.

Example 3.24: Cohen–Macaulay Rings. Let X be the quotient space $\mathbb{CP}^\infty / \mathbb{CP}^{n-1}$. The quotient map $\mathbb{CP}^\infty \to X$ induces an injection $H^*(X; \mathbb{Z}) \to H^*(\mathbb{CP}^\infty; \mathbb{Z})$ embedding $H^*(X; \mathbb{Z})$ in $\mathbb{Z}[\alpha]$ as the subring generated by $1, \alpha^n, \alpha^{n+1}, \cdots$. If we view this subring as a module over $\mathbb{Z}[\alpha^n]$, it is free with basis $\{1, \alpha^{n+1}, \alpha^{n+2}, \cdots, \alpha^{2n-1}\}$. Thus $H^*(X; \mathbb{Z})$ is an example of a *Cohen-Macaulay* ring: a ring containing a polynomial subring over which it is a finitely generated free module. While polynomial cup product rings are rather rare, Cohen–Macauley cup product rings occur much more frequently.

Exercises

1. Assuming as known the cup product structure on the torus $S^1 \times S^1$, compute the cup product structure in $H^*(M_g)$ for M_g the closed orientable surface of genus g by using the quotient map from M_g to a wedge sum of g tori, shown below.

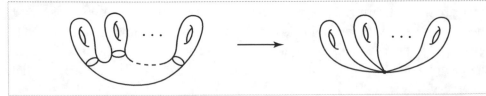

2. Using the cup product $H^k(X, A; R) \times H^\ell(X, B; R) \to H^{k+\ell}(X, A \cup B; R)$, show that if X is the union of contractible open subsets A and B, then all cup products of positive-dimensional classes in $H^*(X; R)$ are zero. This applies in particular if X is a suspension. Generalize to the situation that X is the union of n contractible open subsets, to show that all n-fold cup products of positive-dimensional classes are zero.

3. (a) Using the cup product structure, show there is no map $\mathbb{RP}^n \to \mathbb{RP}^m$ inducing a nontrivial map $H^1(\mathbb{RP}^m; \mathbb{Z}_2) \to H^1(\mathbb{RP}^n; \mathbb{Z}_2)$ if $n > m$. What is the corresponding result for maps $\mathbb{CP}^n \to \mathbb{CP}^m$?

(b) Prove the Borsuk–Ulam theorem by the following argument. Suppose on the contrary that $f: S^n \to \mathbb{R}^n$ satisfies $f(x) \neq f(-x)$ for all x. Then define $g: S^n \to S^{n-1}$ by $g(x) = (f(x) - f(-x))/|f(x) - f(-x)|$, so $g(-x) = -g(x)$ and g induces a map $\mathbb{RP}^n \to \mathbb{RP}^{n-1}$. Show that part (a) applies to this map.

4. Apply the Lefschetz fixed point theorem to show that every map $f: \mathbb{CP}^n \to \mathbb{CP}^n$ has a fixed point if n is even, using the fact that $f^*: H^*(\mathbb{CP}^n; \mathbb{Z}) \to H^*(\mathbb{CP}^n; \mathbb{Z})$ is a ring homomorphism. When n is odd show there is a fixed point unless $f^*(\alpha) = -\alpha$, for α a generator of $H^2(\mathbb{CP}^n; \mathbb{Z})$. [See Exercise 3 in §2.C for an example of a map without fixed points in this exceptional case.]

5. Show the ring $H^*(\mathbb{RP}^\infty; \mathbb{Z}_{2k})$ is isomorphic to $\mathbb{Z}_{2k}[\alpha, \beta]/(2\alpha, 2\beta, \alpha^2 - k\beta)$ where $|\alpha| = 1$ and $|\beta| = 2$. [Use the coefficient map $\mathbb{Z}_{2k} \to \mathbb{Z}_2$ and the proof of Theorem 3.19.]

6. Use cup products to compute the map $H^*(\mathbb{CP}^n; \mathbb{Z}) \to H^*(\mathbb{CP}^n; \mathbb{Z})$ induced by the map $\mathbb{CP}^n \to \mathbb{CP}^n$ that is a quotient of the map $\mathbb{C}^{n+1} \to \mathbb{C}^{n+1}$ raising each coordinate to the dth power, $(z_0, \cdots, z_n) \mapsto (z_0^d, \cdots, z_n^d)$, for a fixed integer $d > 0$. [First do the case $n = 1$.]

7. Use cup products to show that $\mathbb{RP}^3$ is not homotopy equivalent to $\mathbb{RP}^2 \vee S^3$.

8. Let X be $\mathbb{CP}^2$ with a cell e^3 attached by a map $S^2 \to \mathbb{CP}^1 \subset \mathbb{CP}^2$ of degree p, and let $Y = M(\mathbb{Z}_p, 2) \vee S^4$. Thus X and Y have the same 3-skeleton but differ in the way their 4-cells are attached. Show that X and Y have isomorphic cohomology rings with $\mathbb{Z}$ coefficients but not with $\mathbb{Z}_p$ coefficients.

9. Show that if $H_n(X; \mathbb{Z})$ is free for each n, then $H^*(X; \mathbb{Z}_p)$ and $H^*(X; \mathbb{Z}) \otimes \mathbb{Z}_p$ are isomorphic as rings, so in particular the ring structure with $\mathbb{Z}$ coefficients determines the ring structure with $\mathbb{Z}_p$ coefficients.

10. Show that the cross product map $H^*(X; \mathbb{Z}) \otimes H^*(Y; \mathbb{Z}) \to H^*(X \times Y; \mathbb{Z})$ is not an isomorphism if X and Y are infinite discrete sets. [This shows the necessity of the hypothesis of finite generation in Theorem 3.15.]

11. Using cup products, show that every map $S^{k+\ell} \to S^k \times S^\ell$ induces the trivial homomorphism $H_{k+\ell}(S^{k+\ell}) \to H_{k+\ell}(S^k \times S^\ell)$, assuming $k > 0$ and $\ell > 0$.

12. Show that the spaces $(S^1 \times \mathbb{CP}^\infty)/(S^1 \times \{x_0\})$ and $S^3 \times \mathbb{CP}^\infty$ have isomorphic cohomology rings with $\mathbb{Z}$ or any other coefficients. [An exercise for §4.L is to show these two spaces are not homotopy equivalent.]

13. Describe $H^*(\mathbb{C}\mathrm{P}^\infty/\mathbb{C}\mathrm{P}^1;\mathbb{Z})$ as a ring with finitely many multiplicative generators. How does this ring compare with $H^*(S^6 \times \mathbb{H}\mathrm{P}^\infty;\mathbb{Z})$?

14. Let $q:\mathbb{R}\mathrm{P}^\infty \to \mathbb{C}\mathrm{P}^\infty$ be the natural quotient map obtained by regarding both spaces as quotients of S^∞, modulo multiplication by real scalars in one case and complex scalars in the other. Show that the induced map $q^*:H^*(\mathbb{C}\mathrm{P}^\infty;\mathbb{Z}) \to H^*(\mathbb{R}\mathrm{P}^\infty;\mathbb{Z})$ is surjective in even dimensions by showing first by a geometric argument that the restriction $q:\mathbb{R}\mathrm{P}^2 \to \mathbb{C}\mathrm{P}^1$ induces a surjection on H^2 and then appealing to cup product structures. Next, form a quotient space X of $\mathbb{R}\mathrm{P}^\infty \amalg \mathbb{C}\mathrm{P}^n$ by identifying each point $x \in \mathbb{R}\mathrm{P}^{2n}$ with $q(x) \in \mathbb{C}\mathrm{P}^n$. Show there are ring isomorphisms $H^*(X;\mathbb{Z}) \approx \mathbb{Z}[\alpha]/(2\alpha^{n+1})$ and $H^*(X;\mathbb{Z}_2) \approx \mathbb{Z}_2[\alpha,\beta]/(\beta^2 - \alpha^{2n+1})$, where $|\alpha| = 2$ and $|\beta| = 2n + 1$. Make a similar construction and analysis for the quotient map $q:\mathbb{C}\mathrm{P}^\infty \to \mathbb{H}\mathrm{P}^\infty$.

15. For a fixed coefficient field F, define the **Poincaré series** of a space X to be the formal power series $p(t) = \sum_i a_i t^i$ where a_i is the dimension of $H^i(X;F)$ as a vector space over F, assuming this dimension is finite for all i. Show that $p(X \times Y) = p(X)p(Y)$. Compute the Poincaré series for S^n, $\mathbb{R}\mathrm{P}^n$, $\mathbb{R}\mathrm{P}^\infty$, $\mathbb{C}\mathrm{P}^n$, $\mathbb{C}\mathrm{P}^\infty$, and the spaces in the preceding three exercises.

16. Show that if X and Y are finite CW complexes such that $H^*(X;\mathbb{Z})$ and $H^*(Y;\mathbb{Z})$ contain no elements of order a power of a given prime p, then the same is true for $X \times Y$. [Apply Theorem 3.15 with coefficients in various fields.]

17. [This has now been incorporated into Proposition 3.22.]

18. For the closed orientable surface M of genus $g \geq 1$, show that for each nonzero $\alpha \in H^1(M;\mathbb{Z})$ there exists $\beta \in H^1(M;\mathbb{Z})$ with $\alpha\beta \neq 0$. Deduce that M is not homotopy equivalent to a wedge sum $X \vee Y$ of CW complexes with nontrivial reduced homology. Do the same for closed nonorientable surfaces using cohomology with $\mathbb{Z}_2$ coefficients.

3.3 Poincaré Duality

    Algebraic topology is most often concerned with properties of spaces that depend only on homotopy type, so local topological properties do not play much of a role. Digressing somewhat from this viewpoint, we study in this section a class of spaces whose most prominent feature is their local topology, namely manifolds, which are locally homeomorphic to $\mathbb{R}^n$. It is somewhat miraculous that just this local homogeneity property, together with global compactness, is enough to impose a strong symmetry on the homology and cohomology groups of such spaces, as well as strong nontriviality of cup products. This is the Poincaré duality theorem, one of the earliest theorems in the subject. In fact, Poincaré's original work on the duality property came before homology and cohomology had even been properly defined, and it took many years for the concepts of homology and cohomology to be refined sufficiently to put Poincaré duality on a firm footing.

Let us begin with some definitions. A **manifold of dimension** n, or more concisely an **n-manifold**, is a Hausdorff space M in which each point has an open neighborhood homeomorphic to $\mathbb{R}^n$. The dimension of M is intrinsically characterized by the fact that for $x \in M$, the local homology group $H_i(M, M-\{x\}; \mathbb{Z})$ is nonzero only for $i = n$:

$$H_i(M, M-\{x\}; \mathbb{Z}) \approx H_i(\mathbb{R}^n, \mathbb{R}^n - \{0\}; \mathbb{Z}) \qquad \text{by excision}$$
$$\approx \tilde{H}_{i-1}(\mathbb{R}^n - \{0\}; \mathbb{Z}) \qquad \text{since } \mathbb{R}^n \text{ is contractible}$$
$$\approx \tilde{H}_{i-1}(S^{n-1}; \mathbb{Z}) \qquad \text{since } \mathbb{R}^n - \{0\} \simeq S^{n-1}$$

A compact manifold is called **closed**, to distinguish it from the more general notion of a compact manifold with boundary, considered later in this section. For example S^n is a closed manifold, as are $\mathbb{R}\mathrm{P}^n$ and lens spaces since they have S^n as a covering space. Another closed manifold is $\mathbb{C}\mathrm{P}^n$. This is compact since it is a quotient space of S^{2n+1}, and the manifold property is satisfied since there is an open cover by subsets homeomorphic to $\mathbb{R}^{2n}$, the sets $U_i = \{[z_0, \cdots, z_n] \in \mathbb{C}\mathrm{P}^n \mid z_i = 1\}$. The same reasoning applies also for quaternionic projective spaces. Further examples of closed manifolds can be generated from these using the obvious fact that the product of closed manifolds of dimensions m and n is a closed manifold of dimension $m + n$.

Poincaré duality in its most primitive form asserts that for a closed orientable manifold M of dimension n, there are isomorphisms $H_k(M; \mathbb{Z}) \approx H^{n-k}(M; \mathbb{Z})$ for all k. Implicit here is the convention that homology and cohomology groups of negative dimension are zero, so the duality statement includes the fact that all the nontrivial homology and cohomology of M lies in the dimension range from 0 to n. The definition of 'orientable' will be given below. Without the orientability hypothesis there is a weaker statement that $H_k(M; \mathbb{Z}_2) \approx H^{n-k}(M; \mathbb{Z}_2)$ for all k. As we show in Corollaries A.8 and A.9 in the Appendix, the homology groups of a closed manifold are all finitely generated. So via the universal coefficient theorem, Poincaré duality for a closed orientable n-manifold M can be stated just in terms of homology: Modulo their torsion subgroups, $H_k(M; \mathbb{Z})$ and $H_{n-k}(M; \mathbb{Z})$ are isomorphic, and the torsion subgroups of $H_k(M; \mathbb{Z})$ and $H_{n-k-1}(M; \mathbb{Z})$ are isomorphic. However, the statement in terms of cohomology is really more natural.

Poincaré duality thus expresses a certain symmetry in the homology of closed orientable manifolds. For example, consider the n-dimensional torus T^n, the product of n circles. By induction on n it follows from the Künneth formula, or from the easy special case $H_i(X \times S^1; \mathbb{Z}) \approx H_i(X; \mathbb{Z}) \oplus H_{i-1}(X; \mathbb{Z})$ which was an exercise in §2.2, that $H_k(T^n; \mathbb{Z})$ is isomorphic to the direct sum of $\binom{n}{k}$ copies of $\mathbb{Z}$. So Poincaré duality is reflected in the relation $\binom{n}{k} = \binom{n}{n-k}$. The reader might also check that Poincaré duality is consistent with our calculations of the homology of projective spaces and lens spaces, which are all orientable except for $\mathbb{R}\mathrm{P}^n$ with n even.

For many manifolds there is a very nice geometric proof of Poincaré duality using the notion of *dual cell structures*. The germ of this idea can be traced back to the five regular Platonic solids: the tetrahedron, cube, octahedron, dodecahedron, and icosahedron. Each of these polyhedra has a dual polyhedron whose vertices are the center points of the faces of the given polyhedron. Thus the dual of the cube is the octahedron, and vice versa. Similarly the dodecahedron and icosahedron are dual to each other, and the tetrahedron is its own dual. One can regard each of these polyhedra as defining a cell structure C on S^2 with a dual cell structure C^* determined by the dual polyhedron. Each vertex of C lies in a dual 2-cell of C^*, each edge of C crosses a dual edge of C^*, and each 2-cell of C contains a dual vertex of C^*.

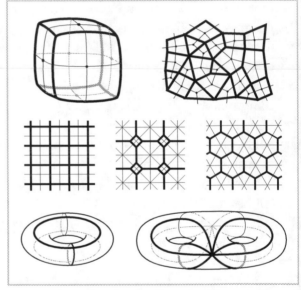

The first figure at the right shows the case of the cube and octahedron. There is no need to restrict to regular polyhedra here, and we can generalize further by replacing S^2 by any surface. A portion of a more-or-less random pair of dual cell structures is shown in the second figure. On the torus, if we lift a dual pair of cell structures to the universal cover $\mathbb{R}^2$, we get a dual pair of periodic tilings of the plane, as in the next three figures. The last two figures show that the standard CW structure on the surface of genus g, obtained from a $4g$-gon by identifying edges via the product of commutators $[a_1, b_1] \cdots [a_g, b_g]$, is homeomorphic to its own dual.

Given a pair of dual cell structures C and C^* on a closed surface M, the pairing of cells with dual cells gives identifications of cellular chain groups $C_0^* = C_2$, $C_1^* = C_1$, and $C_2^* = C_0$. If we use $\mathbb{Z}$ coefficients these identifications are not quite canonical since there is an ambiguity of sign for each cell, the choice of a generator for the corresponding $\mathbb{Z}$ summand of the cellular chain complex. We can avoid this ambiguity by considering the simpler situation of $\mathbb{Z}_2$ coefficients, where the identifications $C_i = C_{2-i}^*$ are completely canonical. The key observation now is that under these identifications, the cellular boundary map $\partial : C_i \to C_{i-1}$ becomes the cellular coboundary map $\delta : C_{2-i}^* \to C_{2-i+1}^*$ since ∂ assigns to a cell the sum of the cells which are faces of it, while δ assigns to a cell the sum of the cells of which it is a face. Thus $H_i(C; \mathbb{Z}_2) \approx H^{2-i}(C^*; \mathbb{Z}_2)$, and hence $H_i(M; \mathbb{Z}_2) \approx H^{2-i}(M; \mathbb{Z}_2)$ since C and C^* are cell structures on the same surface M.

To refine this argument to $\mathbb{Z}$ coefficients the problem of signs must be addressed. After analyzing the situation more closely, one sees that if M is orientable, it is possible to make consistent choices of orientations of all the cells of C and C^* so that the boundary maps in C agree with the coboundary maps in C^*, and therefore one gets $H_i(C;\mathbb{Z}) \approx H^{2-i}(C^*;\mathbb{Z})$, hence $H_i(M;\mathbb{Z}) \approx H^{2-i}(M;\mathbb{Z})$.

For manifolds of higher dimension the situation is entirely analogous. One would consider dual cell structures C and C^* on a closed n-manifold M, each i-cell of C being dual to a unique $(n-i)$-cell of C^* which it intersects in one point 'transversely.' For example on the 3-dimensional torus $S^1 \times S^1 \times S^1$ one could take the standard cell structure lifting to the decomposition of the universal cover $\mathbb{R}^3$ into cubes with vertices at the integer lattice points $\mathbb{Z}^3$, and then the dual cell structure is obtained by translating this by the vector $(1/2, 1/2, 1/2)$. Each edge in either cell structure then has a dual 2-cell which it pierces orthogonally, and each vertex lies in a dual 3-cell.

All the manifolds one commonly meets, for example all differentiable manifolds, have dually paired cell structures with the properties needed to carry out the proof of Poincaré duality we have just sketched. However, to construct these cell structures requires a certain amount of manifold theory. To avoid this, and to get a theorem that applies to all manifolds, we will take a completely different approach, using algebraic topology to replace the geometry of dual cell structures.

Orientations and Homology

Let us consider the question of how one might define orientability for manifolds. First there is the local question: What is an orientation of $\mathbb{R}^n$? Whatever an orientation of $\mathbb{R}^n$ is, it should have the property that it is preserved under rotations and reversed by reflections. For example, in $\mathbb{R}^2$ the notions of 'clockwise' and 'counterclockwise' certainly have this property, as do 'right-handed' and 'left-handed' in $\mathbb{R}^3$. We shall take the viewpoint that this property is what characterizes orientations, so anything satisfying the property can be regarded as an orientation.

With this in mind, we propose the following as an algebraic-topological definition: An orientation of $\mathbb{R}^n$ at a point x is a choice of generator of the infinite cyclic group $H_n(\mathbb{R}^n, \mathbb{R}^n - \{x\})$, where the absence of a coefficient group from the notation means that we take coefficients in $\mathbb{Z}$. To verify that the characteristic property of orientations is satisfied we use the isomorphisms $H_n(\mathbb{R}^n, \mathbb{R}^n - \{x\}) \approx H_{n-1}(\mathbb{R}^n - \{x\}) \approx H_{n-1}(S^{n-1})$ where S^{n-1} is a sphere centered at x. Since these isomorphisms are natural, and rotations of S^{n-1} have degree 1, being homotopic to the identity, while reflections have degree -1, we see that a rotation ρ of $\mathbb{R}^n$ fixing x takes a generator α of $H_n(\mathbb{R}^n, \mathbb{R}^n - \{x\})$ to itself, $\rho_*(\alpha) = \alpha$, while a reflection takes α to $-\alpha$.

Note that with this definition, an orientation of $\mathbb{R}^n$ at a point x determines an orientation at every other point y via the canonical isomorphisms $H_n(\mathbb{R}^n, \mathbb{R}^n - \{x\}) \approx H_n(\mathbb{R}^n, \mathbb{R}^n - B) \approx H_n(\mathbb{R}^n, \mathbb{R}^n - \{y\})$ where B is any ball containing both x and y.

An advantage of this definition of local orientation is that it can be applied to any n-dimensional manifold M: A **local orientation** of M at a point x is a choice of generator μ_x of the infinite cyclic group $H_n(M, M - \{x\})$.

Notational Convention. In what follows we will very often be looking at homology groups of the form $H_n(X, X - A)$. To simplify notation we will write $H_n(X, X - A)$ as $H_n(X \mid A)$, or more generally $H_n(X \mid A; G)$ if a coefficient group G needs to be specified. By excision, $H_n(X \mid A)$ depends only on a neighborhood of the closure of A in X, so it makes sense to view $H_n(X \mid A)$ as *local homology of X at A*.

Having settled what local orientations at points of a manifold are, a global orientation ought to be 'a consistent choice of local orientations at all points.' We make this precise by the following definition. An **orientation** of an n-dimensional manifold M is a function $x \mapsto \mu_x$ assigning to each $x \in M$ a local orientation $\mu_x \in H_n(M \mid x)$, satisfying the 'local consistency' condition that each $x \in M$ has a neighborhood $\mathbb{R}^n \subset M$ containing an open ball B of finite radius about x such that all the local orientations μ_y at points $y \in B$ are the images of one generator μ_B of $H_n(M \mid B) \approx H_n(\mathbb{R}^n \mid B)$ under the natural maps $H_n(M \mid B) \to H_n(M \mid y)$. If an orientation exists for M, then M is called **orientable**.

Every manifold M has an orientable two-sheeted covering space $\widetilde{M}$. For example, $\mathbb{R}P^2$ is covered by S^2, and the Klein bottle has the torus as a two-sheeted covering space. The general construction goes as follows. As a set, let

$$\widetilde{M} = \{ \mu_x \mid x \in M \text{ and } \mu_x \text{ is a local orientation of } M \text{ at } x \}$$

The map $\mu_x \mapsto x$ defines a two-to-one surjection $\widetilde{M} \to M$, and we wish to topologize $\widetilde{M}$ to make this a covering space projection. Given an open ball $B \subset \mathbb{R}^n \subset M$ of finite radius and a generator $\mu_B \in H_n(M \mid B)$, let $U(\mu_B)$ be the set of all $\mu_x \in \widetilde{M}$ such that $x \in B$ and μ_x is the image of μ_B under the natural map $H_n(M \mid B) \to H_n(M \mid x)$. It is easy to check that these sets $U(\mu_B)$ form a basis for a topology on $\widetilde{M}$, and that the projection $\widetilde{M} \to M$ is a covering space. The manifold $\widetilde{M}$ is orientable since each point $\mu_x \in \widetilde{M}$ has a canonical local orientation given by the element $\widetilde{\mu}_x \in H_n(\widetilde{M} \mid \mu_x)$ corresponding to μ_x under the isomorphisms $H_n(\widetilde{M} \mid \mu_x) \approx H_n(U(\mu_B) \mid \mu_x) \approx H_n(B \mid x)$, and by construction these local orientations satisfy the local consistency condition necessary to define a global orientation.

Proposition 3.25. *If M is connected, then M is orientable iff $\widetilde{M}$ has two components. In particular, M is orientable if it is simply-connected, or more generally if $\pi_1(M)$ has no subgroup of index two.*

The first statement is a formulation of the intuitive notion of nonorientability as being able to go around some closed loop and come back with the opposite orientation, since in terms of the covering space $\widetilde{M} \to M$ this corresponds to a loop in M that lifts to a path in $\widetilde{M}$ connecting two distinct points with the same image in M. The existence of such paths is equivalent to $\widetilde{M}$ being connected.

Proof: If M is connected, $\widetilde{M}$ has either one or two components since it is a two-sheeted covering space of M. If it has two components, they are each mapped homeomorphically to M by the covering projection, so M is orientable, being homeomorphic to a component of the orientable manifold $\widetilde{M}$. Conversely, if M is orientable, it has exactly two orientations since it is connected, and each of these orientations defines a component of $\widetilde{M}$. The last statement of the proposition follows since connected two-sheeted covering spaces of M correspond to index-two subgroups of $\pi_1(M)$, by the classification of covering spaces. $\square$

The covering space $\widetilde{M} \to M$ can be embedded in a larger covering space $M_{\mathbb{Z}} \to M$ where $M_{\mathbb{Z}}$ consists of all elements $\alpha_x \in H_n(M|x)$ as x ranges over M. As before, we topologize $M_{\mathbb{Z}}$ via the basis of sets $U(\alpha_B)$ consisting of α_x's with $x \in B$ and α_x the image of an element $\alpha_B \in H_n(M|B)$ under the map $H_n(M|B) \to H_n(M|x)$. The covering space $M_{\mathbb{Z}} \to M$ is infinite-sheeted since for fixed $x \in M$, the α_x's range over the infinite cyclic group $H_n(M|x)$. Restricting α_x to be zero, we get a copy M_0 of M in $M_{\mathbb{Z}}$. The rest of $M_{\mathbb{Z}}$ consists of an infinite sequence of copies M_k of $\widetilde{M}$, $k = 1, 2, \cdots$, where M_k consists of the α_x's that are k times either generator of $H_n(M|x)$.

A continuous map $M \to M_{\mathbb{Z}}$ of the form $x \mapsto \alpha_x \in H_n(M|x)$ is called a **section** of the covering space. An orientation of M is the same thing as a section $x \mapsto \mu_x$ such that μ_x is a generator of $H_n(M|x)$ for each x.

One can generalize the definition of orientation by replacing the coefficient group $\mathbb{Z}$ by any commutative ring R with identity. Then an **R-orientation** of M assigns to each $x \in M$ a generator of $H_n(M|x;R) \approx R$, subject to the corresponding local consistency condition, where a 'generator' of R is an element u such that $Ru = R$. Since we assume R has an identity element, this is equivalent to saying that u is a unit, an invertible element of R. The definition of the covering space $M_{\mathbb{Z}}$ generalizes immediately to a covering space $M_R \to M$, and an R-orientation is a section of this covering space whose value at each $x \in M$ is a generator of $H_n(M|x;R)$.

The structure of M_R is easy to describe. In view of the canonical isomorphism $H_n(M|x;R) \approx H_n(M|x) \otimes R$, each $r \in R$ determines a subcovering space M_r of M_R consisting of the points $\pm \mu_x \otimes r \in H_n(M|x;R)$ for μ_x a generator of $H_n(M|x)$. If r has order 2 in R then $r = -r$ so M_r is just a copy of M, and otherwise M_r is isomorphic to the two-sheeted cover $\widetilde{M}$. The covering space M_R is the union of these M_r's, which are disjoint except for the equality $M_r = M_{-r}$.

In particular we see that an orientable manifold is R-orientable for all R, while a nonorientable manifold is R-orientable iff R contains a unit of order 2, which is equivalent to having $2 = 0$ in R. Thus every manifold is $\mathbb{Z}_2$-orientable. In practice this means that the two most important cases are $R = \mathbb{Z}$ and $R = \mathbb{Z}_2$. In what follows the reader should keep these two cases foremost in mind, but we will usually state results for a general R.

The orientability of a closed manifold is reflected in the structure of its homology, according to the following result.

Theorem 3.26. *Let M be a closed connected n-manifold. Then*:
(a) *If M is R-orientable, the map $H_n(M;R) \to H_n(M|x;R) \approx R$ is an isomorphism for all $x \in M$.*
(b) *If M is not R-orientable, the map $H_n(M;R) \to H_n(M|x;R) \approx R$ is injective with image $\{ r \in R \mid 2r = 0 \}$ for all $x \in M$.*
(c) *$H_i(M;R) = 0$ for $i > n$.*

In particular, $H_n(M;\mathbb{Z})$ is $\mathbb{Z}$ or 0 depending on whether M is orientable or not, and in either case $H_n(M;\mathbb{Z}_2) = \mathbb{Z}_2$.

An element of $H_n(M;R)$ whose image in $H_n(M|x;R)$ is a generator for all x is called a **fundamental class** for M with coefficients in R. By the theorem, a fundamental class exists if M is closed and R-orientable. To show that the converse is also true, let $\mu \in H_n(M;R)$ be a fundamental class and let μ_x denote its image in $H_n(M|x;R)$. The function $x \mapsto \mu_x$ is then an R-orientation since the map $H_n(M;R) \to H_n(M|x;R)$ factors through $H_n(M|B;R)$ for B any open ball in M containing x. Furthermore, M must be compact since μ_x can only be nonzero for x in the image of a cycle representing μ, and this image is compact. In view of these remarks a fundamental class could also be called an **orientation class** for M.

The theorem will follow fairly easily from a more technical statement:

Lemma 3.27. *Let M be a manifold of dimension n and let $A \subset M$ be a compact subset. Then*:
(a) *If $x \mapsto \alpha_x$ is a section of the covering space $M_R \to M$, then there is a unique class $\alpha_A \in H_n(M|A;R)$ whose image in $H_n(M|x;R)$ is α_x for all $x \in A$.*
(b) *$H_i(M|A;R) = 0$ for $i > n$.*

To deduce the theorem from this, choose $A = M$, a compact set by assumption. Part (c) of the theorem is immediate from (b) of the lemma. To obtain (a) and (b) of the theorem, let $\Gamma_R(M)$ be the set of sections of $M_R \to M$. The sum of two sections is a section, and a scalar multiple of a section is a section, so $\Gamma_R(M)$ is an R-module. There is a homomorphism $H_n(M;R) \to \Gamma_R(M)$ sending a class α to the section $x \mapsto \alpha_x$, where α_x is the image of α under the map $H_n(M;R) \to H_n(M|x;R)$. Part (a) of the lemma asserts that this homomorphism is an isomorphism. If M is connected, each section is uniquely determined by its value at one point, so statements (a) and (b) of the theorem are apparent from the earlier discussion of the structure of M_R. $\square$

Proof of 3.27: The coefficient ring R will play no special role in the argument so we shall omit it from the notation. We break the proof up into four steps.

(1) First we observe that if the lemma is true for compact sets A, B, and $A \cap B$, then it is true for $A \cup B$. To see this, consider the Mayer-Vietoris sequence

$$0 \longrightarrow H_n(M \mid A \cup B) \xrightarrow{\Phi} H_n(M \mid A) \oplus H_n(M \mid B) \xrightarrow{\Psi} H_n(M \mid A \cap B)$$

Here the zero on the left comes from the assumption that $H_{n+1}(M \mid A \cap B) = 0$. The map Φ is $\Phi(\alpha) = (\alpha, -\alpha)$ and Ψ is $\Psi(\alpha, \beta) = \alpha + \beta$, where we omit notation for maps on homology induced by inclusion. The terms $H_i(M \mid A \cup B)$ farther to the left in this sequence are sandwiched between groups that are zero by assumption, so $H_i(M \mid A \cup B) = 0$ for $i > n$. This gives (b). For the existence half of (a), if $x \mapsto \alpha_x$ is a section, the hypothesis gives unique classes $\alpha_A \in H_n(M \mid A)$, $\alpha_B \in H_n(M \mid B)$, and $\alpha_{A \cap B} \in H_n(M \mid A \cap B)$ having image α_x for all x in A, B, or $A \cap B$ respectively. The images of α_A and α_B in $H_n(M \mid A \cap B)$ satisfy the defining property of $\alpha_{A \cap B}$, hence must equal $\alpha_{A \cap B}$. Exactness of the sequence then implies that $(\alpha_A, -\alpha_B) = \Phi(\alpha_{A \cup B})$ for some $\alpha_{A \cup B} \in H_n(M \mid A \cup B)$. This means that $\alpha_{A \cup B}$ maps to α_A and α_B, so $\alpha_{A \cup B}$ has image α_x for all $x \in A \cup B$ since α_A and α_B have this property. To see that $\alpha_{A \cup B}$ is unique, observe that if a class $\alpha \in H_n(M \mid A \cup B)$ has image zero in $H_n(M \mid x)$ for all $x \in A \cup B$, then its images in $H_n(M \mid A)$ and $H_n(M \mid B)$ have the same property, hence are zero by hypothesis, so α itself must be zero since Φ is injective. Uniqueness of $\alpha_{A \cup B}$ follows by applying this observation to the difference between two choices for $\alpha_{A \cup B}$.

(2) Next we reduce to the case $M = \mathbb{R}^n$. A compact set $A \subset M$ can be written as the union of finitely many compact sets $A_1, \cdots, A_m$ each contained in an open $\mathbb{R}^n \subset M$. We apply the result in (1) to $A_1 \cup \cdots \cup A_{m-1}$ and A_m. The intersection of these two sets is $(A_1 \cap A_m) \cup \cdots \cup (A_{m-1} \cap A_m)$, a union of $m - 1$ compact sets each contained in an open $\mathbb{R}^n \subset M$. By induction on m this gives a reduction to the case $m = 1$. When $m = 1$, excision allows us to replace M by the neighborhood $\mathbb{R}^n \subset M$.

(3) When $M = \mathbb{R}^n$ and A is a union of convex compact sets $A_1, \cdots, A_m$, an inductive argument as in (2) reduces to the case that A itself is convex. When A is convex the result is evident since the map $H_i(\mathbb{R}^n \mid A) \to H_i(\mathbb{R}^n \mid x)$ is an isomorphism for any $x \in A$, as both $\mathbb{R}^n - A$ and $\mathbb{R}^n - \{x\}$ deformation retract onto a sphere centered at x.

(4) For an arbitrary compact set $A \subset \mathbb{R}^n$ let $\alpha \in H_i(\mathbb{R}^n \mid A)$ be represented by a relative cycle z, and let $C \subset \mathbb{R}^n - A$ be the union of the images of the singular simplices in ∂z. Since C is compact, it has a positive distance δ from A. We can cover A by finitely many closed balls of radius less than δ centered at points of A. Let K be the union of these balls, so K is disjoint from C. The relative cycle z defines an element $\alpha_K \in H_i(\mathbb{R}^n \mid K)$ mapping to the given $\alpha \in H_i(\mathbb{R}^n \mid A)$. If $i > n$ then by (3) we have $H_i(\mathbb{R}^n \mid K) = 0$, so $\alpha_K = 0$, which implies $\alpha = 0$ and hence $H_i(\mathbb{R}^n \mid A) = 0$. If $i = n$ and α_x is zero in $H_n(\mathbb{R}^n \mid x)$ for all $x \in A$, then in fact this holds for all $x \in K$, where α_x in this case means the image of α_K. This is because K is a union of balls B meeting A and $H_n(\mathbb{R}^n \mid B) \to H_n(\mathbb{R}^n \mid x)$ is an isomorphism for all $x \in B$. Since

$\alpha_x = 0$ for all $x \in K$, (3) then says that α_K is zero, hence also α. This finishes the uniqueness statement in (a). The existence statement is easy since we can let α_A be the image of the element α_B associated to any ball $B \supset A$.   □

For a closed n-manifold having the structure of a Δ-complex there is a more explicit construction for a fundamental class. Consider the case of $\mathbb{Z}$ coefficients. In simplicial homology a fundamental class must be represented by some linear combination $\sum_i k_i \sigma_i$ of the n-simplices σ_i of M. The condition that the fundamental class maps to a generator of $H_n(M|x;\mathbb{Z})$ for points x in the interiors of the σ_i's means that each coefficient k_i must be ± 1. The k_i's must also be such that $\sum_i k_i \sigma_i$ is a cycle. This implies that if σ_i and σ_j share a common $(n-1)$-dimensional face, then k_i determines k_j and vice versa. Analyzing the situation more closely, one can show that a choice of signs for the k_i's making $\sum_i k_i \sigma_i$ a cycle is possible iff M is orientable, and if such a choice is possible, then the cycle $\sum_i k_i \sigma_i$ defines a fundamental class. With $\mathbb{Z}_2$ coefficients there is no issue of signs, and $\sum_i \sigma_i$ always defines a fundamental class.

Some information about $H_{n-1}(M)$ can also be squeezed out of the preceding theorem:

Corollary 3.28. *If M is a closed connected n-manifold, the torsion subgroup of $H_{n-1}(M;\mathbb{Z})$ is trivial if M is orientable and $\mathbb{Z}_2$ if M is nonorientable.*

Proof: This is an application of the universal coefficient theorem for homology, using the fact that the homology groups of M are finitely generated, from Corollaries A.8 and A.9 in the Appendix. In the orientable case, if $H_{n-1}(M;\mathbb{Z})$ contained torsion, then for some prime p, $H_n(M;\mathbb{Z}_p)$ would be larger than the $\mathbb{Z}_p$ coming from $H_n(M;\mathbb{Z})$. In the nonorientable case, $H_n(M;\mathbb{Z}_m)$ is either $\mathbb{Z}_2$ or 0 depending on whether m is even or odd. This forces the torsion subgroup of $H_{n-1}(M;\mathbb{Z})$ to be $\mathbb{Z}_2$.   □

The reader who is familiar with Bockstein homomorphisms, which are discussed in §3.E, will recognize that the $\mathbb{Z}_2$ in $H_{n-1}(M;\mathbb{Z})$ in the nonorientable case is the image of the Bockstein homomorphism $H_n(M;\mathbb{Z}_2) \to H_{n-1}(M;\mathbb{Z})$ coming from the short exact sequence of coefficient groups $0 \to \mathbb{Z} \to \mathbb{Z} \to \mathbb{Z}_2 \to 0$.

The structure of $H_n(M;G)$ and $H_{n-1}(M;G)$ for a closed connected n-manifold M can be explained very nicely in terms of cellular homology when M has a CW structure with a single n-cell, which is the case for a large number of manifolds. Note that there can be no cells of higher dimension since a cell of maximal dimension produces nontrivial local homology in that dimension. Consider the cellular boundary map $d: C_n(M) \to C_{n-1}(M)$ with $\mathbb{Z}$ coefficients. Since M has a single n-cell we have $C_n(M) = \mathbb{Z}$. If M is orientable, d must be zero since $H_n(M;\mathbb{Z}) = \mathbb{Z}$. Then since d is zero, $H_{n-1}(M;\mathbb{Z})$ must be free. On the other hand, if M is nonorientable then d

must take a generator of $C_n(M)$ to twice a generator α of a $\mathbb{Z}$ summand of $C_{n-1}(M)$, in order for $H_n(M;\mathbb{Z}_p)$ to be zero for odd primes p and $\mathbb{Z}_2$ for $p = 2$. The cellular chain α must be a cycle since 2α is a boundary and hence a cycle. It follows that the torsion subgroup of $H_{n-1}(M;\mathbb{Z})$ must be a $\mathbb{Z}_2$ generated by α.

Concerning the homology of noncompact manifolds there is the following general statement.

Proposition 3.29. *If M is a connected noncompact n-manifold, then $H_i(M;R) = 0$ for $i \geq n$.*

Proof: Represent an element of $H_i(M;R)$ by a cycle z. This has compact image in M, so there is an open set $U \subset M$ containing the image of z and having compact closure $\overline{U} \subset M$. Let $V = M - \overline{U}$. Part of the long exact sequence of the triple $(M, U \cup V, V)$ fits into a commutative diagram

$$H_{i+1}(M,U\cup V;R) \longrightarrow H_i(U\cup V,V;R) \longrightarrow H_i(M,V;R)$$
$$\uparrow \approx \qquad\qquad \uparrow$$
$$H_i(U;R) \longrightarrow H_i(M;R)$$

When $i > n$, the two groups on either side of $H_i(U \cup V, V;R)$ are zero by Lemma 3.27 since $U \cup V$ and V are the complements of compact sets in M. Hence $H_i(U;R) = 0$, so z is a boundary in U and therefore in M, and we conclude that $H_i(M;R) = 0$.

When $i = n$, the class $[z] \in H_n(M;R)$ defines a section $x \mapsto [z]_x$ of M_R. Since M is connected, this section is determined by its value at a single point, so $[z]_x$ will be zero for all x if it is zero for some x, which it must be since z has compact image and M is noncompact. By Lemma 3.27, z then represents zero in $H_n(M,V;R)$, hence also in $H_n(U;R)$ since the first term in the upper row of the diagram above is zero when $i = n$, by Lemma 3.27 again. So $[z] = 0$ in $H_n(M;R)$, and therefore $H_n(M;R) = 0$ since $[z]$ was an arbitrary element of this group. $\qquad\square$

The Duality Theorem

The form of Poincaré duality we will prove asserts that for an R-orientable closed n-manifold, a certain naturally defined map $H^k(M;R) \to H_{n-k}(M;R)$ is an isomorphism. The definition of this map will be in terms of a more general construction called *cap product*, which has close connections with cup product.

For an arbitrary space X and coefficient ring R, define an R-bilinear cap product $\frown : C_k(X;R) \times C^\ell(X;R) \to C_{k-\ell}(X;R)$ for $k \geq \ell$ by setting

$$\sigma \frown \varphi = \varphi(\sigma|[v_0, \cdots, v_\ell]) \, \sigma|[v_\ell, \cdots, v_k]$$

for $\sigma : \Delta^k \to X$ and $\varphi \in C^\ell(X;R)$. To see that this induces a cap product in homology and cohomology we use the formula

$$\partial(\sigma \frown \varphi) = (-1)^\ell (\partial\sigma \frown \varphi - \sigma \frown \delta\varphi)$$

which is checked by a calculation:

$$\partial\sigma \frown \varphi = \sum_{i=0}^{\ell} (-1)^i \varphi(\sigma|[v_0, \cdots, \hat{v}_i, \cdots, v_{\ell+1}]) \sigma|[v_{\ell+1}, \cdots, v_k]$$

$$+ \sum_{i=\ell+1}^{k} (-1)^i \varphi(\sigma|[v_0, \cdots, v_\ell]) \sigma|[v_\ell, \cdots, \hat{v}_i, \cdots, v_k]$$

$$\sigma \frown \delta\varphi = \sum_{i=0}^{\ell+1} (-1)^i \varphi(\sigma|[v_0, \cdots, \hat{v}_i, \cdots, v_{\ell+1}]) \sigma|[v_{\ell+1}, \cdots, v_k]$$

$$\partial(\sigma \frown \varphi) = \sum_{i=\ell}^{k} (-1)^{i-\ell} \varphi(\sigma|[v_0, \cdots, v_\ell]) \sigma|[v_\ell, \cdots, \hat{v}_i, \cdots, v_k]$$

From the relation $\partial(\sigma \frown \varphi) = \pm(\partial\sigma \frown \varphi - \sigma \frown \delta\varphi)$ it follows that the cap product of a cycle σ and a cocycle φ is a cycle. Further, if $\partial\sigma = 0$ then $\partial(\sigma \frown \varphi) = \pm(\sigma \frown \delta\varphi)$, so the cap product of a cycle and a coboundary is a boundary. And if $\delta\varphi = 0$ then $\partial(\sigma \frown \varphi) = \pm(\partial\sigma \frown \varphi)$, so the cap product of a boundary and a cocycle is a boundary. These facts imply that there is an induced cap product

$$H_k(X;R) \times H^\ell(X;R) \xrightarrow{\frown} H_{k-\ell}(X;R)$$

which is R-linear in each variable.

Using the same formulas, one checks that cap product has the relative forms

$$H_k(X, A;R) \times H^\ell(X;R) \xrightarrow{\frown} H_{k-\ell}(X, A;R)$$
$$H_k(X, A;R) \times H^\ell(X, A;R) \xrightarrow{\frown} H_{k-\ell}(X;R)$$

For example, in the second case the cap product $C_k(X;R) \times C^\ell(X;R) \to C_{k-\ell}(X;R)$ restricts to zero on the submodule $C_k(A;R) \times C^\ell(X, A;R)$, so there is an induced cap product $C_k(X, A;R) \times C^\ell(X, A;R) \to C_{k-\ell}(X;R)$. The formula for $\partial(\sigma \frown \varphi)$ still holds, so we can pass to homology and cohomology groups. There is also a more general relative cap product

$$H_k(X, A \cup B;R) \times H^\ell(X, A;R) \xrightarrow{\frown} H_{k-\ell}(X, B;R),$$

defined when A and B are open sets in X, using the fact that $H_k(X, A \cup B;R)$ can be computed using the chain groups $C_n(X, A + B;R) = C_n(X;R)/C_n(A + B;R)$, as in the derivation of relative Mayer–Vietoris sequences in §2.2.

Cap product satisfies a naturality property that is a little more awkward to state than the corresponding result for cup product since both covariant and contravariant functors are involved. Given a map $f : X \to Y$, the relevant induced maps on homology and cohomology fit into the diagram shown below. It does not quite make sense to say this diagram commutes, but the spirit of commutativity is contained in the formula

$$f_*(\alpha) \frown \varphi = f_*(\alpha \frown f^*(\varphi))$$

$$\begin{array}{ccc} H_k(X) \times H^\ell(X) & \xrightarrow{\ \frown\ } & H_{k-\ell}(X) \\ \downarrow f_* & \uparrow f^* & \downarrow f_* \\ H_k(Y) \times H^\ell(Y) & \xrightarrow{\ \frown\ } & H_{k-\ell}(Y) \end{array}$$

which is obtained by substituting $f\sigma$ for σ in the definition of cap product: $f\sigma \frown \varphi = \varphi(f\sigma|[v_0,\cdots,v_\ell])\,f\sigma|[v_\ell,\cdots,v_k]$. There are evident relative versions as well.

Now we can state Poincaré duality for closed manifolds:

Theorem 3.30 (Poincaré Duality). *If M is a closed R-orientable n-manifold with fundamental class $[M] \in H_n(M;R)$, then the map $D:H^k(M;R) \longrightarrow H_{n-k}(M;R)$ defined by $D(\alpha) = [M] \frown \alpha$ is an isomorphism for all k.*

Recall that a fundamental class for M is an element of $H_n(M;R)$ whose image in $H_n(M|x;R)$ is a generator for each $x \in M$. The existence of such a class was shown in Theorem 3.26.

Example 3.31: Surfaces. Let M be the closed orientable surface of genus g, obtained as usual from a $4g$-gon by identifying pairs of edges according to the word $a_1b_1a_1^{-1}b_1^{-1}\cdots a_gb_ga_g^{-1}b_g^{-1}$. A Δ-complex structure on M is obtained by coning off the $4g$-gon to its center, as indicated in the figure for the case $g - 2$. We can compute cap products using simplicial homology and cohomology since cap products are defined for simplicial homology and cohomology by exactly the same formula as for singular homology and cohomology, so the isomorphism between the simplicial and singular theories respects cap products. A fundamental class $[M]$ generating $H_2(M)$ is represented by the 2-cycle formed by the

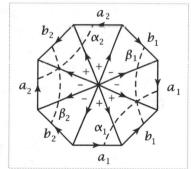

sum of all $4g$ 2-simplices with the signs indicated. The edges a_i and b_i form a basis for $H_1(M)$. Under the isomorphism $H^1(M) \approx \mathrm{Hom}(H_1(M),\mathbb{Z})$, the cohomology class α_i corresponding to a_i assigns the value 1 to a_i and 0 to the other basis elements. This class α_i is represented by the cocycle φ_i assigning the value 1 to the 1-simplices meeting the arc labeled α_i in the figure and 0 to the other 1-simplices. Similarly we have a class β_i corresponding to b_i, represented by the cocycle ψ_i assigning the value 1 to the 1-simplices meeting the arc β_i and 0 to the other 1-simplices. Applying the definition of cap product, we have $[M] \frown \varphi_i = b_i$ and $[M] \frown \psi_i = -a_i$ since in both cases there is just one 2-simplex $[v_0,v_1,v_2]$ where φ_i or ψ_i is nonzero on the edge $[v_0,v_1]$. Thus b_i is the Poincaré dual of α_i and $-a_i$ is the Poincaré dual of β_i. If we interpret Poincaré duality entirely in terms of homology, identifying α_i with its Hom-dual a_i and β_i with b_i, then the classes a_i and b_i are Poincaré duals of each other, up to sign at least. Geometrically, Poincaré duality is reflected in the fact that the loops α_i and b_i are homotopic, as are the loops β_i and a_i.

The closed nonorientable surface N of genus g can be treated in the same way if we use $\mathbb{Z}_2$ coefficients. We view N as obtained from a $2g$-gon by identifying consecutive pairs of edges according to the word $a_1^2 \cdots a_g^2$. We have classes $\alpha_i \in H^1(N; \mathbb{Z}_2)$ represented by cocycles φ_i assigning the value 1 to the edges meeting the arc α_i. Then $[N] \frown \varphi_i = a_i$, so a_i is the Poincaré dual of α_i. In terms of homology, a_i is the Hom-dual of α_i, so a_i is its own Poincaré dual.

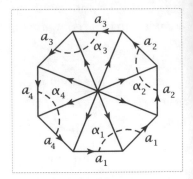

Geometrically, the loops a_i on N are homotopic to their Poincaré dual loops α_i.

Our proof of Poincaré duality, like the construction of fundamental classes, will be by an inductive argument using Mayer–Vietoris sequences. The induction step requires a version of Poincaré duality for open subsets of M, which are noncompact and can satisfy Poincaré duality only when a different kind of cohomology called *cohomology with compact supports* is used.

Cohomology with Compact Supports

Before giving the general definition, let us look at the conceptually simpler notion of simplicial cohomology with compact supports. Here one starts with a Δ-complex X which is locally compact. This is equivalent to saying that every point has a neighborhood that meets only finitely many simplices. Consider the subgroup $\Delta_c^i(X; G)$ of the simplicial cochain group $\Delta^i(X; G)$ consisting of cochains that are compactly supported in the sense that they take nonzero values on only finitely many simplices. The coboundary of such a cochain φ can have a nonzero value only on those $(i+1)$-simplices having a face on which φ is nonzero, and there are only finitely many such simplices by the local compactness assumption, so $\delta\varphi$ lies in $\Delta_c^{i+1}(X; G)$. Thus we have a subcomplex of the simplicial cochain complex. The cohomology groups for this subcomplex will be denoted temporarily by $H_c^i(X; G)$.

Example 3.32. Let us compute these cohomology groups when $X = \mathbb{R}$ with the Δ-complex structure having vertices at the integer points. For a simplicial 0-cochain to be a cocycle it must take the same value on all vertices, but then if the cochain lies in $\Delta_c^0(X)$ it must be identically zero. Thus $H_c^0(\mathbb{R}; G) = 0$. However, $H_c^1(\mathbb{R}; G)$ is nonzero. Namely, consider the map $\Sigma : \Delta_c^1(\mathbb{R}; G) \to G$ sending each cochain to the sum of its values on all the 1-simplices. Note that Σ is not defined on all of $\Delta^1(X)$, just on $\Delta_c^1(X)$. The map Σ vanishes on coboundaries, so it induces a map $H_c^1(\mathbb{R}; G) \to G$. This is surjective since every element of $\Delta_c^1(X)$ is a cocycle. It is an easy exercise to verify that it is also injective, so $H_c^1(\mathbb{R}; G) \approx G$.

Compactly supported cellular cohomology for a locally compact CW complex could be defined in a similar fashion, using cellular cochains that are nonzero on only finitely many cells. However, what we really need is singular cohomology with compact supports for spaces without any simplicial or cellular structure. The quickest definition of this is the following. Let $C_c^i(X;G)$ be the subgroup of $C^i(X;G)$ consisting of cochains $\varphi : C_i(X) \to G$ for which there exists a compact set $K = K_\varphi \subset X$ such that φ is zero on all chains in $X - K$. Note that $\delta\varphi$ is then also zero on chains in $X - K$, so $\delta\varphi$ lies in $C_c^{i+1}(X;G)$ and the $C_c^i(X;G)$'s for varying i form a subcomplex of the singular cochain complex of X. The cohomology groups $H_c^i(X;G)$ of this subcomplex are the **cohomology groups with compact supports**.

Cochains in $C_c^i(X;G)$ have compact support in only a rather weak sense. A stronger and perhaps more natural condition would have been to require cochains to be nonzero only on singular simplices contained in some compact set, depending on the cochain. However, cochains satisfying this condition do not in general form a subcomplex of the singular cochain complex. For example, if $X = \mathbb{R}$ and φ is a 0-cochain assigning a nonzero value to one point of $\mathbb{R}$ and zero to all other points, then $\delta\varphi$ assigns a nonzero value to arbitrarily large 1-simplices.

It will be quite useful to have an alternative definition of $H_c^i(X;G)$ in terms of algebraic limits, which enter the picture in the following way. The cochain group $C_c^i(X;G)$ is the union of its subgroups $C^i(X, X - K;G)$ as K ranges over compact subsets of X. Each inclusion $K \hookrightarrow L$ induces inclusions $C^i(X, X - K;G) \hookrightarrow C^i(X, X - L;G)$ for all i, so there are induced maps $H^i(X, X - K;G) \to H^i(X, X - L;G)$. These need not be injective, but one might still hope that $H_c^i(X;G)$ is somehow describable in terms of the system of groups $H^i(X, X - K;G)$ for varying K. This is indeed the case, and it is algebraic limits that provide the description.

Suppose one has abelian groups G_α indexed by some partially ordered index set I having the property that for each pair $\alpha, \beta \in I$ there exists $\gamma \in I$ with $\alpha \leq \gamma$ and $\beta \leq \gamma$. Such an I is called a **directed set**. Suppose also that for each pair $\alpha \leq \beta$ one has a homomorphism $f_{\alpha\beta} : G_\alpha \to G_\beta$, such that $f_{\alpha\alpha} = \mathbb{1}$ for each α, and if $\alpha \leq \beta \leq \gamma$ then $f_{\alpha\gamma}$ is the composition of $f_{\alpha\beta}$ and $f_{\beta\gamma}$. Given this data, which is called a **directed system** of groups, there are two equivalent ways of defining the **direct limit** group $\varinjlim G_\alpha$. The shorter definition is that $\varinjlim G_\alpha$ is the quotient of the direct sum $\bigoplus_\alpha G_\alpha$ by the subgroup generated by all elements of the form $a - f_{\alpha\beta}(a)$ for $a \in G_\alpha$, where we are viewing each G_α as a subgroup of $\bigoplus_\alpha G_\alpha$. The other definition, which is often more convenient to work with, runs as follows. Define an equivalence relation on the set $\coprod_\alpha G_\alpha$ by $a \sim b$ if $f_{\alpha\gamma}(a) = f_{\beta\gamma}(b)$ for some γ, where $a \in G_\alpha$ and $b \in G_\beta$. This is clearly reflexive and symmetric, and transitivity follows from the directed set property. It could also be described as the equivalence relation generated by setting $a \sim f_{\alpha\beta}(a)$. Any two equivalence classes $[a]$ and $[b]$ have representatives a' and b' lying in the same G_γ, so define $[a] + [b] = [a' + b']$. One checks this is well-defined and gives an abelian group structure to the set of equivalence classes. It is easy to check further that the map sending an equivalence class $[a]$ to the coset of a

in $\varinjlim G_\alpha$ is a homomorphism, with an inverse induced by the map $\sum_i a_i \mapsto \sum_i [a_i]$ for $a_i \in G_{\alpha_i}$. Thus we can identify $\varinjlim G_\alpha$ with the group of equivalence classes $[a]$.

A useful consequence of this is that if we have a subset $J \subset I$ with the property that for each $\alpha \in I$ there exists a $\beta \in J$ with $\alpha \le \beta$, then $\varinjlim G_\alpha$ is the same whether we compute it with α varying over I or just over J. In particular, if I has a maximal element γ, we can take $J = \{\gamma\}$ and then $\varinjlim G_\alpha = G_\gamma$.

Suppose now that we have a space X expressed as the union of a collection of subspaces X_α forming a directed set with respect to the inclusion relation. Then the groups $H_i(X_\alpha; G)$ for fixed i and G form a directed system, using the homomorphisms induced by inclusions. The natural maps $H_i(X_\alpha; G) \to H_i(X; G)$ induce a homomorphism $\varinjlim H_i(X_\alpha; G) \to H_i(X; G)$.

Proposition 3.33. *If a space X is the union of a directed set of subspaces X_α with the property that each compact set in X is contained in some X_α, then the natural map $\varinjlim H_i(X_\alpha; G) \to H_i(X; G)$ is an isomorphism for all i and G.*

Proof: For surjectivity, represent a cycle in X by a finite sum of singular simplices. The union of the images of these singular simplices is compact in X, hence lies in some X_α, so the map $\varinjlim H_i(X_\alpha; G) \to H_i(X; G)$ is surjective. Injectivity is similar: If a cycle in some X_α is a boundary in X, compactness implies it is a boundary in some $X_\beta \supset X_\alpha$, hence represents zero in $\varinjlim H_i(X_\alpha; G)$. $\qquad\qquad\square$

Now we can give the alternative definition of cohomology with compact supports in terms of direct limits. For a space X, the compact subsets $K \subset X$ form a directed set under inclusion since the union of two compact sets is compact. To each compact $K \subset X$ we associate the group $H^i(X, X - K; G)$, with a fixed i and coefficient group G, and to each inclusion $K \subset L$ of compact sets we associate the natural homomorphism $H^i(X, X - K; G) \to H^i(X, X - L; G)$. The resulting limit group $\varinjlim H^i(X, X - K; G)$ is then equal to $H^i_c(X; G)$ since each element of this limit group is represented by a cocycle in $C^i(X, X - K; G)$ for some compact K, and such a cocycle is zero in $\varinjlim H^i(X, X - K; G)$ iff it is the coboundary of a cochain in $C^{i-1}(X, X - L; G)$ for some compact $L \supset K$.

Note that if X is compact, then $H^i_c(X; G) = H^i(X; G)$ since there is a unique maximal compact set $K \subset X$, namely X itself. This is also immediate from the original definition since $C^i_c(X; G) = C^i(X; G)$ if X is compact.

Example 3.34: $H^*_c(\mathbb{R}^n; G)$. To compute $\varinjlim H^i(\mathbb{R}^n, \mathbb{R}^n - K; G)$ it suffices to let K range over balls B_k of integer radius k centered at the origin since every compact set is contained in such a ball. Since $H^i(\mathbb{R}^n, \mathbb{R}^n - B_k; G)$ is nonzero only for $i = n$, when it is G, and the maps $H^n(\mathbb{R}^n, \mathbb{R}^n - B_k; G) \to H^n(\mathbb{R}^n, \mathbb{R}^n - B_{k+1}; G)$ are isomorphisms, we deduce that $H^i_c(\mathbb{R}^n; G) = 0$ for $i \ne n$ and $H^n_c(\mathbb{R}^n; G) \approx G$.

This example shows that cohomology with compact supports is not an invariant of homotopy type. This can be traced to difficulties with induced maps. For example, the constant map from $\mathbb{R}^n$ to a point does not induce a map on cohomology with compact supports. The maps which do induce maps on H_c^* are the *proper* maps, those for which the inverse image of each compact set is compact. In the proof of Poincaré duality, however, we will need induced maps of a different sort going in the opposite direction from what is usual for cohomology, maps $H_c^i(U;G) \to H_c^i(V;G)$ associated to inclusions $U \hookrightarrow V$ of open sets in the fixed manifold M.

The group $H^i(X, X-K;G)$ for K compact depends only on a neighborhood of K in X by excision, assuming X is Hausdorff so that K is closed. As convenient shorthand notation we will write this group as $H^i(X \,|\, K;G)$, in analogy with the similar notation used earlier for local homology. One can think of cohomology with compact supports as the limit of these 'local cohomology groups at compact subsets.'

Duality for Noncompact Manifolds

For M an R-orientable n-manifold, possibly noncompact, we can define a duality map $D_M : H_c^k(M;R) \to H_{n-k}(M;R)$ by a limiting process in the following way. For compact sets $K \subset L \subset M$ we have a diagram

$$
\begin{array}{ccc}
H_n(M|L;R) \times H^k(M|L;R) & & \\
\Big\downarrow i_* \qquad\qquad \Big\uparrow i^* & \overset{\frown}{\searrow} & H_{n-k}(M;R) \\
II_n(M|K;R) \times II^k(M|K;R) & \overset{\frown}{\nearrow} &
\end{array}
$$

where $H_n(M\,|\,A;R) = H_n(M, M-A;R)$ and $H^k(M\,|\,A;R) = H^k(M, M-A;R)$. By Lemma 3.27 there are unique elements $\mu_K \in H_n(M\,|\,K;R)$ and $\mu_L \in H_n(M\,|\,L;R)$ restricting to a given orientation of M at each point of K and L, respectively. From the uniqueness we have $i_*(\mu_L) = \mu_K$. The naturality of cap product implies that $i_*(\mu_L) \frown x = \mu_L \frown i^*(x)$ for all $x \in H^k(M\,|\,K;R)$, so $\mu_K \frown x = \mu_L \frown i^*(x)$. Therefore, letting K vary over compact sets in M, the homomorphisms $H^k(M\,|\,K;R) \to H_{n-k}(M;R)$, $x \mapsto \mu_K \frown x$, induce in the limit a duality homomorphism $D_M : H_c^k(M;R) \to H_{n-k}(M;R)$.

Since $H_c^*(M;R) = H^*(M;R)$ if M is compact, the following theorem generalizes Poincaré duality for closed manifolds:

Theorem 3.35. *The duality map $D_M : H_c^k(M;R) \to H_{n-k}(M;R)$ is an isomorphism for all k whenever M is an R-oriented n-manifold.*

The proof will not be difficult once we establish a technical result stated in the next lemma, concerning the commutativity of a certain diagram. Commutativity statements of this sort are usually routine to prove, but this one seems to be an exception. The reader who consults other books for alternative expositions will find somewhat uneven treatments of this technical point, and the proof we give is also not as simple as one would like.

The coefficient ring R will be fixed throughout the proof, and for simplicity we will omit it from the notation for homology and cohomology.

Lemma 3.36. *If M is the union of two open sets U and V, then there is a diagram of Mayer-Vietoris sequences, commutative up to sign:*

$$\cdots \longrightarrow H_c^k(U \cap V) \longrightarrow H_c^k(U) \oplus H_c^k(V) \longrightarrow H_c^k(M) \longrightarrow H_c^{k+1}(U \cap V) \longrightarrow \cdots$$
$$\downarrow D_{U \cap V} \qquad\qquad \downarrow D_U \oplus -D_V \qquad\qquad \downarrow D_M \qquad\qquad \downarrow D_{U \cap V}$$
$$\cdots \longrightarrow H_{n-k}(U \cap V) \longrightarrow H_{n-k}(U) \oplus H_{n-k}(V) \longrightarrow H_{n-k}(M) \longrightarrow H_{n-k-1}(U \cap V) \longrightarrow \cdots$$

Proof: Compact sets $K \subset U$ and $L \subset V$ give rise to the Mayer-Vietoris sequence in the upper row of the following diagram, whose lower row is also a Mayer-Vietoris sequence:

$$\cdots \longrightarrow H^k(M|K \cap L) \longrightarrow H^k(M|K) \oplus H^k(M|L) \longrightarrow H^k(M|K \cup L) \longrightarrow \cdots$$
$$\downarrow \approx \qquad\qquad\qquad \downarrow \approx$$
$$H^k(U \cap V|K \cap L) \qquad\quad H^k(U|K) \oplus H^k(V|L)$$
$$\downarrow \mu_{K \cap L} \cap \qquad\qquad \downarrow \mu_K \cap \oplus -\mu_L \cap \qquad\qquad \downarrow \mu_{K \cup L} \cap$$
$$\cdots \longrightarrow H_{n-k}(U \cap V) \longrightarrow H_{n-k}(U) \oplus H_{n-k}(V) \longrightarrow H_{n-k}(M) \longrightarrow \cdots$$

The two maps labeled isomorphisms come from excision. Assuming this diagram commutes, consider passing to the limit over compact sets $K \subset U$ and $L \subset V$. Since each compact set in $U \cap V$ is contained in an intersection $K \cap L$ of compact sets $K \subset U$ and $L \subset V$, and similarly for $U \cup V$, the diagram induces a limit diagram having the form stated in the lemma. The first row of this limit diagram is exact since a direct limit of exact sequences is exact; this is an exercise at the end of the section, and follows easily from the definition of direct limits.

It remains to consider the commutativity of the preceding diagram involving K and L. In the two squares shown, not involving boundary or coboundary maps, it is a triviality to check commutativity at the level of cycles and cocycles. Less trivial is the third square, which we rewrite in the following way:

$$H^k(M|K \cup L) \xrightarrow{\delta} H^{k+1}(M|K \cap L) \xrightarrow{\approx} H^{k+1}(U \cap V|K \cap L)$$
$$(*) \qquad \downarrow \mu_{K \cup L} \cap \qquad\qquad\qquad\qquad\qquad \downarrow \mu_{K \cap L} \cap$$
$$H_{n-k}(M) \xrightarrow{\qquad\qquad \partial \qquad\qquad} H_{n-k-1}(U \cap V)$$

Letting $A = M - K$ and $B = M - L$, the map δ is the coboundary map in the Mayer-Vietoris sequence obtained from the short exact sequence of cochain complexes

$$0 \longrightarrow C^*(M, A + B) \longrightarrow C^*(M, A) \oplus C^*(M, B) \longrightarrow C^*(M, A \cap B) \longrightarrow 0$$

where $C^*(M, A + B)$ consists of cochains on M vanishing on chains in A and chains in B. To evaluate the Mayer-Vietoris coboundary map δ on a cohomology class represented by a cocycle $\varphi \in C^*(M, A \cap B)$, the first step is to write $\varphi = \varphi_A - \varphi_B$

for $\varphi_A \in C^*(M,A)$ and $\varphi_B \in C^*(M,B)$. Then $\delta[\varphi]$ is represented by the cocycle $\delta\varphi_A = \delta\varphi_B \in C^*(M, A + B)$, where the equality $\delta\varphi_A = \delta\varphi_B$ comes from the fact that φ is a cocycle, so $\delta\varphi = \delta\varphi_A - \delta\varphi_B = 0$. Similarly, the boundary map ∂ in the homology Mayer–Vietoris sequence is obtained by representing an element of $H_i(M)$ by a cycle z that is a sum of chains $z_U \in C_i(U)$ and $z_V \in C_i(V)$, and then $\partial[z] = [\partial z_U]$.

Via barycentric subdivision, the class $\mu_{K \cup L}$ can be represented by a chain α that is a sum $\alpha_{U-L} + \alpha_{U \cap V} + \alpha_{V-K}$ of chains in $U - L$, $U \cap V$, and $V - K$, respectively, since these three open sets cover M. The chain $\alpha_{U \cap V}$ represents $\mu_{K \cap L}$ since the other two chains α_{U-L} and α_{V-K} lie in the complement of $K \cap L$, hence vanish in $H_n(M \mid K \cap L) \approx H_n(U \cap V \mid K \cap L)$. Similarly, $\alpha_{U-L} + \alpha_{U \cap V}$ represents μ_K.

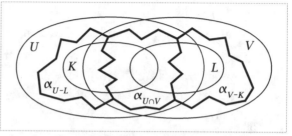

In the square $(*)$ let φ be a cocycle representing an element of $H^k(M \mid K \cup L)$. Under δ this maps to the cohomology class of $\delta\varphi_A$. Continuing on to $H_{n-k-1}(U \cap V)$ we obtain $\alpha_{U \cap V} \frown \delta\varphi_A$, which is in the same homology class as $\partial\alpha_{U \cap V} \frown \varphi_A$ since

$$\partial(\alpha_{U \cap V} \frown \varphi_A) = (-1)^k(\partial\alpha_{U \cap V} \frown \varphi_A - \alpha_{U \cap V} \frown \delta\varphi_A)$$

and $\alpha_{U \cap V} \frown \varphi_A$ is a chain in $U \cap V$.

Going around the square $(*)$ the other way, φ maps first to $\alpha \frown \varphi$. To apply the Mayer–Vietoris boundary map ∂ to this, we first write $\alpha \frown \varphi$ as a sum of a chain in U and a chain in V:

$$\alpha \frown \varphi = (\alpha_{U-L} \frown \varphi) + (\alpha_{U \cap V} \frown \varphi + \alpha_{V-K} \frown \varphi)$$

Then we take the boundary of the first of these two chains, obtaining the homology class $[\partial(\alpha_{U-L} \frown \varphi)] \in H_{n-k-1}(U \cap V)$. To compare this with $[\partial\alpha_{U \cap V} \frown \varphi_A]$, we have

$$\partial(\alpha_{U-L} \frown \varphi) = (-1)^k \partial\alpha_{U-L} \frown \varphi \qquad \text{since } \delta\varphi = 0$$
$$= (-1)^k \partial\alpha_{U-L} \frown \varphi_A \qquad \text{since } \partial\alpha_{U-L} \frown \varphi_B = 0, \ \varphi_B \text{ being}$$
$$\text{zero on chains in } B = M - L$$
$$= (-1)^{k+1} \partial\alpha_{U \cap V} \frown \varphi_A$$

where this last equality comes from the fact that $\partial(\alpha_{U-L} + \alpha_{U \cap V}) \frown \varphi_A = 0$ since $\partial(\alpha_{U-L} + \alpha_{U \cap V})$ is a chain in $U - K$ by the earlier observation that $\alpha_{U-L} + \alpha_{U \cap V}$ represents μ_K, and φ_A vanishes on chains in $A = M - K$.

Thus the square $(*)$ commutes up to a sign depending only on k. $\qquad \square$

Proof of Poincaré Duality: There are two inductive steps, finite and infinite:

(A) If M is the union of open sets U and V and if D_U, D_V, and $D_{U \cap V}$ are isomorphisms, then so is D_M. Via the five-lemma, this is immediate from the preceding lemma.

(B) If M is the union of a sequence of open sets $U_1 \subset U_2 \subset \cdots$ and each duality map $D_{U_i} : H_c^k(U_i) \to H_{n-k}(U_i)$ is an isomorphism, then so is D_M. To show this we notice first that by excision, $H_c^k(U_i)$ can be regarded as the limit of the groups $H^k(M \mid K)$ as K ranges over compact subsets of U_i. Then there are natural maps $H_c^k(U_i) \to H_c^k(U_{i+1})$ since the second of these groups is a limit over a larger collection of K's. Thus we can form $\varinjlim H_c^k(U_i)$ which is obviously isomorphic to $H_c^k(M)$ since the compact sets in M are just the compact sets in all the U_i's. By Proposition 3.33, $H_{n-k}(M) \approx \varinjlim H_{n-k}(U_i)$. The map D_M is thus the limit of the isomorphisms D_{U_i}, hence is an isomorphism.

Now after all these preliminaries we can prove the theorem in three easy steps:

(1) The case $M = \mathbb{R}^n$ can be proved by regarding $\mathbb{R}^n$ as the interior of Δ^n, and then the map D_M can be identified with the map $H^k(\Delta^n, \partial\Delta^n) \to H_{n-k}(\Delta^n)$ given by cap product with a unit times the generator $[\Delta^n] \in H_n(\Delta^n, \partial\Delta^n)$ defined by the identity map of Δ^n, which is a relative cycle. The only nontrivial value of k is $k = n$, when the cap product map is an isomorphism since a generator of $H^n(\Delta^n, \partial\Delta^n) \approx$ $\text{Hom}(H_n(\Delta^n, \partial\Delta^n), R)$ is represented by a cocycle φ taking the value 1 on Δ^n, so by the definition of cap product, $\Delta^n \frown \varphi$ is the last vertex of Δ^n, representing a generator of $H_0(\Delta^n)$.

(2) More generally, D_M is an isomorphism for M an arbitrary open set in $\mathbb{R}^n$. To see this, first write M as a countable union of nonempty bounded convex open sets U_i, for example open balls, and let $V_i = \bigcup_{j<i} U_j$. Both V_i and $U_i \cap V_i$ are unions of $i - 1$ bounded convex open sets, so by induction on the number of such sets in a cover we may assume that D_{V_i} and $D_{U_i \cap V_i}$ are isomorphisms. By (1), D_{U_i} is an isomorphism since U_i is homeomorphic to $\mathbb{R}^n$. Hence $D_{U_i \cup V_i}$ is an isomorphism by (A). Since M is the increasing union of the V_i's and each D_{V_i} is an isomorphism, so is D_M by (B).

(3) If M is a finite or countably infinite union of open sets U_i homeomorphic to $\mathbb{R}^n$, the theorem now follows by the argument in (2), with each appearance of the words 'bounded convex open set' replaced by 'open set in $\mathbb{R}^n$.' Thus the proof is finished for closed manifolds, as well as for all the noncompact manifolds one ever encounters in actual practice.

To handle a completely general noncompact manifold M we use a Zorn's Lemma argument. Consider the collection of open sets $U \subset M$ for which the duality maps D_U are isomorphisms. This collection is partially ordered by inclusion, and the union of every totally ordered subcollection is again in the collection by the argument in (B), which did not really use the hypothesis that the collection $\{U_i\}$ was indexed by the positive integers. Zorn's Lemma then implies that there exists a maximal open set U for which the theorem holds. If $U \neq M$, choose a point $x \in M - U$ and an open neighborhood V of x homeomorphic to $\mathbb{R}^n$. The theorem holds for V and $U \cap V$ by (1) and (2), and it holds for U by assumption, so by (A) it holds for $U \cup V$, contradicting the maximality of U. $\qquad\square$

|| **Corollary 3.37.** *A closed manifold of odd dimension has Euler characteristic zero.*

Proof: Let M be a closed n-manifold. If M is orientable, we have rank $H_i(M;\mathbb{Z}) =$ rank $H^{n-i}(M;\mathbb{Z})$, which equals rank $H_{n-i}(M;\mathbb{Z})$ by the universal coefficient theorem. Thus if n is odd, all the terms of $\sum_i(-1)^i$ rank $H_i(M;\mathbb{Z})$ cancel in pairs.

If M is not orientable we apply the same argument using $\mathbb{Z}_2$ coefficients, with rank $H_i(M;\mathbb{Z})$ replaced by $\dim H_i(M;\mathbb{Z}_2)$, the dimension as a vector space over $\mathbb{Z}_2$, to conclude that $\sum_i(-1)^i \dim H_i(M;\mathbb{Z}_2) = 0$. It remains to check that this alternating sum equals the Euler characteristic $\sum_i(-1)^i$ rank $H_i(M;\mathbb{Z})$. We can do this by using the isomorphisms $H_i(M;\mathbb{Z}_2) \approx H^i(M;\mathbb{Z}_2)$ and applying the universal coefficient theorem for cohomology. Each $\mathbb{Z}$ summand of $H_i(M;\mathbb{Z})$ gives a $\mathbb{Z}_2$ summand of $H^i(M;\mathbb{Z}_2)$. Each $\mathbb{Z}_m$ summand of $H_i(M;\mathbb{Z})$ with m even gives $\mathbb{Z}_2$ summands of $H^i(M;\mathbb{Z}_2)$ and $H^{i+1}(M,\mathbb{Z}_2)$, whose contributions to $\sum_i(-1)^i \dim H_i(M;\mathbb{Z}_2)$ cancel. And $\mathbb{Z}_m$ summands of $H_i(M;\mathbb{Z})$ with m odd contribute nothing to $H^*(M;\mathbb{Z}_2)$.   □

Connection with Cup Product

Cup and cap product are related by the formula

$$(*)\qquad\qquad \psi(\alpha \frown \varphi) = (\varphi \smile \psi)(\alpha)$$

for $\alpha \in C_{k+\ell}(X;R)$, $\varphi \in C^k(X;R)$, and $\psi \in C^\ell(X;R)$. This holds since for a singular $(k+\ell)$-simplex $\sigma : \Delta^{k+\ell} \to X$ we have

$$\psi(\sigma \frown \varphi) = \psi(\varphi(\sigma|[v_0,\cdots,v_k])\sigma|[v_k,\cdots,v_{k+\ell}])$$
$$= \varphi(\sigma|[v_0,\cdots,v_k])\psi(\sigma|[v_k,\cdots,v_{k+\ell}]) = (\varphi \smile \psi)(\sigma)$$

The formula $(*)$ says that the map $\varphi \smile : C^\ell(X;R) \to C^{k+\ell}(X;R)$ is equal to the map $\mathrm{Hom}_R(C_\ell(X;R),R) \to \mathrm{Hom}_R(C_{k+\ell}(X;R),R)$ dual to $\frown \varphi$. Passing to homology and cohomology, we obtain the commutative diagram at the right. When the maps h are isomorphisms, for example when R is a field or when $R = \mathbb{Z}$ and the homology groups of X are free, then the map $\varphi \smile$ is the dual of $\frown \varphi$. Thus in these cases

$$\begin{array}{ccc} H^\ell(X;R) & \xrightarrow{\ h\ } & \mathrm{Hom}_R(H_\ell(X;R),R) \\ \downarrow{\scriptstyle\varphi\smile} & & \downarrow{\scriptstyle(\frown\varphi)^*} \\ H^{k+\ell}(X;R) & \xrightarrow{\ h\ } & \mathrm{Hom}_R(H_{k+\ell}(X;R),R) \end{array}$$

cup and cap product determine each other, at least if one assumes finite generation so that cohomology determines homology as well as vice versa. However, there are examples where cap and cup products are not equivalent when $R = \mathbb{Z}$ and there is torsion in homology.

By means of the formula $(*)$, Poincaré duality has nontrivial implications for the cup product structure of manifolds. For a closed R-orientable n-manifold M, consider the cup product pairing

$$H^k(M;R) \times H^{n-k}(M;R) \longrightarrow R, \qquad (\varphi,\psi) \mapsto (\varphi \smile \psi)[M]$$

Such a bilinear pairing $A \times B \to R$ is said to be **nonsingular** if the maps $A \to \mathrm{Hom}_R(B, R)$ and $B \to \mathrm{Hom}_R(A, R)$, obtained by viewing the pairing as a function of each variable separately, are both isomorphisms.

Proposition 3.38. *The cup product pairing is nonsingular for closed R-orientable manifolds when R is a field, or when $R = \mathbb{Z}$ and torsion in $H^*(M; \mathbb{Z})$ is factored out.*

Proof: Consider the composition

$$H^{n-k}(M; R) \xrightarrow{h} \mathrm{Hom}_R(H_{n-k}(M; R), R) \xrightarrow{D^*} \mathrm{Hom}_R(H^k(M; R), R)$$

where h is the map appearing in the universal coefficient theorem, induced by evaluation of cochains on chains, and D^* is the Hom-dual of the Poincaré duality map $D : H^k \to H_{n-k}$. The composition $D^* h$ sends $\psi \in H^{n-k}(M; R)$ to the homomorphism $\varphi \mapsto \psi([M] \frown \varphi) = (\varphi \smile \psi)[M]$. For field coefficients or for integer coefficients with torsion factored out, h is an isomorphism. Nonsingularity of the pairing in one of its variables is then equivalent to D being an isomorphism. Nonsingularity in the other variable follows by commutativity of cup product. $\qquad \square$

Corollary 3.39. *If M is a closed connected orientable n-manifold, then an element $\alpha \in H^k(M; \mathbb{Z})$ generates an infinite cyclic summand of $H^k(M; \mathbb{Z})$ iff there exists an element $\beta \in H^{n-k}(M; \mathbb{Z})$ such that $\alpha \smile \beta$ is a generator of $H^n(M; \mathbb{Z}) \approx \mathbb{Z}$. With coefficients in a field this holds for any $\alpha \neq 0$.*

Proof: For α to generate a $\mathbb{Z}$ summand of $H^k(M; \mathbb{Z})$ is equivalent to the existence of a homomorphism $\varphi : H^k(M; \mathbb{Z}) \to \mathbb{Z}$ with $\varphi(\alpha) = \pm 1$. By the nonsingularity of the cup product pairing, φ is realized by taking cup product with an element $\beta \in H^{n-k}(M; \mathbb{Z})$ and evaluating on $[M]$, so having a β with $\alpha \smile \beta$ generating $H^n(M; \mathbb{Z})$ is equivalent to having φ with $\varphi(\alpha) = \pm 1$. The case of field coefficients is similar but easier. $\qquad \square$

Example 3.40: Projective Spaces. The cup product structure of $H^*(\mathbb{C}P^n; \mathbb{Z})$ as a truncated polynomial ring $\mathbb{Z}[\alpha]/(\alpha^{n+1})$ with $|\alpha| = 2$ can easily be deduced from this as follows. The inclusion $\mathbb{C}P^{n-1} \hookrightarrow \mathbb{C}P^n$ induces an isomorphism on H^i for $i \leq 2n - 2$, so by induction on n, $H^{2i}(\mathbb{C}P^n; \mathbb{Z})$ is generated by α^i for $i < n$. By the corollary, there is an integer m such that the product $\alpha \smile m\alpha^{n-1} = m\alpha^n$ generates $H^{2n}(\mathbb{C}P^n; \mathbb{Z})$. This can only happen if $m = \pm 1$, and therefore $H^*(\mathbb{C}P^n; \mathbb{Z}) \approx \mathbb{Z}[\alpha]/(\alpha^{n+1})$. The same argument shows $H^*(\mathbb{H}P^n; \mathbb{Z}) \approx \mathbb{Z}[\alpha]/(\alpha^{n+1})$ with $|\alpha| = 4$. For $\mathbb{R}P^n$ one can use the same argument with $\mathbb{Z}_2$ coefficients to deduce that $H^*(\mathbb{R}P^n; \mathbb{Z}_2) \approx \mathbb{Z}_2[\alpha]/(\alpha^{n+1})$ with $|\alpha| = 1$. The cup product structure in infinite-dimensional projective spaces follows from the finite-dimensional case, as we saw in the proof of Theorem 3.19.

Could there be a closed manifold whose cohomology is additively isomorphic to that of $\mathbb{C}P^n$ but with a different cup product structure? For $n = 2$ the answer is no since duality implies that the square of a generator of H^2 must be a generator of

H^4. For $n = 3$, duality says that the product of generators of H^2 and H^4 must be a generator of H^6, but nothing is said about the square of a generator of H^2. Indeed, for $S^2 \times S^4$, whose cohomology has the same additive structure as $\mathbb{CP}^3$, the square of the generator of $H^2(S^2 \times S^4; \mathbb{Z})$ is zero since it is the pullback of a generator of $H^2(S^2; \mathbb{Z})$ under the projection $S^2 \times S^4 \to S^2$, and in $H^*(S^2; \mathbb{Z})$ the square of the generator of H^2 is zero. More generally, an exercise for §4.D describes closed 6-manifolds having the same cohomology groups as $\mathbb{CP}^3$ but where the square of the generator of H^2 is an arbitrary multiple of a generator of H^4.

Example 3.41: **Lens Spaces.** Cup products in lens spaces can be computed in the same way as in projective spaces. For a lens space L^{2n+1} of dimension $2n + 1$ with fundamental group $\mathbb{Z}_m$, we computed $H_i(L^{2n+1}; \mathbb{Z})$ in Example 2.43 to be $\mathbb{Z}$ for $i = 0$ and $2n + 1$, $\mathbb{Z}_m$ for odd $i < 2n + 1$, and 0 otherwise. In particular, this implies that L^{2n+1} is orientable, which can also be deduced from the fact that L^{2n+1} is the orbit space of an action of $\mathbb{Z}_m$ on S^{2n+1} by orientation-preserving homeomorphisms, using an exercise at the end of this section. By the universal coefficient theorem, $H^i(L^{2n+1}; \mathbb{Z}_m)$ is $\mathbb{Z}_m$ for each $i \le 2n + 1$. Let $\alpha \in H^1(L^{2n+1}; \mathbb{Z}_m)$ and $\beta \in H^2(L^{2n+1}; \mathbb{Z}_m)$ be generators. The statement we wish to prove is:

$$H^j(L^{2n+1}; \mathbb{Z}_m) \text{ is generated by } \begin{cases} \beta^i & \text{for } j = 2i \\ \alpha\beta^i & \text{for } j = 2i + 1 \end{cases}$$

By induction on n we may assume this holds for $j \le 2n - 1$ since we have a lens space $L^{2n-1} \subset L^{2n+1}$ with this inclusion inducing an isomorphism on H^j for $j \le 2n - 1$, as one sees by comparing the cellular chain complexes for L^{2n-1} and L^{2n+1}. The preceding corollary does not apply directly for $\mathbb{Z}_m$ coefficients with arbitrary m, but its proof does since the maps $h : H^i(L^{2n+1}; \mathbb{Z}_m) \to \operatorname{Hom}(H_i(L^{2n+1}; \mathbb{Z}_m), \mathbb{Z}_m)$ are isomorphisms. We conclude that $\beta \smile k\alpha\beta^{n-1}$ generates $H^{2n+1}(L^{2n+1}; \mathbb{Z}_m)$ for some integer k. We must have k relatively prime to m, otherwise the product $\beta \smile k\alpha\beta^{n-1} = k\alpha\beta^n$ would have order less than m and so could not generate $H^{2n+1}(L^{2n+1}; \mathbb{Z}_m)$. Then since k is relatively prime to m, $\alpha\beta^n$ is also a generator of $H^{2n+1}(L^{2n+1}; \mathbb{Z}_m)$. From this it follows that β^n must generate $H^{2n}(L^{2n+1}; \mathbb{Z}_m)$, otherwise it would have order less than m and so therefore would $\alpha\beta^n$.

The rest of the cup product structure on $H^*(L^{2n+1}; \mathbb{Z}_m)$ is determined once α^2 is expressed as a multiple of β. When m is odd, the commutativity formula for cup product implies $\alpha^2 = 0$. When m is even, commutativity implies only that α^2 is either zero or the unique element of $H^2(L^{2n+1}; \mathbb{Z}_m) \approx \mathbb{Z}_m$ of order two. In fact it is the latter possibility which holds, since the 2-skeleton L^2 is the circle L^1 with a 2-cell attached by a map of degree m, and we computed the cup product structure in this 2-complex in Example 3.9. It does not seem to be possible to deduce the nontriviality of α^2 from Poincaré duality alone, except when $m = 2$.

The cup product structure for an infinite-dimensional lens space L^∞ follows from the finite-dimensional case since the restriction map $H^j(L^\infty; \mathbb{Z}_m) \to H^j(L^{2n+1}; \mathbb{Z}_m)$ is an isomorphism for $j \leq 2n + 1$. As with $\mathbb{R}P^n$, the ring structure in $H^*(L^{2n+1}; \mathbb{Z})$ is determined by the ring structure in $H^*(L^{2n+1}; \mathbb{Z}_m)$, and likewise for L^∞, where one has the slightly simpler structure $H^*(L^\infty; \mathbb{Z}) \approx \mathbb{Z}[\alpha]/(m\alpha)$ with $|\alpha| = 2$. The case of L^{2n+1} is obtained from this by setting $\alpha^{n+1} = 0$ and adjoining the extra $\mathbb{Z} \approx H^{2n+1}(L^{2n+1}; \mathbb{Z})$.

A different derivation of the cup product structure in lens spaces is given in Example 3E.2.

Using the ad hoc notation $H^k_{free}(M)$ for $H^k(M)$ modulo its torsion subgroup, the preceding proposition implies that for a closed orientable manifold M of dimension $2n$, the middle-dimensional cup product pairing $H^n_{free}(M) \times H^n_{free}(M) \to \mathbb{Z}$ is a nonsingular bilinear form on $H^n_{free}(M)$. This form is symmetric or skew-symmetric according to whether n is even or odd. The algebra in the skew-symmetric case is rather simple: With a suitable choice of basis, the matrix of a skew-symmetric nonsingular bilinear form over $\mathbb{Z}$ can be put into the standard form consisting of 2×2 blocks $\left(\begin{smallmatrix} 0 & -1 \\ 1 & 0 \end{smallmatrix} \right)$ along the diagonal and zeros elsewhere, according to an algebra exercise at the end of the section. In particular, the rank of $H^n(M^{2n})$ must be even when n is odd. We are already familiar with these facts in the case $n = 1$ by the explicit computations of cup products for surfaces in §3.2.

The symmetric case is much more interesting algebraically. There are only finitely many isomorphism classes of symmetric nonsingular bilinear forms over $\mathbb{Z}$ of a fixed rank, but this 'finitely many' grows rather rapidly, for example it is more than 80 million for rank 32; see [Serre 1973] for an exposition of this beautiful chapter of number theory. One can ask whether all these forms actually occur as cup product pairings in closed manifolds M^{4k} for a given k. The answer is yes for $4k = 4, 8, 16$ but seems to be unknown in other dimensions. In dimensions 4, 8, and 16 one can even take M^{4k} to be simply-connected and have the bare minimum of homology: $\mathbb{Z}$'s in dimensions 0 and $4k$ and a free abelian group in dimension $2k$. In dimension 4 there are at most two nonhomeomorphic simply-connected closed 4-manifolds with the same bilinear form. Namely, there are two manifolds with the same form if the square $\alpha \smile \alpha$ of some $\alpha \in H^2(M^4)$ is an odd multiple of a generator of $H^4(M^4)$, for example for $\mathbb{C}P^2$, and otherwise the M^4 is unique, for example for S^4 or $S^2 \times S^2$; see [Freedman & Quinn 1990]. In §4.C we take the first step in this direction by proving a classical result of J. H. C. Whitehead that the homotopy type of a simply-connected closed 4-manifold is uniquely determined by its cup product structure.

Other Forms of Duality

Generalizing the definition of a manifold, an **n-manifold with boundary** is a Hausdorff space M in which each point has an open neighborhood homeomorphic either to $\mathbb{R}^n$ or to the half-space $\mathbb{R}^n_+ = \{ (x_1, \cdots, x_n) \in \mathbb{R}^n \mid x_n \geq 0 \}$. If a point $x \in M$ corresponds under such a homeomorphism to a point $(x_1, \cdots, x_n) \in \mathbb{R}^n_+$ with

$x_n = 0$, then by excision we have $H_n(M, M - \{x\}; \mathbb{Z}) \approx H_n(\mathbb{R}^n_+, \mathbb{R}^n_+ - \{0\}; \mathbb{Z}) = 0$, whereas if x corresponds to a point $(x_1, \cdots, x_n) \in \mathbb{R}^n_+$ with $x_n > 0$ or to a point of $\mathbb{R}^n$, then $H_n(M, M - \{x\}; \mathbb{Z}) \approx H_n(\mathbb{R}^n, \mathbb{R}^n - \{0\}; \mathbb{Z}) \approx \mathbb{Z}$. Thus the points x with $H_n(M, M - \{x\}; \mathbb{Z}) = 0$ form a well-defined subspace, called the **boundary** of M and denoted ∂M. For example, $\partial \mathbb{R}^n_+ = \mathbb{R}^{n-1}$ and $\partial D^n = S^{n-1}$. It is evident that ∂M is an $(n-1)$-dimensional manifold with empty boundary.

If M is a manifold with boundary, then a **collar** neighborhood of ∂M in M is an open neighborhood homeomorphic to $\partial M \times [0, 1)$ by a homeomorphism taking ∂M to $\partial M \times \{0\}$.

Proposition 3.42. *If M is a compact manifold with boundary, then ∂M has a collar neighborhood.*

Proof: Let M' be M with an external collar attached, the quotient of the disjoint union of M and $\partial M \times [0, 1]$ in which $x \in \partial M$ is identified with $(x, 0) \in \partial M \times [0, 1]$. It will suffice to construct a homeomorphism $h : M \to M'$ since $\partial M'$ clearly has a collar neighborhood.

Since M is compact, so is the closed subspace ∂M. This implies that we can choose a finite number of continuous functions $\varphi_i : \partial M \to [0, 1]$ such that the sets $V_i = \varphi_i^{-1}(0, 1]$ form an open cover of ∂M and each V_i has closure contained in an open set $U_i \subset M$ homeomorphic to the half-space $\mathbb{R}^n_+$. After dividing each φ_i by $\sum_j \varphi_j$ we may assume $\sum_i \varphi_i = 1$.

Let $\psi_k = \varphi_1 + \cdots + \varphi_k$ and let $M_k \subset M'$ be the union of M with the points $(x, t) \in \partial M \times [0, 1]$ with $t \le \psi_k(x)$. By definition $\psi_0 = 0$ and $M_0 = M$. We construct a homeomorphism $h_k : M_{k-1} \to M_k$ as follows. The homeomorphism $U_k \approx \mathbb{R}^n_+$ gives a collar neighborhood $\partial U_k \times [-1, 0]$ of ∂U_k in U_k, with $x \in \partial U_k$ corresponding to $(x, 0) \in \partial U_k \times [-1, 0]$. Via the external collar $\partial M \times [0, 1]$ we then have an embedding $\partial U_k \times [-1, 1] \subset M'$. We define h_k to be the identity outside this $\partial U_k \times [-1, 1]$, and for $x \in \partial U_k$ we let h_k stretch the segment $\{x\} \times [-1, \psi_{k-1}(x)]$ linearly onto $\{x\} \times [-1, \psi_k(x)]$. The composition of all the h_k's then gives a homeomorphism $M \approx M'$, finishing the proof. $\qquad\square$

More generally, collars can be constructed for the boundaries of paracompact manifolds in the same way.

A compact manifold M with boundary is defined to be R-orientable if $M - \partial M$ is R-orientable as a manifold without boundary. If $\partial M \times [0, 1)$ is a collar neighborhood of ∂M in M then $H_i(M, \partial M; R)$ is naturally isomorphic to $H_i(M - \partial M, \partial M \times (0, \varepsilon); R)$, so when M is R-orientable, Lemma 3.27 gives a relative fundamental class $[M]$ in $H_n(M, \partial M; R)$ restricting to a given orientation at each point of $M - \partial M$.

It will not be difficult to deduce the following generalization of Poincaré duality to manifolds with boundary from the version we have already proved for noncompact manifolds:

Theorem 3.43. *Suppose M is a compact R-orientable n-manifold whose boundary ∂M is decomposed as the union of two compact $(n-1)$-dimensional manifolds A and B with a common boundary $\partial A = \partial B = A \cap B$. Then cap product with a fundamental class $[M] \in H_n(M, \partial M; R)$ gives isomorphisms $D_M : H^k(M, A; R) \to H_{n-k}(M, B; R)$ for all k.*

The possibility that A, B, or $A \cap B$ is empty is not excluded. The cases $A = \varnothing$ and $B = \varnothing$ are sometimes called Lefschetz duality.

Proof: The cap product map $D_M : H^k(M, A; R) \to H_{n-k}(M, B; R)$ is defined since the existence of collar neighborhoods of $A \cap B$ in A and B and ∂M in M implies that A and B are deformation retracts of open neighborhoods U and V in M such that $U \cup V$ deformation retracts onto $A \cup B = \partial M$ and $U \cap V$ deformation retracts onto $A \cap B$.

The case $B = \varnothing$ is proved by applying Theorem 3.35 to $M - \partial M$. Via a collar neighborhood of ∂M we see that $H^k(M, \partial M; R) \approx H^k_c(M - \partial M; R)$, and there are obvious isomorphisms $H_{n-k}(M; R) \approx H_{n-k}(M - \partial M; R)$.

The general case reduces to the case $B = \varnothing$ by applying the five-lemma to the following diagram, where coefficients in R are implicit:

$$\cdots \longrightarrow H^k(M, \partial M) \longrightarrow H^k(M, A) \longrightarrow H^k(\partial M, A) \longrightarrow H^{k+1}(M, \partial M) \longrightarrow \cdots$$

$$\cdots \longrightarrow H_{n-k}(M) \longrightarrow H_{n-k}(M, B) \longrightarrow H_{n-k-1}(B) \longrightarrow H_{n-k-1}(M) \longrightarrow \cdots$$

with vertical maps $[M]\frown$, $[M]\frown$, and $[M]\frown$, and the middle column $H^k(\partial M, A) \xrightarrow{\approx} H^k(B, \partial B) \xrightarrow{[B]\frown} H_{n-k-1}(B)$.

For commutativity of the middle square one needs to check that the boundary map $H_n(M, \partial M) \to H_{n-1}(\partial M)$ sends a fundamental class for M to a fundamental class for ∂M. We leave this as an exercise at the end of the section. $\qquad \square$

Here is another kind of duality which generalizes the calculation of the local homology groups $H_i(M, M - \{x\}; \mathbb{Z})$:

Theorem 3.44. *If K is a compact, locally contractible subspace of a closed orientable n-manifold M, then $H_i(M, M - K; \mathbb{Z}) \approx H^{n-i}(K; \mathbb{Z})$ for all i.*

Proof: Let U be an open neighborhood of K in M. Consider the following diagram whose rows are long exact sequences of pairs:

$$\cdots \longrightarrow H_i(M-K) \longrightarrow H_i(M) \longrightarrow H_i(M, M-K) \longrightarrow H_{i-1}(M-K) \longrightarrow \cdots$$

$$\cdots \longrightarrow H^{n-i}(M, U) \longrightarrow H^{n-i}(M) \longrightarrow H^{n-i}(U) \longrightarrow H^{n-i+1}(M, U) \longrightarrow \cdots$$

with vertical maps $H^{n-i}(M-K, U-K)$, $H_i(U, U-K)$, and $H^{n-i+1}(M-K, U-K)$.

The second vertical map is the Poincaré duality isomorphism given by cap products with a fundamental class $[M]$. This class can be represented by a cycle which is the sum of a chain in $M - K$ and a chain in U representing elements of $H_n(M - K, U - K)$ and $H_n(U, U - K)$ respectively, and the first and third vertical maps are given by relative cap products with these classes. It is not hard to check that the diagram commutes up to sign, where for the square involving boundary and coboundary maps one uses the formula for the boundary of a cap product.

Passing to the direct limit over decreasing $U \supset K$, the first vertical arrow become the Poincaré duality isomorphism $H_i(M - K) \approx H_c^{n-i}(M - K)$. The five-lemma then gives an isomorphism $H_i(M, M - K) \approx \varinjlim H^{n-i}(U)$. We will show that the natural map from this limit to $H^{n-i}(K)$ is an isomorphism. This is easy when K has a neighborhood that is a mapping cylinder of some map $X \to K$, as in the 'letter examples' at the beginning of Chapter 0, since in this case we can compute the direct limit using neighborhoods U which are segments of the mapping cylinder that deformation retract to K.

For the general case we use Theorem A.7 and Corollary A.9 in the Appendix. The latter says that M can be embedded in some $\mathbb{R}^k$ as a retract of a neighborhood N in $\mathbb{R}^k$, and then Theorem A.7 says that K is a retract of a neighborhood in $\mathbb{R}^k$ and hence, by restriction, of a neighborhood W in M. We can compute $\varinjlim H^{n-i}(U)$ using just neighborhoods U in W, so these also retract to K and hence the map $\varinjlim H^{n-i}(U) \to H^{n-i}(K)$ is surjective. To show that it is injective, note first that the retraction $U \to K$ is homotopic to the identity $U \to U$ through maps $U \to \mathbb{R}^k$, via the standard linear homotopy. Choosing a smaller U if necessary, we may assume this homotopy is through maps $U \to N$ since K is stationary during the homotopy. Applying the retraction $N \to M$ gives a homotopy through maps $U \to M$ fixed on K. Restricting to sufficiently small $V \subset U$, we then obtain a homotopy in U from the inclusion map $V \to U$ to the retraction $V \to K$. Thus the map $H^{n-i}(U) \to H^{n-i}(V)$ factors as $H^{n-i}(U) \to H^{n-i}(K) \to H^{n-i}(V)$ where the first map is induced by inclusion and the second by the retraction. This implies that the kernel of $\varinjlim H^{n-i}(U) \to H^{n-i}(K)$ is trivial. $\square$

From this theorem we can easily deduce **Alexander duality**:

Corollary 3.45. *If K is a compact, locally contractible, nonempty, proper subspace of S^n, then $\tilde{H}_i(S^n - K; \mathbb{Z}) \approx \tilde{H}^{n-i-1}(K; \mathbb{Z})$ for all i.*

Proof: The long exact sequence of reduced homology for the pair $(S^n, S^n - K)$ gives isomorphisms $\tilde{H}_i(S^n - K; \mathbb{Z}) \approx H_{i+1}(S^n, S^n - K; \mathbb{Z})$ for most values of i. The exception is when $i = n - 1$ and we have only a short exact sequence

$$0 \to \tilde{H}_n(S^n; \mathbb{Z}) \to H_n(S^n, S^n - K; \mathbb{Z}) \to \tilde{H}_{n-1}(S^n - K; \mathbb{Z}) \to 0$$

where the initial 0 is $\tilde{H}_n(S^n - K;\mathbb{Z})$ which is zero since the components of $S^n - K$ are noncompact n-manifolds. This short exact sequence splits since we can map it to the corresponding sequence with K replaced by a point in K. Thus $\tilde{H}_{n-1}(S^n - K;\mathbb{Z})$ is $H_n(S^n, S^n - K;\mathbb{Z})$ with a $\mathbb{Z}$ summand canceled, just as $\tilde{H}^0(K;\mathbb{Z})$ is $H^0(K;\mathbb{Z})$ with a $\mathbb{Z}$ summand canceled. $\qquad\square$

The special case of Alexander duality when K is a sphere or disk was treated by more elementary means in Proposition 2B.1. As remarked there, it is interesting that the homology of $S^n - K$ does not depend on the way that K is embedded in S^n. There can be local pathologies as in the case of the Alexander horned sphere, or global complications as with knotted circles in S^3, but these have no effect on the homology of the complement. The only requirement is that K is not too bad a space itself. An example where the theorem fails without the local contractibility assumption is the 'quasi-circle,' defined in an exercise for §1.3. This compact subspace $K \subset \mathbb{R}^2$ can be regarded as a subspace of S^2 by adding a point at infinity. Then we have $\tilde{H}_0(S^2 - K;\mathbb{Z}) \approx \mathbb{Z}$ since $S^2 - K$ has two path-components, but $\tilde{H}^1(K;\mathbb{Z}) = 0$ since K is simply-connected.

Corollary 3.46. *If $X \subset \mathbb{R}^n$ is compact and locally contractible then $H_i(X;\mathbb{Z})$ is 0 for $i \geq n$ and torsionfree for $i = n - 1$ and $n - 2$.*

For example, a closed nonorientable n-manifold M cannot be embedded as a subspace of $\mathbb{R}^{n+1}$ since $H_{n-1}(M;\mathbb{Z})$ contains a $\mathbb{Z}_2$ subgroup, by Corollary 3.28. Thus the Klein bottle cannot be embedded in $\mathbb{R}^3$. More generally, the 2-dimensional complex $X_{m,n}$ studied in Example 1.24, the quotient spaces of $S^1 \times I$ under the identifications $(z,0) \sim (e^{2\pi i/m}z, 0)$ and $(z,1) \sim (e^{2\pi i/n}z, 1)$, cannot be embedded in $\mathbb{R}^3$ if m and n are not relatively prime, since $H_1(X_{m,n}\mathbb{Z})$ is $\mathbb{Z} \times \mathbb{Z}_d$ where d is the greatest common divisor of m and n. The Klein bottle is the case $m = n = 2$.

Proof: Viewing X as a subspace of the one-point compactification S^n, Alexander duality gives isomorphisms $\tilde{H}^i(X;\mathbb{Z}) \approx \tilde{H}_{n-i-1}(S^n - X;\mathbb{Z})$. The latter group is zero for $i \geq n$ and torsionfree for $i = n - 1$, so the result follows from the universal coefficient theorem since X has finitely generated homology groups. $\qquad\square$

There is a way of extending Alexander duality and the duality in Theorem 3.44 to compact sets K that are not locally contractible, by replacing the singular cohomology of K with another kind of cohomology called Čech cohomology. This is defined in the following way. To each open cover $\mathcal{U} = \{U_\alpha\}$ of a given space X we can associate a simplicial complex $N(\mathcal{U})$ called the **nerve** of $\mathcal{U}$. This has a vertex v_α for each U_α, and a set of $k + 1$ vertices spans a k-simplex whenever the $k + 1$ corresponding U_α's have nonempty intersection. When another cover $\mathcal{V} = \{V_\beta\}$ is a refinement of $\mathcal{U}$, so each V_β is contained in some U_α, then these inclusions induce a simplicial map

$N(\mathcal{V}) \to N(\mathcal{U})$ that is well-defined up to homotopy. We can then form the direct limit $\varinjlim H^i(N(\mathcal{U}); G)$ with respect to finer and finer open covers $\mathcal{U}$. This limit group is by definition the **Čech cohomology group** $\check{H}^i(X; G)$. For a full exposition of this cohomology theory see [Eilenberg & Steenrod 1952]. With an analogous definition of relative groups, Čech cohomology turns out to satisfy the same axioms as singular cohomology. For spaces homotopy equivalent to CW complexes, Čech cohomology coincides with singular cohomology, but for spaces with local complexities it often behaves more reasonably. For example, if X is the subspace of $\mathbb{R}^3$ consisting of the spheres of radius $^1/_n$ and center $(^1/_n, 0, 0)$ for $n = 1, 2, \cdots$, then contrary to what one might expect, $H^3(X; \mathbb{Z})$ is nonzero, as shown in [Barratt & Milnor 1962]. But $\check{H}^3(X; \mathbb{Z}) = 0$ and $\check{H}^2(X; \mathbb{Z}) = \mathbb{Z}^\infty$, the direct sum of countably many copies of $\mathbb{Z}$.

Oddly enough, the corresponding Čech homology groups defined using inverse limits are not so well-behaved. This is because the exactness axiom fails due to the algebraic fact that an inverse limit of exact sequences need not be exact, as a direct limit would be; see §3.F. However, there is a way around this problem using a more refined definition. This is Steenrod homology theory, which the reader can learn about in [Milnor 1995].

Exercises

1. Show that there exist nonorientable 1-dimensional manifolds if the Hausdorff condition is dropped from the definition of a manifold.

2. Show that deleting a point from a manifold of dimension greater than 1 does not affect orientability of the manifold.

3. Show that every covering space of an orientable manifold is an orientable manifold.

4. Given a covering space action of a group G on an orientable manifold M by orientation-preserving homeomorphisms, show that M/G is also orientable.

5. Show that $M \times N$ is orientable iff M and N are both orientable.

6. Given two disjoint connected n-manifolds M_1 and M_2, a connected n-manifold $M_1 \# M_2$, their *connected sum*, can be constructed by deleting the interiors of closed n-balls $B_1 \subset M_1$ and $B_2 \subset M_2$ and identifying the resulting boundary spheres ∂B_1 and ∂B_2 via some homeomorphism between them. (Assume that each B_i embeds nicely in a larger ball in M_i.)

(a) Show that if M_1 and M_2 are closed then there are isomorphisms $H_i(M_1 \# M_2; \mathbb{Z}) \approx H_i(M_1; \mathbb{Z}) \oplus H_i(M_2; \mathbb{Z})$ for $0 < i < n$, with one exception: If both M_1 and M_2 are nonorientable, then $H_{n-1}(M_1 \# M_2; \mathbb{Z})$ is obtained from $H_{n-1}(M_1; \mathbb{Z}) \oplus H_{n-1}(M_2; \mathbb{Z})$ by replacing one of the two $\mathbb{Z}_2$ summands by a $\mathbb{Z}$ summand. [Euler characteristics may help in the exceptional case.]

(b) Show that $\chi(M_1 \# M_2) = \chi(M_1) + \chi(M_2) - \chi(S^n)$ if M_1 and M_2 are closed.

7. For a map $f : M \to N$ between connected closed orientable n-manifolds with fundamental classes $[M]$ and $[N]$, the *degree* of f is defined to be the integer d such that $f_*([M]) = d[N]$, so the sign of the degree depends on the choice of fundamental classes. Show that for any connected closed orientable n-manifold M there is a degree 1 map $M \to S^n$.

8. For a map $f : M \to N$ between connected closed orientable n-manifolds, suppose there is a ball $B \subset N$ such that $f^{-1}(B)$ is the disjoint union of balls B_i each mapped homeomorphically by f onto B. Show the degree of f is $\sum_i \varepsilon_i$ where ε_i is $+1$ or -1 according to whether $f : B_i \to B$ preserves or reverses local orientations induced from given fundamental classes $[M]$ and $[N]$.

9. Show that a p-sheeted covering space projection $M \to N$ has degree $\pm p$, when M and N are connected closed orientable manifolds.

10. Show that for a degree 1 map $f : M \to N$ of connected closed orientable manifolds, the induced map $f_* : \pi_1 M \to \pi_1 N$ is surjective, hence also $f_* : H_1(M) \to H_1(N)$. [Lift f to the covering space $\tilde{N} \to N$ corresponding to the subgroup $\text{Im} f_* \subset \pi_1 N$, then consider the two cases that this covering is finite-sheeted or infinite-sheeted.]

11. If M_g denotes the closed orientable surface of genus g, show that degree 1 maps $M_g \to M_h$ exist iff $g \geq h$.

12. As an algebraic application of the preceding problem, show that in a free group F with basis $x_1, \cdots, x_{2k}$, the product of commutators $[x_1, x_2] \cdots [x_{2k-1}, x_{2k}]$ is not equal to a product of fewer than k commutators $[v_i, w_i]$ of elements $v_i, w_i \in F$. [Recall that the 2-cell of M_k is attached by the product $[x_1, x_2] \cdots [x_{2k-1}, x_{2k}]$. From a relation $[x_1, x_2] \cdots [x_{2k-1}, x_{2k}] = [v_1, w_1] \cdots [v_j, w_j]$ in F, construct a degree 1 map $M_j \to M_k$.]

13. Let $M'_h \subset M_g$ be a compact subsurface of genus h with one boundary circle, so M'_h is homeomorphic to M_h with an open disk removed. Show there is no retraction $M_g \to M'_h$ if $h > g/2$. [Apply the previous problem, using the fact that $M_g - M'_h$ has genus $g - h$.]

14. Let X be the shrinking wedge of circles in Example 1.25, the subspace of $\mathbb{R}^2$ consisting of the circles of radius $1/n$ and center $(1/n, 0)$ for $n = 1, 2, \cdots$.

(a) If $f_n : I \to X$ is the loop based at the origin winding once around the nth circle, show that the infinite product of commutators $[f_1, f_2][f_3, f_4] \cdots$ defines a loop in X that is nontrivial in $H_1(X)$. [Use Exercise 12.]

(b) If we view X as the wedge sum of the subspaces A and B consisting of the odd-numbered and even-numbered circles, respectively, use the same loop to show that the map $H_1(X) \to H_1(A) \oplus H_1(B)$ induced by the retractions of X onto A and B is not an isomorphism.

15. For an n-manifold M and a compact subspace $A \subset M$, show that $H_n(M, M - A; R)$ is isomorphic to the group $\Gamma_R(A)$ of sections of the covering space $M_R \to M$ over A, that is, maps $A \to M_R$ whose composition with $M_R \to M$ is the identity.

16. Show that $(\alpha \frown \varphi) \frown \psi = \alpha \frown (\varphi \smile \psi)$ for all $\alpha \in C_k(X; R)$, $\varphi \in C^\ell(X; R)$, and $\psi \in C^m(X; R)$. Deduce that cap product makes $H_*(X; R)$ a right $H^*(X; R)$-module.

17. Show that a direct limit of exact sequences is exact. More generally, show that homology commutes with direct limits: If $\{C_\alpha, f_{\alpha\beta}\}$ is a directed system of chain complexes, with the maps $f_{\alpha\beta} : C_\alpha \to C_\beta$ chain maps, then $H_n(\varinjlim C_\alpha) = \varinjlim H_n(C_\alpha)$.

18. Show that a direct limit $\varinjlim G_\alpha$ of torsionfree abelian groups G_α is torsionfree. More generally, show that any finitely generated subgroup of $\varinjlim G_\alpha$ is realized as a subgroup of some G_α.

19. Show that a direct limit of countable abelian groups over a countable indexing set is countable. Apply this to show that if X is an open set in $\mathbb{R}^n$ then $H_i(X; \mathbb{Z})$ is countable for all i.

20. Show that $H_c^0(X; G) = 0$ if X is path-connected and noncompact.

21. For a space X, let X^+ be the one-point compactification. If the added point, denoted ∞, has a neighborhood in X^+ that is a cone with ∞ the cone point, show that the evident map $H_c^n(X; G) \to H^n(X^+, \infty; G)$ is an isomorphism for all n. [Question: Does this result hold when $X = \mathbb{Z} \times \mathbb{R}$?]

22. Show that $H_c^n(X \times \mathbb{R}; G) \approx H_c^{n-1}(X; G)$ for all n.

23. Show that for a locally compact Δ-complex X the simplicial and singular cohomology groups $H_c^i(X; G)$ are isomorphic. This can be done by showing that $\Delta_c^i(X; G)$ is the union of its subgroups $\Delta^i(X, A; G)$ as A ranges over subcomplexes of X that contain all but finitely many simplices, and likewise $C_c^i(X; G)$ is the union of its subgroups $C^i(X, A; G)$ for the same family of subcomplexes A.

24. Let M be a closed connected 3-manifold, and write $H_1(M; \mathbb{Z})$ as $\mathbb{Z}^r \oplus F$, the direct sum of a free abelian group of rank r and a finite group F. Show that $H_2(M; \mathbb{Z})$ is $\mathbb{Z}^r$ if M is orientable and $\mathbb{Z}^{r-1} \oplus \mathbb{Z}_2$ if M is nonorientable. In particular, $r \geq 1$ when M is nonorientable. Using Exercise 6, construct examples showing there are no other restrictions on the homology groups of closed 3-manifolds. [In the nonorientable case consider the manifold N obtained from $S^2 \times I$ by identifying $S^2 \times \{0\}$ with $S^2 \times \{1\}$ via a reflection of S^2.]

25. Show that if a closed orientable manifold M of dimension $2k$ has $H_{k-1}(M; \mathbb{Z})$ torsionfree, then $H_k(M; \mathbb{Z})$ is also torsionfree.

26. Compute the cup product structure in $H^*(S^2 \times S^8 \sharp S^4 \times S^6; \mathbb{Z})$, and in particular show that the only nontrivial cup products are those dictated by Poincaré duality. [See Exercise 6. The result has an evident generalization to connected sums of $S^i \times S^{n-i}$'s for fixed n and varying i.]

27. Show that after a suitable change of basis, a skew-symmetric nonsingular bilinear form over $\mathbb{Z}$ can be represented by a matrix consisting of 2×2 blocks $\left(\begin{smallmatrix} 0 & -1 \\ 1 & 0 \end{smallmatrix}\right)$ along the diagonal and zeros elsewhere. [For the matrix of a bilinear form, the following operation can be realized by a change of basis: Add an integer multiple of the ith row to the jth row and add the same integer multiple of the ith column to the jth column. Use this to fix up each column in turn. Note that a skew-symmetric matrix must have zeros on the diagonal.]

28. Show that a nonsingular symmetric or skew-symmetric bilinear pairing over a field F, of the form $F^n \times F^n \to F$, cannot be identically zero when restricted to all pairs of vectors v, w in a k-dimensional subspace $V \subset F^n$ if $k > n/2$.

29. Use the preceding problem to show that if the closed orientable surface M_g of genus g retracts onto a graph $X \subset M_g$, then $H_1(X)$ has rank at most g. Deduce an alternative proof of Exercise 13 from this, and construct a retraction of M_g onto a wedge sum of k circles for each $k \le g$.

30. Show that the boundary of an R-orientable manifold is also R-orientable.

31. Show that if M is a compact R-orientable n-manifold, then the boundary map $H_n(M, \partial M; R) \to H_{n-1}(\partial M; R)$ sends a fundamental class for $(M, \partial M)$ to a fundamental class for ∂M.

32. Show that a compact manifold does not retract onto its boundary.

33. Show that if M is a compact contractible n-manifold then ∂M is a homology $(n-1)$-sphere, that is, $H_i(\partial M; \mathbb{Z}) \approx H_i(S^{n-1}; \mathbb{Z})$ for all i.

34. For a compact manifold M verify that the following diagram relating Poincaré duality for M and ∂M is commutative, up to sign at least:

$$
\begin{array}{ccccccc}
H^{k-1}(\partial M; R) & \longrightarrow & H^k(M, \partial M; R) & \longrightarrow & H^k(M; R) & \longrightarrow & H^k(\partial M; R) \\
\downarrow {\scriptstyle [\partial M]\,\cap} & & \downarrow {\scriptstyle [M]\,\cap} & & \downarrow {\scriptstyle [M]\,\cap} & & \downarrow {\scriptstyle [\partial M]\,\cap} \\
H_{n-k}(\partial M; R) & \longrightarrow & H_{n-k}(M; R) & \longrightarrow & H_{n-k}(M, \partial M; R) & \longrightarrow & H_{n-k-1}(\partial M; R)
\end{array}
$$

35. If M is a noncompact R-orientable n-manifold with boundary ∂M having a collar neighborhood in M, show that there are Poincaré duality isomorphisms $H_c^k(M; R) \approx H_{n-k}(M, \partial M; R)$ for all k, using the five-lemma and the following diagram:

$$
\begin{array}{ccccccccc}
\cdots & \longrightarrow & H_c^{k-1}(\partial M; R) & \longrightarrow & H_c^k(M, \partial M; R) & \longrightarrow & H_c^k(M; R) & \longrightarrow & H_c^k(\partial M; R) & \longrightarrow & \cdots \\
& & \downarrow {\scriptstyle D_{\partial M}} & & \downarrow {\scriptstyle D_M} & & \downarrow {\scriptstyle D_M} & & \downarrow {\scriptstyle D_{\partial M}} \\
\cdots & \longrightarrow & H_{n-k}(\partial M; R) & \longrightarrow & H_{n-k}(M; R) & \longrightarrow & H_{n-k}(M, \partial M; R) & \longrightarrow & H_{n-k-1}(\partial M; R) & \longrightarrow & \cdots
\end{array}
$$

Additional Topics

3.A Universal Coefficients for Homology

The main goal in this section is an algebraic formula for computing homology with arbitrary coefficients in terms of homology with $\mathbb{Z}$ coefficients. The theory parallels rather closely the universal coefficient theorem for cohomology in §3.1.

The first step is to formulate the definition of homology with coefficients in terms of tensor products. The chain group $C_n(X;G)$ as defined in §2.2 consists of the finite formal sums $\sum_i g_i \sigma_i$ with $g_i \in G$ and $\sigma_i : \Delta^n \to X$. This means that $C_n(X;G)$ is a direct sum of copies of G, with one copy for each singular n-simplex in X. More generally, the relative chain group $C_n(X, A; G) = C_n(X; G)/C_n(A; G)$ is also a direct sum of copies of G, one for each singular n-simplex in X not contained in A. From the basic properties of tensor products listed in the discussion of the Künneth formula in §3.2 it follows that $C_n(X, A; G)$ is naturally isomorphic to $C_n(X, A) \otimes G$, via the correspondence $\sum_i g_i \sigma_i \mapsto \sum_i \sigma_i \otimes g_i$. Under this isomorphism the boundary map $C_n(X, A; G) \to C_{n-1}(X, A; G)$ becomes the map $\partial \otimes \mathbb{1} : C_n(X, A) \otimes G \to C_{n-1}(X, A) \otimes G$ where $\partial : C_n(X, A) \to C_{n-1}(X, A)$ is the usual boundary map for $\mathbb{Z}$ coefficients. Thus we have the following algebraic problem:

> Given a chain complex $\cdots \to C_n \xrightarrow{\partial_n} C_{n-1} \to \cdots$ of free abelian groups C_n, is it possible to compute the homology groups $H_n(C; G)$ of the associated chain complex $\cdots \to C_n \otimes G \xrightarrow{\partial_n \otimes \mathbb{1}} C_{n-1} \otimes G \to \cdots$ just in terms of G and the homology groups $H_n(C)$ of the original complex?

To approach this problem, the idea will be to compare the chain complex C with two simpler subcomplexes, the subcomplexes consisting of the cycles and the boundaries in C, and see what happens upon tensoring all three complexes with G.

Let $Z_n = \operatorname{Ker} \partial_n \subset C_n$ and $B_n = \operatorname{Im} \partial_{n+1} \subset C_n$. The restrictions of ∂_n to these two subgroups are zero, so they can be regarded as subcomplexes Z and B of C with trivial boundary maps. Thus we have a short exact sequence of chain complexes consisting of the commutative diagrams

(i)
$$
\begin{array}{ccccccccc}
0 & \longrightarrow & Z_n & \longrightarrow & C_n & \xrightarrow{\partial_n} & B_{n-1} & \longrightarrow & 0 \\
& & \downarrow{\partial_n} & & \downarrow{\partial_n} & & \downarrow{\partial_{n-1}} & & \\
0 & \longrightarrow & Z_{n-1} & \longrightarrow & C_{n-1} & \xrightarrow{\partial_{n-1}} & B_{n-2} & \longrightarrow & 0
\end{array}
$$

The rows in this diagram split since each B_n is free, being a subgroup of the free group C_n. Thus $C_n \approx Z_n \oplus B_{n-1}$, but the chain complex C is not the direct sum of the chain complexes Z and B since the latter have trivial boundary maps but the boundary maps in C may be nontrivial. Now tensor with G to get a commutative diagram

(ii)

$$0 \longrightarrow Z_n \otimes G \longrightarrow C_n \otimes G \xrightarrow{\ \partial_n \otimes \mathbb{1}\ } B_{n-1} \otimes G \longrightarrow 0$$

$$\downarrow \partial_n \otimes \mathbb{1} \qquad\qquad \downarrow \partial_n \otimes \mathbb{1} \qquad\qquad \downarrow \partial_{n-1} \otimes \mathbb{1}$$

$$0 \longrightarrow Z_{n-1} \otimes G \longrightarrow C_{n-1} \otimes G \xrightarrow{\ \partial_{n-1} \otimes \mathbb{1}\ } B_{n-2} \otimes G \longrightarrow 0$$

The rows are exact since the rows in (i) split and tensor products satisfy $(A \oplus B) \otimes G \approx A \otimes G \oplus B \otimes G$, so the rows in (ii) are split exact sequences too. Thus we have a short exact sequence of chain complexes $0 \to Z \otimes G \to C \otimes G \to B \otimes G \to 0$. Since the boundary maps are trivial in $Z \otimes G$ and $B \otimes G$, the associated long exact sequence of homology groups has the form

(iii) $\qquad \cdots \longrightarrow B_n \otimes G \longrightarrow Z_n \otimes G \longrightarrow H_n(C; G) \longrightarrow B_{n-1} \otimes G \longrightarrow Z_{n-1} \otimes G \longrightarrow \cdots$

The 'boundary' maps $B_n \otimes G \to Z_n \otimes G$ in this sequence are simply the maps $i_n \otimes \mathbb{1}$ where $i_n : B_n \to Z_n$ is the inclusion. This is evident from the definition of the boundary map in a long exact sequence of homology groups: In diagram (ii) one takes an element of $B_{n-1} \otimes G$, pulls it back via $(\partial_n \otimes \mathbb{1})^{-1}$ to $C_n \otimes G$, then applies $\partial_n \otimes \mathbb{1}$ to get into $C_{n-1} \otimes G$, then pulls back to $Z_{n-1} \otimes G$.

The long exact sequence (iii) can be broken up into short exact sequences

(iv) $\qquad\qquad 0 \longrightarrow \mathrm{Coker}(i_n \otimes \mathbb{1}) \longrightarrow H_n(C; G) \longrightarrow \mathrm{Ker}(i_{n-1} \otimes \mathbb{1}) \longrightarrow 0$

where $\mathrm{Coker}(i_n \otimes \mathbb{1}) = (Z_n \otimes G)/\mathrm{Im}(i_n \otimes \mathbb{1})$. The next lemma shows this cokernel is just $H_n(C) \otimes G$.

Lemma 3A.1. *If the sequence of abelian groups* $A \xrightarrow{\ i\ } B \xrightarrow{\ j\ } C \longrightarrow 0$ *is exact, then so is* $A \otimes G \xrightarrow{\ i \otimes \mathbb{1}\ } B \otimes G \xrightarrow{\ j \otimes \mathbb{1}\ } C \otimes G \longrightarrow 0$.

Proof: Certainly the compositions of two successive maps in the latter sequence are zero. Also, $j \otimes \mathbb{1}$ is clearly surjective since j is. To check exactness at $B \otimes G$ it suffices to show that the map $B \otimes G / \mathrm{Im}(i \otimes \mathbb{1}) \to C \otimes G$ induced by $j \otimes \mathbb{1}$ is an isomorphism, which we do by constructing its inverse. Define a map $\varphi : C \times G \to B \otimes G / \mathrm{Im}(i \otimes \mathbb{1})$ by $\varphi(c, g) = b \otimes g$ where $j(b) = c$. This φ is well-defined since if $j(b) = j(b') = c$ then $b - b' = i(a)$ for some $a \in A$ by exactness, so $b \otimes g - b' \otimes g = (b - b') \otimes g = i(a) \otimes g \in \mathrm{Im}(i \otimes \mathbb{1})$. Since φ is a homomorphism in each variable separately, it induces a homomorphism $C \otimes G \to B \otimes G / \mathrm{Im}(i \otimes \mathbb{1})$. This is clearly an inverse to the map $B \otimes G / \mathrm{Im}(i \otimes \mathbb{1}) \to C \otimes G$. $\qquad \square$

It remains to understand $\mathrm{Ker}(i_{n-1} \otimes \mathbb{1})$, or equivalently $\mathrm{Ker}(i_n \otimes \mathbb{1})$. The situation is that tensoring the short exact sequence

(v) $\qquad\qquad\qquad 0 \longrightarrow B_n \xrightarrow{\ i_n\ } Z_n \longrightarrow H_n(C) \longrightarrow 0$

with G produces a sequence which becomes exact only by insertion of the extra term $\mathrm{Ker}(i_n \otimes \mathbb{1})$:

(vi) $\qquad 0 \longrightarrow \mathrm{Ker}(i_n \otimes \mathbb{1}) \longrightarrow B_n \otimes G \xrightarrow{\ i_n \otimes \mathbb{1}\ } Z_n \otimes G \longrightarrow H_n(C) \otimes G \longrightarrow 0$

What we will show is that $\text{Ker}(i_n \otimes \mathbb{1})$ does not really depend on B_n and Z_n but only on their quotient $H_n(C)$, and of course G.

The sequence (v) is a free resolution of $H_n(C)$, where as in §3.1 a free resolution of an abelian group H is an exact sequence

$$\cdots \longrightarrow F_2 \xrightarrow{f_2} F_1 \xrightarrow{f_1} F_0 \xrightarrow{f_0} H \longrightarrow 0$$

with each F_n free. Tensoring a free resolution of this form with a fixed group G produces a chain complex

$$\cdots \longrightarrow F_1 \otimes G \xrightarrow{f_1 \otimes \mathbb{1}} F_0 \otimes G \xrightarrow{f_0 \otimes \mathbb{1}} H \otimes G \longrightarrow 0$$

By the preceding lemma this is exact at $F_0 \otimes G$ and $H \otimes G$, but to the left of these two terms it may not be exact. For the moment let us write $H_n(F \otimes G)$ for the homology group $\text{Ker}(f_n \otimes \mathbb{1}) / \text{Im}(f_{n+1} \otimes \mathbb{1})$.

Lemma 3A.2. *For any two free resolutions F and F' of H there are canonical isomorphisms $H_n(F \otimes G) \approx H_n(F' \otimes G)$ for all n.*

Proof: We will use Lemma 3.1(a). In the situation described there we have two free resolutions F and F' with a chain map between them. If we tensor the two free resolutions with G we obtain chain complexes $F \otimes G$ and $F' \otimes G$ with the maps $\alpha_n \otimes \mathbb{1}$ forming a chain map between them. Passing to homology, this chain map induces homomorphisms $\alpha_* : H_n(F \otimes G) \to H_n(F' \otimes G)$ which are independent of the choice of α_n's since if α_n and α_n' are chain homotopic via a chain homotopy λ_n then $\alpha_n \otimes \mathbb{1}$ and $\alpha_n' \otimes \mathbb{1}$ are chain homotopic via $\lambda_n \otimes \mathbb{1}$.

For a composition $H \xrightarrow{\alpha} H' \xrightarrow{\beta} H''$ with free resolutions F, F', and F'' of these three groups also given, the induced homomorphisms satisfy $(\beta\alpha)_* = \beta_* \alpha_*$ since we can choose for the chain map $F \to F''$ the composition of chain maps $F \to F' \to F''$. In particular, if we take α to be an isomorphism, with β its inverse and $F'' = F$, then $\beta_* \alpha_* = (\beta\alpha)_* = \mathbb{1}_* = \mathbb{1}$, and similarly with β and α reversed. So α_* is an isomorphism if α is an isomorphism. Specializing further, taking α to be the identity but with two different free resolutions F and F', we get a canonical isomorphism $\mathbb{1}_* : H_n(F \otimes G) \to H_n(F' \otimes G)$. $\square$

The group $H_n(F \otimes G)$, which depends only on H and G, is denoted $\text{Tor}_n(H, G)$. Since a free resolution $0 \to F_1 \to F_0 \to H \to 0$ always exists, as noted in §3.1, it follows that $\text{Tor}_n(H, G) = 0$ for $n > 1$. Usually $\text{Tor}_1(H, G)$ is written simply as $\text{Tor}(H, G)$. As we shall see later, $\text{Tor}(H, G)$ provides a measure of the common torsion of H and G, hence the name 'Tor.'

Is there a group $\text{Tor}_0(H, G)$? With the definition given above it would be zero since Lemma 3A.1 implies that $F_1 \otimes G \to F_0 \otimes G \to H \otimes G \to 0$ is exact. It is probably better to modify the definition of $H_n(F \otimes G)$ to be the homology groups of the sequence

$\cdots \to F_1 \otimes G \to F_0 \otimes G \to 0$, omitting the term $H \otimes G$ which can be regarded as a kind of augmentation. With this revised definition, Lemma 3A.1 then gives an isomorphism $\mathrm{Tor}_0(H, G) \approx H \otimes G$.

We should remark that $\mathrm{Tor}(H, G)$ is a functor of both G and H: Homomorphisms $\alpha: H \to H'$ and $\beta: G \to G'$ induce homomorphisms $\alpha_*: \mathrm{Tor}(H, G) \to \mathrm{Tor}(H', G)$ and $\beta_*: \mathrm{Tor}(H, G) \to \mathrm{Tor}(H, G')$, satisfying $(\alpha \alpha')_* = \alpha_* \alpha'_*$, $(\beta \beta')_* = \beta_* \beta'_*$, and $\mathbb{1}_* = \mathbb{1}$. The induced map α_* was constructed in the proof of Lemma 3A.2, while for β the construction of β_* is obvious.

Before going into calculations of $\mathrm{Tor}(H, G)$ let us finish analyzing the earlier exact sequence (iv). Recall that we have a chain complex C of free abelian groups, with homology groups denoted $H_n(C)$, and tensoring C with G gives another complex $C \otimes G$ whose homology groups are denoted $H_n(C; G)$. The following result is known as the **universal coefficient theorem for homology** since it describes homology with arbitrary coefficients in terms of homology with the 'universal' coefficient group $\mathbb{Z}$.

Theorem 3A.3. *If C is a chain complex of free abelian groups, then there are natural short exact sequences*

$$0 \longrightarrow H_n(C) \otimes G \longrightarrow H_n(C; G) \longrightarrow \mathrm{Tor}(H_{n-1}(C), G) \longrightarrow 0$$

for all n and all G, and these sequences split, though not naturally.

Naturality means that a chain map $C \to C'$ induces a map between the corresponding short exact sequences, with commuting squares.

Proof: The exact sequence in question is (iv) since we have shown that we can identify $\mathrm{Coker}(i_n \otimes \mathbb{1})$ with $H_n(C) \otimes G$ and $\mathrm{Ker}\, i_{n-1}$ with $\mathrm{Tor}(H_{n-1}(C), G)$. Verifying the naturality of this sequence is a mental exercise in definition-checking, left to the reader.

The splitting is obtained as follows. We observed earlier that the short exact sequence $0 \to Z_n \to C_n \to B_{n-1} \to 0$ splits, so there is a projection $p: C_n \to Z_n$ restricting to the identity on Z_n. The map p gives an extension of the quotient map $Z_n \to H_n(C)$ to a homomorphism $C_n \to H_n(C)$. Letting n vary, we then have a chain map $C \to H(C)$ where the groups $H_n(C)$ are regarded as a chain complex with trivial boundary maps, so the chain map condition is automatic. Now tensor with G to get a chain map $C \otimes G \to H(C) \otimes G$. Taking homology groups, we then have induced homomorphisms $H_n(C; G) \to H_n(C) \otimes G$ since the boundary maps in the chain complex $H(C) \otimes G$ are trivial. The homomorphisms $H_n(C; G) \to H_n(C) \otimes G$ give the desired splitting since at the level of chains they are the identity on cycles in C, by the definition of p. □

Corollary 3A.4. *For each pair of spaces (X, A) there are split exact sequences*

$$0 \longrightarrow H_n(X, A) \otimes G \longrightarrow H_n(X, A; G) \longrightarrow \mathrm{Tor}(H_{n-1}(X, A), G) \longrightarrow 0$$

for all n, and these sequences are natural with respect to maps $(X, A) \to (Y, B)$. □

The splitting is not natural, for if it were, a map $X \to Y$ that induced trivial maps $H_n(X) \to H_n(Y)$ and $H_{n-1}(X) \to H_{n-1}(Y)$ would have to induce the trivial map

$H_n(X;G) \to H_n(Y;G)$ for all G, but in Example 2.51 we saw an instance where this fails, namely the quotient map $M(\mathbb{Z}_m, n) \to S^{n+1}$ with $G = \mathbb{Z}_m$.

The basic tools for computing Tor are given by:

Proposition 3A.5.

(1) $\text{Tor}(A, B) \approx \text{Tor}(B, A)$.

(2) $\text{Tor}(\bigoplus_i A_i, B) \approx \bigoplus_i \text{Tor}(A_i, B)$.

(3) $\text{Tor}(A, B) = 0$ *if A or B is free, or more generally torsionfree.*

(4) $\text{Tor}(A, B) \approx \text{Tor}(T(A), B)$ *where $T(A)$ is the torsion subgroup of A.*

(5) $\text{Tor}(\mathbb{Z}_n, A) \approx \text{Ker}(A \xrightarrow{n} A)$.

(6) *For each short exact sequence $0 \to B \to C \to D \to 0$ there is a naturally associated exact sequence*
$$0 \to \text{Tor}(A, B) \to \text{Tor}(A, C) \to \text{Tor}(A, D) \to A \otimes B \to A \otimes C \to A \otimes D \to 0$$

Proof: Statement (2) is easy since one can choose as a free resolution of $\bigoplus_i A_i$ the direct sum of free resolutions of the A_i's. Also easy is (5), which comes from tensoring the free resolution $0 \to \mathbb{Z} \xrightarrow{n} \mathbb{Z} \to \mathbb{Z}_n \to 0$ with A.

For (3), if A is free, it has a free resolution with $F_n = 0$ for $n \geq 1$, so $\text{Tor}(A, B) = 0$ for all B. On the other hand, if B is free, then tensoring a free resolution of A with B preserves exactness, since tensoring a sequence with a direct sum of $\mathbb{Z}$'s produces just a direct sum of copies of the given sequence. So $\text{Tor}(A, B) = 0$ in this case too. The generalization to torsionfree A or B will be given below.

For (6), choose a free resolution $0 \to F_1 \to F_0 \to A \to 0$ and tensor with the given short exact sequence to get a commutative diagram

$$
\begin{array}{ccccccccc}
0 & \longrightarrow & F_1 \otimes B & \longrightarrow & F_1 \otimes C & \longrightarrow & F_1 \otimes D & \longrightarrow & 0 \\
 & & \downarrow & & \downarrow & & \downarrow & & \\
0 & \longrightarrow & F_0 \otimes B & \longrightarrow & F_0 \otimes C & \longrightarrow & F_0 \otimes D & \longrightarrow & 0
\end{array}
$$

The rows are exact since tensoring with a free group preserves exactness. Extending the three columns by zeros above and below, we then have a short exact sequence of chain complexes whose associated long exact sequence of homology groups is the desired six-term exact sequence.

To prove (1) we apply (6) to a free resolution $0 \to F_1 \to F_0 \to B \to 0$. Since $\text{Tor}(A, F_1)$ and $\text{Tor}(A, F_0)$ vanish by the part of (3) which we have proved, the six-term sequence in (6) reduces to the first row of the following diagram:

$$
\begin{array}{ccccccccc}
0 & \longrightarrow & \text{Tor}(A, B) & \longrightarrow & A \otimes F_1 & \longrightarrow & A \otimes F_0 & \longrightarrow & A \otimes B & \longrightarrow & 0 \\
 & & & & \downarrow \approx & & \downarrow \approx & & \downarrow \approx & & \\
0 & \longrightarrow & \text{Tor}(B, A) & \longrightarrow & F_1 \otimes A & \longrightarrow & F_0 \otimes A & \longrightarrow & B \otimes A & \longrightarrow & 0
\end{array}
$$

The second row comes from the definition of $\text{Tor}(B, A)$. The vertical isomorphisms come from the natural commutativity of tensor product. Since the squares commute, there is induced a map $\text{Tor}(A, B) \to \text{Tor}(B, A)$, which is an isomorphism by the five-lemma.

Now we can prove the statement (3) in the torsionfree case. For a free resolution $0 \longrightarrow F_1 \xrightarrow{\varphi} F_0 \longrightarrow A \longrightarrow 0$ we wish to show that $\varphi \otimes \mathbb{1} : F_1 \otimes B \to F_0 \otimes B$ is injective if B is torsionfree. Suppose $\sum_i x_i \otimes b_i$ lies in the kernel of $\varphi \otimes \mathbb{1}$. This means that $\sum_i \varphi(x_i) \otimes b_i$ can be reduced to 0 by a finite number of applications of the defining relations for tensor products. Only a finite number of elements of B are involved in this process. These lie in a finitely generated subgroup $B_0 \subset B$, so $\sum_i x_i \otimes b_i$ lies in the kernel of $\varphi \otimes \mathbb{1} : F_1 \otimes B_0 \to F_0 \otimes B_0$. This kernel is zero since $\mathrm{Tor}(A, B_0) = 0$, as B_0 is finitely generated and torsionfree, hence free.

Finally, we can obtain statement (4) by applying (6) to the short exact sequence $0 \to T(A) \to A \to A/T(A) \to 0$ since $A/T(A)$ is torsionfree. $\qquad\square$

In particular, (5) gives $\mathrm{Tor}(\mathbb{Z}_m, \mathbb{Z}_n) \approx \mathbb{Z}_q$ where q is the greatest common divisor of m and n. Thus $\mathrm{Tor}(\mathbb{Z}_m, \mathbb{Z}_n)$ is isomorphic to $\mathbb{Z}_m \otimes \mathbb{Z}_n$, though somewhat by accident. Combining this isomorphism with (2) and (3) we see that for finitely generated A and B, $\mathrm{Tor}(A, B)$ is isomorphic to the tensor product of the torsion subgroups of A and B, or roughly speaking, the common torsion of A and B. This is one reason for the 'Tor' designation, further justification being (3) and (4).

Homology calculations are often simplified by taking coefficients in a field, usually $\mathbb{Q}$ or $\mathbb{Z}_p$ for p prime. In general this gives less information than taking $\mathbb{Z}$ coefficients, but still some of the essential features are retained, as the following result indicates:

Corollary 3A.6. (a) $H_n(X;\mathbb{Q}) \approx H_n(X;\mathbb{Z}) \otimes \mathbb{Q}$, so when $H_n(X;\mathbb{Z})$ is finitely generated, the dimension of $H_n(X;\mathbb{Q})$ as a vector space over $\mathbb{Q}$ equals the rank of $H_n(X;\mathbb{Z})$.
(b) If $H_n(X;\mathbb{Z})$ and $H_{n-1}(X;\mathbb{Z})$ are finitely generated, then for p prime, $H_n(X;\mathbb{Z}_p)$ consists of
  (i) a $\mathbb{Z}_p$ summand for each $\mathbb{Z}$ summand of $H_n(X;\mathbb{Z})$,
  (ii) a $\mathbb{Z}_p$ summand for each $\mathbb{Z}_{p^k}$ summand in $H_n(X;\mathbb{Z})$, $k \geq 1$,
  (iii) a $\mathbb{Z}_p$ summand for each $\mathbb{Z}_{p^k}$ summand in $H_{n-1}(X;\mathbb{Z})$, $k \geq 1$. $\qquad\square$

Even in the case of nonfinitely generated homology groups, field coefficients still give good qualitative information:

Corollary 3A.7. (a) $\tilde{H}_n(X;\mathbb{Z}) = 0$ for all n iff $\tilde{H}_n(X;\mathbb{Q}) = 0$ and $\tilde{H}_n(X;\mathbb{Z}_p) = 0$ for all n and all primes p.
(b) A map $f : X \to Y$ induces isomorphisms on homology with $\mathbb{Z}$ coefficients iff it induces isomorphisms on homology with $\mathbb{Q}$ and $\mathbb{Z}_p$ coefficients for all primes p.

Proof: Statement (b) follows from (a) by passing to the mapping cone of f. The universal coefficient theorem gives the 'only if' half of (a). For the 'if' implication it suffices to show that if an abelian group A is such that $A \otimes \mathbb{Q} = 0$ and $\mathrm{Tor}(A, \mathbb{Z}_p) = 0$

for all primes p, then $A = 0$. For the short exact sequences $0 \to \mathbb{Z} \xrightarrow{p} \mathbb{Z} \to \mathbb{Z}_p \to 0$ and $0 \to \mathbb{Z} \to \mathbb{Q} \to \mathbb{Q}/\mathbb{Z} \to 0$, the six-term exact sequences in (6) of the proposition become

$$0 \longrightarrow \text{Tor}(A, \mathbb{Z}_p) \longrightarrow A \xrightarrow{p} A \longrightarrow A \otimes \mathbb{Z}_p \longrightarrow 0$$

$$0 \longrightarrow \text{Tor}(A, \mathbb{Q}/\mathbb{Z}) \longrightarrow A \longrightarrow A \otimes \mathbb{Q} \longrightarrow A \otimes \mathbb{Q}/\mathbb{Z} \longrightarrow 0$$

If $\text{Tor}(A, \mathbb{Z}_p) = 0$ for all p, then exactness of the first sequence implies that $A \xrightarrow{p} A$ is injective for all p, so A is torsionfree. Then $\text{Tor}(A, \mathbb{Q}/\mathbb{Z}) = 0$ by (3) or (4) of the proposition, so the second sequence implies that $A \to A \otimes \mathbb{Q}$ is injective, hence $A = 0$ if $A \otimes \mathbb{Q} = 0$. $\qquad\qquad\square$

The algebra by means of which the Tor functor is derived from tensor products has a very natural generalization in which abelian groups are replaced by modules over a fixed ring R with identity, using the definition of tensor product of R-modules given in §3.2. Free resolutions of R-modules are defined in the same way as for abelian groups, using free R-modules, which are direct sums of copies of R. Lemmas 3A.1 and 3A.2 carry over to this context without change, and so one has functors $\text{Tor}_n^R(A, B)$. However, it need not be true that $\text{Tor}_n^R(A, B) = 0$ for $n > 1$. The reason this was true when $R = \mathbb{Z}$ was that subgroups of free groups are free, but submodules of free R-modules need not be free in general. If R is a principal ideal domain, submodules of free R-modules are free, so in this case the rest of the algebra, in particular the universal coefficient theorem, goes through without change. When R is a field F, every module is free and $\text{Tor}_n^F(A, B) = 0$ for $n > 0$ via the free resolution $0 \to A \to A \to 0$. Thus $H_n(C \otimes_F G) \approx H_n(C) \otimes_F G$ if F is a field.

Exercises

1. Use the universal coefficient theorem to show that if $H_*(X; \mathbb{Z})$ is finitely generated, so the Euler characteristic $\chi(X) = \sum_n (-1)^n \text{rank } H_n(X; \mathbb{Z})$ is defined, then for any coefficient field F we have $\chi(X) = \sum_n (-1)^n \dim H_n(X; F)$.

2. Show that $\text{Tor}(A, \mathbb{Q}/\mathbb{Z})$ is isomorphic to the torsion subgroup of A. Deduce that A is torsionfree iff $\text{Tor}(A, B) = 0$ for all B.

3. Show that if $\widetilde{H}^n(X; \mathbb{Q})$ and $\widetilde{H}^n(X; \mathbb{Z}_p)$ are zero for all n and all primes p, then $\widetilde{H}_n(X; \mathbb{Z}) = 0$ for all n, and hence $\widetilde{H}^n(X; G) = 0$ for all G and n.

4. Show that $\otimes$ and Tor commute with direct limits: $(\varinjlim A_\alpha) \otimes B = \varinjlim (A_\alpha \otimes B)$ and $\text{Tor}(\varinjlim A_\alpha, B) = \varinjlim \text{Tor}(A_\alpha, B)$.

5. From the fact that $\text{Tor}(A, B) = 0$ if A is free, deduce that $\text{Tor}(A, B) = 0$ if A is torsionfree by applying the previous problem to the directed system of finitely generated subgroups A_α of A.

6. Show that $\text{Tor}(A, B)$ is always a torsion group, and that $\text{Tor}(A, B)$ contains an element of order n iff both A and B contain elements of order n.

3.B   The General Künneth Formula

Künneth formulas describe the homology or cohomology of a product space in terms of the homology or cohomology of the factors. In nice cases these formulas take the form $H_*(X \times Y; R) \approx H_*(X; R) \otimes H_*(Y; R)$ or $H^*(X \times Y; R) \approx H^*(X; R) \otimes H^*(Y; R)$ for a coefficient ring R. For the case of cohomology, such a formula was given in Theorem 3.15, with hypotheses of finite generation and freeness on the cohomology of one factor. To obtain a completely general formula without these hypotheses it turns out that homology is more natural than cohomology, and the main aim in this section is to derive the general Künneth formula for homology. The new feature of the general case is that an extra Tor term is needed to describe the full homology of a product.

The Cross Product in Homology

A major component of the Künneth formula is a **cross product** map

$$H_i(X; R) \times H_j(Y; R) \xrightarrow{\ \times\ } H_{i+j}(X \times Y; R)$$

There are two ways to define this. One is a direct definition for singular homology, involving explicit simplicial formulas. More enlightening, however, is the definition in terms of cellular homology. This necessitates assuming X and Y are CW complexes, but this hypothesis can later be removed by the technique of CW approximation in §4.1. We shall focus therefore on the cellular definition, leaving the simplicial definition to later in this section for those who are curious to see how it goes.

The key ingredient in the definition of the cellular cross product will be the fact that the cellular boundary map satisfies $d(e^i \times e^j) = de^i \times e^j + (-1)^i e^i \times de^j$. Implicit in the right side of this formula is the convention of treating the symbol $\times$ as a bilinear operation on cellular chains. With this convention we can then say more generally that $d(a \times b) = da \times b + (-1)^i a \times db$ whenever a is a cellular i-chain and b is a cellular j-chain. From this formula it is obvious that the cross product of two cycles is a cycle. Also, the product of a boundary and a cycle is a boundary since $da \times b = d(a \times b)$ if $db = 0$, and similarly $a \times db = (-1)^i d(a \times b)$ if $da = 0$. Hence there is an induced bilinear map $H_i(X; R) \times H_j(Y; R) \to H_{i+j}(X \times Y; R)$, which is by definition the cross product in cellular homology. Since it is bilinear, it could also be viewed as a homomorphism $H_i(X; R) \otimes_R H_j(Y; R) \to H_{i+j}(X \times Y; R)$. In either form, this cross product turns out to be independent of the cell structures on X and Y.

Our task then is to express the boundary maps in the cellular chain complex $C_*(X \times Y)$ for $X \times Y$ in terms of the boundary maps in the cellular chain complexes $C_*(X)$ and $C_*(Y)$. For simplicity we consider homology with $\mathbb{Z}$ coefficients here, but the same formula for arbitrary coefficients follows immediately from this special case. With $\mathbb{Z}$ coefficients, the cellular chain group $C_i(X)$ is free with basis the i-cells of X, but there is a sign ambiguity for the basis element corresponding to each cell e^i, namely the choice of a generator for the $\mathbb{Z}$ summand of $H_i(X^i, X^{i-1})$ corresponding to e^i. Only when $i = 0$ is this choice canonical. We refer to these choices as 'choosing orientations for the cells.' A choice of such orientations allows cellular i-chains to be written unambiguously as linear combinations of i-cells.

The formula $d(e^i \times e^j) = de^i \times e^j + (-1)^i e^i \times de^j$ is not completely canonical since it contains the sign $(-1)^i$ but not $(-1)^j$. Evidently there is some distinction being made between the two factors of $e^i \times e^j$. Since the signs arise from orientations, we need to make explicit how an orientation of cells e^i and e^j determines an orientation of $e^i \times e^j$. Via characteristic maps, orientations can be obtained from orientations of the domain disks of the characteristic maps. It will be convenient to choose these domains to be cubes since the product of two cubes is again a cube. Thus for a cell e_α^i we take a characteristic map $\Phi_\alpha : I^i \to X$ where I^i is the product of i intervals $[0,1]$. An orientation of I^i is a generator of $H_i(I^i, \partial I^i)$, and the image of this generator under $\Phi_{\alpha *}$ gives an orientation of e_α^i. We can identify $H_i(I^i, \partial I^i)$ with $H_i(I^i, I^i - \{x\})$ for any point x in the interior of I^i, and then an orientation is determined by a linear embedding $\Delta^i \to I^i$ with x chosen in the interior of the image of this embedding. The embedding is determined by its sequence of vertices $v_0, \cdots, v_i$. The vectors $v_1 - v_0, \cdots, v_i - v_0$ are linearly independent in I^i, thought of as the unit cube in $\mathbb{R}^i$, so an orientation in our sense is equivalent to an orientation in the sense of linear algebra, that is, an equivalence class of ordered bases, two ordered bases being equivalent if they differ by a linear transformation of positive determinant. (An ordered basis can be continuously deformed to an orthonormal basis, by the Gram–Schmidt process, and two orthonormal bases are related either by a rotation or a rotation followed by a reflection, according to the sign of the determinant of the transformation taking one to the other.)

With this in mind, we adopt the convention that an orientation of $I^i \times I^j = I^{i+j}$ is obtained by choosing an ordered basis consisting of an ordered basis for I^i followed by an ordered basis for I^j. Notice that reversing the orientation for either I^i or I^j then reverses the orientation for I^{i+j}, so all that really matters is the order of the two factors of $I^i \times I^j$.

Proposition 3B.1. *The boundary map in the cellular chain complex $C_*(X \times Y)$ is determined by the boundary maps in the cellular chain complexes $C_*(X)$ and $C_*(Y)$ via the formula $d(e^i \times e^j) = de^i \times e^j + (-1)^i e^i \times de^j$.*

Proof: Let us first consider the special case of the cube I^n. We give I the CW structure with two vertices and one edge, so the ith copy of I has a 1-cell e_i and 0-cells 0_i and 1_i, with $de_i = 1_i - 0_i$. The n-cell in the product I^n is $e_1 \times \cdots \times e_n$, and we claim that the boundary of this cell is given by the formula

$$(*) \qquad d(e_1 \times \cdots \times e_n) = \sum_i (-1)^{i+1} e_1 \times \cdots \times de_i \times \cdots \times e_n$$

This formula is correct modulo the signs of the individual terms $e_1 \times \cdots \times 0_i \times \cdots \times e_n$ and $e_1 \times \cdots \times 1_i \times \cdots \times e_n$ since these are exactly the $(n-1)$-cells in the boundary sphere ∂I^n of I^n. To obtain the signs in $(*)$, note that switching the two ends of an I factor of I^n produces a reflection of ∂I^n, as does a transposition of two adjacent I factors. Since reflections have degree -1, this implies that $(*)$ is correct up to an overall sign. This final sign can be determined by looking at any term, say the term $0_1 \times e_2 \times \cdots \times e_n$, which has a minus sign in $(*)$. To check that this is right, consider the n-simplex $[v_0, \cdots, v_n]$ with v_0 at the origin and v_k the unit vector along the kth coordinate axis for $k > 0$. This simplex defines the 'positive' orientation of I^n as described earlier, and in the usual formula for its boundary the face $[v_0, v_2, \cdots, v_n]$, which defines the positive orientation for the face $0_1 \times e_2 \times \cdots \times e_n$ of I^n, has a minus sign.

If we write $I^n = I^i \times I^j$ with $i + j = n$ and we set $e^i = e_1 \times \cdots \times e_i$ and $e^j = e_{i+1} \times \cdots \times e_n$, then the formula $(*)$ becomes $d(e^i \times e^j) = de^i \times e^j + (-1)^i e^i \times de^j$. We will use naturality to reduce the general case of the boundary formula to this special case. When dealing with cellular homology, the maps $f: X \to Y$ that induce chain maps $f_*: C_*(X) \to C_*(Y)$ of the cellular chain complexes are the *cellular maps*, taking X^n to Y^n for all n, hence (X^n, X^{n-1}) to (Y^n, Y^{n-1}). The naturality statement we want is then:

> **Lemma 3B.2.** *For cellular maps $f: X \to Z$ and $g: Y \to W$, the cellular chain maps $f_*: C_*(X) \to C_*(Z)$, $g_*: C_*(Y) \to C_*(W)$, and $(f \times g)_*: C_*(X \times Y) \to C_*(Z \times W)$ are related by the formula $(f \times g)_* = f_* \times g_*$.*

Proof: The relation $(f \times g)_* = f_* \times g_*$ means that if $f_*(e_\alpha^i) = \sum_\gamma m_{\alpha\gamma} e_\gamma^i$ and if $g_*(e_\beta^j) = \sum_\delta n_{\beta\delta} e_\delta^j$, then $(f \times g)_*(e_\alpha^i \times e_\beta^j) = \sum_{\gamma\delta} m_{\alpha\gamma} n_{\beta\delta}(e_\gamma^i \times e_\delta^j)$. The coefficient $m_{\alpha\gamma}$ is the degree of the composition $f_{\alpha\gamma}: S^i \to X^i/X^{i-1} \to Z^i/Z^{i-1} \to S^i$ where the first and third maps are induced by characteristic maps for the cells e_α^i and e_γ^i, and the middle map is induced by the cellular map f. With the natural choices of basepoints in these quotient spaces, $f_{\alpha\gamma}$ is basepoint-preserving. The $n_{\beta\delta}$'s are obtained similarly from maps $g_{\beta\delta}: S^j \to S^j$. For $f \times g$, the map $(f \times g)_{\alpha\beta, \gamma\delta}: S^{i+j} \to S^{i+j}$ whose degree is the coefficient of $e_\gamma^i \times e_\delta^j$ in $(f \times g)_*(e_\alpha^i \times e_\beta^j)$ is obtained from the product map $f_{\alpha\gamma} \times g_{\beta\delta}: S^i \times S^j \to S^i \times S^j$ by collapsing the $(i + j - 1)$-skeleton of $S^i \times S^j$ to a point. In other words, $(f \times g)_{\alpha\beta, \gamma\delta}$ is the smash product map $f_{\alpha\gamma} \wedge g_{\beta\delta}$. What we need to show is the formula $\deg(f \wedge g) = \deg(f)\deg(g)$ for basepoint-preserving maps $f: S^i \to S^i$ and $g: S^j \to S^j$.

Since $f \wedge g$ is the composition of $f \wedge \mathbb{1}$ and $\mathbb{1} \wedge g$, it suffices to show that $\deg(f \wedge \mathbb{1}) = \deg(f)$ and $\deg(\mathbb{1} \wedge g) = \deg(g)$. We do this by relating smash products to suspension. The smash product $X \wedge S^1$ can be viewed as $X \times I/(X \times \partial I \cup \{x_0\} \times I)$, so it is the reduced suspension ΣX, the quotient of the ordinary suspension SX obtained by collapsing the segment $\{x_0\} \times I$ to a point. If X is a CW complex with x_0 a 0-cell, the quotient map $SX \to X \wedge S^1$ induces an isomorphism on homology since it collapses a contractible subcomplex to a point. Taking $X = S^i$, we have the commutative diagram at the right, and from the induced commutative diagram of homology groups H_{i+1} we deduce that Sf and $f \wedge 1$ have the same degree. Since

$$S(S^i) \xrightarrow{Sf} S(S^i)$$
$$\downarrow \qquad\qquad \downarrow$$
$$S^i \wedge S^1 \xrightarrow{f \wedge 1} S^i \wedge S^1$$

suspension preserves degree by Proposition 2.33, we conclude that $\deg(f \wedge 1) = \deg(f)$. The 1 in this formula is the identity map on S^1, and by iteration we obtain the same result for 1 the identity map on S^j since S^j is the smash product of j copies of S^1. This implies also that $\deg(1 \wedge g) = \deg(g)$ since a permutation of coordinates in S^{i+j} does not affect the degree of maps $S^{i+j} \to S^{i+j}$. $\qquad\square$

Now to finish the proof of the proposition, let $\Phi : I^i \to X^i$ and $\Psi : I^j \to Y^j$ be characteristic maps of cells $e^i_\alpha \subset X$ and $e^j_\beta \subset Y$. The restriction of Φ to ∂I^i is the attaching map of e^i_α. We may perform a preliminary homotopy of this attaching map $\partial I^i \to X^{i-1}$ to make it cellular. There is no need to appeal to the cellular approximation theorem to do this since a direct argument is easy: First deform the attaching map so that it sends all but one face of I^i to a point, which is possible since the union of these faces is contractible, then do a further deformation so that the image point of this union of faces is a 0-cell. A homotopy of the attaching map $\partial I^i \to X^{i-1}$ does not affect the cellular boundary de^i_α, since de^i_α is determined by the induced map $H_{i-1}(\partial I^i) \to H_{i-1}(X^{i-1}) \to H_{i-1}(X^{i-1}, X^{i-2})$. So we may assume Φ is cellular, and likewise Ψ, hence also $\Phi \times \Psi$. The map of cellular chain complexes induced by a cellular map between CW complexes is a chain map, commuting with the cellular boundary maps.

If e^i is the i-cell of I^i and e^j the j-cell of I^j, then $\Phi_*(e^i) = e^i_\alpha$, $\Psi_*(e^j) = e^j_\beta$, and $(\Phi \times \Psi)_*(e^i \times e^j) = e^i_\alpha \times e^j_\beta$, hence

$$
\begin{aligned}
d(e^i_\alpha \times e^j_\beta) &= d\big((\Phi \times \Psi)_*(e^i \times e^j)\big) \\
&= (\Phi \times \Psi)_* d(e^i \times e^j) \qquad \text{since } (\Phi \times \Psi)_* \text{ is a chain map} \\
&= (\Phi \times \Psi)_* \big(de^i \times e^j + (-1)^i e^i \times de^j\big) \qquad \text{by the special case} \\
&= \Phi_*(de^i) \times \Psi_*(e^j) + (-1)^i \Phi_*(e^i) \times \Psi_*(de^j) \qquad \text{by the lemma} \\
&= d\Phi_*(e^i) \times \Psi_*(e^j) + (-1)^i \Phi_*(e^i) \times d\Psi_*(e^j) \qquad \begin{array}{l}\text{since } \Phi_* \text{ and } \Psi_* \text{ are} \\ \text{chain maps}\end{array} \\
&= de^i_\alpha \times e^j_\beta + (-1)^i e^i_\alpha \times de^j_\beta
\end{aligned}
$$

which completes the proof of the proposition. $\qquad\square$

Example 3B.3. Consider $X \times S^k$ where we give S^k its usual CW structure with two cells. The boundary formula in $C_*(X \times S^k)$ takes the form $d(a \times b) = da \times b$ since $d = 0$ in $C_*(S^k)$. So the chain complex $C_*(X \times S^k)$ is just the direct sum of two copies of the chain complex $C_*(X)$, one of the copies having its dimension shifted upward by k. Hence $H_n(X \times S^k; \mathbb{Z}) \approx H_n(X; \mathbb{Z}) \oplus H_{n-k}(X; \mathbb{Z})$ for all n. In particular, we see that all the homology classes in $X \times S^k$ are cross products of homology classes in X and S^k.

Example 3B.4. More subtle things can happen when X and Y both have torsion in their homology. To take the simplest case, let X be S^1 with a cell e^2 attached by a map $S^1 \to S^1$ of degree m, so $H_1(X; \mathbb{Z}) \approx \mathbb{Z}_m$ and $H_i(X; \mathbb{Z}) = 0$ for $i > 1$. Similarly, let Y be obtained from S^1 by attaching a 2-cell by a map of degree n. Thus X and Y each have CW structures with three cells and so $X \times Y$ has nine cells. These are indicated by the dots in the diagram at the right, with X in the horizontal direction and Y in the vertical direction. The arrows denote the nonzero cellular boundary maps. For example the two arrows leaving the dot in the upper right corner indi- cate that $\partial(e^2 \times e^2) = m(e^1 \times e^2) + n(e^2 \times e^1)$. Obviously

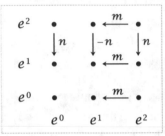

$H_1(X \times Y; \mathbb{Z})$ is $\mathbb{Z}_m \oplus \mathbb{Z}_n$. In dimension 2, $\mathrm{Ker}\,\partial$ is generated by $e^1 \times e^1$, and the image of the boundary map from dimension 3 consists of the multiples $(\ell m - kn)(e^1 \times e^1)$. These form a cyclic group generated by $q(e^1 \times e^1)$ where q is the greatest common divisor of m and n, so $H_2(X \times Y; \mathbb{Z}) \approx \mathbb{Z}_q$. In dimension 3 the cycles are the multiples of $(m/q)(e^1 \times e^2) + (n/q)(e^2 \times e^1)$, and the smallest such multiple that is a boundary is $q[(m/q)(e^1 \times e^2) + (n/q)(e^2 \times e^1)] = m(e^1 \times e^2) + n(e^2 \times e^1)$, so $H_3(X \times Y; \mathbb{Z}) \approx \mathbb{Z}_q$. Since X and Y have no homology above dimension 1, this 3-dimensional homol- ogy of $X \times Y$ cannot be realized by cross products. As the general theory will show, $H_2(X \times Y; \mathbb{Z})$ is $H_1(X; \mathbb{Z}) \otimes H_1(Y; \mathbb{Z})$ and $H_3(X \times Y; \mathbb{Z})$ is $\mathrm{Tor}(H_1(X; \mathbb{Z}), H_1(Y; \mathbb{Z}))$.

This example generalizes easily to higher dimensions, with $X = S^i \cup e^{i+1}$ and $Y = S^j \cup e^{j+1}$, the attaching maps having degrees m and n, respectively. Essentially the same calculation shows that $X \times Y$ has both H_{i+j} and H_{i+j+1} isomorphic to $\mathbb{Z}_q$.

We should say a few words about why the cross product is independent of CW structures. For this we will need a fact proved in the next chapter in Theorem 4.8, that every map between CW complexes is homotopic to a cellular map. As we mentioned earlier, a cellular map induces a chain map between cellular chain complexes. It is easy to see from the equivalence between cellular and singular homology that the map on cellular homology induced by a cellular map is the same as the map induced on singular homology. Now suppose we have cellular maps $f : X \to Z$ and $g : Y \to W$. Then Lemma 3B.2 implies that we have a commutative diagram

$$
\begin{array}{ccc}
H_i(X; \mathbb{Z}) \times H_j(Y; \mathbb{Z}) & \xrightarrow{\times} & H_{i+j}(X \times Y; \mathbb{Z}) \\
\downarrow{\scriptstyle f_* \times g_*} & & \downarrow{\scriptstyle (f \times g)_*} \\
H_i(Z; \mathbb{Z}) \times H_j(W; \mathbb{Z}) & \xrightarrow{\times} & H_{i+j}(Z \times W; \mathbb{Z})
\end{array}
$$

Now take Z and W to be the same spaces as X and Y but with different CW structures, and let f and g be cellular maps homotopic to the identity. The vertical maps in the diagram are then the identity, and commutativity of the diagram says that the cross products defined using the different CW structures coincide.

Cross product is obviously bilinear, or in other words, distributive. It is not hard to check that it is also associative. What about commutativity? If $T:X \times Y \to Y \times X$ is transposition of the factors, then we can ask whether $T_*(a \times b)$ equals $b \times a$. The only effect transposing the factors has on the definition of cross product is in the convention for orienting a product $I^i \times I^j$ by taking an ordered basis in the first factor followed by an ordered basis in the second factor. Switching the two factors can be achieved by moving each of the i coordinates of I^i past each of the coordinates of I^j. This is a total of ij transpositions of adjacent coordinates, each realizable by a reflection, so a sign of $(-1)^{ij}$ is introduced. Thus the correct formula is $T_*(a \times b) = (-1)^{ij} b \times a$ for $a \in H_i(X)$ and $b \in H_j(Y)$.

The Algebraic Künneth Formula

By adding together the various cross products we obtain a map

$$\bigoplus_i (H_i(X;\mathbb{Z}) \otimes H_{n-i}(Y;\mathbb{Z})) \longrightarrow H_n(X \times Y;\mathbb{Z})$$

and it is natural to ask whether this is an isomorphism. Example 3B.4 above shows that this is not always the case, though it is true in Example 3B.3. Our main goal in what follows is to show that the map is always injective, and that its cokernel is $\bigoplus_i \mathrm{Tor}(H_i(X;\mathbb{Z}), H_{n-i-1}(Y;\mathbb{Z}))$. More generally, we consider other coefficients besides $\mathbb{Z}$ and show in particular that with field coefficients the map is an isomorphism.

For CW complexes X and Y, the relationship between the cellular chain complexes $C_*(X)$, $C_*(Y)$, and $C_*(X \times Y)$ can be expressed nicely in terms of tensor products. Since the n-cells of $X \times Y$ are the products of i-cells of X with $(n-i)$-cells of Y, we have $C_n(X \times Y) \approx \bigoplus_i (C_i(X) \otimes C_{n-i}(Y))$, with $e^i \times e^j$ corresponding to $e^i \otimes e^j$. Under this identification the boundary formula of Proposition 3B.1 becomes $d(e^i \otimes e^j) = de^i \otimes e^j + (-1)^i e^i \otimes de^j$. Our task now is purely algebraic, to compute the homology of the chain complex $C_*(X \times Y)$ from the homology of $C_*(X)$ and $C_*(Y)$.

Suppose we are given chain complexes C and C' of abelian groups C_n and C'_n, or more generally R-modules over a commutative ring R. The **tensor product chain complex** $C \otimes_R C'$ is then defined by $(C \otimes_R C')_n = \bigoplus_i (C_i \otimes_R C'_{n-i})$, with boundary maps given by $\partial(c \otimes c') = \partial c \otimes c' + (-1)^i c \otimes \partial c'$ for $c \in C_i$ and $c' \in C'_{n-i}$. The sign $(-1)^i$ guarantees that $\partial^2 = 0$ in $C \otimes_R C'$, since

$$\partial^2(c \otimes c') = \partial(\partial c \otimes c' + (-1)^i c \otimes \partial c')$$
$$= \partial^2 c \otimes c' + (-1)^{i-1} \partial c \otimes \partial c' + (-1)^i \partial c \otimes \partial c' + c \otimes \partial^2 c' = 0$$

From the boundary formula $\partial(c \otimes c') = \partial c \otimes c' + (-1)^i c \otimes \partial c'$ it follows that the tensor product of cycles is a cycle, and the tensor product of a cycle and a boundary, in either order, is a boundary, just as for the cross product defined earlier. So there is induced a natural map on homology groups $H_i(C) \otimes_R H_{n-i}(C') \to H_n(C \otimes_R C')$. Summing over i

then gives a map $\bigoplus_i (H_i(C) \otimes_R H_{n-i}(C')) \to H_n(C \otimes_R C')$. This figures in the following algebraic version of the Künneth formula:

> $\mathbf{Theorem}$ **3B.5.** *If R is a principal ideal domain and the R-modules C_i are free, then for each n there is a natural short exact sequence*
>
> $$0 \to \bigoplus_i (H_i(C) \otimes_R H_{n-i}(C')) \to H_n(C \otimes_R C') \to \bigoplus_i (\mathrm{Tor}_R(H_i(C), H_{n-i-1}(C'))) \to 0$$
>
> *and this sequence splits.*

This is a generalization of the universal coefficient theorem for homology, which is the case that C' consists of just the coefficient group G in dimension zero. The proof will also be a natural generalization of the proof of the universal coefficient theorem.

Proof: First we do the special case that the boundary maps in C are all zero, so $H_i(C) = C_i$. In this case $\partial(c \otimes c') = (-1)^i c \otimes \partial c'$ and the chain complex $C \otimes_R C'$ is simply the direct sum of the complexes $C_i \otimes_R C'$, each of which is a direct sum of copies of C' since C_i is free. Hence $H_n(C_i \otimes_R C') \approx C_i \otimes_R H_{n-i}(C') = H_i(C) \otimes_R H_{n-i}(C')$. Summing over i yields an isomorphism $H_n(C \otimes_R C') \approx \bigoplus_i (H_i(C) \otimes_R H_{n-i}(C'))$, which is the statement of the theorem since there are no Tor terms, $H_i(C) = C_i$ being free.

In the general case, let $Z_i \subset C_i$ and $B_i \subset C_i$ denote kernel and image of the boundary homomorphisms for C. These give subchain complexes Z and B of C with trivial boundary maps. We have a short exact sequence of chain complexes $0 \to Z \to C \to B \to 0$ made up of the short exact sequences $0 \to Z_i \to C_i \xrightarrow{\partial} B_{i-1} \to 0$ each of which splits since B_{i-1} is free, being a submodule of C_{i-1} which is free by assumption. Because of the splitting, when we tensor $0 \to Z \to C \to B \to 0$ with C' we obtain another short exact sequence of chain complexes, and hence a long exact sequence in homology

$$\cdots \to H_n(Z \otimes_R C') \to H_n(C \otimes_R C') \to H_{n-1}(B \otimes_R C') \to H_{n-1}(Z \otimes_R C') \to \cdots$$

where we have $H_{n-1}(B \otimes_R C')$ instead of the expected $H_n(B \otimes_R C')$ since $\partial : C \to B$ decreases dimension by one. Checking definitions, one sees that the 'boundary' map $H_{n-1}(B \otimes_R C') \to H_{n-1}(Z \otimes_R C')$ in the preceding long exact sequence is just the map induced by the natural map $B \otimes_R C' \to Z \otimes_R C'$ coming from the inclusion $B \subset Z$.

Since Z and B are chain complexes with trivial boundary maps, the special case at the beginning of the proof converts the preceding exact sequence into

$$\cdots \xrightarrow{i_n} \bigoplus_i (Z_i \otimes_R H_{n-i}(C')) \to H_n(C \otimes_R C') \to \bigoplus_i (B_i \otimes_R H_{n-i-1}(C')) \xrightarrow{i_{n-1}}$$
$$\bigoplus_i (Z_i \otimes_R H_{n-i-1}(C')) \to \cdots$$

So we have short exact sequences

$$0 \to \mathrm{Coker}\, i_n \to H_n(C \otimes_R C') \to \mathrm{Ker}\, i_{n-1} \to 0$$

where $\operatorname{Coker} i_n = \bigoplus_i (Z_i \otimes_R H_{n-i}(C'))/\operatorname{Im} i_n$, and this equals $\bigoplus_i (H_i(C) \otimes_R H_{n-i}(C'))$ by Lemma 3A.1. It remains to identify $\operatorname{Ker} i_{n-1}$ with $\bigoplus_i \operatorname{Tor}_R(H_i(C), H_{n-i}(C'))$.

By the definition of Tor, tensoring the free resolution $0 \to B_i \to Z_i \to H_i(C) \to 0$ with $H_{n-i}(C')$ yields an exact sequence

$$0 \longrightarrow \operatorname{Tor}_R(H_i(C), H_{n-i}(C')) \longrightarrow B_i \otimes_R H_{n-i}(C') \longrightarrow Z_i \otimes_R H_{n-i}(C') \longrightarrow$$
$$H_i(C) \otimes_R H_{n-i}(C') \longrightarrow 0$$

Hence, summing over i, $\operatorname{Ker} i_n = \bigoplus_i \operatorname{Tor}_R(H_i(C), H_{n-i}(C'))$.

Naturality should be obvious, and we leave it for the reader to fill in the details.

We will show that the short exact sequence in the statement of the theorem splits assuming that both C and C' are free. This suffices for our applications. For the extra argument needed to show splitting when C' is not free, see the exposition in [Hilton & Stammbach 1970].

The splitting is via a homomorphism $H_n(C \otimes_R C') \to \bigoplus_i (H_i(C) \otimes_R H_{n-i}(C'))$ constructed in the following way. As already noted, the sequence $0 \to Z_i \to C_i \to B_{i-1} \to 0$ splits, so the quotient maps $Z_i \to H_i(C)$ extend to homomorphisms $C_i \to H_i(C)$. Similarly we obtain $C_j' \to H_j(C')$ if C' is free. Viewing the sequences of homology groups $H_i(C)$ and $H_j(C')$ as chain complexes $H(C)$ and $H(C')$ with trivial boundary maps, we thus have chain maps $C \to H(C)$ and $C' \to H(C')$, whose tensor product is a chain map $C \otimes_R C' \to H(C) \otimes_R H(C')$. The induced map on homology for this last chain map is the desired splitting map since the chain complex $H(C) \otimes_R H(C')$ equals its own homology, the boundary maps being trivial.   □

The Topological Künneth Formula

Now we can apply the preceding algebra to obtain the topological statement we are looking for:

Theorem 3B.6. *If X and Y are CW complexes and R is a principal ideal domain, then there are natural short exact sequences*

$$0 \longrightarrow \bigoplus_i (H_i(X;R) \otimes_R H_{n-i}(Y;R)) \longrightarrow H_n(X \times Y; R) \longrightarrow$$
$$\bigoplus_i \operatorname{Tor}_R(H_i(X;R), H_{n-i-1}(Y;R)) \longrightarrow 0$$

and these sequences split.

Naturality means that maps $X \to X'$ and $Y \to Y'$ induce a map from the short exact sequence for $X \times Y$ to the corresponding short exact sequence for $X' \times Y'$, with commuting squares. The splitting is not natural, however, as an exercise at the end of this section demonstrates.

Proof: When dealing with products of CW complexes there is always the bothersome fact that the compactly generated CW topology may not be the same as the product topology. However, in the present context this is not a real problem. Since the two topologies have the same compact sets, they have the same singular simplices and hence the same singular homology groups.

Let $C = C_*(X;R)$ and $C' = C_*(Y;R)$, the cellular chain complexes with coefficients in R. Then $C \otimes_R C' = C_*(X \times Y;R)$ by Proposition 3B.1, so the algebraic Künneth formula gives the desired short exact sequences. Their naturality follows from naturality in the algebraic Künneth formula, since we can homotope arbitrary maps $X \to X'$ and $Y \to Y'$ to be cellular by Theorem 4.8, assuring that they induce chain maps of cellular chain complexes. $\qquad\square$

With field coefficients the Künneth formula simplifies because the Tor terms are always zero over a field:

Corollary 3B.7. *If F is a field and X and Y are CW complexes, then the cross product map $h : \bigoplus_i (H_i(X;F) \otimes_F H_{n-i}(Y;F)) \to H_n(X \times Y;F)$ is an isomorphism for all n.* $\quad\square$

There is also a relative version of the Künneth formula for CW pairs (X, A) and (Y, B). This is a split short exact sequence

$$0 \to \bigoplus_i (H_i(X,A;R) \otimes_R H_{n-i}(Y,B;R)) \to H_n(X \times Y, A \times Y \cup X \times B;R) \to$$
$$\bigoplus_i \mathrm{Tor}_R(H_i(X,A;R), H_{n-i-1}(Y,B;R)) \to 0$$

for R a principal ideal domain. This too follows from the algebraic Künneth formula since the isomorphism of cellular chain complexes $C_*(X \times Y) \approx C_*(X) \otimes C_*(Y)$ passes down to a quotient isomorphism

$$C_*(X \times Y)/C_*(A \times Y \cup X \times B) \approx C_*(X)/C_*(A) \otimes C_*(Y)/C_*(B)$$

since bases for these three relative cellular chain complexes correspond bijectively with the cells of $(X - A) \times (Y - B)$, $X - A$, and $Y - B$, respectively.

As a special case, suppose A and B are basepoints $x_0 \in X$ and $y_0 \in Y$. Then the subcomplex $A \times Y \cup X \times B$ can be identified with the wedge sum $X \vee Y$ and the quotient $X \times Y / X \vee Y$ is the smash product $X \wedge Y$. Thus we have a reduced Künneth formula

$$0 \to \bigoplus_i (\tilde{H}_i(X;R) \otimes_R \tilde{H}_{n-i}(Y;R)) \to \tilde{H}_n(X \wedge Y;R) \to$$
$$\bigoplus_i \mathrm{Tor}_R(\tilde{H}_i(X;R), \tilde{H}_{n-i-1}(Y;R)) \to 0$$

If we take $Y = S^k$ for example, then $X \wedge S^k$ is the k-fold reduced suspension of X, and we obtain isomorphisms $\tilde{H}_n(X;R) \approx \tilde{H}_{n+k}(X \wedge S^k;R)$.

The Künneth formula and the universal coefficient theorem can be combined to give a more concise formula $H_n(X \times Y;R) \approx \bigoplus_i H_i(X;H_{n-i}(Y;R))$. The naturality of this isomorphism is somewhat problematic, however, since it uses the splittings in the Künneth formula and universal coefficient theorem. With a little more algebra the formula can be shown to hold more generally for an arbitrary coefficient group G in place of R; see [Hilton & Wylie 1967], p. 227.

There is an analogous formula $\tilde{H}_n(X \wedge Y; R) \approx \bigoplus_i \tilde{H}_i(X; \tilde{H}_{n-i}(Y; R))$. As a special case, when Y is a Moore space $M(G, k)$ we obtain isomorphisms $\tilde{H}_n(X; G) \approx \tilde{H}_{n+k}(X \wedge M(G, k); \mathbb{Z})$. Again naturality is an issue, but in this case there is a natural isomorphism obtainable by applying Theorem 4.59 in §4.3, after verifying that the functors $h_n(X) = \tilde{H}_{n+k}(X \wedge M(G, k); \mathbb{Z})$ define a reduced homology theory, which is not hard. The isomorphism $\tilde{H}_n(X; G) \approx \tilde{H}_{n+k}(X \wedge M(G, k); \mathbb{Z})$ says that homology with arbitrary coefficients can be obtained from homology with $\mathbb{Z}$ coefficients by a topological construction as well as by the algebra of tensor products. For general homology theories this formula can be used as a definition of homology with coefficients.

One might wonder about a cohomology version of the Künneth formula. Taking coefficients in a field F and using the natural isomorphism $\mathrm{Hom}(A \otimes B, C) \approx \mathrm{Hom}(A, \mathrm{Hom}(B, C))$, the Künneth formula for homology and the universal coefficient theorem give isomorphisms

$$H^n(X \times Y; F) \approx \mathrm{Hom}_F(H_n(X \times Y; F), F) \approx \bigoplus_i \mathrm{Hom}_F(H_i(X; F) \otimes H_{n-i}(Y; F), F)$$
$$\approx \bigoplus_i \mathrm{Hom}_F(H_i(X; F), \mathrm{Hom}_F(H_{n-i}(Y; F), F))$$
$$\approx \bigoplus_i \mathrm{Hom}_F(H_i(X; F), H^{n-i}(Y; F))$$
$$\approx \bigoplus_i H^i(X; H^{n-i}(Y; F))$$

More generally, there are isomorphisms $H^n(X \times Y; G) \approx \bigoplus_i H^i(X; H^{n-i}(Y; G))$ for any coefficient group G; see [Hilton & Wylie 1967], p. 227. However, in practice it usually suffices to apply the Künneth formula for homology and the universal coefficient theorem for cohomology separately. Also, Theorem 3.15 shows that with stronger hypotheses one can draw stronger conclusions using cup products.

The Simplicial Cross Product

Let us sketch how the cross product $H_m(X; R) \otimes H_n(Y; R) \to H_{m+n}(X \times Y; R)$ can be defined directly in terms of singular homology. What one wants is a cross product at the level of singular chains, $C_m(X; R) \otimes C_n(Y; R) \to C_{m+n}(X \times Y; R)$. If we are given singular simplices $f : \Delta^m \to X$ and $g : \Delta^n \to Y$, then we have the product map $f \times g : \Delta^m \times \Delta^n \to X \times Y$, and the idea is to subdivide $\Delta^m \times \Delta^n$ into simplices of dimension $m + n$ and then take the sum of the restrictions of $f \times g$ to these simplices, with appropriate signs.

In the special cases that m or n is 1 we have already seen how to subdivide $\Delta^m \times \Delta^n$ into simplices when we constructed prism operators in §2.1. The generalization to $\Delta^m \times \Delta^n$ is not completely obvious, however. Label the vertices of Δ^m as $v_0, v_1, \cdots, v_m$ and the vertices of Δ^n as $w_0, w_1, \cdots, w_n$. Think of the pairs (i, j) with $0 \le i \le m$ and $0 \le j \le n$ as the vertices of an $m \times n$ rectangular grid in $\mathbb{R}^2$. Let σ be a path formed by a sequence of $m + n$ horizontal and vertical edges in this grid starting at $(0, 0)$ and ending at (m, n), always moving either to the right or upward. To such a path σ we associate a linear map $\ell_\sigma : \Delta^{m+n} \to \Delta^m \times \Delta^n$ sending the kth vertex of Δ^{m+n} to (v_{i_k}, w_{j_k}) where (i_k, j_k) is the kth vertex of the edgepath σ. Then we define a simplicial cross product

$$C_m(X;R) \otimes C_n(Y;R) \xrightarrow{\ \times\ } C_{m+n}(X \times Y;R)$$

by the formula

$$f \times g = \sum_\sigma (-1)^{|\sigma|} (f \times g) \ell_\sigma$$

where $|\sigma|$ is the number of squares in the grid lying below the path σ. Note that the symbol '$\times$' means different things on the two sides of the equation. From this definition it is a calculation to show that $\partial(f \times g) = \partial f \times g + (-1)^m f \times \partial g$. This implies that the cross product of two cycles is a cycle, and the cross product of a cycle and a boundary is a boundary, so there is an induced cross product in singular homology.

One can see that the images of the maps ℓ_σ give a simplicial structure on $\Delta^m \times \Delta^n$ in the following way. We can view Δ^m as the subspace of $\mathbb{R}^m$ defined by the inequalities $0 \le x_1 \le \cdots \le x_m \le 1$, with the vertex v_i as the point having coordinates $m - i$ zeros followed by i ones. Similarly we have $\Delta^n \subset \mathbb{R}^n$ with coordinates $0 \le y_1 \le \cdots \le y_n \le 1$. The product $\Delta^m \times \Delta^n$ then consists of $(m + n)$-tuples $(x_1, \cdots, x_m, y_1, \cdots, y_n)$ satisfying both sets of inequalities. The combined inequalities $0 \le x_1 \le \cdots \le x_m \le y_1 \le \cdots \le y_n \le 1$ define a simplex Δ^{m+n} in $\Delta^m \times \Delta^n$, and every other point of $\Delta^m \times \Delta^n$ satisfies a similar set of inequalities obtained from $0 \le x_1 \le \cdots \le x_m \le y_1 \le \cdots \le y_n \le 1$ by a permutation of the variables 'shuffling' the y_j's into the x_i's. Each such shuffle corresponds to an edgepath σ consisting of a rightward edge for each x_i and an upward edge for each y_j in the shuffled sequence. Thus we have $\Delta^m \times \Delta^n$ expressed as the union of simplices Δ_σ^{m+n} indexed by the edgepaths σ. One can check that these simplices fit together nicely to form a Δ-complex structure on $\Delta^m \times \Delta^n$, which is also a simplicial complex structure. See [Eilenberg & Steenrod 1952], p. 68. In fact this construction is sufficiently natural to make the product of any two Δ-complexes into a Δ-complex.

The Cohomology Cross Product

In §3.2 we defined a cross product

$$H^k(X;R) \times H^\ell(Y;R) \xrightarrow{\ \times\ } H^{k+\ell}(X \times Y;R)$$

in terms of the cup product. Let us now describe the alternative approach in which the cross product is defined directly via cellular cohomology, and then cup product is defined in terms of this cross product.

The cellular definition of cohomology cross product is very much like the definition in homology. Given CW complexes X and Y, define a cross product of cellular cochains $\varphi \in C^k(X;R)$ and $\psi \in C^\ell(Y;R)$ by setting

$$(\varphi \times \psi)(e_\alpha^k \times e_\beta^\ell) = \varphi(e_\alpha^k) \psi(e_\beta^\ell)$$

and letting $\varphi \times \psi$ take the value 0 on $(k + \ell)$-cells of $X \times Y$ which are not the product of a k-cell of X with an ℓ-cell of Y. Another way of saying this is to use the convention that a cellular cochain in $C^k(X;R)$ takes the value 0 on cells of dimension different from k, and then we can let $(\varphi \times \psi)(e_\alpha^m \times e_\beta^n) = \varphi(e_\alpha^m)\psi(e_\beta^n)$ for all m and n.

The cellular coboundary formula $\delta(\varphi \times \psi) = \delta\varphi \times \psi + (-1)^k \varphi \times \delta\psi$ for cellular cochains $\varphi \in C^k(X;R)$ and $\psi \in C^\ell(Y;R)$ follows easily from the corresponding boundary formula in Proposition 3B.1, namely

$$\delta(\varphi \times \psi)(e_\alpha^m \times e_\beta^n) = (\varphi \times \psi)(\partial(e_\alpha^m \times e_\beta^n))$$
$$= (\varphi \times \psi)(\partial e_\alpha^m \times e_\beta^n + (-1)^m e_\alpha^m \times \partial e_\beta^n)$$
$$= \delta\varphi(e_\alpha^m)\psi(e_\beta^n) + (-1)^m \varphi(e_\alpha^m)\delta\psi(e_\beta^n)$$
$$= (\delta\varphi \times \psi + (-1)^k \varphi \times \delta\psi)(e_\alpha^m \times e_\beta^n)$$

where the coefficient $(-1)^m$ in the next-to-last line can be replaced by $(-1)^k$ since $\varphi(e_\alpha^m) = 0$ unless $k = m$. From the formula $\delta(\varphi \times \psi) = \delta\varphi \times \psi + (-1)^k \varphi \times \delta\psi$ it follows just as for homology and for cup product that there is an induced cross product in cellular cohomology.

To show this agrees with the earlier definition, we can first reduce to the case that X has trivial $(k-1)$-skeleton and Y has trivial $(\ell-1)$-skeleton via the commutative diagram

$$H^k(X/X^{k-1};R) \times H^\ell(Y/Y^{\ell-1};R) \xrightarrow{\ \times\ } H^{k+\ell}(X/X^{k-1} \times Y/Y^{\ell-1};R)$$
$$\downarrow \qquad\qquad\qquad\qquad\qquad\qquad \downarrow$$
$$H^k(X;R) \times H^\ell(Y;R) \xrightarrow{\qquad\ \times\ \qquad} H^{k+\ell}(X \times Y;R)$$

The left-hand vertical map is surjective, so by commutativity, if the two definitions of cross product agree in the upper row, they agree in the lower row. Next, assuming X^{k-1} and $Y^{\ell-1}$ are trivial, consider the commutative diagram

$$H^k(X;R) \times H^\ell(Y;R) \xrightarrow{\ \times\ } H^{k+\ell}(X \times Y;R)$$
$$\downarrow \qquad\qquad\qquad\qquad\qquad \downarrow$$
$$H^k(X^k;R) \times H^\ell(Y^\ell;R) \xrightarrow{\ \times\ } H^{k+\ell}(X^k \times Y^\ell;R)$$

The vertical maps here are injective, $X^k \times Y^\ell$ being the $(k+\ell)$-skeleton of $X \times Y$, so it suffices to see that the two definitions agree in the lower row. We have $X^k = \bigvee_\alpha S_\alpha^k$ and $Y^\ell = \bigvee_\beta S_\beta^\ell$, so by restriction to these wedge summands the question is reduced finally to the case of a product $S_\alpha^k \times S_\beta^\ell$. In this case, taking $R = \mathbb{Z}$, we showed in Theorem 3.15 that the cross product in question is the map $\mathbb{Z} \times \mathbb{Z} \to \mathbb{Z}$ sending $(1,1)$ to ± 1, with the original definition of cross product. The same is obviously true using the cellular cross product. So for $R = \mathbb{Z}$ the two cross products agree up to sign, and it follows that this is also true for arbitrary R. We leave it to the reader to sort out the matter of signs.

To relate cross product to cup product we use the diagonal map $\Delta : X \to X \times X$, $x \mapsto (x,x)$. If we are given a definition of cross product, we can define cup product as the composition

$$H^k(X;R) \times H^\ell(X;R) \xrightarrow{\ \times\ } H^{k+\ell}(X \times X;R) \xrightarrow{\ \Delta^*\ } H^{k+\ell}(X;R)$$

This agrees with the original definition of cup product since we have $\Delta^*(a \times b) = \Delta^*(p_1^*(a) \smile p_2^*(b)) = \Delta^*(p_1^*(a)) \smile \Delta^*(p_2^*(b)) = a \smile b$, as both compositions $p_1 \Delta$ and $p_2 \Delta$ are the identity map of X.

Unfortunately, the definition of cellular cross product cannot be combined with Δ to give a definition of cup product at the level of cellular cochains. This is because Δ is not a cellular map, so it does not induce a map of cellular cochains. It is possible to homotope Δ to a cellular map by Theorem 4.8, but this involves arbitrary choices. For example, the diagonal of a square can be pushed across either adjacent triangle. In particular cases one might hope to understand the geometry well enough to compute an explicit cellular approximation to the diagonal map, but usually other techniques for computing cup products are preferable.

The cohomology cross product satisfies the same commutativity relation as for homology, namely $T^*(a \times b) = (-1)^{k\ell} b \times a$ for $T : X \times Y \to Y \times X$ the transposition map, $a \in H^k(Y; R)$, and $b \in H^\ell(X; R)$. The proof is the same as for homology. Taking $X = Y$ and noting that $T\Delta = \Delta$, we obtain a new proof of the commutativity property of cup product.

Exercises

1. Compute the groups $H_i(\mathbb{RP}^m \times \mathbb{RP}^n; G)$ and $H^i(\mathbb{RP}^m \times \mathbb{RP}^n; G)$ for $G = \mathbb{Z}$ and $\mathbb{Z}_2$ via the cellular chain and cochain complexes. [See Example 3B.4.]

2. Let C and C' be chain complexes, and let I be the chain complex consisting of $\mathbb{Z}$ in dimension 1 and $\mathbb{Z} \times \mathbb{Z}$ in dimension 0, with the boundary map taking a generator e in dimension 1 to the difference $v_1 - v_0$ of generators v_i of the two $\mathbb{Z}$'s in dimension 0. Show that a chain map $f : I \otimes C \to C'$ is precisely the same as a chain homotopy between the two chain maps $f_i : C \to C'$, $c \mapsto f(v_i \otimes c)$, $i = 0, 1$. [The chain homotopy is $h(c) = f(e \otimes c)$.]

3. Show that the splitting in the topological Künneth formula cannot be natural by considering the map $f \times \mathbb{1} : M(\mathbb{Z}_m, n) \times M(\mathbb{Z}_m, n) \to S^{n+1} \times M(\mathbb{Z}_m, n)$ where f collapses the n-skeleton of $M(\mathbb{Z}_m, n) = S^n \cup e^{n+1}$ to a point.

4. Show that the cross product of fundamental classes for closed R-orientable manifolds M and N is a fundamental class for $M \times N$.

5. Show that **slant products**

$$H_n(X \times Y; R) \times H^j(Y; R) \longrightarrow H_{n-j}(X; R), \quad (e^i \times e^j, \varphi) \mapsto \varphi(e^j) e^i$$

$$H^n(X \times Y; R) \times H_j(Y; R) \longrightarrow H^{n-j}(X; R), \quad (\varphi, e^j) \mapsto (e^i \mapsto \varphi(e^i \times e^j))$$

can be defined via the indicated cellular formulas. [These 'products' are in some ways more like division than multiplication, and this is reflected in the common notation a/b for them, or $a\backslash b$ when the order of the factors is reversed. The first of the two slant products is related to cap product in the same way that the cohomology cross product is related to cup product.]

3.C H-Spaces and Hopf Algebras

Of the three axioms for a group, it would seem that the least subtle is the existence of an identity element. However, we shall see in this section that when topology is added to the picture, the identity axiom becomes much more potent. To give a name to the objects we will be considering, define a space X to be an **H–space**, 'H' standing for 'Hopf,' if there is a continuous multiplication map $\mu : X \times X \to X$ and an 'identity' element $e \in X$ such that the two maps $X \to X$ given by $x \mapsto \mu(x,e)$ and $x \mapsto \mu(e,x)$ are homotopic to the identity through maps $(X,e) \to (X,e)$. In particular, this implies that $\mu(e,e) = e$.

In terms of generality, this definition represents something of a middle ground. One could weaken the definition by dropping the condition that the homotopies preserve the basepoint e, or one could strengthen it by requiring that e be a strict identity, without any homotopies. An exercise at the end of the section is to show the three possible definitions are equivalent if X is a CW complex. An advantage of allowing homotopies in the definition is that a space homotopy equivalent in the basepointed sense to an H–space is again an H–space. Imposing basepoint conditions is fairly standard in homotopy theory, and is usually not a serious restriction.

The most classical examples of H–spaces are **topological groups**, spaces X with a group structure such that both the multiplication map $X \times X \to X$ and the inversion map $X \to X$, $x \mapsto x^{-1}$, are continuous. For example, the group $GL_n(\mathbb{R})$ of invertible $n \times n$ matrices with real entries is a topological group when topologized as a subspace of the n^2-dimensional vector space $M_n(\mathbb{R})$ of all $n \times n$ matrices over $\mathbb{R}$. It is an open subspace since the invertible matrices are those with nonzero determinant, and the determinant function $M_n(\mathbb{R}) \to \mathbb{R}$ is continuous. Matrix multiplication is certainly continuous, being defined by simple algebraic formulas, and it is not hard to see that matrix inversion is also continuous if one thinks for example of the classical adjoint formula for the inverse matrix.

Likewise $GL_n(\mathbb{C})$ is a topological group, as is the quaternionic analog $GL_n(\mathbb{H})$, though in the latter case one needs a somewhat different justification since determinants of quaternionic matrices do not have the good properties one would like. Since these groups GL_n over $\mathbb{R}$, $\mathbb{C}$, and $\mathbb{H}$ are open subsets of Euclidean spaces, they are examples of *Lie groups*, which can be defined as topological groups which are also manifolds. The GL_n groups are noncompact, being open subsets of Euclidean spaces, but they have the homotopy types of compact Lie groups $O(n)$, $U(n)$, and $Sp(n)$. This is explained in §3.D for $GL_n(\mathbb{R})$, and the other two cases are similar.

Among the simplest H–spaces from a topological viewpoint are the unit spheres S^1 in $\mathbb{C}$, S^3 in the quaternions $\mathbb{H}$, and S^7 in the octonions $\mathbb{O}$. These are H–spaces since the multiplications in these division algebras are continuous, being defined by polynomial formulas, and are norm-preserving, $|ab| = |a||b|$, hence restrict to multiplications on the unit spheres, and the identity element of the division algebra lies in the unit sphere in each case. Both S^1 and S^3 are Lie groups since the multiplications in $\mathbb{C}$ and $\mathbb{H}$ are associative and inverses exist since $a\bar{a} = |a|^2 = 1$ if $|a| = 1$. However, S^7 is not a group since multiplication of octonions is not associative. Of course $S^0 = \{\pm 1\}$ is also a topological group, trivially. A famous theorem of J. F. Adams asserts that S^0, S^1, S^3, and S^7 are the only spheres that are H–spaces; see §4.B for a fuller discussion.

Let us describe now some associative H–spaces where inverses fail to exist. Multiplication of polynomials provides an H–space structure on $\mathbb{CP}^\infty$ in the following way. A nonzero polynomial $a_0 + a_1 z + \cdots + a_n z^n$ with coefficients $a_i \in \mathbb{C}$ corresponds to a point $(a_0, \cdots, a_n, 0, \cdots) \in \mathbb{C}^\infty - \{0\}$. Multiplication of two such polynomials determines a multiplication $\mathbb{C}^\infty - \{0\} \times \mathbb{C}^\infty - \{0\} \to \mathbb{C}^\infty - \{0\}$ which is associative, commutative, and has an identity element $(1, 0, \cdots)$. Since $\mathbb{C}$ is commutative we can factor out by scalar multiplication by nonzero constants and get an induced product $\mathbb{CP}^\infty \times \mathbb{CP}^\infty \to \mathbb{CP}^\infty$ with the same properties. Thus $\mathbb{CP}^\infty$ is an associative, commutative H–space with a strict identity. Instead of factoring out by all nonzero scalars, we could factor out only by scalars of the form $\rho e^{2\pi i k/q}$ with ρ an arbitrary positive real, k an arbitrary integer, and q a fixed positive integer. The quotient of $\mathbb{C}^\infty - \{0\}$ under this identification, an infinite-dimensional lens space L^∞ with $\pi_1(L^\infty) \approx \mathbb{Z}_q$, is therefore also an associative, commutative H–space. This includes $\mathbb{RP}^\infty$ in particular.

The spaces $J(X)$ defined in §3.2 are also H–spaces, with the multiplication given by $(x_1, \cdots, x_m)(y_1, \cdots, y_n) = (x_1, \cdots, x_m, y_1, \cdots, y_n)$, which is associative and has an identity element (e) where e is the basepoint of X. One could describe $J(X)$ as the free associative H–space generated by X. There is also a commutative analog of $J(X)$ called the **infinite symmetric product** $SP(X)$ defined in the following way. Let $SP_n(X)$ be the quotient space of the n-fold product X^n obtained by identifying all n-tuples $(x_1, \cdots, x_n)$ that differ only by a permutation of their coordinates. The inclusion $X^n \hookrightarrow X^{n+1}$, $(x_1, \cdots, x_n) \mapsto (x_1, \cdots, x_n, e)$ induces an inclusion $SP_n(X) \hookrightarrow SP_{n+1}(X)$, and $SP(X)$ is defined to be the union of this increasing sequence of $SP_n(X)$'s, with the weak topology. Alternatively, $SP(X)$ is the quotient of $J(X)$ obtained by identifying points that differ only by permutation of coordinates. The H–space structure on $J(X)$ induces an H–space structure on $SP(X)$ which is commutative in addition to being associative and having a strict identity. The spaces $SP(X)$ are studied in more detail in §4.K.

The goal of this section will be to describe the extra structure which the multiplication in an H–space gives to its homology and cohomology. This is of particular interest since many of the most important spaces in algebraic topology turn out to be H–spaces.

Hopf Algebras

Let us look at cohomology first. Choosing a commutative ring R as coefficient ring, we can regard the cohomology ring $H^*(X;R)$ of a space X as an algebra over R rather than merely a ring. Suppose X is an H–space satisfying two conditions:

(1) X is path-connected, hence $H^0(X;R) \approx R$.

(2) $H^n(X;R)$ is a finitely generated free R-module for each n, so the cross product $H^*(X;R) \otimes_R H^*(X;R) \to H^*(X \times X;R)$ is an isomorphism.

The multiplication $\mu : X \times X \to X$ induces a map $\mu^* : H^*(X;R) \to H^*(X \times X;R)$, and when we combine this with the cross product isomorphism in (2) we get a map

$$H^*(X;R) \xrightarrow{\Delta} H^*(X;R) \otimes_R H^*(X;R)$$

which is an algebra homomorphism since both μ^* and the cross product isomorphism are algebra homomorphisms. The key property of Δ turns out to be that for any $\alpha \in H^n(X;R)$, $n > 0$, we have

$$\Delta(\alpha) = \alpha \otimes 1 + 1 \otimes \alpha + \sum_i \alpha_i' \otimes \alpha_i'' \quad \text{where } |\alpha_i'| > 0 \text{ and } |\alpha_i''| > 0$$

To verify this, let $i : X \to X \times X$ be the inclusion $x \mapsto (x, e)$ for e the identity element of X, and consider the commutative diagram

$$
\begin{array}{ccccc}
H^*(X;R) & \xrightarrow{\mu^*} & H^*(X \times X;R) & \xrightarrow{\quad i^* \quad} & H^*(X;R) \\
& {\scriptstyle \Delta} \searrow & {\scriptstyle \times} \uparrow {\scriptstyle \approx} & {\scriptstyle P} \nearrow & {\scriptstyle \times} \uparrow {\scriptstyle \approx} \\
& & H^*(X;R) \otimes_R H^*(X;R) & \xrightarrow{1 \otimes i^*} & H^*(X;R) \otimes_R H^*(e;R)
\end{array}
$$

The map P is defined by commutativity, and by looking at the lower right triangle we see that $P(\alpha \otimes 1) = \alpha$ and $P(\alpha \otimes \beta) = 0$ if $|\beta| > 0$. The H–space property says that $\mu i \simeq \mathbb{1}$, so $P\Delta = \mathbb{1}$. This implies that the component of $\Delta(\alpha)$ in $H^n(X;R) \otimes_R H^0(X;R)$ is $\alpha \otimes 1$. A similar argument shows the component in $H^0(X;R) \otimes_R H^n(X;R)$ is $1 \otimes \alpha$.

We can summarize this situation by saying that $H^*(X;R)$ is a **Hopf algebra**, that is, a graded algebra $A = \bigoplus_{n \geq 0} A^n$ over a commutative base ring R, satisfying the following two conditions:

(1) There is an identity element $1 \in A^0$ such that the map $R \to A^0$, $r \mapsto r \cdot 1$, is an isomorphism. In this case one says A is *connected*.

(2) There is a **diagonal** or **coproduct** $\Delta : A \to A \otimes A$, a homomorphism of graded algebras satisfying $\Delta(\alpha) = \alpha \otimes 1 + 1 \otimes \alpha + \sum_i \alpha_i' \otimes \alpha_i''$ where $|\alpha_i'| > 0$ and $|\alpha_i''| > 0$, for all α with $|\alpha| > 0$.

Here and in what follows we take $\otimes$ to mean $\otimes_R$. The multiplication in $A \otimes A$ is given by the standard formula $(\alpha \otimes \beta)(\gamma \otimes \delta) = (-1)^{|\beta||\gamma|}(\alpha\gamma \otimes \beta\delta)$. For a general Hopf algebra the multiplication is not assumed to be either associative or commutative (in the graded sense), though in the example of $H^*(X;R)$ for X an H–space the algebra structure is of course associative and commutative.

Example 3C.1. One of the simplest Hopf algebras is a polynomial ring $R[\alpha]$. The coproduct $\Delta(\alpha)$ must equal $\alpha \otimes 1 + 1 \otimes \alpha$ since the only elements of $R[\alpha]$ of lower dimension than α are the elements of R in dimension zero, so the terms α_i' and α_i'' in the coproduct formula $\Delta(\alpha) = \alpha \otimes 1 + 1 \otimes \alpha + \sum_i \alpha_i' \otimes \alpha_i''$ must be zero. The requirement that Δ be an algebra homomorphism then determines Δ completely. To describe Δ explicitly we distinguish two cases. If the dimension of α is even or if $2 = 0$ in R, then the multiplication in $R[\alpha] \otimes R[\alpha]$ is strictly commutative and $\Delta(\alpha^n) = (\alpha \otimes 1 + 1 \otimes \alpha)^n = \sum_i \binom{n}{i} \alpha^i \otimes \alpha^{n-i}$. In the opposite case that α is odd-dimensional, then $\Delta(\alpha^2) = (\alpha \otimes 1 + 1 \otimes \alpha)^2 = \alpha^2 \otimes 1 + 1 \otimes \alpha^2$ since $(\alpha \otimes 1)(1 \otimes \alpha) = \alpha \otimes \alpha$ and $(1 \otimes \alpha)(\alpha \otimes 1) = -\alpha \otimes \alpha$ if α has odd dimension. Thus if we set $\beta = \alpha^2$, then β is even-dimensional and we have $\Delta(\alpha^{2n}) = \Delta(\beta^n) = (\beta \otimes 1 + 1 \otimes \beta)^n = \sum_i \binom{n}{i} \beta^i \otimes \beta^{n-i}$ and $\Delta(\alpha^{2n+1}) = \Delta(\alpha\beta^n) = \Delta(\alpha)\Delta(\beta^n) = \sum_i \binom{n}{i} \alpha\beta^i \otimes \beta^{n-i} + \sum_i \binom{n}{i} \beta^i \otimes \alpha\beta^{n-i}$.

Example 3C.2. The exterior algebra $\Lambda_R[\alpha]$ on an odd-dimensional generator α is a Hopf algebra, with $\Delta(\alpha) = \alpha \otimes 1 + 1 \otimes \alpha$. To verify that Δ is an algebra homomorphism we must check that $\Delta(\alpha^2) = \Delta(\alpha)^2$, or in other words, since $\alpha^2 = 0$, we need to see that $\Delta(\alpha)^2 = 0$. As in the preceding example we have $\Delta(\alpha)^2 = (\alpha \otimes 1 + 1 \otimes \alpha)^2 = \alpha^2 \otimes 1 + 1 \otimes \alpha^2$, so $\Delta(\alpha)^2$ is indeed 0. Note that if α were even-dimensional we would instead have $\Delta(\alpha)^2 = \alpha^2 \otimes 1 + 2\alpha \otimes \alpha + 1 \otimes \alpha^2$, which would be 0 in $\Lambda_R[\alpha] \otimes \Lambda_R[\alpha]$ only if $2 = 0$ in R.

An element α of a Hopf algebra is called **primitive** if $\Delta(\alpha) = \alpha \otimes 1 + 1 \otimes \alpha$. As the preceding examples illustrate, if a Hopf algebra is generated as an algebra by primitive elements, then the coproduct Δ is uniquely determined by the product. This happens in a number of interesting special cases, but certainly not in general, as we shall see.

The existence of the coproduct in a Hopf algebra turns out to restrict the multiplicative structure considerably. Here is an important example illustrating this:

Example 3C.3. Suppose that the truncated polynomial algebra $F[\alpha]/(\alpha^n)$ over a field F is a Hopf algebra. Then α is primitive, just as it is in $F[\alpha]$, so if we assume either that α is even-dimensional or that F has characteristic 2, then the relation $\alpha^n = 0$ yields an equation

$$0 = \Delta(\alpha^n) = \alpha^n \otimes 1 + 1 \otimes \alpha^n + \sum_{0 < i < n} \binom{n}{i} \alpha^i \otimes \alpha^{n-i} = \sum_{0 < i < n} \binom{n}{i} \alpha^i \otimes \alpha^{n-i}$$

which implies that $\binom{n}{i} = 0$ in F for each i in the range $0 < i < n$. This is impossible if F has characteristic 0, and if the characteristic of F is $p > 0$ then it happens only when n is a power of p. For $p = 2$ this was shown in the proof of Theorem 3.21, and the argument given there works just as well for odd primes. Conversely, it is easy to check that if F has characteristic p then $F[\alpha]/(\alpha^{p^i})$ is a Hopf algebra, assuming still that α is even-dimensional if p is odd.

The characteristic 0 case of this result implies that $\mathbb{C}P^n$ is not an H–space for finite n, in contrast with $\mathbb{C}P^\infty$ which is an H–space as we saw earlier. Similarly, taking

$F = \mathbb{Z}_2$, we deduce that $\mathbb{R}P^n$ can be an H-space only if $n + 1$ is a power of 2. Indeed, $\mathbb{R}P^1 = S^1/\pm 1$, $\mathbb{R}P^3 = S^3/\pm 1$, and $\mathbb{R}P^7 = S^7/\pm 1$ have quotient H-space structures from S^1, S^3 and S^7 since -1 commutes with all elements of S^1, S^3, or S^7. However, these are the only cases when $\mathbb{R}P^n$ is an H-space since, by an exercise at the end of this section, the universal cover of an H–space is an H-space, and S^1, S^3, and S^7 are the only spheres that are H–spaces, by the theorem of Adams mentioned earlier.

It is an easy exercise to check that the tensor product of Hopf algebras is again a Hopf algebra, with the coproduct $\Delta(\alpha \otimes \beta) = \Delta(\alpha) \otimes \Delta(\beta)$. So the preceding examples yield many other Hopf algebras, tensor products of polynomial, truncated polynomial, and exterior algebras on any number of generators. The following theorem of Hopf is a partial converse:

Theorem 3C.4. *If A is a commutative, associative Hopf algebra over a field F of characteristic 0, and A^n is finite-dimensional over F for each n, then A is isomorphic as an algebra to the tensor product of an exterior algebra on odd-dimensional generators and a polynomial algebra on even-dimensional generators.*

There is an analogous theorem of Borel when F is a finite field of characteristic p. In this case A is again isomorphic to a tensor product of single-generator Hopf algebras, of one of the following types:

- $F[\alpha]$, with α even-dimensional if $p \neq 2$.
- $\Lambda_F[\alpha]$ with α odd-dimensional.
- $F[\alpha]/(\alpha^{p^i})$, with α even-dimensional if $p \neq 2$.

For a proof see [Borel 1953] or [Kane 1988].

Proof of 3C.4: Since A^n is finitely generated over F for each n, we may choose algebra generators $x_1, x_2, \cdots$ for A with $|x_i| \leq |x_{i+1}|$ for all i. Let A_n be the subalgebra generated by $x_1, \cdots, x_n$. This is a Hopf subalgebra of A, that is, $\Delta(A_n) \subset A_n \otimes A_n$, since $\Delta(x_i)$ involves only x_i and terms of smaller dimension. We may assume x_n does not lie in A_{n-1}. Since A is associative and commutative, there is a natural surjection $A_{n-1} \otimes F[x_n] \to A_n$ if $|x_n|$ is even, or $A_{n-1} \otimes \Lambda_F[x_n] \to A_n$ if $|x_n|$ is odd. By induction on n it will suffice to prove these surjections are injective. Thus in the two cases we must rule out nontrivial relations $\sum_i \alpha_i x_n^i = 0$ and $\alpha_0 + \alpha_1 x_n = 0$, respectively, with coefficients $\alpha_i \in A_{n-1}$.

Let I be the ideal in A_n generated by x_n^2 and the positive-dimensional elements of A_{n-1}, so I consists of the polynomials $\sum_i \alpha_i x_n^i$ with coefficients $\alpha_i \in A_{n-1}$, the first two coefficients α_0 and α_1 having trivial components in A^0. Note that $x_n \notin I$ since elements of I having dimension $|x_n|$ must lie in A_{n-1}. Consider the composition

$$A_n \xrightarrow{\;\Delta\;} A_n \otimes A_n \xrightarrow{\;q\;} A_n \otimes (A_n/I)$$

with q the natural quotient map. By the definition of I, this composition $q\Delta$ sends $\alpha \in A_{n-1}$ to $\alpha \otimes 1$ and x_n to $x_n \otimes 1 + 1 \otimes \overline{x}_n$ where $\overline{x}_n$ is the image of x_n in A_n/I.

In case $|x_n|$ is even, applying $q\Delta$ to a nontrivial relation $\sum_i \alpha_i x_n^i = 0$ gives

$$0 = \sum_i (\alpha_i \otimes 1)(x_n \otimes 1 + 1 \otimes \overline{x}_n)^i = (\sum_i \alpha_i x_n^i) \otimes 1 + \sum_i i\alpha_i x_n^{i-1} \otimes \overline{x}_n$$

Since $\sum_i \alpha_i x_n^i = 0$, this implies that $\sum_i i\alpha_i x_n^{i-1} \otimes \overline{x}_n$ is zero in the tensor product $A_n \otimes (A_n/I)$, hence $\sum_i i\alpha_i x_n^{i-1} = 0$ since $x_n \notin I$ implies $\overline{x}_n \neq 0$. The relation $\sum_i i\alpha_i x_n^{i-1} = 0$ has lower degree than the original relation, and is not the trivial relation since F has characteristic 0, $\alpha_i \neq 0$ implying $i\alpha_i \neq 0$ if $i > 0$. Since we could assume the original relation had minimum degree, we have reached a contradiction.

The case $|x_n|$ odd is similar. Applying $q\Delta$ to a relation $\alpha_0 + \alpha_1 x_n = 0$ gives $0 = \alpha_0 \otimes 1 + (\alpha_1 \otimes 1)(x_n \otimes 1 + 1 \otimes \overline{x}_n) = (\alpha_0 + \alpha_1 x_n) \otimes 1 + \alpha_1 \otimes \overline{x}_n$. Since $\alpha_0 + \alpha_1 x_n = 0$, we get $\alpha_1 \otimes \overline{x}_n = 0$, which implies $\alpha_1 = 0$ and hence $\alpha_0 = 0$. $\quad\square$

The structure of Hopf algebras over $\mathbb{Z}$ is much more complicated than over a field. Here is an example that is still fairly simple.

Example 3C.5: Divided Polynomial Algebras. We showed in Proposition 3.22 that the H–space $J(S^n)$ for n even has $H^*(J(S^n); \mathbb{Z})$ a divided polynomial algebra, the algebra $\Gamma_{\mathbb{Z}}[\alpha]$ with additive generators α_i in dimension ni and multiplication given by $\alpha_1^k = k!\alpha_k$, hence $\alpha_i \alpha_j = \binom{i+j}{i}\alpha_{i+j}$. The coproduct in $\Gamma_{\mathbb{Z}}[\alpha]$ is uniquely determined by the multiplicative structure since $\Delta(\alpha_1^k) = (\alpha_1 \otimes 1 + 1 \otimes \alpha_1)^k = \sum_i \binom{k}{i}\alpha_1^i \otimes \alpha_1^{k-i}$, which implies that $\Delta(\alpha_1^k/k!) = \sum_i (\alpha_1^i/i!) \otimes (\alpha_1^{k-i}/(k-i)!)$, that is, $\Delta(\alpha_k) = \sum_i \alpha_i \otimes \alpha_{k-i}$. So in this case the coproduct has a simpler description than the product.

It is interesting to see what happens to the divided polynomial algebra $\Gamma_{\mathbb{Z}}[\alpha]$ when we change to field coefficients. Clearly $\Gamma_{\mathbb{Q}}[\alpha]$ is the same as $\mathbb{Q}[\alpha]$. In contrast with this, $\Gamma_{\mathbb{Z}_p}[\alpha]$, with multiplication defined by $\alpha_i \alpha_j = \binom{i+j}{i}\alpha_{i+j}$, happens to be isomorphic as an algebra to the infinite tensor product $\bigotimes_{i \geq 0} \mathbb{Z}_p[\alpha_{p^i}]/(\alpha_{p^i}^p)$, as we will show in a moment. However, as Hopf algebras these two objects are different since α_{p^i} is primitive in $\bigotimes_{i \geq 0} \mathbb{Z}_p[\alpha_{p^i}]/(\alpha_{p^i}^p)$ but not in $\Gamma_{\mathbb{Z}_p}[\alpha]$ when $i > 0$, since the coproduct in $\Gamma_{\mathbb{Z}_p}[\alpha]$ is given by $\Delta(\alpha_k) = \sum_i \alpha_i \otimes \alpha_{k-i}$.

Now let us show that there is an algebra isomorphism

$$\Gamma_{\mathbb{Z}_p}[\alpha] \approx \bigotimes_{i \geq 0} \mathbb{Z}_p[\alpha_{p^i}]/(\alpha_{p^i}^p)$$

Since $\Gamma_{\mathbb{Z}_p}[\alpha] = \Gamma_{\mathbb{Z}}[\alpha] \otimes \mathbb{Z}_p$, this is equivalent to:

$(*)$ The element $\alpha_1^{n_0} \alpha_p^{n_1} \cdots \alpha_{p^k}^{n_k}$ in $\Gamma_{\mathbb{Z}}[\alpha]$ is divisible by p iff $n_i \geq p$ for some i.

The product $\alpha_1^{n_0} \alpha_p^{n_1} \cdots \alpha_{p^k}^{n_k}$ equals $m\alpha_n$ for $n = n_0 + n_1 p + \cdots + n_k p^k$ and some integer m. The question is whether p divides m. We will show:

$(**)$ $\alpha_n \alpha_{p^k}$ is divisible by p iff $n_k = p - 1$, assuming that $n_i < p$ for each i.

This implies $(*)$ by an inductive argument in which we build up the product in $(*)$ by repeated multiplication on the right by terms α_{p^i}.

To prove $(\ast\ast)$ we recall that $\alpha_n \alpha_{p^k} = \binom{n+p^k}{n}\alpha_{n+p^k}$. The mod p value of this binomial coefficient can be computed using Lemma 3C.6 below. Assuming that $n_i < p$ for each i and that $n_k + 1 < p$, the p-adic representations of $n + p^k$ and n differ only in the coefficient of p^k, so mod p we have $\binom{n+p^k}{n} = \binom{n_k+1}{n_k} = n_k + 1$. This conclusion also holds if $n_k + 1 = p$, when the p-adic representations of $n + p^k$ and n differ also in the coefficient of p^{k+1}. The statement $(\ast\ast)$ then follows.

Lemma 3C.6. *If p is a prime, then $\binom{n}{k} \equiv \prod_i \binom{n_i}{k_i}$ mod p where $n = \sum_i n_i p^i$ and $k = \sum_i k_i p^i$ with $0 \le n_i < p$ and $0 \le k_i < p$ are the p-adic representations of n and k.*

Here the convention is that $\binom{n}{k} = 0$ if $n < k$, and $\binom{n}{0} = 1$ for all $n \ge 0$.

Proof: In $\mathbb{Z}_p[x]$ there is an identity $(1 + x)^p = 1 + x^p$ since p clearly divides $\binom{p}{k} = p!/k!(p - k)!$ for $0 < k < p$. By induction it follows that $(1 + x)^{p^i} = 1 + x^{p^i}$. Hence if $n = \sum_i n_i p^i$ is the p-adic representation of n then:

$$(1 + x)^n = (1 + x)^{n_0}(1 + x^p)^{n_1}(1 + x^{p^2})^{n_2} \cdots$$
$$= \left[1 + \binom{n_0}{1}x + \binom{n_0}{2}x^2 + \cdots + \binom{n_0}{p-1}x^{p-1}\right]$$
$$\times \left[1 + \binom{n_1}{1}x^p + \binom{n_1}{2}x^{2p} + \cdots + \binom{n_1}{p-1}x^{(p-1)p}\right]$$
$$\times \left[1 + \binom{n_2}{1}x^{p^2} + \binom{n_2}{2}x^{2p^2} + \cdots + \binom{n_2}{p-1}x^{(p-1)p^2}\right] \times \cdots$$

When this is multiplied out, one sees that no terms combine, and the coefficient of x^k is just $\prod_i \binom{n_i}{k_i}$ where $k = \sum_i k_i p^i$ is the p-adic representation of k. $\square$

Pontryagin Product

Another special feature of H–spaces is that their homology groups have a product operation, called the **Pontryagin product**. For an H–space X with multiplication $\mu : X \times X \to X$, this is the composition

$$H_*(X;R) \otimes H_*(X;R) \xrightarrow{\times} H_*(X \times X;R) \xrightarrow{\mu_*} H_*(X;R)$$

where the first map is the cross product defined in §3.B. Thus the Pontryagin product consists of bilinear maps $H_i(X;R) \times H_j(X;R) \to H_{i+j}(X;R)$. Unlike cup product, the Pontryagin product is not in general associative unless the multiplication μ is associative or at least associative up to homotopy, in the sense that the maps $X \times X \times X \to X$, $(x, y, z) \mapsto \mu(x, \mu(y, z))$ and $(x, y, z) \mapsto \mu(\mu(x, y), z)$ are homotopic. Fortunately most H–spaces one meets in practice satisfy this associativity property. Nor is the Pontryagin product generally commutative, even in the graded sense, unless μ is commutative or homotopy-commutative, which is relatively rare for H–spaces. We will give examples shortly where the Pontryagin product is not commutative.

In case X is a CW complex and μ is a cellular map the Pontryagin product can be computed using cellular homology via the cellular chain map

$$C_i(X;R) \times C_j(X;R) \xrightarrow{\times} C_{i+j}(X \times X;R) \xrightarrow{\mu_*} C_{i+j}(X;R)$$

where the cross product map sends generators corresponding to cells e^i and e^j to the generator corresponding to the product cell $e^i \times e^j$, and then μ_* is applied to this product cell.

Example 3C.7. Let us compute the Pontryagin product for $J(S^n)$. Here there is one cell e^{in} for each $i \geq 0$, and μ takes the product cell $e^{in} \times e^{jn}$ homeomorphically onto the cell $e^{(i+j)n}$. This means that $H_*(J(S^n);\mathbb{Z})$ is simply the polynomial ring $\mathbb{Z}[x]$ on an n-dimensional generator x. This holds for n odd as well as for n even, so the Pontryagin product need not satisfy the same general commutativity relation as cup product. In this example the Pontryagin product structure is simpler than the cup product structure, though for some H–spaces it is the other way round. In applications it is often convenient to have the choice of which product structure to use.

This calculation immediately generalizes to $J(X)$ where X is any connected CW complex whose cellular boundary maps are all trivial. The cellular boundary maps in the product X^m of m copies of X are then trivial by induction on m using Proposition 3B.1, and therefore the cellular boundary maps in $J(X)$ are all trivial since the quotient map $X^m \to J_m(X)$ is cellular and each cell of $J_m(X)$ is the homeomorphic image of a cell of X^m. Thus $H_*(J(X);\mathbb{Z})$ is free with additive basis the products $e^{n_1} \times \cdots \times e^{n_k}$ of positive-dimensional cells of X, and the multiplicative structure is that of polynomials in noncommuting variables corresponding to the positive-dimensional cells of X.

Another way to describe $H_*(J(X);\mathbb{Z})$ in this example is as the tensor algebra $T\tilde{H}_*(X;\mathbb{Z})$, where for a graded R-module M that is trivial in dimension zero, like the reduced homology of a path-connected space, the tensor algebra TM is the direct sum of the n-fold tensor products of M with itself for all $n \geq 1$, together with a copy of R in dimension zero, with the obvious multiplication coming from tensor product and scalar multiplication.

Generalizing the preceding example, we have:

Proposition 3C.8. *If X is a connected CW complex with $H_*(X;R)$ a free R-module, then $H_*(J(X);R)$ is isomorphic to the tensor algebra $T\tilde{H}_*(X;R)$.*

This can be paraphrased as saying that the homology of the free H–space generated by a space with free homology is the free algebra generated by the homology of the space.

Proof: With coefficients in R, let $\varphi : T\tilde{H}_*(X) \to H_*(J(X))$ be the homomorphism whose restriction to the n-fold tensor product $\tilde{H}_*(X)^{\otimes n}$ is the composition

$$\tilde{H}_*(X)^{\otimes n} \hookrightarrow H_*(X)^{\otimes n} \xrightarrow{\times} H_*(X^n) \to H_*(J_n(X)) \to H_*(J(X))$$

where the next-to-last map is induced by the quotient map $X^n \to J_n(X)$. It is clear that φ is a ring homomorphism since the product in $J(X)$ is induced from the natural map $X^m \times X^n \to X^{m+n}$. To show that φ is an isomorphism, consider the following commutative diagram of short exact sequences:

$$
\begin{array}{ccccccccc}
0 & \longrightarrow & T_{n-1}\widetilde{H}_*(X) & \longrightarrow & T_n\widetilde{H}_*(X) & \longrightarrow & \widetilde{H}_*(X)^{\otimes n} & \longrightarrow & 0 \\
 & & \varphi\downarrow & & \varphi\downarrow & & \times\downarrow\approx & & \\
0 & \longrightarrow & H_*(J_{n-1}(X)) & \longrightarrow & H_*(J_n(X)) & \longrightarrow & \widetilde{H}_*(X^{\wedge n}) & \longrightarrow & 0
\end{array}
$$

In the upper row, $T_m\widetilde{H}_*(X)$ denotes the direct sum of the products $\widetilde{H}_*(X)^{\otimes k}$ for $k \le m$, so this row is exact. The second row is the homology exact sequence for the pair $(J_n(X), J_{n-1}(X))$, with quotient $J_n(X)/J_{n-1}(X)$ the n-fold smash product $X^{\wedge n}$. This long exact sequence breaks up into short exact sequences as indicated, by commutativity of the right-hand square and the fact that the right-hand vertical map is an isomorphism by the Künneth formula, using the hypothesis that $H_*(X)$ is free over the given coefficient ring. By induction on n and the five-lemma we deduce from the diagram that $\varphi : T_n\widetilde{H}_*(X) \to H_*(J_n(X))$ is an isomorphism for all n. Letting n go to ∞, this implies that $\varphi : T\widetilde{H}_*(X) \to H_*(J(X))$ is an isomorphism since in any given dimension $T_n\widetilde{H}_*(X)$ is independent of n when n is sufficiently large, and the same is true of $H_*(J_n(X))$ by the second row of the diagram. $\square$

Dual Hopf Algebras

There is a close connection between the Pontryagin product in homology and the Hopf algebra structure on cohomology. Suppose that X is an H–space such that, with coefficients in a field R, the vector spaces $H_n(X;R)$ are finite-dimensional for all n. Alternatively, we could take $R = \mathbb{Z}$ and assume $H_n(X;\mathbb{Z})$ is finitely generated and free for all n. In either case we have $H^n(X;R) = \mathrm{Hom}_R(H_n(X;R), R)$, and as a consequence the Pontryagin product $H_*(X;R) \otimes H_*(X;R) \to H_*(X;R)$ and the coproduct $\Delta : H^*(X;R) \to H^*(X;R) \otimes H^*(X;R)$ are dual to each other, both being induced by the H–space product $\mu : X \times X \to X$. Therefore the coproduct in cohomology determines the Pontryagin product in homology, and vice versa. Specifically, the component $\Delta_{ij} : H^{i+j}(X;R) \to H^i(X;R) \otimes H^j(X;R)$ of Δ is dual to the product $H_i(X;R) \otimes H_j(X;R) \to H_{i+j}(X;R)$.

Example 3C.9. Consider $J(S^n)$ with n even, so $H^*(J(S^n);\mathbb{Z})$ is the divided polynomial algebra $\Gamma_{\mathbb{Z}}[\alpha]$. In Example 3C.5 we derived the coproduct formula $\Delta(\alpha_k) = \sum_i \alpha_i \otimes \alpha_{k-i}$. Thus Δ_{ij} takes α_{i+j} to $\alpha_i \otimes \alpha_j$, so if x_i is the generator of $H_{in}(J(S^n);\mathbb{Z})$ dual to α_i, then $x_i x_j = x_{i+j}$. This says that $H_*(J(S^n);\mathbb{Z})$ is the polynomial ring $\mathbb{Z}[x]$. We showed this in Example 3C.7 using the cell structure of $J(S^n)$, but the present proof deduces it purely algebraically from the cup product structure.

Now we wish to show that the relation between $H^*(X;R)$ and $H_*(X;R)$ is perfectly symmetric: They are *dual Hopf algebras*. This is a purely algebraic fact:

Proposition 3C.10. *Let A be a Hopf algebra over R that is a finitely generated free R-module in each dimension. Then the product $\pi : A \otimes A \to A$ and coproduct $\Delta : A \to A \otimes A$ have duals $\pi^* : A^* \to A^* \otimes A^*$ and $\Delta^* : A^* \otimes A^* \to A^*$ that give A^* the structure of a Hopf algebra.*

Proof: This will be apparent if we reinterpret the Hopf algebra structure on A formally as a pair of graded R-module homomorphisms $\pi : A \otimes A \to A$ and $\Delta : A \to A \otimes A$ together with an element $1 \in A^0$ satisfying:

(1) The two compositions $A \xrightarrow{i_\ell} A \otimes A \xrightarrow{\pi} A$ and $A \xrightarrow{i_r} A \otimes A \xrightarrow{\pi} A$ are the identity, where $i_\ell(a) = a \otimes 1$ and $i_r(a) = 1 \otimes a$. This says that 1 is a two-sided identity for the multiplication in A.

(2) The two compositions $A \xrightarrow{\Delta} A \otimes A \xrightarrow{p_\ell} A$ and $A \xrightarrow{\Delta} A \otimes A \xrightarrow{p_r} A$ are the identity, where $p_\ell(a \otimes 1) = a = p_r(1 \otimes a)$, $p_\ell(a \otimes b) = 0$ if $|b| > 0$, and $p_r(a \otimes b) = 0$ if $|a| > 0$. This is just the coproduct formula $\Delta(a) = a \otimes 1 + 1 \otimes a + \sum_i a_i' \otimes a_i''$.

(3) The diagram at the right commutes, with
$$\tau(a \otimes b \otimes c \otimes d) = (-1)^{|b||c|} a \otimes c \otimes b \otimes d.$$
This is the condition that Δ is an algebra homomorphism since if we follow

$$
\begin{array}{ccccc}
A \otimes A & \xrightarrow{\pi} & A & \xrightarrow{\Delta} & A \otimes A \\
\downarrow{\scriptstyle \Delta \otimes \Delta} & & & & \uparrow{\scriptstyle \pi \otimes \pi} \\
A \otimes A \otimes A \otimes A & & \xrightarrow{\tau} & & A \otimes A \otimes A \otimes A
\end{array}
$$

an element $a \otimes b$ across the top of the diagram we get $\Delta(ab)$, while the lower route gives first $\Delta(a) \otimes \Delta(b) = (\sum_i a_i' \otimes a_i'') \otimes (\sum_j b_j' \otimes b_j'')$, then after applying τ and $\pi \otimes \pi$ this becomes $\sum_{i,j}(-1)^{|a_i''||b_j'|} a_i' b_j' \otimes a_i'' b_j'' = (\sum_i a_i' \otimes a_i'')(\sum_j b_j' \otimes b_j'')$, which is $\Delta(a)\Delta(b)$.

Condition (1) for A dualizes to (2) for A^*, and similarly (2) for A dualizes to (1) for A^*. Condition (3) for A dualizes to (3) for A^*. $\qquad\square$

Example 3C.11. Let us compute the dual of a polynomial algebra $R[x]$. Suppose first that x has even dimension. Then $\Delta(x^n) = (x \otimes 1 + 1 \otimes x)^n = \sum_i \binom{n}{i} x^i \otimes x^{n-i}$, so if α_i is dual to x^i, the term $\binom{n}{i} x^i \otimes x^{n-i}$ in $\Delta(x^n)$ gives the product relation $\alpha_i \alpha_{n-i} = \binom{n}{i} \alpha_n$. This is the rule for multiplication in a divided polynomial algebra, so the dual of $R[x]$ is $\Gamma_R[\alpha]$ if the dimension of x is even. This also holds if $2 = 0$ in R, since the even-dimensionality of x was used only to deduce that $R[x] \otimes R[x]$ is strictly commutative.

In case x is odd-dimensional, then as we saw in Example 3C.1, if we set $y = x^2$, we have $\Delta(y^n) = (y \otimes 1 + 1 \otimes y)^n = \sum_i \binom{n}{i} y^i \otimes y^{n-i}$ and $\Delta(xy^n) = \Delta(x)\Delta(y^n) = \sum_i \binom{n}{i} xy^i \otimes y^{n-i} + \sum_i \binom{n}{i} y^i \otimes xy^{n-i}$. These formulas for Δ say that the dual of $R[x]$ is $\Lambda_R[\alpha] \otimes \Gamma_R[\beta]$ where α is dual to x and β is dual to y.

This algebra allows us to deduce the cup product structure on $H^*(J(S^n); R)$ from the geometric calculation $H_*(J(S^n); R) \approx R[x]$ in Example 3C.7. As another application, recall from earlier in this section that $\mathbb{R}P^\infty$ and $\mathbb{C}P^\infty$ are H–spaces, so from their cup product structures we can conclude that the Pontryagin rings $H_*(\mathbb{RP}^\infty;\mathbb{Z}_2)$ and $H_*(\mathbb{CP}^\infty;\mathbb{Z})$ are divided polynomial algebras.

In these examples the Hopf algebra is generated as an algebra by primitive elements, so the product determines the coproduct and hence the dual algebra. This is not true in general, however. For example, we have seen that the Hopf algebra $\Gamma_{\mathbb{Z}_p}[\alpha]$ is isomorphic as an algebra to $\bigotimes_{i\geq 0}\mathbb{Z}_p[\alpha_{p^i}]/(\alpha_{p^i}^p)$, but if we regard the latter tensor product as the tensor product of the Hopf algebras $\mathbb{Z}_p[\alpha_{p^i}]/(\alpha_{p^i}^p)$ then the elements α_{p^i} are primitive, though they are not primitive in $\Gamma_{\mathbb{Z}_p}[\alpha]$ for $i > 0$. In fact, the Hopf algebra $\bigotimes_{i\geq 0}\mathbb{Z}_p[\alpha_{p^i}]/(\alpha_{p^i}^p)$ is its own dual, according to one of the exercises below, but the dual of $\Gamma_{\mathbb{Z}_p}[\alpha]$ is $\mathbb{Z}_p[\alpha]$.

Exercises

1. Suppose that X is a CW complex with basepoint $e \in X$ a 0-cell. Show that X is an H–space if there is a map $\mu:X\times X\to X$ such that the maps $X\to X$, $x \mapsto \mu(x,e)$ and $x \mapsto \mu(e,x)$, are homotopic to the identity. [Sometimes this is taken as the definition of an H–space, rather than the more restrictive condition in the definition we have given.] With the same hypotheses, show also that μ can be homotoped so that e is a strict two-sided identity.

2. Show that a retract of an H space is an H-space if it contains the identity element.

3. Show that in a homotopy-associative H–space whose set of path-components is a group with respect to the multiplication induced by the H–space structure, all the path-components must be homotopy equivalent. [Homotopy-associative means associative up to homotopy.]

4. Show that an H–space or topological group structure on a path-connected, locally path-connected space can be lifted to such a structure on its universal cover. [For the group $SO(n)$ considered in the next section, the universal cover for $n > 2$ is a 2-sheeted cover, a group called $Spin(n)$.]

5. Show that if (X,e) is an H–space then $\pi_1(X,e)$ is abelian. [Compare the usual composition $f \cdot g$ of loops with the product $\mu(f(t),g(t))$ coming from the H–space multiplication μ.]

6. Show that S^n is an H–space iff the attaching map of the $2n$-cell of $J_2(S^n)$ is homotopically trivial.

7. What are the primitive elements of the Hopf algebra $\mathbb{Z}_p[x]$ for p prime?

8. Show that the tensor product of two Hopf algebras is a Hopf algebra.

9. Apply the theorems of Hopf and Borel to show that for an H–space X that is a connected finite CW complex with $\tilde{H}_*(X;\mathbb{Z}) \neq 0$, the Euler characteristic $\chi(X)$ is 0.

10. Let X be a path-connected H–space with $H^*(X;R)$ free and finitely generated in each dimension. For maps $f,g:X\to X$, the product $fg:X\to X$ is defined by $(fg)(x) = f(x)g(x)$, using the H–space product.

(a) Show that $(fg)^*(\alpha) = f^*(\alpha) + g^*(\alpha)$ for primitive elements $\alpha \in H^*(X;R)$.

(b) Deduce that the kth-power map $x \mapsto x^k$ induces the map $\alpha \mapsto k\alpha$ on primitive elements α. In particular the quaternionic k^{th}-power map $S^3 \to S^3$ has degree k.

(c) Show that every polynomial $a_n x^n b_n + \cdots + a_1 x b_1 + a_0$ of nonzero degree with coefficients in $\mathbb{H}$ has a root in $\mathbb{H}$. [See Theorem 1.8.]

11. If T^n is the n-dimensional torus, the product of n circles, show that the Pontryagin ring $H_*(T^n;\mathbb{Z})$ is the exterior algebra $\Lambda_{\mathbb{Z}}[x_1, \cdots, x_n]$ with $|x_i| = 1$.

12. Compute the Pontryagin product structure in $H_*(L;\mathbb{Z}_p)$ where L is an infinite-dimensional lens space $S^\infty/\mathbb{Z}_p$, for p an odd prime, using the coproduct in $H^*(L;\mathbb{Z}_p)$.

13. Verify that the Hopf algebras $\Lambda_R[\alpha]$ and $\mathbb{Z}_p[\alpha]/(\alpha^p)$ are self-dual.

14. Show that the coproduct in the Hopf algebra $H_*(X;R)$ dual to $H^*(X;R)$ is induced by the diagonal map $X \to X \times X$, $x \mapsto (x,x)$.

15. Suppose that X is a path-connected H–space such that $H^*(X;\mathbb{Z})$ is free and finitely generated in each dimension, and $H^*(X;\mathbb{Q})$ is a polynomial ring $\mathbb{Q}[\alpha]$. Show that the Pontryagin ring $H_*(X;\mathbb{Z})$ is commutative and associative, with a structure uniquely determined by the ring $H^*(X;\mathbb{Z})$.

16. Classify algebraically the Hopf algebras A over $\mathbb{Z}$ such that A^n is free for each n and $A \otimes \mathbb{Q} \approx \mathbb{Q}[\alpha]$. In particular, determine which Hopf algebras $A \otimes \mathbb{Z}_p$ arise from such A's.

3.D The Cohomology of SO(n)

After the general discussion of homological and cohomological properties of H–spaces in the preceding section, we turn now to a family of quite interesting and subtle examples, the orthogonal groups $O(n)$. We will compute their homology and cohomology by constructing very nice CW structures on them, and the results illustrate the general structure theorems of the last section quite well. After dealing with the orthogonal groups we then describe the straightforward generalization to Stiefel manifolds, which are also fairly basic objects in algebraic and geometric topology.

The orthogonal group $O(n)$ can be defined as the group of isometries of $\mathbb{R}^n$ fixing the origin. Equivalently, this is the group of $n \times n$ matrices A with entries in $\mathbb{R}$ such that $AA^t = I$, where A^t is the transpose of A. From this viewpoint, $O(n)$ is topologized as a subspace of $\mathbb{R}^{n^2}$, with coordinates the n^2 entries of an $n \times n$ matrix. Since the columns of a matrix in $O(n)$ are unit vectors, $O(n)$ can also be regarded as a subspace of the product of n copies of S^{n-1}. It is a closed subspace since the conditions that columns be orthogonal are defined by polynomial equations. Hence

$O(n)$ is compact. The map $O(n) \times O(n) \to O(n)$ given by matrix multiplication is continuous since it is defined by polynomials. The inversion map $A \mapsto A^{-1} = A^t$ is clearly continuous, so $O(n)$ is a topological group, and in particular an H–space.

The determinant map $O(n) \to \{\pm 1\}$ is a surjective homomorphism, so its kernel $SO(n)$, the 'special orthogonal group,' is a subgroup of index two. The two cosets $SO(n)$ and $O(n) - SO(n)$ are homeomorphic to each other since for fixed $B \in O(n)$ of determinant -1, the maps $A \mapsto AB$ and $A \mapsto AB^{-1}$ are inverse homeomorphisms between these two cosets. The subgroup $SO(n)$ is a union of components of $O(n)$ since the image of the map $O(n) \to \{\pm 1\}$ is discrete. In fact, $SO(n)$ is path-connected since by linear algebra, each $A \in SO(n)$ is a rotation, a composition of rotations in a family of orthogonal 2-dimensional subspaces of $\mathbb{R}^n$, with the identity map on the subspace orthogonal to all these planes, and such a rotation can obviously be joined to the identity by a path of rotations of the same planes through decreasing angles. Another reason why $SO(n)$ is connected is that it has a CW structure with a single 0-cell, as we show in Proposition 3D.1. An exercise at the end of the section is to show that a topological group with a finite-dimensional CW structure is an orientable manifold, so $SO(n)$ is a closed orientable manifold. From the CW structure it follows that its dimension is $n(n-1)/2$. These facts can also be proved using fiber bundles.

The group $O(n)$ is a subgroup of $GL_n(\mathbb{R})$, the 'general linear group' of all invertible $n \times n$ matrices with entries in $\mathbb{R}$, discussed near the beginning of §3.C. The Gram–Schmidt orthogonalization process applied to the columns of matrices in $GL_n(\mathbb{R})$ provides a retraction $r : GL_n(\mathbb{R}) \to O(n)$, continuity of r being evident from the explicit formulas for the Gram–Schmidt process. By inserting appropriate scalar factors into these formulas it is easy to see that $O(n)$ is in fact a deformation retract of $GL_n(\mathbb{R})$. Using a bit more linear algebra, namely the polar decomposition, it is possible to show that $GL_n(\mathbb{R})$ is actually homeomorphic to $O(n) \times \mathbb{R}^k$ for $k = n(n+1)/2$.

The topological structure of $SO(n)$ for small values of n can be described in terms of more familiar spaces:

- $SO(1)$ is a point.

- $SO(2)$, the rotations of $\mathbb{R}^2$, is both homeomorphic and isomorphic as a group to S^1, thought of as the unit complex numbers.

- $SO(3)$ is homeomorphic to $\mathbb{R}P^3$. To see this, let $\varphi : D^3 \to SO(3)$ send a nonzero vector x to the rotation through angle $|x|\pi$ about the axis formed by the line through the origin in the direction of x. An orientation convention such as the 'right-hand rule' is needed to make this unambiguous. By continuity, φ then sends 0 to the identity. Antipodal points of $S^2 = \partial D^3$ are sent to the same rotation through angle π, so φ induces a map $\overline{\varphi} : \mathbb{R}P^3 \to SO(3)$, regarding $\mathbb{R}P^3$ as D^3 with antipodal boundary points identified. The map $\overline{\varphi}$ is clearly injective since the axis of a nontrivial rotation is uniquely determined as its fixed point set, and $\overline{\varphi}$ is surjective since by easy linear algebra each nonidentity element of $SO(3)$ is a rotation about some axis. It follows that $\overline{\varphi}$ is a homeomorphism $\mathbb{R}\mathrm{P}^3 \approx SO(3)$.

- $SO(4)$ is homeomorphic to $S^3 \times SO(3)$. Identifying $\mathbb{R}^4$ with the quaternions $\mathbb{H}$ and S^3 with the group of unit quaternions, the quaternion multiplication $v \mapsto vw$ for fixed $w \in S^3$ defines an isometry $\rho_w \in O(4)$ since $|vw| = |v||w| = |v|$ if $|w| = 1$. Points of $O(4)$ are 4-tuples $(v_1, \cdots, v_4)$ of orthonormal vectors $v_i \in \mathbb{H} = \mathbb{R}^4$, and we view $O(3)$ as the subspace with $v_1 = 1$. A homeomorphism $S^3 \times O(3) \to O(4)$ is defined by sending $(v, (1, v_2, v_3, v_4))$ to $(v, v_2 v, v_3 v, v_4 v) = \rho_v(1, v_2, v_3, v_4)$, with inverse $(v, v_2, v_3, v_4) \mapsto (v, (1, v_2 v^{-1}, v_3 v^{-1}, v_4 v^{-1})) = (v, \rho_{v^{-1}}(v, v_2, v_3, v_4))$. Restricting to identity components, we obtain a homeomorphism $S^3 \times SO(3) \approx SO(4)$. This is not a group isomorphism, however. It can be shown, though we will not digress to do so here, that the homomorphism $\psi : S^3 \times S^3 \to SO(4)$ sending a pair (u, v) of unit quaternions to the isometry $w \mapsto uwv^{-1}$ of $\mathbb{H}$ is surjective with kernel $\mathbb{Z}_2 = \{\pm(1,1)\}$, and that ψ is a covering space projection, representing $S^3 \times S^3$ as a 2-sheeted cover of $SO(4)$, the universal cover. Restricting ψ to the diagonal $S^3 = \{(u, u)\} \subset S^3 \times S^3$ gives the universal cover $S^3 \to SO(3)$, so $SO(3)$ is isomorphic to the quotient group of S^3 by the normal subgroup $\{\pm 1\}$.

Using octonions one can construct in the same way a homeomorphism $SO(8) \approx S^7 \times SO(7)$. But in all other cases $SO(n)$ is only a 'twisted product' of $SO(n-1)$ and S^{n-1}; see Example 4.55 and the discussion following Corollary 4D.3.

Cell Structure

Our first task is to construct a CW structure on $SO(n)$. This will come with a very nice cellular map $\rho : \mathbb{R}\mathrm{P}^{n-1} \times \mathbb{R}\mathrm{P}^{n-2} \times \cdots \times \mathbb{R}\mathrm{P}^1 \to SO(n)$. To simplify notation we will write P^i for $\mathbb{R}\mathrm{P}^i$.

To each nonzero vector $v \in \mathbb{R}^n$ we can associate the reflection $r(v) \in O(n)$ across the hyperplane consisting of all vectors orthogonal to v. Since $r(v)$ is a reflection, it has determinant -1, so to get an element of $SO(n)$ we consider the composition $\rho(v) = r(v)r(e_1)$ where e_1 is the first standard basis vector $(1, 0, \cdots, 0)$. Since $\rho(v)$ depends only on the line spanned by v, ρ defines a map $P^{n-1} \to SO(n)$. This map is injective since it is the composition of $v \mapsto r(v)$, which is obviously an injection of P^{n-1} into $O(n) - SO(n)$, with the homeomorphism $O(n) - SO(n) \to SO(n)$ given by right-multiplication by $r(e_1)$. Since ρ is injective and P^{n-1} is compact Hausdorff, we may think of ρ as embedding P^{n-1} as a subspace of $SO(n)$.

More generally, for a sequence $I = (i_1, \cdots, i_m)$ with each $i_j < n$, we define a map $\rho : P^I = P^{i_1} \times \cdots \times P^{i_m} \to SO(n)$ by letting $\rho(v_1, \cdots, v_m)$ be the composition $\rho(v_1) \cdots \rho(v_m)$. If $\varphi^i : D^i \to P^i$ is the standard characteristic map for the i-cell of P^i, restricting to the 2-sheeted covering projection $\partial D^i \to P^{i-1}$, then the product $\varphi^I : D^I \to P^I$ of the appropriate φ^{i_j}'s is a characteristic map for the top-dimensional cell of P^I. We will be especially interested in the sequences $I = (i_1, \cdots, i_m)$ satisfying $n > i_1 > \cdots > i_m > 0$. These sequences will be called *admissible*, as will the sequence consisting of a single 0.

Proposition 3D.1. *The maps $\rho\varphi^I : D^I \to SO(n)$, for I ranging over all admissible sequences, are the characteristic maps of a CW structure on $SO(n)$ for which the map $\rho : P^{n-1} \times P^{n-2} \times \cdots \times P^1 \to SO(n)$ is cellular.*

In particular, there is a single 0-cell $e^0 = \{\mathbb{1}\}$, so $SO(n)$ is path-connected. The other cells $e^I = e^{i_1} \cdots e^{i_m}$ are products, via the group operation in $SO(n)$, of the cells $e^i \subset P^{n-1} \subset SO(n)$.

Proof: According to Proposition A.2 in the Appendix, there are three things to show in order to obtain the CW structure:

(1) For each decreasing sequence I, $\rho\varphi^I$ is a homeomorphism from the interior of D^I onto its image.

(2) The resulting image cells e^I arc all disjoint and cover $SO(n)$.

(3) For each e^I, $\rho\varphi^I(\partial D^I)$ is contained in a union of cells of lower dimension than e^I.

To begin the verification of these properties, define $p : SO(n) \to S^{n-1}$ by evaluation at the vector $e_n = (0, \cdots, 0, 1)$, $p(\alpha) = \alpha(e_n)$. Isometries in $P^{n-2} \subset P^{n-1} \subset SO(n)$ fix e_n, so $p(P^{n-2}) = \{e_n\}$. We claim that p is a homeomorphism from $P^{n-1} - P^{n-2}$ onto $S^{n-1} - \{e_n\}$. This can be seen as follows. Thinking of a point in P^{n-1} as a vector v, the map p takes this to $\rho(v)(e_n) = r(v)r(e_1)(e_n)$, which equals $r(v)(e_n)$ since e_n is in the hyperplane orthogonal to e_1. From the picture at the right it is then clear that p simply stretches the lower half of each meridian circle in S^{n-1} onto

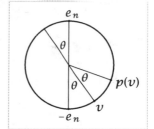

the whole meridian circle, doubling the angle up from the south pole, so $P^{n-1} - P^{n-2}$, represented by vectors whose last coordinate is negative, is taken homeomorphically onto $S^{n-1} - \{e_n\}$.

The next statement is that the map

$$h : (P^{n-1} \times SO(n-1), P^{n-2} \times SO(n-1)) \to (SO(n), SO(n-1)), \quad h(v, \alpha) = \rho(v)\alpha$$

is a homeomorphism from $(P^{n-1} - P^{n-2}) \times SO(n-1)$ onto $SO(n) - SO(n-1)$. Here we view $SO(n-1)$ as the subgroup of $SO(n)$ fixing the vector e_n. To construct an inverse to this homeomorphism, let $\beta \in SO(n) - SO(n-1)$ be given. Then $\beta(e_n) \neq e_n$ so by the preceding paragraph there is a unique $v_\beta \in P^{n-1} - P^{n-2}$ with $\rho(v_\beta)(e_n) = \beta(e_n)$, and v_β depends continuously on β since $\beta(e_n)$ does. The composition $\alpha_\beta = \rho(v_\beta)^{-1}\beta$ then fixes e_n, hence lies in $SO(n-1)$. Since $\rho(v_\beta)\alpha_\beta = \beta$, the map $\beta \mapsto (v_\beta, \alpha_\beta)$ is an inverse to h on $SO(n) - SO(n-1)$.

Statements (1) and (2) can now be proved by induction on n. The map ρ takes P^{n-2} to $SO(n-1)$, so we may assume inductively that the maps $\rho\varphi^I$ for I ranging over admissible sequences with first term $i_1 < n - 1$ are the characteristic maps for a CW structure on $SO(n-1)$, with cells the corresponding products e^I. The admissible sequences I with $i_1 = n - 1$ then give disjoint cells e^I covering $SO(n) - SO(n-1)$ by what was shown in the previous paragraph. So (1) and (2) hold for $SO(n)$.

To prove (3) it suffices to show there is an inclusion $P^i P^i \subset P^i P^{i-1}$ in $SO(n)$ since for an admissible sequence I, the map $\rho : P^I \to SO(n)$ takes the boundary of the top-dimensional cell of P^I to the image of products P^J with J obtained from I by decreasing one term i_j by 1, yielding a sequence which is admissible except perhaps for having two successive terms equal. As a preliminary to showing that $P^i P^i \subset P^i P^{i-1}$, observe that for $\alpha \in O(n)$ we have $r(\alpha(v)) = \alpha r(v) \alpha^{-1}$. Hence $\rho(v)\rho(w) = r(v)r(e_1)r(w)r(e_1) = r(v)r(w')$ where $w' = r(e_1)w$. Thus to show $P^i P^i \subset P^i P^{i-1}$ it suffices to find for each pair $v, w \in \mathbb{R}^{i+1}$ a pair $x \in \mathbb{R}^{i+1}$, $y \in \mathbb{R}^i$ with $r(v)r(w) = r(x)r(y)$.

Let $V \subset \mathbb{R}^{i+1}$ be a 2-dimensional subspace containing v and w. Since $V \cap \mathbb{R}^i$ is at least 1-dimensional, we can choose a unit vector $y \in V \cap \mathbb{R}^i$. Let $\alpha \in O(i+1)$ take V to $\mathbb{R}^2$ and y to e_1. Then the conjugate $\alpha r(v)r(w)\alpha^{-1} = r(\alpha(v))r(\alpha(w))$ lies in $SO(2)$, hence has the form $\rho(z) = r(z)r(e_1)$ for some $z \in \mathbb{R}^2$ by statement (2) for $n = 2$. Therefore

$$r(v)r(w) = \alpha^{-1}r(z)r(e_1)\alpha = r(\alpha^{-1}(z))r(\alpha^{-1}(e_1)) = r(x)r(y)$$

for $x = \alpha^{-1}(z) \in \mathbb{R}^{i+1}$ and $y \in \mathbb{R}^i$.

It remains to show that the map $\rho : P^{n-1} \times P^{n-2} \times \cdots \times P^1 \to SO(n)$ is cellular. This follows from the inclusions $P^i P^i \subset P^i P^{i-1}$ derived above, together with another family of inclusions $P^i P^j \subset P^j P^i$ for $i < j$. To prove the latter we have the formulas

$$
\begin{aligned}
\rho(v)\rho(w) &= r(v)r(w') \quad && \text{where } w' = r(e_1)w, \text{ as earlier} \\
&= r(v)r(w')r(v)r(v) \\
&= r(r(v)w')r(v) \quad && \text{from } r(\alpha(v)) = \alpha r(v)\alpha^{-1} \\
&= r(r(v)r(e_1)w)r(v) = r(\rho(v)w)r(v) \\
&= \rho(\rho(v)w)\rho(v') \quad && \text{where } v' = r(e_1)v, \text{ hence } v = r(e_1)v'
\end{aligned}
$$

In particular, taking $v \in \mathbb{R}^{i+1}$ and $w \in \mathbb{R}^{j+1}$ with $i < j$, we have $\rho(v)w \in \mathbb{R}^{j+1}$, and the product $\rho(v)\rho(w) \in P^i P^j$ equals the product $\rho(\rho(v)w)\rho(v') \in P^j P^i$. $\square$

Mod 2 Homology and Cohomology

Each cell of $SO(n)$ is the homeomorphic image of a cell in $P^{n-1} \times P^{n-2} \times \cdots \times P^1$, so the cellular chain map induced by $\rho : P^{n-1} \times P^{n-2} \times \cdots \times P^1 \to SO(n)$ is surjective. It follows that with $\mathbb{Z}_2$ coefficients the cellular boundary maps for $SO(n)$ are all trivial since this is true in P^i and hence in $P^{n-1} \times P^{n-2} \times \cdots \times P^1$ by Proposition 3B.1. Thus $H_*(SO(n); \mathbb{Z}_2)$ has a $\mathbb{Z}_2$ summand for each cell of $SO(n)$. One can rephrase this as saying that there are isomorphisms $H_i(SO(n);\mathbb{Z}_2) \approx H_i(S^{n-1} \times S^{n-2} \times \cdots \times S^1;\mathbb{Z}_2)$ for all i since this product of spheres also has cells in one-to-one correspondence with admissible sequences. The full structure of the $\mathbb{Z}_2$ homology and cohomology rings is given by:

Theorem 3D.2. **(a)** $H^*(SO(n);\mathbb{Z}_2) \approx \otimes_{i \text{ odd}} \mathbb{Z}_2[\beta_i]/(\beta_i^{p_i})$ *where* $|\beta_i| = i$ *and* p_i *is the smallest power of* 2 *such that* $|\beta_i^{p_i}| \geq n$.

(b) *The Pontryagin ring* $H_*(SO(n);\mathbb{Z}_2)$ *is the exterior algebra* $\Lambda_{\mathbb{Z}_2}[e^1, \cdots, e^{n-1}]$.

Here e^i denotes the cellular homology class of the cell $e^i \subset P^{n-1} \subset SO(n)$, and β_i is the dual class to e^i, represented by the cellular cochain assigning the value 1 to the cell e^i and 0 to all other i-cells.

Proof: As we noted above, ρ induces a surjection on cellular chains. Since the cellular boundary maps with $\mathbb{Z}_2$ coefficients are trivial for both $P^{n-1} \times \cdots \times P^1$ and $SO(n)$, it follows that ρ_* is surjective on $H_*(-;\mathbb{Z}_2)$ and ρ^* is injective on $H^*(-;\mathbb{Z}_2)$. We know that $H^*(P^{n-1} \times \cdots \times P^1;\mathbb{Z}_2)$ is the polynomial ring $\mathbb{Z}_2[\alpha_1, \cdots, \alpha_{n-1}]$ truncated by the relations $\alpha_i^{i+1} = 0$. For $\beta_i \in H^i(SO(n);\mathbb{Z}_2)$ the dual class to e^i, we have $\rho^*(\beta_i) = \sum_j \alpha_j^i$, the class assigning 1 to each i-cell in a factor P^j of $P^{n-1} \times \cdots \times P^1$ and 0 to all other i-cells, which are products of lower-dimensional cells and hence map to cells in $SO(n)$ disjoint from e^i.

First we will show that the monomials $\beta_I = \beta_{i_1} \cdots \beta_{i_m}$ corresponding to admissible sequences I are linearly independent in $H^*(SO(n);\mathbb{Z}_2)$, hence are a vector space basis. Since ρ^* is injective, we may identify each β_i with its image $\sum_j \alpha_j^i$ in the truncated polynomial ring $\mathbb{Z}_2[\alpha_1, \cdots, \alpha_{n-1}]/(\alpha_1^2, \cdots, \alpha_{n-1}^n)$. Suppose we have a linear relation $\sum_I b_I \beta_I = 0$ with $b_I \in \mathbb{Z}_2$ and I ranging over the admissible sequences. Since each β_I is a product of distinct β_i's, we can write the relation in the form $x\beta_1 + y = 0$ where neither x nor y has β_1 as a factor. Since α_1 occurs only in the term β_1 of $x\beta_1 + y$, where it has exponent 1, we have $x\beta_1 + y = x\alpha_1 + z$ where neither x nor z involves α_1. The relation $x\alpha_1 + z = 0$ in $\mathbb{Z}_2[\alpha_1, \cdots, \alpha_{n-1}]/(\alpha_1^2, \cdots, \alpha_{n-1}^n)$ then implies $x = 0$. Thus we may assume the original relation does not involve β_1. Now we repeat the argument for β_2. Write the relation in the form $x\beta_2 + y = 0$ where neither x nor y involves β_2 or β_1. The variable α_2 now occurs only in the term β_2 of $x\beta_2 + y$, where it has exponent 2, so we have $x\beta_2 + y = x\alpha_2^2 + z$ where x and z do not involve α_1 or α_2. Then $x\alpha_2^2 + z = 0$ implies $x = 0$ and we have a relation involving neither β_1 nor β_2. Continuing inductively, we eventually deduce that all coefficients b_I in the original relation $\sum_I b_I \beta_I = 0$ must be zero.

Observe now that $\beta_i^2 = \beta_{2i}$ if $2i < n$ and $\beta_i^2 = 0$ if $2i \geq n$, since $(\sum_j \alpha_j^i)^2 = \sum_j \alpha_j^{2i}$. The quotient Q of the algebra $\mathbb{Z}_2[\beta_1, \beta_2, \cdots]$ by the relations $\beta_i^2 = \beta_{2i}$ and $\beta_j = 0$ for $j \geq n$ then maps onto $H^*(SO(n);\mathbb{Z}_2)$. This map $Q \to H^*(SO(n);\mathbb{Z}_2)$ is also injective since the relations defining Q allow every element of Q to be represented as a linear combination of admissible monomials β_I, and the admissible monomials are linearly independent in $H^*(SO(n);\mathbb{Z}_2)$. The algebra Q can also be described as the tensor product in statement (a) of the theorem since the relations $\beta_i^2 = \beta_{2i}$ allow admissible monomials to be written uniquely as monomials in powers of the β_i's with i odd, and the relation $\beta_j = 0$ for $j \geq n$ becomes $\beta_{ip_i} = \beta_i^{p_i} = 0$ where $j = ip_i$ with i odd and p_i a power of 2. For a given i, this relation holds iff $ip_i \geq n$, or in other words, iff $|\beta_i^{p_i}| \geq n$. This finishes the proof of (a).

For part (b), note first that the group multiplication $SO(n) \times SO(n) \to SO(n)$ is cellular in view of the inclusions $P^i P^i \subset P^i P^{i-1}$ and $P^i P^j \subset P^j P^i$ for $i < j$. So we can compute Pontryagin products at the cellular level. We know that there is at least an additive isomorphism $H_*(SO(n);\mathbb{Z}_2) \approx \Lambda_{\mathbb{Z}_2}[e^1, \cdots, e^{n-1}]$ since the products $e^I = e^{i_1} \cdots e^{i_m}$ with I admissible form a basis for $H_*(SO(n);\mathbb{Z}_2)$. The inclusion $P^i P^i \subset P^i P^{i-1}$ then implies that the Pontryagin product $(e^i)^2$ is 0. It remains only to see the commutativity relation $e^i e^j = e^j e^i$. The inclusion $P^i P^j \subset P^j P^i$ for $i < j$ was obtained from the formula $\rho(v)\rho(w) = \rho(\rho(v)w)\rho(v')$ for $v \in \mathbb{R}^{i+1}$, $w \in \mathbb{R}^{j+1}$, and $v' = r(e_1)v$. The map $f : P^i \times P^j \to P^j \times P^i$, $f(v,w) = (\rho(v)w, v')$, is a homeomorphism since it is the composition of homeomorphisms $(v,w) \mapsto (v, \rho(v)w) \mapsto (v', \rho(v)w) \mapsto (\rho(v)w, v')$. The first of these maps takes $e^i \times e^j$ homeomorphically onto itself since $\rho(v)(e^j) = e^j$ if $i < j$. Obviously the second map also takes $e^i \times e^j$ homeomorphically onto itself, while the third map simply transposes the two factors. Thus f restricts to a homeomorphism from $e^i \times e^j$ onto $e^j \times e^i$, and therefore $e^i e^j = e^j e^i$ in $H_*(SO(n);\mathbb{Z}_2)$. $\qquad\qquad\qquad\qquad\qquad\qquad\qquad\qquad\qquad\qquad$ $\square$

The cup product and Pontryagin product structures in this theorem may seem at first glance to be unrelated, but in fact the relationship is fairly direct. As we saw in the previous section, the dual of a polynomial algebra $\mathbb{Z}_2[x]$ is a divided polynomial algebra $\Gamma_{\mathbb{Z}_2}[\alpha]$, and with $\mathbb{Z}_2$ coefficients the latter is an exterior algebra $\Lambda_{\mathbb{Z}_2}[\alpha_0, \alpha_1, \cdots]$ where $|\alpha_i| = 2^i |x|$. If we truncate the polynomial algebra by a relation $x^{2^n} = 0$, then this just eliminates the generators α_i for $i \geq n$. In view of this, if it were the case that the generators β_i for the algebra $H^*(SO(n);\mathbb{Z}_2)$ happened to be primitive, then $H^*(SO(n);\mathbb{Z}_2)$ would be isomorphic as a Hopf algebra to the tensor product of the single-generator Hopf algebras $\mathbb{Z}_2[\beta_i]/(\beta_i^{p_i})$, $i = 1, 3, \cdots$, hence the dual algebra $H_*(SO(n);\mathbb{Z}_2)$ would be the tensor product of the corresponding truncated divided polynomial algebras, in other words an exterior algebra as just explained. This is in fact the structure of $H_*(SO(n);\mathbb{Z}_2)$, so since the Pontryagin product in $H_*(SO(n);\mathbb{Z}_2)$ determines the coproduct in $H^*(SO(n);\mathbb{Z}_2)$ uniquely, it follows that the β_i's must indeed be primitive.

It is not difficult to give a direct argument that each β_i is primitive. The coproduct $\Delta : H^*(SO(n);\mathbb{Z}_2) \to H^*(SO(n);\mathbb{Z}_2) \otimes H^*(SO(n);\mathbb{Z}_2)$ is induced by the group multiplication $\mu : SO(n) \times SO(n) \to SO(n)$. We need to show that the value of $\Delta(\beta_i)$ on $e^I \otimes e^J$, which we denote $\langle \Delta(\beta_i), e^I \otimes e^J \rangle$, is the same as the value $\langle \beta_i \otimes 1 + 1 \otimes \beta_i, e^I \otimes e^J \rangle$

for all cells e^I and e^J whose dimensions add up to i. Since $\Delta = \mu^*$, we have $\langle \Delta(\beta_i), e^I \otimes e^J \rangle = \langle \beta_i, \mu_*(e^I \otimes e^J) \rangle$. Because μ is the multiplication map, $\mu(e^I \times e^J)$ is contained in $P^I P^J$, and if we use the relations $P^j P^j \subset P^j P^{j-1}$ and $P^j P^k \subset P^k P^j$ for $j < k$ to rearrange the factors P^j of $P^I P^J$ so that their dimensions are in decreasing order, then the only way we will end up with a term P^i is if we start with $P^I P^J$ equal to $P^i P^0$ or $P^0 P^i$. Thus $\langle \beta_i, \mu_*(e^I \otimes e^J) \rangle = 0$ unless $e^I \otimes e^J$ equals $e^i \otimes e^0$ or $e^0 \otimes e^i$. Hence $\Delta(\beta_i)$ contains no other terms besides $\beta_i \otimes 1 + 1 \otimes \beta_i$, and β_i is primitive.

Integer Homology and Cohomology

With $\mathbb{Z}$ coefficients the homology and cohomology of $SO(n)$ turns out to be a good bit more complicated than with $\mathbb{Z}_2$ coefficients. One can see a little of this complexity already for small values of n, where the homeomorphisms $SO(3) \approx \mathbb{R}P^3$ and $SO(4) \approx S^3 \times \mathbb{R}P^3$ would allow one to compute the additive structure as a direct sum of a certain number of $\mathbb{Z}$'s and $\mathbb{Z}_2$'s. For larger values of n the additive structure is qualitatively the same:

|| **Proposition 3D.3.** $H_*(SO(n); \mathbb{Z})$ is a direct sum of $\mathbb{Z}$'s and $\mathbb{Z}_2$'s.

Proof: We compute the cellular chain complex of $SO(n)$, showing that it splits as a tensor product of simpler complexes. For a cell $e^i \subset P^{n-1} \subset SO(n)$ the cellular boundary de^i is $2e^{i-1}$ for even $i > 0$ and 0 for odd i. To compute the cellular boundary of a cell $e^{i_1} \cdots e^{i_m}$ we can pull it back to a cell $e^{i_1} \times \cdots \times e^{i_m}$ of $P^{n-1} \times \cdots \times P^1$ whose cellular boundary, by Proposition 3B.1, is $\sum_j (-1)^{\sigma_j} e^{i_1} \times \cdots \times de^{i_j} \times \cdots \times e^{i_m}$ where $\sigma_j = i_1 + \cdots + i_{j-1}$. Hence $d(e^{i_1} \cdots e^{i_m}) = \sum_j (-1)^{\sigma_j} e^{i_1} \cdots de^{i_j} \cdots e^{i_m}$, where it is understood that $e^{i_1} \cdots de^{i_j} \cdots e^{i_m}$ is zero if $i_j = i_{j+1} + 1$ since $P^{i_j-1} P^{i_j-1} \subset P^{i_j-1} P^{i_j-2}$, in a lower-dimensional skeleton.

To split the cellular chain complex $C_*(SO(n))$ as a tensor product of smaller chain complexes, let C^{2i} be the subcomplex of $C_*(SO(n))$ with basis the cells e^0, e^{2i}, e^{2i-1}, and $e^{2i} e^{2i-1}$. This is a subcomplex since $de^{2i-1} = 0$, $de^{2i} = 2e^{2i-1}$, and, in $P^{2i} \times P^{2i-1}$, $d(e^{2i} \times e^{2i-1}) = de^{2i} \times e^{2i-1} + e^{2i} \times de^{2i-1} = 2e^{2i-1} \times e^{2i-1}$, hence $d(e^{2i} e^{2i-1}) = 0$ since $P^{2i-1} P^{2i-1} \subset P^{2i-1} P^{2i-2}$. The claim is that there are chain complex isomorphisms

$$C_*(SO(2k+1)) \approx C^2 \otimes C^4 \otimes \cdots \otimes C^{2k}$$
$$C_*(SO(2k+2)) \approx C^2 \otimes C^4 \otimes \cdots \otimes C^{2k} \otimes C^{2k+1}$$

where C^{2k+1} has basis e^0 and e^{2k+1}. Certainly these isomorphisms hold for the chain groups themselves, so it is only a matter of checking that the boundary maps agree. For the case of $C_*(SO(2k+1))$ this can be seen by induction on k, as the reader can easily verify. Then the case of $C_*(SO(2k+2))$ reduces to the first case by a similar argument.

Since $H_*(C^{2i})$ consists of $\mathbb{Z}$'s in dimensions 0 and $4i-1$ and a $\mathbb{Z}_2$ in dimension $2i-1$, while $H_*(C^{2k+1})$ consists of $\mathbb{Z}$'s in dimensions 0 and $2k+1$, we conclude from the algebraic Künneth formula that $H_*(SO(n);\mathbb{Z})$ is a direct sum of $\mathbb{Z}$'s and $\mathbb{Z}_2$'s. □

Note that the calculation shows that $SO(2k)$ and $SO(2k-1) \times S^{2k-1}$ have isomorphic homology groups in all dimensions.

In view of the preceding proposition, one can get rather complete information about $H_*(SO(n);\mathbb{Z})$ by considering the natural maps to $H_*(SO(n);\mathbb{Z}_2)$ and to the quotient of $H_*(SO(n);\mathbb{Z})$ by its torsion subgroup. Let us denote this quotient by $H_*^{free}(SO(n);\mathbb{Z})$. The same strategy applies equally well to cohomology, and the universal coefficient theorem gives an isomorphism $H_{free}^*(SO(n);\mathbb{Z}) \approx H_*^{free}(SO(n);\mathbb{Z})$.

The proof of the proposition shows that the additive structure of $H_*^{free}(SO(n);\mathbb{Z})$ is fairly simple:

$$H_*^{free}(SO(2k+1);\mathbb{Z}) \approx H_*(S^3 \times S^7 \times \cdots \times S^{4k-1})$$
$$H_*^{free}(SO(2k+2);\mathbb{Z}) \approx H_*(S^3 \times S^7 \times \cdots \times S^{4k-1} \times S^{2k+1})$$

The multiplicative structure is also as simple as it could be:

Proposition 3D.4. *The Pontryagin ring $H_*^{free}(SO(n);\mathbb{Z})$ is an exterior algebra,*

$$H_*^{free}(SO(2k+1);\mathbb{Z}) \approx \Lambda_{\mathbb{Z}}[a_3, a_7, \cdots, a_{4k-1}] \quad \text{where } |a_i| = i$$
$$H_*^{free}(SO(2k+2);\mathbb{Z}) \approx \Lambda_{\mathbb{Z}}[a_3, a_7, \cdots, a_{4k-1}, a'_{2k+1}]$$

The generators a_i and a'_{2k+1} are primitive, so the dual Hopf algebra $H_{free}^*(SO(n);\mathbb{Z})$ is an exterior algebra on the dual generators α_i and α'_{2k+1}.

Proof: As in the case of $\mathbb{Z}_2$ coefficients we can work at the level of cellular chains since the multiplication in $SO(n)$ is cellular. Consider first the case $n = 2k+1$. Let E^i be the cycle $e^{2i}e^{2i-1}$ generating a $\mathbb{Z}$ summand of $H_*(SO(n);\mathbb{Z})$. By what we have shown above, the products $E^{i_1} \cdots E^{i_m}$ with $i_1 > \cdots > i_m$ form an additive basis for $H_*^{free}(SO(n);\mathbb{Z})$, so we need only verify that the multiplication is as in an exterior algebra on the classes E^i. The map f in the proof of Theorem 3D.2 gives a homeomorphism $e^i \times e^j \approx e^j \times e^i$ if $i < j$, and this homeomorphism has local degree $(-1)^{ij+1}$ since it is the composition $(v,w) \mapsto (v,\rho(v)w) \mapsto (v',\rho(v)w) \mapsto (\rho(v)w, v')$ of homeomorphisms with local degrees $+1, -1$, and $(-1)^{ij}$. Applying this four times to commute $E^iE^j = e^{2i}e^{2i-1}e^{2j}e^{2j-1}$ to $E^jE^i = e^{2j}e^{2j-1}e^{2i}e^{2i-1}$, three of the four applications give a sign of -1 and the fourth gives a $+1$, so we conclude that $E^iE^j = -E^jE^i$ if $i < j$. When $i = j$ we have $(E^i)^2 = 0$ since $e^{2i}e^{2i-1}e^{2i}e^{2i-1} = e^{2i}e^{2i}e^{2i-1}e^{2i-1}$, which lies in a lower-dimensional skeleton because of the relation $P^{2i}P^{2i} \subset P^{2i}P^{2i-1}$.

Thus we have shown that $H_*(SO(2k+1);\mathbb{Z})$ contains $\Lambda_{\mathbb{Z}}[E^1, \cdots, E^k]$ as a subalgebra. The same reasoning shows that $H_*(SO(2k+2);\mathbb{Z})$ contains the subalgebra $\Lambda_{\mathbb{Z}}[E^1, \cdots, E^k, e^{2k+1}]$. These exterior subalgebras account for all the nontorsion in $H_*(SO(n);\mathbb{Z})$, so the product structure in $H_*^{free}(SO(n);\mathbb{Z})$ is as stated.

Now we show that the generators E^i and e^{2k+1} are primitive in $H_*^{free}(SO(n);\mathbb{Z})$. Looking at the formula for the boundary maps in the cellular chain complex of $SO(n)$, we see that this chain complex is the direct sum of the subcomplexes $C(m)$ with basis the m-fold products $e^{i_1} \cdots e^{i_m}$ with $i_1 > \cdots > i_m > 0$. We allow $m = 0$ here, with $C(0)$ having basis the 0-cell of $SO(n)$. The direct sum $C(0) \oplus \cdots \oplus C(m)$ is the cellular chain complex of the subcomplex of $SO(n)$ consisting of cells that are products of m or fewer cells e^i. In particular, taking $m = 2$ we have a subcomplex $X \subset SO(n)$ whose homology, mod torsion, consists of the $\mathbb{Z}$ in dimension zero and the $\mathbb{Z}$'s generated by the cells E^i, together with the cell e^{2k+1} when $n = 2k + 2$. The inclusion $X \hookrightarrow SO(n)$ induces a commutative diagram

$$
\begin{array}{ccc}
H_*^{free}(X;\mathbb{Z}) & \xrightarrow{\Delta} & H_*^{free}(X;\mathbb{Z}) \otimes H_*^{free}(X;\mathbb{Z}) \\
\downarrow & & \downarrow \\
H_*^{free}(SO(n);\mathbb{Z}) & \xrightarrow{\Delta} & H_*^{free}(SO(n);\mathbb{Z}) \otimes H_*^{free}(SO(n);\mathbb{Z})
\end{array}
$$

where the lower Δ is the coproduct in $H_*^{free}(SO(n);\mathbb{Z})$ and the upper Δ is its analog for X, coming from the diagonal map $X \to X \times X$ and the Künneth formula. The classes E^i in the lower left group pull back to elements we label $\tilde{E}^i$ in the upper left group. Since these have odd dimension and $H_*^{free}(X;\mathbb{Z})$ vanishes in even positive dimensions, the images $\Delta(\tilde{E}^i)$ can have no components $a \otimes b$ with both a and b positive-dimensional. The same is therefore true for $\Delta(E^i)$ by commutativity of the diagram, so the classes E^i are primitive. This argument also works for e^{2k+1} when $n = 2k + 2$.

Since the exterior algebra generators of $H_*^{free}(SO(n);\mathbb{Z})$ are primitive, this algebra splits as a Hopf algebra into a tensor product of single-generator exterior algebras $\Lambda_{\mathbb{Z}}[a_i]$ (and $\Lambda_{\mathbb{Z}}[a'_{2k+1}]$). The dual Hopf algebra $H^*_{free}(SO(n);\mathbb{Z})$ therefore splits as the tensor product of the dual exterior algebras $\Lambda_{\mathbb{Z}}[\alpha_i]$ (and $\Lambda_{\mathbb{Z}}[\alpha'_{2k+1}]$), hence $H^*_{free}(SO(n);\mathbb{Z})$ is also an exterior algebra. $\qquad\square$

The exact ring structure of $H^*(SO(n);\mathbb{Z})$ can be deduced from these results via Bockstein homomorphisms, as we show in Example 3E.7, though the process is somewhat laborious and the answer not very neat.

Stiefel Manifolds

Consider the **Stiefel manifold** $V_{n,k}$, whose points are the *orthonormal k-frames* in $\mathbb{R}^n$, that is, orthonormal k-tuples of vectors. Thus $V_{n,k}$ is a subset of the product of k copies of S^{n-1}, and it is given the subspace topology. As special cases, $V_{n,n} = O(n)$ and $V_{n,1} = S^{n-1}$. Also, $V_{n,2}$ can be identified with the space of unit tangent vectors to S^{n-1} since a vector v at the point $x \in S^{n-1}$ is tangent to S^{n-1} iff it is orthogonal to x. We can also identify $V_{n,n-1}$ with $SO(n)$ since there is a unique way of extending an orthonormal $(n-1)$-frame to a positively oriented orthonormal n-frame.

There is a natural projection $p : O(n) \to V_{n,k}$ sending $\alpha \in O(n)$ to the k-frame consisting of the last k columns of α, which are the images under α of the last k standard basis vectors in $\mathbb{R}^n$. This projection is onto, and the preimages of points are precisely the cosets $\alpha O(n-k)$, where we embed $O(n-k)$ in $O(n)$ as the orthogonal transformations of the first $n-k$ coordinates of $\mathbb{R}^n$. Thus $V_{n,k}$ can be viewed as the space $O(n)/O(n-k)$ of such cosets, with the quotient topology from $O(n)$. This is the same as the previously defined topology on $V_{n,k}$ since the projection $O(n) \to V_{n,k}$ is a surjection of compact Hausdorff spaces.

When $k < n$ the projection $p : SO(n) \to V_{n,k}$ is surjective, and $V_{n,k}$ can also be viewed as the coset space $SO(n)/SO(n-k)$. We can use this to induce a CW structure on $V_{n,k}$ from the CW structure on $SO(n)$. The cells are the sets of cosets of the form $e^I SO(n-k) = e^{i_1} \cdots e^{i_m} SO(n-k)$ for $n > i_1 > \cdots > i_m \geq n-k$, together with the coset $SO(n-k)$ itself as a 0-cell of $V_{n,k}$. These sets of cosets are unions of cells of $SO(n)$ since $SO(n-k)$ consists of the cells $e^J = e^{j_1} \cdots e^{j_\ell}$ with $n-k > j_1 > \cdots > j_\ell$. This implies that $V_{n,k}$ is the disjoint union of its cells, and the boundary of each cell is contained in cells of lower dimension, so we do have a CW structure.

Since the projection $SO(n) \to V_{n,k}$ is a cellular map, the structure of the cellular chain complex of $V_{n,k}$ can easily be deduced from that of $SO(n)$. For example, the cellular chain complex of $V_{2k+1,2}$ is just the complex C^{2k} defined earlier, while for $V_{2k,2}$ the cellular boundary maps are all trivial. Hence the nonzero homology groups of $V_{n,2}$ are

$$H_i(V_{2k+1,2}; \mathbb{Z}) = \begin{cases} \mathbb{Z} & \text{for } i = 0,\, 4k-1 \\ \mathbb{Z}_2 & \text{for } i = 2k-1 \end{cases}$$

$$H_i(V_{2k,2}; \mathbb{Z}) = \mathbb{Z} \quad \text{for } i = 0,\, 2k-2,\, 2k-1,\, 4k-3$$

Thus $SO(n)$ has the same homology and cohomology groups as the product space $V_{3,2} \times V_{5,2} \times \cdots \times V_{2k+1,2}$ when $n = 2k+1$, or as $V_{3,2} \times V_{5,2} \times \cdots \times V_{2k+1,2} \times S^{2k+1}$ when $n = 2k+2$. However, our calculations show that $SO(n)$ is distinguished from these products by its cup product structure with $\mathbb{Z}_2$ coefficients, at least when $n \geq 5$, since β_1^4 is nonzero in $H^4(SO(n); \mathbb{Z}_2)$ if $n \geq 5$, while for the product spaces the nontrivial element of $H^1(-; \mathbb{Z}_2)$ must lie in the factor $V_{3,2}$, and $H^4(V_{3,2}; \mathbb{Z}_2) = 0$. When $n = 4$ we have $SO(4)$ homeomorphic to $SO(3) \times S^3 = V_{3,2} \times S^3$ as we noted at the beginning of this section. Also $SO(3) = V_{3,2}$ and $SO(2) = S^1$.

Exercises

1. Show that a topological group with a finite-dimensional CW structure is an orientable manifold. [Consider the homeomorphisms $x \mapsto gx$ or $x \mapsto xg$ for fixed g and varying x in the group.]

2. Using the CW structure on $SO(n)$, show that $\pi_1 SO(n) \approx \mathbb{Z}_2$ for $n \geq 3$. Find a loop representing a generator, and describe how twice this loop is nullhomotopic.

3. Compute the Pontryagin ring structure in $H_*(SO(5); \mathbb{Z})$.

3.E Bockstein Homomorphisms

Homology and cohomology with coefficients in a field, particularly $\mathbb{Z}_p$ with p prime, often have more structure and are easier to compute than with $\mathbb{Z}$ coefficients. Of course, passing from $\mathbb{Z}$ to $\mathbb{Z}_p$ coefficients can involve a certain loss of information, a blurring of finer distinctions. For example, a $\mathbb{Z}_{p^n}$ in integer homology becomes a pair of $\mathbb{Z}_p$'s in $\mathbb{Z}_p$ homology or cohomology, so the exponent n is lost with $\mathbb{Z}_p$ coefficients. In this section we introduce Bockstein homomorphisms, which in many interesting cases allow one to recover $\mathbb{Z}$ coefficient information from $\mathbb{Z}_p$ coefficients. Bockstein homomorphisms also provide a small piece of extra internal structure to $\mathbb{Z}_p$ homology or cohomology itself, which can be quite useful.

We will concentrate on cohomology in order to have cup products available, but the basic constructions work equally well for homology. If we take a short exact sequence $0 \to G \to H \to K \to 0$ of abelian groups and apply the covariant functor $\mathrm{Hom}(C_n(X), -)$, we obtain

$$0 \to C^n(X;G) \to C^n(X;H) \to C^n(X;K) \to 0$$

which is exact since $C_n(X)$ is free. Letting n vary, we have a short exact sequence of chain complexes, so there is an associated long exact sequence

$$\cdots \to H^n(X;G) \to H^n(X;H) \to H^n(X;K) \to H^{n+1}(X;G) \to \cdots$$

whose 'boundary' map $H^n(X;K) \to H^{n+1}(X;G)$ is called a **Bockstein homomorphism**.

We shall be interested primarily in the Bockstein $\beta : H^n(X;\mathbb{Z}_m) \to H^{n+1}(X;\mathbb{Z}_m)$ associated to the coefficient sequence $0 \to \mathbb{Z}_m \xrightarrow{m} \mathbb{Z}_{m^2} \to \mathbb{Z}_m \to 0$, especially when m is prime, but for the moment we do not need this assumption. Closely related to β is the Bockstein $\tilde{\beta} : H^n(X;\mathbb{Z}_m) \to H^{n+1}(X;\mathbb{Z})$ associated to $0 \to \mathbb{Z} \xrightarrow{m} \mathbb{Z} \to \mathbb{Z}_m \to 0$. From the natural map of the latter short exact sequence onto the former one, we obtain the relationship $\beta = \rho\tilde{\beta}$ where $\rho : H^*(X;\mathbb{Z}) \to H^*(X;\mathbb{Z}_m)$ is the homomorphism induced by the map $\mathbb{Z} \to \mathbb{Z}_m$ reducing coefficients mod m. Thus we have a commutative triangle in the following diagram, whose upper row is the exact sequence containing $\tilde{\beta}$.

$$H^n(X;\mathbb{Z}) \xrightarrow{\rho} H^n(X;\mathbb{Z}_m) \xrightarrow{\tilde{\beta}} H^{n+1}(X;\mathbb{Z}) \xrightarrow{m} H^{n+1}(X;\mathbb{Z})$$
$$\searrow_{\beta} \qquad \downarrow_{\rho}$$
$$H^{n+1}(X;\mathbb{Z}_m)$$

Example 3E.1. Let X be a $K(\mathbb{Z}_m, 1)$, for example $\mathbb{RP}^\infty$ when $m = 2$ or an infinite-dimensional lens space with fundamental group $\mathbb{Z}_m$ for arbitrary m. From the homology calculations in Examples 2.42 and 2.43 together with the universal coefficient theorem or cellular cohomology we have $H^n(X;\mathbb{Z}_m) \approx \mathbb{Z}_m$ for all n. Let us show that $\beta : H^n(X;\mathbb{Z}_m) \to H^{n+1}(X;\mathbb{Z}_m)$ is an isomorphism for n odd and zero for n even. If n is odd the vertical map ρ in the diagram above is surjective for $X = K(\mathbb{Z}_m, 1)$, as is $\tilde{\beta}$ since the map m is trivial, so β is surjective, hence an isomorphism. On the other hand, when n is even the first map ρ in the diagram is surjective, so $\tilde{\beta} = 0$ by exactness, hence $\beta = 0$.

A useful fact about β is that it satisfies the derivation property

$$(*) \qquad \beta(a \smile b) = \beta(a) \smile b + (-1)^{|a|} a \smile \beta(b)$$

which comes from the corresponding formula for ordinary coboundary. Namely, let φ and ψ be $\mathbb{Z}_m$ cocycles representing a and b, and let $\tilde{\varphi}$ and $\tilde{\psi}$ be lifts of these to $\mathbb{Z}_{m^2}$ cochains. Concretely, one can view φ and ψ as functions on singular simplices with values in $\{0, 1, \cdots, m-1\}$, and then $\tilde{\varphi}$ and $\tilde{\psi}$ can be taken to be the same functions, but with $\{0, 1, \cdots, m-1\}$ regarded as a subset of $\mathbb{Z}_{m^2}$. Then $\delta\tilde{\varphi} = m\eta$ and $\delta\tilde{\psi} = m\mu$ for $\mathbb{Z}_m$ cocycles η and μ representing $\beta(a)$ and $\beta(b)$. Taking cup products, $\tilde{\varphi} \smile \tilde{\psi}$ is a $\mathbb{Z}_{m^2}$ cochain lifting the $\mathbb{Z}_m$ cocycle $\varphi \smile \psi$, and

$$\delta(\tilde{\varphi} \smile \tilde{\psi}) = \delta\tilde{\varphi} \smile \tilde{\psi} \pm \tilde{\varphi} \smile \delta\tilde{\psi} = m\eta \smile \tilde{\psi} \pm \tilde{\varphi} \smile m\mu = m(\eta \smile \psi \pm \varphi \smile \mu)$$

where the sign $\pm$ is $(-1)^{|a|}$. Hence $\eta \smile \psi + (-1)^{|a|}\varphi \smile \mu$ represents $\beta(a \smile b)$, giving the formula $(*)$.

Example 3E.2: Cup Products in Lens Spaces. The cup product structure for lens spaces was computed in Example 3.41 via Poincaré duality, but using Bocksteins we can deduce it from the cup product structure in $\mathbb{CP}^{\infty}$, which was computed in Theorem 3.19 without Poincaré duality. Consider first the infinite-dimensional lens space $L = S^{\infty}/\mathbb{Z}_m$ where $\mathbb{Z}_m$ acts on the unit sphere S^{∞} in $\mathbb{C}^{\infty}$ by scalar multiplication, so the action is generated by the rotation $v \mapsto e^{2\pi i/m}v$. The quotient map $S^{\infty} \to \mathbb{CP}^{\infty}$ factors through L, so we have a projection $L \to \mathbb{CP}^{\infty}$. Looking at the cell structure on L described in Example 2.43, we see that each even-dimensional cell of L projects homeomorphically onto the corresponding cell of $\mathbb{CP}^{\infty}$. Namely, the $2n$-cell of L is the homeomorphic image of the $2n$-cell in $S^{2n+1} \subset \mathbb{C}^{n+1}$ formed by the points $\cos\theta(z_1, \cdots, z_n, 0) + \sin\theta(0, \cdots, 0, 1)$ with $\sum_i z_i^2 = 1$ and $0 < \theta \leq \pi$, and the same is true for the $2n$-cell of $\mathbb{CP}^{\infty}$. From cellular cohomology it then follows that the map $L \to \mathbb{CP}^{\infty}$ induces isomorphisms on even-dimensional cohomology with $\mathbb{Z}_m$ coefficients. Since $H^*(\mathbb{CP}^{\infty}; \mathbb{Z}_m)$ is a polynomial ring, we deduce that if $y \in H^2(L; \mathbb{Z}_m)$ is a generator, then y^k generates $H^{2k}(L; \mathbb{Z}_m)$ for all k.

By Example 3E.1 there is a generator $x \in H^1(L; \mathbb{Z}_m)$ with $\beta(x) = y$. The product formula $(*)$ gives $\beta(xy^k) = \beta(x)y^k - x\beta(y^k) = y^{k+1}$. Thus β takes xy^k to a generator, hence xy^k must be a generator of $H^{2k+1}(L; \mathbb{Z}_m)$. This completely determines the cup product structure in $H^*(L; \mathbb{Z}_m)$ if m is odd since the commutativity property of cup product implies that $x^2 = 0$ in this case. The result is that $H^*(L; \mathbb{Z}_m) \approx \Lambda_{\mathbb{Z}_m}[x] \otimes \mathbb{Z}_m[y]$ for odd m. When m is even this statement needs to be modified slightly by inserting the relation that x^2 is the unique element of order

2 in $H^2(L;\mathbb{Z}_m) \approx \mathbb{Z}_m$, as we showed in Example 3.9 by an explicit calculation in the 2-skeleton of L.

The cup product structure in finite-dimensional lens spaces follows from this since a finite-dimensional lens space embeds as a skeleton in an infinite-dimensional lens space, and the homotopy type of an infinite-dimensional lens space is determined by its fundamental group since it is a $K(\pi, 1)$. It follows that the cup product structure on a lens space $S^{2n+1}/\mathbb{Z}_m$ with $\mathbb{Z}_m$ coefficients is obtained from the preceding calculation by truncating via the relation $y^{n+1} = 0$.

The relation $\beta = \rho\tilde{\beta}$ implies that $\beta^2 = \rho\tilde{\beta}\rho\tilde{\beta} = 0$ since $\tilde{\beta}\rho = 0$ in the long exact sequence containing $\tilde{\beta}$. Because $\beta^2 = 0$, the groups $H^n(X;\mathbb{Z}_m)$ form a chain complex with the Bockstein homomorphisms β as the 'boundary' maps. We can then form the associated *Bockstein cohomology groups* $\mathrm{Ker}\,\beta/\mathrm{Im}\,\beta$, which we denote $BH^n(X;\mathbb{Z}_m)$ in dimension n. The most interesting case is when m is a prime p, so we shall assume this from now on.

Proposition 3E.3. *If $H_n(X;\mathbb{Z})$ is finitely generated for all n, then the Bockstein cohomology groups $BH^n(X;\mathbb{Z}_p)$ are determined by the following rules:*
(a) *Each $\mathbb{Z}$ summand of $H^n(X;\mathbb{Z})$ contributes a $\mathbb{Z}_p$ summand to $BH^n(X;\mathbb{Z}_p)$.*
(b) *Each $\mathbb{Z}_{p^k}$ summand of $H^n(X;\mathbb{Z})$ with $k > 1$ contributes $\mathbb{Z}_p$ summands to both $BH^{n-1}(X;\mathbb{Z}_p)$ and $BH^n(X;\mathbb{Z}_p)$.*
(c) *A $\mathbb{Z}_p$ summand of $H^n(X;\mathbb{Z})$ gives $\mathbb{Z}_p$ summands of $H^{n-1}(X;\mathbb{Z}_p)$ and $H^n(X;\mathbb{Z}_p)$ with β an isomorphism between these two summands, hence there is no contribution to $BH^*(X;\mathbb{Z}_p)$.*

Proof: We will use the algebraic notion of *minimal chain complexes*. Suppose that C is a chain complex of free abelian groups for which the homology groups $H_n(C)$ are finitely generated for each n. Choose a splitting of each $H_n(C)$ as a direct sum of cyclic groups. There are countably many of these cyclic groups, so we can list them as $G_1, G_2, \cdots$. For each G_i choose a generator g_i and define a corresponding chain complex $M(g_i)$ by the following prescription. If g_i has infinite order in $G_i \subset H_{n_i}(C)$, let $M(g_i)$ consist of just a $\mathbb{Z}$ in dimension n_i, with generator z_i. On the other hand, if g_i has finite order k in $H_{n_i}(C)$, let $M(g_i)$ consist of $\mathbb{Z}$'s in dimensions n_i and $n_i + 1$, generated by x_i and y_i respectively, with $\partial y_i = kx_i$. Let M be the direct sum of the chain complexes $M(g_i)$. Define a chain map $\sigma : M \to C$ by sending z_i and x_i to cycles ζ_i and ξ_i representing the corresponding homology classes g_i, and y_i to a chain η_i with $\partial\eta_i = k\xi_i$. The chain map σ induces an isomorphism on homology, hence also on cohomology with any coefficients, by Corollary 3.4. The dual cochain complex M^* obtained by applying $\mathrm{Hom}(-,\mathbb{Z})$ splits as the direct sum of the dual complexes $M^*(g_i)$. So in cohomology with $\mathbb{Z}$ coefficients the dual basis element z_i^* generates a $\mathbb{Z}$ summand in dimension n_i, while y_i^* generates a $\mathbb{Z}_k$ summand in dimension $n_i + 1$ since $\delta x_i^* = ky_i^*$. With $\mathbb{Z}_p$ coefficients, p prime, z_i^* gives a $\mathbb{Z}_p$ summand of

$H^{n_i}(M;\mathbb{Z}_p)$, while x_i^* and y_i^* give $\mathbb{Z}_p$ summands of $H^{n_i}(M;\mathbb{Z}_p)$ and $H^{n_i+1}(M;\mathbb{Z}_p)$ if p divides k and otherwise they give nothing.

The map σ induces an isomorphism between the associated Bockstein long exact sequences of cohomology groups, with commuting squares, so we can use M^* to compute β and $\tilde{\beta}$, and we can do the calculation separately on each summand $M^*(g_i)$. Obviously β and $\tilde{\beta}$ are zero on y_i^* and z_i^*. When p divides k we have the class $x_i^* \in H^{n_i}(M;\mathbb{Z}_p)$, and from the definition of Bockstein homomorphisms it follows that $\tilde{\beta}(x_i^*) = (k/p)y_i^* \in H^{n_i+1}(M;\mathbb{Z})$ and $\beta(x_i^*) = (k/p)y_i^* \in H^{n_i+1}(M;\mathbb{Z}_p)$. The latter element is nonzero iff k is not divisible by p^2. $\square$

Corollary 3E.4. *In the situation of the preceding proposition, $H^*(X;\mathbb{Z})$ contains no elements of order p^2 iff the dimension of $BH^n(X;\mathbb{Z}_p)$ as a vector space over $\mathbb{Z}_p$ equals the rank of $H^n(X;\mathbb{Z})$ for all n. In this case $\rho:H^*(X;\mathbb{Z}) \to H^*(X;\mathbb{Z}_p)$ is injective on the p-torsion, and the image of this p-torsion under ρ is equal to $\operatorname{Im}\beta$.*

Proof: The first statement is evident from the proposition. The injectivity of ρ on p-torsion is in fact equivalent to there being no elements of order p^2. The equality $\operatorname{Im}\rho = \operatorname{Im}\beta$ follows from the fact that $\operatorname{Im}\beta = \rho(\operatorname{Im}\tilde{\beta}) = \rho(\operatorname{Ker} m)$ in the commutative diagram near the beginning of this section, and the fact that for $m = p$ the kernel of m is exactly the p-torsion when there are no elements of order p^2. $\square$

Example 3E.5. Let us use Bocksteins to compute $H^*(\mathbb{RP}^\infty \times \mathbb{RP}^\infty;\mathbb{Z})$. This could instead be done by first computing the homology via the general Künneth formula, then applying the universal coefficient theorem, but with Bocksteins we will only need the simpler Künneth formula for field coefficients in Theorem 3.15. The cup product structure in $H^*(\mathbb{RP}^\infty \times \mathbb{RP}^\infty;\mathbb{Z})$ will also be easy to determine via Bocksteins.

For p an odd prime we have $\tilde{H}^*(\mathbb{RP}^\infty;\mathbb{Z}_p) = 0$, hence $\tilde{H}^*(\mathbb{RP}^\infty \times \mathbb{RP}^\infty;\mathbb{Z}_p) = 0$ by Theorem 3.15. The universal coefficient theorem then implies that $\tilde{H}^*(\mathbb{RP}^\infty \times \mathbb{RP}^\infty;\mathbb{Z})$ consists entirely of elements of order a power of 2. From Example 3E.1 we know that Bockstein homomorphisms in $H^*(\mathbb{RP}^\infty;\mathbb{Z}_2) \approx \mathbb{Z}_2[x]$ are given by $\beta(x^{2k-1}) = x^{2k}$ and $\beta(x^{2k}) = 0$. In $H^*(\mathbb{RP}^\infty \times \mathbb{RP}^\infty;\mathbb{Z}_2) \approx \mathbb{Z}_2[x,y]$ we can then compute β via the product formula $\beta(x^m y^n) = (\beta x^m)y^n + x^m(\beta y^n)$. The answer can be represented graphically by the figure to the right. Here the dot, diamond, or circle in the (m,n) position represents the monomial $x^m y^n$ and line segments indicate nontrivial Bocksteins. For example, the lower left square records the formulas $\beta(xy) = x^2 y + xy^2$, $\beta(x^2 y) = x^2 y^2 = \beta(xy^2)$, and $\beta(x^2 y^2) = 0$. Thus in this square we see that $\operatorname{Ker}\beta = \operatorname{Im}\beta$, with generators the 'diagonal' sum $x^2 y + xy^2$ and $x^2 y^2$. The

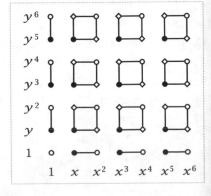

same thing happens in all the other squares, so it is apparent that $\text{Ker}\,\beta = \text{Im}\,\beta$ except for the zero-dimensional class '1.' By the preceding corollary this says that all nontrivial elements of $\widetilde{H}^*(\mathbb{RP}^\infty \times \mathbb{RP}^\infty; \mathbb{Z})$ have order 2. Furthermore, $\text{Im}\,\beta$ consists of the subring $\mathbb{Z}_2[x^2, y^2]$, indicated by the circles in the figure, together with the multiples of $x^2y + xy^2$ by elements of $\mathbb{Z}_2[x^2, y^2]$. It follows that there is a ring isomorphism

$$H^*(\mathbb{RP}^\infty \times \mathbb{RP}^\infty; \mathbb{Z}) \approx \mathbb{Z}[\lambda, \mu, \nu]/(2\lambda, 2\mu, 2\nu, \nu^2 + \lambda^2\mu + \lambda\mu^2)$$

where $\rho(\lambda) = x^2$, $\rho(\mu) = y^2$, $\rho(\nu) = x^2y + xy^2$, and the relation $\nu^2 + \lambda^2\mu + \lambda\mu^2 = 0$ holds since $(x^2y + xy^2)^2 = x^4y^2 + x^2y^4$.

This calculation illustrates the general principle that cup product structures with $\mathbb{Z}$ coefficients tend to be considerably more complicated than with field coefficients. One can see even more striking evidence of this by computing $H^*(\mathbb{RP}^\infty \times \mathbb{RP}^\infty \times \mathbb{RP}^\infty; \mathbb{Z})$ by the same technique.

Example 3E.6. Let us construct finite CW complexes X_1, X_2, and Y such that the rings $H^*(X_1; \mathbb{Z})$ and $H^*(X_2; \mathbb{Z})$ are isomorphic but $H^*(X_1 \times Y; \mathbb{Z})$ and $H^*(X_2 \times Y; \mathbb{Z})$ are isomorphic only as groups, not as rings. According to Theorem 3.15 this can happen only if all three of X_1, X_2, and Y have torsion in their $\mathbb{Z}$-cohomology. The space X_1 is obtained from $S^2 \times S^2$ by attaching a 3-cell e^3 to the second S^2 factor by a map of degree 2. Thus X_1 has a CW structure with cells e^0, e_1^2, e_2^2, e^3, e^4 with e^3 attached to the 2-sphere $e_0 \cup e_2^2$. The space X_2 is obtained from $S^2 \vee S^2 \vee S^4$ by attaching a 3-cell to the second S^2 summand by a map of degree 2, so it has a CW structure with the same collection of five cells, the only difference being that in X_2 the 4-cell is attached trivially. For the space Y we choose a Moore space $M(\mathbb{Z}_2, 2)$, with cells labeled f^0, f^2, f^3, the 3-cell being attached by a map of degree 2.

From cellular cohomology we see that both $H^*(X_1; \mathbb{Z})$ and $H^*(X_2; \mathbb{Z})$ consist of $\mathbb{Z}$'s in dimensions 0, 2, and 4, and a $\mathbb{Z}_2$ in dimension 3. In both cases all cup products of positive-dimensional classes are zero since for dimension reasons the only possible nontrivial product is the square of the 2-dimensional class, but this is zero as one sees by restricting to the subcomplex $S^2 \times S^2$ or $S^2 \vee S^2 \vee S^4$. For the space Y we have $H^*(Y; \mathbb{Z})$ consisting of a $\mathbb{Z}$ in dimension 0 and a $\mathbb{Z}_2$ in dimension 3, so the cup product structure here is trivial as well.

With $\mathbb{Z}_2$ coefficients the cellular cochain complexes for X_i, Y, and $X_i \times Y$ are all trivial, so we can identify the cells with a basis for $\mathbb{Z}_2$ cohomology. In X_i and Y the only nontrivial $\mathbb{Z}_2$ Bocksteins are $\beta(e_2^2) = e^3$ and $\beta(f^2) = f^3$. The Bocksteins in $X_i \times Y$ can then be computed using the product formula for β, which applies to cross product as well as cup product since cross product is defined in terms of cup product. The results are shown in the following table, where an arrow denotes a nontrivial Bockstein.

$$e^0 \times f^0 \quad e^2_1 \times f^0 \quad \nearrow \quad e^3 \times f^0 \quad e^4 \times f^0 \quad e^2_1 \times f^3 \quad e^4 \times f^2 \longrightarrow e^4 \times f^3$$

$$e^2_2 \times f^0 \quad \nearrow \quad e^0 \times f^3 \quad e^2_1 \times f^2 \quad e^3 \times f^2 \longrightarrow e^3 \times f^3 \quad \nearrow$$

$$e^0 \times f^2 \quad\quad\quad\quad e^2_2 \times f^2 \longrightarrow e^2_2 \times f^3$$

The two arrows from $e^2_2 \times f^2$ mean that $\beta(e^2_2 \times f^2) = e^3 \times f^2 + e^2_2 \times f^3$. It is evident that $BH^*(X_i \times Y; \mathbb{Z}_2)$ consists of $\mathbb{Z}_2$'s in dimensions 0, 2, and 4, so Proposition 3E.3 implies that the nontorsion in $H^*(X_i \times Y; \mathbb{Z})$ consists of $\mathbb{Z}$'s in these dimensions. Furthermore, by Corollary 3E.4 the 2-torsion in $H^*(X_i \times Y; \mathbb{Z})$ corresponds to the image of β and consists of $\mathbb{Z}_2 \times \mathbb{Z}_2$'s in dimensions 3 and 5 together with $\mathbb{Z}_2$'s in dimensions 6 and 7. In particular, there is a $\mathbb{Z}_2$ corresponding to $e^3 \times f^2 + e^2_2 \times f^3$ in dimension 5. There is no p-torsion for odd primes p since $H^*(X_i \times Y; \mathbb{Z}_p) \approx H^*(X_i; \mathbb{Z}_p) \otimes H^*(Y; \mathbb{Z}_p)$ is nonzero only in even dimensions.

We can see now that with $\mathbb{Z}$ coefficients, the cup product $H^2 \times H^5 \to H^7$ is nontrivial for $X_1 \times Y$ but trivial for $X_2 \times Y$. For in $H^*(X_i \times Y; \mathbb{Z}_2)$ we have, using the relation $(a \times b) \smile (c \times d) = (a \smile c) \times (b \smile d)$ which follows immediately from the definition of cross product,

(1) $e^2_1 \times f^0 \smile e^2_1 \times f^3 = (e^2_1 \smile e^2_1) \times (f^0 \smile f^3) = 0$ since $e^2_1 \smile e^2_1 = 0$

(2) $e^2_1 \times f^0 \smile (e^3 \times f^2 + e^2_2 \times f^3) = (e^2_1 \smile e^3) \times (f^0 \smile f^2) + (e^2_1 \smile e^2_2) \times (f^0 \smile f^3) = (e^2_1 \smile e^2_2) \times f^3$ since $e^2_1 \smile e^3 = 0$

and in $H^7(X_i \times Y; \mathbb{Z}_2) \approx H^7(X_i \times Y; \mathbb{Z})$ we have $(e^2_1 \smile e^2_2) \times f^3 = e^4 \times f^3 \neq 0$ for $i = 1$ but $(e^2_1 \smile e^2_2) \times f^3 = 0 \times f^3 = 0$ for $i = 2$.

Thus the cohomology ring of a product space is not always determined by the cohomology rings of the factors.

Example 3E.7. Bockstein homomorphisms can be used to get a more complete picture of the structure of $H^*(SO(n); \mathbb{Z})$ than we obtained in the preceding section. Continuing the notation employed there, we know from the calculation for $\mathbb{RP}^\infty$ in Example 3E.1 that $\beta(\sum_j \alpha^{2i-1}_j) = \sum_j \alpha^{2i}_j$ and $\beta(\sum_j \alpha^{2i}_j) = 0$, hence $\beta(\beta_{2i-1}) = \beta_{2i}$ and $\beta(\beta_{2i}) = 0$. Taking the case $n = 5$ as an example, we have $H^*(SO(5); \mathbb{Z}_2) \approx \mathbb{Z}_2[\beta_1, \beta_3]/(\beta^8_1, \beta^2_3)$. The upper part of the table at the top of the next page shows the nontrivial Bocksteins. Once again two arrows from an element mean 'sum,' for example $\beta(\beta_1 \beta_3) = \beta(\beta_1)\beta_3 + \beta_1\beta(\beta_3) = \beta_2\beta_3 + \beta_1\beta_4 = \beta^2_1\beta_3 + \beta^5_1$. This Bockstein data allows us to calculate $H^i(SO(5); \mathbb{Z})$ modulo odd torsion, with the results indicated in the remainder of the table, where the vertical arrows denote the map ρ. As we showed in Proposition 3D.3, there is no odd torsion, so this in fact gives the full calculation of $H^i(SO(5); \mathbb{Z})$.

$$
\begin{array}{cccccccc}
1 & \beta_1 \to \beta_1^2 & \beta_1^3 \to \beta_1^4 & \beta_1^5 \to \beta_1^6 & \beta_1^7 \\
& & \beta_3 \nearrow \beta_1\beta_3 \to \beta_1^2\beta_3 & \beta_1^3\beta_3 \to \beta_1^4\beta_3 & \beta_1^5\beta_3 \to \beta_1^6\beta_3 & \beta_1^7\beta_3
\end{array}
$$

| $\mathbb{Z}$ | 0 | $\mathbb{Z}_2$ | $\mathbb{Z}$ | $\mathbb{Z}_2$ | $\mathbb{Z}_2$ | $\mathbb{Z}_2$ | $\mathbb{Z}_2\times\mathbb{Z}$ | 0 | $\mathbb{Z}_2$ | $\mathbb{Z}$ |

$$
\begin{array}{cccccccc}
1 & x & y & x^2 & xy & x^3=y^2 & x^2y,\,z & x^3y=y^3 & yz \\
 & \downarrow & \downarrow & \downarrow & \downarrow & \downarrow & \downarrow & \downarrow & \downarrow \\
 & \beta_1^2 & \beta_1^3+\beta_3 & \beta_1^4 & \beta_1^5+\beta_1^2\beta_3 & \beta_1^6 & \beta_1^7+\beta_1^4\beta_3 & \beta_1^6\beta_3 & \beta_1^7\beta_3
\end{array}
$$

It is interesting that the generator $y \in H^3(SO(5);\mathbb{Z}) \approx \mathbb{Z}$ has y^2 nontrivial, since this implies that the ring structures of $H^*(SO(5);\mathbb{Z})$ and $H^*(\mathbb{RP}^7 \times S^3;\mathbb{Z})$ are not isomorphic, even though the cohomology groups and the $\mathbb{Z}_2$ cohomology rings of these two spaces are the same. An exercise at the end of the section is to show that in fact $SO(5)$ is not homotopy equivalent to the product of any two CW complexes with nontrivial cohomology.

A natural way to describe $H^*(SO(5);\mathbb{Z})$ would be as a quotient of a free graded commutative associative algebra $F[x,y,z]$ over $\mathbb{Z}$ with $|x| = 2$, $|y| = 3$, and $|z| = 7$. Elements of $F[x,y,z]$ are representable as polynomials $p(x,y,z)$, subject only to the relations imposed by commutativity. In particular, since y and z are odd-dimensional we have $yz = -zy$, and y^2 and z^2 are nonzero elements of order 2 in $F[x,y,z]$. Any monomial containing y^2 or z^2 as a factor also has order 2. In these terms, the calculation of $H^*(SO(5);\mathbb{Z})$ can be written

$$
H^*(SO(5);\mathbb{Z}) \approx F[x,y,z]/(2x, x^4, y^4, z^2, xz, x^3 - y^2)
$$

The next figure shows the nontrivial Bocksteins for $H^*(SO(7);\mathbb{Z}_2)$. Here the numbers across the top indicate dimension, stopping with 21, the dimension of $SO(7)$. The labels on the dots refer to the basis of products of distinct β_i's. For example, the dot labeled 135 is $\beta_1\beta_3\beta_5$.

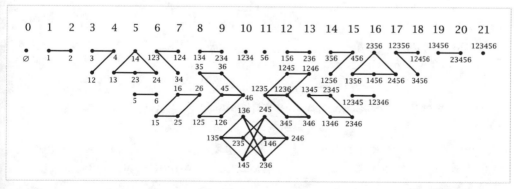

The left-right symmetry of the figure displays Poincaré duality quite graphically. Note that the corresponding diagram for $SO(5)$, drawn in a slightly different way from the preceding figure, occurs in the upper left corner as the subdiagram with labels 1 through 4. This subdiagram has the symmetry of Poincaré duality as well.

From the diagram one can with some effort work out the cup product structure in $H^*(SO(7);\mathbb{Z})$, but the answer is rather complicated, just as the diagram is:

$$F[x,y,z,v,w]/(2x,2v,x^4,y^4,z^2,v^2,w^2,xz,vz,vw,y^2w,x^3y^2v,$$
$$y^2z-x^3v,xw-y^2v-x^3v)$$

where x, y, z, v, w have dimensions 2, 3, 7, 7, 11, respectively. It is curious that the relation $x^3 = y^2$ in $H^*(SO(5);\mathbb{Z})$ no longer holds in $H^*(SO(7);\mathbb{Z})$.

Exercises

1. Show that $H^*(K(\mathbb{Z}_m,1);\mathbb{Z}_k)$ is isomorphic as a ring to $H^*(K(\mathbb{Z}_m,1);\mathbb{Z}_m)\otimes\mathbb{Z}_k$ if k divides m. In particular, if m/k is even, this is $\Lambda_{\mathbb{Z}_k}[x]\otimes\mathbb{Z}_k[y]$.

2. In this problem we will derive one half of the classification of lens spaces up to homotopy equivalence, by showing that if $L_m(\ell_1,\cdots,\ell_n) \simeq L_m(\ell_1',\cdots,\ell_n')$ then $\ell_1\cdots\ell_n \equiv \pm\ell_1'\cdots\ell_n'k^n$ mod m for some integer k. The converse is Exercise 29 for §4.2.

(a) Let $L = L_m(\ell_1,\cdots,\ell_n)$ and let $\mathbb{Z}_m^*$ be the multiplicative group of invertible elements of $\mathbb{Z}_m$. Define $t \in \mathbb{Z}_m^*$ by the equation $xy^{n-1} = tz$ where x is a generator of $H^1(L;\mathbb{Z}_m)$, $y = \beta(x)$, and $z \in H^{2n-1}(L;\mathbb{Z}_m)$ is the image of a generator of $H^{2n-1}(L;\mathbb{Z})$. Show that the image $\tau(L)$ of t in the quotient group $\mathbb{Z}_m^*/\pm(\mathbb{Z}_m^*)^n$ depends only on the homotopy type of L.

(b) Given nonzero integers $k_1,\cdots,k_n$, define a map $\tilde{f}:S^{2n-1}\to S^{2n-1}$ sending the unit vector $(r_1e^{i\theta_1},\cdots,r_ne^{i\theta_n})$ in $\mathbb{C}^n$ to $(r_1e^{ik_1\theta_1},\cdots,r_ne^{ik_n\theta_n})$. Show:

   (i) $\tilde{f}$ has degree $k_1\cdots k_n$.

  (ii) $\tilde{f}$ induces a quotient map $f:L\to L'$ for $L' = L_m(\ell_1',\cdots,\ell_n')$ provided that $k_j\ell_j \equiv \ell_j'$ mod m for each j.

 (iii) f induces an isomorphism on π_1, hence on $H^1(-;\mathbb{Z}_m)$.

 (iv) f has degree $k_1\cdots k_n$, i.e., f_* is multiplication by $k_1\cdots k_n$ on $H_{2n-1}(-;\mathbb{Z})$.

(c) Using the f in (b), show that $\tau(L) = k_1\cdots k_n\tau(L')$.

(d) Deduce that if $L_m(\ell_1,\cdots,\ell_n) \simeq L_m(\ell_1',\cdots,\ell_n')$, then $\ell_1\cdots\ell_n \equiv \pm\ell_1'\cdots\ell_n'k^n$ mod m for some integer k.

3. Let X be the smash product of k copies of a Moore space $M(\mathbb{Z}_p,n)$ with p prime. Compute the Bockstein homomorphisms in $H^*(X;\mathbb{Z}_p)$ and use this to describe $H^*(X;\mathbb{Z})$.

4. Using the cup product structure in $H^*(SO(5);\mathbb{Z})$, show that $SO(5)$ is not homotopy equivalent to the product of any two CW complexes with nontrivial cohomology.

3.F Limits and Ext

It often happens that one has a CW complex X expressed as a union of an increasing sequence of subcomplexes $X_0 \subset X_1 \subset X_2 \subset \cdots$. For example, X_i could be the i-skeleton of X, or the X_i's could be finite complexes whose union is X. In situations of this sort, Proposition 3.33 says that $H_n(X; G)$ is the direct limit $\varinjlim H_n(X_i; G)$. Our goal in this section is to show this holds more generally for any homology theory, and to derive the corresponding formula for cohomology theories, which is a bit more complicated even for ordinary cohomology with $\mathbb{Z}$ coefficients. For ordinary homology and cohomology the results apply somewhat more generally than just to CW complexes, since if a space X is the union of an increasing sequence of subspaces X_i with the property that each compact set in X is contained in some X_i, then the singular complex of X is the union of the singular complexes of the X_i's, and so this gives a reduction to the CW case.

Passing to limits can often result in nonfinitely generated homology and cohomology groups. At the end of this section we describe some of the rather subtle behavior of Ext for nonfinitely generated groups.

Direct and Inverse Limits

As a special case of the general definition in §3.3, the direct limit $\varinjlim G_i$ of a sequence of homomorphisms of abelian groups $G_1 \xrightarrow{\alpha_1} G_2 \xrightarrow{\alpha_2} G_3 \longrightarrow \cdots$ is defined to be the quotient of the direct sum $\bigoplus_i G_i$ by the subgroup consisting of elements of the form $(g_1, g_2 - \alpha_1(g_1), g_3 - \alpha_2(g_2), \cdots)$. It is easy to see from this definition that every element of $\varinjlim G_i$ is represented by an element $g_i \in G_i$ for some i, and two such representatives $g_i \in G_i$ and $g_j \in G_j$ define the same element of $\varinjlim G_i$ iff they have the same image in some G_k under the appropriate composition of α_ℓ's. If all the α_i's are injective and are viewed as inclusions of subgroups, $\varinjlim G_i$ is just $\bigcup_i G_i$.

Example 3F.1. For a prime p, consider the sequence $\mathbb{Z} \xrightarrow{p} \mathbb{Z} \xrightarrow{p} \mathbb{Z} \longrightarrow \cdots$ with all maps multiplication by p. Then $\varinjlim G_i$ can be identified with the subgroup $\mathbb{Z}[1/p]$ of $\mathbb{Q}$ consisting of rational numbers with denominator a power of p. More generally, we can realize any subgroup of $\mathbb{Q}$ as the direct limit of a sequence $\mathbb{Z} \longrightarrow \mathbb{Z} \longrightarrow \mathbb{Z} \longrightarrow \cdots$ with an appropriate choice of maps. For example, if the nth map is multiplication by n, then the direct limit is $\mathbb{Q}$ itself.

Example 3F.2. The sequence of injections $\mathbb{Z}_p \xrightarrow{p} \mathbb{Z}_{p^2} \xrightarrow{p} \mathbb{Z}_{p^3} \longrightarrow \cdots$, with p prime, has direct limit a group we denote $\mathbb{Z}_{p^\infty}$. This is isomorphic to $\mathbb{Z}[1/p]/\mathbb{Z}$, the subgroup of $\mathbb{Q}/\mathbb{Z}$ represented by fractions with denominator a power of p. In fact $\mathbb{Q}/\mathbb{Z}$ is isomorphic to the direct sum of the subgroups $\mathbb{Z}[1/p]/\mathbb{Z} \approx \mathbb{Z}_{p^\infty}$ for all primes p. It is not hard to determine all the subgroups of $\mathbb{Q}/\mathbb{Z}$ and see that each one can be realized as a direct limit of finite cyclic groups with injective maps between them. Conversely, every such direct limit is isomorphic to a subgroup of $\mathbb{Q}/\mathbb{Z}$.

We can realize these algebraic examples topologically by the following construction. The **mapping telescope** of a sequence of maps $X_0 \xrightarrow{f_0} X_1 \xrightarrow{f_1} X_2 \rightarrow \cdots$ is the union of the mapping cylinders M_{f_i} with the copies of X_i in M_{f_i} and $M_{f_{i-1}}$ identified for all i. Thus the mapping tele-

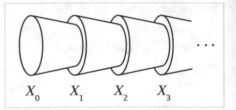

scope is the quotient space of the disjoint union $\coprod_i (X_i \times [i, i+1])$ in which each point $(x_i, i+1) \in X_i \times [i, i+1]$ is identified with $(f_i(x_i), i+1) \in X_{i+1} \times [i+1, i+2]$. In the mapping telescope T, let T_i be the union of the first i mapping cylinders. This deformation retracts onto X_i by deformation retracting each mapping cylinder onto its right end in turn. If the maps f_i are cellular, each mapping cylinder is a CW complex and the telescope T is the increasing union of the subcomplexes $T_i \simeq X_i$. Then Proposition 3.33, or Theorem 3F.8 below, implies that $H_n(T; G) \approx \varinjlim H_n(X_i; G)$.

Example 3F.3. Suppose each f_i is a map $S^n \rightarrow S^n$ of degree p for a fixed prime p. Then $H_n(T)$ is the direct limit of the sequence $\mathbb{Z} \xrightarrow{p} \mathbb{Z} \xrightarrow{p} \mathbb{Z} \rightarrow \cdots$ considered in Example 3F.1 above, and $\widetilde{H}_k(T) = 0$ for $k \neq n$, so T is a Moore space $M(\mathbb{Z}[1/p], n)$.

Example 3F.4. In the preceding example, if we attach a cell e^{n+1} to the first S^n in T via the identity map of S^n, we obtain a space X which is a Moore space $M(\mathbb{Z}_{p^\infty}, n)$ since X is the union of its subspaces $X_i = T_i \cup e^{n+1}$, which are $M(\mathbb{Z}_{p^i}, n)$'s, and the inclusion $X_i \subset X_{i+1}$ induces the inclusion $\mathbb{Z}_{p^i} \subset \mathbb{Z}_{p^{i+1}}$ on H_n.

Generalizing these two examples, we can obtain Moore spaces $M(G, n)$ for arbitrary subgroups G of $\mathbb{Q}$ or $\mathbb{Q}/\mathbb{Z}$ by choosing maps $f_i : S^n \rightarrow S^n$ of suitable degrees.

The behavior of cohomology groups is more complicated. If X is the increasing union of subcomplexes X_i, then the cohomology groups $H^n(X_i; G)$, for fixed n and G, form a sequence of homomorphisms

$$\cdots \longrightarrow G_2 \xrightarrow{\alpha_2} G_1 \xrightarrow{\alpha_1} G_0$$

Given such a sequence of group homomorphisms, the **inverse limit** $\varprojlim G_i$ is defined to be the subgroup of $\prod_i G_i$ consisting of sequences (g_i) with $\alpha_i(g_i) = g_{i-1}$ for all i. There is a natural map $\lambda : H^n(X; G) \rightarrow \varprojlim H^n(X_i; G)$ sending an element of $H^n(X; G)$ to its sequence of images in $H^n(X_i; G)$ under the maps $H^n(X; G) \rightarrow H^n(X_i; G)$ induced by inclusion. One might hope that λ is an isomorphism, but this is not true in general, as we shall see. However, for some choices of G it is:

Proposition 3F.5. *If the CW complex X is the union of an increasing sequence of subcomplexes X_i and if G is one of the fields $\mathbb{Q}$ or $\mathbb{Z}_p$, then $\lambda : H^n(X; G) \rightarrow \varprojlim H^n(X_i; G)$ is an isomorphism for all n.*

Proof: First we have an easy algebraic fact: Given a sequence of homomorphisms of abelian groups $G_1 \xrightarrow{\alpha_1} G_2 \xrightarrow{\alpha_2} G_3 \rightarrow \cdots$, then $\mathrm{Hom}(\varinjlim G_i, G) = \varprojlim \mathrm{Hom}(G_i, G)$

for any G. Namely, it follows from the definition of $\varinjlim G_i$ that a homomorphism $\varphi : \varinjlim G_i \to G$ is the same thing as a sequence of homomorphisms $\varphi_i : G_i \to G$ with $\varphi_i = \varphi_{i+1}\alpha_i$ for all i. Such a sequence (φ_i) is exactly an element of $\varprojlim \mathrm{Hom}(G_i, G)$.

Now if G is a field $\mathbb{Q}$ or $\mathbb{Z}_p$ we have

$$H^n(X; G) = \mathrm{Hom}(H_n(X; G), G)$$
$$= \mathrm{Hom}(\varinjlim H_n(X_i; G), G)$$
$$= \varprojlim \mathrm{Hom}(H_n(X_i; G), G)$$
$$= \varprojlim H^n(X_i; G) \qquad\qquad \square$$

Let us analyze what happens for cohomology with an arbitrary coefficient group, or more generally for any cohomology theory. Given a sequence of homomorphisms of abelian groups

$$\cdots \longrightarrow G_2 \xrightarrow{\alpha_2} G_1 \xrightarrow{\alpha_1} G_0$$

define a map $\delta : \prod_i G_i \to \prod_i G_i$ by $\delta(\cdots, g_i, \cdots) = (\cdots, g_i - \alpha_{i+1}(g_{i+1}), \cdots)$, so that $\varprojlim G_i$ is the kernel of δ. Denoting the cokernel of δ by $\varprojlim^1 G_i$, we have then an exact sequence

$$0 \to \varprojlim G_i \to \prod_i G_i \xrightarrow{\delta} \prod_i G_i \to \varprojlim^1 G_i \to 0$$

This may be compared with the corresponding situation for the direct limit of a sequence $G_1 \xrightarrow{\alpha_1} G_2 \xrightarrow{\alpha_2} G_3 \longrightarrow \cdots$. In this case one has a short exact sequence

$$0 \to \bigoplus_i G_i \xrightarrow{\delta} \bigoplus_i G_i \to \varinjlim G_i \to 0$$

where $\delta(\cdots, g_i, \cdots) = (\cdots, g_i - \alpha_{i-1}(g_{i-1}), \cdots)$, so δ is injective and there is no term $\varinjlim^1 G_i$ analogous to $\varprojlim^1 G_i$.

Here are a few simple observations about $\varprojlim$ and $\varprojlim^1$:

- If all the α_i's are isomorphisms then $\varprojlim G_i \approx G_0$ and $\varprojlim^1 G_i = 0$. In fact, $\varprojlim^1 G_i = 0$ if each α_i is surjective, for to realize a given element $(h_i) \in \prod_i G_i$ as $\delta(g_i)$ we can take $g_0 = 0$ and then solve $\alpha_1(g_1) = -h_0$, $\alpha_2(g_2) = g_1 - h_1$, $\cdots$.
- If all the α_i's are zero then $\varprojlim G_i = \varprojlim^1 G_i = 0$.
- Deleting a finite number of terms from the end of the sequence $\cdots \to G_1 \to G_0$ does not affect $\varprojlim G_i$ or $\varprojlim^1 G_i$. More generally, $\varprojlim G_i$ and $\varprojlim^1 G_i$ are unchanged if we replace the sequence $\cdots \to G_1 \to G_0$ by a subsequence, with the appropriate compositions of α_j's as the maps.

Example 3F.6. Consider the sequence of natural surjections $\cdots \to \mathbb{Z}_{p^3} \to \mathbb{Z}_{p^2} \to \mathbb{Z}_p$ with p a prime. The inverse limit of this sequence is a famous object in number theory, called the **p-adic integers**. Our notation for it will be $\widehat{\mathbb{Z}}_p$. It is actually a commutative ring, not just a group, since the projections $\mathbb{Z}_{p^{i+1}} \to \mathbb{Z}_{p^i}$ are ring homomorphisms, but we will be interested only in the additive group structure. Elements of $\widehat{\mathbb{Z}}_p$ are infinite sequences $(\cdots, a_2, a_1)$ with $a_i \in \mathbb{Z}_{p^i}$ such that a_i is the mod p^i reduction of a_{i+1}.

For each choice of a_i there are exactly p choices for a_{i+1}, so $\hat{\mathbb{Z}}_p$ is uncountable. There is a natural inclusion $\mathbb{Z} \subset \hat{\mathbb{Z}}_p$ as the constant sequences $a_i = n \in \mathbb{Z}$. It is easy to see that $\hat{\mathbb{Z}}_p$ is torsionfree by checking that it has no elements of prime order.

There is another way of looking at $\hat{\mathbb{Z}}_p$. An element of $\hat{\mathbb{Z}}_p$ has a unique representation as a sequence $(\cdots, a_2, a_1)$ of integers a_i with $0 \le a_i < p^i$ for each i. We can write each a_i uniquely in the form $b_{i-1} p^{i-1} + \cdots + b_1 p + b_0$ with $0 \le b_j < p$. The fact that a_{i+1} reduces mod p^i to a_i means that the numbers b_j depend only on the element $(\cdots, a_2, a_1) \in \hat{\mathbb{Z}}_p$, so we can view the elements of $\hat{\mathbb{Z}}_p$ as the 'base p infinite numbers' $\cdots b_1 b_0$ with $0 \le b_i < p$ for all i, with the familiar rule for addition in base p notation. The finite expressions $b_n \cdots b_1 b_0$ represent the nonnegative integers, but negative integers have infinite expansions. For example, -1 has $b_i = p - 1$ for all i, as one can see by adding 1 to this number.

Since the maps $\mathbb{Z}_{p^{i+1}} \to \mathbb{Z}_{p^i}$ are surjective, $\varprojlim^1 \mathbb{Z}_{p^i} = 0$. The next example shows how p-adic integers can also give rise to a nonvanishing $\varprojlim^1$ term.

Example 3F.7. Consider the sequence $\cdots \to \mathbb{Z} \xrightarrow{p} \mathbb{Z} \xrightarrow{p} \mathbb{Z}$ for p prime. In this case the inverse limit is zero since a nonzero integer can only be divided by p finitely often. The $\varprojlim^1$ term is the cokernel of the map $\delta : \prod_\infty \mathbb{Z} \to \prod_\infty \mathbb{Z}$ given by $\delta(y_1, y_2, \cdots) = (y_1 - p y_2, y_2 - p y_3, \cdots)$. We claim that the map $\hat{\mathbb{Z}}_p / \mathbb{Z} \to \text{Coker}\,\delta$ sending a p-adic number $\cdots b_1 b_0$ as in the preceding example to $(b_0, b_1, \cdots)$ is an isomorphism. To see this, note that the image of δ consists of the sums $y_1(1, 0, \cdots) + y_2(-p, 1, 0, \cdots) + y_3(0, -p, 1, 0, \cdots) + \cdots$. The terms after $y_1(1, 0, \cdots)$ give exactly the relations that hold among the p-adic numbers $\cdots b_1 b_0$, and in particular allow one to reduce an arbitrary sequence $(b_0, b_1, \cdots)$ to a unique sequence with $0 \le b_i < p$ for all i. The term $y_1(1, 0, \cdots)$ corresponds to the subgroup $\mathbb{Z} \subset \hat{\mathbb{Z}}_p$.

We come now to the main result of this section:

Theorem 3F.8. *For a CW complex X which is the union of an increasing sequence of subcomplexes $X_0 \subset X_1 \subset \cdots$ there is an exact sequence*

$$0 \longrightarrow \varprojlim{}^1 h^{n-1}(X_i) \longrightarrow h^n(X) \xrightarrow{\ \lambda\ } \varprojlim h^n(X_i) \longrightarrow 0$$

where h^* is any reduced or unreduced cohomology theory. For any homology theory h_*, reduced or unreduced, the natural maps $\varinjlim h_n(X_i) \to h_n(X)$ are isomorphisms.

Proof: Let T be the mapping telescope of the inclusion sequence $X_0 \hookrightarrow X_1 \hookrightarrow \cdots$. This is a subcomplex of $X \times [0, \infty)$ when $[0, \infty)$ is given the CW structure with the integer points as 0-cells. We have $T \simeq X$ since T is a deformation retract of $X \times [0, \infty)$, as we showed in the proof of Lemma 2.34 in the special case that X_i is the i-skeleton of X, but the argument works just as well for arbitrary subcomplexes X_i.

Let $T_1 \subset T$ be the union of the products $X_i \times [i, i+1]$ for i odd, and let T_2 be the corresponding union for i even. Thus $T_1 \cap T_2 = \coprod_i X_i$ and $T_1 \cup T_2 = T$. For an unreduced cohomology theory h^* we have then a Mayer–Vietoris sequence

$$h^{n-1}(T_1) \oplus h^{n-1}(T_2) \longrightarrow h^{n-1}(T_1 \cap T_2) \longrightarrow h^n(T) \longrightarrow h^n(T_1) \oplus h^n(T_2) \longrightarrow h^n(T_1 \cap T_2)$$

$$\wr\wr \qquad\qquad\qquad \wr\wr \qquad\qquad \wr\wr \qquad\qquad \wr\wr \qquad\qquad\qquad \wr\wr$$

$$\textstyle\prod_i h^{n-1}(X_i) \xrightarrow{\;\varphi\;} \prod_i h^{n-1}(X_i) \longrightarrow h^n(X) \longrightarrow \prod_i h^n(X_i) \xrightarrow{\;\varphi\;} \prod_i h^n(X_i)$$

The maps φ making the diagram commute are given by the formula $\varphi(\cdots, g_i, \cdots) = (\cdots, (-1)^{i-1}(g_i - \rho(g_{i+1})), \cdots)$, the ρ's being the appropriate restriction maps. This differs from δ only in the sign of its even coordinates, so if we change the isomorphism $h^k(T_1 \cap T_2) \approx \prod_i h^k(X_i)$ by inserting a minus sign in the even coordinates, we can replace φ by δ in the second row of the diagram. This row then yields a short exact sequence $0 \to \operatorname{Coker} \delta \to h^n(X; G) \to \operatorname{Ker} \delta \to 0$, finishing the proof for unreduced cohomology.

The same argument works for reduced cohomology if we use the reduced telescope obtained from T by collapsing $\{x_0\} \times [0, \infty)$ to a point, for x_0 a basepoint 0-cell of X_0. Then $T_1 \cap T_2 = \bigvee_i X_i$ rather than $\coprod_i X_i$, and the rest of the argument goes through unchanged. The proof also applies for homology theories, with direct products replaced by direct sums in the second row of the diagram. As we noted earlier, $\operatorname{Ker} \delta = 0$ in the direct limit case, and $\operatorname{Coker} \delta = \varinjlim$. $\qquad\square$

Example 3F.9. As in Example 3F.3, consider the mapping telescope T for the sequence of degree p maps $S^n \to S^n \to \cdots$. Letting T_i be the union of the first i mapping cylinders in the telescope, the inclusions $T_1 \hookrightarrow T_2 \hookrightarrow \cdots$ induce on $H^n(-; \mathbb{Z})$ the sequence $\cdots \to \mathbb{Z} \xrightarrow{p} \mathbb{Z}$ in Example 3F.7. From the theorem we deduce that $H^{n+1}(T; \mathbb{Z}) \approx \widehat{\mathbb{Z}}_p / \mathbb{Z}$ and $\tilde{H}^k(T; \mathbb{Z}) = 0$ for $k \neq n+1$. Thus we have the rather strange situation that the CW complex T is the union of subcomplexes T_i each having cohomology consisting only of a $\mathbb{Z}$ in dimension n, but T itself has no cohomology in dimension n and instead has a huge uncountable group $\widehat{\mathbb{Z}}_p / \mathbb{Z}$ in dimension $n+1$. This contrasts sharply with what happens for homology, where the groups $H_n(T_i) \approx \mathbb{Z}$ fit together nicely to give $H_n(T) \approx \mathbb{Z}[1/p]$.

Example 3F.10. A more reasonable behavior is exhibited if we consider the space $X = M(\mathbb{Z}_{p^\infty}, n)$ in Example 3F.4 expressed as the union of its subspaces X_i. By the universal coefficient theorem, the reduced cohomology of X_i with $\mathbb{Z}$ coefficients consists of a $\mathbb{Z}_{p^i} = \operatorname{Ext}(\mathbb{Z}_{p^i}, \mathbb{Z})$ in dimension $n+1$. The inclusion $X_i \hookrightarrow X_{i+1}$ induces the inclusion $\mathbb{Z}_{p^i} \hookrightarrow \mathbb{Z}_{p^{i+1}}$ on H_n, and on Ext this induced map is a surjection $\mathbb{Z}_{p^{i+1}} \to \mathbb{Z}_{p^i}$ as one can see by looking at the diagram of free resolutions on the left:

$$
\begin{array}{ccccccc}
0 \longrightarrow \mathbb{Z} & \xrightarrow{p^i} & \mathbb{Z} & \longrightarrow & \mathbb{Z}_{p^i} & \longrightarrow & 0 \\
\downarrow{\scriptstyle 1} & & \downarrow{\scriptstyle p} & & \downarrow{\scriptstyle p} & & \\
0 \longrightarrow \mathbb{Z} & \xrightarrow{p^{i+1}} & \mathbb{Z} & \longrightarrow & \mathbb{Z}_{p^{i+1}} & \longrightarrow & 0
\end{array}
\qquad
\begin{array}{ccccc}
0 \longleftarrow \operatorname{Ext}(\mathbb{Z}_{p^i}, \mathbb{Z}) & \longleftarrow & \operatorname{Hom}(\mathbb{Z}, \mathbb{Z}) & \longleftarrow & \cdots \\
\uparrow & & \uparrow{\scriptstyle 1} & & \\
0 \longleftarrow \operatorname{Ext}(\mathbb{Z}_{p^{i+1}}, \mathbb{Z}) & \longleftarrow & \operatorname{Hom}(\mathbb{Z}, \mathbb{Z}) & \longleftarrow & \cdots
\end{array}
$$

Applying $\operatorname{Hom}(-, \mathbb{Z})$ to this diagram, we get the diagram on the right, with exact rows, and the left-hand vertical map is a surjection since the vertical map to the right of it is surjective. Thus the sequence $\cdots \to H^{n+1}(X_2; \mathbb{Z}) \to H^{n+1}(X_1; \mathbb{Z})$ is the sequence in Example 3F.6, and we deduce that $H^{n+1}(X;\mathbb{Z}) \approx \hat{\mathbb{Z}}_p$, the p-adic integers, and $\tilde{H}^k(X;\mathbb{Z}) = 0$ for $k \neq n + 1$.

This example can be related to the preceding one. If we view X as the mapping cone of the inclusion $S^n \hookrightarrow T$ of one end of the telescope, then the long exact sequences of homology and cohomology groups for the pair (T, S^n) reduce to the short exact sequences at the right.

$$0 \longrightarrow H_n(S^n) \longrightarrow H_n(T) \longrightarrow H_n(X) \longrightarrow 0$$
$$\| \qquad\qquad \| \qquad\qquad \|$$
$$\mathbb{Z} \qquad\quad \mathbb{Z}[1/p] \qquad \mathbb{Z}_{p^\infty}$$

$$0 \longrightarrow H^n(S^n) \longrightarrow H^{n+1}(X) \longrightarrow H^{n+1}(T) \longrightarrow 0$$
$$\| \qquad\qquad\quad \| \qquad\qquad \|$$
$$\mathbb{Z} \qquad\qquad \hat{\mathbb{Z}}_p \qquad\quad \hat{\mathbb{Z}}_p/\mathbb{Z}$$

From these examples and the universal coefficient theorem we obtain isomorphisms $\text{Ext}(\mathbb{Z}_{p^\infty}, \mathbb{Z}) \approx \hat{\mathbb{Z}}_p$ and $\text{Ext}(\mathbb{Z}[1/p], \mathbb{Z}) \approx \hat{\mathbb{Z}}_p/\mathbb{Z}$. These can also be derived directly from the definition of Ext. A free resolution of $\mathbb{Z}_{p^\infty}$ is

$$0 \longrightarrow \mathbb{Z}^\infty \overset{\varphi}{\longrightarrow} \mathbb{Z}^\infty \longrightarrow \mathbb{Z}_{p^\infty} \longrightarrow 0$$

where $\mathbb{Z}^\infty$ is the direct sum of an infinite number of $\mathbb{Z}$'s, the sequences $(x_1, x_2, \cdots)$ of integers all but finitely many of which are zero, and φ sends $(x_1, x_2, \cdots)$ to $(px_1 - x_2, px_2 - x_3, \cdots)$. We can view φ as the linear map corresponding to the infinite matrix with p's on the diagonal, -1's just above the diagonal, and 0's everywhere else. Clearly $\text{Ker}\,\varphi = 0$ since integers cannot be divided by p infinitely often. The image of φ is generated by the vectors $(p, 0, \cdots), (-1, p, 0, \cdots), (0, -1, p, 0, \cdots), \cdots$ so $\text{Coker}\,\varphi \approx \mathbb{Z}_{p^\infty}$. Dualizing by taking $\text{Hom}(-, \mathbb{Z})$, we have $\text{Hom}(\mathbb{Z}^\infty, \mathbb{Z})$ the infinite direct product of $\mathbb{Z}$'s, and $\varphi^*(y_1, y_2, \cdots) = (py_1, py_2 - y_1, py_3 - y_2, \cdots)$, corresponding to the transpose of the matrix of φ. By definition, $\text{Ext}(\mathbb{Z}_{p^\infty}, \mathbb{Z}) = \text{Coker}\,\varphi^*$. The image of φ^* consists of the infinite sums $y_1(p, -1, 0 \cdots) + y_2(0, p, -1, 0, \cdots) + \cdots$, so $\text{Coker}\,\varphi^*$ can be identified with $\hat{\mathbb{Z}}_p$ by rewriting a sequence $(z_1, z_2, \cdots)$ as the p-adic number $\cdots z_2 z_1$.

The calculation $\text{Ext}(\mathbb{Z}[1/p], \mathbb{Z}) \approx \hat{\mathbb{Z}}_p/\mathbb{Z}$ is quite similar. A free resolution of $\mathbb{Z}[1/p]$ can be obtained from the free resolution of $\mathbb{Z}_{p^\infty}$ by omitting the first column of the matrix of φ and, for convenience, changing sign. This gives the formula $\varphi(x_1, x_2, \cdots) = (x_1, x_2 - px_1, x_3 - px_2, \cdots)$, with the image of φ generated by the elements $(1, -p, 0, \cdots), (0, 1, -p, 0, \cdots), \cdots$. The dual map φ^* is given by $\varphi^*(y_1, y_2, \cdots) = (y_1 - py_2, y_2 - py_3, \cdots)$, and this has image consisting of the sums $y_1(1, 0 \cdots) + y_2(-p, 1, 0, \cdots) + y_3(0, -p, 1, 0, \cdots) + \cdots$, so we get $\text{Ext}(\mathbb{Z}[1/p], \mathbb{Z}) = \text{Coker}\,\varphi^* \approx \hat{\mathbb{Z}}_p/\mathbb{Z}$. Note that φ^* is exactly the map δ in Example 3F.7.

It is interesting to note also that the map $\varphi : \mathbb{Z}^\infty \to \mathbb{Z}^\infty$ in the two cases $\mathbb{Z}_{p^\infty}$ and $\mathbb{Z}[1/p]$ is precisely the cellular boundary map $H_{n+1}(X^{n+1}, X^n) \to H_n(X^n, X^{n-1})$ for the Moore space $M(\mathbb{Z}_{p^\infty}, n)$ or $M(\mathbb{Z}[1/p], n)$ constructed as the mapping telescope of the sequence of degree p maps $S^n \to S^n \to \cdots$, with a cell e^{n+1} attached to the first S^n in the case of $\mathbb{Z}_{p^\infty}$.

More About Ext

The functors Hom and Ext behave fairly simply for finitely generated groups, when cohomology and homology are essentially the same except for a dimension shift in the torsion. But matters are more complicated in the nonfinitely generated case. A useful tool for getting a handle on this complication is the following:

Proposition 3F.11. *Given an abelian group G and a short exact sequence of abelian groups $0 \to A \to B \to C \to 0$, there are exact sequences*

$$0 \to \text{Hom}(G,A) \to \text{Hom}(G,B) \to \text{Hom}(G,C) \to \text{Ext}(G,A) \to \text{Ext}(G,B) \to \text{Ext}(G,C) \to 0$$

$$0 \to \text{Hom}(C,G) \to \text{Hom}(B,G) \to \text{Hom}(A,G) \to \text{Ext}(C,G) \to \text{Ext}(B,G) \to \text{Ext}(A,G) \to 0$$

Proof: A free resolution $0 \to F_1 \to F_0 \to G \to 0$ gives rise to a commutative diagram

$$
\begin{array}{ccccccccc}
0 & \longrightarrow & \text{Hom}(F_0,A) & \longrightarrow & \text{Hom}(F_0,B) & \longrightarrow & \text{Hom}(F_0,C) & \longrightarrow & 0 \\
& & \downarrow & & \downarrow & & \downarrow & & \\
0 & \longrightarrow & \text{Hom}(F_1,A) & \longrightarrow & \text{Hom}(F_1,B) & \longrightarrow & \text{Hom}(F_1,C) & \longrightarrow & 0
\end{array}
$$

Since F_0 and F_1 are free, the two rows are exact, as they are simply direct products of copies of the exact sequence $0 \to A \to B \to C \to 0$, in view of the general fact that $\text{Hom}(\bigoplus_i G_i, H) = \prod_i \text{Hom}(G_i, H)$. Enlarging the diagram by zeros above and below, it becomes a short exact sequence of chain complexes, and the associated long exact sequence of homology groups is the first of the two six-term exact sequences in the proposition.

To obtain the other exact sequence we will construct the commutative diagram at the right, where the columns are free resolutions and the rows are exact. To start, let $F_0 \to A$ and $F_0'' \to C$ be surjections from free abelian groups onto A and C. Then let $F_0' = F_0 \oplus F_0''$, with the obvious

$$
\begin{array}{ccccccccc}
& & 0 & & 0 & & 0 & & \\
& & \downarrow & & \downarrow & & \downarrow & & \\
0 & \longrightarrow & F_1 & \longrightarrow & F_1' & \longrightarrow & F_1'' & \longrightarrow & 0 \\
& & \downarrow & & \downarrow & & \downarrow & & \\
0 & \longrightarrow & F_0 & \longrightarrow & F_0' & \longrightarrow & F_0'' & \longrightarrow & 0 \\
& & \downarrow & & \downarrow & & \downarrow & & \\
0 & \longrightarrow & A & \longrightarrow & B & \longrightarrow & C & \longrightarrow & 0 \\
& & \downarrow & & \downarrow & & \downarrow & & \\
& & 0 & & 0 & & 0 & &
\end{array}
$$

maps in the second row, inclusion and projection. The map $F_0' \to B$ is defined on the summand F_0 to make the lower left square commute, and on the summand F_0'' it is defined by sending basis elements of F_0'' to elements of B mapping to the images of these basis elements in C, so the lower right square also commutes. Now we have the bottom two rows of the diagram, and we can regard these two rows as a short exact sequence of two-term chain complexes. The associated long exact sequence of homology groups has six terms, the first three being the kernels of the three vertical maps to A, B, and C, and the last three being the cokernels of these maps. Since the vertical maps to A and C are surjective, the fourth and sixth of the six homology groups vanish, hence also the fifth, which says the vertical map to B is surjective. The first three of the original six homology groups form a short exact sequence, and we let this be the top row of the diagram, formed by the kernels of the vertical maps to A, B, and C. These kernels are subgroups of free abelian groups, hence are also free.

Thus the three columns are free resolutions. The upper two squares automatically commute, so the construction of the diagram is complete.

The first two rows of the diagram split by freeness, so applying $\text{Hom}(-, G)$ yields a diagram

$$
\begin{array}{ccccccccc}
0 & \longrightarrow & \text{Hom}(F_0'', G) & \longrightarrow & \text{Hom}(F_0', G) & \longrightarrow & \text{Hom}(F_0, G) & \longrightarrow & 0 \\
& & \downarrow & & \downarrow & & \downarrow & & \\
0 & \longrightarrow & \text{Hom}(F_1'', G) & \longrightarrow & \text{Hom}(F_1', G) & \longrightarrow & \text{Hom}(F_1, G) & \longrightarrow & 0
\end{array}
$$

with exact rows. Again viewing this as a short exact sequence of chain complexes, the associated long exact sequence of homology groups is the second six-term exact sequence in the statement of the proposition. $\square$

The second sequence in the proposition says in particular that an injection $A \to B$ induces a surjection $\text{Ext}(B, C) \to \text{Ext}(A, C)$ for any C. For example, if A has torsion, this says $\text{Ext}(A, \mathbb{Z})$ is nonzero since it maps onto $\text{Ext}(\mathbb{Z}_n, \mathbb{Z}) \approx \mathbb{Z}_n$ for some $n > 1$. The calculation $\text{Ext}(\mathbb{Z}_{p^\infty}, \mathbb{Z}) \approx \hat{\mathbb{Z}}_p$ earlier in this section shows that torsion in A does not necessarily yield torsion in $\text{Ext}(A, \mathbb{Z})$, however.

Two other useful formulas whose proofs we leave as exercises are:

$$\text{Ext}(\oplus_i A_i, B) \approx \prod_i \text{Ext}(A_i, B) \qquad \text{Ext}(A, \oplus_i B_i) \approx \oplus_i \text{Ext}(A, B_i)$$

For example, since $\mathbb{Q}/\mathbb{Z} = \oplus_p \mathbb{Z}_{p^\infty}$ we obtain $\text{Ext}(\mathbb{Q}/\mathbb{Z}, \mathbb{Z}) \approx \prod_p \hat{\mathbb{Z}}_p$ from the calculation $\text{Ext}(\mathbb{Z}_{p^\infty}, \mathbb{Z}) \approx \hat{\mathbb{Z}}_p$. Then from the exact sequence $0 \to \mathbb{Z} \to \mathbb{Q} \to \mathbb{Q}/\mathbb{Z} \to 0$ we get $\text{Ext}(\mathbb{Q}, \mathbb{Z}) \approx (\prod_p \hat{\mathbb{Z}}_p)/\mathbb{Z}$ using the second exact sequence in the proposition.

In these examples the groups $\text{Ext}(A, \mathbb{Z})$ are rather large, and the next result says this is part of a general pattern:

Proposition 3F.12. *If A is not finitely generated then either* $\text{Hom}(A, \mathbb{Z})$ *or* $\text{Ext}(A, \mathbb{Z})$ *is uncountable. Hence if* $H_n(X; \mathbb{Z})$ *is not finitely generated then either* $H^n(X; \mathbb{Z})$ *or* $H^{n+1}(X; \mathbb{Z})$ *is uncountable.*

Both possibilities can occur, as we see from the examples $\text{Hom}(\oplus_\infty \mathbb{Z}, \mathbb{Z}) \approx \prod_\infty \mathbb{Z}$ and $\text{Ext}(\mathbb{Z}_{p^\infty}, \mathbb{Z}) \approx \hat{\mathbb{Z}}_p$.

This proposition has some interesting topological consequences. First, it implies that if a space X has $\tilde{H}^*(X; \mathbb{Z}) = 0$, then $\tilde{H}_*(X; \mathbb{Z}) = 0$, since the case of finitely generated homology groups follows from our earlier results. And second, it says that one cannot always construct a space X with prescribed cohomology groups $H^n(X; \mathbb{Z})$, as one can for homology. For example there is no space whose only nonvanishing $\tilde{H}^n(X; \mathbb{Z})$ is a countable nonfinitely generated group such as $\mathbb{Q}$ or $\mathbb{Q}/\mathbb{Z}$. Even in the finitely generated case the dimension $n = 1$ is somewhat special since the group $H^1(X; \mathbb{Z}) \approx \text{Hom}(H_1(X), \mathbb{Z})$ is always torsionfree.

Proof: We begin with two consequences of Proposition 3F.11:

(a) An inclusion $B \hookrightarrow A$ induces a surjection $\text{Ext}(A, \mathbb{Z}) \to \text{Ext}(B, \mathbb{Z})$. Hence $\text{Ext}(A, \mathbb{Z})$ is uncountable if $\text{Ext}(B, \mathbb{Z})$ is.

(b) If $A \to A/B$ is a quotient map with B finitely generated, then the first term in the exact sequence $\mathrm{Hom}(B, \mathbb{Z}) \to \mathrm{Ext}(A/B, \mathbb{Z}) \to \mathrm{Ext}(A, \mathbb{Z})$ is countable, so $\mathrm{Ext}(A, \mathbb{Z})$ is uncountable if $\mathrm{Ext}(A/B, \mathbb{Z})$ is.

There are two explicit calculations that will be used in the proof:

(c) If A is a direct sum of infinitely many nontrivial finite cyclic groups, then $\mathrm{Ext}(A, \mathbb{Z})$ is uncountable, the product of infinitely many nontrivial groups $\mathrm{Ext}(\mathbb{Z}_n, \mathbb{Z}) \approx \mathbb{Z}_n$.

(d) For p prime, Example 3F.10 gives $\mathrm{Ext}(\mathbb{Z}_{p^\infty}, \mathbb{Z}) \approx \hat{\mathbb{Z}}_p$ which is uncountable.

Consider now the map $A \to A$ given by $a \mapsto pa$ for a fixed prime p. Denote the kernel, image, and cokernel of this map by $_pA$, pA, and A_p, respectively. The functor $A \mapsto A_p$ is the same as $A \mapsto A \otimes \mathbb{Z}_p$. We call the dimension of A_p as a vector space over $\mathbb{Z}_p$ the p-rank of A.

Suppose the p-rank of A is infinite. Then $\mathrm{Ext}(A_p, \mathbb{Z})$ is uncountable by (c). There is an exact sequence $0 \to pA \to A \to A_p \to 0$, so $\mathrm{Hom}(pA, \mathbb{Z}) \to \mathrm{Ext}(A_p, \mathbb{Z}) \to \mathrm{Ext}(A, \mathbb{Z})$ is exact, hence either $\mathrm{Hom}(pA, \mathbb{Z})$ or $\mathrm{Ext}(A, \mathbb{Z})$ is uncountable. Also, we have an isomorphism $\mathrm{Hom}(pA, \mathbb{Z}) \approx \mathrm{Hom}(A, \mathbb{Z})$ since the exact sequence $0 \to {}_pA \to A \to pA \to 0$ gives an exact sequence $0 \to \mathrm{Hom}(pA, \mathbb{Z}) \to \mathrm{Hom}(A, \mathbb{Z}) \to \mathrm{Hom}({}_pA, \mathbb{Z})$ whose last term is 0 since $_pA$ is a torsion group. Thus we have shown that either $\mathrm{Hom}(A, \mathbb{Z})$ or $\mathrm{Ext}(A, \mathbb{Z})$ is uncountable if A has infinite p-rank for some p.

In the remainder of the proof we will show that $\mathrm{Ext}(A, \mathbb{Z})$ is uncountable if A has finite p-rank for all p and A is not finitely generated.

Let C be a nontrivial cyclic subgroup of A, either finite or infinite. If there is no maximal cyclic subgroup of A containing C then there is an infinite ascending chain of cyclic subgroups $C = C_1 \subset C_2 \subset \cdots$. If the indices $[C_i : C_{i-1}]$ involve infinitely many distinct prime factors p then A/C contains an infinite sum $\bigoplus_\infty \mathbb{Z}_p$ for these p so $\mathrm{Ext}(A/C, \mathbb{Z})$ is uncountable by (a) and (c) and hence also $\mathrm{Ext}(A, \mathbb{Z})$ by (b). If only finitely many primes are factors of the indices $[C_i : C_{i-1}]$ then A/C contains a subgroup $\mathbb{Z}_{p^\infty}$ so $\mathrm{Ext}(A/C, \mathbb{Z})$ and hence $\mathrm{Ext}(A, \mathbb{Z})$ is uncountable in this case as well by (a), (b), and (d). Thus we may assume that each nonzero element of A lies in a maximal cyclic subgroup.

If A has positive finite p-rank we can choose a cyclic subgroup mapping nontrivially to A_p and then a maximal cyclic subgroup C containing this one will also map nontrivially to A_p. The quotient A/C has smaller p-rank since $C \to A \to A/C \to 0$ exact implies $C_p \to A_p \to (A/C)_p \to 0$ exact, as tensoring with $\mathbb{Z}_p$ preserves exactness to this extent. By (b) and induction on p-rank this gives a reduction to the case $A_p = 0$, so $A = pA$.

If A is torsionfree, the maximality of the cyclic subgroup C in the preceding paragraph implies that A/C is also torsionfree, so by induction on p-rank we reduce to the case that A is torsionfree and $A = pA$. But in this case A has no maximal cyclic subgroups so this case has already been covered.

If A has torsion, its torsion subgroup T is the direct sum of the p-torsion subgroups $T(p)$ for all primes p. Only finitely many of these $T(p)$'s can be nonzero, otherwise A contains finite cyclic subgroups not contained in maximal cyclic subgroups. If some $T(p)$ is not finitely generated then by (a) we can assume $A = T(p)$. In this case the reduction from finite p-rank to p-rank 0 given above stays within the realm of p-torsion groups. But if $A = pA$ we again have no maximal cyclic subgroups, so we are done in the case that T is not finitely generated. Finally, when T is finitely generated then we can use (b) to reduce to the torsionfree case by passing from A to A/T. $\square$

Exercises

1. Given maps $f_i : X_i \to X_{i+1}$ for integers $i < 0$, show that the 'reverse mapping telescope' obtained by glueing together the mapping cylinders of the f_i's in the obvious way deformation retracts onto X_0. Similarly, if maps $f_i : X_i \to X_{i+1}$ are given for all $i \in \mathbb{Z}$, show that the resulting 'double mapping telescope' deformation retracts onto any of the ordinary mapping telescopes contained in it, the union of the mapping cylinders of the f_i's for i greater than a given number n.

2. Show that $\varprojlim^1 G_i = 0$ if the sequence $\cdots \longrightarrow G_2 \xrightarrow{\alpha_2} G_1 \xrightarrow{\alpha_1} G_0$ satisfies the *Mittag-Leffler condition* that for each i the images of the maps $G_{i+n} \to G_i$ are independent of n for sufficiently large n.

3. Show that $\mathrm{Ext}(A, \mathbb{Q}) = 0$ for all A. [Consider the homology with $\mathbb{Q}$ coefficients of a Moore space $M(A, n)$.]

4. An abelian group G is defined to be *divisible* if the map $G \xrightarrow{n} G$, $g \mapsto ng$, is surjective for all $n > 1$. Show that a group is divisible iff it is a quotient of a direct sum of $\mathbb{Q}$'s. Deduce from the previous problem that if G is divisible then $\mathrm{Ext}(A, G) = 0$ for all A.

5. Show that $\mathrm{Ext}(A, \mathbb{Z})$ is isomorphic to the cokernel of $\mathrm{Hom}(A, \mathbb{Q}) \to \mathrm{Hom}(A, \mathbb{Q}/\mathbb{Z})$, the map induced by the quotient map $\mathbb{Q} \to \mathbb{Q}/\mathbb{Z}$. Use this to get another proof that $\mathrm{Ext}(\mathbb{Z}_{p^\infty}, \mathbb{Z}) \approx \hat{\mathbb{Z}}_p$ for p prime.

6. Show that $\mathrm{Ext}(\mathbb{Z}_{p^\infty}, \mathbb{Z}_p) \approx \mathbb{Z}_p$.

7. Show that for a short exact sequence of abelian groups $0 \to A \to B \to C \to 0$, a Moore space $M(C, n)$ can be realized as a quotient $M(B, n)/M(A, n)$. Applying the long exact sequence of cohomology for the pair $(M(B, n), M(A, n))$ with any coefficient group G, deduce an exact sequence

$$0 \to \mathrm{Hom}(C, G) \to \mathrm{Hom}(B, G) \to \mathrm{Hom}(A, G) \to \mathrm{Ext}(C, G) \to \mathrm{Ext}(B, G) \to \mathrm{Ext}(A, G) \to 0$$

8. Show that for a Moore space $M(G, n)$ the Bockstein long exact sequence in cohomology associated to the short exact sequence of coefficient groups $0 \to A \to B \to C \to 0$ reduces to an exact sequence

$$0 \to \mathrm{Hom}(G, A) \to \mathrm{Hom}(G, B) \to \mathrm{Hom}(G, C) \to \mathrm{Ext}(G, A) \to \mathrm{Ext}(G, B) \to \mathrm{Ext}(G, C) \to 0$$

9. For an abelian group A let $p : A \to A$ be multiplication by p, and let $_pA = \mathrm{Ker}\, p$, $pA = \mathrm{Im}\, p$, and $A_p = \mathrm{Coker}\, p$ as in the proof of Proposition 3F.12. Show that the six-term exact sequences involving $\mathrm{Hom}(-, \mathbb{Z})$ and $\mathrm{Ext}(-, \mathbb{Z})$ associated to the short exact sequences $0 \to {_pA} \to A \to pA \to 0$ and $0 \to pA \to A \to A_p \to 0$ can be spliced together to yield the exact sequence across the top of the following diagram

$$\text{Hom}(pA,\mathbb{Z}) \longrightarrow \text{Ext}(A_p,\mathbb{Z}) \longrightarrow \text{Ext}(A,\mathbb{Z}) \xrightarrow{\;p\;} \text{Ext}(A,\mathbb{Z}) \longrightarrow \text{Ext}(_pA,\mathbb{Z}) \longrightarrow 0$$

$$\text{Ext}(pA,\mathbb{Z})$$

$$0 \longrightarrow \text{Hom}(pA,\mathbb{Z}) \xrightarrow{\;\approx\;} \text{Hom}(A,\mathbb{Z}) \longrightarrow 0 \qquad\qquad 0$$

where the map labeled 'p' is multiplication by p. Use this to show:

(a) $\text{Ext}(A,\mathbb{Z})$ is divisible iff A is torsionfree.

(b) $\text{Ext}(A,\mathbb{Z})$ is torsionfree if A is divisible, and the converse holds if $\text{Hom}(A,\mathbb{Z}) = 0$.

3.G Transfer Homomorphisms

There is a simple construction called 'transfer' that provides very useful information about homology and cohomology of finite-sheeted covering spaces. After giving the definition and proving a few elementary properties, we will use the transfer in the construction of a number of spaces whose $\mathbb{Z}_p$ cohomology is a polynomial ring.

Let $\pi : \widetilde{X} \to X$ be an n-sheeted covering space, for some finite n. In addition to the induced map on singular chains $\pi_\# : C_k(\widetilde{X}) \to C_k(X)$ there is also a homomorphism in the opposite direction $\tau . C_k(X) \to C_k(\widetilde{X})$ which assigns to a singular simplex $\sigma : \Delta^k \to X$ the sum of the n distinct lifts $\widetilde{\sigma} : \Delta^k \to \widetilde{X}$. This is obviously a chain map, commuting with boundary homomorphisms, so it induces **transfer homomorphisms** $\tau_* : H_k(X;G) \to H_k(\widetilde{X};G)$ and $\tau^* : H^k(\widetilde{X};G) \to H^k(X;G)$ for any coefficient group G. We focus on cohomology in what follows, but similar statements hold for homology as well.

The composition $\pi_\# \tau$ is clearly multiplication by n, hence $\tau^* \pi^* = n$. This has the consequence that the kernel of $\pi^* : H^k(X;G) \to H^k(\widetilde{X};G)$ consists of torsion elements of order dividing n, since $\pi^*(\alpha) = 0$ implies $\tau^* \pi^*(\alpha) = n\alpha = 0$. Thus the cohomology of $\widetilde{X}$ must be 'larger' than that of X except possibly for torsion of order dividing n. This can be a genuine exception as one sees from the examples of S^m covering $\mathbb{RP}^m$ and lens spaces. More generally, if $S^m \to X$ is any n-sheeted covering space, then the relation $\tau^* \pi^* = n$ implies that $\widetilde{H}^*(X;\mathbb{Z})$ consists entirely of torsion elements of order dividing n, apart from a possible $\mathbb{Z}$ in dimension m. (Since X is a closed manifold, its homology groups are finitely generated by Corollaries A.8 and A.9 in the Appendix.)

By studying the other composition $\pi^* \tau^*$ we will prove:

Proposition 3G.1. *Let $\pi : \widetilde{X} \to X$ be an n-sheeted covering space defined by an action of a group Γ on $\widetilde{X}$. Then with coefficients in a field F whose characteristic is 0 or a prime not dividing n, the map $\pi^* : H^k(X;F) \to H^k(\widetilde{X};F)$ is injective with image the subgroup $H^*(\widetilde{X};F)^\Gamma$ consisting of classes α such that $\gamma^*(\alpha) = \alpha$ for all $\gamma \in \Gamma$.*

Proof: We have already seen that elements of the kernel of π^* have finite order dividing n, so π^* is injective for the coefficient fields we are considering here. It remains to describe the image of π^*. Note first that $\tau \pi_\#$ sends a singular simplex $\Delta^k \to \widetilde{X}$ to the sum of all its images under the Γ-action. Hence $\pi^* \tau^*(\alpha) = \sum_{\gamma \in \Gamma} \gamma^*(\alpha)$ for $\alpha \in H^k(\widetilde{X}; F)$. If α is fixed under the action of Γ on $H^k(\widetilde{X}; F)$, the sum $\sum_{\gamma \in \Gamma} \gamma^*(\alpha)$ equals $n\alpha$, so if the coefficient field F has characteristic 0 or a prime not dividing n, we can write $\alpha = \pi^* \tau^*(\alpha/n)$ and thus α lies in the image of π^*. Conversely, since $\pi\gamma = \pi$ for all $\gamma \in \Gamma$, we have $\gamma^* \pi^*(\alpha) = \pi^*(\alpha)$ for all α, and so the image of π^* is contained in $H^*(\widetilde{X}; F)^\Gamma$. $\qquad\qquad\square$

Example 3G.2. Let $X = S^1 \vee S^k$, $k > 1$, with $\widetilde{X}$ the n-sheeted cover corresponding to the index n subgroup of $\pi_1(X)$, so $\widetilde{X}$ is a circle with n S^k's attached at equally spaced points around the circle. The deck transformation group $\mathbb{Z}_n$ acts by rotating the circle, permuting the S^k's cyclically. Hence for any coefficient group G, the invariant cohomology $H^*(\widetilde{X}; G)^{\mathbb{Z}_n}$ is all of H^0 and H^1, plus a copy of G in dimension k, the cellular cohomology classes assigning the same element of G to each S^k. Thus $H^i(\widetilde{X}; G)^{\mathbb{Z}_n}$ is exactly the image of π^* for $i = 0$ and k, while the image of π^* in dimension 1 is the subgroup $nH^1(\widetilde{X}; G)$. Whether this equals $H^1(\widetilde{X}; G)^{\mathbb{Z}_n}$ or not depends on G. For $G = \mathbb{Q}$ or $\mathbb{Z}_p$ with p not dividing n, we have equality, but not for $G = \mathbb{Z}$ or $\mathbb{Z}_p$ with p dividing n. In this last case the map π^* is not injective on H^1.

Spaces with Polynomial mod p Cohomology

An interesting special case of the general problem of realizing graded commutative rings as cup product rings of spaces is the case of polynomial rings $\mathbb{Z}_p[x_1, \cdots, x_n]$ over the coefficient field $\mathbb{Z}_p$, p prime. The basic question here is, which sets of numbers $d_1, \cdots, d_n$ are realizable as the dimensions $|x_i|$ of the generators x_i? From §3.2 we have the examples of products of $\mathbb{C}\mathrm{P}^\infty$'s and $\mathbb{H}\mathrm{P}^\infty$'s with d_i's equal to 2 or 4, for arbitrary p, and when $p = 2$ we can also take $\mathbb{R}\mathrm{P}^\infty$'s with d_i's equal to 1.

As an application of transfer homomorphisms we will construct some examples with larger d_i's. In the case of polynomials in one variable, it turns out that these examples realize everything that can be realized. But for two or more variables, more sophisticated techniques are necessary to realize all the realizable cases; see the end of this section for further remarks on this.

The construction can be outlined as follows. Start with a space Y already known to have polynomial cohomology $H^*(Y; \mathbb{Z}_p) = \mathbb{Z}_p[y_1, \cdots, y_n]$, and suppose there is an action of a finite group Γ on Y. A simple trick called the Borel construction shows that without loss of generality we may assume the action is free, defining a covering space $Y \to Y/\Gamma$. Then by Proposition 3G.1 above, if p does not divide the order of Γ, $H^*(Y/\Gamma; \mathbb{Z}_p)$ is isomorphic to the subring of $\mathbb{Z}_p[y_1, \cdots, y_n]$ consisting of polynomials that are invariant under the induced action of Γ on $H^*(Y; \mathbb{Z}_p)$. And in some cases this subring is itself a polynomial ring.

For example, if Y is the product of n copies of $\mathbb{CP}^\infty$ then the symmetric group Σ_n acts on Y by permuting the factors, with the induced action on $H^*(Y; Z_p) \approx Z_p[y_1, \cdots, y_n]$ permuting the y_i's. A standard theorem in algebra says that the invariant polynomials form a polynomial ring $Z_p[\sigma_1, \cdots, \sigma_n]$ where σ_i is the ith elementary symmetric polynomial, the sum of all products of i distinct y_j's. Thus σ_i is a homogeneous polynomial of degree i. The order of Σ_n is $n!$ so the condition that p not divide the order of Γ amounts to $p > n$. Thus we realize the polynomial ring $Z_p[x_1, \cdots, x_n]$ with $|x_i| = 2i$, provided that $p > n$.

This example is less than optimal since there happens to be another space, the Grassmann manifold of n-dimensional linear subspaces of $\mathbb{C}^\infty$, whose cohomology with any coefficient ring R is $R[x_1, \cdots, x_n]$ with $|x_i| = 2i$, as we show in §4.D, so the restriction $p > n$ is not really necessary.

To get further examples the idea is to replace $\mathbb{CP}^\infty$ by a space with the same Z_p cohomology but with 'more symmetry,' allowing for larger groups Γ to act. The constructions will be made using $K(\pi, 1)$ spaces, which were introduced in §1.B. For a group π we constructed there a Δ-complex $B\pi$ with contractible universal cover $E\pi$. The construction is functorial: A homomorphism $\varphi : \pi \to \pi'$ induces a map $B\varphi : B\pi \to B\pi'$, $B\varphi([g_1 | \cdots | g_n]) = [\varphi(g_1) | \cdots | \varphi(g_n)]$, satisfying the functor properties $B(\varphi\psi) = B\varphi B\psi$ and $B\mathbb{1} = \mathbb{1}$. In particular, if Γ is a group of automorphisms of π, then Γ acts on $B\pi$.

The other ingredient we shall need is the **Borel construction**, which converts an action of a group Γ on a space Y into a free action of Γ on a homotopy equivalent space Y'. Namely, take $Y' = Y \times E\Gamma$ with the diagonal action of Γ, $\gamma(y, z) = (\gamma y, \gamma z)$ where Γ acts on $E\Gamma$ as deck transformations. The diagonal action is free, in fact a covering space action, since this is true for the action in the second coordinate. The orbit space of this diagonal action is denoted $Y \times_\Gamma E\Gamma$.

Example 3G.3. Let $\pi = Z_p$ and let Γ be the full automorphism group $\mathrm{Aut}(Z_p)$. Automorphisms of Z_p have the form $x \mapsto mx$ for $(m, p) = 1$, so Γ is the multiplicative group of invertible elements in the field Z_p. By elementary field theory this is a cyclic group, of order $p - 1$. The preceding constructions then give a covering space $K(Z_p, 1) \to K(Z_p, 1)/\Gamma$ with $H^*(K(Z_p, 1)/\Gamma; Z_p) \approx H^*(K(Z_p, 1); Z_p)^\Gamma$. We may assume we are in the nontrivial case $p > 2$. From the calculation of the cup product structure of lens spaces in Example 3.41 or Example 3E.2 we have $H^*(K(Z_p, 1); Z_p) \approx \Lambda_{Z_p}[\alpha] \otimes Z_p[\beta]$ with $|\alpha| = 1$ and $|\beta| = 2$, and we need to figure out how Γ acts on this cohomology ring.

Let $\gamma \in \Gamma$ be a generator, say $\gamma(x) = mx$. The induced action of γ on $\pi_1 K(Z_p, 1)$ is also multiplication by m since we have taken $K(Z_p, 1) = BZ_p \times E\Gamma$ and γ takes an edge loop $[g]$ in BZ_p to $[\gamma(g)] = [mg]$. Hence γ acts on $H_1(K(Z_p, 1); Z)$ by multiplication by m. It follows that $\gamma(\alpha) = m\alpha$ and $\gamma(\beta) = m\beta$ since $H^1(K(Z_p, 1); Z_p) \approx \mathrm{Hom}(H_1(K(Z_p, 1)), Z_p)$ and $H^2(K(Z_p, 1); Z_p) \approx \mathrm{Ext}(H_1(K(Z_p, 1)), Z_p)$, and it is a general fact, following easily from the definitions, that multiplication by an integer m in an abelian group H induces multiplication by m in $\mathrm{Hom}(H,G)$ and $\mathrm{Ext}(H,G)$.

Thus $\gamma(\beta^k) = m^k\beta^k$ and $\gamma(\alpha\beta^k) = m^{k+1}\alpha\beta^k$. Since m was chosen to be a generator of the multiplicative group of invertible elements of $\mathbb{Z}_p$, it follows that the only elements of $H^*(K(\mathbb{Z}_p,1);\mathbb{Z}_p)$ fixed by γ, hence by Γ, are the scalar multiples of $\beta^{i(p-1)}$ and $\alpha\beta^{i(p-1)-1}$. Thus $H^*(K(\mathbb{Z}_p,1);\mathbb{Z}_p)^\Gamma = \Lambda_{\mathbb{Z}_p}[\alpha\beta^{p-2}]\otimes\mathbb{Z}_p[\beta^{p-1}]$, so we have produced a space whose $\mathbb{Z}_p$ cohomology ring is $\Lambda_{\mathbb{Z}_p}[x_{2p-3}]\otimes\mathbb{Z}_p[y_{2p-2}]$, subscripts indicating dimension.

Example 3G.4. As an easy generalization of the preceding example, replace the group Γ there by a subgroup of $\mathrm{Aut}(\mathbb{Z}_p)$ of order d, where d is any divisor of $p-1$. The new Γ is generated by the automorphism $x \mapsto m^{(p-1)/d}x$, and the same analysis shows that we obtain a space with $\mathbb{Z}_p$ cohomology $\Lambda_{\mathbb{Z}_p}[x_{2d-1}]\otimes\mathbb{Z}_p[y_{2d}]$, subscripts again denoting dimension. For a given choice of d the condition that d divides $p-1$ says $p \equiv 1 \bmod d$, which is satisfied by infinitely many p's, according to a classical theorem of Dirichlet.

Example 3G.5. The two preceding examples can be modified so as to eliminate the exterior algebra factors, by replacing $\mathbb{Z}_p$ by $\mathbb{Z}_{p^\infty}$, the union of the increasing sequence $\mathbb{Z}_p \subset \mathbb{Z}_{p^2} \subset \mathbb{Z}_{p^3} \subset \cdots$. The first step is to show that $H^*(K(\mathbb{Z}_{p^\infty},1);\mathbb{Z}_p) \approx \mathbb{Z}_p[\beta]$ with $|\beta| = 2$. We know that $\tilde{H}_*(K(\mathbb{Z}_{p^i},1);\mathbb{Z})$ consists of $\mathbb{Z}_{p^i}$'s in odd dimensions. The inclusion $\mathbb{Z}_{p^i} \hookrightarrow \mathbb{Z}_{p^{i+1}}$ induces a map $K(\mathbb{Z}_{p^i},1) \to K(\mathbb{Z}_{p^{i+1}},1)$ that is unique up to homotopy. We can take this map to be a p-sheeted covering space since the covering space of a $K(\mathbb{Z}_{p^{i+1}},1)$ corresponding to the unique index p subgroup of $\pi_1 K(\mathbb{Z}_{p^{i+1}},1)$ is a $K(\mathbb{Z}_{p^i},1)$. The homology transfer formula $\pi_*\tau_* = p$ shows that the image of the induced map $H_n(K(\mathbb{Z}_{p^i},1);\mathbb{Z}) \to H_n(K(\mathbb{Z}_{p^{i+1}},1);\mathbb{Z})$ for n odd contains the multiples of p, hence this map is the inclusion $\mathbb{Z}_{p^i} \hookrightarrow \mathbb{Z}_{p^{i+1}}$. We can use the universal coefficient theorem to compute the induced map $H^*(K(\mathbb{Z}_{p^{i+1}},1);\mathbb{Z}_p) \to H^*(K(\mathbb{Z}_{p^i},1);\mathbb{Z}_p)$. Namely, the inclusion $\mathbb{Z}_{p^i} \hookrightarrow \mathbb{Z}_{p^{i+1}}$ induces the trivial map $\mathrm{Hom}(\mathbb{Z}_{p^{i+1}},\mathbb{Z}_p) \to \mathrm{Hom}(\mathbb{Z}_{p^i},\mathbb{Z}_p)$, so on odd-dimensional cohomology the induced map is trivial. On the other hand, the induced map on even-dimensional cohomology is an isomorphism since the map of free resolutions

$$\begin{array}{ccccccccc} 0 & \longrightarrow & \mathbb{Z} & \xrightarrow{\ p^i\ } & \mathbb{Z} & \longrightarrow & \mathbb{Z}_{p^i} & \longrightarrow & 0 \\ & & \downarrow{\mathbb{1}} & & \downarrow{p} & & \downarrow{p} & & \\ 0 & \longrightarrow & \mathbb{Z} & \xrightarrow{\ p^{i+1}\ } & \mathbb{Z} & \longrightarrow & \mathbb{Z}_{p^{i+1}} & \longrightarrow & 0 \end{array}$$

dualizes to

$$\begin{array}{ccccccc} 0 & \longleftarrow & \mathrm{Ext}(\mathbb{Z}_{p^i},\mathbb{Z}_p) & \longleftarrow & \mathrm{Hom}(\mathbb{Z},\mathbb{Z}_p) & \xleftarrow{\ 0\ } & \mathrm{Hom}(\mathbb{Z},\mathbb{Z}_p) \\ & & \uparrow & & \uparrow{\mathbb{1}} & & \uparrow \\ 0 & \longleftarrow & \mathrm{Ext}(\mathbb{Z}_{p^{i+1}},\mathbb{Z}_p) & \longleftarrow & \mathrm{Hom}(\mathbb{Z},\mathbb{Z}_p) & \xleftarrow{\ 0\ } & \mathrm{Hom}(\mathbb{Z},\mathbb{Z}_p) \end{array}$$

Since $\mathbb{Z}_{p^\infty}$ is the union of the increasing sequence of subgroups $\mathbb{Z}_{p^i}$, the space $B\mathbb{Z}_{p^\infty}$ is the union of the increasing sequence of subcomplexes $B\mathbb{Z}_{p^i}$. We can therefore apply

Proposition 3F.5 to conclude that $H^*(K(\mathbb{Z}_{p^\infty}, 1); \mathbb{Z}_p)$ is zero in odd dimensions, while in even dimensions the map $H^*(K(\mathbb{Z}_{p^\infty}, 1); \mathbb{Z}_p) \to H^*(K(\mathbb{Z}_p, 1); \mathbb{Z}_p)$ induced by the inclusion $\mathbb{Z}_p \hookrightarrow \mathbb{Z}_{p^\infty}$ is an isomorphism. Thus $H^*(K(\mathbb{Z}_{p^\infty}, 1); \mathbb{Z}_p) \approx \mathbb{Z}_p[\beta]$ as claimed.

Next we show that the map $\mathrm{Aut}(\mathbb{Z}_{p^\infty}) \to \mathrm{Aut}(\mathbb{Z}_p)$ obtained by restriction to the subgroup $\mathbb{Z}_p \subset \mathbb{Z}_{p^\infty}$ is a split surjection. Automorphisms of $\mathbb{Z}_{p^i}$ are the maps $x \mapsto mx$ for $(m, p) = 1$, so the restriction map $\mathrm{Aut}(\mathbb{Z}_{p^{i+1}}) \to \mathrm{Aut}(\mathbb{Z}_{p^i})$ is surjective. Since $\mathrm{Aut}(\mathbb{Z}_{p^\infty}) = \varprojlim \mathrm{Aut}(\mathbb{Z}_{p^i})$, the restriction map $\mathrm{Aut}(\mathbb{Z}_{p^\infty}) \to \mathrm{Aut}(\mathbb{Z}_p)$ is also surjective. The order of $\mathrm{Aut}(\mathbb{Z}_{p^i})$, the multiplicative group of invertible elements of $\mathbb{Z}_{p^i}$, is $p^i - p^{i-1} = p^{i-1}(p - 1)$ and $p - 1$ is relatively prime to p^{i-1}, so the abelian group $\mathrm{Aut}(\mathbb{Z}_{p^i})$ contains a subgroup of order $p - 1$. This subgroup maps onto the cyclic group $\mathrm{Aut}(\mathbb{Z}_p)$ of the same order, so $\mathrm{Aut}(\mathbb{Z}_{p^i}) \to \mathrm{Aut}(\mathbb{Z}_p)$ is a split surjection, hence so is $\mathrm{Aut}(\mathbb{Z}_{p^\infty}) \to \mathrm{Aut}(\mathbb{Z}_p)$.

Thus we have an action of $\Gamma = \mathrm{Aut}(\mathbb{Z}_p)$ on $B\mathbb{Z}_{p^\infty}$ extending its natural action on $B\mathbb{Z}_p$. The Borel construction then gives an inclusion $B\mathbb{Z}_p \times_\Gamma E\Gamma \hookrightarrow B\mathbb{Z}_{p^\infty} \times_\Gamma E\Gamma$ inducing an isomorphism of $H^*(B\mathbb{Z}_{p^\infty} \times_\Gamma E\Gamma; \mathbb{Z}_p)$ onto the even-dimensional part of $H^*(B\mathbb{Z}_p \times_\Gamma E\Gamma; \mathbb{Z}_p)$, a polynomial algebra $\mathbb{Z}_p[y_{2p-2}]$. Similarly, if d is any divisor of $p - 1$, then taking Γ to be the subgroup of $\mathrm{Aut}(\mathbb{Z}_p)$ of order d yields a space with $\mathbb{Z}_p$ cohomology the polynomial ring $\mathbb{Z}_p[y_{2d}]$.

Example 3G.6. Now we enlarge the preceding example by taking products and bringing in the permutation group to produce a space with $\mathbb{Z}_p$ cohomology the polynomial ring $\mathbb{Z}_p[y_{2d}, y_{4d}, \cdots, y_{2nd}]$ where d is any divisor of $p - 1$ and $p > n$. Let X be the product of n copies of $B\mathbb{Z}_{p^\infty}$ and let Γ be the group of homeomorphisms of X generated by permutations of the factors together with the actions of $\mathbb{Z}_d$ in each factor constructed in the preceding example. We can view Γ as a group of $n \times n$ matrices with entries in $\mathbb{Z}_p$, the matrices obtained by replacing some of the 1's in a permutation matrix by elements of $\mathbb{Z}_p$ of multiplicative order a divisor of d. Thus there is a split short exact sequence $0 \to (\mathbb{Z}_d)^n \to \Gamma \to \Sigma_n \to 0$, and the order of Γ is $d^n n!$. The product space X has $H^*(X; \mathbb{Z}_p) \approx \mathbb{Z}_p[\beta_1, \cdots, \beta_n]$ with $|\beta_i| = 2$, so $H^*(X \times_\Gamma E\Gamma; \mathbb{Z}_p) \approx \mathbb{Z}_p[\beta_1, \cdots, \beta_n]^\Gamma$ provided that p does not divide the order of Γ, which means $p > n$. For a polynomial to be invariant under the $\mathbb{Z}_d$ action in each factor it must be a polynomial in the powers β_i^d, and to be invariant under permutations of the variables it must be a symmetric polynomial in these powers. Since symmetric polynomials are exactly the polynomials in the elementary symmetric functions, the polynomials in the β_i's invariant under Γ form a polynomial ring $\mathbb{Z}_p[y_{2d}, y_{4d}, \cdots, y_{2nd}]$ with y_{2k} the sum of all products of k distinct powers β_i^d.

Example 3G.7. As a further variant on the preceding example, choose a divisor q of d and replace Γ by its subgroup consisting of matrices for which the product of the qth powers of the nonzero entries is 1. This has the effect of enlarging the ring of polynomials invariant under the action, and it can be shown that the invariant polynomials form a polynomial ring $\mathbb{Z}_p[y_{2d}, y_{4d}, \cdots, y_{2(n-1)d}, y_{2nq}]$, with the last generator y_{2nd} replaced by $y_{2nq} = \prod_i \beta_i^q$. For example, if $n = 2$ and $q = 1$ we obtain $\mathbb{Z}_p[y_4, y_{2d}]$ with $y_4 = \beta_1 \beta_2$ and $y_{2d} = \beta_1^d + \beta_2^d$. The group Γ in this case happens to be isomorphic to the dihedral group of order $2d$.

General Remarks

The problem of realizing graded polynomial rings $\mathbb{Z}_p[y]$ in one variable as cup product rings of spaces was discussed in §3.2, and Example 3G.5 provides the remaining examples, showing that $|y|$ can be any even divisor of $2(p-1)$. In more variables the problem of realizing $\mathbb{Z}_p[y_1, \cdots, y_n]$ with specified dimensions $|y_i|$ is more difficult, but has been solved for odd primes p. Here is a sketch of the answer.

Assuming that p is odd, the dimensions $|y_i|$ are even. Call the number $d_i = |y_i|/2$ the *degree* of y_i. In the examples above this was in fact the degree of y_i as a polynomial in the 2-dimensional classes β_j invariant under the action of Γ. It was proved in [Dwyer, Miller, & Wilkerson 1992] that every realizable polynomial algebra $\mathbb{Z}_p[y_1, \cdots, y_n]$ is the ring of invariant polynomials $\mathbb{Z}_p[\beta_1, \cdots, \beta_n]^\Gamma$ for an action of some finite group Γ on $\mathbb{Z}_p[\beta_1, \cdots, \beta_n]$, where $|\beta_i| = 2$. The basic examples, whose products yield all realizable polynomial algebras, can be divided into two categories. First there are classifying spaces of Lie groups, each of which realizes a polynomial algebra for all but finitely many primes p. These are listed in the following table.

Lie group	degrees	primes
S^1	1	all
$SU(n)$	$2, 3, \cdots, n$	all
$Sp(n)$	$2, 4, \cdots, 2n$	all
$SO(2k)$	$2, 4, \cdots, 2k-2, k$	$p > 2$
G_2	$2, 6$	$p > 2$
F_4	$2, 6, 8, 12$	$p > 3$
E_6	$2, 5, 6, 8, 9, 12$	$p > 3$
E_7	$2, 6, 8, 10, 12, 14$	$p > 3$
E_8	$2, 8, 12, 14, 18, 20, 24, 30$	$p > 5$

The remaining examples form two infinite families plus 30 sporadic exceptions shown in the table on the next page. The first row is the examples we have constructed, though our construction needed the extra condition that p not divide the order of the group Γ. For all entries in both tables the order of Γ, the group such that $\mathbb{Z}_p[y_1, \cdots, y_n] = \mathbb{Z}_p[\beta_1, \cdots, \beta_n]^\Gamma$, turns out to equal the product of the degrees. When p does not divide this order, the method we used for the first row can also be applied to give examples for all the other rows. In some cases the congruence conditions on p, which are needed in order for Γ to be a subgroup of $\mathrm{Aut}(\mathbb{Z}_p^n) = GL_n(\mathbb{Z}_p)$, automatically imply that p does not divide the order of Γ. But when this is not the case a different construction of a space with the desired cohomology is needed. To find out more about this the reader can begin by consulting [Kane 1988] and [Notbohm 1999].

degrees	primes
$d, 2d, \cdots, (n-1)d, nq$ with $q \mid d$	$p \equiv 1 \bmod d$
$2, d$	$p \equiv -1 \bmod d$

degrees	primes		degrees	primes
4, 6	$p \equiv 1 \bmod 3$		60, 60	$p \equiv 1 \bmod 60$
6, 12	$p \equiv 1 \bmod 3$		12, 30	$p \equiv 1, 4 \bmod 15$
4, 12	$p \equiv 1 \bmod 12$		12, 60	$p \equiv 1, 49 \bmod 60$
12, 12	$p \equiv 1 \bmod 12$		12, 20	$p \equiv 1, 9 \bmod 20$
8, 12	$p \equiv 1 \bmod 4$		2, 6, 10	$p \equiv 1, 4 \bmod 5$
8, 24	$p \equiv 1 \bmod 8$		4, 6, 14	$p \equiv 1, 2, 4 \bmod 7$
12, 24	$p \equiv 1 \bmod 12$		6, 9, 12	$p \equiv 1 \bmod 3$
24, 24	$p \equiv 1 \bmod 24$		6, 12, 18	$p \equiv 1 \bmod 3$
6, 8	$p \equiv 1, 3 \bmod 8$		6, 12, 30	$p \equiv 1, 4 \bmod 15$
8, 12	$p \equiv 1 \bmod 8$		4, 8, 12, 20	$p \equiv 1 \bmod 4$
6, 24	$p \equiv 1, 19 \bmod 24$		2, 12, 20, 30	$p \equiv 1, 4 \bmod 5$
12, 24	$p \equiv 1 \bmod 24$		8, 12, 20, 24	$p \equiv 1 \bmod 4$
20, 30	$p \equiv 1 \bmod 5$		12, 18, 24, 30	$p \equiv 1 \bmod 3$
20, 60	$p \equiv 1 \bmod 20$		4, 6, 10, 12, 18	$p \equiv 1 \bmod 3$
30, 60	$p \equiv 1 \bmod 15$		6, 12, 18, 24, 30, 42	$p \equiv 1 \bmod 3$

For the prime 2 the realization problem has not yet been completely solved. Among the known examples are those in the table at the right. The construction for the last entry, which does not arise from a Lie group, is in [Dwyer & Wilkerson 1993]. (For $p = 2$ 'degree' means the actual cohomological dimension.)

Lie group	degrees
$O(1)$	1
$SO(n)$	$2, 3, \cdots, n$
$SU(n)$	$4, 6, \cdots, 2n$
$Sp(n)$	$4, 8, \cdots, 4n$
$PSp(2n+1)$	$2, 3, 8, 12, \cdots, 8n+4$
G_2	4, 6, 7
$Spin(7)$	4, 6, 7, 8
$Spin(8)$	4, 6, 7, 8, 8
$Spin(9)$	4, 6, 7, 8, 16
F_4	4, 6, 7, 16, 24
—	8, 12, 14, 15

3.H Local Coefficients

Homology and cohomology with local coefficients are fancier versions of ordinary homology and cohomology that can be defined for nonsimply-connected spaces. In various situations these more refined homology and cohomology theories arise naturally and inevitably. For example, the only way to extend Poincaré duality with $\mathbb{Z}$ coefficients to nonorientable manifolds is to use local coefficients. In the overall scheme of algebraic topology, however, the role played by local coefficients is fairly small. Local coefficients bring an extra level of complication that one tries to avoid whenever possible. With this in mind, the goal of this section will not be to give a full exposition but rather just to sketch the main ideas, leaving the technical details for the interested reader to fill in.

The plan for this section is first to give the quick algebraic definition of homology and cohomology with local coefficients, and then to reinterpret this definition more geometrically in a way that looks more like ordinary homology and cohomology. The reinterpretation also allows the familiar properties of homology and cohomology to be extended to the local coefficient case with very little effort.

Local Coefficients via Modules

Let X be a path-connected space having a universal cover $\tilde{X}$ and fundamental group π, so that X is the quotient of $\tilde{X}$ by the action of π by deck transformations $\tilde{x} \mapsto \gamma \cdot \tilde{x}$ for $\gamma \in \pi$ and $\tilde{x} \in \tilde{X}$. The action of π on $\tilde{X}$ induces an action of π on the group $C_n(\tilde{X})$ of singular n-chains in $\tilde{X}$, by sending a singular n-simplex $\sigma : \Delta^n \to \tilde{X}$ to the composition $\Delta^n \xrightarrow{\sigma} \tilde{X} \xrightarrow{\gamma} \tilde{X}$. The action of π on $C_n(\tilde{X})$ makes $C_n(\tilde{X})$ a module over the group ring $\mathbb{Z}[\pi]$, which consists of the finite formal sums $\sum_i m_i \gamma_i$ with $m_i \in \mathbb{Z}$ and $\gamma_i \in \pi$, with the natural addition $\sum_i m_i \gamma_i + \sum_i n_i \gamma_i = \sum_i (m_i + n_i) \gamma_i$ and multiplication $(\sum_i m_i \gamma_i)(\sum_j n_j \gamma_j) = \sum_{i,j} m_i n_j \gamma_i \gamma_j$. The boundary maps $\partial : C_n(\tilde{X}) \to C_{n-1}(\tilde{X})$ are $\mathbb{Z}[\pi]$-module homomorphisms since the action of π on these groups comes from an action on $\tilde{X}$.

If M is an arbitrary module over $\mathbb{Z}[\pi]$, we would like to define $C_n(X;M)$ to be $C_n(\tilde{X}) \otimes_{\mathbb{Z}[\pi]} M$, but for tensor products over a noncommutative ring one has to be a little careful with left and right module structures. In general, if R is a ring, possibly noncommutative, one defines the tensor product $A \otimes_R B$ of a right R-module A and a left R-module B to be the abelian group with generators $a \otimes b$ for $a \in A$ and $b \in B$, subject to distributivity and associativity relations:

(i) $(a_1 + a_2) \otimes b = a_1 \otimes b + a_2 \otimes b$ and $a \otimes (b_1 + b_2) = a \otimes b_1 + a \otimes b_2$.

(ii) $ar \otimes b = a \otimes rb$.

In case $R = \mathbb{Z}[\pi]$, a left $\mathbb{Z}[\pi]$-module A can be regarded as a right $\mathbb{Z}[\pi]$-module by setting $a\gamma = \gamma^{-1}a$ for $\gamma \in \pi$. So the tensor product of two left $\mathbb{Z}[\pi]$-modules A and B is defined, and the relation $a\gamma \otimes b = a \otimes \gamma b$ becomes $\gamma^{-1}a \otimes b = a \otimes \gamma b$, or equivalently $a' \otimes b = \gamma a' \otimes \gamma b$ where $a' = \gamma^{-1}a$. Thus tensoring over $\mathbb{Z}[\pi]$ has the effect of factoring out the action of π. To simplify notation we shall write $A \otimes_{\mathbb{Z}[\pi]} B$ as $A \otimes_\pi B$, emphasizing the fact that the essential part of a $\mathbb{Z}[\pi]$-module structure is the action of π.

In particular, $C_n(\tilde{X}) \otimes_\pi M$ is defined if M is a left $\mathbb{Z}[\pi]$-module. These chain groups $C_n(X;M) = C_n(\tilde{X}) \otimes_\pi M$ form a chain complex with the boundary maps $\partial \otimes \mathbb{1}$. The homology groups $H_n(X;M)$ of this chain complex are by definition **homology groups with local coefficients**.

For cohomology one can set $C^n(X;M) = \mathrm{Hom}_{\mathbb{Z}[\pi]}(C_n(\tilde{X}), M)$, the $\mathbb{Z}[\pi]$-module homomorphisms $C_n(\tilde{X}) \to M$. These groups $C^n(X;M)$ form a cochain complex whose cohomology groups $H^n(X;M)$ are **cohomology groups with local coefficients**.

Example 3H.1. Let us check that when M is a trivial $\mathbb{Z}[\pi]$-module, with $\gamma m = m$ for all $\gamma \in \pi$ and $m \in M$, then $H_n(X; M)$ is just ordinary homology with coefficients in the abelian group M. For a singular n-simplex $\sigma : \Delta^n \to X$, the various lifts $\tilde{\sigma} : \Delta^n \to \tilde{X}$ form an orbit of the action of π on $C_n(\tilde{X})$. In $C_n(\tilde{X}) \otimes_\pi M$ all these lifts are identified via the relation $\tilde{\sigma} \otimes m = \gamma\tilde{\sigma} \otimes \gamma m = \gamma\tilde{\sigma} \otimes m$. Thus we can identify $C_n(\tilde{X}) \otimes_\pi M$ with $C_n(X) \otimes M$, the chain group denoted $C_n(X; M)$ in ordinary homology theory, so $H_n(X; M)$ reduces to ordinary homology with coefficients in M. The analogous statement for cohomology is also true since elements of $\mathrm{Hom}_{\mathbb{Z}[\pi]}(C_n(\tilde{X}), M)$ are functions from singular n-simplices $\tilde{\sigma} : \Delta^n \to \tilde{X}$ to M taking the same value on all elements of a π-orbit since the action of π on M is trivial, so $\mathrm{Hom}_{\mathbb{Z}[\pi]}(C_n(\tilde{X}), M)$ is identifiable with $\mathrm{Hom}(C_n(X), M)$, ordinary cochains with coefficients in M.

Example 3H.2. Suppose we take $M = \mathbb{Z}[\pi]$, viewed as a module over itself via its ring structure. For a ring R with identity element, $A \otimes_R R$ is naturally isomorphic to A via the correspondence $a \otimes r \mapsto ar$. So we have a natural identification of $C_n(\tilde{X}) \otimes_\pi \mathbb{Z}[\pi]$ with $C_n(\tilde{X})$, and hence an isomorphism $H_n(X; \mathbb{Z}[\pi]) \approx H_n(\tilde{X})$. Generalizing this, let $X' \to X$ be the cover corresponding to a subgroup $\pi' \subset \pi$. Then the free abelian group $\mathbb{Z}[\pi/\pi']$ with basis the cosets $\gamma\pi'$ is a $\mathbb{Z}[\pi]$-module and $C_n(\tilde{X}) \otimes_{\mathbb{Z}[\pi]} \mathbb{Z}[\pi/\pi'] \approx C_n(X')$, so $H_n(X; \mathbb{Z}[\pi/\pi']) \approx H_n(X')$. More generally, if A is an abelian group then $A[\pi/\pi']$ is a $\mathbb{Z}[\pi]$-module and $H_n(X; A[\pi/\pi']) \approx H_n(X'; A)$. So homology of covering spaces is a special case of homology with local coefficients. The corresponding assertions for cohomology are not true, however, as we shall see later in the section.

For a $\mathbb{Z}[\pi]$-module M, let π' be the kernel of the homomorphism $\rho : \pi \to \mathrm{Aut}(M)$ defining the module structure, given by $\rho(\gamma)(m) = \gamma m$, where $\mathrm{Aut}(M)$ is the group of automorphisms of the abelian group M. If $X' \to X$ is the cover corresponding to the normal subgroup π' of π, then $C_n(\tilde{X}) \otimes_\pi M \approx C_n(X') \otimes_\pi M \approx C_n(X') \otimes_{\mathbb{Z}[\pi/\pi']} M$. This gives a more efficient description of $H_n(X; M)$.

Example 3H.3. As a special case, suppose that we take $M = \mathbb{Z}$, so $\mathrm{Aut}(\mathbb{Z}) \approx \mathbb{Z}_2 = \{\pm 1\}$. For a nontrivial $\mathbb{Z}[\pi]$-module structure on M, π' is a subgroup of index 2 and $X' \to X$ is a 2-sheeted covering space. If τ is the nontrivial deck transformation of X', let $C_n^+(X') = \{\alpha \in C_n(X') \mid \tau_\#(\alpha) = \alpha\}$ and $C_n^-(X') = \{\alpha \in C_n(X') \mid \tau_\#(\alpha) = -\alpha\}$. It follows easily that $C_n^\pm(X')$ has basis the chains $\sigma \pm \tau\sigma$ for $\sigma : \Delta^n \to X'$, and we have short exact sequences

$$0 \longrightarrow C_n^-(X') \hookrightarrow C_n(X') \xrightarrow{\Sigma} C_n^+(X') \longrightarrow 0$$

$$0 \longrightarrow C_n^+(X') \hookrightarrow C_n(X') \xrightarrow{\Delta} C_n^-(X') \longrightarrow 0$$

where $\Sigma(\alpha) = \alpha + \tau_\#(\alpha)$ and $\Delta(\alpha) = \alpha - \tau_\#(\alpha)$. The homomorphism $C_n(X) \to C_n^+(X')$ sending a singular simplex in X to the sum of its two lifts to X' is an isomorphism. The quotient map $C_n(X') \to C_n(X') \otimes_\pi \mathbb{Z}$ has kernel $C_n^+(X')$, so the second short exact sequence gives an isomorphism $C_n^-(X') \approx C_n(X') \otimes_\pi \mathbb{Z}$. These isomorphisms are isomorphisms of chain complexes and the short exact sequences are short exact sequence of chain complexes, so from the first short exact sequence we get a long exact sequence of homology groups

$$\cdots \longrightarrow H_n(X; \widetilde{\mathbb{Z}}) \longrightarrow H_n(X') \xrightarrow{\ p_* \ } H_n(X) \longrightarrow H_{n-1}(X; \widetilde{\mathbb{Z}}) \longrightarrow \cdots$$

where the symbol $\widetilde{\mathbb{Z}}$ indicates local coefficients in the module $\mathbb{Z}$ and p_* is induced by the covering projection $p : X' \to X$.

Let us apply this exact sequence when X is a nonorientable n-manifold M which is closed and connected. We shall use terminology and notation from §3.3. We can view $\mathbb{Z}$ as a $\mathbb{Z}[\pi_1 M]$-module by letting a loop γ in M act on $\mathbb{Z}$ by multiplication by $+1$ or -1 according to whether γ preserves or reverses local orientations of M. The double cover $X' \to X$ is then the 2-sheeted cover $\widetilde{M} \to M$ with $\widetilde{M}$ orientable. The nonorientability of M implies that $H_n(M) = 0$. Since $H_{n+1}(M) = 0$, the exact sequence above then gives $H_n(M; \widetilde{\mathbb{Z}}) \approx H_n(\widetilde{M}) \approx \mathbb{Z}$. This can be interpreted as saying that by taking homology with local coefficients we obtain a fundamental class for a nonorientable manifold.

Local Coefficients via Bundles of Groups

Now we wish to reinterpret homology and cohomology with local coefficients in more geometric terms, making it look more like ordinary homology and cohomology.

Let us first define a special kind of covering space with extra algebraic structure. A **bundle of groups** is a map $p : E \to X$ together with a group structure on each subset $p^{-1}(x)$, such that all these groups $p^{-1}(x)$ are isomorphic to a fixed group G in the following special way: Each point of X has a neighborhood U for which there exists a homeomorphism $h_U : p^{-1}(U) \to U \times G$ taking each $p^{-1}(x)$ to $\{x\} \times G$ by a group isomorphism. Since G is given the discrete topology, the projection p is a covering space. Borrowing terminology from the theory of fiber bundles, the subsets $p^{-1}(x)$ are called the **fibers** of $p : E \to X$, and one speaks of E as a bundle of groups with fiber G. It may be worth remarking that if we modify the definition by replacing the word 'group' with 'vector space' throughout, then we obtain the much more common notion of a vector bundle; see [VBKT].

Trivial examples are provided by products $E = X \times G$. Nontrivial examples we have considered are the covering spaces $M_{\mathbb{Z}} \to M$ of nonorientable manifolds M defined in §3.3. Here the group G is the homology coefficient group $\mathbb{Z}$, though one could equally well define a bundle of groups $M_G \to M$ for any abelian coefficient group G.

Homology groups of X with coefficients in a bundle E of abelian groups may be defined as follows. Consider finite sums $\sum_i n_i \sigma_i$ where each $\sigma_i : \Delta^n \to X$ is a singular n-simplex in X and $n_i : \Delta^n \to E$ is a lifting of σ_i. The sum of two lifts n_i and m_i of the same σ_i is defined by $(n_i + m_i)(s) = n_i(s) + m_i(s)$, and is also a lift of σ_i. In this way the finite sums $\sum_i n_i \sigma_i$ form an abelian group $C_n(X; E)$, provided we allow the deletion of terms $n_i \sigma_i$ when n_i is the zero-valued lift. A boundary homomorphism $\partial : C_n(X;E) \to C_{n-1}(X;E)$ is defined by the formula $\partial(\sum_i n_i \sigma_i) = \sum_{i,j}(-1)^j n_i \sigma_i | [v_0, \cdots, \hat{v}_j, \cdots, v_n]$ where 'n_i' in the right side of the equation means the restricted lifting $n_i | [v_0, \cdots, \hat{v}_j, \cdots, v_n]$. The proof that the usual boundary homomorphism ∂ satisfies $\partial^2 = 0$ still works in the present context, so the groups $C_n(X;E)$ form a chain complex. We denote the homology groups of this chain complex by $H_n(X;E)$.

In case E is the product bundle $X \times G$, lifts n_i are simply elements of G, so $H_n(X;E) = H_n(X;G)$, ordinary homology. In the general case, lifts $n_i : \Delta^n \to E$ are uniquely determined by their value at one point $s \in \Delta^n$, and these values can be specified arbitrarily since Δ^n is simply-connected, so the n_i's can be thought of as elements of $p^{-1}(\sigma_i(s))$, a group isomorphic to G. However if E is not a product, there is no canonical isomorphism between different fibers $p^{-1}(x)$, so one cannot identify $H_n(X;E)$ with ordinary homology.

An alternative approach would be to take the coefficients n_i to be elements of the fiber group over a specific point of $\sigma_i(\Delta^n)$, say $\sigma_i(v_0)$. However, with such a definition the formula for the boundary operator ∂ becomes more complicated since there is no point of Δ^n that lies in all the faces.

Our task now is to relate the homology groups $H_n(X;E)$ to homology groups with coefficients in a module, as defined earlier. In §1.3 we described how covering spaces of X with a given fiber F can be classified in terms of actions of $\pi_1(X)$ on F, assuming X is path-connected and has the local properties guaranteeing the existence of a universal cover. It is easy to check that covering spaces that are bundles of groups with fiber a group G are equivalent to actions of $\pi_1(X)$ on G by automorphisms of G, that is, homomorphisms from $\pi_1(X)$ to $\mathrm{Aut}(G)$.

For example, for the bundle $M_{\mathbb{Z}} \to M$ the action of a loop y on the fiber $\mathbb{Z}$ is multiplication by ± 1 according to whether y preserves or reverses orientation in M, that is, whether y lifts to a closed loop in the orientable double cover $\widetilde{M} \to M$ or not. As another example, the action of $\pi_1(X)$ on itself by inner automorphisms corresponds to a bundle of groups $p : E \to X$ with fibers $p^{-1}(x) = \pi_1(X,x)$. This example is rather similar in spirit to the examples $M_{\mathbb{Z}} \to M$. In both cases one has a functor associating a group to each point of a space, and all the groups at different points are isomorphic, but not canonically so. Different choices of isomorphisms are obtained by choosing different paths between two points, and loops give rise to an action of π_1 on the fibers.

In the case of bundles of groups $p : E \to X$ whose fiber G is abelian, an action of $\pi_1(X)$ on G by automorphisms is the same as a $\mathbb{Z}[\pi_1 X]$-module structure on G.

Proposition 3H.4. *If X is a path-connected space having a universal covering space, then the groups $H_n(X;E)$ are naturally isomorphic to the homology groups $H_n(X;G)$ with local coefficients in the $\mathbb{Z}[\pi]$-module G associated to E, where $\pi = \pi_1(X)$.*

Proof: As noted earlier, a bundle of groups $E \to X$ with fiber G is equivalent to an action of π on G. In more explicit terms this means that if $\widetilde{X}$ is the universal cover of X, then E is identifiable with the quotient of $\widetilde{X} \times G$ by the diagonal action of π, $\gamma(\widetilde{x}, g) = (\gamma \widetilde{x}, \gamma g)$ where the action in the first coordinate is by deck transformations of $\widetilde{X}$. For a chain $\sum_i n_i \sigma_i \in C_n(X; E)$, the coefficient n_i gives a lift of σ_i to E, and n_i in turn has various lifts to $\widetilde{X} \times G$. Thus we have natural surjections $C_n(\widetilde{X} \times G) \to C_n(E) \to C_n(X; E)$ expressing each of these groups as a quotient of the preceding one. More precisely, identifying $C_n(\widetilde{X} \times G)$ with $C_n(\widetilde{X}) \otimes \mathbb{Z}[G]$ in the obvious way, then $C_n(E)$ is the quotient of $C_n(\widetilde{X}) \otimes \mathbb{Z}[G]$ under the identifications $\widetilde{\sigma} \otimes g \sim \gamma \cdot \widetilde{\sigma} \otimes \gamma \cdot g$. This quotient is the tensor product $C_n(\widetilde{X}) \otimes_\pi \mathbb{Z}[G]$. To pass to the quotient $C_n(X; E)$ of $C_n(E) = C_n(\widetilde{X}) \otimes_\pi \mathbb{Z}[G]$ we need to take into account the sum operation in $C_n(X; E)$, addition of lifts $n_i : \Delta^n \to E$. This means that in sums $\widetilde{\sigma} \otimes g_1 + \widetilde{\sigma} \otimes g_2 = \widetilde{\sigma} \otimes (g_1 + g_2)$, the term $g_1 + g_2$ should be interpreted not in $\mathbb{Z}[G]$ but in the natural quotient G of $\mathbb{Z}[G]$. Hence $C_n(X; E)$ is identified with the quotient $C_n(\widetilde{X}) \otimes_\pi G$ of $C_n(\widetilde{X}) \otimes_\pi \mathbb{Z}[G]$. This natural identification commutes with the boundary homomorphisms, so the homology groups are also identified. $\square$

More generally, if X has a number of path-components X_α with universal covers $\widetilde{X}_\alpha$, then $C_n(X; E) = \bigoplus_\alpha (C_n(\widetilde{X}_\alpha) \otimes_{\mathbb{Z}[\pi_1(X_\alpha)]} G)$, so $H_n(X; E)$ splits accordingly as a direct sum of the local coefficient homology groups for the path-components X_α.

We turn now to the question of whether homology with local coefficients satisfies axioms similar to those for ordinary homology. The main novelty is with the behavior of induced homomorphisms. In order for a map $f : X \to X'$ to induce a map on homology with local coefficients we must have bundles of groups $E \to X$ and $E' \to X'$ that are related in some way. The natural assumption to make is that there is a commutative diagram as at the right, such that $\widetilde{f}$ restricts to a homomorphism in each fiber. With this hypothesis there is then a chain homomorphism $f_\# : C_n(X; E) \to C_n(X'; E')$ obtained by composing singular simplices with f and their lifts with $\widetilde{f}$, hence there is an induced homomorphism $f_* : H_n(X; E) \to H_n(X'; E')$. The fibers of E and E' need not be isomorphic groups, so change-of-coefficient homomorphisms $H_n(X; G_1) \to H_n(X; G_2)$ for ordinary homology are a special case. To avoid this extra complication we shall consider only the case that $\widetilde{f}$ restricts to an isomorphism on each fiber. With this condition, a commutative diagram as above will be called a **bundle map**.

$$
\begin{array}{ccc}
E & \xrightarrow{\widetilde{f}} & E' \\
\downarrow{\scriptstyle p} & & \downarrow{\scriptstyle p'} \\
X & \xrightarrow{f} & X'
\end{array}
$$

Here is a method for constructing bundle maps. Starting with a map $f : X \to X'$ and a bundle of groups $p' : E' \to X'$, let

$$E = \{ (x, e') \in X \times E' \mid f(x) = p'(e') \}.$$

This fits into a commutative diagram as above if we define $p(x, e') = x$ and $\widetilde{f}(x, e') = e'$. In particular, the fiber $p^{-1}(x)$ consists of pairs (x, e') with $p'(e') = f(x)$, so $\widetilde{f}$ is a bijection of this fiber with the fiber of $E' \to X'$ over $f(x)$. We use this bijection to give $p^{-1}(x)$ a group structure. To check that $p : E \to X$ is a bundle of groups, let $h' : (p')^{-1}(U') \to U' \times G$ be an isomorphism as in the definition of a bundle of groups. Define $h : p^{-1}(U) \to U \times G$ over $U = f^{-1}(U')$ by $h(x, e') = (x, h_2'(e'))$ where h_2' is the second coordinate of h'. An inverse for h is $(x, g) \in (x, (h')^{-1}(f(x), g))$, and h is clearly an isomorphism on each fiber. Thus $p : E \to X$ is a bundle of groups, called the **pullback** of $E' \to X'$ via f, or the **induced bundle**. The notation $f^*(E')$ is often used for the pullback bundle.

Given any bundle map $E \to E'$ as in the diagram above, it is routine to check that the map $E \to f^*(E')$, $e \mapsto (p(e), \tilde{f}(e))$, is an isomorphism of bundles over X, so the pullback construction produces all bundle maps. Thus we see one reason why homology with local coefficients is somewhat complicated: $H_n(X; E)$ is really a functor of two variables, covariant in X and contravariant in E.

Viewing bundles of abelian groups over X as $\mathbb{Z}[\pi_1 X]$-modules, the pullback construction corresponds to making a $\mathbb{Z}[\pi_1 X']$-module into a $\mathbb{Z}[\pi_1 X]$-module by defining $\gamma g = f_*(\gamma)g$ for $f_* : \pi_1(X) \to \pi_1(X')$. This follows easily from the definitions. In particular, this implies that homotopic maps $f_0, f_1 : X \to X'$ induce isomorphic pullback bundles $f_0^*(E'), f_1^*(E')$. Hence the map $f_* : H_n(X; E) \to H_n(X'; E')$ induced by a bundle map depends only on the homotopy class of f.

Generalizing the definition of $H_n(X; E)$ to pairs (X, A) is straightforward, starting with the definition of $H_n(X, A; E)$ as the nth homology group of the chain complex of quotients $C_n(X; E)/C_n(A; E)$ where $p : E \to X$ becomes a bundle of groups over A by restriction to $p^{-1}(A)$. Associated to the pair (X, A) there is then a long exact sequence of homology groups with local coefficients in the bundle E. The excision property is proved just as for ordinary homology, via iterated barycentric subdivision. The final axiom for homology, involving disjoint unions, extends trivially to homology with local coefficients. Simplicial and cellular homology also extend without difficulty to the case of local coefficients, as do the proofs that these forms of homology agree with singular homology for Δ-complexes and CW complexes, respectively. We leave the verifications of all these statements to the energetic reader.

Now we turn to cohomology. One might try defining $H^n(X; E)$ by simply dualizing, taking $\mathrm{Hom}(C_n(X), E)$, but this makes no sense since E is not a group. Instead, the cochain group $C^n(X; E)$ is defined to consist of all functions φ assigning to each singular simplex $\sigma : \Delta^n \to X$ a lift $\varphi(\sigma) : \Delta^n \to E$. In case E is the product $X \times G$, this amounts to assigning an element of G to each σ, so this definition generalizes ordinary cohomology. Coboundary maps $\delta : C^n(X; E) \to C^{n+1}(X; E)$ are defined just as with ordinary cohomology, and satisfy $\delta^2 = 0$, so we have cohomology groups $H^n(X; E)$, and in the relative case, $H^n(X, A; E)$, defined via relative cochains $C^n(X, A; E) = \mathrm{Ker}(C^n(X; E) \to C^n(A; E))$.

For a path-connected space X with universal cover $\tilde{X}$ and fundamental group π, we can identify $H^n(X; E)$ with $H^n(X; G)$, cohomology with local coefficients in the

$\mathbb{Z}[\pi]$-module G corresponding to E, by identifying $C^n(X;E)$ with $\mathrm{Hom}_{\mathbb{Z}[\pi]}(C_n(\widetilde{X}),G)$ in the following way. An element $\varphi \in C^n(X;E)$ assigns to each $\sigma:\Delta^n \to X$ a lift to E. Regarding E as the quotient of $\widetilde{X} \times G$ under the diagonal action of π, a lift of σ to E is the same as an orbit of a lift to $\widetilde{X} \times G$. Such an orbit is a function f assigning to each lift $\widetilde{\sigma}:\Delta^n \to \widetilde{X}$ an element $f(\widetilde{\sigma}) \in G$ such that $f(\gamma\widetilde{\sigma}) = \gamma f(\widetilde{\sigma})$ for all $\gamma \in \pi$, that is, an element of $\mathrm{Hom}_{\mathbb{Z}[\pi]}(C_n(\widetilde{X}),G)$.

The basic properties of ordinary cohomology in §3.1 extend without great difficulty to cohomology groups with local coefficients. In order to define the map $f^*:H^n(X';E') \to H^n(X;E)$ induced by a bundle map as before, it suffices to observe that a singular simplex $\sigma:\Delta^n \to X$ and a lift $\widetilde{\sigma}':\Delta^n \to E'$ of $f\sigma$ define a lift $\widetilde{\sigma} = (\sigma,\widetilde{\sigma}'):\Delta^n \to f^*(E)$ of σ. To show that $f \simeq g$ implies $f^* = g^*$ requires some modification of the proof of the corresponding result for ordinary cohomology in §3.1, which proceeded by dualizing the proof for homology. In the local coefficient case one constructs a chain homotopy P^* satisfying $g^\sharp - f^\sharp = P^*\delta + \delta P^*$ directly from the subdivision of $\Delta^n \times I$ used in the proof of the homology result. Similar remarks apply to proving excision and Mayer–Vietoris sequences for cohomology with local coefficients. To prove the equivalence of simplicial and cellular cohomology with singular cohomology in the local coefficient context, one should use the telescope argument from the proof of Lemma 2.34 to show that $H^n(X^k;E) \approx H^n(X;E)$ for $k > n$. Once again details will be left to the reader.

The difference between homology with local coefficients and cohomology with local coefficients is illuminated by comparing the following proposition with our earlier identification of $H_*(X;\mathbb{Z}[\pi_1 X])$ with the ordinary homology of the universal cover of X.

> **Proposition 3H.5.** *If X is a finite CW complex with universal cover $\widetilde{X}$ and fundamental group π, then for all n, $H^n(X;\mathbb{Z}[\pi])$ is isomorphic to $H_c^n(\widetilde{X};\mathbb{Z})$, cohomology of $\widetilde{X}$ with compact supports and ordinary integer coefficients.*

For example, consider the n-dimensional torus T^n, the product of n circles, with fundamental group $\pi = \mathbb{Z}^n$ and universal cover $\mathbb{R}^n$. We have $H_i(T^n;\mathbb{Z}[\pi]) \approx H_i(\mathbb{R}^n)$, which is zero except for a $\mathbb{Z}$ in dimension 0, but $H^i(T^n;\mathbb{Z}[\pi]) \approx H_c^i(\mathbb{R}^n)$ vanishes except for a $\mathbb{Z}$ in dimension n, as we saw in Example 3.34.

To prove the proposition we shall use a few general facts about cohomology with compact supports. One significant difference between ordinary cohomology and cohomology with compact supports is in induced maps. A map $f:X \to Y$ induces $f^\sharp:C_c^n(Y;G) \to C_c^n(X;G)$ and hence $f^*:H_c^n(Y;G) \to H_c^n(X;G)$ provided that f is **proper**: The preimage $f^{-1}(K)$ of each compact set K in Y is compact in X. Thus if $\varphi \in C^n(Y;G)$ vanishes on chains in $Y - K$ then $f^\sharp(\varphi) \in C^n(X;G)$ vanishes on chains in $X - f^{-1}(K)$. Further, to guarantee that $f \simeq g$ implies $f^* = g^*$ we should restrict attention to homotopies that are proper as maps $X \times I \to Y$. Relative groups

$H_c^n(X,A;G)$ are defined when A is a closed subset of X, which guarantees that the inclusion $A \hookrightarrow X$ is a proper map. With these constraints the basic theory of §3.1 translates without difficulty to cohomology with compact supports.

In particular, for a locally compact CW complex X one can compute $H_c^*(X;G)$ using *finite cellular cochains*, the cellular cochains vanishing on all but finitely many cells. Namely, to compute $H_c^n(X^n, X^{n-1}; G)$ using excision one first has to identify this group with $H_c^n(X^n, N(X^{n-1}); G)$ where $N(X^{n-1})$ is a closed neighborhood of X^{n-1} in X^n obtained by deleting an open n-disk from the interior of each n-cell. If X is locally compact, the obvious deformation retraction of $N(X^{n-1})$ onto X^{n-1} is a proper homotopy equivalence. Hence via long exact sequences and the five-lemma we obtain isomorphisms $H_c^n(X^n, X^{n-1}; G) \approx H_c^n(X^n, N(X^{n-1}); G)$, and by excision the latter group can be identified with the finite cochains.

Proof of 3H.5: As noted above, we can compute $H_c^*(\widetilde{X};\mathbb{Z})$ using the groups $C_f^n(\widetilde{X};\mathbb{Z})$ of finite cellular cochains $\varphi : C_n \to \mathbb{Z}$, where $C_n = H_n(\widetilde{X}^n, \widetilde{X}^{n-1})$. Giving $\widetilde{X}$ the CW structure lifting the CW structure on X, then since X is compact, finite cellular cochains are exactly homomorphisms $\varphi : C_n \to \mathbb{Z}$ such that for each cell e^n of $\widetilde{X}$, $\varphi(\gamma e^n)$ is nonzero for only finitely many covering transformations $\gamma \in \pi$. Such a φ determines a map $\hat{\varphi} : C_n \to \mathbb{Z}[\pi]$ by setting $\hat{\varphi}(e^n) = \sum_\gamma \varphi(\gamma^{-1} e^n) \gamma$. The map $\hat{\varphi}$ is a $\mathbb{Z}[\pi]$-homomorphism since if we replace the summation index γ in the right side of $\varphi(\eta e^n) = \sum_\gamma \varphi(\gamma^{-1}\eta e^n)\gamma$ by $\eta\gamma$, we get $\sum_\gamma \varphi(\gamma^{-1} e^n)\eta\gamma$. The function $\varphi \mapsto \hat{\varphi}$ defines a homomorphism $C_f^n(\widetilde{X};\mathbb{Z}) \to \mathrm{Hom}_{\mathbb{Z}[\pi]}(C_n, \mathbb{Z}[\pi])$ which is injective since φ is recoverable from $\hat{\varphi}$ as the coefficient of $\gamma = 1$. Furthermore, this homomorphism is surjective since a $\mathbb{Z}[\pi]$-homomorphism $\psi : M \to \mathbb{Z}[\pi]$ has the form $\psi(x) = \sum_\gamma \psi_\gamma(x)\gamma$ with $\psi_\gamma \in \mathrm{Hom}_{\mathbb{Z}}(M, \mathbb{Z})$ satisfying $\psi_\gamma(x) = \psi_1(\gamma^{-1}x)$, so ψ_1 determines ψ. The isomorphisms $C_f^n(\widetilde{X};\mathbb{Z}) \approx \mathrm{Hom}_{\mathbb{Z}[\pi]}(C_n, \mathbb{Z}[\pi])$ are isomorphisms of cochain complexes, so the respective cohomology groups $H_c^n(\widetilde{X};\mathbb{Z})$ and $H^n(X;\mathbb{Z}[\pi])$ are isomorphic. $\qquad\square$

Cup and cap product work easily with local coefficients in a bundle of rings, the latter concept being defined in the obvious way. The cap product can be used to give a version of Poincaré duality for a closed n-manifold M using coefficients in a bundle of rings E under the same assumption as with ordinary coefficients that there exists a fundamental class $[M] \in H_n(M;E)$ restricting to a generator of $H_n(M, M - \{x\}; E)$ for all $x \in M$. By excision the latter group is isomorphic to the fiber ring R of E. The same proof as for ordinary coefficients then shows that $[M] \frown : H^k(M;E) \to H_{n-k}(M;E)$ is an isomorphism for all k.

Taking R to be one of the standard rings $\mathbb{Z}$, $\mathbb{Q}$, or $\mathbb{Z}_p$ does not give anything new since the only ring automorphism these rings have is the identity, so the bundle of rings E must be the product $M \times R$. To get something more interesting, suppose we take R to be the ring $\mathbb{Z}[i]$ of Gaussian integers, the complex numbers $a + bi$ with

$a, b \in \mathbb{Z}$. This has complex conjugation $a + bi \mapsto a - bi$ as a ring isomorphism. If M is nonorientable and connected we can use the homomorphism $\omega : \pi_1(M) \to \{\pm 1\}$ that defines the bundle of groups $M_{\mathbb{Z}}$ to build a bundle of rings E corresponding to the action of $\pi_1(M)$ on $\mathbb{Z}[i]$ given by $\gamma(a + bi) = a + \omega(\gamma)bi$. The homology and cohomology groups of M with coefficients in E depend only on the additive structure of $\mathbb{Z}[i]$ so they split as the direct sum of their real and imaginary parts, which are just the homology or cohomology groups with ordinary coefficients $\mathbb{Z}$ and twisted coefficients $\widetilde{\mathbb{Z}}$, respectively. The fundamental class in $H_n(M; \widetilde{\mathbb{Z}})$ constructed in Example 3H.3 can be viewed as a pure imaginary fundamental class $[M] \in H_n(M; E)$. Since cap product with $[M]$ interchanges real and imaginary parts, we obtain:

Theorem 3H.6. *If M is a nonorientable closed connected n-manifold then cap product with the pure imaginary fundamental class $[M]$ gives isomorphisms $H^k(M; \mathbb{Z}) \approx H_{n-k}(M; \widetilde{\mathbb{Z}})$ and $H^k(M; \widetilde{\mathbb{Z}}) \approx H_{n-k}(M; \mathbb{Z})$.*   □

More generally this holds with $\mathbb{Z}$ replaced by other rings such as $\mathbb{Q}$ or $\mathbb{Z}_p$. There is also a version for noncompact manifolds using cohomology with compact supports.

Exercises

1. Compute $H_*(S^1; E)$ and $H^*(S^1; E)$ for $E \to S^1$ the nontrivial bundle with fiber $\mathbb{Z}$.

2. Compute the homology groups with local coefficients $H_n(M; M_{\mathbb{Z}})$ for a closed nonorientable surface M.

3. Let $\mathcal{B}(X; G)$ be the set of isomorphism classes of bundles of groups $E \to X$ with fiber G, and let $E_0 \to B\mathrm{Aut}(G)$ be the bundle corresponding to the 'identity' action $\rho : \mathrm{Aut}(G) \to \mathrm{Aut}(G)$. Show that the map $[X, B\mathrm{Aut}(G)] \to \mathcal{B}(X, G)$, $[f] \mapsto f^*(E_0)$, is a bijection if X is a CW complex, where $[X, Y]$ denotes the set of homotopy classes of maps $X \to Y$.

4. Show that if finite connected CW complexes X and Y are homotopy equivalent, then their universal covers $\widetilde{X}$ and $\widetilde{Y}$ are proper homotopy equivalent.

5. If X is a finite connected graph with $\pi_1(X)$ free on $g > 0$ generators, show that $H^n(X; \mathbb{Z}[\pi_1 X])$ is zero unless $n = 1$, when it is $\mathbb{Z}$ when $g = 1$ and the direct sum of a countably infinite number of $\mathbb{Z}$'s when $g > 1$. [Use Proposition 3H.5 and compute $H_c^n(\widetilde{X})$ as $\varinjlim H^n(\widetilde{X}, \widetilde{X} - T_i)$ for a suitable sequence of finite subtrees $T_1 \subset T_2 \subset \cdots$ of $\widetilde{X}$ with $\bigcup_i T_i = \widetilde{X}$.]

6. Show that homology groups $H_n^{\ell f}(X; G)$ can be defined using **locally finite** chains, which are formal sums $\sum_\sigma g_\sigma \sigma$ of singular simplices $\sigma : \Delta^n \to X$ with coefficients $g_\sigma \in G$, such that each $x \in X$ has a neighborhood meeting the images of only finitely many σ's with $g_\sigma \neq 0$. Develop this version of homology far enough to show that for a finite-dimensional locally compact CW complex X, $H_n^{\ell f}(X; G)$ can be computed using infinite cellular chains $\sum_\alpha g_\alpha e_\alpha^n$.

Chapter 4

Homotopy Theory

Homotopy theory begins with the homotopy groups $\pi_n(X)$, which are the natural higher-dimensional analogs of the fundamental group. These higher homotopy groups have certain formal similarities with homology groups. For example, $\pi_n(X)$ turns out to be always abelian for $n \geq 2$, and there are relative homotopy groups fitting into a long exact sequence just like the long exact sequence of homology groups. However, the higher homotopy groups are much harder to compute than either homology groups or the fundamental group, due to the fact that neither the excision property for homology nor van Kampen's theorem for π_1 holds for higher homotopy groups.

In spite of these computational difficulties, homotopy groups are of great theoretical significance. One reason for this is Whitehead's theorem that a map between CW complexes which induces isomorphisms on all homotopy groups is a homotopy equivalence. The stronger statement that two CW complexes with isomorphic homotopy groups are homotopy equivalent is usually false, however. One of the rare cases when a CW complex does have its homotopy type uniquely determined by its homotopy groups is when it has just a single nontrivial homotopy group. Such spaces, known as Eilenberg–MacLane spaces, turn out to play a fundamental role in algebraic topology for a variety of reasons. Perhaps the most important is their close connection with cohomology: Cohomology classes in a CW complex correspond bijectively with homotopy classes of maps from the complex into an Eilenberg–MacLane space.

Thus cohomology has a strictly homotopy-theoretic interpretation, and there is an analogous but more subtle homotopy-theoretic interpretation of homology, explained in §4.F.

A more elementary and direct connection between homotopy and homology is the Hurewicz theorem, asserting that the first nonzero homotopy group $\pi_n(X)$ of a simply-connected space X is isomorphic to the first nonzero homology group $\tilde{H}_n(X)$. This result, along with its relative version, is one of the cornerstones of algebraic topology.

Though the excision property does not always hold for homotopy groups, in some important special cases there is a range of dimensions in which it does hold. This leads to the idea of stable homotopy groups, the beginning of stable homotopy theory. Perhaps the major unsolved problem in algebraic topology is the computation of the stable homotopy groups of spheres. Near the end of §4.2 we give some tables of known calculations that show quite clearly the complexity of the problem.

Included in §4.2 is a brief introduction to fiber bundles, which generalize covering spaces and play a somewhat analogous role for higher homotopy groups. It would easily be possible to devote a whole book to the subject of fiber bundles, even the special case of vector bundles, but here we use fiber bundles only to provide a few basic examples and to motivate their more flexible homotopy-theoretic generalization, fibrations, which play a large role in §4.3. Among other things, fibrations allow one to describe, in theory at least, how the homotopy type of an arbitrary CW complex is built up from its homotopy groups by an inductive procedure of forming 'twisted products' of Eilenberg–MacLane spaces. This is the notion of a Postnikov tower. In favorable cases, including all simply-connected CW complexes, the additional data beyond homotopy groups needed to determine a homotopy type can also be described, in the form of a sequence of cohomology classes called the k-invariants of a space. If these are all zero, the space is homotopy equivalent to a product of Eilenberg–MacLane spaces, and otherwise not. Unfortunately the k-invariants are cohomology classes in rather complicated spaces in general, so this is not a practical way of classifying homotopy types, but it is useful for various more theoretical purposes.

This chapter is arranged so that it begins with purely homotopy-theoretic notions, largely independent of homology and cohomology theory, whose roles gradually increase in later sections of the chapter. It should therefore be possible to read a good portion of this chapter immediately after reading Chapter 1, with just an occasional glimpse at Chapter 2 for algebraic definitions, particularly the notion of an exact sequence which is just as important in homotopy theory as in homology and cohomology theory.

4.1 Homotopy Groups

Perhaps the simplest noncontractible spaces are spheres, so to get a glimpse of the subtlety inherent in homotopy groups let us look at some of the calculations of the groups $\pi_i(S^n)$ that have been made. A small sample is shown in the table below, extracted from [Toda 1962].

$$\pi_i(S^n)$$

$i \rightarrow$	1	2	3	4	5	6	7	8	9	10	11	12
n 1	$\mathbb{Z}$	0	0	0	0	0	0	0	0	0	0	0
$\downarrow$ 2	0	$\mathbb{Z}$	$\mathbb{Z}$	$\mathbb{Z}_2$	$\mathbb{Z}_2$	$\mathbb{Z}_{12}$	$\mathbb{Z}_2$	$\mathbb{Z}_2$	$\mathbb{Z}_3$	$\mathbb{Z}_{15}$	$\mathbb{Z}_2$	$\mathbb{Z}_2 \times \mathbb{Z}_2$
3	0	0	$\mathbb{Z}$	$\mathbb{Z}_2$	$\mathbb{Z}_2$	$\mathbb{Z}_{12}$	$\mathbb{Z}_2$	$\mathbb{Z}_2$	$\mathbb{Z}_3$	$\mathbb{Z}_{15}$	$\mathbb{Z}_2$	$\mathbb{Z}_2 \times \mathbb{Z}_2$
4	0	0	0	$\mathbb{Z}$	$\mathbb{Z}_2$	$\mathbb{Z}_2$	$\mathbb{Z} \times \mathbb{Z}_{12}$	$\mathbb{Z}_2 \times \mathbb{Z}_2$	$\mathbb{Z}_2 \times \mathbb{Z}_2$	$\mathbb{Z}_{24} \times \mathbb{Z}_3$	$\mathbb{Z}_{15}$	$\mathbb{Z}_2$
5	0	0	0	0	$\mathbb{Z}$	$\mathbb{Z}_2$	$\mathbb{Z}_2$	$\mathbb{Z}_{24}$	$\mathbb{Z}_2$	$\mathbb{Z}_2$	$\mathbb{Z}_2$	$\mathbb{Z}_{30}$
6	0	0	0	0	0	$\mathbb{Z}$	$\mathbb{Z}_2$	$\mathbb{Z}_2$	$\mathbb{Z}_{24}$	0	$\mathbb{Z}$	$\mathbb{Z}_2$
7	0	0	0	0	0	0	$\mathbb{Z}$	$\mathbb{Z}_2$	$\mathbb{Z}_2$	$\mathbb{Z}_{24}$	0	0
8	0	0	0	0	0	0	0	$\mathbb{Z}$	$\mathbb{Z}_2$	$\mathbb{Z}_2$	$\mathbb{Z}_{24}$	0

This is an intriguing mixture of pattern and chaos. The most obvious feature is the large region of zeros below the diagonal, and indeed $\pi_i(S^n) = 0$ for all $i < n$ as we show in Corollary 4.9. There is also the sequence of zeros in the first row, suggesting that $\pi_i(S^1) = 0$ for all $i > 1$. This too is a fairly elementary fact, a special case of Proposition 4.1, following easily from covering space theory.

The coincidences in the second and third rows can hardly be overlooked. These are the case $n = 1$ of isomorphisms $\pi_i(S^{2n}) \approx \pi_{i-1}(S^{2n-1}) \times \pi_i(S^{4n-1})$ that hold for $n = 1, 2, 4$ and all i. The next case $n = 2$ says that each entry in the fourth row is the product of the entry diagonally above it to the left and the entry three units below it. Actually, these isomorphisms $\pi_i(S^{2n}) \approx \pi_{i-1}(S^{2n-1}) \times \pi_i(S^{4n-1})$ hold for all n if one factors out 2-torsion, the elements of order a power of 2. This is a theorem of James that will be proved in [SSAT].

The next regular feature in the table is the sequence of $\mathbb{Z}$'s down the diagonal. This is an illustration of the Hurewicz theorem, which asserts that for a simply-connected space X, the first nonzero homotopy group $\pi_n(X)$ is isomorphic to the first nonzero homology group $H_n(X)$.

One may observe that all the groups above the diagonal are finite except for $\pi_3(S^2)$, $\pi_7(S^4)$, and $\pi_{11}(S^6)$. In §4.B we use cup products in cohomology to show that $\pi_{4k-1}(S^{2k})$ contains a $\mathbb{Z}$ direct summand for all $k \geq 1$. It is a theorem of Serre proved in [SSAT] that $\pi_i(S^n)$ is finite for $i > n$ except for $\pi_{4k-1}(S^{2k})$, which is the direct sum of $\mathbb{Z}$ with a finite group. So all the complexity of the homotopy groups of spheres resides in finite abelian groups. The problem thus reduces to computing the p-torsion in $\pi_i(S^n)$ for each prime p.

An especially interesting feature of the table is that along each diagonal the groups $\pi_{n+k}(S^n)$ with k fixed and varying n eventually become independent of n for large enough n. This stability property is the Freudenthal suspension theorem, proved in §4.2 where we give more extensive tables of these stable homotopy groups of spheres.

Definitions and Basic Constructions

Let I^n be the n-dimensional unit cube, the product of n copies of the interval $[0, 1]$. The boundary ∂I^n of I^n is the subspace consisting of points with at least one coordinate equal to 0 or 1. For a space X with basepoint $x_0 \in X$, define $\pi_n(X, x_0)$ to be the set of homotopy classes of maps $f : (I^n, \partial I^n) \to (X, x_0)$, where homotopies f_t are required to satisfy $f_t(\partial I^n) = x_0$ for all t. The definition extends to the case $n = 0$ by taking I^0 to be a point and ∂I^0 to be empty, so $\pi_0(X, x_0)$ is just the set of path-components of X.

When $n \geq 2$, a sum operation in $\pi_n(X, x_0)$, generalizing the composition operation in π_1, is defined by

$$(f + g)(s_1, s_2, \cdots, s_n) = \begin{cases} f(2s_1, s_2, \cdots, s_n), & s_1 \in [0, 1/2] \\ g(2s_1 - 1, s_2, \cdots, s_n), & s_1 \in [1/2, 1] \end{cases}$$

It is evident that this sum is well-defined on homotopy classes. Since only the first coordinate is involved in the sum operation, the same arguments as for π_1 show that $\pi_n(X, x_0)$ is a group, with identity element the constant map sending I^n to x_0 and with inverses given by $-f(s_1, s_2, \cdots, s_n) = f(1 - s_1, s_2, \cdots, s_n)$.

The additive notation for the group operation is used because $\pi_n(X, x_0)$ is abelian for $n \geq 2$. Namely, $f + g \simeq g + f$ via the homotopy indicated in the following figures.

The homotopy begins by shrinking the domains of f and g to smaller subcubes of I^n, with the region outside these subcubes mapping to the basepoint. After this has been done, there is room to slide the two subcubes around anywhere in I^n as long as they stay disjoint, so if $n \geq 2$ they can be slid past each other, interchanging their positions. Then to finish the homotopy, the domains of f and g can be enlarged back to their original size. If one likes, the whole process can be done using just the coordinates s_1 and s_2, keeping the other coordinates fixed.

Maps $(I^n, \partial I^n) \to (X, x_0)$ are the same as maps of the quotient $I^n / \partial I^n = S^n$ to X taking the basepoint $s_0 = \partial I^n / \partial I^n$ to x_0. This means that we can also view $\pi_n(X, x_0)$ as homotopy classes of maps $(S^n, s_0) \to (X, x_0)$, where homotopies are through maps of the same form $(S^n, s_0) \to (X, x_0)$. In this interpretation of $\pi_n(X, x_0)$, the sum $f + g$ is the composition $S^n \xrightarrow{c} S^n \vee S^n \xrightarrow{f \vee g} X$ where c collapses the equator S^{n-1} in S^n to a point and we choose the basepoint s_0 to lie in this S^{n-1}.

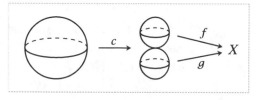

We will show next that if X is path-connected, different choices of the basepoint x_0 always produce isomorphic groups $\pi_n(X, x_0)$, just as for π_1, so one is justified in writing $\pi_n(X)$ for $\pi_n(X, x_0)$ in these cases. Given a path $\gamma : I \to X$ from $x_0 = \gamma(0)$ to another basepoint $x_1 = \gamma(1)$, we may associate to each map $f : (I^n, \partial I^n) \to (X, x_1)$ a new map $\gamma f : (I^n, \partial I^n) \to (X, x_0)$ by shrinking the domain of f to a smaller concentric cube in I^n, then inserting the path γ on each radial segment in the shell between this smaller cube and ∂I^n. When

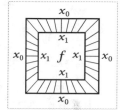

$n = 1$ the map γf is the composition of the three paths γ, f, and the inverse of γ, so the notation γf conflicts with the notation for composition of paths. Since we are mainly interested in the cases $n > 1$, we leave it to the reader to make the necessary notational adjustments when $n = 1$.

A homotopy of γ or f through maps fixing ∂I or ∂I^n, respectively, yields a homotopy of γf through maps $(I^n, \partial I^n) \to (X, x_0)$. Here are three other basic properties:

(1) $\gamma(f + g) \simeq \gamma f + \gamma g$.

(2) $(\gamma \eta) f \simeq \gamma(\eta f)$.

(3) $1 f \simeq f$, where 1 denotes the constant path.

The homotopies in (2) and (3) are obvious. For (1), we first deform f and g to be constant on the right and left halves of I^n, respectively, producing maps we may call $f + 0$ and $0 + g$, then we excise a progressively wider symmetric middle slab of $\gamma(f + 0) + \gamma(0 + g)$ until it becomes $\gamma(f + g)$:

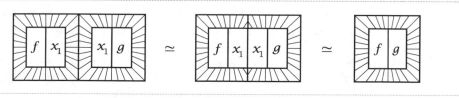

An explicit formula for this homotopy is

$$h_t(s_1, s_2, \cdots, s_n) = \begin{cases} \gamma(f + 0)((2 - t)s_1, s_2, \cdots, s_n), & s_1 \in [0, 1/2] \\ \gamma(0 + g)((2 - t)s_1 + t - 1, s_2, \cdots, s_n), & s_1 \in [1/2, 1] \end{cases}$$

Thus we have $\gamma(f + g) \simeq \gamma(f + 0) + \gamma(0 + g) \simeq \gamma f + \gamma g$.

If we define a change-of-basepoint transformation $\beta_\gamma : \pi_n(X, x_1) \to \pi_n(X, x_0)$ by $\beta_\gamma([f]) = [\gamma f]$, then (1) shows that β_γ is a homomorphism, while (2) and (3) imply that β_γ is an isomorphism with inverse $\beta_{\overline{\gamma}}$ where $\overline{\gamma}$ is the inverse path of γ,

$\overline{y}(s) = y(1-s)$. Thus if X is path-connected, different choices of basepoint x_0 yield isomorphic groups $\pi_n(X, x_0)$, which may then be written simply as $\pi_n(X)$.

Now let us restrict attention to loops y at the basepoint x_0. Since $\beta_{y\eta} = \beta_y \beta_\eta$, the association $[y] \mapsto \beta_y$ defines a homomorphism from $\pi_1(X, x_0)$ to $\mathrm{Aut}(\pi_n(X, x_0))$, the group of automorphisms of $\pi_n(X, x_0)$. This is called the **action of π_1 on π_n**, each element of π_1 acting as an automorphism $[f] \mapsto [yf]$ of π_n. When $n = 1$ this is the action of π_1 on itself by inner automorphisms. When $n > 1$, the action makes the abelian group $\pi_n(X, x_0)$ into a module over the group ring $\mathbb{Z}[\pi_1(X, x_0)]$. Elements of $\mathbb{Z}[\pi_1]$ are finite sums $\sum_i n_i y_i$ with $n_i \in \mathbb{Z}$ and $y_i \in \pi_1$, multiplication being defined by distributivity and the multiplication in π_1. The module structure on π_n is given by $(\sum_i n_i y_i)\alpha = \sum_i n_i(y_i \alpha)$ for $\alpha \in \pi_n$. For brevity one sometimes says π_n is a π_1-module rather than a $\mathbb{Z}[\pi_1]$-module.

In the literature, a space with trivial π_1 action on π_n is called 'n-simple,' and 'simple' means 'n-simple for all n.' In this book we will call a space **abelian** if it has trivial action of π_1 on all homotopy groups π_n, since when $n = 1$ this is the condition that π_1 be abelian. This terminology is consistent with a long-established usage of the term 'nilpotent' to refer to spaces with nilpotent π_1 and nilpotent action of π_1 on all higher homotopy groups; see [Hilton, Mislin, & Roitberg 1975]. An important class of abelian spaces is H-spaces, as we show in Example 4A.3.

We next observe that π_n is a functor. Namely, a map $\varphi: (X, x_0) \to (Y, y_0)$ induces $\varphi_*: \pi_n(X, x_0) \to \pi_n(Y, y_0)$ defined by $\varphi_*([f]) = [\varphi f]$. It is immediate from the definitions that φ_* is well-defined and a homomorphism for $n \geq 1$. The functor properties $(\varphi\psi)_* = \varphi_* \psi_*$ and $\mathbb{1}_* = \mathbb{1}$ are also evident, as is the fact that if $\varphi_t: (X, x_0) \to (Y, y_0)$ is a homotopy then $\varphi_{0*} = \varphi_{1*}$.

In particular, a homotopy equivalence $(X, x_0) \simeq (Y, y_0)$ in the basepointed sense induces isomorphisms on all homotopy groups π_n. This is true even if basepoints are not required to be stationary during homotopies. We showed this for π_1 in Proposition 1.18, and the generalization to higher n's is an exercise at the end of this section.

Homotopy groups behave very nicely with respect to covering spaces:

Proposition 4.1. *A covering space projection $p: (\widetilde{X}, \widetilde{x}_0) \to (X, x_0)$ induces isomorphisms $p_*: \pi_n(\widetilde{X}, \widetilde{x}_0) \to \pi_n(X, x_0)$ for all $n \geq 2$.*

Proof: For surjectivity of p_* we apply the lifting criterion in Proposition 1.33, which implies that every map $(S^n, s_0) \to (X, x_0)$ lifts to $(\widetilde{X}, \widetilde{x}_0)$ provided that $n \geq 2$ so that S^n is simply-connected. Injectivity of p_* is immediate from the covering homotopy property, just as in Proposition 1.31 which treated the case $n = 1$. $\square$

In particular, $\pi_n(X, x_0) = 0$ for $n \geq 2$ whenever X has a contractible universal cover. This applies for example to S^1, so we obtain the first row of the table of homotopy groups of spheres shown earlier. More generally, the n-dimensional torus T^n, the product of n circles, has universal cover $\mathbb{R}^n$, so $\pi_i(T^n) = 0$ for $i > 1$. This is in marked contrast to the homology groups $H_i(T^n)$ which are nonzero for all $i \leq n$. Spaces with $\pi_n = 0$ for all $n \geq 2$ are sometimes called **aspherical**.

The behavior of homotopy groups with respect to products is very simple:

Proposition 4.2. *For a product $\prod_\alpha X_\alpha$ of an arbitrary collection of path-connected spaces X_α there are isomorphisms $\pi_n(\prod_\alpha X_\alpha) \approx \prod_\alpha \pi_n(X_\alpha)$ for all n.*

Proof: A map $f : Y \to \prod_\alpha X_\alpha$ is the same thing as a collection of maps $f_\alpha : Y \to X_\alpha$. Taking Y to be S^n and $S^n \times I$ gives the result.   □

Very useful generalizations of the homotopy groups $\pi_n(X, x_0)$ are the **relative homotopy groups** $\pi_n(X, A, x_0)$ for a pair (X, A) with a basepoint $x_0 \in A$. To define these, regard I^{n-1} as the face of I^n with the last coordinate $s_n = 0$ and let J^{n-1} be the closure of $\partial I^n - I^{n-1}$, the union of the remaining faces of I^n. Then $\pi_n(X, A, x_0)$ for $n \geq 1$ is defined to be the set of homotopy classes of maps $(I^n, \partial I^n, J^{n-1}) \to (X, A, x_0)$, with homotopies through maps of the same form. There does not seem to be a completely satisfactory way of defining $\pi_0(X, A, x_0)$, so we shall leave this undefined (but see the exercises for one possible definition). Note that $\pi_n(X, x_0, x_0) = \pi_n(X, x_0)$, so absolute homotopy groups are a special case of relative homotopy groups.

A sum operation is defined in $\pi_n(X, A, x_0)$ by the same formulas as for $\pi_n(X, x_0)$, except that the coordinate s_n now plays a special role and is no longer available for the sum operation. Thus $\pi_n(X, A, x_0)$ is a group for $n \geq 2$, and this group is abelian for $n \geq 3$. For $n = 1$ we have $I^1 = [0, 1]$, $I^0 = \{0\}$, and $J^0 = \{1\}$, so $\pi_1(X, A, x_0)$ is the set of homotopy classes of paths in X from a varying point in A to the fixed basepoint $x_0 \in A$. In general this is not a group in any natural way.

Just as elements of $\pi_n(X, x_0)$ can be regarded as homotopy classes of maps $(S^n, s_0) \to (X, x_0)$, there is an alternative definition of $\pi_n(X, A, x_0)$ as the set of homotopy classes of maps $(D^n, S^{n-1}, s_0) \to (X, A, x_0)$, since collapsing J^{n-1} to a point converts $(I^n, \partial I^n, J^{n-1})$ into (D^n, S^{n-1}, s_0). From this viewpoint, addition is done via the map $c : D^n \to D^n \vee D^n$ collapsing $D^{n-1} \subset D^n$ to a point.

A useful and conceptually enlightening reformulation of what it means for an element of $\pi_n(X, A, x_0)$ to be trivial is given by the following *compression criterion*:

- A map $f : (D^n, S^{n-1}, s_0) \to (X, A, x_0)$ represents zero in $\pi_n(X, A, x_0)$ iff it is homotopic rel S^{n-1} to a map with image contained in A.

For if we have such a homotopy to a map g, then $[f] = [g]$ in $\pi_n(X, A, x_0)$, and $[g] = 0$ via the homotopy obtained by composing g with a deformation retraction of D^n onto s_0. Conversely, if $[f] = 0$ via a homotopy $F : D^n \times I \to X$, then by restricting F to a family of n-disks in $D^n \times I$ starting with $D^n \times \{0\}$ and ending with the disk $D^n \times \{1\} \cup S^{n-1} \times I$, all the disks in the family having the same boundary, then we get a homotopy from f to a map into A, stationary on S^{n-1}.

A map $\varphi : (X, A, x_0) \to (Y, B, y_0)$ induces maps $\varphi_* : \pi_n(X, A, x_0) \to \pi_n(Y, B, y_0)$ which are homomorphisms for $n \geq 2$ and have properties analogous to those in the absolute case: $(\varphi \psi)_* = \varphi_* \psi_*$, $\mathbb{1}_* = \mathbb{1}$, and $\varphi_* = \psi_*$ if $\varphi \simeq \psi$ through maps $(X, A, x_0) \to (Y, B, y_0)$.

Probably the most useful feature of the relative groups $\pi_n(X, A, x_0)$ is that they fit into a long exact sequence

$$\cdots \longrightarrow \pi_n(A, x_0) \xrightarrow{\ i_*\ } \pi_n(X, x_0) \xrightarrow{\ j_*\ } \pi_n(X, A, x_0) \xrightarrow{\ \partial\ } \pi_{n-1}(A, x_0) \longrightarrow \cdots \longrightarrow \pi_0(X, x_0)$$

Here i and j are the inclusions $(A, x_0) \hookrightarrow (X, x_0)$ and $(X, x_0, x_0) \hookrightarrow (X, A, x_0)$. The map ∂ comes from restricting maps $(I^n, \partial I^n, J^{n-1}) \to (X, A, x_0)$ to I^{n-1}, or by restricting maps $(D^n, S^{n-1}, s_0) \to (X, A, x_0)$ to S^{n-1}. The map ∂, called the *boundary map*, is a homomorphism when $n > 1$.

‖ Theorem 4.3. *This sequence is exact.*

Near the end of the sequence, where group structures are not defined, exactness still makes sense: The image of one map is the kernel of the next, those elements mapping to the homotopy class of the constant map.

Proof: With only a little more effort we can derive the long exact sequence of a triple (X, A, B, x_0) with $x_0 \in B \subset A \subset X$:

$$\cdots \longrightarrow \pi_n(A, B, x_0) \xrightarrow{\ i_*\ } \pi_n(X, B, x_0) \xrightarrow{\ j_*\ } \pi_n(X, A, x_0) \xrightarrow{\ \partial\ } \pi_{n-1}(A, B, x_0) \longrightarrow \cdots$$
$$\longrightarrow \pi_1(X, A, x_0)$$

When $B = x_0$ this reduces to the exact sequence for the pair (X, A, x_0), though the latter sequence continues on two more steps to $\pi_0(X, x_0)$. The verification of exactness at these last two steps is left as a simple exercise.

Exactness at $\pi_n(X, B, x_0)$: First note that the composition $j_* i_*$ is zero since every map $(I^n, \partial I^n, J^{n-1}) \to (A, B, x_0)$ represents zero in $\pi_n(X, A, x_0)$ by the compression criterion. To see that $\text{Ker}\, j_* \subset \text{Im}\, i_*$, let $f : (I^n, \partial I^n, J^{n-1}) \to (X, B, x_0)$ represent zero in $\pi_n(X, A, x_0)$. Then by the compression criterion again, f is homotopic rel ∂I^n to a map with image in A, hence the class $[f] \in \pi_n(X, B, x_0)$ is in the image of i_*.

Exactness at $\pi_n(X, A, x_0)$: The composition ∂j_* is zero since the restriction of a map $(I^n, \partial I^n, J^{n-1}) \to (X, B, x_0)$ to I^{n-1} has image lying in B, and hence represents zero in $\pi_{n-1}(A, B, x_0)$. Conversely, suppose the restriction of $f : (I^n, \partial I^n, J^{n-1}) \to (X, A, x_0)$ to I^{n-1} represents zero in $\pi_{n-1}(A, B, x_0)$. Then $f \vert I^{n-1}$ is homotopic to a map with image in B via a homotopy $F : I^{n-1} \times I \to A$ rel ∂I^{n-1}. We can tack F onto f to get a new map $(I^n, \partial I^n, J^{n-1}) \to (X, B, x_0)$ which, as a map $(I^n, \partial I^n, J^{n-1}) \to (X, A, x_0)$, is homotopic to f by the homotopy that tacks on increasingly longer initial segments of F. So $[f] \in \text{Im}\, j_*$.

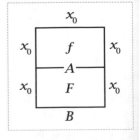

Exactness at $\pi_n(A, B, x_0)$: The composition $i_* \partial$ is zero since the restriction of a map $f : (I^{n+1}, \partial I^{n+1}, J^n) \to (X, A, x_0)$ to I^n is homotopic rel ∂I^n to a constant map via f itself. The converse is easy if B is a point, since a nullhomotopy $f_t : (I^n, \partial I^n) \to (X, x_0)$ of $f_0 : (I^n, \partial I^n) \to (A, x_0)$ gives a map $F : (I^{n+1}, \partial I^{n+1}, J^n) \to (X, A, x_0)$ with $\partial([F]) = [f_0]$. Thus the proof is finished in this case. For a general B, let F be a nullhomotopy of $f : (I^n, \partial I^n, J^{n-1}) \to (A, B, x_0)$ through maps $(I^n, \partial I^n, J^{n-1}) \to (X, B, x_0)$, and let g be the restriction of F to $I^{n-1} \times I$, as in the first of the two pictures below. Reparametrizing the nth and $(n+1)$st coordinates as shown in the second picture, we see that f with g tacked on is in the image of ∂. But as we noted in the preceding paragraph, tacking g onto f gives the same element of $\pi_n(A, B, x_0)$. $\square$

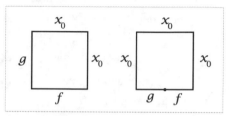

Example 4.4. Let CX be the cone on a path-connected space X, the quotient space of $X \times I$ obtained by collapsing $X \times \{0\}$ to a point. We can view X as the subspace $X \times \{1\} \subset CX$. Since CX is contractible, the long exact sequence of homotopy groups for the pair (CX, X) gives isomorphisms $\pi_n(CX, X, x_0) \approx \pi_{n-1}(X, x_0)$ for all $n \geq 1$. Taking $n = 2$, we can thus realize any group G, abelian or not, as a relative π_2 by choosing X to have $\pi_1(X) \approx G$.

The long exact sequence of homotopy groups is clearly natural: A map of basepointed triples $(X, A, B, x_0) \to (Y, C, D, y_0)$ induces a map between the associated long exact sequences, with commuting squares.

There are change-of-basepoint isomorphisms β_γ for relative homotopy groups analogous to those in the absolute case. One starts with a path γ in $A \subset X$ from x_0 to x_1, and this induces $\beta_\gamma : \pi_n(X, A, x_1) \to \pi_n(X, A, x_0)$ by setting $\beta_\gamma([f]) = [\gamma f]$ where γf is defined as in the picture, by placing a copy of f in a smaller cube with its face I^{n-1} centered in the corresponding face of the larger cube. This construction satisfies the same basic properties as in the absolute case, with very similar

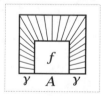

proofs that we leave to the exercises. Separate proofs must be given in the two cases since the definition of γf in the relative case does not specialize to the definition of γf in the absolute case.

The isomorphisms β_γ show that $\pi_n(X, A, x_0)$ is independent of x_0 when A is path-connected. In this case $\pi_n(X, A, x_0)$ is often written simply as $\pi_n(X, A)$.

Restricting to loops at the basepoint, the association $\gamma \mapsto \beta_\gamma$ defines an action of $\pi_1(A, x_0)$ on $\pi_n(X, A, x_0)$ analogous to the action of $\pi_1(X, x_0)$ on $\pi_n(X, x_0)$ in the absolute case. In fact, it is clear from the definitions that $\pi_1(A, x_0)$ acts on the whole long exact sequence of homotopy groups for (X, A, x_0), the action commuting with the various maps in the sequence.

A space X with basepoint x_0 is said to be **n-connected** if $\pi_i(X, x_0) = 0$ for $i \leq n$. Thus 0-connected means path-connected and 1-connected means simply-connected. Since n-connected implies 0-connected, the choice of the basepoint x_0 is not significant. The condition of being n-connected can be expressed without mention of a basepoint since it is an easy exercise to check that the following three conditions are equivalent.

(1) Every map $S^i \to X$ is homotopic to a constant map.
(2) Every map $S^i \to X$ extends to a map $D^{i+1} \to X$.
(3) $\pi_i(X, x_0) = 0$ for all $x_0 \in X$.

Thus X is n-connected if any one of these three conditions holds for all $i \leq n$. Similarly, in the relative case it is not hard to see that the following four conditions are equivalent, for $i > 0$:

(1) Every map $(D^i, \partial D^i) \to (X, A)$ is homotopic rel ∂D^i to a map $D^i \to A$.
(2) Every map $(D^i, \partial D^i) \to (X, A)$ is homotopic through such maps to a map $D^i \to A$.
(3) Every map $(D^i, \partial D^i) \to (X, A)$ is homotopic through such maps to a constant map $D^i \to A$.
(4) $\pi_i(X, A, x_0) = 0$ for all $x_0 \in A$.

When $i = 0$ we did not define the relative π_0, and (1)–(3) are each equivalent to saying that each path-component of X contains points in A since D^0 is a point and ∂D^0 is empty. The pair (X, A) is called **n-connected** if (1)–(4) hold for all $i \leq n$, $i > 0$, and (1)–(3) hold for $i = 0$.

Note that X is n-connected iff (X, x_0) is n-connected for some x_0 and hence for all x_0.

Whitehead's Theorem

Since CW complexes are built using attaching maps whose domains are spheres, it is perhaps not too surprising that homotopy groups of CW complexes carry a lot of information. Whitehead's theorem makes this explicit:

Theorem 4.5. *If a map $f : X \to Y$ between connected CW complexes induces isomorphisms $f_* : \pi_n(X) \to \pi_n(Y)$ for all n, then f is a homotopy equivalence. In case f is the inclusion of a subcomplex $X \hookrightarrow Y$, the conclusion is stronger: X is a deformation retract of Y.*

The proof will follow rather easily from a more technical result that turns out to be very useful in quite a number of arguments. For convenient reference we call this the **compression lemma**.

Lemma 4.6. *Let (X, A) be a CW pair and let (Y, B) be any pair with $B \neq \varnothing$. For each n such that $X - A$ has cells of dimension n, assume that $\pi_n(Y, B, y_0) = 0$ for all $y_0 \in B$. Then every map $f : (X, A) \to (Y, B)$ is homotopic rel A to a map $X \to B$.*

When $n = 0$ the condition that $\pi_n(Y, B, y_0) = 0$ for all $y_0 \in B$ is to be regarded as saying that (Y, B) is 0-connected.

Proof: Assume inductively that f has already been homotoped to take the skeleton X^{k-1} to B. If Φ is the characteristic map of a cell e^k of $X - A$, the composition $f\Phi : (D^k, \partial D^k) \to (Y, B)$ can be homotoped into B rel ∂D^k in view of the hypothesis that $\pi_k(Y, B, y_0) = 0$ if $k > 0$, or that (Y, B) is 0-connected if $k = 0$. This homotopy of $f\Phi$ induces a homotopy of f on the quotient space $X^{k-1} \cup e^k$ of $X^{k-1} \amalg D^k$, a homotopy rel X^{k-1}. Doing this for all k-cells of $X - A$ simultaneously, and taking the constant homotopy on A, we obtain a homotopy of $f \mid X^k \cup A$ to a map into B. By the homotopy extension property in Proposition 0.16, this homotopy extends to a homotopy defined on all of X, and the induction step is completed.

Finitely many applications of the induction step finish the proof if the cells of $X - A$ are of bounded dimension. In the general case we perform the homotopy of the induction step during the t-interval $[1 - 1/2^k, 1 - 1/2^{k+1}]$. Any finite skeleton X^k is eventually stationary under these homotopies, hence we have a well-defined homotopy f_t, $t \in [0, 1]$, with $f_1(X) \subset B$. $\qquad\square$

Proof of Whitehead's Theorem: In the special case that f is the inclusion of a subcomplex, consider the long exact sequence of homotopy groups for the pair (Y, X). Since f induces isomorphisms on all homotopy groups, the relative groups $\pi_n(Y, X)$ are zero. Applying the lemma to the identity map $(Y, X) \to (Y, X)$ then yields a deformation retraction of Y onto X.

The general case can be proved using mapping cylinders. Recall that the mapping cylinder M_f of a map $f : X \to Y$ is the quotient space of the disjoint union of $X \times I$ and Y under the identifications $(x, 1) \sim f(x)$. Thus M_f contains both $X = X \times \{0\}$ and Y as subspaces, and M_f deformation retracts onto Y. The map f becomes the composition of the inclusion $X \hookrightarrow M_f$ with the retraction $M_f \to Y$. Since this retraction is a homotopy equivalence, it suffices to show that M_f deformation retracts onto X if f induces isomorphisms on homotopy groups, or equivalently, if the relative groups $\pi_n(M_f, X)$ are all zero.

If the map f happens to be cellular, taking the n-skeleton of X to the n-skeleton of Y for all n, then (M_f, X) is a CW pair and so we are done by the first paragraph of the proof. If f is not cellular, we can either appeal to Theorem 4.8 which says that f is homotopic to a cellular map, or we can use the following argument. First apply the preceding lemma to obtain a homotopy rel X of the inclusion $(X \cup Y, X) \hookrightarrow (M_f, X)$ to a map into X. Since the pair $(M_f, X \cup Y)$ obviously satisfies the homotopy extension property, this homotopy extends to a homotopy from the identity map of M_f to a map $g : M_f \to M_f$ taking $X \cup Y$ into X. Then apply the lemma again to the composition $(X \times I \amalg Y, X \times \partial I \amalg Y) \to (M_f, X \cup Y) \xrightarrow{g} (M_f, X)$ to finish the construction of a deformation retraction of M_f onto X. $\qquad\square$

Whitehead's theorem does not say that two CW complexes X and Y with isomorphic homotopy groups are homotopy equivalent, since there is a big difference between saying that X and Y have isomorphic homotopy groups and saying that there is a map $X \rightarrow Y$ inducing isomorphisms on homotopy groups. For example, consider $X = \mathbb{R}P^2$ and $Y = S^2 \times \mathbb{R}P^\infty$. These both have fundamental group $\mathbb{Z}_2$, and Proposition 4.1 implies that their higher homotopy groups are isomorphic since their universal covers S^2 and $S^2 \times S^\infty$ are homotopy equivalent, S^∞ being contractible. But $\mathbb{R}P^2$ and $S^2 \times \mathbb{R}P^\infty$ are not homotopy equivalent since their homology groups are vastly different, $S^2 \times \mathbb{R}P^\infty$ having nonvanishing homology in infinitely many dimensions since it retracts onto $\mathbb{R}P^\infty$. Another pair of CW complexes that are not homotopy equivalent but have isomorphic homotopy groups is S^2 and $S^3 \times \mathbb{C}P^\infty$, as we shall see in Example 4.51.

One very special case when the homotopy type of a CW complex is determined by its homotopy groups is when all the homotopy groups are trivial, for then the inclusion map of a 0-cell into the complex induces an isomorphism on homotopy groups, so the complex deformation retracts to the 0-cell.

Somewhat similar in spirit to the compression lemma is the following rather basic **extension lemma**:

Lemma 4.7. *Given a CW pair (X, A) and a map $f : A \rightarrow Y$ with Y path-connected, then f can be extended to a map $X \rightarrow Y$ if $\pi_{n-1}(Y) = 0$ for all n such that $X - A$ has cells of dimension n.*

Proof: Assume inductively that f has been extended over the $(n-1)$-skeleton. Then an extension over an n-cell exists iff the composition of the cell's attaching map $S^{n-1} \rightarrow X^{n-1}$ with $f : X^{n-1} \rightarrow Y$ is nullhomotopic. $\square$

Cellular Approximation

When we showed that $\pi_1(S^k) = 0$ for $k > 1$ in Proposition 1.14, we first showed that every loop in S^k can be deformed to miss at least one point if $k > 1$, then we used the fact that the complement of a point in S^k is contractible to finish the proof. The same strategy could be used to show that $\pi_n(S^k) = 0$ for $n < k$ if we could do the first step of deforming a map $S^n \rightarrow S^k$ to be nonsurjective. One might at first think this step was unnecessary, that no continuous map $S^n \rightarrow S^k$ could be surjective when $n < k$, but it is not hard to use space-filling curves from point-set topology to produce such maps. Some work must then be done to construct homotopies eliminating this rather strange behavior.

For maps between CW complexes it turns out to be sufficient for this and many other purposes in homotopy theory to require just that cells map to cells of the same or lower dimension. Such a map $f : X \rightarrow Y$, satisfying $f(X^n) \subset Y^n$ for all n, is called a **cellular map**. It is a fundamental fact that arbitrary maps can always be deformed to be cellular. This is the **cellular approximation theorem**:

Theorem 4.8. *Every map $f : X \to Y$ of CW complexes is homotopic to a cellular map. If f is already cellular on a subcomplex $A \subset X$, the homotopy may be taken to be stationary on A.*

Corollary 4.9. $\pi_n(S^k) = 0$ *for* $n < k$.

Proof: If S^n and S^k are given their usual CW structures, with the 0-cells as basepoints, then every basepoint-preserving map $S^n \to S^k$ can be homotoped, fixing the basepoint, to be cellular, and hence constant if $n < k$. $\square$

Linear maps cannot increase dimension, so one might try to prove the theorem by showing that arbitrary maps between CW complexes can be homotoped to have some sort of linearity properties. For simplicial complexes the simplicial approximation theorem, Theorem 2C.1, achieves this, and cellular approximation can be regarded as a CW analog of simplicial approximation since simplicial maps are cellular. However, simplicial maps are much more rigid than cellular maps, which perhaps explains why subdivision of the domain is required for simplicial approximation but not for cellular approximation. The core of the proof of cellular approximation will be a weak form of simplicial approximation that can be proved by a rather elementary direct argument.

Proof of 4.8: Suppose inductively that $f : X \to Y$ is already cellular on the skeleton X^{n-1}, and let e^n be an n-cell of X. The closure of e^n in X is compact, being the image of a characteristic map for e^n, so f takes the closure of e^n to a compact set in Y. Since a compact set in a CW complex can meet only finitely many cells by Proposition A.1 in the Appendix, it follows that $f(e^n)$ meets only finitely many cells of Y. Let $e^k \subset Y$ be a cell of highest dimension meeting $f(e^n)$. We may assume $k > n$, otherwise f is already cellular on e^n. We will show below that it is possible to deform $f | X^{n-1} \cup e^n$, staying fixed on X^{n-1}, so that $f(e^n)$ misses some point $p \in e^k$. Then we can deform $f | X^{n-1} \cup e^n$ rel X^{n-1} so that $f(e^n)$ misses the whole cell e^k by composing with a deformation retraction of $Y^k - \{p\}$ onto $Y^k - e^k$. By finitely many iterations of this process we eventually make $f(e^n)$ miss all cells of dimension greater than n. Doing this for all n-cells, staying fixed on n-cells in A where f is already cellular, we obtain a homotopy of $f | X^n$ rel $X^{n-1} \cup A^n$ to a cellular map. The induction step is then completed by appealing to the homotopy extension property in Proposition 0.16 to extend this homotopy, together with the constant homotopy on A, to a homotopy defined on all of X. Letting n go to ∞, the resulting possibly infinite string of homotopies can be realized as a single homotopy by performing the nth homotopy during the t-interval $[1 - 1/2^n, 1 - 1/2^{n+1}]$. This makes sense since each point of X lies in some X^n, which is eventually stationary in the infinite chain of homotopies.

To fill in the missing step in this argument we will need a technical lemma about deforming maps to create some linearity. Define a **polyhedron** in $\mathbb{R}^n$ to be a subspace that is the union of finitely many convex polyhedra, each of which is a compact set obtained by intersecting finitely many half-spaces defined by linear inequalities of the form $\sum_i a_i x_i \leq b$. By a **PL (piecewise linear) map** from a polyhedron to $\mathbb{R}^k$ we shall mean a map which is linear when restricted to each convex polyhedron in some such decomposition of the polyhedron into convex polyhedra.

Lemma 4.10. *Let $f : I^n \to Z$ be a map, where Z is obtained from a subspace W by attaching a cell e^k. Then there is a homotopy $f_t : (I^n, f^{-1}(e^k)) \to (Z, e^k)$ rel $f^{-1}(W)$ from $f = f_0$ to a map f_1 for which there is a polyhedron $K \subset I^n$ such that:*
  (a) *$f_1(K) \subset e^k$ and $f_1 | K$ is PL with respect to some identification of e^k with $\mathbb{R}^k$.*
  (b) *$K \supset f_1^{-1}(U)$ for some nonempty open set U in e^k.*

Before proving the lemma, let us see how it finishes the proof of the cellular approximation theorem. Composing the given map $f : X^{n-1} \cup e^n \to Y^k$ with a characteristic map $I^n \to X$ for e^n, we obtain a map f as in the lemma, with $Z = Y^k$ and $W = Y^k - e^k$. The homotopy given by the lemma is fixed on ∂I^n, hence induces a homotopy f_t of $f | X^{n-1} \cup e^n$ fixed on X^{n-1}. The image of the resulting map f_1 intersects the open set U in e^k in a set contained in the union of finitely many hyperplanes of dimension at most n, so if $n < k$ there will be points $p \in U$ not in the image of f_1. $\qquad\square$

Proof of 4.10: Identifying e^k with $\mathbb{R}^k$, let $B_1, B_2 \subset e^k$ be the closed balls of radius 1 and 2 centered at the origin. Since $f^{-1}(B_2)$ is closed and therefore compact in I^n, it follows that f is uniformly continuous on $f^{-1}(B_2)$. Thus there exists $\varepsilon > 0$ such that $|x - y| < \varepsilon$ implies $|f(x) - f(y)| < 1/2$ for all $x, y \in f^{-1}(B_2)$. Subdivide the interval I so that the induced subdivision of I^n into cubes has each cube lying in a ball of diameter less than ε. Let K_1 be the union of all the cubes meeting $f^{-1}(B_1)$, and let K_2 be the union of all the cubes meeting K_1. We may assume ε is chosen smaller than half the distance between the compact sets $f^{-1}(B_1)$ and $I^n - f^{-1}(\text{int}(B_2))$, and then we will have $K_2 \subset f^{-1}(B_2)$.

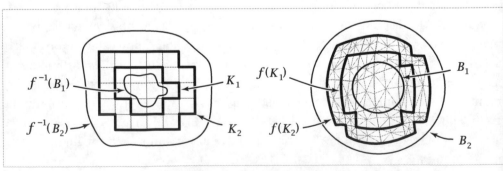

Now we subdivide all the cubes of K_2 into simplices. This can be done inductively. The boundary of each cube is a union of cubes of one lower dimension, so assuming these lower-dimensional cubes have already been subdivided into simplices, we obtain a subdivision of the cube itself by taking its center point as a new vertex and joining this by a cone to each simplex in the boundary of the cube.

Let $g : K_2 \to e^k = \mathbb{R}^k$ be the map that equals f on all vertices of simplices of the subdivision and is linear on each simplex. Let $\varphi : K_2 \to [0, 1]$ be the map that is linear on simplices and has the value 1 on vertices in K_1 and 0 on vertices in $K_2 - K_1$. Thus $\varphi(K_1) = 1$. Define a homotopy $f_t : K_2 \to e^k$ by the formula $(1 - t\varphi)f + (t\varphi)g$, so $f_0 = f$ and $f_1 | K_1 = g | K_1$. Since f_t is the constant homotopy on simplices in K_2 disjoint from K_1, and in particular on simplices in the closure of $I^n - K_2$, we may extend f_t to be the constant homotopy of f on $I^n - K_2$.

The map f_1 takes the closure of $I^n - K_1$ to a compact set C which, we claim, is disjoint from the centerpoint 0 of B_1 and hence from a neighborhood U of 0. This will prove the lemma, with $K = K_1$, since we will then have $f_1^{-1}(U) \subset K_1$ with f_1 PL on K_1 where it is equal to g.

The verification of the claim has two steps:

(1) On $I^n - K_2$ we have $f_1 = f$, and f takes $I^n - K_2$ outside B_1 since $f^{-1}(B_1) \subset K_2$ by construction.

(2) For a simplex σ of K_2 not in K_1 we have $f(\sigma)$ contained in some ball B_σ of radius $1/2$ by the choice of ε and the fact that $K_2 \subset f^{-1}(B_2)$. Since $f(\sigma) \subset B_\sigma$ and B_σ is convex, we must have $g(\sigma) \subset B_\sigma$, hence also $f_t(\sigma) \subset B_\sigma$ for all t, and in particular $f_1(\sigma) \subset B_\sigma$. We know that B_σ is not contained in B_1 since σ contains points outside K_1 hence outside $f^{-1}(B_1)$. The radius of B_σ is half that of B_1, so it follows that 0 is not in B_σ, and hence 0 is not in $f_1(\sigma)$.   □

Example 4.11: Cellular Approximation for Pairs. Every map $f : (X, A) \to (Y, B)$ of CW pairs can be deformed through maps $(X, A) \to (Y, B)$ to a cellular map. This follows from the theorem by first deforming the restriction $f : A \to B$ to be cellular, then extending this to a homotopy of f on all of X, then deforming the resulting map to be cellular staying fixed on A. As a further refinement, the homotopy of f can be taken to be stationary on any subcomplex of X where f is already cellular.

An easy consequence of this is:

Corollary 4.12. *A CW pair (X, A) is n-connected if all the cells in $X - A$ have dimension greater than n. In particular the pair (X, X^n) is n-connected, hence the inclusion $X^n \hookrightarrow X$ induces isomorphisms on π_i for $i < n$ and a surjection on π_n.*

Proof: Applying cellular approximation to maps $(D^i, \partial D^i) \to (X, A)$ with $i \le n$ gives the first statement. The last statement comes from the long exact sequence of the pair (X, X^n).   □

CW Approximation

A map $f: X \to Y$ is called a **weak homotopy equivalence** if it induces isomorphisms $\pi_n(X, x_0) \to \pi_n(Y, f(x_0))$ for all $n \geq 0$ and all choices of basepoint x_0. Whitehead's theorem can be restated as saying that a weak homotopy equivalence between CW complexes is a homotopy equivalence. It follows easily that this holds also for spaces homotopy equivalent to CW complexes. In general, however, weak homotopy equivalence is strictly weaker than homotopy equivalence. For example, there exist noncontractible spaces whose homotopy groups are all trivial, such as the 'quasi-circle' according to an exercise at the end of this section, and for such spaces a map to a point is a weak homotopy equivalence that is not a homotopy equivalence.

We will show that for every space X there is a CW complex Z and a weak homotopy equivalence $f: Z \to X$. Such a map $f: Z \to X$ is called a **CW approximation** to X. A weak homotopy equivalence induces isomorphisms on all homology and cohomology groups, as we will see, so CW approximations allow many general statements in algebraic topology to be proved using cell-by-cell arguments for CW complexes.

The construction of a CW approximation $f: Z \to X$ for a space X is inductive, so let us describe the induction step. Suppose given a CW complex A with a map $f: A \to X$ and suppose we have chosen a basepoint 0-cell a_y in each component of A. Then for an integer $k \geq 0$ we will attach k-cells to A to form a CW complex B with a map $f: B \to X$ extending the given f, such that:

(∗)   The induced map $f_*: \pi_i(B, a_y) \to \pi_i(X, f(a_y))$ is injective for $i = k - 1$ and surjective for $i = k$, for all a_y.

There are two steps to the construction:

(1) Choose maps $\varphi_\alpha: (S^{k-1}, s_0) \to (A, a_y)$ representing a set of generators for the kernel of $f_*: \pi_{k-1}(A, a_y) \to \pi_{k-1}(X, f(a_y))$ for all the basepoints a_y. We may assume the maps φ_α are cellular, where S^{k-1} has its standard CW structure with s_0 as 0-cell. Attaching cells e_α^k to A via the maps φ_α then produces a CW complex, and the map f extends over these cells using nullhomotopies of the compositions $f\varphi_\alpha$, which exist by the choice of the φ_α's.

(2) Choose maps $f_\beta: S^k \to X$ representing generators for the groups $\pi_k(X, f(a_y))$, attach cells e_β^k to A via the constant maps at the appropriate basepoints a_y, and extend f over the resulting spheres S_β^k via the f_β's.

The surjectivity condition in (∗) then holds by construction. For the injectivity condition, an element of the kernel of $f_*: \pi_{k-1}(B, a_y) \to \pi_{k-1}(X, f(a_y))$ can be represented by a cellular map $h: S^{k-1} \to B$. This has image in A, so is in the kernel of $f_*: \pi_{k-1}(A, a_y) \to \pi_{k-1}(X, f(a_y))$ and hence is homotopic to a linear combination of the φ_α's, which are nullhomotopic in B, so h is nullhomotopic as well. When $k = 1$ there is no group structure on π_{k-1} so injectivity on π_0 does not follow from having a trivial kernel, and we modify the construction by choosing the cells e_α^1 to join each pair of basepoints a_γ that map by f to the same path-component of X. The map f can then be extended over these 1-cells e_α^1.

Note that if the given map $f : A \rightarrow X$ happened to be injective or surjective on π_i for some $i < k - 1$ or $i < k$, respectively, then this remains true after attaching the k-cells. This is because attaching k-cells does not affect π_i if $i < k - 1$, by cellular approximation, nor does it destroy surjectivity on π_{k-1} or indeed any π_i, obviously.

Now to construct a CW approximation $f : Z \rightarrow X$ one can start with A consisting of one point for each path-component of X, with $f : A \rightarrow X$ mapping each of these points to the corresponding path-component. Having now a bijection on π_0, attach 1-cells to A to create a surjection on π_1 for each path-component, then 2-cells to improve this to an isomorphism on π_1 and a surjection on π_2, and so on for each successive π_i in turn. After all cells have been attached one has a CW complex Z with a weak homotopy equivalence $f : Z \rightarrow X$. This proves:

Proposition 4.13. *Every space X has a CW approximation $f : Z \rightarrow X$. If X is path-connected, Z can be chosen to have a single 0-cell, with all other cells attached by basepoint-preserving maps. Thus every connected CW complex is homotopy equivalent to a CW complex with these additional properties.* □

Example 4.14. One can also apply this technique to produce a CW approximation to a pair (X, X_0). First construct a CW approximation $f_0 : Z_0 \rightarrow X_0$, then starting with the composition $Z_0 \rightarrow X_0 \hookrightarrow X$, attach cells to Z_0 to create a weak homotopy equivalence $f : Z \rightarrow X$ extending f_0. By the five-lemma, the map $f : (Z, Z_0) \rightarrow (X, X_0)$ induces isomorphisms on relative as well as absolute homotopy groups.

Here is another application of the technique, giving a more geometric interpretation to the homotopy-theoretic notion of n-connectedness:

Proposition 4.15. *If (X, A) is an n-connected CW pair, then there exists a CW pair $(Z, A) \simeq (X, A)$ rel A such that all cells of $Z - A$ have dimension greater than n.*

Proof: Starting with the inclusion $A \hookrightarrow X$, attach cells to A of dimension $n + 1$ and higher to produce a CW complex Z and a map $f : Z \rightarrow X$ that is the identity on A and induces an injection on π_n and isomorphisms on all higher homotopy groups. The induced map on π_n is also surjective since this is true for the composition $A \hookrightarrow Z \xrightarrow{f} X$ by the hypothesis that (X, A) is n-connected. In dimensions below n, f induces isomorphisms on homotopy groups since both inclusions $A \hookrightarrow Z$ and $A \hookrightarrow X$ induce isomorphisms in these dimensions. Thus f is a weak homotopy equivalence, and hence a homotopy equivalence by Whitehead's theorem.

To see that f is a homotopy equivalence rel A, form a quotient space W of the mapping cylinder M_f by collapsing each segment $\{a\} \times I$ to a point, for $a \in A$. Assuming f has been made cellular, W is a CW complex containing X and Z as subcomplexes, and W deformation retracts to X just as M_f does. Also, $\pi_i(W, Z) = 0$ for all

i since f induces isomorphisms on all homotopy groups, so W deformation retracts onto Z. These two deformation retractions of W onto X and Z are stationary on A, hence give a homotopy equivalence $X \simeq Z$ rel A. $\square$

Example 4.16: Postnikov Towers. We can also apply the technique to construct, for each connected CW complex X and each integer $n \geq 1$, a CW complex X_n containing X as a subcomplex such that:

(a) $\pi_i(X_n) = 0$ for $i > n$.

(b) The inclusion $X \hookrightarrow X_n$ induces an isomorphism on π_i for $i \leq n$.

To do this, all we have to do is apply the general construction to the constant map of X to a point, starting at the stage of attaching cells of dimension $n + 2$. Thus we attach $(n + 2)$-cells to X using cellular maps $S^{n+1} \to X$ that generate $\pi_{n+1}(X)$ to form a space with π_{n+1} trivial, then for this space we attach $(n + 3)$-cells to make π_{n+2} trivial, and so on. The result is a CW complex X_n with the desired properties.

The inclusion $X \hookrightarrow X_n$ extends to a map $X_{n+1} \to X_n$ since X_{n+1} is obtained from X by attaching cells of dimension $n + 3$ and greater, and $\pi_i(X_n) = 0$ for $i > n$ so we can apply Lemma 4.7, the extension lemma. Thus we have a commutative diagram as at the right. This is a called a *Postnikov tower* for X. One can regard the spaces X_n as truncations of X which provide successively better approximations to X as n increases. Postnikov towers turn out to be quite powerful tools for proving general theorems, and we will study them further in §4.3.

Now that we have seen several varied applications of the technique of attaching cells to make a map $f : A \to X$ more nearly a weak homotopy equivalence, it might be useful to give a name to the properties that the construction can achieve. To simplify the description, we may assume without loss of generality that the given f is an inclusion $A \hookrightarrow X$ by replacing X by the mapping cylinder of f. Thus, starting with a pair (X, A) where the subspace $A \subset X$ is a nonempty CW complex, we define an **n-connected CW model for** (X, A) to be an n-connected CW pair (Z, A) and a map $f : Z \to X$ with $f | A$ the identity, such that $f_* : \pi_i(Z) \to \pi_i(X)$ is an isomorphism for $i > n$ and an injection for $i = n$, for all choices of basepoint. Since (Z, A) is n-connected, the map $\pi_i(A) \to \pi_i(Z)$ is an isomorphism for $i < n$ and a surjection for $i = n$. In the critical dimension n, the maps $A \hookrightarrow Z \xrightarrow{f} X$ induce a composition $\pi_n(A) \to \pi_n(Z) \to \pi_n(X)$ factoring the map $\pi_n(A) \to \pi_n(X)$ as a surjection followed by an injection, just as any homomorphism $\varphi : G \to H$ can be factored (uniquely) as a surjection $\varphi : G \to \operatorname{Im} \varphi$ followed by an injection $\operatorname{Im} \varphi \hookrightarrow H$. One can think of Z as a sort of homotopy-theoretic hybrid of A and X. As n increases, the hybrid looks more and more like A, and less and less like X.

Our earlier construction shows:

Proposition 4.17. *For every pair* (X, A) *with* A *a nonempty CW complex there exist* n-*connected CW models* $f : (Z, A) \rightarrow (X, A)$ *for all* $n \geq 0$, *and these models can be chosen to have the additional property that* Z *is obtained from* A *by attaching cells of dimension greater than* n. $\square$

The construction of n-connected CW models involves many arbitrary choices, so it may be somewhat surprising that they turn out to be unique up to homotopy equivalence. This will follow easily from the next proposition. Another application of the proposition will be to build a tower like the Postnikov tower from the various n-connected CW models for a given pair (X, A).

Proposition 4.18. *Suppose we are given*:

   (i) *an* n-*connected CW model* $f : (Z, A) \rightarrow (X, A)$,
   (ii) *an* n'-*connected CW model* $f' : (Z', A') \rightarrow (X', A')$,
   (iii) *a map* $g : (X, A) \rightarrow (X', A')$.

$$\begin{array}{ccc} Z & \xrightarrow{\ f\ } & X \\ h \downarrow & & \downarrow g \\ Z' & \xrightarrow{\ f'\ } & X' \end{array}$$

Then if $n \geq n'$, *there is a map* $h : Z \rightarrow Z'$ *such that* $h \mid A = g$ *and* $gf \simeq f'h$ *rel* A, *so the diagram above is commutative up to homotopy* rel A. *Furthermore, such a map* h *is unique up to homotopy* rel A.

Proof: By Proposition 4.15 we may assume all cells of $Z - A$ have dimension greater than n. Let W be the quotient space of the mapping cylinder of f' obtained by collapsing each line segment $\{a'\} \times I$ to a point, for $a' \in A'$. We can think of W as a relative mapping cylinder, and like the ordinary mapping cylinder, W contains copies of Z' and X', the latter as a deformation retract. The assumption that (Z', A') is an n'-connected CW model for (X', A') implies that the relative groups $\pi_i(W, Z')$ are zero for $i > n'$.

Via the inclusion $X' \hookrightarrow W$ we can view gf as a map $Z \rightarrow W$. As a map of pairs $(Z, A) \rightarrow (W, Z')$, gf is homotopic rel A to a map h with image in Z', by the compression lemma and the hypothesis $n \geq n'$. This proves the first assertion. For the second, suppose h_0 and h_1 are two maps $Z \rightarrow Z'$ whose compositions with f' are homotopic to gf rel A. Thus if we regard h_0 and h_1 as maps to W, they are homotopic rel A. Such a homotopy gives a map $(Z \times I, Z \times \partial I \cup A \times I) \rightarrow (W, Z')$, and by the compression lemma again this map can be deformed rel $Z \times \partial I \cup A \times I$ to a map with image in Z', which gives the desired homotopy $h_0 \simeq h_1$ rel A. $\square$

Corollary 4.19. *An* n-*connected CW model for* (X, A) *is unique up to homotopy equivalence* rel A. *In particular, CW approximations to spaces are unique up to homotopy equivalence.*

Proof: Given two n-connected CW models (Z, A) and (Z', A) for (X, A), we apply the proposition twice with g the identity map to obtain maps $h : Z \rightarrow Z'$ and $h' : Z' \rightarrow Z$. The uniqueness statement gives homotopies $hh' \simeq \mathbb{1}$ and $h'h \simeq \mathbb{1}$ rel A. $\square$

Taking $n = n'$ in the proposition, we obtain also a functoriality property for n-connected CW models. For example, a map $X \to X'$ induces a map of CW approximations $Z \to Z'$, which is unique up to homotopy.

The proposition allows us to relate n-connected CW models (Z_n, A) for (X, A) for varying n, by means of maps $Z_n \to Z_{n-1}$ that form a tower as shown in the diagram, with commutative triangles on the left and homotopy-commutative triangles on the right. We can make the triangles on the right strictly commutative by replacing the maps $Z_n \to X$ by the compositions through Z_0.

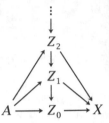

Example 4.20: Whitehead Towers . If we take X to be an arbitrary CW complex with the subspace A a point, then the resulting tower of n-connected CW models amounts to a sequence of maps

$$\cdots \to Z_2 \to Z_1 \to Z_0 \to X$$

with Z_n n-connected and the map $Z_n \to X$ inducing an isomorphism on all homotopy groups π_i with $i > n$. The space Z_0 is path-connected and homotopy equivalent to the component of X containing A, so one may as well assume Z_0 equals this component. The next space Z_1 is simply-connected, and the map $Z_1 \to X$ has the homotopy properties of the universal cover of the component Z_0 of X. For larger values of n one can by analogy view the map $Z_n \to X$ as an 'n-connected cover' of X. For $n > 1$ these do not seem to arise so frequently in nature as in the case $n = 1$. A rare exception is the Hopf map $S^3 \to S^2$ defined in Example 4.45, which is a 2-connected cover.

Now let us show that CW approximations behave well with respect to homology and cohomology:

Proposition 4.21. *A weak homotopy equivalence $f : X \to Y$ induces isomorphisms $f_* : H_n(X; G) \to H_n(Y; G)$ and $f^* : H^n(Y; G) \to H^n(X; G)$ for all n and all coefficient groups G.*

Proof: Replacing Y by the mapping cylinder M_f and looking at the long exact sequences of homotopy, homology, and cohomology groups for (M_f, X), we see that it suffices to show:

- If (Z, X) is an n-connected pair of path-connected spaces, then $H_i(Z, X; G) = 0$ and $H^i(Z, X; G) = 0$ for all $i \leq n$ and all G.

Let $\alpha = \sum_j n_j \sigma_j$ be a relative cycle representing an element of $H_k(Z, X; G)$, for singular k-simplices $\sigma_j : \Delta^k \to Z$. Build a finite Δ-complex K from a disjoint union of k-simplices, one for each σ_j, by identifying all $(k-1)$-dimensional faces of these k-simplices for which the corresponding restrictions of the σ_j's are equal. Thus the σ_j's induce a map $\sigma : K \to Z$. Since α is a relative cycle, $\partial \alpha$ is a chain in X. Let

$L \subset K$ be the subcomplex consisting of $(k-1)$-simplices corresponding to the singular $(k-1)$-simplices in $\partial \alpha$, so $\sigma(L) \subset X$. The chain α is the image under the chain map $\sigma_\sharp$ of a chain $\tilde{\alpha}$ in K, with $\partial \tilde{\alpha}$ a chain in L. In relative homology we then have $\sigma_*[\tilde{\alpha}] = [\alpha]$. If we assume $\pi_i(Z,X) = 0$ for $i \le k$, then $\sigma : (K,L) \to (Z,X)$ is homotopic rel L to a map with image in X, by the compression lemma. Hence $\sigma_*[\tilde{\alpha}]$ is in the image of the map $H_k(X,X;G) \to H_k(Z,X;G)$, and since $H_k(X,X;G) = 0$ we conclude that $[\alpha] = \sigma_*[\tilde{\alpha}] = 0$. This proves the result for homology, and the result for cohomology then follows by the universal coefficient theorem. $\qquad \square$

CW approximations can be used to reduce many statements about general spaces to the special case of CW complexes. For example, the cup product version of the Künneth formula in Theorem 3.15, asserting that $H^*(X \times Y;R) \approx H^*(X;R) \otimes H^*(Y;R)$ under certain conditions, can now be extended to non-CW spaces since if X and Y are CW approximations to spaces Z and W, respectively, then $X \times Y$ is a CW approximation to $Z \times W$. Here we are giving $X \times Y$ the CW topology rather than the product topology, but this has no effect on homotopy groups since the two topologies have the same compact sets, as explained in the Appendix. Similarly, the general Künneth formula for homology in §3.B holds for arbitrary products $X \times Y$.

The condition for a map $Y \to Z$ to be a weak homotopy equivalence involves only maps of spheres into Y and Z, but in fact weak homotopy equivalences $Y \to Z$ behave nicely with respect to maps of arbitrary CW complexes into Y and Z, not just spheres. The following proposition gives a precise statement, using the notations $[X,Y]$ for the set of homotopy classes of maps $X \to Y$ and $\langle X,Y \rangle$ for the set of basepoint-preserving-homotopy classes of basepoint-preserving maps $X \to Y$. (The notation $\langle X,Y \rangle$ is not standard, but is intended to suggest 'pointed homotopy classes.')

Proposition 4.22. *A weak homotopy equivalence $f : Y \to Z$ induces bijections $[X,Y] \to [X,Z]$ and $\langle X,Y \rangle \to \langle X,Z \rangle$ for all CW complexes X.*

Proof: Consider first $[X,Y] \to [X,Z]$. We may assume f is an inclusion by replacing Z by the mapping cylinder M_f as usual. The groups $\pi_n(Z,Y,y_0)$ are then zero for all n and all basepoints $y_0 \in Y$, so the compression lemma implies that any map $X \to Z$ can be homotoped to have image in Y. This gives surjectivity of $[X,Y] \to [X,Z]$. A relative version of this argument shows injectivity since we can deform a homotopy $(X \times I, X \times \partial I) \to (Z,Y)$ to have image in Y.

In the case of $\langle X,Y \rangle \to \langle X,Z \rangle$ the same argument applies if M_f is replaced by the reduced mapping cylinder, the quotient of M_f obtained by collapsing the segment $\{y_0\} \times I$ to a point, for y_0 the basepoint of Y. This collapsed segment then serves as the common basepoint of Y, Z, and the reduced mapping cylinder. The reduced mapping cylinder deformation retracts to Z just as the unreduced one does, but with the advantage that the basepoint does not move. $\qquad \square$

Exercises

1. Suppose a sum $f +' g$ of maps $f, g : (I^n, \partial I^n) \to (X, x_0)$ is defined using a coordinate of I^n other than the first coordinate as in the usual sum $f + g$. Verify the formula $(f + g) +' (h + k) = (f +' h) + (g +' k)$, and deduce that $f +' k \simeq f + k$ so the two sums agree on $\pi_n(X, x_0)$, and also that $g +' h \simeq h + g$ so the addition is abelian.

2. Show that if $\varphi : X \to Y$ is a homotopy equivalence, then the induced homomorphisms $\varphi_* : \pi_n(X, x_0) \to \pi_n(Y, \varphi(x_0))$ are isomorphisms for all n. [The case $n = 1$ is Proposition 1.18.]

3. For an H–space (X, x_0) with multiplication $\mu : X \times X \to X$, show that the group operation in $\pi_n(X, x_0)$ can also be defined by the rule $(f + g)(x) = \mu(f(x), g(x))$.

4. Let $p : \tilde{X} \to X$ be the universal cover of a path-connected space X. Show that under the isomorphism $\pi_n(X) \approx \pi_n(\tilde{X})$, which holds for $n \geq 2$, the action of $\pi_1(X)$ on $\pi_n(X)$ corresponds to the action of $\pi_1(X)$ on $\pi_n(\tilde{X})$ induced by the action of $\pi_1(X)$ on $\tilde{X}$ as deck transformations. More precisely, prove a formula like $\gamma p_*(\alpha) = p_*(\beta_{\tilde{\gamma}}(\gamma_*(\alpha)))$ where $\gamma \in \pi_1(X, x_0)$, $\alpha \in \pi_n(\tilde{X}, \tilde{x}_0)$, and γ_* denotes the homomorphism induced by the action of γ on $\tilde{X}$.

5. For a pair (X, A) of path-connected spaces, show that $\pi_1(X, A, x_0)$ can be identified in a natural way with the set of cosets αH of the subgroup $H \subset \pi_1(X, x_0)$ represented by loops in A at x_0.

6. If $p : (\tilde{X}, \tilde{A}, \tilde{x}_0) \to (X, A, x_0)$ is a covering space with $\tilde{A} = p^{-1}(A)$, show that the map $p_* : \pi_n(\tilde{X}, \tilde{A}, \tilde{x}_0) \to \pi_n(X, A, x_0)$ is an isomorphism for all $n > 1$.

7. Extend the results proved near the beginning of this section for the change-of-basepoint maps β_γ to the case of relative homotopy groups.

8. Show the sequence $\pi_1(X, x_0) \to \pi_1(X, A, x_0) \xrightarrow{\partial} \pi_0(A, x_0) \to \pi_0(X, x_0)$ is exact.

9. Suppose we define $\pi_0(X, A, x_0)$ to be the quotient set $\pi_0(X, x_0)/i^*(\pi_0(A, x_0))$, so that the long exact sequence of homotopy groups for the pair (X, A) extends to $\cdots \to \pi_0(A, x_0) \xrightarrow{i_*} \pi_0(X, x_0) \to \pi_0(X, A, x_0) \to 0$.

  (a) Show that with this extension, the five-lemma holds for the map of long exact sequences induced by a map $(X, A, x_0) \to (Y, B, y_0)$, in the following form: One of the maps between the two sequences is a bijection if the four surrounding maps are bijections for all choices of x_0.

  (b) Show that the long exact sequence of a triple (X, A, B, x_0) can be extended only to the term $\pi_0(A, B, x_0)$ in general, and that the five-lemma holds for this extension.

10. Show the 'quasi-circle' described in Exercise 7 in §1.3 has trivial homotopy groups but is not contractible, hence does not have the homotopy type of a CW complex.

11. Show that a CW complex is contractible if it is the union of an increasing sequence of subcomplexes $X_1 \subset X_2 \subset \cdots$ such that each inclusion $X_i \hookrightarrow X_{i+1}$ is nullhomotopic, a condition sometimes expressed by saying X_i is contractible in X_{i+1}. An example is

S^∞, or more generally the infinite suspension $S^\infty X$ of any CW complex X, the union of the iterated suspensions $S^n X$.

12. Show that an n-connected, n-dimensional CW complex is contractible.

13. Use the extension lemma to show that a CW complex retracts onto any contractible subcomplex.

14. Use cellular approximation to show that the n-skeletons of homotopy equivalent CW complexes without cells of dimension $n + 1$ are also homotopy equivalent.

15. Show that every map $f : S^n \to S^n$ is homotopic to a multiple of the identity map by the following steps.

(a) Use Lemma 4.10 (or simplicial approximation, Theorem 2C.1) to reduce to the case that there exists a point $q \in S^n$ with $f^{-1}(q) = \{p_1, \cdots, p_k\}$ and f is an invertible linear map near each p_i.

(b) For f as in (a), consider the composition gf where $g : S^n \to S^n$ collapses the complement of a small ball about q to the basepoint. Use this to reduce (a) further to the case $k = 1$.

(c) Finish the argument by showing that an invertible $n \times n$ matrix can be joined by a path of such matrices to either the identity matrix or the matrix of a reflection. (Use Gaussian elimination, for example.)

16. Show that a map $f : X \to Y$ between connected CW complexes factors as a composition $X \to Z_n \to Y$ where the first map induces isomorphisms on π_i for $i \leq n$ and the second map induces isomorphisms on π_i for $i \geq n + 1$.

17. Show that if X and Y are CW complexes with X m-connected and Y n-connected, then $(X \times Y, X \vee Y)$ is $(m + n + 1)$-connected, as is the smash product $X \wedge Y$.

18. Give an example of a weak homotopy equivalence $X \to Y$ for which there does not exist a weak homotopy equivalence $Y \to X$.

19. Consider the equivalence relation $\simeq_w$ generated by weak homotopy equivalence: $X \simeq_w Y$ if there are spaces $X = X_1, X_2, \cdots, X_n = Y$ with weak homotopy equivalences $X_i \to X_{i+1}$ or $X_i \leftarrow X_{i+1}$ for each i. Show that $X \simeq_w Y$ iff X and Y have a common CW approximation.

20. Show that $[X, Y]$ is finite if X is a finite connected CW complex and $\pi_i(Y)$ is finite for $i \leq \dim X$.

21. For this problem it is convenient to use the notations X^n for the nth stage in a Postnikov tower for X and X_m for an $(m - 1)$-connected covering of X, where X is a connected CW complex. Show that $(X^n)_m \simeq (X_m)^n$, so the notation X_m^n is unambiguous. Thus $\pi_i(X_m^n) \approx \pi_i(X)$ for $m \leq i \leq n$ and all other homotopy groups of X_m^n are zero.

$$\begin{array}{ccc} X_m & \longrightarrow & X_m^n \\ \downarrow & & \downarrow \\ X & \longrightarrow & X^n \end{array}$$

22. Show that a path-connected space X has a CW approximation with countably many cells iff $\pi_n(X)$ is countable for all n. [Use the results on simplicial approximations to maps and spaces in §2.C.]

23. If $f : X \to Y$ is a map with X and Y homotopy equivalent to CW complexes, show that the pair (M_f, X) is homotopy equivalent to a CW pair, where M_f is the mapping cylinder. Deduce that the mapping cone C_f has the homotopy type of a CW complex.

4.2 Elementary Methods of Calculation

We have not yet computed any nonzero homotopy groups $\pi_n(X)$ with $n \geq 2$. In Chapter 1 the two main tools we used for computing fundamental groups were van Kampen's theorem and covering spaces. In the present section we will study the higher-dimensional analogs of these: the excision theorem for homotopy groups, and fiber bundles. Both of these are quite a bit weaker than their fundamental group analogs, in that they do not directly compute homotopy groups but only give relations between the homotopy groups of different spaces. Their applicability is thus more limited, but suffices for a number of interesting calculations, such as $\pi_n(S^n)$ and more generally the Hurewicz theorem relating the first nonzero homotopy and homology groups of a space. Another noteworthy application is the Freudenthal suspension theorem, which leads to stable homotopy groups and in fact the whole subject of stable homotopy theory.

Excision for Homotopy Groups

What makes homotopy groups so much harder to compute than homology groups is the failure of the excision property. However, there is a certain dimension range, depending on connectivities, in which excision does hold for homotopy groups:

Theorem 4.23. *Let X be a CW complex decomposed as the union of subcomplexes A and B with nonempty connected intersection $C = A \cap B$. If (A, C) is m-connected and (B, C) is n-connected, $m, n \geq 0$, then the map $\pi_i(A, C) \to \pi_i(X, B)$ induced by inclusion is an isomorphism for $i < m + n$ and a surjection for $i = m + n$.*

This yields the **Freudenthal suspension theorem**:

Corollary 4.24. *The suspension map $\pi_i(S^n) \to \pi_{i+1}(S^{n+1})$ is an isomorphism for $i < 2n - 1$ and a surjection for $i = 2n - 1$. More generally this holds for the suspension $\pi_i(X) \to \pi_{i+1}(SX)$ whenever X is an $(n-1)$-connected CW complex.*

Proof: Decompose the suspension SX as the union of two cones C_+X and C_-X intersecting in a copy of X. The suspension map is the same as the map

$$\pi_i(X) \approx \pi_{i+1}(C_+X, X) \longrightarrow \pi_{i+1}(SX, C_-X) \approx \pi_{i+1}(SX)$$

where the two isomorphisms come from long exact sequences of pairs and the middle map is induced by inclusion. From the long exact sequence of the pair $(C_{\pm}X, X)$ we see that this pair is n-connected if X is $(n-1)$-connected. The preceding theorem then says that the middle map is an isomorphism for $i + 1 < 2n$ and surjective for $i + 1 = 2n$. $\qquad\square$

Corollary 4.25. $\pi_n(S^n) \approx \mathbb{Z}$, *generated by the identity map, for all* $n \geq 1$. *In particular, the degree map* $\pi_n(S^n) \to \mathbb{Z}$ *is an isomorphism.*

Proof: From the preceding corollary we know that in the sequence of suspension maps $\pi_1(S^1) \to \pi_2(S^2) \to \pi_3(S^3) \to \cdots$ the first map is surjective and all the subsequent maps are isomorphisms. Since $\pi_1(S^1)$ is $\mathbb{Z}$ generated by the identity map, it follows that $\pi_n(S^n)$ for $n \geq 2$ is a finite or infinite cyclic group independent of n, generated by the identity map. The fact that this cyclic group is infinite can be deduced from homology theory since there exist basepoint-preserving maps $S^n \to S^n$ of arbitrary degree, and degree is a homotopy invariant. Alternatively, if one wants to avoid appealing to homology theory one can use the Hopf bundle $S^1 \to S^3 \to S^2$ described in Example 4.45, whose long exact sequence of homotopy groups gives an isomorphism $\pi_1(S^1) \approx \pi_2(S^2)$.

The degree map $\pi_n(S^n) \to \mathbb{Z}$ is an isomorphism since the map $z \mapsto z^k$ of S^1 has degree k, as do its iterated suspensions by Proposition 2.33. $\qquad\square$

Proof of 4.23: We proceed by proving successively more general cases. The first case contains the heart of the argument, and suffices for the calculation of $\pi_n(S^n)$.

Case 1: A is obtained from C by attaching cells e_α^{m+1} and B is obtained from C by attaching a cell e^{n+1}. To show surjectivity of $\pi_i(A, C) \to \pi_i(X, B)$ we start with a map $f : (I^i, \partial I^i, J^{i-1}) \to (X, B, x_0)$. This has compact image, meeting only finitely many of the cells e_α^{m+1} and e^{n+1}. By Lemma 4.10 we may homotope f through maps $(I^i, \partial I^i, J^{i-1}) \to (X, B, x_0)$ so that there are simplices $\Delta_\alpha^{m+1} \subset e_\alpha^{m+1}$ and $\Delta^{n+1} \subset e^{n+1}$ for which $f^{-1}(\Delta_\alpha^{m+1})$ and $f^{-1}(\Delta^{n+1})$ are finite unions of convex polyhedra, on each of which f is the restriction of a linear map from $\mathbb{R}^i$ to $\mathbb{R}^{m+1}$ or $\mathbb{R}^{n+1}$. We may assume these linear maps are surjections by rechoosing smaller simplices Δ_α^{m+1} and Δ^{n+1} in the complement of the images of the nonsurjective linear maps.

Claim: If $i \leq m + n$, then there exist points $p_\alpha \in \Delta_\alpha^{m+1}$, $q \in \Delta^{n+1}$, and a map $\varphi : I^{i-1} \to [0, 1)$ such that:

(a) $f^{-1}(q)$ lies below the graph of φ in $I^{i-1} \times I = I^i$.

(b) $f^{-1}(p_\alpha)$ lies above the graph of φ for each α.

(c) $\varphi = 0$ on ∂I^{i-1}.

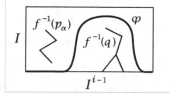

Granting this, let f_t be a homotopy of f excising the region under the graph of φ by restricting f to the region above the graph of $t\varphi$ for $0 \leq t \leq 1$. By (b), $f_t(I^{i-1})$ is disjoint from $P = \bigcup_\alpha \{p_\alpha\}$ for all t, and by (a), $f_1(I^i)$ is disjoint from $Q = \{q\}$. This means that in the commutative diagram at the right the given element $[f]$ in the upper-right group, when regarded as an element of the lower-right group,

$$\begin{array}{ccc} \pi_i(A,C) & \longrightarrow & \pi_i(X,B) \\ \downarrow \approx & & \downarrow \approx \\ \pi_i(X-Q,X-Q-P) & \longrightarrow & \pi_i(X,X-P) \end{array}$$

is equal to the element $[f_1]$ in the image of the lower horizontal map. Since the vertical maps are isomorphisms, this proves the surjectivity statement.

Now we prove the Claim. For any $q \in \Delta^{n+1}$, $f^{-1}(q)$ is a finite union of convex polyhedra of dimension $\leq i - n - 1$ since $f^{-1}(\Delta^{n+1})$ is a finite union of convex polyhedra on each of which f is the restriction of a linear surjection $\mathbb{R}^i \to \mathbb{R}^{n+1}$. We wish to choose the points $p_\alpha \in \Delta_\alpha^{m+1}$ so that not only is $f^{-1}(q)$ disjoint from $f^{-1}(p_\alpha)$ for each α, but also so that $f^{-1}(q)$ and $f^{-1}(p_\alpha)$ have disjoint images under the projection $\pi: I^i \to I^{i-1}$. This is equivalent to saying that $f^{-1}(p_\alpha)$ is disjoint from $T = \pi^{-1}(\pi(f^{-1}(q)))$, the union of all segments $\{x\} \times I$ meeting $f^{-1}(q)$. This set T is a finite union of convex polyhedra of dimension $\leq i - n$ since $f^{-1}(q)$ is a finite union of convex polyhedra of dimension $\leq i - n - 1$. Since linear maps cannot increase dimension, $f(T) \cap \Delta_\alpha^{m+1}$ is also a finite union of convex polyhedra of dimension $\leq i - n$. Thus if $m + 1 > i - n$, there is a point $p_\alpha \in \Delta_\alpha^{m+1}$ not in $f(T)$. This gives $f^{-1}(p_\alpha) \cap T = \varnothing$ if $i \leq m + n$. Hence we can choose a neighborhood U of $\pi(f^{-1}(q))$ in I^{i-1} disjoint from $\pi(f^{-1}(p_\alpha))$ for all α. Then there exists $\varphi: I^{i-1} \to [0, 1)$ having support in U, with $f^{-1}(q)$ lying under the graph of φ. This verifies the Claim, and so finishes the proof of surjectivity in Case 1.

For injectivity in Case 1 the argument is very similar. Suppose we have two maps $f_0, f_1: (I^i, \partial I^i, J^{i-1}) \to (A, C, x_0)$ representing elements of $\pi_i(A, C, x_0)$ having the same image in $\pi_i(X, B, x_0)$. Thus there is a homotopy from f_0 to f_1 in the form of a map $F: (I^i, \partial I^i, J^{i-1}) \times [0, 1] \to (X, B, x_0)$. After a preliminary deformation of F via Lemma 4.10, we construct a function $\varphi: I^{i-1} \times I \to [0, 1)$ separating $F^{-1}(q)$ from the sets $F^{-1}(p_\alpha)$ as before. This allows us to excise $F^{-1}(q)$ from the domain of F, from which it follows that f_0 and f_1 represent the same element of $\pi_i(A, C, x_0)$. Since $I^i \times I$ now plays the role of I^i, the dimension i is replaced by $i + 1$ and the dimension restriction $i \leq m + n$ becomes $i + 1 \leq m + n$, or $i < m + n$.

Case 2: A is obtained from C by attaching $(m + 1)$-cells as in Case 1 and B is obtained from C by attaching cells of dimension $\geq n + 1$. To show surjectivity of $\pi_i(A, C) \to \pi_i(X, B)$, consider a map $f: (I^i, \partial I^i, J^{i-1}) \to (X, B, x_0)$ representing an element of $\pi_i(X, B)$. The image of f is compact, meeting only finitely many cells, and by repeated applications of Case 1 we can push f off the cells of $B - C$ one at a time, in order of decreasing dimension. Injectivity is quite similar, starting with a homotopy $F: (I^i, \partial I^i, J^{i-1}) \times [0, 1] \to (X, B, x_0)$ and pushing this off cells of $B - C$.

Case 3: A is obtained from C by attaching cells of dimension $\geq m + 1$ and B is as in Case 2. We may assume all cells of $A - C$ have dimension $\leq m + n + 1$ since higher-dimensional cells have no effect on π_i for $i \leq m + n$, by cellular approximation. Let

$A_k \subset A$ be the union of C with the cells of A of dimension $\leq k$ and let $X_k = A_k \cup B$. We prove the result for $\pi_i(A_k, C) \to \pi_i(X_k, B)$ by induction on k. The induction starts with $k = m + 1$, which is Case 2. For the induction step consider the following commutative diagram formed by the exact sequences of the triples (A_k, A_{k-1}, C) and (X_k, X_{k-1}, B):

$$\begin{array}{ccccccccc}
\pi_{i+1}(A_k, A_{k-1}) & \longrightarrow & \pi_i(A_{k-1}, C) & \longrightarrow & \pi_i(A_k, C) & \longrightarrow & \pi_i(A_k, A_{k-1}) & \longrightarrow & \pi_{i-1}(A_{k-1}, C) \\
\downarrow & & \downarrow & & \downarrow & & \downarrow & & \downarrow \\
\pi_{i+1}(X_k, X_{k-1}) & \longrightarrow & \pi_i(X_{k-1}, B) & \longrightarrow & \pi_i(X_k, B) & \longrightarrow & \pi_i(X_k, X_{k-1}) & \longrightarrow & \pi_{i-1}(X_{k-1}, B)
\end{array}$$

When $i < m + n$ the first and fourth vertical maps are isomorphisms by Case 2, while by induction the second and fifth maps are isomorphisms, so the middle map is an isomorphism by the five-lemma. Similarly, when $i = m + n$ the second and fourth maps are surjective and the fifth map is injective, which is enough to imply the middle map is surjective by one half of the five-lemma. When $i = 2$ the diagram may contain nonabelian groups and the two terms on the right may not be groups, but the five-lemma remains valid in this generality, with trivial modifications to the proof in §2.1. When $i = 1$ the assertion about $\pi_1(A, C) \to \pi_1(X, B)$ follows by a direct argument: If $m \geq 1$ then both terms are trivial, while if $m = 0$ then $n \geq 1$ and the result follows by cellular approximation.

After these special cases we can now easily deal with the general case. The connectivity assumptions on the pairs (A, C) and (B, C) imply by Proposition 4.15 that they are homotopy equivalent to pairs (A', C) and (B', C) as in Case 3, via homotopy equivalences fixed on C, so these homotopy equivalences fit together to give a homotopy equivalence $A \cup B \simeq A' \cup B'$. Thus the general case reduces to Case 3. $\qquad\square$

Example 4.26. The calculation of $\pi_n(S^n)$ can be extended to show that $\pi_n(\bigvee_\alpha S_\alpha^n)$ for $n \geq 2$ is free abelian with basis the homotopy classes of the inclusions $S_\alpha^n \hookrightarrow \bigvee_\alpha S_\alpha^n$. Suppose first that there are only finitely many summands S_α^n. We can regard $\bigvee_\alpha S_\alpha^n$ as the n-skeleton of the product $\prod_\alpha S_\alpha^n$, where S_α^n is given its usual CW structure and $\prod_\alpha S_\alpha^n$ has the product CW structure. Since $\prod_\alpha S_\alpha^n$ has cells only in dimensions a multiple of n, the pair $(\prod_\alpha S_\alpha^n, \bigvee_\alpha S_\alpha^n)$ is $(2n - 1)$-connected. Hence from the long exact sequence of homotopy groups for this pair we see that the inclusion $\bigvee_\alpha S_\alpha^n \hookrightarrow \prod_\alpha S_\alpha^n$ induces an isomorphism on π_n if $n \geq 2$. By Proposition 4.2 we have $\pi_n(\prod_\alpha S_\alpha^n) \approx \bigoplus_\alpha \pi_n(S_\alpha^n)$, a free abelian group with basis the inclusions $S_\alpha^n \hookrightarrow \prod_\alpha S_\alpha^n$, so the same is true for $\bigvee_\alpha S_\alpha^n$. This takes care of the case of finitely many S_α^n's.

To reduce the case of infinitely many summands S_α^n to the finite case, consider the homomorphism $\Phi : \bigoplus_\alpha \pi_n(S_\alpha^n) \to \pi_n(\bigvee_\alpha S_\alpha^n)$ induced by the inclusions $S_\alpha^n \hookrightarrow \bigvee_\alpha S_\alpha^n$. Then Φ is surjective since any map $f : S^n \to \bigvee_\alpha S_\alpha^n$ has compact image contained in the wedge sum of finitely many S_α^n's, so by the finite case already proved, $[f]$ is in the image of Φ. Similarly, a nullhomotopy of f has compact image contained in a finite wedge sum of S_α^n's, so the finite case also implies that Φ is injective.

Example 4.27. Let us show that $\pi_n(S^1 \vee S^n)$ for $n \geq 2$ is free abelian on a countably infinite number of generators. By Proposition 4.1 we may compute $\pi_i(S^1 \vee S^n)$ for $i \geq 2$ by passing to the universal cover. This consists of a copy of $\mathbb{R}$ with a sphere S_k^n attached at each integer point $k \in \mathbb{R}$, so it is homotopy equivalent to $\bigvee_k S_k^n$. The preceding Example 4.26 says that $\pi_n(\bigvee_k S_k^n)$ is free abelian with basis represented by the inclusions of the wedge summands. So a basis for π_n of the universal cover of $S^1 \vee S^n$ is represented by maps that lift the maps obtained from the inclusion $S^n \hookrightarrow S^1 \vee S^n$ by the action of the various elements of $\pi_1(S^1 \vee S^n) \approx \mathbb{Z}$. This means that $\pi_n(S^1 \vee S^n)$ is a free $\mathbb{Z}[\pi_1(S^1 \vee S^n)]$-module on a single basis element, the homotopy class of the inclusion $S^n \hookrightarrow S^1 \vee S^n$. Writing a generator of $\pi_1(S^1 \vee S^n)$ as t, the group ring $\mathbb{Z}[\pi_1(S^1 \vee S^n)]$ becomes $\mathbb{Z}[t, t^{-1}]$, the Laurent polynomials in t and t^{-1} with $\mathbb{Z}$ coefficients, and we have $\pi_n(S^1 \vee S^n) \approx \mathbb{Z}[t, t^{-1}]$.

This example shows that the homotopy groups of a finite CW complex need not be finitely generated, in contrast to the homology groups. However, if we restrict attention to spaces with trivial action of π_1 on all π_n's, then a theorem of Serre, proved in [SSAT], says that the homotopy groups of such a space are finitely generated iff the homology groups are finitely generated.

In this example, $\pi_n(S^1 \vee S^n)$ is finitely generated as a $\mathbb{Z}[\pi_1]$-module, but there are finite CW complexes where even this fails. This happens in fact for $\pi_3(S^1 \vee S^2)$, according to Exercise 38 at the end of this section. In §4.A we construct more complicated examples for each π_n with $n > 1$, in particular for π_2.

A useful tool for more complicated calculations is the following general result:

Proposition 4.28. *If a CW pair (X, A) is r-connected and A is s-connected, with $r, s \geq 0$, then the map $\pi_i(X, A) \to \pi_i(X/A)$ induced by the quotient map $X \to X/A$ is an isomorphism for $i \leq r + s$ and a surjection for $i = r + s + 1$.*

Proof: Consider $X \cup CA$, the complex obtained from X by attaching a cone CA along $A \subset X$. Since CA is a contractible subcomplex of $X \cup CA$, the quotient map $X \cup CA \to (X \cup CA)/CA = X/A$ is a homotopy equivalence by Proposition 0.17. So we have a commutative diagram

$$\pi_i(X, A) \longrightarrow \pi_i(X \cup CA, CA) \longrightarrow \pi_i(X \cup CA/CA) = \pi_i(X/A)$$
$$\uparrow \approx \qquad \nearrow \approx$$
$$\pi_i(X \cup CA)$$

where the vertical isomorphism comes from a long exact sequence. Now apply the excision theorem to the first map in the diagram, using the fact that (CA, A) is $(s + 1)$-connected if A is s-connected, which comes from the exact sequence for the pair (CA, A).   □

Example 4.29. Suppose X is obtained from a wedge of spheres $\bigvee_\alpha S_\alpha^n$ by attaching cells e_β^{n+1} via basepoint-preserving maps $\varphi_\beta : S^n \to \bigvee_\alpha S_\alpha^n$, with $n \geq 2$. By cellular approximation we know that $\pi_i(X) = 0$ for $i < n$, and we shall show that $\pi_n(X)$ is the quotient of the free abelian group $\pi_n(\bigvee_\alpha S^n_\alpha) \approx \bigoplus_\alpha \mathbb{Z}$ by the subgroup generated by the classes $[\varphi_\beta]$. Any subgroup can be realized in this way, by choosing maps φ_β to represent a set of generators for the subgroup, so it follows that every abelian group can be realized as $\pi_n(X)$ for such a space $X = (\bigvee_\alpha S^n_\alpha) \bigcup_\beta e^{n+1}_\beta$. This is the higher-dimensional analog of the construction in Corollary 1.28 of a 2-dimensional CW complex with prescribed fundamental group.

To see that $\pi_n(X)$ is as claimed, consider the following portion of the long exact sequence of the pair $(X, \bigvee_\alpha S^n_\alpha)$:

$$\pi_{n+1}(X, \textstyle\bigvee_\alpha S^n_\alpha) \xrightarrow{\ \partial\ } \pi_n(\textstyle\bigvee_\alpha S^n_\alpha) \longrightarrow \pi_n(X) \longrightarrow 0$$

The quotient $X/\bigvee_\alpha S^n_\alpha$ is a wedge of spheres S^{n+1}_β, so the preceding proposition and Example 4.26 imply that $\pi_{n+1}(X, \bigvee_\alpha S^n_\alpha)$ is free with basis the characteristic maps of the cells e^{n+1}_β. The boundary map ∂ takes these to the classes $[\varphi_\beta]$, and the result follows.

Eilenberg–MacLane Spaces

A space X having just one nontrivial homotopy group $\pi_n(X) \approx G$ is called an **Eilenberg–MacLane space** $K(G, n)$. The case $n = 1$ was considered in §1.B, where the condition that $\pi_i(X) = 0$ for $i > 1$ was replaced by the condition that X have a contractible universal cover, which is equivalent for spaces that have a universal cover of the homotopy type of a CW complex.

We can build a CW complex $K(G, n)$ for arbitrary G and n, assuming G is abelian if $n > 1$, in the following way. To begin, let X be an $(n - 1)$-connected CW complex of dimension $n + 1$ such that $\pi_n(X) \approx G$, as was constructed in Example 4.29 above when $n > 1$ and in Corollary 1.28 when $n = 1$. Then we showed in Example 4.16 how to attach higher-dimensional cells to X to make π_i trivial for $i > n$ without affecting π_n or the lower homotopy groups.

By taking products of $K(G, n)$'s for varying n we can then realize any sequence of groups G_n, abelian for $n > 1$, as the homotopy groups π_n of a space.

A fair number of $K(G, 1)$'s arise naturally in a variety of contexts, and a few of these are mentioned in §1.B. By contrast, naturally occurring $K(G, n)$'s for $n \geq 2$ are rare. It seems the only real example is $\mathbb{C}P^\infty$, which is a $K(\mathbb{Z}, 2)$ as we shall see in Example 4.50. One could of course trivially generalize this example by taking a product of $\mathbb{C}P^\infty$'s to get a $K(G, 2)$ with G a product of $\mathbb{Z}$'s.

Actually there is a fairly natural construction of a $K(\mathbb{Z}, n)$ for arbitrary n, the infinite symmetric product $SP(S^n)$ defined in §3.C. In §4.K we prove that the functor SP has the surprising property of converting homology groups into homotopy groups, namely $\pi_i(SP(X)) \approx H_i(X; \mathbb{Z})$ for all $i > 0$ and all connected CW complexes X. Taking X to be a sphere, we deduce that $SP(S^n)$ is a $K(\mathbb{Z}, n)$. More generally, $SP(M(G, n))$ is a $K(G, n)$ for each Moore space $M(G, n)$.

Having shown the existence of $K(G,n)$'s, we now consider the uniqueness question, which has the nicest possible answer:

Proposition 4.30. *The homotopy type of a CW complex $K(G,n)$ is uniquely determined by G and n.*

The proof will be based on a more technical statement:

Lemma 4.31. *Let X be a CW complex of the form $(\bigvee_\alpha S_\alpha^n) \bigcup_\beta e_\beta^{n+1}$ for some $n \geq 1$. Then for every homomorphism $\psi : \pi_n(X) \to \pi_n(Y)$ with Y path-connected there exists a map $f : X \to Y$ with $f_* = \psi$.*

Proof: To begin, let f send the natural basepoint of $\bigvee_\alpha S_\alpha^n$ to a chosen basepoint $y_0 \in Y$. Extend f over each sphere S_α^n via a map representing $\psi([i_\alpha])$ where i_α is the inclusion $S_\alpha^n \hookrightarrow X$. Thus for the map $f : X^n \to Y$ constructed so far we have $f_*([i_\alpha]) = \psi([i_\alpha])$ for all α, hence $f_*([\varphi]) = \psi([\varphi])$ for all basepoint-preserving maps $\varphi : S^n \to X^n$ since the i_α's generate $\pi_n(X^n)$. To extend f over a cell e_β^{n+1} all we need is that the composition of the attaching map $\varphi_\beta : S^n \to X^n$ for this cell with f be nullhomotopic in Y. But this composition $f\varphi_\beta$ represents $f_*([\varphi_\beta]) = \psi([\varphi_\beta])$, and $\psi([\varphi_\beta]) = 0$ because $[\varphi_\beta]$ is zero in $\pi_n(X)$ since φ_β is nullhomotopic in X via the characteristic map of e_β^{n+1}. Thus we obtain an extension $f : X \to Y$. This has $f_* = \psi$ since the elements $[i_\alpha]$ generate $\pi_n(X^n)$ and hence also $\pi_n(X)$ by cellular approximation. $\square$

Proof of 4.30: Suppose K and K' are $K(G,n)$ CW complexes. Since homotopy equivalence is an equivalence relation, there is no loss of generality if we assume K is a particular $K(G,n)$, namely one constructed from a space X as in the lemma by attaching cells of dimension $n+2$ and greater. By the lemma there is a map $f : X \to K'$ inducing an isomorphism on π_n. To extend this f over K we proceed inductively. For each cell e^{n+2}, the composition of its attaching map with f is nullhomotopic in K' since $\pi_{n+1}(K') = 0$, so f extends over this cell. The same argument applies for all the higher-dimensional cells in turn. The resulting $f : K \to K'$ is a homotopy equivalence since it induces isomorphisms on all homotopy groups. $\square$

The Hurewicz Theorem

Using the calculations of homotopy groups done above we can easily prove the simplest and most often used cases of the Hurewicz theorem:

Theorem 4.32. *If a space X is $(n-1)$-connected, $n \geq 2$, then $\tilde{H}_i(X) = 0$ for $i < n$ and $\pi_n(X) \approx H_n(X)$. If a pair (X,A) is $(n-1)$-connected, $n \geq 2$, with A simply-connected and nonempty, then $H_i(X,A) = 0$ for $i < n$ and $\pi_n(X,A) \approx H_n(X,A)$.*

Thus the first nonzero homotopy and homology groups of a simply-connected space occur in the same dimension and are isomorphic. One cannot expect any nice relationship between $\pi_i(X)$ and $H_i(X)$ beyond this. For example, S^n has trivial homology groups above dimension n but many nontrivial homotopy groups in this range when $n \geq 2$. In the other direction, Eilenberg–MacLane spaces such as $\mathbb{C}P^\infty$ have trivial higher homotopy groups but many nontrivial homology groups.

The theorem can sometimes be used to compute $\pi_2(X)$ if X is a path-connected space that is nice enough to have a universal cover. For if $\widetilde{X}$ is the universal cover, then $\pi_2(X) \approx \pi_2(\widetilde{X})$ and the latter group is isomorphic to $H_2(\widetilde{X})$ by the Hurewicz theorem. So if one can understand $\widetilde{X}$ well enough to compute $H_2(\widetilde{X})$, one can compute $\pi_2(X)$.

In the part of the theorem dealing with relative groups, notice that X must be simply-connected as well as A since (X,A) is 1-connected by hypothesis. There is a more general version of the relative Hurewicz theorem given later in Theorem 4.37 that allows A and X to be nonsimply-connected, but this requires $\pi_n(X,A)$ to be replaced by a certain quotient group.

Proof: We may assume X is a CW complex and (X,A) is a CW pair by taking CW approximations to X and (X,A). For CW pairs the relative case then reduces to the absolute case since $\pi_i(X,A) \approx \pi_i(X/A)$ for $i \leq n$ by Proposition 4.28, while $H_i(X,A) \approx \widetilde{H}_i(X/A)$ for all i by Proposition 2.22.

In the absolute case we can apply Proposition 4.15 to replace X by a homotopy equivalent CW complex with $(n-1)$-skeleton a point, hence $\widetilde{H}_i(X) = 0$ for $i < n$. To show $\pi_n(X) \approx H_n(X)$, we can further simplify by throwing away cells of dimension greater than $n+1$ since these have no effect on π_n or H_n. Thus X has the form $(\bigvee_\alpha S_\alpha^n) \bigcup_\beta e_\beta^{n+1}$. We may assume the attaching maps φ_β of the cells e_β^{n+1} are basepoint-preserving since this is what the proof of Proposition 4.15 gives. Example 4.29 then applies to compute $\pi_n(X)$ as the cokernel of the boundary map $\pi_{n+1}(X,X^n) \to \pi_n(X^n)$, a map $\bigoplus_\beta \mathbb{Z} \to \bigoplus_\alpha \mathbb{Z}$. This is the same as the cellular boundary map $d : H_{n+1}(X^{n+1}, X^n) \to H_n(X^n, X^{n-1})$ since for a cell e_β^{n+1}, the coefficients of $d e_\beta^{n+1}$ are the degrees of the compositions $q_\alpha \varphi_\beta$ where q_α collapses all n-cells except e_α^n to a point, and the isomorphism $\pi_n(S^n) \approx \mathbb{Z}$ in Corollary 4.25 is given by degree. Since there are no $(n-1)$-cells, we have $H_n(X) \approx \operatorname{Coker} d$. $\quad\square$

Since homology groups are usually more computable than homotopy groups, the following version of Whitehead's theorem is often easier to apply:

Corollary 4.33. *A map $f : X \to Y$ between simply-connected CW complexes is a homotopy equivalence if $f_* : H_n(X) \to H_n(Y)$ is an isomorphism for each n.*

Proof: After replacing Y by the mapping cylinder M_f we may take f to be an inclusion $X \hookrightarrow Y$. Since X and Y are simply-connected, we have $\pi_1(Y,X) = 0$. The relative Hurewicz theorem then says that the first nonzero $\pi_n(Y,X)$ is isomorphic to the first nonzero $H_n(Y,X)$. All the groups $H_n(Y,X)$ are zero from the long exact sequence of homology, so all the groups $\pi_n(Y,X)$ also vanish. This means that the inclusion

$X \hookrightarrow Y$ induces isomorphisms on all homotopy groups, and therefore this inclusion is a homotopy equivalence. $\square$

Example 4.34: Uniqueness of Moore Spaces. Let us show that the homotopy type of a CW complex Moore space $M(G, n)$ is uniquely determined by G and n if $n > 1$, so $M(G, n)$ is simply-connected. Let X be an $M(G, n)$ as constructed in Example 2.40 by attaching $(n + 1)$-cells to a wedge sum of n-spheres, and let Y be any other $M(G, n)$ CW complex. By Lemma 4.31 there is a map $f : X \rightarrow Y$ inducing an isomorphism on π_n. If we can show that f also induces an isomorphism on H_n, then the preceding corollary will imply the result.

One way to show that f induces an isomorphism on H_n would be to use a more refined version of the Hurewicz theorem giving an isomorphism between π_n and H_n that is natural with respect to maps between spaces, as in Theorem 4.37 below. However, here is a direct argument which avoids naturality questions. For the mapping cylinder M_f we know that $\pi_i(M_f, X) = 0$ for $i \le n$. If this held also for $i = n + 1$ then the relative Hurewicz theorem would say that $H_i(M_f, X) = 0$ for $i \le n + 1$ and hence that f_* would be an isomorphism on H_n. To make this argument work, let us temporarily enlarge Y by attaching $(n + 2)$-cells to make π_{n+1} zero. The new mapping cylinder M_f then has $\pi_{n+1}(M_f, X) = 0$ from the long exact sequence of the pair. So for the enlarged Y the map f induces an isomorphism on H_n. But attaching $(n + 2)$-cells has no effect on H_n, so the original $f : X \rightarrow Y$ had to be an isomorphism on H_n.

It is certainly possible for a map of nonsimply-connected spaces to induce isomorphisms on all homology groups but not on homotopy groups. Nonsimply-connected acyclic spaces, for which the inclusion of a point induces an isomorphism on homology, exhibit this phenomenon in its purest form. Perhaps the simplest nontrivial acyclic space is the 2-dimensional complex constructed in Example 2.38 with fundamental group $\langle a, b \mid a^5 = b^3 = (ab)^2 \rangle$ of order 120.

It is also possible for a map between spaces with abelian fundamental groups to induce isomorphisms on homology but not on higher homotopy groups, as the next example shows.

Example 4.35. We construct a space $X = (S^1 \vee S^n) \cup e^{n+1}$, for arbitrary $n > 1$, such that the inclusion $S^1 \hookrightarrow X$ induces an isomorphism on all homology groups and on π_i for $i < n$, but not on π_n. From Example 4.27 we have $\pi_n(S^1 \vee S^n) \approx \mathbb{Z}[t, t^{-1}]$. Let X be obtained from $S^1 \vee S^n$ by attaching a cell e^{n+1} via a map $S^n \rightarrow S^1 \vee S^n$ corresponding to $2t - 1 \in \mathbb{Z}[t, t^{-1}]$. By looking in the universal cover we see that $\pi_n(X) \approx \mathbb{Z}[t, t^{-1}]/(2t - 1)$, where $(2t - 1)$ denotes the ideal in $\mathbb{Z}[t, t^{-1}]$ generated by $2t - 1$. Note that setting $t = 1/2$ embeds $\mathbb{Z}[t, t^{-1}]/(2t - 1)$ in $\mathbb{Q}$ as the subring $\mathbb{Z}[1/2]$ consisting of rationals with denominator a power of 2. From the long exact sequence of homotopy groups for the $(n - 1)$-connected pair (X, S^1) we see that the inclusion

$S^1 \hookrightarrow X$ induces an isomorphism on π_i for $i < n$. The fact that this inclusion also induces isomorphisms on all homology groups can be deduced from cellular homology. The key point is that the cellular boundary map $H_{n+1}(X^{n+1}, X^n) \to H_n(X^n, X^{n-1})$ is an isomorphism since the degree of the composition of the attaching map $S^n \to S^1 \vee S^n$ of e^{n+1} with the collapse $S^1 \vee S^n \to S^n$ is $2 - 1 = 1$.

This example relies heavily on the nontriviality of the action of $\pi_1(X)$ on $\pi_n(X)$, so one might ask whether the simple-connectivity assumption in Corollary 4.33 can be weakened to trivial action of π_1 on all π_n's. This is indeed the case, as we will show in Proposition 4.74.

The form of the Hurewicz theorem given above asserts merely the existence of an isomorphism between homotopy and homology groups, but one might want a more precise statement which says that a particular map is an isomorphism. In fact, there are always natural maps from homotopy groups to homology groups, defined in the following way. Thinking of $\pi_n(X, A, x_0)$ for $n > 0$ as homotopy classes of maps $f : (D^n, \partial D^n, s_0) \to (X, A, x_0)$, the **Hurewicz map** $h : \pi_n(X, A, x_0) \to H_n(X, A)$ is defined by $h([f]) = f_*(\alpha)$ where α is a fixed generator of $H_n(D^n, \partial D^n) \approx \mathbb{Z}$ and $f_* : H_n(D^n, \partial D^n) \to H_n(X, A)$ is induced by f. If we have a homotopy $f \simeq g$ through maps $(D^n, \partial D^n, s_0) \to (X, A, x_0)$, or even through maps $(D^n, \partial D^n) \to (X, A)$ not preserving the basepoint, then $f_* = g_*$, so h is well-defined.

Proposition 4.36. *The Hurewicz map* $h : \pi_n(X, A, x_0) \to H_n(X, A)$ *is a homomorphism, assuming* $n > 1$ *so that* $\pi_n(X, A, x_0)$ *is a group.*

Proof: It suffices to show that for maps $f, g : (D^n, \partial D^n) \to (X, A)$, the induced maps on homology satisfy $(f + g)_* = f_* + g_*$, for if this is the case then $h([f + g]) = (f + g)_*(\alpha) = f_*(\alpha) + g_*(\alpha) = h([f]) + h([g])$. Our proof that $(f + g)_* = f_* + g_*$ will in fact work for any homology theory.

Let $c : D^n \to D^n \vee D^n$ be the map collapsing the equatorial D^{n-1} to a point, and let $q_1, q_2 : D^n \vee D^n \to D^n$ be the quotient maps onto the two summands, collapsing the other summand to a point. We then have a diagram

$$H_n(D^n, \partial D^n) \xrightarrow{c_*} H_n(D^n \vee D^n, \partial D^n \vee \partial D^n) \xrightarrow{(f \vee g)_*} H_n(X, A)$$
$$q_{1*} \oplus q_{2*} \downarrow \approx$$
$$H_n(D^n, \partial D^n) \oplus H_n(D^n, \partial D^n)$$

The map $q_{1*} \oplus q_{2*}$ is an isomorphism with inverse $i_{1*} + i_{2*}$ where i_1 and i_2 are the inclusions of the two summands $D^n \hookrightarrow D^n \vee D^n$. Since $q_1 c$ and $q_2 c$ are homotopic to the identity through maps $(D^n, \partial D^n) \to (D^n, \partial D^n)$, the composition $(q_{1*} \oplus q_{2*})c_*$ is the diagonal map $x \mapsto (x, x)$. From the equalities $(f \vee g)i_1 = f$ and $(f \vee g)i_2 = g$ we deduce that $(f \vee g)_*(i_{1*} + i_{2*})$ sends $(x, 0)$ to $f_*(x)$ and $(0, x)$ to $g_*(x)$, hence (x, x) to $f_*(x) + g_*(x)$. Thus the composition across the top of the diagram is

$x \mapsto f_*(x) + g_*(x)$. On the other hand, $f + g = (f \vee g)c$, so this composition is also $(f + g)_*$. $\square$

There is also an absolute Hurewicz map $h : \pi_n(X, x_0) \to H_n(X)$ defined in a similar way by setting $h([f]) = f_*(\alpha)$ for $f : (S^n, s_0) \to (X, x_0)$ and α a chosen generator of $H_n(S^n)$. For example, if $X = S^n$ then $f_*(\alpha)$ is $(\deg f)\alpha$ by the definition of degree, so we can view h in this case as the degree map $\pi_n(S^n) \to \mathbb{Z}$, which we know is an isomorphism by Corollary 4.25. The proof of the preceding proposition is readily modified to show that the absolute h is a homomorphism for $n \geq 1$.

The absolute and relative Hurewicz maps can be combined in a diagram of long exact sequences

$$
\begin{array}{ccccccccc}
\cdots & \longrightarrow & \pi_n(A, x_0) & \longrightarrow & \pi_n(X, x_0) & \longrightarrow & \pi_n(X, A, x_0) & \longrightarrow & \pi_{n-1}(A, x_0) & \longrightarrow & \cdots \\
 & & \downarrow h & & \downarrow h & & \downarrow h & & \downarrow h & & \\
\cdots & \longrightarrow & H_n(A) & \longrightarrow & H_n(X) & \longrightarrow & H_n(X, A) & \longrightarrow & H_{n-1}(A) & \longrightarrow & \cdots
\end{array}
$$

An easy definition check which we leave to the reader shows that this diagram commutes up to sign at least. With more care in the choice of the generators α it can be made to commute exactly.

Another elementary property of Hurewicz maps is that they are natural: A map $f : (X, x_0) \to (Y, y_0)$ induces a commutative diagram as at the right, and similarly in the relative case.

$$
\begin{array}{ccc}
\pi_n(X, x_0) & \xrightarrow{\ f_*\ } & \pi_n(Y, y_0) \\
\downarrow h & & \downarrow h \\
H_n(X) & \xrightarrow{\ f_*\ } & H_n(Y)
\end{array}
$$

It is easy to construct nontrivial elements of the kernel of the Hurewicz homomorphism $h : \pi_n(X, x_0) \to H_n(X)$ if $\pi_1(X, x_0)$ acts nontrivially on $\pi_n(X, x_0)$, namely elements of the form $[\gamma][f] - [f]$. This is because γf and f, viewed as maps $S^n \to X$, are homotopic if we do not require the basepoint to be fixed during the homotopy, so $(\gamma f)_*(\alpha) = f_*(\alpha)$ for α a generator of $H_n(S^n)$.

Similarly in the relative case the kernel of $h : \pi_n(X, A, x_0) \to H_n(X, A)$ contains the elements of the form $[\gamma][f] - [f]$ for $[\gamma] \in \pi_1(A, x_0)$. For example the Hurewicz map $\pi_n(S^1 \vee S^n, S^1) \to H_n(S^1 \vee S^n, S^1)$ is the homomorphism $\mathbb{Z}[t, t^{-1}] \to \mathbb{Z}$ sending all powers of t to 1. Since the pair $(S^1 \vee S^n, S^1)$ is $(n-1)$-connected, this example shows that the hypothesis $\pi_1(A, x_0) = 0$ in the relative form of the Hurewicz theorem proved earlier cannot be dropped.

If we define $\pi'_n(X, A, x_0)$ to be the quotient group of $\pi_n(X, A, x_0)$ obtained by factoring out the subgroup generated by all elements of the form $[\gamma][f] - [f]$, or the normal subgroup generated by such elements in the case $n = 2$ when $\pi_2(X, A, x_0)$ may not be abelian, then h induces a homomorphism $h' : \pi'_n(X, A, x_0) \to H_n(X, A)$. The general form of the Hurewicz theorem deals with this homomorphism:

‖ **Theorem 4.37.** *If (X, A) is an $(n - 1)$-connected pair of path-connected spaces with $n \geq 2$ and $A \neq \varnothing$, then $h' : \pi_n'(X, A, x_0) \to H_n(X, A)$ is an isomorphism and $H_i(X, A) = 0$ for $i < n$.*

Note that this statement includes the absolute form of the theorem by taking A to be the basepoint.

Before starting the proof of this general Hurewicz theorem we have a preliminary step:

‖ **Lemma 4.38.** *If X is a connected CW complex to which cells e_α^n are attached for a fixed $n \geq 2$, forming a CW complex $W = X \bigcup_\alpha e_\alpha^n$, then $\pi_n(W, X)$ is a free $\pi_1(X)$-module with basis the homotopy classes of the characteristic maps Φ_α of the cells e_α^n, provided that the map $\pi_1(X) \to \pi_1(W)$ induced by inclusion is an isomorphism. In particular, this is always the case if $n \geq 3$. In the general $n = 2$ case, $\pi_2(W, X)$ is generated by the classes of the characteristic maps of the cells e_α^2 together with their images under the action of $\pi_1(X)$.*

If the characteristic maps $\Phi_\alpha : (D^n, \partial D^n) \to (W, X)$ do not take a basepoint s_0 in ∂D^n to the basepoint x_0 in X, then they will define elements of $\pi_n(W, X, x_0)$ only after we choose change-of-basepoint paths from the points $\Phi_\alpha(s_0)$ to x_0. Different choices of such paths yield elements of $\pi_n(W, X, x_0)$ related by the action of $\pi_1(X, x_0)$, so the basis for $\pi_n(W, X, x_0)$ is well-defined up to multiplication by invertible elements of $\mathbb{Z}[\pi_1(X)]$.

The situation when $n = 2$ and the map $\pi_1(X) \to \pi_1(W)$ is not an isomorphism is more complicated because the relative π_2 can be nonabelian in this case. Whitehead analyzed what happens here and showed that $\pi_2(W, X)$ has the structure of a 'free crossed $\pi_1(X)$-module.' See [Whitehead 1949] or [Sieradski 1993].

Proof: Since $W/X = \bigvee_\alpha S_\alpha^n$, we have $\pi_n(W, X) \approx \pi_n(\bigvee_\alpha S_\alpha^n)$ when X is simply-connected, by Proposition 4.28. The conclusion of the lemma in this case is then immediate from Example 4.26.

When X is not simply-connected but the inclusion $X \hookrightarrow W$ induces an isomorphism on π_1, then the universal cover of W is obtained from the universal cover of X by attaching n-cells lifting the cells e_α^n. If we choose one such lift $\widetilde{e}_\alpha^n$ of e_α^n, then all the other lifts are the images $y\widetilde{e}_\alpha^n$ of $\widetilde{e}_\alpha^n$ under the deck transformations $y \in \pi_1(X)$. The special case proved in the preceding paragraph shows that the relative π_n for the universal cover is the free abelian group with basis corresponding to the cells $y\widetilde{e}_\alpha^n$. By the relative version of Proposition 4.1, the projection of the universal cover of W onto W induces an isomorphism on relative π_n's, so $\pi_n(W, X)$ is free abelian with basis the classes $[ye_\alpha^n]$ as y ranges over $\pi_1(X)$, or in other words the free $\pi_1(X)$-module with basis the cells e_α^n.

It remains to consider the $n = 2$ case in general. Since both of the pairs (W, X) and $(X^1 \bigcup_\alpha e_\alpha^2, X^1)$ are 1-connected, the homotopy excision theorem implies that the map $\pi_2(X^1 \bigcup_\alpha e_\alpha^2, X^1) \to \pi_2(W, X)$ is surjective. This gives a reduction to the case that X is 1-dimensional. We may also assume the 2-cells e_α^2 are attached along loops passing through the basepoint 0-cell x_0, since this can be achieved by homotopy of the attaching maps, which does not affect the homotopy type of the pair (W, X).

In the closure of each 2-cell e_α^2 choose an embedded disk D_α^2 which contains x_0 but is otherwise contained entirely in the interior of e_α^2. Let $Y = X \bigcup_\alpha D_\alpha^2$, the wedge sum of X with the disks D_α^2, and let $Z = W - \bigcup_\alpha \mathrm{int}(D_\alpha^2)$, so Y and Z are 2-dimensional CW complexes with a common 1-skeleton $Y^1 = Z^1 = Y \cap Z = X \bigvee_\alpha \partial D_\alpha^2$. The inclusion $(W, X) \hookrightarrow (W, Z)$ is a homotopy equivalence of pairs. Homotopy excision gives a surjection $\pi_2(Y, Y^1) \to \pi_2(W, Z)$. The universal cover $\tilde{Y}$ of Y is obtained from the universal cover $\tilde{X}$ of X by taking the wedge sum with lifts $\tilde{D}_{\alpha\beta}^2$ of the disks D_α^2. Hence we have isomorphisms

$$\pi_2(Y, Y^1) \approx \pi_2(\tilde{Y}, \tilde{Y}^1) \qquad \text{where } \tilde{Y}^1 \text{ is the 1-skeleton of } \tilde{Y}$$
$$\approx \pi_2\bigl(\textstyle\bigvee_{\alpha\beta} \tilde{D}_{\alpha\beta}^2, \bigvee_{\alpha\beta} \partial \tilde{D}_{\alpha\beta}^2\bigr) \quad \text{since } \tilde{X} \text{ is contractible}$$
$$\approx \pi_1\bigl(\textstyle\bigvee_{\alpha\beta} \partial \tilde{D}_{\alpha\beta}^2\bigr) \qquad \text{since } \bigvee_{\alpha\beta} \tilde{D}_{\alpha\beta}^2 \text{ is contractible}$$

This last group is free with basis the loops $\partial \tilde{D}_{\alpha\beta}^2$, so the inclusions $\tilde{D}_{\alpha\beta}^2 \hookrightarrow \bigvee_{\alpha\beta} \tilde{D}_{\alpha\beta}^2$ form a basis for $\pi_2(\bigvee_{\alpha\beta} \tilde{D}_{\alpha\beta}^2, \bigvee_{\alpha\beta} \partial \tilde{D}_{\alpha\beta}^2)$. This implies that $\pi_2(Y, Y^1)$ is generated by the inclusions $D_\alpha^2 \hookrightarrow Y$ and their images under the action of loops in X. The same is true for $\pi_2(W, Z)$ via the surjection $\pi_2(Y, Y^1) \to \pi_2(W, Z)$. Using the isomorphism $\pi_2(W, Z) \approx \pi_2(W, X)$, we conclude that $\pi_2(W, X)$ is generated by the characteristic maps of the cells e_α^2 and their images under the action of $\pi_1(X)$. $\qquad\square$

Proof of the general Hurewicz Theorem: As in the earlier form of the theorem we may assume (X, A) is a CW pair such that the cells of $X - A$ have dimension $\geq n$.

We first prove the theorem assuming that $\pi_1(A) \to \pi_1(X)$ is an isomorphism. This is always the case if $n \geq 3$, so this case will finish the proof except when $n = 2$. We may assume also that $X = X^{n+1}$ since higher-dimensional cells have no effect on π_n or H_n. Consider the commutative diagram

$$
\begin{array}{ccccccc}
\pi_{n+1}(X, X^n \cup A) & \xrightarrow{\partial} & \pi_n(X^n \cup A, A) & \xrightarrow{i_*} & \pi_n(X, A) & \longrightarrow & 0 \\
\downarrow & & \downarrow & & \downarrow & & \\
\pi'_{n+1}(X, X^n \cup A) & \xrightarrow{\partial'} & \pi'_n(X^n \cup A, A) & \xrightarrow{i'_*} & \pi'_n(X, A) & \longrightarrow & 0 \\
\downarrow h' & & \downarrow h' & & \downarrow h' & & \\
H_{n+1}(X, X^n \cup A) & \longrightarrow & H_n(X^n \cup A, A) & \longrightarrow & H_n(X, A) & \longrightarrow & 0
\end{array}
$$

The first and third rows are exact sequences for the triple $(X, X^n \cup A, A)$. The left-hand h' is an isomorphism since by the preceding lemma, $\pi_{n+1}(X, X^n \cup A)$ is a free π_1-module with basis the characteristic maps of the $(n+1)$-cells of $X - A$, so $\pi'_{n+1}(X, X^n \cup A)$ is a free abelian group with the same basis, and $H_{n+1}(X, X^n \cup A)$ is also free with basis the $(n+1)$-cells of $X - A$. Similarly, the lemma implies that the middle h' is an isomorphism since the assumption that $\pi_1(A) \to \pi_1(X)$ is an isomorphism implies that $\pi_1(A) \to \pi_1(X^n \cup A)$ is injective, hence an isomorphism if $n \geq 2$.

A simple diagram chase now shows that the right-hand h' is an isomorphism. Namely, surjectivity follows since $H_n(X^n \cup A, A) \to H_n(X, A)$ is surjective and the middle h' is an isomorphism. For injectivity, take an element $x \in \pi'_n(X, A)$ with $h'(x) = 0$. The map i_* is surjective since i_* is, so $x = i'_*(y)$ for some element $y \in \pi'_n(X^n \cup A, A)$. Since the first two maps h' are isomorphisms and the bottom row is exact, there is a $z \in \pi'_{n+1}(X, X^n \cup A)$ with $\partial'(z) = y$. Hence $x = 0$ since $i_* \partial = 0$ implies $i'_* \partial' = 0$.

It remains to prove the theorem when $n = 2$ and $\pi_1(A) \to \pi_1(X)$ is not an isomorphism. The proof above will apply once we show that the middle h' in the diagram is an isomorphism. The preceding lemma implies that $\pi'_2(X^2 \cup A, A)$ is generated by characteristic maps of the 2-cells of $X - A$. The images of these generators under h' form a basis for $H_2(X^2 \cup A, A)$. Thus h' is a homomorphism from a group which, by the lemma below, is abelian to a free abelian group taking a set of generators to a basis, hence h' is an isomorphism. $\square$

Lemma 4.39. *For any (X, A, x_0), the formula $a + b - a = (\partial a)b$ holds for all $a, b \in \pi_2(X, A, x_0)$, where $\partial : \pi_2(X, A, x_0) \to \pi_1(A, x_0)$ is the usual boundary map and $(\partial a)b$ denotes the action of ∂a on b. Hence $\pi'_2(X, A, x_0)$ is abelian.*

Here the '$+$' and '$-$' in $a + b - a$ refer to the group operation in the nonabelian group $\pi_2(X, A, x_0)$.

Proof: The formula is obtained by constructing a homotopy from $a + b - a$ to $(\partial a)b$ as indicated in the pictures below. $\square$

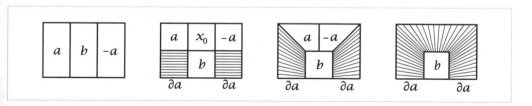

The Plus Construction

There are quite a few situations in algebraic topology where having a nontrivial fundamental group complicates matters considerably. We shall next describe a construction which in certain circumstances allows one to modify a space so as to eliminate its fundamental group, or at least simplify it, without affecting homology or cohomology. Here is the simplest case:

Proposition 4.40. *Let X be a connected CW complex with $H_1(X) = 0$. Then there is a simply-connected CW complex X^+ and a map $X \to X^+$ inducing isomorphisms on all homology groups.*

Proof: Choose loops $\varphi_\alpha : S^1 \to X^1$ generating $\pi_1(X)$ and use these to attach cells e_α^2 to X to form a simply-connected CW complex X'. The homology exact sequence

$$0 \longrightarrow H_2(X) \longrightarrow H_2(X') \longrightarrow H_2(X', X) \longrightarrow 0 = H_1(X)$$

splits since $H_2(X', X)$ is free with basis the cells e_α^2. Thus we have an isomorphism $H_2(X') \approx H_2(X) \oplus H_2(X', X)$. Since X' is simply-connected, the Hurewicz theorem gives an isomorphism $H_2(X') \approx \pi_2(X')$, and so we may represent a basis for the free summand $H_2(X', X)$ by maps $\psi_\alpha : S^2 \to X'$. We may assume these are cellular maps, and then use them to attach cells e_α^3 to X' forming a simply-connected CW complex X^+, with the inclusion $X \hookrightarrow X^+$ an isomorphism on all homology groups. $\qquad\square$

In the preceding proposition, the condition $H_1(X) = 0$ means that $\pi_1(X)$ is equal to its commutator subgroup, that is, $\pi_1(X)$ is a **perfect** group. Suppose more generally that X is a connected CW complex and $H \subset \pi_1(X)$ is a perfect subgroup. Let $p : \widetilde{X} \to X$ be the covering space corresponding to H, so $\pi_1(\widetilde{X}) \approx H$ is perfect and $H_1(\widetilde{X}) = 0$. From the previous proposition we get an inclusion $\widetilde{X} \hookrightarrow \widetilde{X}^+$. Let X^+ be obtained from the disjoint union of $\widetilde{X}^+$ and the mapping cylinder M_p by identifying the copies of $\widetilde{X}$ in these two spaces. Thus we have the commutative diagram of inclusion maps shown at the right. From the van Kampen theorem, the induced map $\pi_1(X) \to \pi_1(X^+)$ is surjective with kernel the normal

$$\begin{array}{ccc} \widetilde{X} & \longrightarrow & \widetilde{X}^+ \\ \downarrow & & \downarrow \\ X \simeq M_p & \longrightarrow & X^+ \end{array}$$

subgroup generated by H. Further, since X^+/M_p is homeomorphic to $\widetilde{X}^+/\widetilde{X}$ we have $H_*(X^+, M_p) = H_*(\widetilde{X}^+, \widetilde{X}) = 0$, so the map $X \to X^+$ induces an isomorphism on homology.

This construction $X \to X^+$, killing a perfect subgroup of $\pi_1(X)$ while preserving homology, is known as the Quillen **plus construction**. In some of the main applications X is a $K(G, 1)$ where G has perfect commutator subgroup, so the map $X \to X^+$ abelianizes π_1 while preserving homology. The space X^+ need no longer be a $K(\pi, 1)$, and in fact its homotopy groups can be quite interesting. The most striking example is $G = \Sigma_\infty$, the infinite symmetric group consisting of permutations of $1, 2, \cdots$ fixing all but finitely many n's, with commutator subgroup the infinite alternating group A_∞, which is perfect. In this case a famous theorem of Barratt, Priddy, and Quillen says that the homotopy groups $\pi_i(K(\Sigma_\infty, 1)^+)$ are the stable homotopy groups of spheres!

There are limits, however, on which subgroups of $\pi_1(X)$ can be killed without affecting the homology of X. For example, for $X = S^1 \vee S^1$ it is impossible to kill the commutator subgroup of $\pi_1(X)$ while preserving homology. In fact, by Exercise 23 at the end of this section every space with fundamental group $\mathbb{Z} \times \mathbb{Z}$ must have H_2 nontrivial.

Fiber Bundles

A 'short exact sequence of spaces' $A \hookrightarrow X \to X/A$ gives rise to a long exact sequence of homology groups, but not to a long exact sequence of homotopy groups due to the failure of excision. However, there is a different sort of 'short exact sequence of spaces' that does give a long exact sequence of homotopy groups. This sort of short exact sequence $F \to E \xrightarrow{p} B$, called a *fiber bundle*, is distinguished from the type $A \hookrightarrow X \to X/A$ in that it has more homogeneity: All the subspaces $p^{-1}(b) \subset E$, which are called **fibers**, are homeomorphic. For example, E could be the product $F \times B$ with $p : E \to B$ the projection. General fiber bundles can be thought of as twisted products. Familiar examples are the Möbius band, which is a twisted annulus with line segments as fibers, and the Klein bottle, which is a twisted torus with circles as fibers.

The topological homogeneity of all the fibers of a fiber bundle is rather like the algebraic homogeneity in a short exact sequence of groups $0 \to K \to G \xrightarrow{p} H \to 0$ where the 'fibers' $p^{-1}(h)$ are the cosets of K in G. In a few fiber bundles $F \to E \to B$ the space E is actually a group, F is a subgroup (though seldom a normal subgroup), and B is the space of left or right cosets. One of the nicest such examples is the Hopf bundle $S^1 \to S^3 \to S^2$ where S^3 is the group of quaternions of unit norm and S^1 is the subgroup of unit complex numbers. For this bundle, the long exact sequence of homotopy groups takes the form

$$\cdots \to \pi_i(S^1) \to \pi_i(S^3) \to \pi_i(S^2) \to \pi_{i-1}(S^1) \to \pi_{i-1}(S^3) \to \cdots$$

In particular, the exact sequence gives an isomorphism $\pi_2(S^2) \approx \pi_1(S^1)$ since the two adjacent terms $\pi_2(S^3)$ and $\pi_1(S^3)$ are zero by cellular approximation. Thus we have a direct homotopy-theoretic proof that $\pi_2(S^2) \approx \mathbb{Z}$. Also, since $\pi_i(S^1) = 0$ for $i > 1$ by Proposition 4.1, the exact sequence implies that there are isomorphisms $\pi_i(S^3) \approx \pi_i(S^2)$ for all $i \geq 3$, so in particular $\pi_3(S^2) \approx \pi_3(S^3)$, and by Corollary 4.25 the latter group is $\mathbb{Z}$.

After these preliminary remarks, let us begin by defining the property that leads to a long exact sequence of homotopy groups. A map $p : E \to B$ is said to have the **homotopy lifting property** with respect to a space X if, given a homotopy $g_t : X \to B$ and a map $\tilde{g}_0 : X \to E$ lifting g_0, so $p\tilde{g}_0 = g_0$, then there exists a homotopy $\tilde{g}_t : X \to E$ lifting g_t. From a formal point of view, this can be regarded as a special case of the *lift extension property for a pair* (Z, A), which asserts that every map $Z \to B$ has a lift $Z \to E$ extending a given lift defined on the subspace $A \subset Z$. The case $(Z, A) = (X \times I, X \times \{0\})$ is the homotopy lifting property.

A **fibration** is a map $p : E \to B$ having the homotopy lifting property with respect to all spaces X. For example, a projection $B \times F \to B$ is a fibration since we can choose lifts of the form $\tilde{g}_t(x) = (g_t(x), h(x))$ where $\tilde{g}_0(x) = (g_0(x), h(x))$.

Theorem 4.41. *Suppose $p : E \to B$ has the homotopy lifting property with respect to disks D^k for all $k \geq 0$. Choose basepoints $b_0 \in B$ and $x_0 \in F = p^{-1}(b_0)$. Then the map $p_* : \pi_n(E, F, x_0) \to \pi_n(B, b_0)$ is an isomorphism for all $n \geq 1$. Hence if B is path-connected, there is a long exact sequence*

$$\cdots \to \pi_n(F, x_0) \to \pi_n(E, x_0) \xrightarrow{p_*} \pi_n(B, b_0) \to \pi_{n-1}(F, x_0) \to \cdots \to \pi_0(E, x_0) \to 0$$

The proof will use a relative form of the homotopy lifting property. The map $p : E \to B$ is said to have the **homotopy lifting property for a pair** (X, A) if each homotopy $f_t : X \to B$ lifts to a homotopy $\tilde{g}_t : X \to E$ starting with a given lift $\tilde{g}_0$ and extending a given lift $\tilde{g}_t : A \to E$. In other words, the homotopy lifting property for (X, A) is the lift extension property for $(X \times I, X \times \{0\} \cup A \times I)$.

The homotopy lifting property for D^k is equivalent to the homotopy lifting property for $(D^k, \partial D^k)$ since the pairs $(D^k \times I, D^k \times \{0\})$ and $(D^k \times I, D^k \times \{0\} \cup \partial D^k \times I)$ are homeomorphic. This implies that the homotopy lifting property for disks is equivalent to the homotopy lifting property for all CW pairs (X, A). For by induction over the skeleta of X it suffices to construct a lifting $\tilde{g}_t$ one cell of $X - A$ at a time. Composing with the characteristic map $\Phi : D^k \to X$ of a cell then gives a reduction to the case $(X, A) = (D^k, \partial D^k)$. A map $p : E \to B$ satisfying the homotopy lifting property for disks is sometimes called a *Serre fibration*.

Proof: First we show that p_* is onto. Represent an element of $\pi_n(B, b_0)$ by a map $f : (I^n, \partial I^n) \to (B, b_0)$. The constant map to x_0 provides a lift of f to E over the subspace $J^{n-1} \subset I^n$, so the relative homotopy lifting property for $(I^{n-1}, \partial I^{n-1})$ extends this to a lift $\tilde{f} : I^n \to E$, and this lift satisfies $\tilde{f}(\partial I^n) \subset F$ since $f(\partial I^n) = b_0$. Then $\tilde{f}$ represents an element of $\pi_n(E, F, x_0)$ with $p_*([\tilde{f}]) = [f]$ since $p\tilde{f} = f$.

Injectivity of p_* is similar. Given $\tilde{f}_0, \tilde{f}_1 : (I^n, \partial I^n, J^{n-1}) \to (E, F, x_0)$ such that $p_*([\tilde{f}_0]) = p_*([\tilde{f}_1])$, let $G : (I^n \times I, \partial I^n \times I) \to (B, b_0)$ be a homotopy from $p\tilde{f}_0$ to $p\tilde{f}_1$. We have a partial lift $\tilde{G}$ given by $\tilde{f}_0$ on $I^n \times \{0\}$, $\tilde{f}_1$ on $I^n \times \{1\}$, and the constant map to x_0 on $J^{n-1} \times I$. After permuting the last two coordinates of $I^n \times I$, the relative homotopy lifting property gives an extension of this partial lift to a full lift $\tilde{G} : I^n \times I \to E$. This is a homotopy $\tilde{f}_t : (I^n, \partial I^n, J^{n-1}) \to (E, F, x_0)$ from $\tilde{f}_0$ to $\tilde{f}_1$. So p_* is injective.

For the last statement of the theorem we plug $\pi_n(B, b_0)$ in for $\pi_n(E, F, x_0)$ in the long exact sequence for the pair (E, F). The map $\pi_n(E, x_0) \to \pi_n(E, F, x_0)$ in the exact sequence then becomes the composition $\pi_n(E, x_0) \to \pi_n(E, F, x_0) \xrightarrow{p_*} \pi_n(B, b_0)$, which is just $p_* : \pi_n(E, x_0) \to \pi_n(B, b_0)$. The 0 at the end of the sequence, surjectivity of $\pi_0(F, x_0) \to \pi_0(E, x_0)$, comes from the hypothesis that B is path-connected since a path in E from an arbitrary point $x \in E$ to F can be obtained by lifting a path in B from $p(x)$ to b_0. $\qquad \square$

A **fiber bundle** structure on a space E, with fiber F, consists of a projection map $p : E \to B$ such that each point of B has a neighborhood U for which there is a

homeomorphism $h : p^{-1}(U) \to U \times F$ making the diagram at
the right commute, where the unlabeled map is projection
onto the first factor. Commutativity of the diagram means
that h carries each fiber $F_b = p^{-1}(b)$ homeomorphically

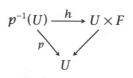

onto the copy $\{b\} \times F$ of F. Thus the fibers F_b are arranged locally as in the product
$B \times F$, though not necessarily globally. An h as above is called a **local trivialization**
of the bundle. Since the first coordinate of h is just p, h is determined by its second
coordinate, a map $p^{-1}(U) \to F$ which is a homeomorphism on each fiber F_b.

The fiber bundle structure is determined by the projection map $p : E \to B$, but to
indicate what the fiber is we sometimes write a fiber bundle as $F \to E \to B$, a 'short
exact sequence of spaces.' The space B is called the **base space** of the bundle, and E
is the **total space**.

Example 4.42. A fiber bundle with fiber a discrete space is a covering space. Conversely, a covering space whose fibers all have the same cardinality, for example a
covering space over a connected base space, is a fiber bundle with discrete fiber.

Example 4.43. One of the simplest nontrivial fiber bundles is the Möbius band, which
is a bundle over S^1 with fiber an interval. Specifically, take E to be the quotient of
$I \times [-1, 1]$ under the identifications $(0, v) \sim (1, -v)$, with $p : E \to S^1$ induced by the
projection $I \times [-1, 1] \to I$, so the fiber is $[-1, 1]$. Glueing two copies of E together
by the identity map between their boundary circles produces a Klein bottle, a bundle
over S^1 with fiber S^1.

Example 4.44. Projective spaces yield interesting fiber bundles. In the real case we
have the familiar covering spaces $S^n \to \mathbb{R}P^n$ with fiber S^0. Over the complex numbers the analog of this is a fiber bundle $S^1 \to S^{2n+1} \to \mathbb{C}P^n$. Here S^{2n+1} is the unit
sphere in $\mathbb{C}^{n+1}$ and $\mathbb{C}P^n$ is viewed as the quotient space of S^{2n+1} under the equivalence relation $(z_0, \cdots, z_n) \sim \lambda(z_0, \cdots, z_n)$ for $\lambda \in S^1$, the unit circle in $\mathbb{C}$. The
projection $p : S^{2n+1} \to \mathbb{C}P^n$ sends $(z_0, \cdots, z_n)$ to its equivalence class $[z_0, \cdots, z_n]$,
so the fibers are copies of S^1. To see that the local triviality condition for fiber bundles is satisfied, let $U_i \subset \mathbb{C}P^n$ be the open set of equivalence classes $[z_0, \cdots, z_n]$
with $z_i \neq 0$. Define $h_i : p^{-1}(U_i) \to U_i \times S^1$ by $h_i(z_0, \cdots, z_n) = ([z_0, \cdots, z_n], z_i/|z_i|)$.
This takes fibers to fibers, and is a homeomorphism since its inverse is the map
$([z_0, \cdots, z_n], \lambda) \mapsto \lambda |z_i| z_i^{-1}(z_0, \cdots, z_n)$, as one checks by calculation.

The construction of the bundle $S^1 \to S^{2n+1} \to \mathbb{C}P^n$ also works when $n = \infty$, so
there is a fiber bundle $S^1 \to S^\infty \to \mathbb{C}P^\infty$.

Example 4.45. The case $n = 1$ is particularly interesting since $\mathbb{C}P^1 = S^2$ and the
bundle becomes $S^1 \to S^3 \to S^2$ with fiber, total space, and base all spheres. This is
known as the **Hopf bundle**, and is of low enough dimension to be seen explicitly. The
projection $S^3 \to S^2$ can be taken to be $(z_0, z_1) \mapsto z_0/z_1 \in \mathbb{C} \cup \{\infty\} = S^2$. In polar
coordinates we have $p(r_0 e^{i\theta_0}, r_1 e^{i\theta_1}) = (r_0/r_1)e^{i(\theta_0 - \theta_1)}$ where $r_0^2 + r_1^2 = 1$. For a

fixed ratio $\rho = r_0/r_1 \in (0, \infty)$ the angles θ_0 and θ_1 vary independently over S^1, so the points $(r_0 e^{i\theta_0}, r_1 e^{i\theta_1})$ form a torus $T_\rho \subset S^3$. Letting ρ vary, these disjoint tori T_ρ fill up S^3, if we include the limiting cases T_0 and T_∞ where the radii r_0 and r_1 are zero, making the tori T_0 and T_∞ degenerate to circles. These two circles are the unit circles in the two $\mathbb{C}$ factors of $\mathbb{C}^2$, so under stereographic projection of S^3 from the point $(0, 1)$ onto $\mathbb{R}^3$ they correspond to the unit circle in the xy-plane and the z-axis. The concentric tori T_ρ are then arranged as in the following figure.

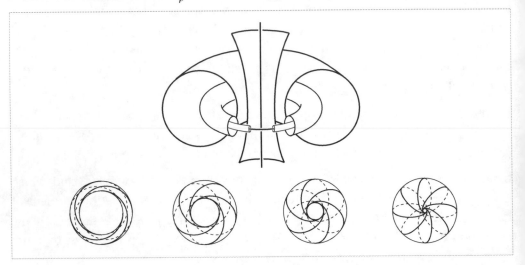

Each torus T_ρ is a union of circle fibers, the pairs (θ_0, θ_1) with $\theta_0 - \theta_1$ constant. These fiber circles have slope 1 on the torus, winding around once longitudinally and once meridionally. With respect to the ambient space it might be more accurate to say they have slope ρ. As ρ goes to 0 or ∞ the fiber circles approach the circles T_0 and T_∞, which are also fibers. The figure shows four of the tori decomposed into fibers.

Example 4.46. Replacing the field $\mathbb{C}$ by the quaternions $\mathbb{H}$, the same constructions yield fiber bundles $S^3 \to S^{4n+3} \to \mathbb{HP}^n$ over quaternionic projective spaces $\mathbb{HP}^n$. Here the fiber S^3 is the unit quaternions, and S^{4n+3} is the unit sphere in $\mathbb{H}^{n+1}$. Taking $n = 1$ gives a second Hopf bundle $S^3 \to S^7 \to S^4 = \mathbb{HP}^1$.

Example 4.47. Another Hopf bundle $S^7 \to S^{15} \to S^8$ can be defined using the octonion algebra $\mathbb{O}$. Elements of $\mathbb{O}$ are pairs of quaternions (a_1, a_2) with multiplication given by $(a_1, a_2)(b_1, b_2) = (a_1 b_1 - \overline{b}_2 a_2, a_2 \overline{b}_1 + b_2 a_1)$. Regarding S^{15} as the unit sphere in the 16-dimensional vector space $\mathbb{O}^2$, the projection map $p : S^{15} \to S^8 = \mathbb{O} \cup \{\infty\}$ is $(z_0, z_1) \mapsto z_0 z_1^{-1}$, just as for the other Hopf bundles, but because $\mathbb{O}$ is not associative, a little care is needed to show this is a fiber bundle with fiber S^7, the unit octonions. Let U_0 and U_1 be the complements of ∞ and 0 in the base space $\mathbb{O} \cup \{\infty\}$. Define $h_i : p^{-1}(U_i) \to U_i \times S^7$ and $g_i : U_i \times S^7 \to p^{-1}(U_i)$ by

$$h_0(z_0, z_1) = (z_0 z_1^{-1}, z_1/|z_1|), \qquad g_0(z, w) = (zw, w)/|(zw, w)|$$
$$h_1(z_0, z_1) = (z_0 z_1^{-1}, z_0/|z_0|), \qquad g_1(z, w) = (w, z^{-1} w)/|(w, z^{-1} w)|$$

If one assumes the known fact that any subalgebra of $\mathbb{O}$ generated by two elements is associative, then it is a simple matter to check that g_i and h_i are inverse homeomorphisms, so we have a fiber bundle $S^7 \to S^{15} \to S^8$. Actually, the calculation that g_i and h_i are inverses needs only the following more elementary facts about octonions z, w, where the conjugate $\bar{z}$ of $z = (a_1, a_2)$ is defined by the expected formula $\bar{z} = (\bar{a}_1, -a_2)$:

(1) $rz = zr$ for all $r \in \mathbb{R}$ and $z \in \mathbb{O}$, where $\mathbb{R} \subset \mathbb{O}$ as the pairs $(r, 0)$.

(2) $|z|^2 = z\bar{z} = \bar{z}z$, hence $z^{-1} = \bar{z}/|z|^2$.

(3) $|zw| = |z||w|$.

(4) $\overline{zw} = \bar{w}\,\bar{z}$, hence $(zw)^{-1} = w^{-1}z^{-1}$.

(5) $z(\bar{z}w) = (z\bar{z})w$ and $(zw)\bar{w} = z(w\bar{w})$, hence $z(z^{-1}w) = w$ and $(zw)w^{-1} = z$.

These facts can be checked by somewhat tedious direct calculation. More elegant derivations can be found in Chapter 8 of [Ebbinghaus 1991].

There is an octonion projective plane $\mathbb{O}P^2$ obtained by attaching a cell e^{16} to S^8 via the Hopf map $S^{15} \to S^8$, just as $\mathbb{C}P^2$ and $\mathbb{H}P^2$ are obtained from the other Hopf maps. However, there is no octonion analog of $\mathbb{R}P^n$, $\mathbb{C}P^n$, and $\mathbb{H}P^n$ for $n > 2$ since associativity of multiplication is needed for the relation $(z_0, \cdots, z_n) \sim \lambda(z_0, \cdots, z_n)$ to be an equivalence relation.

There are no fiber bundles with fiber, total space, and base space spheres of other dimensions than in these Hopf bundle examples. This is discussed in an exercise for §4.D, which reduces the question to the famous 'Hopf invariant one' problem.

Proposition 4.48. *A fiber bundle $p : E \to B$ has the homotopy lifting property with respect to all CW pairs (X, A).*

A theorem of Huebsch and Hurewicz proved in §2.7 of [Spanier 1966] says that fiber bundles over paracompact base spaces are fibrations, having the homotopy lifting property with respect to all spaces. This stronger result is not often needed in algebraic topology, however.

Proof: As noted earlier, the homotopy lifting property for CW pairs is equivalent to the homotopy lifting property for disks, or equivalently, cubes. Let $G : I^n \times I \to B$, $G(x, t) = g_t(x)$, be a homotopy we wish to lift, starting with a given lift $\tilde{g}_0$ of g_0. Choose an open cover $\{U_\alpha\}$ of B with local trivializations $h_\alpha : p^{-1}(U_\alpha) \to U_\alpha \times F$. Using compactness of $I^n \times I$, we may subdivide I^n into small cubes C and I into intervals $I_j = [t_j, t_{j+1}]$ so that each product $C \times I_j$ is mapped by G into a single U_α. We may assume by induction on n that $\tilde{g}_t$ has already been constructed over ∂C for each of the subcubes C. To extend this $\tilde{g}_t$ over a cube C we may proceed in stages, constructing $\tilde{g}_t$ for t in each successive interval I_j. This in effect reduces us to the case that no subdivision of $I^n \times I$ is necessary, so G maps all of $I^n \times I$ to a single U_α. Then we have $\tilde{G}(I^n \times \{0\} \cup \partial I^n \times I) \subset p^{-1}(U_\alpha)$, and composing $\tilde{G}$ with the local trivialization

h_α reduces us to the case of a product bundle $U_\alpha \times F$. In this case the first coordinate of a lift $\tilde{g}_t$ is just the given g_t, so only the second coordinate needs to be constructed. This can be obtained as a composition $I^n \times I \to I^n \times \{0\} \cup \partial I^n \times I \to F$ where the first map is a retraction and the second map is what we are given. $\square$

Example 4.49. Applying this theorem to a covering space $p : E \to B$ with E and B path-connected, and discrete fiber F, the resulting long exact sequence of homotopy groups yields Proposition 4.1 that $p_* : \pi_n(E) \to \pi_n(B)$ is an isomorphism for $n \geq 2$. We also obtain a short exact sequence $0 \to \pi_1(E) \to \pi_1(B) \to \pi_0(F) \to 0$, consistent with the covering space theory facts that $p_* : \pi_1(E) \to \pi_1(B)$ is injective and that the fiber F can be identified, via path-lifting, with the set of cosets of $p_* \pi_1(E)$ in $\pi_1(B)$.

Example 4.50. From the bundle $S^1 \to S^\infty \to \mathbb{CP}^\infty$ we obtain $\pi_i(\mathbb{CP}^\infty) \approx \pi_{i-1}(S^1)$ for all i since S^∞ is contractible. Thus $\mathbb{CP}^\infty$ is a $K(\mathbb{Z}, 2)$. In similar fashion the bundle $S^3 \to S^\infty \to \mathbb{HP}^\infty$ gives $\pi_i(\mathbb{HP}^\infty) \approx \pi_{i-1}(S^3)$ for all i, but these homotopy groups are far more complicated than for $\mathbb{CP}^\infty$ and S^1. In particular, $\mathbb{HP}^\infty$ is not a $K(\mathbb{Z}, 4)$.

Example 4.51. The long exact sequence for the Hopf bundle $S^1 \to S^3 \to S^2$ gives isomorphisms $\pi_2(S^2) \approx \pi_1(S^1)$ and $\pi_n(S^3) \approx \pi_n(S^2)$ for all $n \geq 3$. Taking $n = 3$, we see that $\pi_3(S^2)$ is infinite cyclic, generated by the Hopf map $S^3 \to S^2$.

From this example and the preceding one we see that S^2 and $S^3 \times \mathbb{CP}^\infty$ are simply-connected CW complexes with isomorphic homotopy groups, though they are not homotopy equivalent since they have quite different homology groups.

Example 4.52: Whitehead Products. Let us compute $\pi_3(\bigvee_\alpha S^2_\alpha)$, showing that it is free abelian with basis consisting of the Hopf maps $S^3 \to S^2_\alpha \subset \bigvee_\alpha S^2_\alpha$ together with the attaching maps $S^3 \to S^2_\alpha \vee S^2_\beta \subset \bigvee_\alpha S^2_\alpha$ of the cells $e^2_\alpha \times e^2_\beta$ in the products $S^2_\alpha \times S^2_\beta$ for all unordered pairs $\alpha \neq \beta$.

Suppose first that there are only finitely many summands S^2_α. For a finite product $\prod_\alpha X_\alpha$ of path-connected spaces, the map $\pi_n(\bigvee_\alpha X_\alpha) \to \pi_n(\prod_\alpha X_\alpha)$ induced by inclusion is surjective since the group $\pi_n(\prod_\alpha X_\alpha) \approx \bigoplus_\alpha \pi_n(X_\alpha)$ is generated by the subgroups $\pi_n(X_\alpha)$. Thus the long exact sequence of homotopy groups for the pair $(\prod_\alpha X_\alpha, \bigvee_\alpha X_\alpha)$ breaks up into short exact sequences

$$0 \longrightarrow \pi_{n+1}(\prod_\alpha X_\alpha, \bigvee_\alpha X_\alpha) \longrightarrow \pi_n(\bigvee_\alpha X_\alpha) \longrightarrow \pi_n(\prod_\alpha X_\alpha) \longrightarrow 0$$

These short exact sequences split since the inclusions $X_\alpha \hookrightarrow \bigvee_\alpha X_\alpha$ induce maps $\pi_n(X_\alpha) \to \pi_n(\bigvee_\alpha X_\alpha)$ and hence a splitting homomorphism $\bigoplus_\alpha \pi_n(X_\alpha) \to \pi_n(\bigvee_\alpha X_\alpha)$. Taking $X_\alpha = S^2_\alpha$ and $n = 3$, we get an isomorphism

$$\pi_3(\bigvee_\alpha S^2_\alpha) \approx \pi_4(\prod_\alpha S^2_\alpha, \bigvee_\alpha S^2_\alpha) \oplus (\bigoplus_\alpha \pi_3(S^2_\alpha))$$

The factor $\bigoplus_\alpha \pi_3(S^2_\alpha)$ is free with basis the Hopf maps $S^3 \to S^2_\alpha$ by the preceding example. For the other factor we have $\pi_4(\prod_\alpha S^2_\alpha, \bigvee_\alpha S^2_\alpha) \approx \pi_4(\prod_\alpha S^2_\alpha / \bigvee_\alpha S^2_\alpha)$ by Proposition 4.28. The quotient $\prod_\alpha S^2_\alpha / \bigvee_\alpha S^2_\alpha$ has 5-skeleton a wedge of spheres $S^4_{\alpha\beta}$ for $\alpha \neq \beta$, so $\pi_4(\prod_\alpha S_\alpha^2 / \bigvee_\alpha S_\alpha^2) \approx \pi_4(\bigvee_{\alpha\beta} S_{\alpha\beta}^4)$ is free with basis the inclusions $S_{\alpha\beta}^4 \hookrightarrow \bigvee_{\alpha\beta} S_{\alpha\beta}^4$. Hence $\pi_4(\prod_\alpha S_\alpha^2, \bigvee_\alpha S_\alpha^2)$ is free with basis the characteristic maps of the 4-cells $e_\alpha^2 \times e_\beta^2$. Via the injection $\partial : \pi_4(\prod_\alpha S_\alpha^2, \bigvee_\alpha S_\alpha^2) \to \pi_3(\bigvee_\alpha S_\alpha^2)$ this means that the attaching maps of the cells $e_\alpha^2 \times e_\beta^2$ form a basis for the summand $\mathrm{Im}\,\partial$ of $\pi_3(\bigvee_\alpha S_\alpha^2)$. This finishes the proof for the case of finitely many summands S_α^2. The case of infinitely many S_α^2's follows immediately since any map $S^3 \to \bigvee_\alpha S_\alpha^2$ has compact image, lying in a finite union of summands, and similarly for any homotopy between such maps.

The maps $S^3 \to S_\alpha^2 \vee S_\beta^2$ in this example are expressible in terms of a product in homotopy groups called the **Whitehead product**, defined as follows. Given basepoint-preserving maps $f : S^k \to X$ and $g : S^\ell \to X$, let $[f, g] : S^{k+\ell-1} \to X$ be the composition $S^{k+\ell-1} \to S^k \vee S^\ell \xrightarrow{f \vee g} X$ where the first map is the attaching map of the $(k + \ell)$-cell of $S^k \times S^\ell$ with its usual CW structure. Since homotopies of f or g give rise to homotopies of $[f, g]$, we have a well-defined product $\pi_k(X) \times \pi_\ell(X) \to \pi_{k+\ell-1}(X)$. The notation $[f, g]$ is used since for $k = \ell = 1$ this is just the commutator product in $\pi_1(X)$. It is an exercise to show that when $k = 1$ and $\ell > 1$, $[f, g]$ is the difference between g and its image under the π_1-action of f.

In these terms the map $S^3 \to S_\alpha^2 \vee S_\beta^2$ in the preceding example is the Whitehead product $[i_\alpha, i_\beta]$ of the two inclusions of S^2 into $S_\alpha^2 \vee S_\beta^2$. Another example of a Whitehead product we have encountered previously is $[\mathbb{1}, \mathbb{1}] : S^{2n-1} \to S^n$, which is the attaching map of the $2n$-cell of the space $J(S^n)$ considered in §3.2.

The calculation of $\pi_3(\bigvee_\alpha S_\alpha^2)$ is the first nontrivial case of a more general theorem of Hilton calculating all the homotopy groups of any wedge sum of spheres in terms of homotopy groups of spheres, using Whitehead products. A further generalization by Milnor extends this to wedge sums of suspensions of arbitrary connected CW complexes. See [Whitehead 1978] for an exposition of these results and further information on Whitehead products.

Example 4.53: Stiefel and Grassmann Manifolds. The fiber bundles with total space a sphere and base space a projective space considered above are the cases $n = 1$ of families of fiber bundles in each of the real, complex, and quaternionic cases:

$$
\begin{array}{ll}
O(n) \to V_n(\mathbb{R}^k) \to G_n(\mathbb{R}^k) & \qquad O(n) \to V_n(\mathbb{R}^\infty) \to G_n(\mathbb{R}^\infty) \\
U(n) \to V_n(\mathbb{C}^k) \to G_n(\mathbb{C}^k) & \qquad U(n) \to V_n(\mathbb{C}^\infty) \to G_n(\mathbb{C}^\infty) \\
Sp(n) \to V_n(\mathbb{H}^k) \to G_n(\mathbb{H}^k) & \qquad Sp(n) \to V_n(\mathbb{H}^\infty) \to G_n(\mathbb{H}^\infty)
\end{array}
$$

Taking the real case first, the Stiefel manifold $V_n(\mathbb{R}^k)$ is the space of n-frames in $\mathbb{R}^k$, that is, n-tuples of orthonormal vectors in $\mathbb{R}^k$. This is topologized as a subspace of the product of n copies of the unit sphere in $\mathbb{R}^k$. The Grassmann manifold $G_n(\mathbb{R}^k)$ is the space of n-dimensional vector subspaces of $\mathbb{R}^k$. There is a natural surjection $p : V_n(\mathbb{R}^k) \to G_n(\mathbb{R}^k)$ sending an n-frame to the subspace it spans, and $G_n(\mathbb{R}^k)$ is topologized as a quotient space of $V_n(\mathbb{R}^k)$ via this projection. The fibers of the map

p are the spaces of n-frames in a fixed n-plane in $\mathbb{R}^k$ and so are homeomorphic to $V_n(\mathbb{R}^n)$. An n-frame in $\mathbb{R}^n$ is the same as an orthogonal $n \times n$ matrix, regarding the columns of the matrix as an n-frame, so the fiber can also be described as the orthogonal group $O(n)$. There is no difficulty in allowing $k = \infty$ in these definitions, and in fact $V_n(\mathbb{R}^\infty) = \bigcup_k V_n(\mathbb{R}^k)$ and $G_n(\mathbb{R}^\infty) = \bigcup_k G_n(\mathbb{R}^k)$.

The complex and quaternionic Stiefel manifolds and Grassmann manifolds are defined in the same way using the usual Hermitian inner products in $\mathbb{C}^k$ and $\mathbb{H}^k$. The unitary group $U(n)$ consists of $n \times n$ matrices whose columns form orthonormal bases for $\mathbb{C}^n$, and the symplectic group $Sp(n)$ is the quaternionic analog of this.

We should explain why the various projection maps $V_n \to G_n$ are fiber bundles. Let us take the real case for concreteness, though the argument is the same in all cases. If we fix an n-plane $P \in G_n(\mathbb{R}^k)$ and choose an orthonormal basis for P, then we obtain continuously varying orthonormal bases for all n-planes P' in a neighborhood U of P by projecting the basis for P orthogonally onto P' to obtain a nonorthonormal basis for P', then applying the Gram–Schmidt process to this basis to make it orthonormal. The formulas for the Gram–Schmidt process show that it is continuous. Having orthonormal bases for all n-planes in U, we can use these to identify these n-planes with $\mathbb{R}^n$, hence n-frames in these n-planes are identified with n-frames in $\mathbb{R}^n$, and so $p^{-1}(U)$ is identified with $U \times V_n(\mathbb{R}^n)$. This argument works for $k = \infty$ as well as for finite k.

In the case $n = 1$ the total spaces V_1 are spheres, which are highly connected, and the same is true in general:

- $V_n(\mathbb{R}^k)$ is $(k - n - 1)$-connected.
- $V_n(\mathbb{C}^k)$ is $(2k - 2n)$-connected.
- $V_n(\mathbb{H}^k)$ is $(4k - 4n + 2)$-connected.
- $V_n(\mathbb{R}^\infty)$, $V_n(\mathbb{C}^\infty)$, and $V_n(\mathbb{H}^\infty)$ are contractible.

The first three statements will be proved in the next example. For the last statement the argument is the same in the three cases, so let us consider the real case. Define a homotopy $h_t : \mathbb{R}^\infty \to \mathbb{R}^\infty$ by $h_t(x_1, x_2, \cdots) = (1 - t)(x_1, x_2, \cdots) + t(0, x_1, x_2, \cdots)$. This is linear for each t, and its kernel is easily checked to be trivial. So if we apply h_t to an n-frame we get an n-tuple of independent vectors, which can be made orthonormal by the Gram–Schmidt formulas. Thus we have a deformation retraction, in the weak sense, of $V_n(\mathbb{R}^\infty)$ onto the subspace of n-frames with first coordinate zero. Iterating this n times, we deform into the subspace of n-frames with first n coordinates zero. For such an n-frame $(v_1, \cdots, v_n)$ define a homotopy $(1 - t)(v_1, \cdots, v_n) + t(e_1, \cdots, e_n)$ where e_i is the ith standard basis vector in $\mathbb{R}^\infty$. This homotopy preserves linear independence, so after again applying Gram–Schmidt we have a deformation through n-frames, which finishes the construction of a contraction of $V_n(\mathbb{R}^\infty)$.

Since $V_n(\mathbb{R}^\infty)$ is contractible, we obtain isomorphisms $\pi_i O(n) \approx \pi_{i+1} G_n(\mathbb{R}^\infty)$ for all i and n, and similarly in the complex and quaternionic cases.

Example 4.54. For $m < n \leq k$ there are fiber bundles

$$V_{n-m}(\mathbb{R}^{k-m}) \longrightarrow V_n(\mathbb{R}^k) \xrightarrow{p} V_m(\mathbb{R}^k)$$

where the projection p sends an n-frame onto the m-frame formed by its first m vectors, so the fiber consists of $(n - m)$-frames in the $(k - m)$-plane orthogonal to a given m-frame. Local trivializations can be constructed as follows. For an m-frame F, choose an orthonormal basis for the $(k - m)$-plane orthogonal to F. This determines orthonormal bases for the $(k - m)$-planes orthogonal to all nearby m-frames by orthogonal projection and Gram–Schmidt, as in the preceding example. This allows us to identify these $(k-m)$-planes with $\mathbb{R}^{k-m}$, and in particular the fibers near $p^{-1}(F)$ are identified with $V_{n-m}(\mathbb{R}^{k-m})$, giving a local trivialization.

There are analogous bundles in the complex and quaternionic cases as well, with local triviality shown in the same way.

Restricting to the case $m = 1$, we have bundles $V_{n-1}(\mathbb{R}^{k-1}) \to V_n(\mathbb{R}^k) \to S^{k-1}$ whose associated long exact sequence of homotopy groups allows us to deduce that $V_n(\mathbb{R}^k)$ is $(k - n - 1)$-connected by induction on n. In the complex and quaternionic cases the same argument yields the other connectivity statements in the preceding example.

Taking $k - n$ we obtain fiber bundles $O(k - m) \to O(k) \to V_m(\mathbb{R}^k)$. The fibers are in fact just the cosets $\alpha O(k - m)$ for $\alpha \in O(k)$, where $O(k - m)$ is regarded as the subgroup of $O(k)$ fixing the first m standard basis vectors. So we see that $V_m(\mathbb{R}^k)$ is identifiable with the coset space $O(k)/O(k - m)$, or in other words the orbit space for the free action of $O(k-m)$ on $O(k)$ by right-multiplication. In similar fashion one can see that $G_m(\mathbb{R}^k)$ is the coset space $O(k)/(O(m) \times O(k - m))$ where the subgroup $O(m) \times O(k - m) \subset O(k)$ consists of the orthogonal transformations taking the m-plane spanned by the first m standard basis vectors to itself. The corresponding observations apply also in the complex and quaternionic cases, with the unitary and symplectic groups.

Example 4.55: Bott Periodicity. Specializing the preceding example by taking $m = 1$ and $k = n$ we obtain bundles

$$O(n - 1) \longrightarrow O(n) \xrightarrow{p} S^{n-1}$$
$$U(n - 1) \longrightarrow U(n) \xrightarrow{p} S^{2n-1}$$
$$Sp(n - 1) \longrightarrow Sp(n) \xrightarrow{p} S^{4n-1}$$

The map p can be described as evaluation of an orthogonal, unitary, or symplectic transformation on a fixed unit vector. These bundles show that computing homotopy groups of $O(n)$, $U(n)$, and $Sp(n)$ should be at least as difficult as computing homotopy groups of spheres. For example, if one knew the homotopy groups of $O(n)$ and $O(n - 1)$, then from the long exact sequence of homotopy groups for the first bundle one could say quite a bit about the homotopy groups of S^{n-1}.

The bundles above imply a very interesting stability property. In the real case, the inclusion $O(n-1) \hookrightarrow O(n)$ induces an isomorphism on π_i for $i < n-2$, from the long exact sequence of the first bundle. Hence the groups $\pi_i O(n)$ are independent of n if n is sufficiently large, and the same is true for the groups $\pi_i U(n)$ and $\pi_i Sp(n)$ via the other two bundles. One of the most surprising results in all of algebraic topology is the Bott Periodicity Theorem which asserts that these stable groups repeat periodically, with a period of eight for O and Sp and a period of two for U. Their values are given in the following table:

$i \bmod 8$	0	1	2	3	4	5	6	7
$\pi_i O(n)$	$\mathbb{Z}_2$	$\mathbb{Z}_2$	0	$\mathbb{Z}$	0	0	0	$\mathbb{Z}$
$\pi_i U(n)$	0	$\mathbb{Z}$	0	$\mathbb{Z}$	0	$\mathbb{Z}$	0	$\mathbb{Z}$
$\pi_i Sp(n)$	0	0	0	$\mathbb{Z}$	$\mathbb{Z}_2$	$\mathbb{Z}_2$	0	$\mathbb{Z}$

Stable Homotopy Groups

We showed in Corollary 4.24 that for an n-connected CW complex X, the suspension map $\pi_i(X) \to \pi_{i+1}(SX)$ is an isomorphism for $i < 2n+1$. In particular this holds for $i \le n$ so SX is $(n+1)$-connected. This implies that in the sequence of iterated suspensions

$$\pi_i(X) \longrightarrow \pi_{i+1}(SX) \longrightarrow \pi_{i+2}(S^2 X) \longrightarrow \cdots$$

all maps are eventually isomorphisms, even without any connectivity assumption on X itself. The resulting **stable homotopy group** is denoted $\pi_i^s(X)$.

An especially interesting case is the group $\pi_i^s(S^0)$, which equals $\pi_{i+n}(S^n)$ for $n > i + 1$. This stable homotopy group is often abbreviated to π_i^s and called the **stable i-stem**. It is a theorem of Serre which we prove in [SSAT] that π_i^s is always finite for $i > 0$.

These stable homotopy groups of spheres are among the most fundamental objects in topology, and much effort has gone into their calculation. At the present time, complete calculations are known only for i up to around 60 or so. Here is a table for $i \le 19$, taken from [Toda 1962]:

i	0	1	2	3	4	5	6	7	8	9	10	11	12
π_i^s	$\mathbb{Z}$	$\mathbb{Z}_2$	$\mathbb{Z}_2$	$\mathbb{Z}_{24}$	0	0	$\mathbb{Z}_2$	$\mathbb{Z}_{240}$	$\mathbb{Z}_2 \times \mathbb{Z}_2$	$\mathbb{Z}_2 \times \mathbb{Z}_2 \times \mathbb{Z}_2$	$\mathbb{Z}_6$	$\mathbb{Z}_{504}$	0

13	14	15	16	17	18	19
$\mathbb{Z}_3$	$\mathbb{Z}_2 \times \mathbb{Z}_2$	$\mathbb{Z}_{480} \times \mathbb{Z}_2$	$\mathbb{Z}_2 \times \mathbb{Z}_2$	$\mathbb{Z}_2 \times \mathbb{Z}_2 \times \mathbb{Z}_2 \times \mathbb{Z}_2$	$\mathbb{Z}_8 \times \mathbb{Z}_2$	$\mathbb{Z}_{264} \times \mathbb{Z}_2$

Patterns in this apparent chaos begin to emerge only when one projects π_i^s onto its **p-components,** the quotient groups obtained by factoring out all elements of order relatively prime to the prime p. For $i > 0$ the p-component $_p\pi_i^s$ is of course isomorphic to the subgroup of π_i^s consisting of elements of order a power of p, but the quotient viewpoint is in some ways preferable.

The figure below is a schematic diagram of the 2-components of π_i^s for $i \le 60$. A vertical chain of n dots in the ith column represents a $\mathbb{Z}_{2^n}$ summand of π_i^s. The bottom dot of such a chain denotes a generator of this summand, and the vertical segments denote multiplication by 2, so the second dot up is twice a generator, the next dot is four times a generator, and so on. The three generators η, ν, and σ in dimensions 1, 3, and 7 are represented by the Hopf bundle maps $S^3 \to S^2$, $S^7 \to S^4$, $S^{15} \to S^8$ defined in Examples 4.45, 4.46, and 4.47.

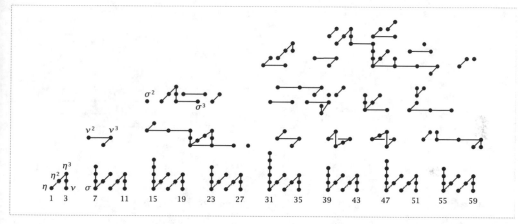

The horizontal and diagonal lines in the diagram provide some information about compositions of maps between spheres. Namely, there are products $\pi_i^s \times \pi_j^s \to \pi_{i+j}^s$ defined by compositions $S^{i+j+k} \to S^{j+k} \to S^k$.

Proposition 4.56. *The composition products $\pi_i^s \times \pi_j^s \to \pi_{i+j}^s$ induce a graded ring structure on $\pi_*^s = \bigoplus_i \pi_i^s$ satisfying the commutativity relation $\alpha\beta = (-1)^{ij}\beta\alpha$ for $\alpha \in \pi_i^s$ and $\beta \in \pi_j^s$.*

This will be proved at the end of this subsection. It follows that $_p\pi_*^s$, the direct sum of the p-components $_p\pi_i^s$, is also a graded ring satisfying the same commutativity property. In $_2\pi_i^s$ many of the compositions with suspensions of the Hopf maps η and ν are nontrivial, and these nontrivial compositions are indicated in the diagram by segments extending 1 or 3 units to the right, diagonally for η and horizontally for ν. Thus for example we see the relation $\eta^3 = 4\nu$ in $_2\pi_3^s$. Remember that $_2\pi_3^s \approx \mathbb{Z}_8$ is a quotient of $\pi_3^s \approx \mathbb{Z}_{24}$, where the actual relation is $\eta^3 = 12\nu$ since $2\eta = 0$ implies $2\eta^3 = 0$, so η^3 is the unique element of order two in this $\mathbb{Z}_{24}$.

Across the bottom of the diagram there is a repeated pattern of pairs of 'teeth.' This pattern continues to infinity, though with the spikes in dimensions $8k - 1$ not all of the same height, namely, the spike in dimension $2^m(2n + 1) - 1$ has height $m + 1$. In the upper part of the diagram, however, there is considerably less regularity, and this complexity seems to persist in higher dimensions as well.

The next diagram shows the 3-components of π_i^s for $i \le 100$, and the increase in regularity is quite noticeable. Here vertical segments denote multiplication by 3 and the other solid segments denote composition with elements $\alpha_1 \in {}_3\pi_3^s$ and $\beta_1 \in {}_3\pi_{10}^s$. The meaning of the dashed lines will be explained below. The most regular part of the diagram is the 'triadic ruler' across the bottom. This continues in the same pattern forever, with spikes of height $m + 1$ in dimension $4k - 1$ for 3^m the highest power of 3 dividing $4k$. Looking back at the $p = 2$ diagram, one can see that the vertical segments of the 'teeth' form a 'dyadic ruler.'

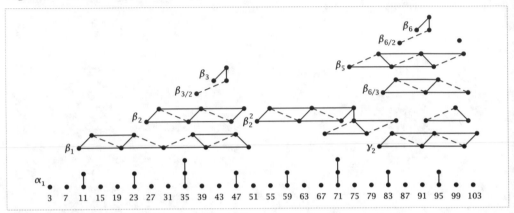

Even more regularity appears with larger primes, beginning with the case $p = 5$ shown in the next diagram. Again one has the infinite ruler, this time a 'pentadic' ruler, but there is also much regularity in the rest of the diagram. The four dots with question marks below them near the right edge of the diagram are hypothetical: The calculations in [Ravenel 1986] do not decide whether these potential elements of ${}_5\pi_i^s$ for $i = 932, 933, 970$, and 971 actually exist.

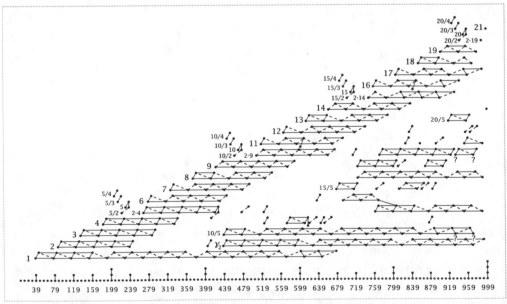

These three diagrams are drawn from tables published in [Kochman 1990] and [Kochman & Mahowald 1995] for $p = 2$ and [Ravenel 1986] for $p = 3, 5$.

For each p there is a similar infinite 'p-adic ruler,' corresponding to cyclic subgroups of order p^{m+1} in $_p\pi^s_{2j(p-1)-1}$ for all j, where p^m is the highest power of p dividing j. These subgroups are the p-components of a certain cyclic subgroup of π^s_{4k-1} known as Im J, the image of a homomorphism $J:\pi_{4k-1}(O)\to\pi^s_{4k-1}$. There are also $\mathbb{Z}_2$ subgroups of π^s_i for $i=8k,8k+1$ forming Im J in these dimensions. In the diagram of $_2\pi^s_*$ these are the parts of the teeth connected to the spike in dimension $8k-1$. The J-homomorphism will be studied in some detail in [VBKT].

There are a few other known infinite families in π^s_*, notably a family of elements $\beta_n\in{}_p\pi^s_{2(p^2-1)n-2p}$ for $p\ge 5$ and a family $\gamma_n\in{}_p\pi^s_{2(p^3-1)n-2p^2-2p+1}$ for $p\ge 7$. The element β_n appears in the diagram for $p=5$ as the element in the upper part of the diagram labeled by the number n. These β_n's generate the strips along the upward diagonal, except when n is a multiple of 5 and the strip is generated by $\beta_2\beta_{n-1}$ rather than β_n. There are also elements β_n for certain fractional values of n. The element γ_2 generates the long strip starting in dimension 437, but $\gamma_3=0$. The element γ_4 in dimension 933 is one of the question marks. The theory behind these families β_n and γ_n and possible generalizations, as explained in [Ravenel 1986 & 1992], is one of the more esoteric branches of algebraic topology.

In π^s_* there are many compositions which are zero. One can get some idea of this from the diagrams above, where all sequences of edges break off after a short time. As a special instance of the vanishing of products, the commutativity formula in Proposition 4.56 implies that the square of an odd-dimensional element of odd order is zero. More generally, a theorem of Nishida says that every positive-dimensional element $\alpha\in\pi^s_*$ is nilpotent, with $\alpha^n=0$ for some n. For example, for the element $\beta_1\in{}_5\pi^s_{38}$ the smallest such n is 18.

The widespread vanishing of products in π^s_* can be seen as limiting their usefulness in describing the structure of π^s_*. But it can also be used to construct new elements of π^s_*. Suppose one has maps $W\xrightarrow{f}X\xrightarrow{g}Y\xrightarrow{h}Z$ such that the compositions gf and hg are both homotopic to constant maps. A nullhomotopy of gf gives an extension of gf to a map $F:CW\to Y$, and a nullhomotopy of hg gives an extension of hg to a map $G:CX\to Z$. Regarding the suspension SW as the union of two cones CW, define the **Toda bracket** $\langle f,g,h\rangle:SW\to Z$ to be the composition $G(Cf)$ on one cone and hF on the other.

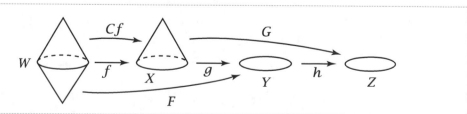

The map $\langle f,g,h\rangle$ is not uniquely determined by f, g, and h since it depends on the choices of the nullhomotopies. In the case of π^s_*, the various choices of $\langle f,g,h\rangle$

range over a coset of a certain subgroup, described in an exercise at the end of the section.

There are also higher-order Toda brackets $\langle f_1, \cdots, f_n \rangle$ defined in somewhat similar fashion. The dashed lines in the diagrams of $_3\pi_*^s$ and $_5\pi_*^s$ join an element x to a bracket element $\langle \alpha_1, \cdots, \alpha_1, x \rangle$. Most of the unlabeled elements above the rulers in all three diagrams are obtained from the labeled elements by compositions and brackets. For example, in $_2\pi_*^s$ the 8-dimensional element is $\langle \nu, \eta, \nu \rangle$ and the 14-dimensional elements are σ^2 and $\langle \nu, \langle \nu, \eta, \nu \rangle, 2, \eta \rangle$.

Proof of 4.56: Only distributivity and commutativity need to be checked. One distributivity law is easy: Given $f, g : S^{i+j+k} \to S^{j+k}$ and $h : S^{j+k} \to S^k$, then $h(f + g) = hf + hg$ since both expressions equal hf and hg on the two hemispheres of S^{i+j+k}. The other distributivity law will follow from this one and the commutativity relation.

To prove the commutativity relation it will be convenient to express suspension in terms of smash product. The smash product $S^n \wedge S^1$ can be regarded as the quotient space of $S^n \times I$ with $S^n \times \partial I \cup \{x_0\} \times I$ collapsed to a point. This is the same as the quotient of the suspension S^{n+1} of S^n obtained by collapsing to a point the suspension of x_0. Collapsing this arc in S^{n+1} to a point again yields S^{n+1}, so we obtain in this way a homeomorphism identifying $S^n \wedge S^1$ with S^{n+1}. Under this identification the suspension Sf of a basepoint-preserving map $f : S^n \to S^n$ becomes the smash product $f \wedge \mathbb{1} : S^n \wedge S^1 \to S^n \wedge S^1$. By iteration, the k-fold suspension $S^k f$ then corresponds to $f \wedge \mathbb{1} : S^n \wedge S^k \to S^n \wedge S^k$.

Now we verify the commutativity relation. Let $f : S^{i+k} \to S^k$ and $g : S^{j+k} \to S^k$ be given. We may assume k is even. Consider the commutative diagram below, where σ and τ transpose the two factors. Thinking of S^{j+k} and S^k as smash products of circles, σ is the composition of $k(j+k)$ transpositions of adjacent circle

$$
\begin{array}{ccccc}
S^{i+k} \wedge S^{j+k} & \xrightarrow{f \wedge \mathbb{1}} & S^k \wedge S^{j+k} & \xrightarrow{\mathbb{1} \wedge g} & S^k \wedge S^k \\
\downarrow \sigma & & & & \downarrow \tau \\
S^{j+k} \wedge S^k & & \xrightarrow{g \wedge \mathbb{1}} & & S^k \wedge S^k
\end{array}
$$

factors. Such a transposition has degree -1 since it is realized as a reflection of the $S^2 = S^1 \wedge S^1$ involved. Hence σ has degree $(-1)^{k(j+k)}$, which is $+1$ since k is even. Thus σ is homotopic to the identity. Similarly, τ is homotopic to the identity. Hence $f \wedge g = (\mathbb{1} \wedge g)(f \wedge \mathbb{1})$ is homotopic to the composition $(g \wedge \mathbb{1})(f \wedge \mathbb{1})$, which is stably equivalent to the composition gf. Symmetrically, fg is stably homotopic to $g \wedge f$. So it suffices to show $f \wedge g \simeq (-1)^{ij} g \wedge f$. This we do by the commutative diagram at the right, where σ and τ are again the transpositions of the two factors. As before, τ is homotopic to the identity, but now σ has

$$
\begin{array}{ccc}
S^{i+k} \wedge S^{j+k} & \xrightarrow{f \wedge g} & S^k \wedge S^k \\
\downarrow \sigma & & \downarrow \tau \\
S^{j+k} \wedge S^{i+k} & \xrightarrow{g \wedge f} & S^k \wedge S^k
\end{array}
$$

degree $(-1)^{(i+k)(j+k)}$, which equals $(-1)^{ij}$ since k is even. The composition $(g \wedge f)\sigma$ is homotopic to $(-1)^{ij}(g \wedge f)$ since additive inverses in homotopy groups are obtained by precomposing with a reflection, of degree -1. Thus from the commutativity of the diagram we obtain the relation $f \wedge g \simeq (-1)^{ij} g \wedge f$. $\qquad\square$

Exercises

1. Use homotopy groups to show there is no retraction $\mathbb{R}P^n \to \mathbb{R}P^k$ if $n > k > 0$.

2. Show the action of $\pi_1(\mathbb{R}P^n)$ on $\pi_n(\mathbb{R}P^n) \approx \mathbb{Z}$ is trivial for n odd and nontrivial for n even.

3. Let X be obtained from a lens space of dimension $2n + 1$ by deleting a point. Compute $\pi_{2n}(X)$ as a module over $\mathbb{Z}[\pi_1(X)]$.

4. Let $X \subset \mathbb{R}^{n+1}$ be the union of the infinite sequence of spheres S_k^n of radius $1/k$ and center $(1/k, 0, \cdots, 0)$. Show that $\pi_i(X) = 0$ for $i < n$ and construct a homomorphism from $\pi_n(X)$ onto $\prod_k \pi_n(S_k^n)$.

5. Let $f : S_\alpha^2 \vee S_\beta^2 \to S_\alpha^2 \vee S_\beta^2$ be the map which is the identity on the S_α^2 summand and which on the S_β^2 summand is the sum of the identity map and a homeomorphism $S_\beta^2 \to S_\alpha^2$. Let X be the mapping torus of f, the quotient space of $(S_\alpha^2 \vee S_\beta^2) \times I$ under the identifications $(x, 0) \sim (f(x), 1)$. The mapping torus of the restriction of f to S_α^2 forms a subspace $A = S^1 \times S_\alpha^2 \subset X$. Show that the maps $\pi_2(A) \to \pi_2(X) \to \pi_2(X, A)$ form a short exact sequence $0 \to \mathbb{Z} \to \mathbb{Z} \oplus \mathbb{Z} \to \mathbb{Z} \to 0$, and compute the action of $\pi_1(A)$ on these three groups. In particular, show the action of $\pi_1(A)$ is trivial on $\pi_2(A)$ and $\pi_2(X, A)$ but is nontrivial on $\pi_2(X)$.

6. Show that the relative form of the Hurewicz theorem in dimension n implies the absolute form in dimension $n - 1$ by considering the pair (CX, X) where CX is the cone on X.

7. Construct a CW complex X with prescribed homotopy groups $\pi_i(X)$ and prescribed actions of $\pi_1(X)$ on the $\pi_i(X)$'s.

8. Show the suspension of an acyclic CW complex is contractible.

9. Show that a map between simply-connected CW complexes is a homotopy equivalence if its mapping cone is contractible. Use the preceding exercise to give an example where this fails in the nonsimply-connected case.

10. Let the CW complex X be obtained from $S^1 \vee S^n$, $n \geq 2$, by attaching a cell e^{n+1} by a map representing the polynomial $p(t) \in \mathbb{Z}[t, t^{-1}] \approx \pi_n(S^1 \vee S^n)$, so $\pi_n(X) \approx \mathbb{Z}[t, t^{-1}]/(p(t))$. Show $\pi_n'(X)$ is cyclic and compute its order in terms of $p(t)$. Give examples showing that the group $\pi_n(X)$ can be finitely generated or not, independently of whether $\pi_n'(X)$ is finite or infinite.

11. Let X be a connected CW complex with 1-skeleton X^1. Show that $\pi_2(X, X^1) \approx \pi_2(X) \times K$ where K is the kernel of $\pi_1(X^1) \to \pi_1(X)$, a free group. Show also that the map $\pi_2'(X) \to \pi_2'(X, X^1)$ need not be injective by considering the case $X = \mathbb{R}P^2$ with its standard CW structure.

12. Show that a map $f : X \to Y$ of connected CW complexes is a homotopy equivalence if it induces an isomorphism on π_1 and if a lift $\tilde{f} : \tilde{X} \to \tilde{Y}$ to the universal covers induces an isomorphism on homology. [The latter condition can be restated in terms of homology with local coefficients as saying that $f_* : H_*(X; \mathbb{Z}[\pi_1 X]) \to H_*(Y; \mathbb{Z}[\pi_1 Y])$ is an isomorphism; see §3.H.]

13. Show that a map between connected n-dimensional CW complexes is a homotopy equivalence if it induces an isomorphism on π_i for $i \leq n$. [Pass to universal covers and use homology.]

14. If an n-dimensional CW complex X contains a subcomplex Y homotopy equivalent to S^n, show that the map $\pi_n(Y) \to \pi_n(X)$ induced by inclusion is injective. [Use the Hurewicz homomorphism.]

15. Show that a closed simply-connected 3-manifold is homotopy equivalent to S^3. [Use Poincaré duality, and also the fact that closed manifolds are homotopy equivalent to CW complexes, from Corollary A.12 in the Appendix. The stronger statement that a closed simply-connected 3-manifold is homeomorphic to S^3 is the *Poincaré conjecture*, finally proved by Perelman. The higher-dimensional analog, that a closed n-manifold homotopy equivalent to S^n is homeomorphic to S^n, had been proved earlier for all $n \geq 4$.]

16. Show that the closed surfaces with infinite fundamental group are $K(\pi, 1)$'s by showing that their universal covers are contractible, via the Hurewicz theorem and results of §3.3.

17. Show that the map $\langle X, Y \rangle \to \mathrm{Hom}(\pi_n(X), \pi_n(Y))$, $[f] \mapsto f_*$, is a bijection if X is an $(n-1)$-connected CW complex and Y is a path-connected space with $\pi_i(Y) = 0$ for $i > n$. Deduce that CW complex $K(G, n)$'s are uniquely determined, up to homotopy type, by G and n.

18. If X and Y are simply-connected CW complexes such that $\tilde{H}_i(X)$ and $\tilde{H}_j(Y)$ are finite and of relatively prime orders for all pairs (i, j), show that the inclusion $X \vee Y \hookrightarrow X \times Y$ is a homotopy equivalence and $X \wedge Y$ is contractible. [Use the Künneth formula.]

19. If X is a $K(G, 1)$ CW complex, show that $\pi_n(X^n)$ is free abelian for $n \geq 2$.

20. Let G be a group and X a simply-connected space. Show that for the product $K(G, 1) \times X$ the action of π_1 on π_n is trivial for all $n > 1$.

21. Given a sequence of CW complexes $K(G_n, n)$, $n = 1, 2, \cdots$, let X_n be the CW complex formed by the product of the first n of these $K(G_n, n)$'s. Via the inclusions $X_{n-1} \subset X_n$ coming from regarding X_{n-1} as the subcomplex of X_n with n^{th} coordinate equal to a basepoint 0-cell of $K(G_n, n)$, we can then form the union of all the X_n's, a CW complex X. Show $\pi_n(X) \approx G_n$ for all n.

22. Show that $H_{n+1}(K(G, n); \mathbb{Z}) = 0$ if $n > 1$. [Build a $K(G, n)$ from a Moore space $M(G, n)$ by attaching cells of dimension $> n + 1$.]

23. Extend the Hurewicz theorem by showing that if X is an $(n-1)$-connected CW complex, then the Hurewicz homomorphism $h : \pi_{n+1}(X) \to H_{n+1}(X)$ is surjective when $n > 1$, and when $n = 1$ show there is an isomorphism $H_2(X)/h(\pi_2(X)) \approx H_2(K(\pi_1(X),1))$. [Build a $K(\pi_n(X),n)$ from X by attaching cells of dimension $n+2$ and greater, and then consider the homology sequence of the pair (Y,X) where Y is X with the $(n+2)$-cells of $K(\pi_n(X),n)$ attached. Note that the image of the boundary map $H_{n+2}(Y,X) \to H_{n+1}(X)$ coincides with the image of h, and $H_{n+1}(Y) \approx H_{n+1}(K(\pi_n(X),n))$. The previous exercise is needed for the case $n > 1$.]

24. Show there is a Moore space $M(G,1)$ with $\pi_1(M(G,1)) \approx G$ iff $H_2(K(G,1);\mathbb{Z}) = 0$. [Use the preceding problem. Build such an $M(G,1)$ from the 2-skeleton K^2 of a $K(G,1)$ by attaching 3-cells according to a basis for the free group $H_2(K^2;\mathbb{Z})$.] In particular, there is no $M(\mathbb{Z}^n,1)$ with fundamental group $\mathbb{Z}^n$, free abelian of rank n, if $n \geq 2$.

25. For X a connected CW complex with $\pi_i(X) = 0$ for $1 < i < n$ for some $n \geq 2$, show that $H_n(X)/h(\pi_n(X)) \approx H_n(K(\pi_1(X),1))$, where h is the Hurewicz map.

26. Generalizing the example of $\mathbb{R}P^2$ and $S^2 \times \mathbb{R}P^\infty$, show that if X is a connected finite-dimensional CW complex with universal cover $\tilde{X}$, then X and $\tilde{X} \times K(\pi_1(X),1)$ have isomorphic homotopy groups but are not homotopy equivalent if $\pi_1(X)$ contains elements of finite order.

27. Show that the image of the map $\pi_2(X,x_0) \to \pi_2(X,A,x_0)$ lies in the center of $\pi_2(X,A,x_0)$. (This exercise should be in §4.1.)

28. Show that the group $\mathbb{Z}_p \times \mathbb{Z}_p$ with p prime cannot act freely on any sphere S^n, by filling in details of the following argument. Such an action would define a covering space $S^n \to M$ with M a closed manifold. When $n > 1$, build a $K(\mathbb{Z}_p \times \mathbb{Z}_p,1)$ from M by attaching a single $(n+1)$-cell and then cells of higher dimension. Deduce that $H^{n+1}(K(\mathbb{Z}_p \times \mathbb{Z}_p,1);\mathbb{Z}_p)$ is $\mathbb{Z}_p$ or 0, a contradiction. (The case $n = 1$ is more elementary.)

29. Finish the homotopy classification of lens spaces begun in Exercise 2 of §3.E by showing that two lens spaces $L_m(\ell_1,\cdots,\ell_n)$ and $L_m(\ell'_1,\cdots,\ell'_n)$ are homotopy equivalent if $\ell_1 \cdots \ell_n \equiv \pm k^n \ell'_1 \cdots \ell'_n \bmod m$ for some integer k, via the following steps:

(a) Reduce to the case $k = 1$ by showing that $L_m(\ell'_1,\cdots,\ell'_n) = L_m(k\ell'_1,\cdots,k\ell'_n)$ if k is relatively prime to m. [Rechoose the generator of the $\mathbb{Z}_m$ action on S^{2n-1}.]

(b) Let $f:L \to L'$ be a map constructed as in part (b) of the exercise in §3.E. Construct a map $g:L \to L'$ as a composition $L \to L \vee S^{2n-1} \to L \vee S^{2n-1} \to L'$ where the first map collapses the boundary of a small ball to a point, the second map is the wedge of the identity on L and a map of some degree d on S^{2n-1}, and the third map is f on L and the projection $S^{2n-1} \to L'$ on S^{2n-1}. Show that g has degree $k_1 \cdots k_n + dm$, that is, g induces multiplication by $k_1 \cdots k_n + dm$ on $H_{2n-1}(-;\mathbb{Z})$. [Show first that a lift of g to the universal cover S^{2n-1} has this degree.]

(c) If $\ell_1 \cdots \ell_n \equiv \pm \ell_1' \cdots \ell_n'$ mod m, choose d so that $k_1 \cdots k_n + dm = \pm 1$ and show this implies that g induces an isomorphism on all homotopy groups, hence is a homotopy equivalence. [For π_i with $i > 1$, consider a lift of g to the universal cover.]

30. Let E be a subspace of $\mathbb{R}^2$ obtained by deleting a subspace of $\{0\} \times \mathbb{R}$. For which such spaces E is the projection $E \to \mathbb{R}$, $(x, y) \mapsto x$, a fiber bundle?

31. For a fiber bundle $F \to E \to B$ such that the inclusion $F \hookrightarrow E$ is homotopic to a constant map, show that the long exact sequence of homotopy groups breaks up into split short exact sequences giving isomorphisms $\pi_n(B) \approx \pi_n(E) \oplus \pi_{n-1}(F)$. In particular, for the Hopf bundles $S^3 \to S^7 \to S^4$ and $S^7 \to S^{15} \to S^8$ this yields isomorphisms

$$\pi_n(S^4) \approx \pi_n(S^7) \oplus \pi_{n-1}(S^3)$$
$$\pi_n(S^8) \approx \pi_n(S^{15}) \oplus \pi_{n-1}(S^7)$$

Thus $\pi_7(S^4)$ and $\pi_{15}(S^8)$ contain $\mathbb{Z}$ summands.

32. Show that if $S^k \to S^m \to S^n$ is a fiber bundle, then $k = n - 1$ and $m = 2n - 1$. [Look at the long exact sequence of homotopy groups.]

33. Show that if there were fiber bundles $S^{n-1} \to S^{2n-1} \to S^n$ for all n, then the groups $\pi_i(S^n)$ would be finitely generated free abelian groups computable by induction, and nonzero for $i \geq n \geq 2$.

34. Let $p : S^3 \to S^2$ be the Hopf bundle and let $q : T^3 \to S^3$ be the quotient map collapsing the complement of a ball in the 3-dimensional torus $T^3 = S^1 \times S^1 \times S^1$ to a point. Show that $pq : T^3 \to S^2$ induces the trivial map on π_* and $\tilde{H}_*$, but is not homotopic to a constant map.

35. Show that the fiber bundle $S^3 \to S^{4n+3} \to \mathbb{HP}^n$ gives rise to a quotient fiber bundle $S^2 \to \mathbb{CP}^{2n+1} \to \mathbb{HP}^n$ by factoring out the action of S^1 on S^{4n+3} by complex scalar multiplication.

36. For basepoint-preserving maps $f : S^1 \to X$ and $g : S^n \to X$ with $n > 1$, show that the Whitehead product $[f, g]$ is $\pm(g - fg)$, where fg denotes the action of f on g.

37. Show that all Whitehead products in a path-connected H–space are trivial.

38. Show $\pi_3(S^1 \vee S^2)$ is not finitely generated as a module over $\mathbb{Z}[\pi_1(S^1 \vee S^2)]$ by considering Whitehead products in the universal cover, using the results in Example 4.52. Generalize this to $\pi_{i+j-1}(S^1 \vee S^i \vee S^j)$ for $i, j > 1$.

39. Show that the indeterminacy of a Toda bracket $\langle f, g, h \rangle$ with $f \in \pi_i^s$, $g \in \pi_j^s$, $h \in \pi_k^s$ is the subgroup $f \cdot \pi_{j+k+1}^s + h \cdot \pi_{i+j+1}^s$ of $\pi_{i+j+k+1}^s$.

4.3 Connections with Cohomology

The Hurewicz theorem provides a strong link between homotopy groups and homology, and hence also an indirect relation with cohomology. But there is a more direct connection with cohomology of a quite different sort. We will show that for every CW complex X there is a natural bijection between $H^n(X;G)$ and the set $\langle X, K(G,n) \rangle$ of basepoint-preserving homotopy classes of maps from X to a $K(G,n)$. We will also define a natural group structure on $\langle X, K(G,n) \rangle$ that makes the bijection a group isomorphism. The mere fact that there is any connection at all between cohomology and homotopy classes of maps is the first surprise here, and the second is that Eilenberg–MacLane spaces are involved, since their definition is entirely in terms of homotopy groups, which on the face of it have nothing to do with cohomology.

After proving this basic isomorphism $H^n(X;G) \approx \langle X, K(G,n) \rangle$ and describing a few of its immediate applications, the later parts of this section aim toward a further study of Postnikov towers, which were introduced briefly in §4.1. These provide a general theoretical method for realizing an arbitrary CW complex as a sort of twisted product of Eilenberg–MacLane spaces, up to homotopy equivalence. The most geometric interpretation of the phrase 'twisted product' is the notion of fiber bundle introduced in the previous section, but here we need the more homotopy-theoretic notion of a *fibration*, so before we begin the discussion of Postnikov towers we first take a few pages to present some basic constructions and results about fibrations.

As we shall see, Postnikov towers can be expressed as sequences of fibrations with fibers Eilenberg–MacLane spaces, so we can again expect close connections with cohomology. One such connection is provided by *k-invariants*, which describe, at least in principle, how Postnikov towers for a broad class of spaces are determined by a sequence of cohomology classes. Another application of these ideas, described at the end of the section, is a technique for factoring basic extension and lifting problems in homotopy theory into a sequence of smaller problems whose solutions are equivalent to the vanishing of certain cohomology classes. This technique goes under the somewhat grandiose title of Obstruction Theory, though it is really quite a simple idea when expressed in terms of Postnikov towers.

The Homotopy Construction of Cohomology

The main result of this subsection is the following fundamental relationship between singular cohomology and Eilenberg–MacLane spaces:

Theorem 4.57. *There are natural bijections* $T : \langle X, K(G,n) \rangle \to H^n(X;G)$ *for all CW complexes X and all $n > 0$, with G any abelian group. Such a T has the form* $T([f]) = f^*(\alpha)$ *for a certain distinguished class* $\alpha \in H^n(K(G,n);G)$.

In the course of the proof we will define a natural group structure on $\langle X, K(G,n) \rangle$ such that the transformation T is an isomorphism.

A class $\alpha \in H^n(K(G,n);G)$ with the property stated in the theorem is called a **fundamental class**. The proof of the theorem will yield an explicit fundamental class, namely the element of $H^n(K(G,n);G) \approx \mathrm{Hom}(H_n(K;\mathbb{Z}),G)$ given by the inverse of the Hurewicz isomorphism $G = \pi_n(K(G,n)) \to H_n(K;\mathbb{Z})$. Concretely, if we choose $K(G,n)$ to be a CW complex with $(n-1)$-skeleton a point, then a fundamental class is represented by the cellular cochain assigning to each n-cell of $K(G,n)$ the element of $\pi_n(K(G,n))$ defined by a characteristic map for the n-cell.

The theorem also holds with $\langle X, K(G,n) \rangle$ replaced by $[X, K(G,n)]$, the non-basepointed homotopy classes. This is easy to see when $n > 1$ since every map $X \to K(G,n)$ can be homotoped to take basepoint to basepoint, and every homotopy between basepoint-preserving maps can be homotoped to be basepoint-preserving since the target space $K(G,n)$ is simply-connected. When $n = 1$ it is still true that $[X, K(G,n)] = \langle X, K(G,n) \rangle$ for abelian G according to an exercise for §4.A. For $n = 0$ it is elementary that $H^0(X;G) = [X, K(G,0)]$ and $\tilde{H}^0(X;G) = \langle X, K(G,0) \rangle$.

It is possible to give a direct proof of the theorem, constructing maps and homotopies cell by cell. This provides geometric insight into why the result is true, but unfortunately the technical details of this proof are rather tedious. So we shall take a different approach, one that has the advantage of placing the result in its natural context via general machinery that turns out to be quite useful in other situations as well. The two main steps will be the following assertions.

(1) The functors $h^n(X) = \langle X, K(G,n) \rangle$ define a reduced cohomology theory on the category of basepointed CW complexes.

(2) If a reduced cohomology theory h^* defined on CW complexes has coefficient groups $h^n(S^0)$ which are zero for $n \neq 0$, then there are natural isomorphisms $h^n(X) \approx \tilde{H}^n(X; h^0(S^0))$ for all CW complexes X and all n.

Towards proving (1) we will study a more general question: When does a sequence of spaces K_n define a cohomology theory by setting $h^n(X) = \langle X, K_n \rangle$? Note that this will be a reduced cohomology theory since $\langle X, K_n \rangle$ is trivial when X is a point.

The first question to address is putting a group structure on the set $\langle X, K \rangle$. This requires that either X or K have some special structure. When $X = S^n$ we have $\langle S^n, K \rangle = \pi_n(K)$, which has a group structure when $n > 0$. The definition of this group structure works more generally whenever S^n is replaced by a suspension SX, with the sum of maps $f, g : SX \to K$ defined as the composition $SX \to SX \vee SX \to K$ where the first map collapses an 'equatorial' $X \subset SX$ to a point and the second map consists of f and g on the two summands. However, for this to make sense we must be talking about basepoint-preserving maps, and there is a problem with where to choose the basepoint in SX. If x_0 is a basepoint of X, the basepoint of SX should be somewhere along the segment $\{x_0\} \times I \subset SX$, most likely either an endpoint or the midpoint, but no single choice of such a basepoint gives a well-defined sum. The sum would be well-defined if we restricted attention to maps sending the whole segment $\{x_0\} \times I$ to the basepoint. This is equivalent to considering basepoint-preserving maps $\Sigma X \to K$ where $\Sigma X = SX/(\{x_0\} \times I)$ and the image of $\{x_0\} \times I$ in ΣX is taken to be the basepoint. If X is a CW complex with x_0 a 0-cell, the quotient map $SX \to \Sigma X$ is a homotopy equivalence since it collapses a contractible subcomplex of SX to a point, so we can identify $\langle SX, K \rangle$ with $\langle \Sigma X, K \rangle$. The space ΣX is called the **reduced suspension** of X when we want to distinguish it from the ordinary suspension SX.

It is easy to check that $\langle \Sigma X, K \rangle$ is a group with respect to the sum defined above, inverses being obtained by reflecting the I coordinate in the suspension. However, what we would really like to have is a group structure on $\langle X, K \rangle$ arising from a special structure on K rather than on X. This can be obtained using the following basic **adjoint relation**:

- $\langle \Sigma X, K \rangle = \langle X, \Omega K \rangle$ where ΩK is the space of loops in K at its chosen basepoint and the constant loop is taken as the basepoint of ΩK.

The space ΩK, called the **loopspace** of K, is topologized as a subspace of the space K^I of all maps $I \to K$, where K^I is given the compact-open topology; see the Appendix for the definition and basic properties of this topology. The adjoint relation $\langle \Sigma X, K \rangle = \langle X, \Omega K \rangle$ holds because basepoint-preserving maps $\Sigma X \to K$ are exactly the same as basepoint-preserving maps $X \to \Omega K$, the correspondence being given by associating to $f : \Sigma X \to K$ the family of loops obtained by restricting f to the images of the segments $\{x\} \times I$ in ΣX.

Taking $X = S^n$ in the adjoint relation, we see that $\pi_{n+1}(K) = \pi_n(\Omega K)$ for all $n \geq 0$. Thus passing from a space to its loopspace has the effect of shifting homotopy groups down a dimension. In particular we see that $\Omega K(G, n)$ is a $K(G, n-1)$. This fact will turn out to be important in what follows.

Note that the association $X \mapsto \Omega X$ is a functor: A basepoint-preserving map $f : X \to Y$ induces a map $\Omega f : \Omega X \to \Omega Y$ by composition with f. A homotopy $f \simeq g$ induces a homotopy $\Omega f \simeq \Omega g$, so it follows formally that $X \simeq Y$ implies $\Omega X \simeq \Omega Y$.

It is a theorem of [Milnor 1959] that the loopspace of a CW complex has the homotopy type of a CW complex. This may be a bit surprising since loopspaces are usually quite large spaces, though of course CW complexes can be quite large too, in terms of the number of cells. What often happens in practice is that if a CW complex X has only finitely many cells in each dimension, then ΩX is homotopy equivalent to a CW complex with the same property. We will see explicitly how this happens for $X = S^n$ in §4.J.

Composition of loops defines a map $\Omega K \times \Omega K \to \Omega K$, and this gives a sum operation in $\langle X, \Omega K \rangle$ by setting $(f + g)(x) = f(x) \cdot g(x)$, the composition of the loops $f(x)$ and $g(x)$. Under the adjoint relation this is the same as the sum in $\langle \Sigma X, K \rangle$ defined previously. If we take the composition of loops as the sum operation then it is perhaps somewhat easier to see that $\langle X, \Omega K \rangle$ is a group since the same reasoning which shows that $\pi_1(K)$ is a group can be applied.

Since cohomology groups are abelian, we would like the group $\langle X, \Omega K \rangle$ to be abelian. This can be achieved by iterating the operation of forming loopspaces. One has a double loopspace $\Omega^2 K = \Omega(\Omega K)$ and inductively an n-fold loopspace $\Omega^n K = \Omega(\Omega^{n-1} K)$. The evident bijection $K^{Y \times Z} \approx (K^Y)^Z$ is a homeomorphism for locally compact Hausdorff spaces Y and Z, as shown in Proposition A.16 in the Appendix, and from this it follows by induction that $\Omega^n K$ can be regarded as the space of maps $I^n \to K$ sending ∂I^n to the basepoint. Taking $n = 2$, we see that the argument that $\pi_2(K)$ is abelian shows more generally that $\langle X, \Omega^2 K \rangle$ is an abelian group. Iterating the adjoint relation gives $\langle \Sigma^n X, K \rangle = \langle X, \Omega^n K \rangle$, so this is an abelian group for all $n \geq 2$.

Thus for a sequence of spaces K_n to define a cohomology theory $h^n(X) = \langle X, K_n \rangle$ we have been led to the assumption that each K_n should be a loopspace and in fact a double loopspace. Actually we do not need K_n to be literally a loopspace since it would suffice for it to be homotopy equivalent to a loopspace, as $\langle X, K_n \rangle$ depends only on the homotopy type of K_n. In fact it would suffice to have just a weak homotopy equivalence $K_n \to \Omega L_n$ for some space L_n since this would induce a bijection $\langle X, K_n \rangle = \langle X, \Omega L_n \rangle$ by Proposition 4.22. In the special case that $K_n = K(G, n)$ for all n, we can take $L_n = K_{n+1} = K(G, n+1)$ by the earlier observation that $\Omega K(G, n+1)$ is a $K(G, n)$. Thus if we take the $K(G, n)$'s to be CW complexes, the map $K_n \to \Omega K_{n+1}$ is just a CW approximation $K(G, n) \to \Omega K(G, n+1)$.

There is another reason to look for weak homotopy equivalences $K_n \to \Omega K_{n+1}$. For a reduced cohomology theory $h^n(X)$ there are natural isomorphisms $h^n(X) \approx h^{n+1}(\Sigma X)$ coming from the long exact sequence of the pair (CX, X) with CX the cone on X, so if $h^n(X) = \langle X, K_n \rangle$ for all n then the isomorphism $h^n(X) \approx h^{n+1}(\Sigma X)$ translates into a bijection $\langle X, K_n \rangle \approx \langle \Sigma X, K_{n+1} \rangle = \langle X, \Omega K_{n+1} \rangle$ and the most natural thing would be for this to come from a weak equivalence $K_n \to \Omega K_{n+1}$. Weak equivalences of this form would give also weak equivalences $K_n \to \Omega K_{n+1} \to \Omega^2 K_{n+2}$ and so we would automatically obtain an abelian group structure on $\langle X, K_n \rangle \approx \langle X, \Omega^2 K_{n+2} \rangle$.

These observations lead to the following definition. An Ω-**spectrum** is a sequence of CW complexes $K_1, K_2, \cdots$ together with weak homotopy equivalences $K_n \to \Omega K_{n+1}$ for all n. By using the theorem of Milnor mentioned above it would be possible to replace 'weak homotopy equivalence' by 'homotopy equivalence' in this definition. However it does not noticeably simplify matters to do this, except perhaps psychologically.

Notice that if we discard a finite number of spaces K_n from the beginning of an Ω-spectrum $K_1, K_2, \cdots$, then these omitted terms can be reconstructed from the remaining K_n's since each K_n determines K_{n-1} as a CW approximation to ΩK_n. So it is not important that the sequence start with K_1. By the same token, this allows us to extend the sequence of K_n's to all negative values of n. This is significant because a general cohomology theory $h^n(X)$ need not vanish for negative n.

Theorem 4.58. *If $\{K_n\}$ is an Ω-spectrum, then the functors $X \mapsto h^n(X) = \langle X, K_n \rangle$, $n \in \mathbb{Z}$, define a reduced cohomology theory on the category of basepointed CW complexes and basepoint-preserving maps.*

Rather amazingly, the converse is also true: Every reduced cohomology theory on CW complexes arises from an Ω-spectrum in this way. This is the Brown representability theorem which will be proved in §4.E.

A space K_n in an Ω-spectrum is sometimes called an *infinite loopspace* since there are weak homotopy equivalences $K_n \to \Omega^k K_{n+k}$ for all k. A number of important spaces in algebraic topology turn out to be infinite loopspaces. Besides Eilenberg–MacLane spaces, two other examples are the infinite-dimensional orthogonal and unitary groups O and U, for which there are weak homotopy equivalences $O \to \Omega^8 O$ and $U \to \Omega^2 U$ by a strong form of the Bott periodicity theorem, as we will show in [VBKT]. So O and U give periodic Ω-spectra, hence periodic cohomology theories known as real and complex K–theory. For a more in-depth introduction to the theory of infinite loopspaces, the book [Adams 1978] can be much recommended.

Proof: Two of the three axioms for a cohomology theory, the homotopy axiom and the wedge sum axiom, are quite easy to check. For the homotopy axiom, a basepoint-preserving map $f : X \to Y$ induces $f^* : \langle Y, K_n \rangle \to \langle X, K_n \rangle$ by composition, sending a map $Y \to K_n$ to $X \xrightarrow{f} Y \to K_n$. Clearly f^* depends only on the basepoint-preserving homotopy class of f, and it is obvious that f^* is a homomorphism if we replace K_n by ΩK_{n+1} and use the composition of loops to define the group structure. The wedge sum axiom holds since in the realm of basepoint-preserving maps, a map $\bigvee_\alpha X_\alpha \to K_n$ is the same as a collection of maps $X_\alpha \to K_n$.

The bulk of the proof involves associating a long exact sequence to each CW pair (X, A). As a first step we build the following diagram:

$$(1) \quad \begin{array}{ccccccccc}
A & \hookrightarrow & X & \hookrightarrow & X \cup CA & \hookrightarrow & (X \cup CA) \cup CX & \hookrightarrow & ((X \cup CA) \cup CX) \cup C(X \cup CA) \\
\| & & \| & & \approx\downarrow & & \approx\downarrow\uparrow & & \approx\downarrow\uparrow \\
A & \hookrightarrow & X & \longrightarrow & X/A & \longrightarrow & SA & \xrightarrow{\ c\ } & SX
\end{array}$$

The first row is obtained from the inclusion $A \hookrightarrow X$ by iterating the rule, 'attach a cone on the preceding subspace,' as shown in the pictures below.

The three downward arrows in the diagram (1) are quotient maps collapsing the most recently attached cone to a point. Since cones are contractible, these downward maps are homotopy equivalences. The second and third of them have homotopy inverses the evident inclusion maps, indicated by the upward arrows. In the lower row of the diagram the maps are the obvious ones, except for the map $X/A \to SA$ which is the composition of a homotopy inverse of the quotient map $X \cup CA \to X/A$ followed by the maps $X \cup CA \to (X \cup CA) \cup CX \to SA$. Thus the square containing this map commutes up to homotopy. It is easy to check that the same is true of the right-hand square as well.

The whole construction can now be repeated with $SA \hookrightarrow SX$ in place of $A \hookrightarrow X$, then with double suspensions, and so on. The resulting infinite sequence can be written in either of the following two forms:

$$A \to X \to X \cup CA \to SA \to SX \to S(X \cup CA) \to S^2A \to S^2X \to \cdots$$

$$A \to X \to X/A \to SA \to SX \to SX/SA \to S^2A \to S^2X \to \cdots$$

In the first version we use the obvious equality $SX \cup CSA = S(X \cup CA)$. The first version has the advantage that the map $X \cup CA \to SA$ is easily described and canonical, whereas in the second version the corresponding map $X/A \to SA$ is only defined up to homotopy since it depends on choosing a homotopy inverse to the quotient map $X \cup CA \to X/A$. The second version does have the advantage of conciseness, however.

When basepoints are important it is generally more convenient to use reduced cones and reduced suspensions, obtained from ordinary cones and suspensions by collapsing the segment $\{x_0\} \times I$ where x_0 is the basepoint. The image point of this segment in the reduced cone or suspension then serves as a natural basepoint in the quotient. Assuming x_0 is a 0-cell, these collapses of $\{x_0\} \times I$ are homotopy equivalences. Using reduced cones and suspensions in the preceding construction yields a sequence

$$(2) \qquad A \hookrightarrow X \to X/A \to \Sigma A \hookrightarrow \Sigma X \to \Sigma(X/A) \to \Sigma^2 A \hookrightarrow \Sigma^2 X \to \cdots$$

where we identify $\Sigma X/\Sigma A$ with $\Sigma(X/A)$, and all the later maps in the sequence are suspensions of the first three maps. This sequence, or its unreduced version, is called the **cofibration sequence** or **Puppe sequence** of the pair (X,A). It has an evident naturality property, namely, a map $(X,A) \to (Y,B)$ induces a map between the cofibration sequences of these two pairs, with homotopy-commutative squares:

$$
\begin{array}{ccccccccccc}
A & \longrightarrow & X & \longrightarrow & X/A & \longrightarrow & \Sigma A & \longrightarrow & \Sigma X & \longrightarrow & \Sigma(X/A) & \longrightarrow & \Sigma^2 A & \longrightarrow & \cdots \\
\downarrow & & \downarrow & & \downarrow & & \downarrow & & \downarrow & & \downarrow & & \downarrow & & \\
B & \longrightarrow & Y & \longrightarrow & Y/B & \longrightarrow & \Sigma B & \longrightarrow & \Sigma Y & \longrightarrow & \Sigma(Y/B) & \longrightarrow & \Sigma^2 B & \longrightarrow & \cdots
\end{array}
$$

Taking basepoint-preserving homotopy classes of maps from the spaces in (2) to a fixed space K gives a sequence

$$(3) \qquad \langle A, K \rangle \leftarrow \langle X, K \rangle \leftarrow \langle X/A, K \rangle \leftarrow \langle \Sigma A, K \rangle \leftarrow \langle \Sigma X, K \rangle \leftarrow \cdots$$

whose maps are defined by composition with those in (2). For example, the map $\langle X, K \rangle \to \langle A, K \rangle$ sends a map $X \to K$ to $A \to X \to K$. The sets in (3) are groups starting with $\langle \Sigma A, K \rangle$, and abelian groups from $\langle \Sigma^2 A, K \rangle$ onward. It is easy to see that the maps between these groups are homomorphisms since the maps in (2) are suspensions from $\Sigma A \to \Sigma X$ onward. In general the first three terms of (3) are only sets with distinguished 'zero' elements, the constant maps.

A key observation is that the sequence (3) is exact. To see this, note first that the diagram (1) shows that, up to homotopy equivalence, each term in (2) is obtained from its two predecessors by the same procedure of forming a mapping cone, so it suffices to show that $\langle A, K \rangle \leftarrow \langle X, K \rangle \leftarrow \langle X \cup CA, K \rangle$ is exact. This is easy: A map $f : X \to K$ goes to zero in $\langle A, K \rangle$ iff its restriction to A is nullhomotopic, fixing the basepoint, and this is equivalent to f extending to a map $X \cup CA \to K$.

If we have a weak homotopy equivalence $K \to \Omega K'$ for some space K', then the sequence (3) can be continued three steps to the left via the commutative diagram

$$
\begin{array}{ccccccc}
\langle A, K \rangle & \longleftarrow & \langle X, K \rangle & \longleftarrow & \langle X/A, K \rangle & \longleftarrow & \cdots \\
\downarrow{\approx} & & \downarrow{\approx} & & \downarrow{\approx} & & \\
\langle A, \Omega K' \rangle & \longleftarrow & \langle X, \Omega K' \rangle & \longleftarrow & \langle X/A, \Omega K' \rangle & \longleftarrow & \cdots \\
\downarrow{\approx} & & \downarrow{\approx} & & \downarrow{\approx} & & \\
\langle A, K' \rangle \longleftarrow \langle X, K' \rangle \longleftarrow \langle X/A, K' \rangle & \longleftarrow & \langle \Sigma A, K' \rangle \longleftarrow \langle \Sigma X, K' \rangle \longleftarrow \langle \Sigma(X/A), K' \rangle \longleftarrow \cdots
\end{array}
$$

Thus if we have a sequence of spaces K_n together with weak homotopy equivalences $K_n \to \Omega K_{n+1}$, we can extend the sequence (3) to the left indefinitely, producing a long exact sequence

$$\cdots \leftarrow \langle A, K_n \rangle \leftarrow \langle X, K_n \rangle \leftarrow \langle X/A, K_n \rangle \leftarrow \langle A, K_{n-1} \rangle \leftarrow \langle X, K_{n-1} \rangle \leftarrow \cdots$$

All the terms here are abelian groups and the maps homomorphisms. This long exact sequence is natural with respect to maps $(X, A) \to (Y, B)$ since cofibration sequences are natural. $\qquad\square$

There is no essential difference between cohomology theories on basepointed CW complexes and cohomology theories on nonbasepointed CW complexes. Given a reduced basepointed cohomology theory $\tilde{h}^*$, one gets an unreduced theory by setting $h^n(X, A) = \tilde{h}^n(X/A)$, where $X/\varnothing = X_+$, the union of X with a disjoint basepoint. This is a nonbasepointed theory since an arbitrary map $X \to Y$ induces a basepoint-preserving map $X_+ \to Y_+$. Furthermore, a nonbasepointed unreduced theory h^* gives a nonbasepointed reduced theory by setting $\tilde{h}^n(X) = \mathrm{Coker}(h^n(point) \to h^n(X))$, where the map is induced by the constant map $X \to point$. One could also give an argument using suspension, which is always an isomorphism for reduced theories, and which takes one from the nonbasepointed to the basepointed category.

Theorem 4.59. *If h^* is an unreduced cohomology theory on the category of CW pairs and $h^n(point) = 0$ for $n \neq 0$, then there are natural isomorphisms $h^n(X, A) \approx H^n(X, A; h^0(point))$ for all CW pairs (X, A) and all n. The corresponding statement for homology theories is also true.*

Proof: The case of homology is slightly simpler, so let us consider this first. For CW complexes, relative homology groups reduce to absolute groups, so it suffices to deal with the latter. For a CW complex X the long exact sequences of h_* homology groups for the pairs (X^n, X^{n-1}) give rise to a cellular chain complex

$$\cdots \longrightarrow h_{n+1}(X^{n+1}, X^n) \xrightarrow{d_{n+1}} h_n(X^n, X^{n-1}) \xrightarrow{d_n} h_{n-1}(X^{n-1}, X^{n-2}) \longrightarrow \cdots$$

just as for ordinary homology. The hypothesis that $h_n(point) = 0$ for $n \neq 0$ implies that this chain complex has homology groups $h_n(X)$ by the same argument as for ordinary homology. The main thing to verify now is that this cellular chain complex is isomorphic to the cellular chain complex in ordinary homology with coefficients in the group $G = h_0(point)$. Certainly the cellular chain groups in the two cases are isomorphic, being direct sums of copies of G with one copy for each cell, so we have only to check that the cellular boundary maps are the same.

It is not really necessary to treat the cellular boundary map d_1 from 1-chains to 0-chains since one can always pass from X to ΣX, suspension being a natural isomorphism in any homology theory, and the double suspension $\Sigma^2 X$ has no 1-cells.

The calculation of cellular boundary maps d_n for $n > 1$ in terms of degrees of certain maps between spheres works equally well for the homology theory h_*, where 'degree' now means degree with respect to the h_* theory, so what is needed is the fact that a map $S^n \to S^n$ of degree m in the usual sense induces multiplication by m on $h_n(S^n) \approx G$. This is obviously true for degrees 0 and 1, represented by a constant map and the identity map. Since $\pi_n(S^n) \approx \mathbb{Z}$, every map $S^n \to S^n$ is homotopic to some multiple of the identity, so the general case will follow if we know that degree in the h_* theory is additive with respect to the sum operation in $\pi_n(S^n)$. This is a special case of the following more general assertion:

Lemma 4.60. *If a functor h from basepointed CW complexes to abelian groups satisfies the homotopy and wedge axioms, then for any two basepoint-preserving maps $f, g : \Sigma X \to K$, we have $(f + g)_* = f_* + g_*$ if h is covariant and $(f + g)^* = f^* + g^*$ if h is contravariant.*

Proof: The map $f + g$ is the composition $\Sigma X \xrightarrow{c} \Sigma X \vee \Sigma X \xrightarrow{f \vee g} K$ where c is the quotient map collapsing an equatorial copy of X. In the covariant case consider the diagram at the right, where i_1 and i_2 are the inclusions $\Sigma X \hookrightarrow \Sigma X \vee \Sigma X$. Let $q_1, q_2 : \Sigma X \vee \Sigma X \to \Sigma X$ be the quotient maps restricting to the identity on the summand indicated by the subscript and collapsing the other summand to a point.

$$h(\Sigma X) \xrightarrow{c_*} h(\Sigma X \vee \Sigma X) \xrightarrow{(f \vee g)_*} h(K)$$
$$i_{1*} \oplus i_{2*} \Big\uparrow \approx$$
$$h(\Sigma X) \oplus h(\Sigma X)$$

Then $q_{1*} \oplus q_{2*}$ is an inverse to $i_{1*} \oplus i_{2*}$ since $q_j i_k$ is the identity map for $j = k$ and the constant map for $j \neq k$.

An element x in the left-hand group $h(\Sigma X)$ in the diagram is sent by the composition $(q_{1*} \oplus q_{2*})c_*$ to the element (x, x) in the lower group $h(\Sigma X) \oplus h(\Sigma X)$ since

q_1c and q_2c are homotopic to the identity. The composition $(f \vee g)_*(i_{1*} \oplus i_{2*})$ sends $(x,0)$ to $f_*(x)$ and $(0,y)$ to $g_*(y)$ since $(f \vee g)i_1 = f$ and $(f \vee g)i_2 = g$. Hence (x,y) is sent to $f_*(x) + g_*(y)$. Combining these facts, we see that the composition across the top of the diagram is $x \mapsto f_*(x) + g_*(x)$. But this composition is also $(f + g)_*$ since $f + g = (f \vee g)c$. This finishes the proof in the covariant case.

The contravariant case is similar, using the corresponding diagram with arrows reversed. The inverse of $i_1^* \oplus i_2^*$ is $q_1^* \oplus q_2^*$ by the same reasoning. An element u in the right-hand group $h(K)$ maps to the element $(f^*(u), g^*(u))$ in the lower group $h(\Sigma X) \oplus h(\Sigma X)$ since $(f \vee g)i_1 = f$ and $(f \vee g)i_2 = g$. An element $(x,0)$ in the lower group in the diagram maps to the element x in the left-hand group since q_1c is homotopic to the identity, and similarly $(0,y)$ maps to y. Hence (x,y) maps to $x + y$ in the left-hand group. We conclude that $u \in h(K)$ maps by the composition across the top of the diagram to $f^*(u) + g^*(u)$ in $h(\Sigma X)$. But this composition is $(f + g)^*$ by definition. $\square$

Returning to the proof of the theorem, we see that the cellular chain complexes for $h_*(X)$ and $H_*(X; G)$ are isomorphic, so we obtain isomorphisms $h_n(X) \approx H_n(X; G)$ for all n. To verify that these isomorphisms are natural with respect to maps $f : X \to Y$ we may first deform such a map f to be cellular. Then f takes each pair (X^n, X^{n-1}) to the pair (Y^n, Y^{n-1}), hence f induces a chain map of cellular chain complexes in the h_* theory, as well as for $H_*(-; G)$. To compute these chain maps we may pass to the quotient maps $X^n/X^{n-1} \to Y^n/Y^{n-1}$. These are maps of the form $\bigvee_\alpha S_\alpha^n \to \bigvee_\beta S_\beta^n$, so the induced maps f_* on h_n are determined by their component maps $f_* : S_\alpha^n \to S_\beta^n$. This is exactly the same situation as with the cellular boundary maps before, where we saw that the degree of a map $S^n \to S^n$ determines the induced map on h_n. We conclude that the cellular chain map induced by f in the h_* theory agrees exactly with the cellular chain map for $H_*(-; G)$. This implies that the isomorphism between the two theories is natural.

The situation for cohomology is quite similar, but there is one point in the argument where a few more words are needed. For cohomology theories the cellular cochain groups are the direct product, rather than the direct sum, of copies of the coefficient group $G = h^0(\textit{point})$, with one copy per cell. This means that when there are infinitely many cells in a given dimension, it is not automatically true that the cellular coboundary maps are uniquely determined by how they map factors of one direct product to factors of the other direct product. To be precise, consider the cellular coboundary map $d_n : h^n(X^n, X^{n-1}) \to h^{n+1}(X^{n+1}, X^n)$. Decomposing the latter group as a product of copies of G for the $(n+1)$-cells, we see that d_n is determined by the maps $h^n(X^n/X^{n-1}) \to h^n(S_\alpha^n)$ associated to the attaching maps φ_α of the cells e_α^{n+1}. The thing to observe is that since φ_α has compact image, meeting only finitely many n-cells, this map $h^n(X^n/X^{n-1}) \to h^n(S_\alpha^n)$ is *finitely supported* in the sense that there is a splitting of the domain into a product of finitely many factors and a product of the remaining possibly infinite number of factors, such that the map is zero on the latter product. Finitely supported maps have the good property that they are determined by their restrictions to the G factors of $h^n(X^n/X^{n-1})$. From this we deduce, using the lemma, that the cellular coboundary maps in the h^* theory agree with those in ordinary cohomology with G coefficients. This extra argument is also needed to prove naturality of the isomorphisms $h^n(X) \approx H^n(X; G)$.

This completes the proof of Theorem 4.59. $\qquad\square$

Proof of Theorem 4.57: The functors $h^n(X) = \langle X, K(G, n) \rangle$ define a reduced cohomology theory, and the coefficient groups $h^n(S^i) = \pi_i(K(G, n))$ are the same as $\tilde{H}^n(S^i; G)$, so Theorem 4.59, translated into reduced cohomology, gives natural isomorphisms $T : \langle X, K(G, n) \rangle \to \tilde{H}^n(X; G)$ for all CW complexes X.

It remains to see that $T([f]) = f^*(\alpha)$ for some $\alpha \in \tilde{H}^n(K(G, n); G)$, independent of f. This is purely formal: Take $\alpha = T(\mathbb{1})$ for $\mathbb{1}$ the identity map of $K(G, n)$, and then naturality gives $T([f]) = T(f^*(\mathbb{1})) = f^*T(\mathbb{1}) = f^*(\alpha)$, where the first f^* refers to induced homomorphisms for the functor h^n, which means composition with f. $\qquad\square$

The fundamental class $\alpha = T(\mathbb{1})$ can be made more explicit if we choose for $K(G, n)$ a CW complex K with $(n-1)$-skeleton a point. Denoting $\langle X, K(G, n) \rangle$ by $h^n(X)$, then we have

$$h^n(K) \approx h^n(K^{n+1}) \approx \operatorname{Ker} d : h^n(K^n) \to h^{n+1}(K^{n+1}, K^n)$$

The map d is the cellular coboundary in h^* cohomology since we have $h^n(K^n) = h^n(K^n, K^{n-1})$ because K^{n-1} is a point and h^* is a reduced theory. The isomorphism of $h^n(K)$ with $\operatorname{Ker} d$ is given by restriction of maps $K \to K$ to K^n, so the element $\mathbb{1} \in h^n(K)$ defining the fundamental class $T(\mathbb{1})$ corresponds, under the isomorphism $h^n(K) \approx \operatorname{Ker} d$, to the inclusion $K^n \hookrightarrow K$ viewed as an element of $h^n(K^n)$. As a cellular cocycle this element assigns to each n-cell of K the element of the coefficient group $G = \pi_n(K)$ given by the inclusion of the closure of this cell into K. This means that the fundamental class $\alpha \in H^n(K; G)$ is represented by the cellular cocycle assigning to each n-cell the element of $\pi_n(K)$ given by a characteristic map for the cell.

By naturality of T it follows that for a cellular map $f : X \to K$, the corresponding element of $H^n(X; G)$ is represented by the cellular cocycle sending each n-cell of X to the element of $G = \pi_n(K)$ represented by the composition of f with a characteristic map for the cell.

The natural isomorphism $H^n(X; G) \approx \langle X, K(G, n) \rangle$ leads to a basic principle which reappears many places in algebraic topology, the idea that the occurrence or nonoccurrence of a certain phenomenon is governed by what happens in a single special case, the *universal example*. To illustrate, let us prove the following special fact:

- The map $H^1(X;\mathbb{Z}) \to H^2(X;\mathbb{Z})$, $\alpha \mapsto \alpha^2$, is identically zero for all spaces X.

By taking a CW approximation to X we are reduced to the case that X is a CW complex. Then every element of $H^1(X;\mathbb{Z})$ has the form $f^*(\alpha)$ for some $f:X \to K(\mathbb{Z},1)$, with α a fundamental class in $H^1(K(\mathbb{Z},1);\mathbb{Z})$, further reducing us to verifying the result for this single α, the 'universal example.' And for this universal α it is evident that $\alpha^2 = 0$ since S^1 is a $K(\mathbb{Z},1)$ and $H^2(S^1;\mathbb{Z}) = 0$.

Does this fact generalize? It certainly does not hold if we replace the coefficient ring $\mathbb{Z}$ by $\mathbb{Z}_2$ since $H^*(\mathbb{RP}^\infty;\mathbb{Z}_2) = \mathbb{Z}_2[x]$. Indeed, the example of $\mathbb{RP}^\infty$ shows more generally that the fundamental class $\alpha \in H^n(K(\mathbb{Z}_2,n);\mathbb{Z}_2)$ generates a polynomial subalgebra $\mathbb{Z}_2[\alpha] \subset H^*(K(\mathbb{Z}_2,n);\mathbb{Z}_2)$ for each $n \geq 1$, since there is a map $f:\mathbb{RP}^\infty \to K(\mathbb{Z}_2,n)$ with $f^*(\alpha) = x^n$ and all the powers of x^n are nonzero, hence also all the powers of α. By the same reasoning, the example of $\mathbb{CP}^\infty$ shows that the fundamental class $\alpha \in H^{2n}(K(\mathbb{Z},2n);\mathbb{Z})$ generates a polynomial subalgebra $\mathbb{Z}[\alpha]$ in $H^*(K(\mathbb{Z},2n);\mathbb{Z})$. As we shall see in [SSAT], $H^*(K(\mathbb{Z},2n);\mathbb{Z})/torsion$ is exactly this polynomial algebra $\mathbb{Z}[\alpha]$.

A little more subtle is the question of identifying the subalgebra of $H^*(K(\mathbb{Z},n);\mathbb{Z})$ generated by the fundamental class α for odd $n \geq 3$. By the commutativity property of cup products we know that α^2 is either zero or of order two. To see that α^2 is nonzero it suffices to find a single space X with an element $y \in H^n(X;\mathbb{Z})$ such that $y^2 \neq 0$. The first place to look might be $\mathbb{RP}^\infty$, but its cohomology with $\mathbb{Z}$ coefficients is concentrated in even dimensions. Instead, consider $X = \mathbb{RP}^\infty \times \mathbb{RP}^\infty$. This has $\mathbb{Z}_2$ cohomology $\mathbb{Z}_2[x,y]$ and Example 3E.5 shows that its $\mathbb{Z}$ cohomology is the $\mathbb{Z}_2[x^2,y^2]$-submodule generated by 1 and $x^2 y + xy^2$, except in dimension zero of course, where 1 generates a $\mathbb{Z}$ rather than a $\mathbb{Z}_2$. In particular we can take $y = x^{2k}(x^2 y + xy^2)$ for any $k \geq 0$, and then all powers y^m are nonzero since we are inside the polynomial ring $\mathbb{Z}_2[x,y]$. It follows that the subalgebra of $H^*(K(\mathbb{Z},n);\mathbb{Z})$ generated by α is $\mathbb{Z}[\alpha]/(2\alpha^2)$ for odd $n \geq 3$.

These examples lead one to wonder just how complicated the cohomology of $K(G,n)$'s is. The general construction of a $K(G,n)$ is not very helpful in answering this question. Consider the case $G = \mathbb{Z}$ for example. Here one would start with S^n and attach $(n+2)$-cells to kill $\pi_{n+1}(S^n)$. Since $\pi_{n+1}(S^n)$ happens to be cyclic, only one $(n+2)$-cell is needed. To continue, one would have to compute generators for π_{n+2} of the resulting space $S^n \cup e^{n+2}$, use these to attach $(n+3)$-cells, then compute the resulting π_{n+3}, and so on for each successive dimension. When $n = 2$ this procedure happens to work out very neatly, and the resulting $K(\mathbb{Z},2)$ is $\mathbb{CP}^\infty$ with its usual CW structure having one cell in each even dimension, according to an exercise at the end of the section. However, for larger n it quickly becomes impractical to make this procedure explicit since homotopy groups are so hard to compute. One can get some idea of the difficulties of the next case $n = 3$ by considering the homology groups of $K(\mathbb{Z},3)$. Using techniques in [SSAT], the groups $H_i(K(\mathbb{Z},3);\mathbb{Z})$ for $0 \leq i \leq 12$ can be computed to be

$$\mathbb{Z}, \ 0, \ 0, \ \mathbb{Z}, \ 0, \ \mathbb{Z}_2, \ 0, \ \mathbb{Z}_3, \ \mathbb{Z}_2, \ \mathbb{Z}_2, \ \mathbb{Z}_3, \ \mathbb{Z}_{10}, \ \mathbb{Z}_2$$

To get this sequence of homology groups would require quite a few cells, and the situation only gets worse in higher dimensions, where the homology groups are not always cyclic.

Indeed, one might guess that computing the homology groups of $K(\mathbb{Z}, n)$'s would be of the same order of difficulty as computing the homotopy groups of spheres, but by some miracle this is not the case. The calculations are indeed complicated, but they were completely done by Serre and Cartan in the 1950s, not just for $K(\mathbb{Z}, n)$'s, but for all $K(G, n)$'s with G finitely generated abelian. For example, $H^*(K(\mathbb{Z}, 3); \mathbb{Z}_2)$ is the polynomial algebra $\mathbb{Z}_2[x_3, x_5, x_9, x_{17}, x_{33}, \cdots]$ with generators of dimensions $2^i + 1$, indicated by the subscripts. And in general, for G finitely generated abelian, $H^*(K(G, n); \mathbb{Z}_p)$ is a polynomial algebra on generators of specified dimensions if p is 2, while for p an odd prime one gets the tensor product of a polynomial ring on generators of specified even dimensions and an exterior algebra on generators of specified odd dimensions. With $\mathbb{Z}$ coefficients the description of the cohomology is not nearly so neat, however. We will study these questions in some detail in [SSAT].

There is a good reason for being interested in the cohomology of $K(G, n)$'s, arising from the equivalence $H^n(X; G) \approx \langle X, K(G, n) \rangle$. Taking $\mathbb{Z}$ coefficients for simplicity, an element of $H^m(K(\mathbb{Z}, n); \mathbb{Z})$ corresponds to a map $\theta : K(\mathbb{Z}, n) \to K(\mathbb{Z}, m)$. We can compose θ with any map $f : X \to K(\mathbb{Z}, n)$ to get a map $\theta f : X \to K(\mathbb{Z}, m)$. Letting f vary and keeping θ fixed, this gives a function $H^n(X; \mathbb{Z}) \to H^m(X; \mathbb{Z})$, depending only on θ. This is the idea of *cohomology operations*, which we study in more detail in §4.L.

The equivalence $H^n(X; G) \approx \langle X, K(G, n) \rangle$ also leads to a new viewpoint toward cup products. Taking G to be a ring R and setting $K_n = K(R, n)$, then if we are given maps $f : X \to K_m$ and $g : Y \to K_n$, we can define the cross product of the corresponding cohomology classes by the composition

$$X \times Y \xrightarrow{\ f \times g\ } K_m \times K_n \longrightarrow K_m \wedge K_n \xrightarrow{\ \mu\ } K_{m+n}$$

where the middle map is the quotient map and μ can be defined in the following way. The space $K_m \wedge K_n$ is $(m + n - 1)$-connected, so by the Hurewicz theorem and the Künneth formula for reduced homology we have isomorphisms $\pi_{m+n}(K_m \wedge K_n) \approx H_{m+n}(K_m \wedge K_n) \approx H_m(K_m) \otimes H_n(K_n) \approx R \otimes R$. By Lemmas 4.7 and 4.31 there is then a map $\mu : K_m \wedge K_n \to K_{m+n}$ inducing the multiplication map $R \otimes R \to R$ on π_{m+n}. Or we could use the isomorphism $H^{m+n}(K_m \wedge K_n; R) \approx \mathrm{Hom}(H_{m+n}(K_m \wedge K_n), R)$ and let μ be the map corresponding to the cohomology class given by the multiplication homomorphism $R \otimes R \to R$.

The case $R = \mathbb{Z}$ is particularly simple. We can take S^m as the $(m + 1)$-skeleton of K_m, and similarly for K_n, so $K_m \wedge K_n$ has $S^m \wedge S^n$ as its $(m + n + 1)$-skeleton and we can obtain μ by extending the inclusion $S^m \wedge S^n = S^{m+n} \hookrightarrow K_{m+n}$.

It is not hard to prove the basic properties of cup product using this definition, and in particular the commutativity property becomes somewhat more transparent from this viewpoint. For example, when $R = \mathbb{Z}$, commutativity just comes down to the fact that the map $S^m \wedge S^n \to S^n \wedge S^m$ switching the factors has degree $(-1)^{mn}$ when regarded as a map of S^{m+n}.

Fibrations

Recall from §4.2 that a fibration is a map $p : E \to B$ having the homotopy lifting property with respect to all spaces. In a fiber bundle all the fibers are homeomorphic by definition, but this need not be true for fibrations. An example is the linear projection of a 2-simplex onto one of its edges, which is a fibration according to an exercise at the end of the section. The following result gives some evidence that fibrations should be thought of as a homotopy-theoretic analog of fiber bundles:

Proposition 4.61. *For a fibration $p : E \to B$, the fibers $F_b = p^{-1}(b)$ over each path component of B are all homotopy equivalent.*

Proof: A path $\gamma : I \to B$ gives rise to a homotopy $g_t : F_{\gamma(0)} \to B$ with $g_t(F_{\gamma(0)}) = \gamma(t)$. The inclusion $F_{\gamma(0)} \hookrightarrow E$ provides a lift $\tilde{g}_0$, so by the homotopy lifting property we have a homotopy $\tilde{g}_t : F_{\gamma(0)} \to E$ with $\tilde{g}_t(F_{\gamma(0)}) \subset F_{\gamma(t)}$ for all t. In particular, $\tilde{g}_1$ gives a map $L_\gamma : F_{\gamma(0)} \to F_{\gamma(1)}$. The association $\gamma \mapsto L_\gamma$ has the following basic properties:

(a) If $\gamma \simeq \gamma'$ rel ∂I, then $L_\gamma \simeq L_{\gamma'}$. In particular the homotopy class of L_γ is independent of the choice of the lifting $\tilde{g}_t$ of g_t.

(b) For a composition of paths $\gamma\gamma'$, $L_{\gamma\gamma'}$ is homotopic to the composition $L_{\gamma'}L_\gamma$.

From these statements it follows that L_γ is a homotopy equivalence with homotopy inverse $L_{\bar{\gamma}}$, where $\bar{\gamma}$ is the inverse path of γ.

Before proving (a), note that a fibration has the homotopy lifting property for pairs $(X \times I, X \times \partial I)$ since the pairs $(I \times I, I \times \{0\} \cup \partial I \times I)$ and $(I \times I, I \times \{0\})$ are homeomorphic, hence the same is true after taking products with X.

To prove (a), let $\gamma(s, t)$ be a homotopy from $\gamma(t)$ to $\gamma'(t)$, $(s, t) \in I \times I$. This determines a family $g_{st} : F_{\gamma(0)} \to B$ with $g_{st}(F_{\gamma(0)}) = \gamma(s, t)$. Let $\tilde{g}_{0,t}$ and $\tilde{g}_{1,t}$ be lifts defining L_γ and $L_{\gamma'}$, and let $\tilde{g}_{s,0}$ be the inclusion $F_{\gamma(0)} \hookrightarrow E$ for all s. Using the homotopy lifting property for the pair $(F_{\gamma(0)} \times I, F_{\gamma(0)} \times \partial I)$, we can extend these lifts to lifts $\tilde{g}_{st}$ for $(s, t) \in I \times I$. Restricting to $t = 1$ then gives a homotopy $L_\gamma \simeq L_{\gamma'}$.

Property (b) holds since for lifts $\tilde{g}_t$ and $\tilde{g}'_t$ defining L_γ and $L_{\gamma'}$ we obtain a lift defining $L_{\gamma\gamma'}$ by taking $\tilde{g}_{2t}$ for $0 \le t \le 1/2$ and $\tilde{g}'_{2t-1}L_\gamma$ for $1/2 \le t \le 1$. $\square$

One may ask whether fibrations satisfy a homotopy analog of the local triviality property of fiber bundles. Observe first that for a fibration $p : E \to B$, the restriction

$p : p^{-1}(A) \to A$ is a fibration for any subspace $A \subset B$. So we can ask whether every point of B has a neighborhood U for which the fibration $p^{-1}(U) \to U$ is equivalent in some homotopy-theoretic sense to a projection $U \times F \to U$. The natural notion of equivalence for fibrations is defined in the following way. Given fibrations $p_1 : E_1 \to B$ and $p_2 : E_2 \to B$, a map $f : E_1 \to E_2$ is called **fiber-preserving** if $p_1 = p_2 f$, or in other words, $f(p_1^{-1}(b)) \subset p_2^{-1}(b)$ for all $b \in B$. A fiber-preserving map $f : E_1 \to E_2$ is a **fiber homotopy equivalence** if there is a fiber-preserving map $g : E_2 \to E_1$ such that both compositions fg and gf are homotopic to the identity through fiber-preserving maps. A fiber homotopy equivalence can be thought of as a family of homotopy equivalences between corresponding fibers of E_1 and E_2. An interesting fact is that a fiber-preserving map that is a homotopy equivalence is a fiber homotopy equivalence; this is an exercise for §4.H.

We will show that a fibration $p : E \to B$ is locally fiber-homotopically trivial in the sense described above if B is locally contractible. In order to do this we first digress to introduce another basic concept.

Given a fibration $p : E \to B$ and a map $f : A \to B$, there is a **pullback** or **induced** fibration $f^*(E) \to A$ obtained by setting $f^*(E) = \{(a, e) \in A \times E \mid f(a) = p(e)\}$, with the projections of $f^*(E)$ onto A and E giving a commutative diagram as shown at the right. The homotopy lifting property holds for $f^*(E) \to A$ since a homotopy $g_t : X \to A$ gives the first coordinate of a lift $\tilde{g}_t : X \to f^*(E)$, the second coordinate being a lifting to E of the composed homotopy fg_t.

$$
\begin{array}{ccc}
f^*(E) & \longrightarrow & E \\
\downarrow & & \downarrow{p} \\
A & \xrightarrow{\ f\ } & B
\end{array}
$$

Proposition 4.62. *Given a fibration $p : E \to B$ and a homotopy $f_t : A \to B$, the pullback fibrations $f_0^*(E) \to A$ and $f_1^*(E) \to A$ are fiber homotopy equivalent.*

Proof: Let $F : A \times I \to B$ be the homotopy f_t. The fibration $F^*(E) \to A \times I$ contains $f_0^*(E)$ and $f_1^*(E)$ over $A \times \{0\}$ and $A \times \{1\}$. So it suffices to prove the following: For a fibration $p : E \to B \times I$, the restricted fibrations $E_s = p^{-1}(B \times \{s\}) \to B$ are all fiber homotopy equivalent for $s \in [0, 1]$.

To prove this assertion the idea is to imitate the construction of the homotopy equivalences L_γ in the proof of Proposition 4.61. A path $\gamma : [0, 1] \to I$ gives rise to a fiber-preserving map $L_\gamma : E_{\gamma(0)} \to E_{\gamma(1)}$ by lifting the homotopy $g_t : E_{\gamma(0)} \to B \times I$, $g_t(x) = (p(x), \gamma(t))$, starting with the inclusion $E_{\gamma(0)} \hookrightarrow E$. As before, one shows the two basic properties (a) and (b), noting that in (a) the homotopy $L_\gamma \simeq L_{\gamma'}$ is fiber-preserving since it is obtained by lifting a homotopy $h_t : E_{\gamma(0)} \times [0, 1] \to B \times I$ of the form $h_t(x, u) = (p(x), -)$. From (a) and (b) it follows that L_γ is a fiber homotopy equivalence with inverse $L_{\bar{\gamma}}$.   □

Corollary 4.63. *A fibration $E \to B$ over a contractible base B is fiber homotopy equivalent to a product fibration $B \times F \to B$.*

Proof: The pullback of E by the identity map $B \to B$ is E itself, while the pullback by a constant map $B \to B$ is a product $B \times F$. $\qquad\square$

Thus we see that if B is locally contractible then any fibration over B is locally fiber homotopy equivalent to a product fibration.

Pathspace Constructions

There is a simple but extremely useful way to turn arbitrary mappings into fibrations. Given a map $f : A \to B$, let E_f be the space of pairs (a, γ) where $a \in A$ and $\gamma : I \to B$ is a path in B with $\gamma(0) = f(a)$. We topologize E_f as a subspace of $A \times B^I$, where B^I is the space of mappings $I \to B$ with the compact-open topology; see the Appendix for the definition and basic properties of this topology, in particular Proposition A.14 which we will be using shortly.

‖ **Proposition 4.64.** *The map* $p : E_f \to B$, $p(a, \gamma) = \gamma(1)$, *is a fibration.*

Proof: Continuity of p follows from (a) of Proposition A.14 in the Appendix which says that the evaluation map $B^I \times I \to B$, $(\gamma, s) \mapsto \gamma(s)$, is continuous.

To verify the fibration property, let a homotopy $g_t : X \to B$ and a lift $\tilde{g}_0 : X \to E_f$ of g_0 be given. Write $\tilde{g}_0(x) = (h(x), \gamma_x)$ for $h : X \to A$ and $\gamma_x : I \to B$. Define a lift $\tilde{g}_t : X \to E_f$ by $\tilde{g}_t(x) = (h(x), \gamma_x \cdot g_{[0,t]}(x))$, the second coordinate being the path γ_x followed by the path traced out by $g_s(x)$ for $0 \le s \le t$. This composition of paths is defined since $g_0(x) = p \tilde{g}_0(x) = \gamma_x(1)$. To check that $\tilde{g}_t$ is a continuous homotopy we regard it as a map $X \times I \to E_f \subset A \times B^I$ and then apply (b) of Proposition A.14 which in the current context asserts that continuity of a map $X \times I \to A \times B^I$ is equivalent to continuity of the associated map $X \times I \times I \to A \times B$. $\qquad\square$

We can regard A as the subspace of E_f consisting of pairs (a, γ) with γ the constant path at $f(a)$, and E_f deformation retracts onto this subspace by restricting all the paths γ to shorter and shorter initial segments. The map $p : E_f \to B$ restricts to f on the subspace A, so we have factored an arbitrary map $f : A \to B$ as the composition $A \hookrightarrow E_f \to B$ of a homotopy equivalence and a fibration. We can also think of this construction as extending f to a fibration $E_f \to B$ by enlarging its domain to a homotopy equivalent space. The fiber F_f of $E_f \to B$ is called the **homotopy fiber** of f. It consists of all pairs (a, γ) with $a \in A$ and γ a path in B from $f(a)$ to a basepoint $b_0 \in B$.

If $f : A \to B$ is the inclusion of a subspace, then E_f is the space of paths in B starting at points of A. In this case a map $(I^{i+1}, \partial I^{i+1}, J^i) \to (B, A, x_0)$ is the same as a map $(I^i, \partial I^i) \to (F_f, \gamma_0)$ where γ_0 is the constant path at x_0 and F_f is the fiber of E_f over x_0. This means that $\pi_{i+1}(B, A, x_0)$ can be identified with $\pi_i(F_f, \gamma_0)$, hence the long exact sequences of homotopy groups of the pair (B, A) and of the fibration $E_f \to B$ can be identified.

An important special case is when f is the inclusion of the basepoint b_0 into B. Then E_f is the space PB of paths in B starting at b_0, and $p : PB \to B$ sends each path to its endpoint. The fiber $p^{-1}(b_0)$ is the loopspace ΩB consisting of all loops in B based at b_0. Since PB is contractible by progressively truncating paths, the long exact sequence of homotopy groups for the path fibration $PB \to B$ yields another proof that $\pi_n(X, x_0) \approx \pi_{n-1}(\Omega X, x_0)$ for all n.

As we mentioned in the discussion of loopspaces earlier in this section, it is a theorem of [Milnor 1959] that the loopspace of a CW complex is homotopy equivalent to a CW complex. Milnor's theorem is actually quite a bit more general than this, and implies in particular that the homotopy fiber of an arbitrary map between CW complexes has the homotopy type of a CW complex. One can usually avoid quoting these results by using CW approximations, though it is reassuring to know they are available if needed, or if one does not want to bother with CW approximations.

If the fibration construction $f \mapsto E_f$ is applied to a map $p : E \to B$ that is already a fibration, one might expect the resulting fibration $E_p \to B$ to be closely related to the original fibration $E \to B$. This is indeed the case:

Proposition 4.65. *If $p : E \to B$ is a fibration, then the inclusion $E \hookrightarrow E_p$ is a fiber homotopy equivalence. In particular, the homotopy fibers of p are homotopy equivalent to the actual fibers.*

Proof: We apply the homotopy lifting property to the homotopy $g_t : E_p \to B$, $g_t(e, \gamma) = \gamma(t)$, with initial lift $\tilde{g}_0 : E_p \to E$, $\tilde{g}_0(e, \gamma) = e$. The lifting $\tilde{g}_t : E_p \to E$ is then the first coordinate of a homotopy $h_t : E_p \to E_p$ whose second coordinate is the restriction of the paths γ to the interval $[t, 1]$. Since the endpoints of the paths γ are unchanged, h_t is fiber-preserving. We have $h_0 = \mathbb{1}$, $h_1(E_p) \subset E$, and $h_t(E) \subset E$ for all t. If we let i denote the inclusion $E \hookrightarrow E_p$, then $ih_1 \simeq \mathbb{1}$ via h_t and $h_1 i \simeq \mathbb{1}$ via $h_t|E$, so i is a fiber homotopy equivalence. $\square$

We have seen that loopspaces occur as fibers of fibrations $PB \to B$ with contractible total space PB. Here is something of a converse:

Proposition 4.66. *If $F \to E \to B$ is a fibration or fiber bundle with E contractible, then there is a weak homotopy equivalence $F \to \Omega B$.*

Proof: If we compose a contraction of E with the projection $p : E \to B$ then we have for each point $x \in E$ a path γ_x in B from $p(x)$ to a basepoint $b_0 = p(x_0)$, where x_0 is the point to which E contracts. This yields a map $E \to PB$, $x \mapsto \overline{\gamma}_x$, whose composition with the fibration $PB \to B$ is p. By restriction this gives a map $F \to \Omega B$ where $F = p^{-1}(b_0)$, and the long exact sequence of homotopy groups for $F \to E \to B$ maps to the long exact sequence for $\Omega B \to PB \to B$. Since E and PB are contractible, the five-lemma implies that the map $F \to \Omega B$ is a weak homotopy equivalence.

$$
\begin{array}{ccccc}
F & \longrightarrow & E & \xrightarrow{\ p\ } & B \\
\downarrow & & \downarrow & & \| \\
\Omega B & \longrightarrow & PB & \longrightarrow & B
\end{array}
$$

$\square$

Examples arising from fiber bundles constructed earlier in the chapter are $O(n) \simeq \Omega G_n(\mathbb{R}^\infty)$, $U(n) \simeq \Omega G_n(\mathbb{C}^\infty)$, and $Sp(n) \simeq \Omega G_n(\mathbb{H}^\infty)$. In particular, taking $n = 1$ in the latter two examples, we have $S^1 \simeq \Omega \mathbb{CP}^\infty$ and $S^3 \simeq \Omega \mathbb{HP}^\infty$. Note that in all these examples it is a topological group that is homotopy equivalent to a loopspace. In [Milnor 1956] this is shown to hold in general: For each topological group G there is a fiber bundle $G \to EG \to BG$ with EG contractible, hence by the proposition there is a weak equivalence $G \simeq \Omega BG$. There is also a converse statement: The loopspace of a CW complex is homotopy equivalent to a topological group.

The relationship between X and ΩX has been much studied, particularly the case that ΩX has the homotopy type of a finite CW complex, which is of special interest because of the examples of the classical Lie groups such as $O(n)$, $U(n)$, and $Sp(n)$. See [Kane 1988] for an introduction to this subject.

It is interesting to see what happens when the process of forming homotopy fibers is iterated. Given a fibration $p : E \to B$ with fiber $F = p^{-1}(b_0)$, we know that the inclusion of F into the homotopy fiber F_p is a homotopy equivalence. Recall that F_p consists of pairs (e, γ) with $e \in E$ and γ a path in B from $p(e)$ to b_0. The inclusion $F \hookrightarrow E$ extends to a map $i : F_p \to E$, $i(e, \gamma) = e$, and this map is obviously a fibration. In fact it is the pullback via p of the path fibration $PB \to B$. This allows us to iterate, taking the homotopy fiber F_i with its map to F_p, and so on, as in the first row of the following diagram:

$$
\begin{array}{ccccccccccc}
\cdots & \longrightarrow & F_j & \longrightarrow & F_i & \xrightarrow{\ j\ } & F_p & \xrightarrow{\ i\ } & E & \xrightarrow{\ p\ } & B \\
& & \simeq \Big\updownarrow & & \simeq \Big\updownarrow & & \simeq \Big\uparrow & & \Big\| & & \Big\| \\
\cdots & \longrightarrow & \Omega E & \xrightarrow{\ \Omega p\ } & \Omega B & \longrightarrow & F & \longrightarrow & E & \xrightarrow{\ p\ } & B
\end{array}
$$

The actual fiber of i over a point $e_0 \in p^{-1}(b_0)$ consists of pairs (e_0, γ) with γ a loop in B at the basepoint b_0, so this fiber is just ΩB, and the inclusion $\Omega B \hookrightarrow F_i$ is a homotopy equivalence. In the second row of the diagram the map $\Omega B \to F$ is the composition $\Omega B \hookrightarrow F_i \to F_p \to F$ where the last map is a homotopy inverse to the inclusion $F \hookrightarrow F_p$, so the square in the diagram containing these maps commutes up to homotopy. The homotopy fiber F_i consists of pairs (γ, η) where η is a path in E ending at e_0 and γ is a path in B from $p(\eta(0))$ to b_0. A homotopy inverse to the inclusion $\Omega B \hookrightarrow F_i$ is the retraction $F_i \to \Omega B$ sending (γ, η) to the loop obtained by composing the inverse path of $p\eta$ with γ. These constructions can now be iterated indefinitely.

Thus we produce a sequence

$$ \cdots \to \Omega^2 B \to \Omega F \to \Omega E \to \Omega B \to F \to E \to B $$

where any two consecutive maps form a fibration, up to homotopy equivalence, and all the maps to the left of ΩB are obtained by applying the functor Ω to the later maps. The long exact sequence of homotopy groups for any fibration in the sequence coincides with the long exact sequence for $F \to E \to B$, as the reader can check.

Postnikov Towers

A Postnikov tower for a path-connected space X is a commutative diagram as at the right, such that:

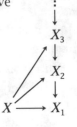

(1) The map $X \to X_n$ induces an isomorphism on π_i for $i \le n$.

(2) $\pi_i(X_n) = 0$ for $i > n$.

As we saw in Example 4.16, every connected CW complex X has a Postnikov tower, and this is unique up to homotopy equivalence by Corollary 4.19.

If we convert the map $X_n \to X_{n-1}$ into a fibration, its fiber F_n is a $K(\pi_n X, n)$, as is apparent from a brief inspection of the long exact sequence of homotopy groups for the fibration:

$$\pi_{i+1}(X_n) \to \pi_{i+1}(X_{n-1}) \to \pi_i(F_n) \to \pi_i(X_n) \to \pi_i(X_{n-1})$$

We can replace each map $X_n \to X_{n-1}$ by a fibration $X'_n \to X'_{n-1}$ in succession, starting with $X_2 \to X_1$ and working upward. For the inductive step we convert the composition $X_n \to X_{n-1} \hookrightarrow X'_{n-1}$ into a fibration $X'_n \to X'_{n-1}$ fitting into the commutative diagram at the right. Thus we obtain a Postnikov tower satisfying also the condition

$$
\begin{array}{ccc}
X_n & \hookrightarrow & X'_n \\
\downarrow & & \downarrow \\
X_{n-1} & \hookrightarrow & X'_{n-1}
\end{array}
$$

(3) The map $X_n \to X_{n-1}$ is a fibration with fiber a $K(\pi_n X, n)$.

To the extent that fibrations can be regarded as twisted products, up to homotopy equivalence, the spaces X_n in a Postnikov tower for X can be thought of as twisted products of Eilenberg-MacLane spaces $K(\pi_n X, n)$.

For many purposes, a CW complex X can be replaced by one of the stages X_n in a Postnikov tower for X, for example if one is interested in homotopy or homology groups in only a finite range of dimensions. However, to determine the full homotopy type of X from its Postnikov tower, some sort of limit process is needed. Let us investigate this question is somewhat greater generality.

Given a sequence of maps $\cdots \to X_2 \to X_1$, define their **inverse limit** $\varprojlim X_n$ to be the subspace of the product $\prod_n X_n$ consisting of sequences of points $x_n \in X_n$ with x_n mapping to x_{n-1} under the map $X_n \to X_{n-1}$. The corresponding algebraic notion is the inverse limit $\varprojlim X_n$ of a sequence of group homomorphisms $\cdots \to G_2 \to G_1$, which is the subgroup of $\prod_n G_n$ consisting of sequences of elements $g_n \in G_n$ with g_n mapping to g_{n-1} under the homomorphism $G_n \to G_{n-1}$.

Proposition 4.67. *For an arbitrary sequence of fibrations* $\cdots \to X_2 \to X_1$ *the natural map* $\lambda : \pi_i(\varprojlim X_n) \to \varprojlim \pi_i(X_n)$ *is surjective, and* λ *is injective if the maps* $\pi_{i+1}(X_n) \to \pi_{i+1}(X_{n-1})$ *are surjective for n sufficiently large.*

Proof: Represent an element of $\varprojlim \pi_i(X_n)$ by maps $f_n : (S^i, s_0) \to (X_n, x_n)$. Since the projection $p_n : X_n \to X_{n-1}$ takes $[f_n]$ to $[f_{n-1}]$, by applying the homotopy lifting property for the pair (S^i, s_0) we can homotope f_n, fixing s_0, so that $p_n f_n = f_{n-1}$. Doing this inductively for $n = 2, 3, \cdots$, we get $p_n f_n = f_{n-1}$ for all n simultaneously, which gives surjectivity of λ.

For injectivity, note first that inverse limits are unaffected by throwing away a finite number of terms at the end of the sequence of spaces or groups, so we may assume the maps $\pi_{i+1}(X_n) \to \pi_{i+1}(X_{n-1})$ are surjective for all n. Given a map $f : S^i \to \varprojlim X_n$, suppose we have nullhomotopies $F_n : D^{i+1} \to X_n$ of the coordinate functions $f_n : S^i \to X_n$ of f. We have $p_n F_n = F_{n-1}$ on S^i, so $p_n F_n$ and F_{n-1} are the restrictions to the two hemispheres of S^{i+1} of a map $g_{n-1} : S^{i+1} \to X_{n-1}$. If the map $\pi_{i+1}(X_n) \to \pi_{i+1}(X_{n-1})$ is surjective, we can rechoose F_n so that the new g_{n-1} is nullhomotopic, that is, so that $p_n F_n \simeq F_{n-1}$ rel S^i. Applying the homotopy lifting property for (D^{i+1}, S^i), we can make $p_n F_n = F_{n-1}$. Doing this inductively for $n = 2, 3, \cdots$, we see that $f : S^i \to \varprojlim X_n$ is nullhomotopic and λ is injective. $\qquad\square$

One might wish to have a description of the kernel of λ in the case of an arbitrary sequence of fibrations $\cdots \to X_2 \to X_1$, though for our present purposes this question is not relevant. In fact, $\operatorname{Ker}\lambda$ is naturally isomorphic to $\varprojlim^1 \pi_{i+1}(X_n)$, where $\varprojlim^1$ is the functor defined in §3.F. Namely, if $f : S^i \to \varprojlim X_n$ determines an element of $\operatorname{Ker}\lambda$, then the sequence of maps $g_n : S^{i+1} \to X_n$ constructed above gives an element of $\prod_n \pi_{i+1}(X_n)$, well-defined up to the choice of the nullhomotopies F_n. Any new choice of F_n is obtained by adding a map $G_n : S^{i+1} \to X_n$ to F_n. The effect of this is to change g_n to $g_n + G_n$ and g_{n-1} to $g_{n-1} - p_n G_n$. Since $\varprojlim^1 \pi_{i+1}(X_n)$ is the quotient of $\prod_n \pi_{i+1}(X_n)$ under exactly these identifications, we get $\operatorname{Ker}\lambda \approx \varprojlim^1 \pi_{i+1}(X_n)$. Thus for each $i > 0$ there is a natural exact sequence

$$0 \longrightarrow \varprojlim{}^1 \pi_{i+1}(X_n) \longrightarrow \pi_i(\varprojlim X_n) \longrightarrow \varprojlim \pi_i(X_n) \longrightarrow 0$$

The proposition says that the $\varprojlim^1$ term vanishes if the maps $\pi_{i+1}(X_n) \to \pi_{i+1}(X_{n-1})$ are surjective for sufficiently large n.

Corollary 4.68. *For the Postnikov tower of a connected CW complex X the natural map $X \to \varprojlim X_n$ is a weak homotopy equivalence, so X is a CW approximation to $\varprojlim X_n$.*

Proof: The composition $\pi_i(X) \to \pi_i(\varprojlim X_n) \overset{\lambda}{\longrightarrow} \varprojlim \pi_i(X_n)$ is an isomorphism since $\pi_i(X) \to \pi_i(X_n)$ is an isomorphism for large n. $\qquad\square$

Having seen how to decompose a space X into the terms in its Postnikov tower, we consider now the inverse process of building a Postnikov tower, starting with X_1 as a $K(\pi, 1)$ and inductively constructing X_n from X_{n-1}. It would be very nice if the fibration $K(\pi, n) \to X_n \to X_{n-1}$ could be extended another term to the right, to form a fibration sequence

$$K(\pi, n) \to X_n \to X_{n-1} \to K(\pi, n+1)$$

for this would say that X_n is the homotopy fiber of a map $X_{n-1} \to K(\pi, n+1)$, and homotopy classes of such maps are in one-to-one correspondence with elements of $H^{n+1}(X_{n-1}; \pi)$ by Theorem 4.57. Since the homotopy fiber of $X_{n-1} \to K(\pi, n+1)$ is the same as the pullback of the path fibration $PK(\pi, n+1) \to K(\pi, n+1)$, its homotopy type depends only on the homotopy class of the map $X_{n-1} \to K(\pi, n+1)$, by Proposition 4.62. Note that the last term $K(\pi, n+1)$ in the fibration sequence above cannot be anything else but a $K(\pi, n+1)$ since its loopspace must be homotopy equivalent to the first term in the sequence, a $K(\pi, n)$.

In general, a fibration $F \to E \to B$ is called **principal** if there is a commutative diagram

$$
\begin{array}{ccccc}
F & \longrightarrow & E & \longrightarrow & B \\
\downarrow & & \downarrow & & \downarrow \\
\Omega B' & \longrightarrow & F' & \longrightarrow & E' & \longrightarrow & B'
\end{array}
$$

where the second row is a fibration sequence and the vertical maps are weak homotopy equivalences. Thus if all the fibrations in a Postnikov tower for X happen to be principal, we have a diagram as at the right, where each X_{n+1} is, up to weak homotopy equivalence, the homotopy fiber of the map $k_n : X_n \to K(\pi_{n+1}X, n+2)$. The map k_n is equivalent to a class in $H^{n+2}(X_n; \pi_{n+1}X)$ called the nth **k-invariant** of X. These classes specify how to construct X inductively from

$$
\begin{array}{ccc}
& \vdots & \\
& \downarrow & \\
K(\pi_3 X, 3) \longrightarrow & X_3 & \xrightarrow{\ k_3\ } K(\pi_4 X, 5) \\
& \downarrow & \\
K(\pi_2 X, 2) \longrightarrow & X_2 & \xrightarrow{\ k_2\ } K(\pi_3 X, 4) \\
& \downarrow & \\
K(\pi_1 X, 1) = & X_1 & \xrightarrow{\ k_1\ } K(\pi_2 X, 3)
\end{array}
$$

Eilenberg–MacLane spaces. For example, if all the k_n's are zero, X is just the product of the spaces $K(\pi_n X, n)$, and in the general case X is some sort of twisted product of $K(\pi_n X, n)$'s.

To actually build a space from its k-invariants is usually too unwieldy a procedure to be carried out in practice, but as a theoretical tool this procedure can be quite useful. The next result tells us when this tool is available:

Theorem 4.69. *A connected CW complex X has a Postnikov tower of principal fibrations iff $\pi_1(X)$ acts trivially on $\pi_n(X)$ for all $n > 1$.*

Notice that in the definition of a principal fibration, the map $F \to \Omega B'$ automatically exists and is a homotopy weak equivalence once one has the right-hand square of the commutative diagram with its vertical maps weak homotopy equivalences. Thus the question of whether a fibration is principal can be rephrased in the following way: Given a map $A \to X$, which one can always replace by an equivalent fibration if one likes, does there exist a fibration $F \to E \to B$ and a commutative square as at the right, with the vertical maps weak homotopy equivalences? By replacing A and X with CW approximations and

$$
\begin{array}{ccc}
A & \longrightarrow & X \\
\downarrow & & \downarrow \\
F & \longrightarrow & E
\end{array}
$$

converting the resulting map $A \to X$ into an inclusion via a mapping cylinder, the question becomes whether a CW pair (X, A) is equivalent to a fibration pair (E, F), that is, whether there is a fibration $F \to E \to B$ and a map $(X, A) \to (E, F)$ for which both $X \to E$ and $A \to F$ are weak homotopy equivalences. In general the answer will rarely be yes, since the homotopy fiber of $A \hookrightarrow X$ would have to have the weak homotopy type of a loopspace, which is a rather severe restriction. However, in the situation of Postnikov towers, the homotopy fiber is a $K(\pi, n)$ with π abelian since $n \geq 2$, so it is a loopspace. But there is another requirement: The action of $\pi_1(A)$ on $\pi_n(X, A)$ must be trivial for all $n \geq 1$. This is equivalent to the action of $\pi_1(F)$ on $\pi_n(E, F)$ being trivial, which is always the case in a fibration since under the isomorphism $p_* : \pi_n(E, F) \to \pi_n(B, x_0)$ an element $y\alpha - \alpha$, with $y \in \pi_1(F)$ and $\alpha \in \pi_n(E, F)$, maps to $p_*(y)p_*(\alpha) - p_*(\alpha)$ which is zero since $p_*(y)$ lies in the trivial group $\pi_1(x_0)$.

The relative group $\pi_n(X, A)$ is always isomorphic to π_{n-1} of the homotopy fiber of the inclusion $A \hookrightarrow X$, so in the case at hand when the homotopy fiber is a $K(\pi, n)$, the only nontrivial relative homotopy group is $\pi_{n+1}(X, A) \approx \pi$. In this case the necessary condition of trivial action is also sufficient:

Lemma 4.70. *Let (X, A) be a CW pair with both X and A connected, such that the homotopy fiber of the inclusion $A \hookrightarrow X$ is a $K(\pi, n)$, $n \geq 1$. Then there exists a fibration $F \to E \to B$ and a map $(X, A) \to (E, F)$ inducing weak homotopy equivalences $X \to E$ and $A \to F$ iff the action of $\pi_1(A)$ on $\pi_{n+1}(X, A)$ is trivial.*

Proof: It remains only to prove the 'if' implication. As we noted just before the statement of the lemma, the groups $\pi_i(X, A)$ are zero except for $\pi_{n+1}(X, A) \approx \pi$. If the action of $\pi_1(A)$ on $\pi_{n+1}(X, A)$ is trivial, the relative Hurewicz theorem gives an isomorphism $\pi_{n+1}(X, A) \approx H_{n+1}(X, A)$. Since (X, A) is n-connected, we may assume A contains the n-skeleton of X, so X/A is n-connected and the absolute Hurewicz theorem gives $\pi_{n+1}(X/A) \approx H_{n+1}(X/A)$. Hence the quotient map $X \to X/A$ induces an isomorphism $\pi_{n+1}(X, A) \approx \pi_{n+1}(X/A)$ since the analogous statement for homology is certainly true.

Since $\pi_{n+1}(X/A) \approx \pi$, we can build a $K(\pi, n+1)$ from X/A by attaching cells of dimension $n + 3$ and greater. This leads to the commutative diagram at the right, where the vertical maps are inclusions and the lower row is obtained by converting the map k into a fibration. The map $A \to F_k$

$$
\begin{array}{ccccc}
A & \longrightarrow & X & \longrightarrow & X/A \\
\downarrow & & \downarrow{\scriptstyle\simeq} & \searrow{\scriptstyle k} & \downarrow \\
F_k & \longrightarrow & E_k & \longrightarrow & K(\pi, n+1)
\end{array}
$$

is a weak homotopy equivalence by the five-lemma applied to the map between the long exact sequences of homotopy groups for the pairs (X, A) and (E_k, F_k), since the only nontrivial relative groups are π_{n+1}, both of which map isomorphically to $\pi_{n+1}(K(\pi, n+1))$. $\square$

Proof of 4.69: In view of the lemma, all that needs to be done is identify the action of $\pi_1(X)$ on $\pi_n(X)$ with the action of $\pi_1(X_n)$ on $\pi_{n+1}(X_{n-1}, X_n)$ for $n \geq 2$, thinking of the map $X_n \to X_{n-1}$ as an inclusion. From the exact sequence

$$0 = \pi_{n+1}(X_{n-1}) \longrightarrow \pi_{n+1}(X_{n-1}, X_n) \overset{\partial}{\longrightarrow} \pi_n(X_n) \longrightarrow \pi_n(X_{n-1}) = 0$$

we have an isomorphism $\pi_{n+1}(X_{n-1}, X_n) \approx \pi_n(X_n)$ respecting the action of $\pi_1(X_n)$. And the map $X \to X_n$ induces isomorphisms on π_1 and π_n, so we are done. $\qquad\square$

Let us consider now a natural generalization of Postnikov towers, in which one starts with a map $f : X \to Y$ between path-connected spaces rather than just a single space X. A **Moore–Postnikov tower** for f is a commutative diagram as shown at the right, with each composition $X \to Z_n \to Y$ homotopic to f, and such that:

(1) The map $X \to Z_n$ induces an isomorphism on π_i for $i < n$ and a surjection for $i = n$.

(2) The map $Z_n \to Y$ induces an isomorphism on π_i for $i > n$ and an injection for $i = n$.

(3) The map $Z_{n+1} \to Z_n$ is a fibration with fiber a $K(\pi_n F, n)$ where F is the homotopy fiber of f.

A Moore–Postnikov tower specializes to a Postnikov tower by taking Y to be a point and then setting $X_n = Z_{n+1}$, discarding the space Z_1 which has trivial homotopy groups.

Theorem 4.71. *Every map $f : X \to Y$ between connected CW complexes has a Moore–Postnikov tower, which is unique up to homotopy equivalence. A Moore–Postnikov tower of principal fibrations exists iff $\pi_1(X)$ acts trivially on $\pi_n(M_f, X)$ for all $n > 1$, where M_f is the mapping cylinder of f.*

Proof: The existence and uniqueness of a diagram satisfying (1) and (2) and commutative at least up to homotopy follows from Propositions 4.17 and 4.18 applied to the pair (M_f, X) with M_f the mapping cylinder of f. Having such a diagram, we proceed as in the earlier case of Postnikov towers, replacing each map $Z_n \to Z_{n-1}$ by a homotopy equivalent fibration, starting with $Z_2 \to Z_1$ and working upward. We can then apply the homotopy lifting property to make all the triangles in the left half of the tower strictly commutative. After these steps the triangles in the right half of the diagram commute up to homotopy, and to make them strictly commute we can just replace each map to Y by the composition through Z_1.

To see that the fibers of the maps $Z_{n+1} \to Z_n$ are Eilenberg–MacLane spaces as in condition (3), consider two successive levels of the tower. We may arrange that the maps $X \to Z_{n+1} \to Z_n \to Y$ are inclusions by taking mapping cylinders, first of $X \to Z_{n+1}$, then of the new $Z_{n+1} \to Z_n$, and then of the new $Z_n \to Y$. From the left-hand triangle we see that $Z_{n+1} \to Z_n$ induces an isomorphism on π_i for $i < n$ and a surjection for $i = n$, hence $\pi_i(Z_n, Z_{n+1}) = 0$ for $i < n + 1$. Similarly, the other triangle gives $\pi_i(Z_n, Z_{n+1}) = 0$ for $i > n + 1$. To show that $\pi_{n+1}(Z_n, Z_{n+1}) \approx \pi_{n+1}(Y, X)$ we use the following diagram:

$$\begin{array}{ccccccccc}
\pi_{n+1}(Z_{n+1}) & \longrightarrow & \pi_{n+1}(Z_n) & \longrightarrow & \pi_{n+1}(Z_n, Z_{n+1}) & \longrightarrow & \pi_n(Z_{n+1}) & \longrightarrow & \pi_n(Z_n) \\
\downarrow{=} & & \downarrow{\approx} & & \downarrow & & \downarrow{=} & & \downarrow \\
\pi_{n+1}(Z_{n+1}) & \longrightarrow & \pi_{n+1}(Y) & \longrightarrow & \pi_{n+1}(Y, Z_{n+1}) & \longrightarrow & \pi_n(Z_{n+1}) & \longrightarrow & \pi_n(Y) \\
\uparrow & & \uparrow{=} & & \uparrow & & \uparrow{\approx} & & \uparrow{=} \\
\pi_{n+1}(X) & \longrightarrow & \pi_{n+1}(Y) & \longrightarrow & \pi_{n+1}(Y, X) & \longrightarrow & \pi_n(X) & \longrightarrow & \pi_n(Y)
\end{array}$$

The upper-right vertical map is injective and the lower-left vertical map is surjective, so the five-lemma implies that the two middle vertical maps are isomorphisms. Since the homotopy fiber of an inclusion $A \hookrightarrow B$ has π_i equal to $\pi_{i+1}(B, A)$, we see that condition (3) is satisfied.

The statement about a tower of principal fibrations can be obtained as an application of Lemma 4.70. As we saw in the previous paragraph, there are isomorphisms $\pi_{n+1}(Y, X) \approx \pi_{n+1}(Z_n, Z_{n+1})$, and these respect the action of $\pi_1(X) \approx \pi_1(Z_{n+1})$, so Lemma 4.70 gives the result. $\square$

Besides the case that Y is a point, which yields Postnikov towers, another interesting special case of Moore–Postnikov towers is when X is a point. In this case the space Z_n is an n-connected covering of Y, as in Example 4.20. The n-connected covering of Y can also be obtained as the homotopy fiber of the nth stage $Y \to Y_n$ of a Postnikov tower for Y. The tower of n-connected coverings of Y can be realized by principal fibrations by taking Z_n to be the homotopy fiber of the map $Z_{n-1} \to K(\pi_n Y, n)$ that is the first nontrivial stage in a Postnikov tower for Z_{n-1}.

$$\begin{array}{l}
\vdots \\
\downarrow \\
Z_2 \to K(\pi_3 Y, 3) \\
\downarrow \\
Z_1 \to K(\pi_2 Y, 2) \\
\downarrow \\
Y \to K(\pi_1 Y, 1)
\end{array}$$

A generalization of the preceding theory allowing nontrivial actions of π_1 can be found in [Robinson 1972].

Obstruction Theory

It is very common in algebraic topology to encounter situations where one would like to extend or lift a given map. Obvious examples are the homotopy extension and homotopy lifting properties. In their simplest forms, extension and lifting questions can often be phrased in one of the following two ways:

The Extension Problem. Given a CW pair (W, A) and a map $A \to X$, does this extend to a map $W \to X$?

The Lifting Problem. Given a fibration $X \to Y$ and a map $W \to Y$, is there a lift $W \to X$?

In order for the lifting problem to include things like the homotopy lifting property, it should be generalized to a relative form:

The Relative Lifting Problem. Given a CW pair (W, A), a fibration $X \to Y$, and a map $W \to Y$, does there exist a lift $W \to X$ extending a given lift on A?

$$
\begin{array}{ccc}
A & \longrightarrow & X \\
\downarrow & \nearrow & \downarrow \\
W & \Longrightarrow & Y
\end{array}
$$

Besides reducing to the absolute lifting problem when $A = \varnothing$, this includes the extension problem by taking Y to be a point. Of course, one could broaden these questions by dropping the requirements that (W, A) be a CW pair and that the map $X \to Y$ be a fibration. However, these conditions are often satisfied in cases of interest, and they make the task of finding solutions much easier.

The term 'obstruction theory' refers to a procedure for defining a sequence of cohomology classes that are the obstructions to finding a solution to the extension, lifting, or relative lifting problem. In the most favorable cases these obstructions lie in cohomology groups that are all zero, so the problem has a solution. But even when the obstructions are nonzero it can be very useful to have the problem expressed in cohomological terms.

There are two ways of developing obstruction theory, which produce essentially the same result in the end. In the more elementary approach one tries to construct the extension or lifting one cell of W at a time, proceeding inductively over skeleta of W. This approach has an appealing directness, but the technical details of working at the level of cochains are perhaps a little tedious. Instead of pursuing this direct line we shall follow the second approach, which is slightly more sophisticated but has the advantage that the theory becomes an almost trivial application of Postnikov towers for the extension problem, or Moore–Postnikov towers for the lifting problem. The cellular viewpoint is explained in [VBKT], where it appears in the study of characteristic classes of vector bundles.

Let us consider the extension problem first, where we wish to extend a map $A \to X$ to the larger complex W. Suppose that X has a Postnikov tower of principal fibrations. Then we have a commutative diagram as shown below, where we have enlarged the tower by adjoining the space X_0, which is just a point, at the bottom. The map $X_1 \to X_0$ is then a fibration, and to say it is principal says that X_1, which in any case is a $K(\pi_1 X, 1)$, is the loopspace of $K(\pi_1 X, 2)$, hence $\pi_1(X)$ must be abelian. Conversely, if $\pi_1(X)$ is abelian and acts trivially on all the higher homotopy groups of X, then there is an extended Postnikov tower of principal fibrations as shown.

$$
\begin{array}{ccc}
& & \vdots \\
& & \downarrow \\
X_3 & \xrightarrow{\ k_3\ } & K(\pi_4 X, 5) \\
\downarrow & & \\
X_2 & \xrightarrow{\ k_2\ } & K(\pi_3 X, 4) \\
\downarrow & & \\
A \longrightarrow X \longrightarrow X_1 & \xrightarrow{\ k_1\ } & K(\pi_2 X, 3) \\
\downarrow \qquad\qquad\qquad \downarrow & & \\
W \longrightarrow\qquad\qquad X_0 & \xrightarrow{\ k_0\ } & K(\pi_1 X, 2)
\end{array}
$$

Our strategy will be to try to lift the constant map $W \to X_0$ to maps $W \to X_n$ for $n = 1, 2, \cdots$ in succession, extending the given maps $A \to X_n$. If we are able to find all these lifts $W \to X_n$, there will then be no difficulty in constructing the desired extension $W \to X$.

For the inductive step we have a commutative diagram as at the right. Since X_n
is the pullback, its points are pairs consist-

$$\begin{array}{ccc} A & \longrightarrow X_n & \longrightarrow PK \\ \downarrow & \downarrow & \downarrow \\ W & \longrightarrow X_{n-1} & \longrightarrow K = K(\pi_n X, n+1) \end{array}$$

ing of a point in X_{n-1} and a path from its image in K to the basepoint. A lift $W \to X_n$
therefore amounts to a nullhomotopy of the composition $W \to X_{n-1} \to K$. We already
have such a lift defined on A, hence a nullhomotopy of $A \to K$, and we want a nullhomotopy of $W \to K$ extending this nullhomotopy on A.

The map $W \to K$ together with the nullhomotopy on A gives a map $W \cup CA \to K$,
where CA is the cone on A. Since K is a $K(\pi_n X, n+1)$, the map $W \cup CA \to K$
determines an **obstruction** class $\omega_n \in H^{n+1}(W \cup CA; \pi_n X) \approx H^{n+1}(W, A; \pi_n X)$.

‖ **Proposition 4.72.** *A lift $W \to X_n$ extending the given $A \to X_n$ exists iff $\omega_n = 0$.*

Proof: We need to show that the map $W \cup CA \to K$ extends to a map $CW \to K$ iff
$\omega_n = 0$, or in other words, iff $W \cup CA \to K$ is homotopic to a constant map.

Suppose that $g_t : W \cup CA \to K$ is such a homotopy. The constant map g_1 then
extends to the constant map $g_1 : CW \to K$, so by the homotopy extension property for
the pair $(CW, W \cup CA)$, applied to the reversed homotopy g_{1-t}, we have a homotopy
$y_t . CW \dashrightarrow K$ extending the previous homotopy $g_t : W \cup CA \to K$. The map $g_0 : CW \to K$
then extends the given map $W \cup CA \to K$.

Conversely, if we have an extension $CW \to K$, then this is nullhomotopic since the
cone CW is contractible, and we may restrict such a nullhomotopy to $W \cup CA$.   □

If we succeed in extending the lifts $A \to X_n$ to lifts $W \to X_n$ for all n, then we obtain a map $W \to \varprojlim X_n$ extending the given $A \to X \to \varprojlim X_n$. Let M be the mapping
cylinder of $X \to \varprojlim X_n$. Since the restriction of $W \to \varprojlim X_n \subset M$ to A factors through
X, this gives a homotopy of this restriction to the map $A \to X \subset M$. Extend this to a
homotopy of $W \to M$, producing a map $(W, A) \to (M, X)$. Since the map $X \to \varprojlim X_n$ is
a weak homotopy equivalence, $\pi_i(M, X) = 0$ for all i, so by Lemma 4.6, the compression lemma, the map $(W, A) \to (M, X)$ can be homotoped to a map $W \to X$ extending
the given $A \to X$, and we have solved the extension problem.

Thus if it happens that at each stage of the inductive process of constructing
lifts $W \to X_n$ the obstruction $\omega_n \in H^{n+1}(W, A; \pi_n X)$ vanishes, then the extension
problem has a solution. In particular, this yields:

‖ **Corollary 4.73.** *If X is a connected abelian CW complex and (W, A) is a CW pair
such that $H^{n+1}(W, A; \pi_n X) = 0$ for all n, then every map $A \to X$ can be extended to
a map $W \to X$.*   □

This is a considerable improvement on the more elementary result that extensions
exist if $\pi_n(X) = 0$ for all n such that $W - A$ has cells of dimension $n + 1$, which is
Lemma 4.7.

We can apply the Hurewicz theorem and obstruction theory to extend the homology version of Whitehead's theorem to CW complexes with trivial action of π_1 on all homotopy groups:

Proposition 4.74. *If X and Y are connected abelian CW complexes, then a map $f : X \to Y$ inducing isomorphisms on all homology groups is a homotopy equivalence.*

Proof: Taking the mapping cylinder of f reduces us to the case of an inclusion $X \hookrightarrow Y$ of a subcomplex. If we can show that $\pi_1(X)$ acts trivially on $\pi_n(Y, X)$ for all n, then the relative Hurewicz theorem will imply that $\pi_n(Y, X) = 0$ for all n, so $X \to Y$ will be a weak homotopy equivalence. The assumptions guarantee that $\pi_1(X) \to \pi_1(Y)$ is an isomorphism, so we know at least that $\pi_1(Y, X) = 0$.

We can use obstruction theory to extend the identity map $X \to X$ to a retraction $Y \to X$. To apply the theory we need $\pi_1(X)$ acting trivially on $\pi_n(X)$, which holds by hypothesis. Since the inclusion $X \hookrightarrow Y$ induces isomorphisms on homology, we have $H_*(Y, X) = 0$, hence $H^{n+1}(Y, X; \pi_n(X)) = 0$ for all n by the universal coefficient theorem. So there are no obstructions, and a retraction $Y \to X$ exists. This implies that the maps $\pi_n(Y) \to \pi_n(Y, X)$ are onto, so trivial action of $\pi_1(X)$ on $\pi_n(Y)$ implies trivial action on $\pi_n(Y, X)$ by naturality of the action.　　□

The generalization of the preceding analysis of the extension problem to the relative lifting problem is straightforward. Assuming the fibration $p : X \to Y$ in the statement of the relative lifting problem has a Moore–Postnikov tower of principal fibrations, we have the diagram at the right, where F is the fiber of the fibration $X \to Y$. The first step is to lift the map $W \to Y$ to Z_1, extending the given lift on A. We may take Z_1 to be the covering space of Y corresponding to the subgroup $p_*(\pi_1(X))$ of $\pi_1(Y)$, so covering space theory tells us when we can lift $W \to Y$ to Z_1, and the unique lifting property for covering spaces can be used to see whether a lift can be chosen to agree with the lift on A given by the diagram; this could only be a problem when A has more than one component.

Having a lift to Z_1, the analysis proceeds exactly as before. One finds a sequence of obstructions $\omega_n \in H^{n+1}(W, A; \pi_n F)$, assuming $\pi_1 F$ is abelian in the case $n = 1$. A lift to X exists, extending the given lift on A, if each successive ω_n is zero.

One can ask the converse question: If a lift exists, must the obstructions ω_n all be zero? Since Proposition 4.72 is an if and only if statement, one might expect the answer to be yes, but upon closer inspection the matter becomes less clear. The difficulty is that, even if at some stage the obstruction ω_n is zero, so a lift to Z_{n+1} exists, there may be many choices of such a lift, and different choices could lead to different ω_{n+1}'s, some zero and others nonzero. Examples of such ambiguities are not hard to produce, for both the lifting and the extension problems, and the ambiguities only become worse with each subsequent choice of a lift. So it is only in rather special circumstances that one can say that there are well-defined obstructions. A simple case is when $\pi_i(F) = 0$ for $i < n$, so the Moore-Postnikov factorization begins with Z_n as in the diagram at the right. In this case the composition across the bottom of the diagram gives a well-defined **primary obstruction** $\omega_n \in H^{n+1}(W, A; \pi_n F)$.

$$
\begin{array}{ccc}
& & \vdots \\
& & \downarrow \\
& Z_{n+2} & \longrightarrow K(\pi_{n+2}F, n+3) \\
& \nearrow \quad \downarrow & \\
A \longrightarrow X \longrightarrow Z_{n+1} & \longrightarrow K(\pi_{n+1}F, n+2) \\
\downarrow \qquad \searrow{\scriptstyle p} \quad \downarrow & \\
W \longrightarrow Y = Z_n & \longrightarrow K(\pi_n F, n+1)
\end{array}
$$

Exercises

1. Show there is a map $\mathbb{RP}^\infty \to \mathbb{CP}^\infty = K(\mathbb{Z}, 2)$ which induces the trivial map on $\tilde{H}_*(-;\mathbb{Z})$ but a nontrivial map on $\tilde{H}^*(-;\mathbb{Z})$. How is this consistent with the universal coefficient theorem?

2. Show that the group structure on S^1 coming from multiplication in $\mathbb{C}$ induces a group structure on $\langle X, S^1 \rangle$ such that the bijection $\langle X, S^1 \rangle \to H^1(X; \mathbb{Z})$ of Theorem 4.57 is an isomorphism.

3. Suppose that a CW complex X contains a subcomplex S^1 such that the inclusion $S^1 \hookrightarrow X$ induces an injection $H_1(S^1; \mathbb{Z}) \to H_1(X; \mathbb{Z})$ with image a direct summand of $H_1(X; \mathbb{Z})$. Show that S^1 is a retract of X.

4. Given abelian groups G and H and CW complexes $K(G, n)$ and $K(H, n)$, show that the map $\langle K(G, n), K(H, n) \rangle \to \text{Hom}(G, H)$ sending a homotopy class $[f]$ to the induced homomorphism $f_* : \pi_n(K(G, n)) \to \pi_n(K(H, n))$ is a bijection.

5. Show that $[X, S^n] \approx H^n(X; \mathbb{Z})$ if X is an n-dimensional CW complex. [Build a $K(\mathbb{Z}, n)$ from S^n by attaching cells of dimension $\geq n + 2$.]

6. Use Exercise 4 to construct a multiplication map $\mu : K(G, n) \times K(G, n) \to K(G, n)$ for any abelian group G, making a CW complex $K(G, n)$ into an H-space whose multiplication is commutative and associative up to homotopy and has a homotopy inverse. Show also that the H-space multiplication μ is unique up to homotopy.

7. Using an H-space multiplication μ on $K(G, n)$, define an addition in $\langle X, K(G, n) \rangle$ by $[f] + [g] = [\mu(f, g)]$ and show that under the bijection $H^n(X; G) \approx \langle X, K(G, n) \rangle$ this addition corresponds to the usual addition in cohomology.

8. Show that a map $p : E \to B$ is a fibration iff the map $\pi : E^I \to E_p$, $\pi(\gamma) = (\gamma(0), p\gamma)$, has a section, that is, a map $s : E_p \to E^I$ such that $\pi s = \mathbb{1}$.

9. Show that a linear projection of a 2-simplex onto one of its edges is a fibration but not a fiber bundle. [Use the preceding problem.]

10. Given a fibration $F \to E \to B$, use the homotopy lifting property to define an action of $\pi_1(E)$ on $\pi_n(F)$, a homomorphism $\pi_1(E) \to \text{Aut}(\pi_n(F))$, such that the composition $\pi_1(F) \to \pi_1(E) \to \text{Aut}(\pi_n(F))$ is the usual action of $\pi_1(F)$ on $\pi_n(F)$. Deduce that if $\pi_1(E) = 0$, then the action of $\pi_1(F)$ on $\pi_n(F)$ is trivial.

11. For a space B, let $\mathcal{F}(B)$ be the set of fiber homotopy equivalence classes of fibrations $E \to B$. Show that a map $f : B_1 \to B_2$ induces $f^* : \mathcal{F}(B_2) \to \mathcal{F}(B_1)$ depending only on the homotopy class of f, with f^* a bijection if f is a homotopy equivalence.

12. Show that for homotopic maps $f, g : A \to B$ the fibrations $E_f \to B$ and $E_g \to B$ are fiber homotopy equivalent.

13. Given a map $f : A \to B$ and a homotopy equivalence $g : C \to A$, show that the fibrations $E_f \to B$ and $E_{fg} \to B$ are fiber homotopy equivalent. [One approach is to use Corollary 0.21 to reduce to the case of deformation retractions.]

14. For a space B, let $\mathcal{M}(B)$ denote the set of equivalence classes of maps $f : A \to B$ where $f_1 : A_1 \to B$ is equivalent to $f_2 : A_2 \to B$ if there exists a homotopy equivalence $g : A_1 \to A_2$ such that $f_1 \simeq f_2 g$. Show the natural map $\mathcal{F}(B) \to \mathcal{M}(B)$ is a bijection. [See Exercises 11 and 13.]

15. If the fibration $p : E \to B$ is a homotopy equivalence, show that p is a fiber homotopy equivalence of E with the trivial fibration $\mathbb{1} : B \to B$.

16. Show that a map $f : X \to Y$ of connected CW complexes is a homotopy equivalence if it induces an isomorphism on π_1 and its homotopy fiber F_f has $\widetilde{H}_*(F_f; \mathbb{Z}) = 0$.

17. Show that ΩX is an H–space with multiplication the composition of loops.

18. Show that a fibration sequence $\cdots \to \Omega B \to F \to E \to B$ induces a long exact sequence $\cdots \to \langle X, \Omega B \rangle \to \langle X, F \rangle \to \langle X, E \rangle \to \langle X, B \rangle$, with groups and group homomorphisms except for the last three terms, abelian groups except for the last six terms.

19. Given a fibration $F \longrightarrow E \xrightarrow{\ p\ } B$, define a natural action of ΩB on the homotopy fiber F_p and use this to show that exactness at $\langle X, F \rangle$ in the long exact sequence in the preceding problem can be improved to the statement that two elements of $\langle X, F \rangle$ have the same image in $\langle X, E \rangle$ iff they are in the same orbit of the induced action of $\langle X, \Omega B \rangle$ on $\langle X, F \rangle$.

20. Show that by applying the loopspace functor to a Postnikov tower for X one obtains a Postnikov tower of principal fibrations for ΩX.

21. Show that in the Postnikov tower of an H–space, all the spaces are H–spaces and the maps are H–maps, commuting with the multiplication, up to homotopy.

22. Show that a principal fibration $\Omega C \longrightarrow E \xrightarrow{\ p\ } B$ is fiber homotopy equivalent to the product $\Omega C \times B$ iff it has a section, a map $s : B \to E$ with $ps = \mathbb{1}$.

23. Prove the following uniqueness result for the Quillen plus construction: Given a connected CW complex X, if there is an abelian CW complex Y and a map $X \to Y$ inducing an isomorphism $H_*(X; \mathbb{Z}) \approx H_*(Y; \mathbb{Z})$, then such a Y is unique up to homotopy equivalence. [Use Corollary 4.73 with W the mapping cylinder of $X \to Y$.]

24. In the situation of the relative lifting problem, suppose one has two different lifts $W \to X$ that agree on the subspace $A \subset W$. Show that the obstructions to finding a homotopy rel A between these two lifts lie in the groups $H^n(W, A; \pi_n F)$.

Additional Topics

4.A Basepoints and Homotopy

In the first part of this section we will use the action of π_1 on π_n to describe the difference between $\pi_n(X, x_0)$ and the set of homotopy classes of maps $S^n \to X$ without conditions on basepoints. More generally, we will compare the set $\langle Z, X \rangle$ of basepoint-preserving homotopy classes of maps $(Z, z_0) \to (X, x_0)$ with the set $[Z, X]$ of unrestricted homotopy classes of maps $Z \to X$, for Z any CW complex with basepoint z_0 a 0-cell. Then the section concludes with an extended example exhibiting some rather subtle nonfinite generation phenomena in homotopy and homology groups.

We begin by constructing an action of $\pi_1(X, x_0)$ on $\langle Z, X \rangle$ when Z is a CW complex with basepoint 0-cell z_0. Given a loop y in X based at x_0 and a map $f_0 : (Z, z_0) \to (X, x_0)$, then by the homotopy extension property there is a homotopy $f_s : Z \to X$ of f_0 such that $f_s(z_0)$ is the loop y. We might try to define an action of $\pi_1(X, x_0)$ on $\langle Z, X \rangle$ by $[y][f_0] = [f_1]$, but this definition encounters a small problem when we compose loops. For if η is another loop at x_0, then by applying the homotopy extension property a second time we get a homotopy of f_1 restricting to η on x_0, and the two homotopies together give the relation $([y][\eta])[f_0] = [\eta]([y][f_0])$, in view of our convention that the product $y\eta$ means first y, then η. This is not quite the relation we want, but the problem is easily corrected by letting the action be an action on the right rather than on the left. Thus we set $[f_0][y] = [f_1]$, and then $[f_0]([y][\eta]) = ([f_0][y])[\eta]$.

Let us check that this right action is well-defined. Suppose we start with maps $f_0, g_0 : (Z, z_0) \to (X, x_0)$ representing the same class in $\langle Z, X \rangle$, together with homotopies f_s and g_s of f_0 and g_0 such that $f_s(z_0)$ and $g_s(z_0)$ are homotopic loops. These various homotopies define a map $H : Z \times I \times \partial I \cup Z \times \{0\} \times I \cup \{z_0\} \times I \times I \to X$ which is f_s on $Z \times I \times \{0\}$, g_s on $Z \times I \times \{1\}$, the basepoint-preserving homotopy from f_0 to g_0 on $Z \times \{0\} \times I$, and the homotopy from $f_s(z_0)$ to $g_s(z_0)$ on $\{z_0\} \times I \times I$. We would like to extend H over $Z \times I \times I$. The pair $(I \times I, I \times \partial I \cup \{0\} \times I)$ is homeomorphic to $(I \times I, I \times \{0\})$, and via this homeomorphism we can view H as a map $Z \times I \times \{0\} \cup \{z_0\} \times I \times I \to X$, that is, a map $Z \times I \to X$ with a homotopy on the subcomplex $\{z_0\} \times I$. This means the homotopy extension property can be applied to produce an extension of the original H to $Z \times I \times I$. Restricting this extended H to $Z \times \{1\} \times I$ gives a basepoint-preserving homotopy $f_1 \simeq g_1$, which shows that $[f_0][y]$ is well-defined.

Note that in this argument we did not have to assume the homotopies f_s and g_s were constructed by applying the homotopy extension property. Thus we have proved the following result:

Proposition 4A.1. *There is a right action of $\pi_1(X, x_0)$ on $\langle Z, X \rangle$ defined by setting $[f_0][\gamma] = [f_1]$ whenever there exists a homotopy $f_s : Z \to X$ from f_0 to f_1 such that $f_s(z_0)$ is the loop γ, or any loop homotopic to γ.* $\qquad\square$

It is easy to convert this right action into a left action, by defining $[\gamma][f_0] = [f_0][\gamma]^{-1}$. This just amounts to choosing the homotopy f_s so that $f_s(z_0)$ is the inverse path of γ.

When $Z = S^n$ this action reduces to the usual action of $\pi_1(X, x_0)$ on $\pi_n(X, x_0)$ since in the original definition of γf in terms of maps $(I^n, \partial I^n) \to (X, x_0)$, a homotopy from γf to f is obtained by restricting γf to smaller and smaller concentric cubes, and on the 'basepoint' ∂I^n this homotopy traces out the loop γ.

Proposition 4A.2. *If (Z, z_0) is a CW pair and X is a path-connected space, then the natural map $\langle Z, X \rangle \to [Z, X]$ induces a bijection of the orbit set $\langle Z, X \rangle / \pi_1(X, x_0)$ onto $[Z, X]$.*

In particular, this implies that $[Z, X] = \langle Z, X \rangle$ if X is simply-connected.

Proof: Since X is path-connected, every $f : Z \to X$ can be homotoped to take z_0 to the basepoint x_0, via homotopy extension, so the map $\langle Z, X \rangle \to [Z, X]$ is onto. If f_0 and f_1 are basepoint-preserving maps that are homotopic via the homotopy $f_s : Z \to X$, then by definition $[f_1] = [f_0][\gamma]$ for the loop $\gamma(s) = f_s(z_0)$, so $[f_0]$ and $[f_1]$ are in the same orbit under the action of $\pi_1(X, x_0)$. Conversely, two basepoint-preserving maps in the same orbit are obviously homotopic. $\qquad\square$

Example 4A.3. If X is an H–space with identity element x_0, then the action of $\pi_1(X, x_0)$ on $\langle Z, X \rangle$ is trivial since for a map $f : (Z, z_0) \to (X, x_0)$ and a loop γ in X based at x_0, the multiplication in X defines a homotopy $f_s(z) = f(z)\gamma(s)$. This starts and ends with a map homotopic to f, and the loop $f_s(z_0)$ is homotopic to γ, both these homotopies being basepoint-preserving by the definition of an H–space.

The set of orbits of the π_1 action on π_n does not generally inherit a group structure from π_n. For example, when $n = 1$ the orbits are just the conjugacy classes in π_1, and these form a group only when π_1 is abelian. Basepoints are thus a necessary technical device for producing the group structure in homotopy groups, though as we have shown, they can be ignored in simply-connected spaces.

For a set of maps $S^n \to X$ to generate $\pi_n(X)$ as a module over $\mathbb{Z}[\pi_1(X)]$ means that all elements of $\pi_n(X)$ can be represented by sums of these maps along arbitrary paths in X, where we allow reversing orientations to get negatives and repetitions to get arbitrary integer multiples. Examples of finite CW complexes X for which $\pi_n(X)$ is not finitely generated as a module over $\mathbb{Z}[\pi_1(X)]$ were given in Exercise 38 in §4.2, provided $n \geq 3$. Finding such an example for $n = 2$ seems to be more difficult. The rest of this section will be devoted to a somewhat complicated construction which does this, and is interesting for other reasons as well.

An Example of Nonfinite Generation

We will construct a finite CW complex having π_n not finitely generated as a $\mathbb{Z}[\pi_1]$-module, for a given integer $n \geq 2$. The complex will be a subcomplex of a $K(\pi, 1)$ having interesting homological properties: It is an $(n + 1)$-dimensional CW complex with H_{n+1} nonfinitely generated, but its n-skeleton is finite so H_i is finitely generated for $i \leq n$ and π is finitely presented if $n > 1$. The first such example was found in [Stallings 1963] for $n = 2$. Our construction will be essentially the n-dimensional generalization of this, but described in a more geometric way as in [Bestvina & Brady 1997], which provides a general technique for constructing many examples of this sort.

To begin, let X be the product of n copies of $S^1 \vee S^1$. Since $S^1 \vee S^1$ is the 1-skeleton of the torus $T^2 = S^1 \times S^1$ in its usual CW structure, X can be regarded as a subcomplex of the $2n$-dimensional torus T^{2n}, the product of $2n$ circles. Define $f : T^{2n} \to S^1$ by $f(\theta_1, \cdots, \theta_{2n}) = \theta_1 + \cdots + \theta_{2n}$ where the coordinates $\theta_i \subset S^1$ are viewed as angles measured in radians. The space $Z = X \cap f^{-1}(0)$ will provide the example we are looking for. As we shall see, Z is a finite CW complex of dimension $n - 1$, with $\pi_{n-1}(Z)$ nonfinitely generated as a module over $\pi_1(Z)$ if $n \geq 3$. We will also see that $\pi_i(Z) = 0$ for $1 < i < n - 1$.

The induced homomorphism $f_* : \pi_1(T^{2n}) \to \pi_1(S^1) = \mathbb{Z}$ sends each generator coming from an S^1 factor to 1. Let $\widetilde{T}^{2n} \to T^{2n}$ be the covering space corresponding to the kernel of f_*. This is a normal covering space since it corresponds to a normal subgroup, and the deck transformation group is $\mathbb{Z}$. The subcomplex of $\widetilde{T}^{2n}$ projecting to X is a normal covering space $\widetilde{X} \to X$ with the same group of deck transformations. Since $\pi_1(X)$ is the product of n free groups on two generators, $\widetilde{X}$ is the covering space of X corresponding to the kernel of the homomorphism $\pi_1(X) \to \mathbb{Z}$ sending each of the two generators of each free factor to 1. Since X is a $K(\pi, 1)$, so is $\widetilde{X}$. For example, when $n = 1$, $\widetilde{X}$ is the union of two helices on the infinite cylinder $\widetilde{T}^2$:

The map f lifts to a map $\widetilde{f} : \widetilde{T}^{2n} \to \mathbb{R}$, and Z lifts homeomorphically to a subspace $Z \subset \widetilde{X}$, namely $\widetilde{f}^{-1}(0) \cap \widetilde{X}$. We will show:

(∗) $\widetilde{X}$ is homotopy equivalent to a space Y obtained from Z by attaching an infinite sequence of n-cells.

Assuming this is true, it follows that $H_n(Y)$ is not finitely generated since in the exact sequence $H_n(Z) \to H_n(Y) \to H_n(Y, Z) \to H_{n-1}(Z)$ the first term is zero and the last term is finitely generated, Z being a finite CW complex of dimension $n - 1$, while the third term is an infinite sum of $\mathbb{Z}$'s, one for each n-cell of Y. If $\pi_{n-1}(Z)$ were finitely generated as a $\pi_1(Z)$-module, then by attaching finitely many n-cells to Z we could make it $(n-1)$-connected since it is already $(n-2)$-connected as the $(n-1)$-skeleton of the $K(\pi, 1)$ Y. Then by attaching cells of dimension greater than n we could build a $K(\pi, 1)$ with finite n-skeleton. But this contradicts the fact that $H_n(Y)$ is not finitely generated.

To begin the verification of $(*)$, consider the torus T^m. The standard cell structure on T^m lifts to a cubical cell structure on the universal cover $\mathbb{R}^m$, with vertices the integer lattice points $\mathbb{Z}^m$. The function f lifts to a linear projection $L : \mathbb{R}^m \to \mathbb{R}$, $L(x_1, \cdots, x_m) = x_1 + \cdots + x_m$. The planes in $L^{-1}(\mathbb{Z})$ cut the cubes of $\mathbb{R}^m$ into convex polyhedra which we call *slabs*. There are m slabs in each m-dimensional cube. The boundary of a slab in $L^{-1}[i, i+1]$ consists of lateral faces that are slabs for lower-dimensional cubes, together with a lower face in $L^{-1}(i)$ and an upper face in $L^{-1}(i+1)$. In each cube there are two exceptional slabs whose lower or upper face degenerates to a point. These are the slabs containing the 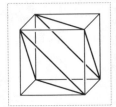 vertices of the cube where L has its maximum and minimum values. A slab deformation retracts onto the union of its lower and lateral faces, provided that the slab has an upper face that is not just a point. Slabs of the latter type are m-simplices, and we will refer to them as *cones* in what follows. These are the slabs containing the vertex of a cube on which L takes its maximal value. The lateral faces of a cone are also cones, of lower dimension.

The slabs, together with all their lower-dimensional faces, give a CW structure on $\mathbb{R}^m$ with the planes of $L^{-1}(\mathbb{Z})$ as subcomplexes. These structures are preserved by the deck transformations of the cover $\mathbb{R}^m \to T^m$ so there is an induced CW structure in the quotient T^m, with $f^{-1}(0)$ as a subcomplex.

If X is any subcomplex of T^m in its original cubical cell structure, then the slab CW structure on T^m restricts to a CW structure on X. In particular, we obtain a CW structure on $Z = X \cap f^{-1}(0)$. Likewise we get a lifted CW structure on the cover $\widetilde{X} \subset \widetilde{T}^m$. Let $\widetilde{X}[i, j] = \widetilde{X} \cap \widetilde{f}^{-1}[i, j]$. The deformation retractions of noncone slabs onto their lateral and lower faces give rise to a deformation retraction of $\widetilde{X}[i, i+1]$ onto $\widetilde{X}[i] \cup C_i$ where C_i consists of all the cones in $\widetilde{X}[i, i+1]$. These cones are attached along their lower faces, and they all have the same vertex in $\widetilde{X}[i+1]$, so C_i is itself a cone in the usual sense, attached to $\widetilde{X}[i]$ along its base.

For the particular X we are interested in, we claim that each C_i is an n-disk attached along its boundary sphere. When $n = 1$ this is evident from the earlier picture of $\widetilde{X}$ as the union of two helices on a cylinder. For larger n we argue by induction. Passing from n to $n+1$ replaces X by two copies of $X \times S^1$ intersecting in X, one copy for each of the additional S^1 factors of T^{2n+2}. Replacing X by $X \times S^1$ changes C_i to its join with a point in the base of the new C_i. The two copies of this join then yield the suspension of C_i attached along the suspension of the base.

The same argument shows that $\widetilde{X}[-i-1,-i]$ deformation retracts onto $\widetilde{X}[-i]$ with an n-cell attached. We build the space Y and a homotopy equivalence $g:Y\to\widetilde{X}$ by an inductive procedure, starting with $Y_0 = Z$. Assuming that Y_i and a homotopy equivalence $g_i:Y_i\to\widetilde{X}[-i,i]$ have already been defined, we form Y_{i+1} by attaching two n-cells by the maps obtained from the attaching maps of the two n-cells in $\widetilde{X}[-i-1,i+1] - \widetilde{X}[-i,i]$ by composing with a homotopy inverse to g_i. This allows g_i to be extended to a homotopy equivalence $g_{i+1}:Y_{i+1}\to\widetilde{X}[-i-1,i+1]$. Taking the union over i gives $g:Y\to\widetilde{X}$. One can check this is a homotopy equivalence by seeing that it induces isomorphisms on all homotopy groups, using the standard compactness argument. This finishes the verification of $(*)$.

It is interesting to see what the complex Z looks like in the case $n = 3$, when Z is 2-dimensional and has π_2 nonfinitely generated over $\mathbb{Z}[\pi_1(Z)]$. In this case X is the product of three $S^1 \vee S^1$'s, so X is the union of the eight 3-tori obtained by choosing one of the two S^1 summands in each $S^1 \vee S^1$ factor. We denote these 3-tori $S^1_\pm \times S^1_\pm \times S^1_\pm$. Viewing each of these 3-tori as the cube in the previous figure with opposite faces identified, we see that Z is the union of the eight 2-tori formed by the two sloping triangles in each cube. Two of these 2-tori intersect along a circle when the corresponding 3-tori of X intersect along a 2-torus. This happens when the triples of $\pm$'s for the two 3-tori differ in exactly one entry. The pattern of intersection of the eight 2-tori of Z can thus be described combinatorially via the 1-skeleton of the cube, with vertices $(\pm1,\pm1,\pm1)$. There is a torus of Z for each vertex of the cube, and two tori intersect along a circle when the corresponding vertices of the cube are the endpoints of an edge of the cube. All eight tori contain the single 0-cell of Z.

To obtain a model of Z itself, consider a regular octahedron inscribed in the cube with vertices $(\pm1,\pm1,\pm1)$. If we identify each pair of oppo-  site edges of the octahedron, each pair of opposite triangular faces becomes a torus. However, there are only four pairs of opposite faces, so we get only four tori this way, not eight. To correct this problem, regard each triangular face of the octahedron as two copies of the same triangle, distinguished from each other by a choice of normal direction, an arrow attached to the triangle pointing either inside the octahedron or outside it, that is, either toward the nearest vertex of the surrounding cube or toward the opposite vertex of the cube. Then each pair of opposite triangles of the octahedron having normal vectors pointing toward the same vertex of the cube determines a torus, when opposite edges are identified as before. Each edge of the original octahedron is also replaced by two edges oriented either toward the interior or exterior of the octahedron. The vertices of the octahedron may be left unduplicated since they will all be identified to a single point anyway. With this scheme, the two tori corresponding to the vertices at the ends of an edge of the cube then intersect along a circle, as they should, and other pairs of tori intersect only at the 0-cell of Z.

This model of Z has the advantage of displaying the symmetry group of the cube, a group of order 48, as a symmetry group of Z, corresponding to the symmetries of X permuting the three $S^1 \vee S^1$ factors and the two S^1's of each $S^1 \vee S^1$. Undoubtedly Z would be very pretty to look at if we lived in a space with enough dimensions to see all of it at one glance.

It might be interesting to see an explicit set of maps $S^2 \to Z$ generating $\pi_2(Z)$ as a $\mathbb{Z}[\pi_1]$-module. One might also ask whether there are simpler examples of these nonfinite generation phenomena.

Exercises

1. Show directly that if X is a topological group with identity element x_0, then any two maps $f, g : (Z, z_0) \to (X, x_0)$ which are homotopic are homotopic through basepoint-preserving maps.

2. Show that under the map $\langle X, Y \rangle \to \mathrm{Hom}(\pi_n(X, x_0), \pi_n(Y, y_0))$, $[f] \mapsto f_*$, the action of $\pi_1(Y, y_0)$ on $\langle X, Y \rangle$ corresponds to composing with the action on $\pi_n(Y, y_0)$, that is, $(\gamma f)_* = \beta_\gamma f_*$. Deduce a bijection of $[X, K(\pi, 1)]$ with the set of orbits of $\mathrm{Hom}(\pi_1(X), \pi)$ under composition with inner automorphisms of π. In particular, if π is abelian then $[X, K(\pi, 1)] = \langle X, K(\pi, 1) \rangle = \mathrm{Hom}(\pi_1(X), \pi)$.

3. For a space X let $\mathrm{Aut}(X)$ denote the group of homotopy classes of homotopy equivalences $X \to X$. Show that for a CW complex $K(\pi, 1)$, $\mathrm{Aut}(K(\pi, 1))$ is isomorphic to the group of outer automorphisms of π, that is, automorphisms modulo inner automorphisms.

4. With the notation of the preceding problem, show that $\mathrm{Aut}(\bigvee_n S^k) \approx \mathrm{GL}_n(\mathbb{Z})$ for $k > 1$, where $\bigvee_n S^k$ denotes the wedge sum of n copies of S^k and $\mathrm{GL}_n(\mathbb{Z})$ is the group of $n \times n$ matrices with entries in $\mathbb{Z}$ having an inverse matrix of the same form. [$\mathrm{GL}_n(\mathbb{Z})$ is the automorphism group of $\mathbb{Z}^n \approx \pi_k(\bigvee_n S^k) \approx H_k(\bigvee_n S^k)$.]

5. This problem involves the spaces constructed in the latter part of this section.
(a) Compute the homology groups of the complex Z in the case $n = 3$, when Z is 2-dimensional.
(b) Letting $\widetilde{X}_n$ denote the n-dimensional complex $\widetilde{X}$, show that $\widetilde{X}_n$ can be obtained inductively from $\widetilde{X}_{n-1}$ as the union of two copies of the mapping torus of the generating deck transformation $\widetilde{X}_{n-1} \to \widetilde{X}_{n-1}$, with copies of $\widetilde{X}_{n-1}$ in these two mapping tori identified. Thus there is a fiber bundle $\widetilde{X}_n \to S^1 \vee S^1$ with fiber $\widetilde{X}_{n-1}$.
(c) Use part (b) to find a presentation for $\pi_1(\widetilde{X}_n)$, and show this presentation reduces to a finite presentation if $n > 2$ and a presentation with a finite number of generators if $n = 2$. In the latter case, deduce that $\pi_1(\widetilde{X}_2)$ has no finite presentation from the fact that $H_2(\widetilde{X}_2)$ is not finitely generated.

4.B The Hopf Invariant

In §2.2 we used homology to distinguish different homotopy classes of maps $S^n \to S^n$ via the notion of degree. We will show here that cup product can be used to do something similar for maps $S^{2n-1} \to S^n$. Originally this was done by Hopf using more geometric constructions, before the invention of cohomology and cup products.

In general, given a map $f: S^m \to S^n$ with $m \geq n$, we can form a CW complex C_f by attaching a cell e^{m+1} to S^n via f. The homotopy type of C_f depends only on the homotopy class of f, by Proposition 0.18. Thus for maps $f, g: S^m \to S^n$, any invariant of homotopy type that distinguishes C_f from C_g will show that f is not homotopic to g. For example, if $m = n$ and f has degree d, then from the cellular chain complex of C_f we see that $H_n(C_f) \approx \mathbb{Z}_{|d|}$, so the homology of C_f detects the degree of f, up to sign. When $m > n$, however, the homology of C_f consists of $\mathbb{Z}$'s in dimensions 0, n, and $m + 1$, independent of f. The same is true of cohomology groups, but cup products have a chance of being nontrivial in $H^*(C_f)$ when $m = 2n - 1$. In this case, if we choose generators $\alpha \in H^n(C_f)$ and $\beta \in H^{2n}(C_f)$, then the multiplicative structure of $H^*(C_f)$ is determined by a relation $\alpha^2 = H(f)\beta$ for an integer $H(f)$ called the **Hopf invariant** of f. The sign of $H(f)$ depends on the choice of the generator β, but this can be specified by requiring β to correspond to a fixed generator of $H^{2n}(D^{2n}, \partial D^{2n})$ under the map $H^{2n}(C_f) \approx H^{2n}(C_f, S^n) \to H^{2n}(D^{2n}, \partial D^{2n})$ induced by the characteristic map of the cell e^{2n}, which is determined by f. We can then change the sign of $H(f)$ by composing f with a reflection of S^{2n-1}, of degree -1. If $f \simeq g$, then under the homotopy equivalence $C_f \simeq C_g$ the chosen generators β for $H^{2n}(C_f)$ and $H^{2n}(C_g)$ correspond, so $H(f)$ depends only on the homotopy class of f.

If f is a constant map then $C_f = S^n \vee S^{2n}$ and $H(f) = 0$ since C_f retracts onto S^n. Also, $H(f)$ is always zero for odd n since in this case $\alpha^2 = -\alpha^2$ by the commutativity property of cup product, hence $\alpha^2 = 0$.

Three basic examples of maps with nonzero Hopf invariant are the maps defining the three Hopf bundles in Examples 4.45, 4.46, and 4.47. The first of these Hopf maps is the attaching map $f: S^3 \to S^2$ for the 4-cell of $\mathbb{C}P^2$. This has $H(f) = 1$ since $H^*(\mathbb{C}P^2; \mathbb{Z}) \approx \mathbb{Z}[\alpha]/(\alpha^3)$ by Theorem 3.19. Similarly, $\mathbb{H}P^2$ gives rise to a map $S^7 \to S^4$ of Hopf invariant 1. In the case of the octonionic projective plane $\mathbb{O}P^2$, which is built from the map $S^{15} \to S^8$ defined in Example 4.47, we can deduce that $H^*(\mathbb{O}P^2; \mathbb{Z}) \approx \mathbb{Z}[\alpha]/(\alpha^3)$ either from Poincaré duality as in Example 3.40 or from Exercise 5 for §4.D.

It is a fundamental theorem of [Adams 1960] that a map $f: S^{2n-1} \to S^n$ of Hopf invariant 1 exists only when $n = 2, 4, 8$. This has a number of very interesting consequences, for example:

- $\mathbb{R}^n$ is a division algebra only for $n = 1,\ 2,\ 4,\ 8$.
- S^n is an H–space only for $n = 0,\ 1,\ 3,\ 7$.
- S^n has n linearly independent tangent vector fields only for $n = 0,\ 1,\ 3,\ 7$.
- The only fiber bundles $S^p \to S^q \to S^r$ occur when $(p, q, r) = (0, 1, 1)$, $(1, 3, 2)$, $(3, 7, 4)$, and $(7, 15, 8)$.

The first and third assertions were in fact proved shortly before Adams' theorem in [Kervaire 1958] and [Milnor 1958] as applications of a theorem of Bott that $\pi_{2n}U(n) \approx \mathbb{Z}_{n!}$. A full discussion of all this, and a proof of Adams' theorem, is given in [VBKT].

Though maps of Hopf invariant 1 are rare, there are maps $S^{2n-1} \to S^n$ of Hopf invariant 2 for all even n. Namely, consider the space $J_2(S^n)$ constructed in §3.2. This has a CW structure with three cells, of dimensions 0, n, and $2n$, so $J_2(S^n)$ has the form C_f for some $f : S^{2n-1} \to S^n$. We showed that if n is even, the square of a generator of $H^n(J_2(S^n); \mathbb{Z})$ is twice a generator of $H^{2n}(J_2(S^n); \mathbb{Z})$, so $H(f) = \pm 2$.

From this example we can get maps of any even Hopf invariant when n is even via the following fact.

‖ Proposition 4B.1. *The Hopf invariant* $H : \pi_{2n-1}(S^n) \to \mathbb{Z}$ *is a homomorphism.*

Proof: For $f, g : S^{2n-1} \to S^n$, let us compare C_{f+g} with the space $C_{f \vee g}$ obtained from S^n by attaching two $2n$-cells via f and g. There is a natural quotient map $q : C_{f+g} \to C_{f \vee g}$ collapsing the equatorial disk of the $2n$-cell of C_{f+g} to a point. The induced cellular chain map q_* sends e_{f+g}^{2n} to $e_f^{2n} + e_g^{2n}$. In cohomology this implies that $q^*(\beta_f) = q^*(\beta_g) = \beta_{f+g}$ where β_f, β_g, and β_{f+g} are the cohomology classes dual to the $2n$-cells. Letting α_{f+g} and $\alpha_{f \vee g}$ be the cohomology classes corresponding to the n-cells, we have $q^*(\alpha_{f \vee g}) = \alpha_{f+g}$ since q is a homeomorphism on the n-cells. By restricting to the subspaces C_f and C_g of $C_{f \vee g}$ we see that $\alpha_{f \vee g}^2 = H(f)\beta_f + H(g)\beta_g$. Thus $\alpha_{f+g}^2 = q^*(\alpha_{f \vee g}^2) = H(f)q^*(\beta_f) + H(g)q^*(\beta_g) = (H(f) + H(g))\beta_{f+g}$.   □

‖ Corollary 4B.2. $\pi_{2n-1}(S^n)$ *contains a* $\mathbb{Z}$ *direct summand when* n *is even.*

Proof: Either H or $H/2$ is a surjective homomorphism $\pi_{2n-1}(S^n) \to \mathbb{Z}$.   □

Exercises

1. Show that the Hopf invariant of a composition $S^{2n-1} \xrightarrow{f} S^{2n-1} \xrightarrow{g} S^n$ is given by $H(gf) = (\deg f)H(g)$, and for a composition $S^{2n-1} \xrightarrow{f} S^n \xrightarrow{g} S^n$ the Hopf invariant satisfies $H(gf) = (\deg g)^2 H(f)$.

2. Show that if $S^k \to S^m \xrightarrow{p} S^n$ is a fiber bundle, then $m = 2n - 1$, $k = n - 1$, and, when $n > 1$, $H(p) = \pm 1$. [Show that C_p is a manifold and apply Poincaré duality.]

4.C Minimal Cell Structures

We can apply the homology version of Whitehead's theorem, Corollary 4.33, to show that a simply-connected CW complex with finitely generated homology groups is always homotopy equivalent to a CW complex having the minimum number of cells consistent with its homology, namely, one n-cell for each $\mathbb{Z}$ summand of H_n and a pair of cells of dimension n and $n+1$ for each $\mathbb{Z}_k$ summand of H_n.

Proposition 4C.1. *Given a simply-connected CW complex X and a decomposition of each of its homology groups $H_n(X)$ as a direct sum of cyclic groups with specified generators, then there is a CW complex Z and a cellular homotopy equivalence $f : Z \to X$ such that each cell of Z is either:*

(a) *a 'generator' n-cell e_α^n, which is a cycle in cellular homology mapped by f to a cellular cycle representing the specified generator α of one of the cyclic summands of $H_n(X)$; or*

(b) *a 'relator' $(n+1)$-cell e_α^{n+1}, with cellular boundary equal to a multiple of the generator n-cell e_α^n, in the case that α has finite order.*

In the nonsimply-connected case this result can easily be false, counterexamples being provided by acyclic spaces and the space $X = (S^1 \vee S^n) \cup e^{n+1}$ constructed in Example 4.35, which has the same homology as S^1 but which must have cells of dimension greater than 1 in order to have π_n nontrivial.

Proof: We build Z inductively over skeleta, starting with Z^1 a point since X is simply-connected. For the inductive step, suppose we have constructed $f : Z^n \to X$ inducing an isomorphism on H_i for $i < n$ and a surjection on H_n. For the mapping cylinder M_f we then have $H_i(M_f, Z^n) = 0$ for $i \le n$ and $H_{n+1}(M_f, Z^n) \approx \pi_{n+1}(M_f, Z^n)$ by the Hurewicz theorem. To construct Z^{n+1} we use the following diagram:

$$
\begin{array}{ccccccc}
H_{n+1}(X) & & \pi_{n+1}(M_f, Z^n) & & & & H_n(X) \\
\wr & & \wr & & & & \wr \\
H_{n+1}(M_f) \longrightarrow & H_{n+1}(M_f, Z^n) & \longrightarrow & H_n(Z^n) & \longrightarrow & H_n(M_f) \longrightarrow 0 \\
\uparrow & \uparrow & & \| & & \uparrow \\
H_{n+1}(Z^{n+1}) \longrightarrow & H_{n+1}(Z^{n+1}, Z^n) & \longrightarrow & H_n(Z^n) & \longrightarrow & H_n(Z^{n+1}) \longrightarrow 0
\end{array}
$$

By induction we know the map $H_n(Z^n) \to H_n(M_f) \approx H_n(X)$ exactly, namely, Z^n has generator n-cells, which are cellular cycles mapping to the given generators of $H_n(X)$, along with relator n-cells that do not contribute to $H_n(Z^n)$. Thus $H_n(Z^n)$ is free with basis the generator n-cells, and the kernel of $H_n(Z^n) \to H_n(X)$ is free with basis given by certain multiples of some of the generator n-cells. Choose 'relator' elements ρ_i in $H_{n+1}(M_f, Z^n)$ mapping to this basis for the kernel, and let the 'generator' elements $\gamma_i \in H_{n+1}(M_f, Z^n)$ be the images of the chosen generators of $H_{n+1}(M_f) \approx H_{n+1}(X)$.

Via the Hurewicz isomorphism $H_{n+1}(M_f, Z^n) \approx \pi_{n+1}(M_f, Z^n)$, the homology classes ρ_i and γ_i are represented by maps $r_i, g_i : (D^{n+1}, S^n) \to (M_f, Z^n)$. We form

Z^{n+1} from Z^n by attaching $(n + 1)$-cells via the restrictions of the maps r_i and g_i to S^n. The maps r_i and g_i themselves then give an extension of the inclusion $Z^n \hookrightarrow M_f$ to a map $Z^{n+1} \rightarrow M_f$, whose composition with the retraction $M_f \rightarrow X$ is the extended map $f : Z^{n+1} \rightarrow X$. This gives us the lower row of the preceding diagram, with commutative squares. By construction, the subgroup of $H_{n+1}(Z^{n+1}, Z^n)$ generated by the relator $(n + 1)$-cells maps injectively to $H_n(Z^n)$, with image the kernel of $H_n(Z^n) \rightarrow H_n(X)$, so $f_* : H_n(Z^{n+1}) \rightarrow H_n(X)$ is an isomorphism. The elements of $H_{n+1}(Z^{n+1}, Z^n)$ represented by the generator $(n+1)$-cells map to the y_i's, hence map to zero in $H_n(Z^n)$, so by exactness of the second row these generator $(n + 1)$-cells are cellular cycles representing elements of $H_{n+1}(Z^{n+1})$ mapped by f_* to the given generators of $H_{n+1}(X)$. In particular, $f_* : H_{n+1}(Z^{n+1}) \rightarrow H_{n+1}(X)$ is surjective, and the induction step is finished.

Doing this for all n, we produce a CW complex Z and a map $f : Z \rightarrow X$ with the desired properties. $\square$

Example 4C.2. Suppose X is a simply-connected CW complex such that for some $n \geq 2$, the only nonzero reduced homology groups of X are $H_n(X)$, which is finitely generated, and $H_{n+1}(X)$, which is finitely generated and free. Then the proposition says that X is homotopy equivalent to a CW complex Z obtained from a wedge sum of n-spheres by attaching $(n + 1)$-cells. The attaching maps of these cells are determined up to homotopy by the cellular boundary map $H_{n+1}(Z^{n+1}, Z^n) \rightarrow H_n(Z^n)$ since $\pi_n(Z^n) \approx H_n(Z^n)$. So the attaching maps are either trivial, in the case of generator $(n + 1)$-cells, or they represent some multiple of an inclusion of one of the wedge summands, in the case of a relator $(n + 1)$-cell. Hence Z is the wedge sum of spheres S^n and S^{n+1} together with Moore spaces $M(\mathbb{Z}_m, n)$ of the form $S^n \cup e^{n+1}$. In particular, the homotopy type of X is uniquely determined by its homology groups.

Proposition 4C.3. *Let X be a simply-connected space homotopy equivalent to a CW complex, such that the only nontrivial reduced homology groups of X are $H_2(X) \approx \mathbb{Z}^m$ and $H_4(X) \approx \mathbb{Z}$. Then the homotopy type of X is uniquely determined by the cup product ring $H^*(X; \mathbb{Z})$. In particular, this applies to any simply-connected closed 4-manifold.*

Proof: By the previous proposition we may assume X is a complex X_φ obtained from a wedge sum $\bigvee_j S_j^2$ of m 2-spheres S_j^2 by attaching a cell e^4 via a map $\varphi : S^3 \rightarrow \bigvee_j S_j^2$. As shown in Example 4.52, $\pi_3(\bigvee_j S_j^2)$ is free with basis the Hopf maps $\eta_j : S^3 \rightarrow S_j^2$ and the Whitehead products $[i_j, i_k]$, $j < k$, where i_j is the inclusion $S_j^2 \hookrightarrow \bigvee_j S_j^2$. Since a homotopy of φ does not change the homotopy type of X_φ, we may assume φ is a linear combination $\sum_j a_j \eta_j + \sum_{j<k} a_{jk}[i_j, i_k]$. We need to see how the coefficients a_j and a_{jk} determine the cup product $H^2(X; \mathbb{Z}) \times H^2(X; \mathbb{Z}) \rightarrow H^4(X; \mathbb{Z})$.

This cup product can be represented by an $m \times m$ symmetric matrix (b_{jk}) where the cup product of the cohomology classes dual to the jth and k^{th} 2-cells is b_{jk}

times the class dual to the 4-cell. We claim that $b_{jk} = a_{jk}$ for $j < k$ and $b_{jj} = a_j$. If φ is one of the generators η_i or $[i_j, i_k]$ this is clear, since if $\varphi = \eta_j$ then X_φ is the wedge sum of $\mathbb{C}P^2$ with $m - 1$ 2-spheres, while if $\varphi = [i_j, i_k]$ then X_φ is the wedge sum of $S_j^2 \times S_k^2$ with $m - 2$ 2-spheres. The claim is also true when φ is $-\eta_j$ or $-[i_j, i_k]$ since changing the sign of φ amounts to composing φ with a reflection of S^3, and this changes the generator of $H^4(X_\varphi; \mathbb{Z})$ to its negative. The general case now follows by induction from the assertion that the matrix (b_{jk}) for $X_{\varphi+\psi}$ is the sum of the corresponding matrices for X_φ and X_ψ. This assertion can be proved as follows. By attaching two 4-cells to $\bigvee_j S_j^2$ by φ and ψ we obtain a complex $X_{\varphi,\psi}$ which we can view as $X_\varphi \cup X_\psi$. There is a quotient map $q : X_{\varphi+\psi} \to X_{\varphi,\psi}$ that is a homeomorphism on the 2-skeleton and collapses the closure of an equatorial 3-disk in the 4-cell of $X_{\varphi+\psi}$ to a point. The induced map $q^* : H^4(X_{\varphi,\psi}) \to H^4(X_{\varphi+\psi})$ sends each of the two generators corresponding to the 4-cells of $X_{\varphi,\psi}$ to a generator, and the assertion follows.

Now suppose X_φ and X_ψ have isomorphic cup product rings. This means bases for $H^*(X_\varphi; \mathbb{Z})$ and $H^*(X_\psi; \mathbb{Z})$ can be chosen so that the matrices specifying the cup product $H^2 \times H^2 \to H^4$ with respect to these bases are the same. The preceding proposition says that any choice of basis can be realized as the dual basis to a cell structure on a CW complex homotopy equivalent to the given complex. Therefore we may assume the matrices (b_{jk}) for X_φ and X_ψ are the same. By what we have shown in the preceding paragraph, this means φ and ψ are homotopic, hence X_φ and X_ψ are homotopy equivalent.

For the statement about simply-connected closed 4-manifolds, Corollaries A.8 and A.9 and Proposition A.11 in the Appendix say that such a manifold M has the homotopy type of a CW complex with finitely generated homology groups. Then Poincaré duality and the universal coefficient theorem imply that the only nontrivial homology groups $H_i(M)$ are $\mathbb{Z}$ for $i = 0, 4$ and $\mathbb{Z}^m$ for $i = 2$, for some $m \geq 0$. $\square$

This result and the example preceding it are special cases of a homotopy classification by Whitehead of simply-connected CW complexes with positive-dimensional cells in three adjacent dimensions n, $n + 1$, and $n + 2$; see [Baues 1996] for a full treatment of this.

4.D Cohomology of Fiber Bundles

While the homotopy groups of the three spaces in a fiber bundle fit into a long exact sequence, the relation between their homology or cohomology groups is much more complicated. The Künneth formula shows that there are some subtleties even for a product bundle, and for general bundles the machinery of spectral sequences, developed in [SSAT], is required. In this section we will describe a few special sorts of fiber bundles where more elementary techniques suffice. As applications we calculate the cohomology rings of some important spaces closely related to Lie groups. In particular we find a number of spaces with exterior and polynomial cohomology rings.

The Leray–Hirsch Theorem

This theorem will be the basis for all the other results in this section. It gives hypotheses sufficient to guarantee that a fiber bundle has cohomology very much like that of a product bundle.

Theorem 4D.1. *Let $F \to E \xrightarrow{p} B$ be a fiber bundle such that, for some commutative coefficient ring R :*
(a) *$H^n(F; R)$ is a finitely generated free R-module for each n.*
(b) *There exist classes $c_j \in H^{k_j}(E; R)$ whose restrictions $i^*(c_j)$ form a basis for $H^*(F; R)$ in each fiber F, where $i : F \to E$ is the inclusion.*
Then the map $\Phi : H^*(B; R) \otimes_R H^*(F; R) \to H^*(E; R)$, $\sum_{ij} b_i \otimes i^*(c_j) \mapsto \sum_{ij} p^*(b_i) \smile c_j$, is an isomorphism.

In other words, $H^*(E; R)$ is a free $H^*(B; R)$-module with basis $\{c_j\}$, where we view $H^*(E; R)$ as a module over the ring $H^*(B; R)$ by defining scalar multiplication by $bc = p^*(b) \smile c$ for $b \in H^*(B; R)$ and $c \in H^*(E; R)$.

In the case of a product $E = B \times F$ with $H^*(F; R)$ free over R, we can pull back a basis for $H^*(F; R)$ via the projection $E \to F$ to obtain the classes c_j. Thus the Leray-Hirsch theorem generalizes the version of the Künneth formula involving cup products, Theorem 3.15, at least as far as the additive structure and the module structure over $H^*(B; R)$ are concerned. However, the Leray-Hirsch theorem does not assert that the isomorphism $H^*(E; R) \approx H^*(B; R) \otimes_R H^*(F; R)$ is a ring isomorphism, and in fact this need not be true, as we shall see by an example later.

An example of a bundle where the classes c_j do not exist is the Hopf bundle $S^1 \to S^3 \to S^2$, since $H^*(S^3) \not\approx H^*(S^2) \otimes H^*(S^1)$.

Proof: We first prove the result for finite-dimensional CW complexes B by induction on their dimension. The case that B is 0-dimensional is trivial. For the induction step, suppose B has dimension n, and let $B' \subset B$ be the subspace obtained by deleting a point x_α from the interior of each n-cell e_α^n of B. Let $E' = p^{-1}(B')$. Then we have a commutative diagram, with coefficients in R understood:

$$\cdots \longrightarrow H^*(B, B') \otimes_R H^*(F) \longrightarrow H^*(B) \otimes_R H^*(F) \longrightarrow H^*(B') \otimes_R H^*(F) \longrightarrow \cdots$$
$$\Big\downarrow \Phi \qquad\qquad\qquad\quad \Big\downarrow \Phi \qquad\qquad\qquad\quad \Big\downarrow \Phi$$
$$\cdots \longrightarrow H^*(E, E') \longrightarrow\qquad\qquad H^*(E) \longrightarrow\qquad\qquad H^*(E') \longrightarrow \cdots$$

The map Φ on the left is defined exactly as in the absolute case, using the relative cup product $H^*(E, E') \otimes_R H^*(E) \to H^*(E, E')$. The first row of the diagram is exact since tensoring with a free module preserves exactness. The second row is of course exact also. The commutativity of the diagram follows from the evident naturality of Φ in the case of the two squares shown. For the other square involving coboundary maps, if we start with an element $b \otimes i^*(c_j) \in H^*(B') \otimes_R H^*(F)$ and map this horizontally we get $\delta b \otimes i^*(c_j)$ which maps vertically to $p^*(\delta b) \smile c_j$, whereas if we first map vertically we get $p^*(b) \smile c_j$ which maps horizontally to $\delta(p^*(b) \smile c_j) = \delta p^*(b) \smile c_j = p^*(\delta b) \smile c_j$ since $\delta c_j = 0$.

The space B' deformation retracts onto the skeleton B^{n-1}, and the following lemma implies that the inclusion $p^{-1}(B^{n-1}) \hookrightarrow E'$ is a weak homotopy equivalence, hence induces an isomorphism on all cohomology groups:

Lemma 4D.2. *Given a fiber bundle $p : E \to B$ and a subspace $A \subset B$ such that (B, A) is k-connected, then $(E, p^{-1}(A))$ is also k-connected.*

Proof: For a map $g : (D^i, \partial D^i) \to (E, p^{-1}(A))$ with $i \le k$, there is by hypothesis a homotopy $f_t : (D^i, \partial D^i) \to (B, A)$ of $f_0 = pg$ to a map f_1 with image in A. The homotopy lifting property then gives a homotopy $g_t : (D^i, \partial D^i) \to (E, p^{-1}(A))$ of g to a map with image in $p^{-1}(A)$. $\qquad\square$

The theorem for finite-dimensional B will now follow by induction on n and the five-lemma once we show that the left-hand Φ in the diagram is an isomorphism.

By the fiber bundle property there are open disk neighborhoods $U_\alpha \subset e_\alpha^n$ of the points x_α such that the bundle is a product over each U_α. Let $U = \bigcup_\alpha U_\alpha$ and let $U' = U \cap B'$. By excision we have $H^*(B, B') \approx H^*(U, U')$, and $H^*(E, E') \approx H^*(p^{-1}(U), p^{-1}(U'))$. This gives a reduction to the problem of showing that the map $\Phi : H^*(U, U') \otimes_R H^*(F) \to H^*(U \times F, U' \times F)$ is an isomorphism. For this we can either appeal to the relative Künneth formula in Theorem 3.18 or we can argue again by induction, applying the five-lemma to the diagram with (B, B') replaced by (U, U'), induction implying that the theorem holds for U and U' since they deformation retract onto complexes of dimensions 0 and $n - 1$, respectively, and by the lemma we can restrict to the bundles over these complexes.

Next there is the case that B is an infinite-dimensional CW complex. Since (B, B^n) is n-connected, the lemma implies that the same is true of $(E, p^{-1}(B^n))$. Hence in the commutative diagram at the right the horizontal maps are isomorphisms below dimension n. Then the fact that the right-hand Φ is an isomorphism, as we have al-

$$
\begin{array}{ccc}
H^*(B) \otimes_R H^*(F) & \longrightarrow & H^*(B^n) \otimes_R H^*(F) \\
\downarrow{\scriptstyle\Phi} & & \downarrow{\scriptstyle\Phi} \\
H^*(E) & \longrightarrow & H^*(p^{-1}(B^n))
\end{array}
$$

ready shown, implies that the left-hand Φ is an isomorphism below dimension n. Since n is arbitrary, this gives the theorem for all CW complexes B.

To extend to the case of arbitrary base spaces B we need the notion of a **pullback bundle** which is used quite frequently in bundle theory. Given a fiber bundle

$p : E \to B$ and a map $f : A \to B$, let $f^*(E) = \{(a, e) \in A \times E \mid f(a) = p(e)\}$, so there is a commutative diagram as at the right, where the two maps from $f^*(E)$ are $(a, e) \mapsto a$ and $(a, e) \mapsto e$. It is a simple exercise to verify that the projection $f^*(E) \to A$ is a fiber bundle with the same fiber as $E \to B$, since a local trivialization of $E \to B$ over $U \subset B$ gives rise to a local trivialization of $f^*(E) \to A$ over $f^{-1}(U)$.

$$\begin{array}{ccc} f^*(E) & \longrightarrow & E \\ \downarrow & & \downarrow{\scriptstyle p} \\ A & \xrightarrow{\ f\ } & B \end{array}$$

If $f : A \to B$ is a CW approximation to an arbitrary base space B, then $f^*(E) \to E$ induces an isomorphism on homotopy groups by the five-lemma applied to the long exact sequences of homotopy groups for the two bundles $E \to B$ and $f^*(E) \to A$ with fiber F. Hence $f^*(E) \to E$ is also an isomorphism on cohomology. The classes c_j pull back to classes in $H^*(f^*(E); R)$ which still restrict to a basis in each fiber, and so the naturality of Φ reduces the theorem for $E \to B$ to the case of $f^*(E) \to A$. $\square$

Corollary 4D.3. (a) $H^*(U(n); \mathbb{Z}) \approx \Lambda_{\mathbb{Z}}[x_1, x_3, \cdots, x_{2n-1}]$, the exterior algebra on generators x_i of odd dimension i.

(b) $H^*(SU(n); \mathbb{Z}) \approx \Lambda_{\mathbb{Z}}[x_3, x_5, \cdots, x_{2n-1}]$.

(c) $H^*(Sp(n); \mathbb{Z}) \approx \Lambda_{\mathbb{Z}}[x_3, x_7, \cdots, x_{4n-1}]$.

This rather simple structure is in marked contrast with the cohomology of $O(n)$ and $SO(n)$ which is considerably more complicated, as shown in §3.D.

Proof: For (a), assume inductively that the result holds for $U(n-1)$. By considering the bundle $U(n-1) \to U(n) \to S^{2n-1}$ we see that the pair $(U(n), U(n-1))$ is $(2n-2)$-connected, so $H^i(U(n); \mathbb{Z}) \to H^i(U(n-1); \mathbb{Z})$ is onto for $i \leq 2n-3$ and the classes $x_1, \cdots, x_{2n-3} \in H^*(U(n-1); \mathbb{Z})$ given by induction are the restrictions of classes $c_1, \cdots, c_{2n-3} \in H^*(U(n); \mathbb{Z})$. The products of distinct x_i's form a basis for $H^*(U(n-1); \mathbb{Z}) \approx \Lambda_{\mathbb{Z}}[x_1, \cdots, x_{2n-3}]$, and these products are restrictions of the corresponding products of c_i's, so the Leray–Hirsch theorem applies, yielding $H^*(U(n); \mathbb{Z}) \approx H^*(U(n-1); \mathbb{Z}) \otimes H^*(S^{2n-1}; \mathbb{Z})$. In view of the commutativity property of cup product, this tensor product is the exterior algebra on odd-dimensional generators $x_1, \cdots, x_{2n-1}$.

The same proof works for $Sp(n)$ using the bundle $Sp(n-1) \to Sp(n) \to S^{4n-1}$. In the case of $SU(n)$ one uses the bundle $SU(n-1) \to SU(n) \to S^{2n-1}$. Since $SU(1)$ is the trivial group, the bundle $SU(1) \to SU(2) \to S^3$ shows that $SU(2) = S^3$, so the first generator is x_3. $\square$

It is illuminating to look more closely at how the homology and cohomology of $O(n)$, $U(n)$, and $Sp(n)$ are related to their bundle structures. For $U(n)$ one has the sequence of bundles

$$\begin{array}{ccccccccc} S^1 = U(1) & \hookrightarrow & U(2) & \hookrightarrow & U(3) & \hookrightarrow & \cdots & \hookrightarrow & U(n-1) & \hookrightarrow & U(n) \\ & & \downarrow & & \downarrow & & & & \downarrow & & \downarrow \\ & & S^3 & & S^5 & & \cdots & & S^{2n-3} & & S^{2n-1} \end{array}$$

If all these were product bundles, $U(n)$ would be homeomorphic to the product $S^1 \times S^3 \times \cdots \times S^{2n-1}$. In actuality the bundles are nontrivial, but the homology and cohomology of $U(n)$ are the same as for this product of spheres, including the cup product structure. For $Sp(n)$ the situation is quite similar, with the corresponding product of spheres $S^3 \times S^7 \times \cdots \times S^{4n-1}$. For $O(n)$ the corresponding sequence of bundles is

$$S^0 = O(1) \hookrightarrow O(2) \hookrightarrow O(3) \hookrightarrow \cdots \hookrightarrow O(n-1) \hookrightarrow O(n)$$
$$\downarrow \qquad\quad \downarrow \qquad\quad \downarrow \qquad\qquad\quad \downarrow \qquad\qquad \downarrow$$
$$S^1 \qquad S^2 \qquad \cdots \qquad S^{n-2} \qquad S^{n-1}$$

The calculations in §3.D show that $H_*(O(n); \mathbb{Z}_2) \approx H_*(S^0 \times S^1 \times \cdots \times S^{n-1}; \mathbb{Z}_2)$, but with $\mathbb{Z}$ coefficients this no longer holds. Instead, consider the coarser sequence of bundles

$$S^0 = O(1) \hookrightarrow O(3) \hookrightarrow O(5) \hookrightarrow \cdots \hookrightarrow O(2k-1) \hookrightarrow O(2k)$$
$$\downarrow \qquad\quad \downarrow \qquad\qquad\quad \downarrow \qquad\qquad \downarrow$$
$$V_2(\mathbb{R}^3) \quad V_2(\mathbb{R}^5) \quad \cdots \quad V_2(\mathbb{R}^{2k-1}) \qquad S^{2k-1}$$

where the last bundle $O(2k) \to S^{2k-1}$ is omitted if $n = 2k - 1$. As we remarked at the end of §3.D in the case of $SO(n)$, $O(n)$ has the same integral homology and cohomology groups as if these bundles were products, but the cup product structure for $O(n)$ with $\mathbb{Z}_2$ coefficients is not the same as in this product.

Cohomology of Grassmannians

Here is an important application of the Leray–Hirsch theorem, generalizing the calculation of the cohomology rings of projective spaces:

Theorem 4D.4. *If $G_n(\mathbb{C}^\infty)$ is the Grassmann manifold of n-dimensional vector subspaces of $\mathbb{C}^\infty$, then $H^*(G_n(\mathbb{C}^\infty); \mathbb{Z})$ is a polynomial ring $\mathbb{Z}[c_1, \cdots, c_n]$ on generators c_i of dimension $2i$. Similarly, $H^*(G_n(\mathbb{R}^\infty); \mathbb{Z}_2)$ is a polynomial ring $\mathbb{Z}_2[w_1, \cdots, w_n]$ on generators w_i of dimension i, and $H^*(G_n(\mathbb{H}^\infty); \mathbb{Z}) \approx \mathbb{Z}[q_1, \cdots, q_n]$ with q_i of dimension $4i$.*

The plan of the proof is to apply the Leray–Hirsch theorem to a fiber bundle $F \to E \xrightarrow{p} G_n(\mathbb{C}^\infty)$ where E has the same cohomology ring as the product of n copies of $\mathbb{C}P^\infty$, a polynomial ring $\mathbb{Z}[x_1, \cdots, x_n]$ with each x_i 2-dimensional. The induced map $p^* : H^*(G_n(\mathbb{C}^\infty); \mathbb{Z}) \to H^*(E; \mathbb{Z})$ will be injective, and we will show that its image consists of the symmetric polynomials in $\mathbb{Z}[x_1, \cdots, x_n]$, the polynomials invariant under permutations of the variables x_i. It is a classical theorem in algebra that the symmetric polynomials themselves form a polynomial ring $\mathbb{Z}[\sigma_1, \cdots, \sigma_n]$ where σ_i is a certain symmetric polynomial of degree i, namely the sum of all products of i distinct x_j's. This gives the result for $G_n(\mathbb{C}^\infty)$, and the same argument will also apply in the real and quaternionic cases.

Proof: Define an **n-flag** in $\mathbb{C}^k$ to be an ordered n-tuple of orthogonal 1-dimensional vector subspaces of $\mathbb{C}^k$. Equivalently, an n-flag could be defined as a chain of vector subspaces $V_1 \subset \cdots \subset V_n$ of $\mathbb{C}^k$ where V_i has dimension i. Why either of these objects should be called a 'flag' is not exactly clear, but that is the traditional name. The set of all n-flags in $\mathbb{C}^k$ forms a subspace $F_n(\mathbb{C}^k)$ of the product of n copies of $\mathbb{C}\mathrm{P}^{k-1}$. There is a natural fiber bundle

$$ F_n(\mathbb{C}^n) \longrightarrow F_n(\mathbb{C}^k) \xrightarrow{\ p\ } G_n(\mathbb{C}^k) $$

where p sends an n-tuple of orthogonal lines to the n-plane it spans. The local triviality property can be verified just as was done for the analogous Stiefel bundle $V_n(\mathbb{C}^n) \to V_n(\mathbb{C}^k) \to G_n(\mathbb{C}^k)$ in Example 4.53. The case $k = \infty$ is covered by the same argument, and this case will be the bundle $F \to E \to G_n(\mathbb{C}^\infty)$ alluded to in the paragraph preceding the proof.

The first step in the proof is to show that $H^*(F_n(\mathbb{C}^\infty); \mathbb{Z}) \approx \mathbb{Z}[x_1, \cdots, x_n]$ where x_i is the pullback of a generator of $H^2(\mathbb{C}\mathrm{P}^\infty; \mathbb{Z})$ under the map $F_n(\mathbb{C}^\infty) \to \mathbb{C}\mathrm{P}^\infty$ projecting an n-flag onto its ith line. This can be seen by considering the fiber bundle

$$ \mathbb{C}\mathrm{P}^\infty \longrightarrow F_n(\mathbb{C}^\infty) \xrightarrow{\ p\ } F_{n-1}(\mathbb{C}^\infty) $$

where p projects an n-flag onto the $(n-1)$-flag obtained by ignoring its last line. The local triviality property can be verified by the argument in Example 4.54. The Leray–Hirsch theorem applies since the powers of x_n restrict to a basis for $H^*(\mathbb{C}\mathrm{P}^\infty; \mathbb{Z})$ in the fibers $\mathbb{C}\mathrm{P}^\infty$, each fiber being the space of lines in a vector subspace $\mathbb{C}^\infty$ of the standard $\mathbb{C}^\infty$. The elements x_i for $i < n$ are the pullbacks via p of elements of $H^*(F_{n-1}(\mathbb{C}^\infty); \mathbb{Z})$ defined in the same way. By induction $H^*(F_{n-1}(\mathbb{C}^\infty); \mathbb{Z})$ is a polynomial ring on these elements. From the Leray–Hirsch theorem we conclude that the products of powers of the x_i's for $1 \le i \le n$ form an additive basis for $H^*(F_n(\mathbb{C}^\infty); \mathbb{Z})$, hence this ring is the polynomial ring on the x_i's.

There is a corresponding result for $F_n(\mathbb{C}^k)$, that $H^*(F_n(\mathbb{C}^k); \mathbb{Z})$ is free with basis the monomials $x_1^{i_1} \cdots x_n^{i_n}$ with $i_j \le k - j$ for each j. This is proved in exactly the same way, using induction on n and the fiber bundle $\mathbb{C}\mathrm{P}^{k-n} \to F_n(\mathbb{C}^k) \to F_{n-1}(\mathbb{C}^k)$. Thus the cohomology groups of $F_n(\mathbb{C}^k)$ are isomorphic to those of $\mathbb{C}\mathrm{P}^{k-1} \times \cdots \times \mathbb{C}\mathrm{P}^{k-n}$.

After these preliminaries we can start the main argument, using the fiber bundle $F_n(\mathbb{C}^n) \to F_n(\mathbb{C}^\infty) \xrightarrow{\ p\ } G_n(\mathbb{C}^\infty)$. The preceding calculations show that the Leray–Hirsch theorem applies, so $H^*(F_n(\mathbb{C}^\infty); \mathbb{Z})$ is a free module over $H^*(G_n(\mathbb{C}^\infty); \mathbb{Z})$ with basis the monomials $x_1^{i_1} \cdots x_n^{i_n}$ with $i_j \le n - j$ for each j. In particular, since 1 is among the basis elements, the homomorphism p^* is injective and its image is a direct summand of $H^*(F_n(\mathbb{C}^\infty); \mathbb{Z})$. It remains to show that the image of p^* is exactly the symmetric polynomials.

To show that the image of p^* is contained in the symmetric polynomials, consider a map $\pi : F_n(\mathbb{C}^\infty) \to F_n(\mathbb{C}^\infty)$ permuting the lines in each n-flag according to a given permutation of the numbers $1, \cdots, n$. The induced map π^* on $H^*(F_n(\mathbb{C}^\infty); \mathbb{Z}) \approx \mathbb{Z}[x_1, \cdots, x_n]$ is the corresponding permutation of the variables x_i. Since permuting the lines in an n-flag has no effect on the n-plane they span, we have $p\pi = p$, hence $\pi^* p^* = p^*$, which says that polynomials in the image of p^* are invariant under permutations of the variables.

As remarked earlier, the symmetric polynomials in $\mathbb{Z}[x_1, \cdots, x_n]$ form a polynomial ring $\mathbb{Z}[\sigma_1, \cdots, \sigma_n]$ where σ_i has degree i. We have shown that the image of p^* is a direct summand, so to show that p^* maps onto the symmetric polynomials it will suffice to show that the graded rings $H^*(G_n(\mathbb{C}^\infty); \mathbb{Z})$ and $\mathbb{Z}[\sigma_1, \cdots, \sigma_n]$ have the same rank in each dimension, where the rank of a finitely generated free abelian group is the number of $\mathbb{Z}$ summands.

For a graded free $\mathbb{Z}$-module $A = \bigoplus_i A_i$, define its **Poincaré series** to be the formal power series $p_A(t) = \sum_i a_i t^i$ where a_i is the rank of A_i, which we assume to be finite for all i. The basic formula we need is that $p_{A \otimes B}(t) = p_A(t) \, p_B(t)$, which is immediate from the definition of the graded tensor product.

In the case at hand all nonzero cohomology is in even dimensions, so let us simplify notation by taking A_i to be the $2i$-dimensional cohomology of the space in question. Since the Poincaré series of $\mathbb{Z}[x]$ is $\sum_i t^i = (1 - t)^{-1}$, the Poincaré series of $H^*(F_n(\mathbb{C}^\infty); \mathbb{Z})$ is $(1 - t)^{-n}$. For $H^*(F_n(\mathbb{C}^n); \mathbb{Z})$ the Poincaré series is

$$(1 + t)(1 + t + t^2) \cdots (1 + t + \cdots + t^{n-1}) = \prod_{i=1}^{n} \frac{1 - t^i}{1 - t} = (1 - t)^{-n} \prod_{i=1}^{n} (1 - t^i)$$

From the additive isomorphism $H^*(F_n(\mathbb{C}^\infty); \mathbb{Z}) \approx H^*(G_n(\mathbb{C}^\infty); \mathbb{Z}) \otimes H^*(F_n(\mathbb{C}^n); \mathbb{Z})$ we see that the Poincaré series $p(t)$ of $H^*(G_n(\mathbb{C}^\infty); \mathbb{Z})$ satisfies

$$p(t)(1 - t)^{-n} \prod_{i=1}^{n} (1 - t^i) = (1 - t)^{-n} \qquad \text{and hence} \qquad p(t) = \prod_{i=1}^{n} (1 - t^i)^{-1}$$

This is exactly the Poincaré series of $\mathbb{Z}[\sigma_1, \cdots, \sigma_n]$ since σ_i has degree i. As noted before, this implies that the image of p^* is all the symmetric polynomials.

This finishes the proof for $G_n(\mathbb{C}^\infty)$. The same arguments apply in the other two cases, using $\mathbb{Z}_2$ coefficients throughout in the real case and replacing 'rank' by 'dimension' for $\mathbb{Z}_2$ vector spaces. $\square$

These calculations show that the isomorphism $H^*(E; R) \approx H^*(B; R) \otimes_R H^*(F; R)$ of the Leray-Hirsch theorem is not generally a ring isomorphism, for if it were, then the polynomial ring $H^*(F_n(\mathbb{C}^\infty); \mathbb{Z})$ would contain a copy of $H^*(F_n(\mathbb{C}^n); \mathbb{Z})$ as a subring, but in the latter ring some power of every positive-dimensional element is zero since $H^k(F_n(\mathbb{C}^n); \mathbb{Z}) = 0$ for sufficiently large k.

The Gysin Sequence

Besides the Leray-Hirsch theorem, which deals with fiber bundles that are cohomologically like products, there is another special class of fiber bundles for which an elementary analysis of their cohomology structure is possible. These are fiber bundles $S^{n-1} \longrightarrow E \xrightarrow{p} B$ satisfying an orientability hypothesis that will always hold if B is simply-connected or if we take cohomology with $\mathbb{Z}_2$ coefficients. For such bundles we will show there is an exact sequence, called the **Gysin sequence**,

$$\cdots \longrightarrow H^{i-n}(B;R) \xrightarrow{\smile e} H^i(B;R) \xrightarrow{p^*} H^i(E;R) \longrightarrow H^{i-n+1}(B;R) \longrightarrow \cdots$$

where e is a certain 'Euler class' in $H^n(B;R)$. Since $H^i(B;R) = 0$ for $i < 0$, the initial portion of the Gysin sequence gives isomorphisms $p^*:H^i(B;R) \xrightarrow{\approx} H^i(E;R)$ for $i < n - 1$, and the more interesting part of the sequence begins

$$0 \longrightarrow H^{n-1}(B;R) \xrightarrow{p^*} H^{n-1}(E;R) \longrightarrow H^0(B;R) \xrightarrow{\smile e} H^n(B;R) \xrightarrow{p^*} H^n(E;R) \longrightarrow \cdots$$

In the case of a product bundle $E = S^{n-1} \times B$ there is a section, a map $s:B \to E$ with $ps = \mathbb{1}$, so the Gysin sequence breaks up into split short exact sequences

$$0 \longrightarrow H^i(B;R) \xrightarrow{p^*} H^i(S^{n-1} \times B;R) \longrightarrow H^{i-n+1}(B;R) \longrightarrow 0$$

which agrees with the Künneth formula $H^*(S^{n-1} \times B;R) \approx H^*(S^{n-1};R) \otimes_R H^*(B;R)$. The splitting holds whenever the bundle has a section, even if it is not a product.

For example, consider the bundle $S^{n-1} \to V_2(\mathbb{R}^{n+1}) \xrightarrow{p} S^n$. Points of $V_2(\mathbb{R}^{n+1})$ are pairs (v_1, v_2) of orthogonal unit vectors in $\mathbb{R}^{n+1}$, and $p(v_1, v_2) = v_1$. If we think of v_1 as a point of S^n and v_2 as a unit vector tangent to S^n at v_1, then $V_2(\mathbb{R}^{n+1})$ is exactly the bundle of unit tangent vectors to S^n. A section of this bundle is a field of unit tangent vectors to S^n, and such a vector field exists iff n is odd by Theorem 2.28. The fact that the Gysin sequence splits when there is a section then says that $V_2(\mathbb{R}^{n+1})$ has the same cohomology as the product $S^{n-1} \times S^n$ if n is odd, at least when $n > 1$ so that the base space S^n is simply-connected and the orientability hypothesis is satisfied. When n is even, the calculations at the end of §3.D show that $H^*(V_2(\mathbb{R}^{n+1});\mathbb{Z})$ consists of $\mathbb{Z}$'s in dimensions 0 and $2n - 1$ and a $\mathbb{Z}_2$ in dimension n. The latter group appears in the Gysin sequence as

$$H^0(S^n) \xrightarrow{\smile e} H^n(S^n) \longrightarrow H^n(V_2(\mathbb{R}^{n+1})) \longrightarrow H^1(S^n)$$
$$\| \qquad\qquad \| \qquad\qquad\qquad \| \qquad\qquad\qquad \|$$
$$\mathbb{Z} \qquad\qquad \mathbb{Z} \qquad\qquad\qquad \mathbb{Z}_2 \qquad\qquad\qquad 0$$

hence the Euler class e must be twice a generator of $H^n(S^n)$ in the case that n is even. When n is odd it must be zero in order for the Gysin sequence to split.

This example illustrates a theorem in differential topology that explains why the Euler class has this name: The Euler class of the unit tangent bundle of a closed orientable smooth n-manifold M is equal to the Euler characteristic $\chi(M)$ times a generator of $H^n(M;\mathbb{Z})$.

Whenever a bundle $S^{n-1} \to E \xrightarrow{p} B$ has a section, the Euler class e must be zero from exactness of $H^0(B) \xrightarrow{\smile e} H^n(B) \xrightarrow{p^*} H^n(E)$ since p^* is injective if there is a section. Thus the Euler class can be viewed as an obstruction to the existence of a section: If the Euler class is nonzero, there can be no section. This qualitative statement can be made more precise by bringing in the machinery of obstruction theory, as explained in [Milnor & Stasheff 1974] or [VBKT].

Before deriving the Gysin sequence let us look at some examples of how it can be used to compute cup products.

Example 4D.5. Consider a bundle $S^{n-1} \to E \xrightarrow{p} B$ with E contractible, for example the bundle $S^1 \to S^\infty \to \mathbb{CP}^\infty$ or its real or quaternionic analogs. The long exact sequence of homotopy groups for the bundle shows that B is $(n-1)$-connected. Thus if $n > 1$, B is simply-connected and we have a Gysin sequence for cohomology with $\mathbb{Z}$ coefficients. For $n = 1$ we take $\mathbb{Z}_2$ coefficients. If $n > 1$ then since E is contractible, the Gysin sequence implies that $H^i(B;\mathbb{Z}) = 0$ for $0 < i < n$ and that $\smile e : H^i(B;\mathbb{Z}) \to H^{i+n}(B;\mathbb{Z})$ is an isomorphism for $i \geq 0$. It follows that $H^*(B;\mathbb{Z})$ is the polynomial ring $\mathbb{Z}[e]$. When $n = 1$ the map $p^* : H^{n-1}(B;\mathbb{Z}_2) \to H^{n-1}(E;\mathbb{Z}_2)$ in the Gysin sequence is surjective, so we see that $\smile e : H^i(B;\mathbb{Z}_2) \to H^{i+n}(B;\mathbb{Z}_2)$ is again an isomorphism for all $i \geq 0$, and hence $H^*(B;\mathbb{Z}_2) \approx \mathbb{Z}_2[e]$. Thus the Gysin sequence gives a new derivation of the cup product structure in projective spaces. Also, since polynomial rings $\mathbb{Z}[e]$ are realizable as $H^*(X;\mathbb{Z})$ only when e has dimension 2 or 4, as we show in Corollary 4L.10, we can conclude that there exist bundles $S^{n-1} \to E \to B$ with E contractible only when n is 1, 2, or 4.

Example 4D.6. For the Grassmann manifold $G_n = G_n(\mathbb{R}^\infty)$ we have $\pi_1(G_n) \approx \pi_0 O(n) \approx \mathbb{Z}_2$, so the universal cover of G_n gives a bundle $S^0 \to \widetilde{G}_n \to G_n$. One can view $\widetilde{G}_n$ as the space of oriented n-planes in $\mathbb{R}^\infty$, which is obviously a 2-sheeted covering space of G_n, hence the universal cover since it is path-connected, being the quotient $V_n(\mathbb{R}^\infty)/SO(n)$ of the contractible space $V_n(\mathbb{R}^\infty)$. A portion of the Gysin sequence for the bundle $S^0 \to \widetilde{G}_n \to G_n$ is $H^0(G_n;\mathbb{Z}_2) \xrightarrow{\smile e} H^1(G_n;\mathbb{Z}_2) \to H^1(\widetilde{G}_n;\mathbb{Z}_2)$. This last group is zero since $\widetilde{G}_n$ is simply-connected, and $H^1(G_n;\mathbb{Z}_2) \approx \mathbb{Z}_2$ since $H^*(G_n;\mathbb{Z}_2) \approx \mathbb{Z}_2[w_1, \cdots, w_n]$ as we showed earlier in this section, so $e = w_1$ and the map $\smile e : H^*(G_n;\mathbb{Z}_2) \to H^*(G_n;\mathbb{Z}_2)$ is injective. The Gysin sequence then breaks up into short exact sequences $0 \to H^i(G_n;\mathbb{Z}_2) \xrightarrow{\smile e} H^{i+1}(G_n;\mathbb{Z}_2) \to H^{i+1}(\widetilde{G}_n;\mathbb{Z}_2) \to 0$, from which it follows that $H^*(\widetilde{G}_n;\mathbb{Z}_2)$ is the quotient ring $\mathbb{Z}_2[w_1, \cdots, w_n]/(w_1) \approx \mathbb{Z}_2[w_2, \cdots, w_n]$.

Example 4D.7. The complex analog of the bundle in the preceding example is a bundle $S^1 \to \widetilde{G}_n(\mathbb{C}^\infty) \to G_n(\mathbb{C}^\infty)$ with $\widetilde{G}_n(\mathbb{C}^\infty)$ 2-connected. This can be constructed in the following way. There is a determinant homomorphism $U(n) \to S^1$ with kernel $SU(n)$, the unitary matrices of determinant 1, so S^1 is the coset space $U(n)/SU(n)$, and by restricting the action of $U(n)$ on $V_n(\mathbb{C}^\infty)$ to $SU(n)$ we obtain the second row of the commutative diagram at the right. The second row is a fiber bundle by the usual

$$
\begin{array}{ccccc}
U(n) & \longrightarrow & V_n(\mathbb{C}^\infty) & \longrightarrow & G_n(\mathbb{C}^\infty) \\
\downarrow & & \downarrow & & \| \\
S^1 & \longrightarrow & V_n(\mathbb{C}^\infty)/SU(n) & \longrightarrow & G_n(\mathbb{C}^\infty)
\end{array}
$$

argument of choosing continuously varying orthonormal bases in n-planes near a given n-plane. One sees that the space $\widetilde{G}_n(\mathbb{C}^\infty) = V_n(\mathbb{C}^\infty)/SU(n)$ is 2-connected by looking at the relevant portion of the diagram of homotopy groups associated to these two bundles:

$$\begin{array}{ccccccccc}
0 & \longrightarrow & \pi_2(G_n) & \xrightarrow[\approx]{\partial} & \pi_1(U(n)) & \longrightarrow & 0 \\
& & \| & & \downarrow{\approx} & & \\
0 & \longrightarrow & \pi_2(\widetilde{G}_n) & \longrightarrow & \pi_2(G_n) & \xrightarrow{\partial} & \pi_1(S^1) & \longrightarrow & \pi_1(\widetilde{G}_n) & \longrightarrow & 0
\end{array}$$

The second vertical map is an isomorphism since S^1 embeds in $U(n)$ as the subgroup $U(1)$. Since the boundary map in the upper row is an isomorphism, so also is the boundary map in the lower row, and then exactness implies that $\widetilde{G}_n$ is 2-connected.

The Gysin sequence for $S^1 \to \widetilde{G}_n(\mathbb{C}^\infty) \to G_n(\mathbb{C}^\infty)$ can be analyzed just as in the preceding example. Part of the sequence is $H^0(G_n; \mathbb{Z}) \xrightarrow{\smile e} H^2(G_n; \mathbb{Z}) \to H^2(\widetilde{G}_n; \mathbb{Z})$, and this last group is zero since $\widetilde{G}_n$ is 2-connected, so e must be a generator of $H^2(G_n; \mathbb{Z}) \approx \mathbb{Z}$. Since $H^*(G_n; \mathbb{Z})$ is a polynomial algebra $\mathbb{Z}[c_1, \cdots, c_n]$, we must have $e = \pm c_1$, so the map $\smile e : H^*(G_n; \mathbb{Z}) \to H^*(G_n; \mathbb{Z})$ is injective, the Gysin sequence breaks up into short exact sequences, and $H^*(\widetilde{G}_n; \mathbb{Z})$ is the quotient ring $\mathbb{Z}[c_1, \cdots, c_n]/(c_1) \approx \mathbb{Z}[c_2, \cdots, c_n]$.

The spaces $\widetilde{G}_n$ in the last two examples are often denoted $BSO(n)$ and $BSU(n)$, expressing the fact that they are related to the groups $SO(n)$ and $SU(n)$ via bundles $SO(n) \to V_n(\mathbb{R}^\infty) \to BSO(n)$ and $SU(n) \to V_n(\mathbb{C}^\infty) \to BSU(n)$ with contractible total spaces V_n. There is no quaternion analog of $BSO(n)$ and $BSU(n)$ since for $n = 2$ this would give a space with cohomology ring $\mathbb{Z}[x]$ on an 8-dimensional generator, which is impossible by Corollary 4L.10.

Now we turn to the derivation of the Gysin sequence, which follows a rather roundabout route:

(1) Deduce a relative version of the Leray–Hirsch theorem from the absolute case.

(2) Specialize this to the case of bundles with fiber a disk, yielding a basic result called the Thom isomorphism.

(3) Show this applies to all orientable disk bundles.

(4) Deduce the Gysin sequence by plugging the Thom isomorphism into the long exact sequence of cohomology groups for the pair consisting of a disk bundle and its boundary sphere bundle.

(1) A **fiber bundle pair** consists of a fiber bundle $p : E \to B$ with fiber F, together with a subspace $E' \subset E$ such that $p : E' \to B$ is a bundle with fiber a subspace $F' \subset F$, with local trivializations for E' obtained by restricting local trivializations for E. For example, if $E \to B$ is a bundle with fiber D^n and $E' \subset E$ is the union of the boundary spheres of the fibers, then (E, E') is a fiber bundle pair since local trivializations of E restrict to local trivializations of E', in view of the fact that homeomorphisms from an n-disk to an n-disk restrict to homeomorphisms between their boundary spheres, boundary and interior points of D^n being distinguished by the local homology groups $H_n(D^n, D^n - \{x\}; \mathbb{Z})$.

Theorem 4D.8. *Suppose that* $(F, F') \to (E, E') \xrightarrow{p} B$ *is a fiber bundle pair such that* $H^*(F, F'; R)$ *is a free* R*-module, finitely generated in each dimension. If there exist classes* $c_j \in H^*(E, E'; R)$ *whose restrictions form a basis for* $H^*(F, F'; R)$ *in each fiber* (F, F'), *then* $H^*(E, E'; R)$, *as a module over* $H^*(B; R)$, *is free with basis* $\{c_j\}$.

The module structure is defined just as in the absolute case by $bc = p^*(b) \smile c$, but now we use the relative cup product $H^*(E; R) \times H^*(E, E'; R) \to H^*(E, E'; R)$.

Proof: Construct a bundle $\hat{E} \to B$ from E by attaching the mapping cylinder M of $p : E' \to B$ to E by identifying the subspaces $E' \subset E$ and $E' \subset M$. Thus the fibers $\hat{F}$ of $\hat{E}$ are obtained from the fibers F by attaching cones CF' on the subspaces $F' \subset F$. Regarding B as the subspace of $\hat{E}$ at one end of the mapping cylinder M, we have $H^*(\hat{E}, M; R) \approx H^*(\hat{E} - B, M - B; R) \approx H^*(E, E'; R)$ via excision and the obvious deformation retraction of $\hat{E} - B$ onto E. The long exact sequence of a triple gives $H^*(\hat{E}, M; R) \approx H^*(\hat{E}, B; R)$ since M deformation retracts to B. All these isomorphisms are $H^*(B; R)$-module isomorphisms. Since B is a retract of $\hat{E}$ via the bundle projection $\hat{E} \to B$, we have a splitting $H^*(\hat{E}; R) \approx H^*(\hat{E}, B; R) \oplus H^*(B; R)$ as $H^*(B; R)$-modules. Let $\hat{c}_j \in H^*(\hat{E}; R)$ correspond to $c_j \in H^*(E, E'; R) \approx H^*(\hat{E}, B; R)$ in this splitting. The classes $\hat{c}_j$ together with 1 restrict to a basis for $H^*(\hat{F}; R)$ in each fiber $\hat{F} = F \cup CF'$, so the absolute form of the Leray-Hirsch theorem implies that $H^*(\hat{E}; R)$ is a free $H^*(B; R)$-module with basis $\{1, \hat{c}_j\}$. It follows that $\{c_j\}$ is a basis for the free $H^*(B; R)$-module $H^*(E, E'; R)$. $\qquad\square$

(2) Now we specialize to the case of a fiber bundle pair $(D^n, S^{n-1}) \to (E, E') \xrightarrow{p} B$. An element $c \in H^n(E, E'; R)$ whose restriction to each fiber (D^n, S^{n-1}) is a generator of $H^n(D^n, S^{n-1}; R) \approx R$ is called a **Thom class** for the bundle. We are mainly interested in the cases $R = \mathbb{Z}$ and $\mathbb{Z}_2$, but R could be any commutative ring with identity, in which case a 'generator' is an element with a multiplicative inverse, so all elements of R are multiples of the generator. A Thom class with $\mathbb{Z}$ coefficients gives rise to a Thom class with any other coefficient ring R under the homomorphism $H^n(E, E'; \mathbb{Z}) \to H^n(E, E'; R)$ induced by the homomorphism $\mathbb{Z} \to R$ sending 1 to the identity element of R.

Corollary 4D.9. *If the disk bundle* $(D^n, S^{n-1}) \to (E, E') \xrightarrow{p} B$ *has a Thom class* $c \in H^n(E, E'; R)$, *then the map* $\Phi : H^i(B; R) \to H^{i+n}(E, E'; R)$, $\Phi(b) = p^*(b) \smile c$, *is an isomorphism for all* $i \geq 0$, *and* $H^i(E, E'; R) = 0$ *for* $i < n$. $\qquad\square$

The isomorphism Φ is called the **Thom isomorphism**. The corollary can be made into a statement about absolute cohomology by defining the **Thom space** $T(E)$ to be the quotient E/E'. Each disk fiber D^n of E becomes a sphere S^n in $T(E)$, and all these spheres coming from different fibers are disjoint except for the common basepoint $x_0 = E'/E'$. A Thom class can be regarded as an element of $H^n(T(E), x_0; R) \approx$

$H^n(T(E); R)$ that restricts to a generator of $H^n(S^n; R)$ in each 'fiber' S^n in $T(E)$, and the Thom isomorphism becomes $H^i(B; R) \approx \tilde{H}^{n+i}(T(E); R)$.

(3) The major remaining step in the derivation of the Gysin sequence is to relate the existence of a Thom class for a disk bundle $D^n \to E \to B$ to a notion of orientability of the bundle. First we define orientability for a sphere bundle $S^{n-1} \to E' \to B$. In the proof of Proposition 4.61 we described a procedure for lifting paths y in B to homotopy equivalences L_y between the fibers above the endpoints of y. We did this for fibrations rather than fiber bundles, but the method applies equally well to fiber bundles whose fiber is a CW complex since the homotopy lifting property was used only for the fiber and for the product of the fiber with I. In the case of a sphere bundle $S^{n-1} \to E' \to B$, if y is a loop in B then L_y is a homotopy equivalence from the fiber S^{n-1} over the basepoint of y to itself, and we define the sphere bundle to be **orientable** if L_y induces the identity map on $H^{n-1}(S^{n-1}; \mathbb{Z})$ for each loop y in B.

For example, the Klein bottle, regarded as a bundle over S^1 with fiber S^1, is nonorientable since as we follow a path looping once around the base circle, the corresponding fiber circles sweep out the full Klein bottle, ending up where they started but with orientation reversed. The same reasoning shows that the torus, viewed as a circle bundle over S^1, is orientable. More generally, any sphere bundle that is a product is orientable since the maps L_y can be taken to be the identity for all loops y. Also, sphere bundles over simply-connected base spaces are orientable since $y \simeq \eta$ implies $L_y \simeq L_\eta$, hence all L_y's are homotopic to the identity when all loops y are nullhomotopic.

One could define orientability for a disk bundle $D^n \to E \to B$ by relativizing the previous definition, constructing lifts L_y which are homotopy equivalences of the fiber pairs (D^n, S^{n-1}). However, since $H^n(D^n, S^{n-1}; \mathbb{Z})$ is canonically isomorphic to $H^{n-1}(S^{n-1}; \mathbb{Z})$ via the coboundary map in the long exact sequence of the pair, it is simpler and amounts to the same thing just to define E to be orientable if its boundary sphere subbundle E' is orientable.

Theorem 4D.10. *Every disk bundle has a Thom class with $\mathbb{Z}_2$ coefficients, and orientable disk bundles have Thom classes with $\mathbb{Z}$ coefficients.*

An exercise at the end of the section is to show that the converse of the last statement is also true: A disk bundle is orientable if it has a Thom class with $\mathbb{Z}$ coefficients.

Proof: The case of a non-CW base space B reduces to the CW case by pulling back over a CW approximation to B, as in the Leray–Hirsch theorem, applying the five-lemma to say that the pullback bundle has isomorphic homotopy groups, hence isomorphic absolute and relative cohomology groups. From the definition of the pullback bundle it is immediate that the pullback of an orientable sphere bundle is orientable. There is also no harm in assuming the base CW complex B is connected. We will show:

If the disk bundle $D^n \to E \to B$ is orientable and B is a connected CW complex,

($*$)   then the restriction map $H^i(E, E'; \mathbb{Z}) \to H^i(D_x^n, S_x^{n-1}; \mathbb{Z})$ is an isomorphism for all fibers D_x^n, $x \in B$, and for all $i \leq n$.

For $\mathbb{Z}_2$ coefficients we will see that ($*$) holds without any orientability hypothesis. Hence with either $\mathbb{Z}$ or $\mathbb{Z}_2$ coefficients, a generator of $H^n(E, E') \approx H^n(D_x^n, S_x^{n-1})$ is a Thom class.

If the disk bundle $D^n \to E \to B$ is orientable, then if we choose an isomorphism $H^n(D_x^n, S_x^{n-1}; \mathbb{Z}) \approx \mathbb{Z}$ for one fiber D_x^n, this determines such isomorphisms for all fibers by composing with the isomorphisms L_γ^*, which depend only on the endpoints of γ. Having made such a choice, then if ($*$) is true, we have a preferred isomorphism $H^n(E, E'; \mathbb{Z}) \approx \mathbb{Z}$ which restricts to the chosen isomorphism $H^n(D_x^n, S_x^{n-1}; \mathbb{Z}) \approx \mathbb{Z}$ for each fiber. This is because for a path γ from x to y, the inclusion $(D_x^n, S_x^{n-1}) \hookrightarrow (E, E')$ is homotopic to the composition of L_γ with the inclusion $(D_y^n, S_y^{n-1}) \hookrightarrow (E, E')$. We will use this preferred isomorphism $H^n(E, E'; \mathbb{Z}) \approx \mathbb{Z}$ in the inductive proof of ($*$) given below. In the case of $\mathbb{Z}_2$ coefficients, there can be only one isomorphism of a group with $\mathbb{Z}_2$ so no choices are necessary and orientability is irrelevant. We will prove ($*$) in the $\mathbb{Z}$ coefficient case, leaving it to the reader to replace all $\mathbb{Z}$'s in the proof by $\mathbb{Z}_2$'s to obtain a proof in the $\mathbb{Z}_2$ case.

Suppose first that the CW complex B has finite dimension k. Let $U \subset B$ be the subspace obtained by deleting one point from the interior of each k-cell of B, and let $V \subset B$ be the union of the open k-cells. Thus $B = U \cup V$. For a subspace $A \subset B$ let $E_A \to A$ and $E'_A \to A$ be the disk and sphere bundles obtained by taking the subspaces of E and E' projecting to A. Consider the following portion of a Mayer–Vietoris sequence, with $\mathbb{Z}$ coefficients implicit from now on:

$$H^n(E, E') \longrightarrow H^n(E_U, E'_U) \oplus H^n(E_V, E'_V) \xrightarrow{\ \Psi\ } H^n(E_{U \cap V}, E'_{U \cap V})$$

The first map is injective since the preceding term in the sequence is zero by induction on k, since $U \cap V$ deformation retracts onto a disjoint union of $(k - 1)$-spheres and we can apply Lemma 4D.2 to replace $E_{U \cap V}$ by the part of E over this union of $(k - 1)$-spheres. By exactness we then have an isomorphism $H^n(E, E') \approx \text{Ker}\, \Psi$. Similarly, by Lemma 4D.2 and induction each of the terms $H^n(E_U, E'_U)$, $H^n(E_V, E'_V)$, and $H^n(E_{U \cap V}, E'_{U \cap V})$ is a product of $\mathbb{Z}$'s, with one $\mathbb{Z}$ factor for each component of the spaces involved, projection onto the $\mathbb{Z}$ factor being given by restriction to any fiber in the component. Elements of $\text{Ker}\, \Psi$ are pairs $(\alpha, \beta) \in H^n(E_U, E'_U) \oplus H^n(E_V, E'_V)$ having the same restriction to $H^n(E_{U \cap V}, E'_{U \cap V})$. Since B is connected, this means that all the $\mathbb{Z}$ coordinates of α and β in the previous direct product decompositions must be equal, since between any two components of U or V one can interpolate a finite sequence of components of U and V alternately, each component in the sequence having nontrivial intersection with its neighbors. Thus $\text{Ker}\, \Psi$ is a copy of $\mathbb{Z}$, with restriction to a fiber being the isomorphism $H^n(E, E') \approx \mathbb{Z}$.

To finish proving $(*)$ for finite-dimensional B it remains to see that $H^i(E, E') = 0$ for $i < n$, but this follows immediately by looking at an earlier stage of the Mayer–Vietoris sequence, where the two terms adjacent to $H^i(E, E')$ vanish by induction.

Proving $(*)$ for an infinite-dimensional CW complex B reduces to the finite-dimensional case as in the Leray–Hirsch theorem since we are only interested in cohomology in a finite range of dimensions. $\square$

(4) Now we can derive the Gysin sequence for a sphere bundle $S^{n-1} \to E \xrightarrow{p} B$. Consider the mapping cylinder M_p, which is a disk bundle $D^n \to M_p \xrightarrow{p} B$ with E as its boundary sphere bundle. Assuming that a Thom class $c \in H^n(M_p, E; R)$ exists, as is the case if E is orientable or if $R = \mathbb{Z}_2$, then the long exact sequence of cohomology groups for the pair (M_p, E) gives the first row of the following commutative diagram, with R coefficients implicit:

$$\cdots \longrightarrow H^i(M_p, E) \xrightarrow{j^*} H^i(M_p) \longrightarrow H^i(E) \longrightarrow H^{i+1}(M_p, E) \longrightarrow \cdots$$
$$ \approx \uparrow \Phi \approx \uparrow p^* \| \approx \uparrow \Phi$$
$$\cdots \longrightarrow H^{i-n}(B) \xrightarrow{\smile e} H^i(B) \xrightarrow{p^*} H^i(E) \longrightarrow H^{i-n+1}(B) \longrightarrow \cdots$$

The maps Φ are the Thom isomorphism, and the vertical map p^* is an isomorphism since M_p deformation retracts onto B. The **Euler class** $e \in H^n(B; R)$ is defined to be $(p^*)^{-1} j^*(c)$, c being a Thom class. The square containing the map $\smile e$ commutes since for $b \in H^{i-n}(B; R)$ we have $j^* \Phi(b) = j^*(p^*(b) \smile c) = p^*(b) \smile j^*(c)$, which equals $p^*(b \smile e) = p^*(b) \smile p^*(e)$ since $p^*(e) = j^*(c)$. Another way of defining e is as the class corresponding to $c \smile c$ under the Thom isomorphism, since $\Phi(e) = p^*(e) \smile c = j^*(c) \smile c = c \smile c$.

Finally, the lower row of the diagram is by definition the Gysin sequence. $\square$

To conclude this section we will use the following rather specialized application of the Gysin sequence to compute a few more examples of spaces with polynomial cohomology.

Proposition 4D.11. *Suppose that $S^{2k-1} \to E \xrightarrow{p} B$ is an orientable sphere bundle such that $H^*(E; R)$ is a polynomial ring $R[x_1, \cdots, x_\ell]$ on even-dimensional generators x_i. Then $H^*(B; R) = R[y_1, \cdots, y_\ell, e]$ where e is the Euler class of the bundle and $p^*(y_i) = x_i$ for each i.*

Proof: Consider the three terms $H^i(B; R) \xrightarrow{\smile e} H^{i+2k}(B; R) \longrightarrow H^{i+2k}(E; R)$ of the Gysin sequence. If i is odd, the third term is zero since E has no odd-dimensional cohomology. Hence the map $\smile e$ is surjective, and by induction on dimension this implies that $H^*(B; R)$ is zero in odd dimensions. This means the Gysin sequence reduces to short exact sequences

$$0 \longrightarrow H^{2i}(B; R) \xrightarrow{\smile e} H^{2i+2k}(B; R) \xrightarrow{p^*} H^{2i+2k}(E; R) \longrightarrow 0$$

Since p^* is surjective, we can choose elements $y_j \in H^*(B;R)$ with $p^*(y_j) = x_j$. It remains to check that $H^*(B;R) = R[y_1, \cdots, y_\ell, e]$, which is elementary algebra: Given $b \in H^*(B;R)$, $p^*(b)$ must be a polynomial $f(x_1, \cdots, x_\ell)$, so $b - f(y_1, \cdots, y_\ell)$ is in the kernel of p^* and exactness gives an equation $b - f(y_1, \cdots, y_\ell) = b' \smile e$ for some $b' \in H^*(B;R)$. Since b' has lower dimension than b, we may assume by induction that b' is a polynomial in $y_1, \cdots, y_\ell, e$. Hence $b = f(y_1, \cdots, y_\ell) + b' \smile e$ is also a polynomial in $y_1, \cdots, y_\ell, e$. Thus the natural map $R[y_1, \cdots, y_\ell, e] \to H^*(B;R)$ is surjective. To see that it is injective, suppose there is a polynomial relation $f(y_1, \cdots, y_\ell, e) = 0$ in $H^*(B;R)$. Applying p^*, we get $f(x_1, \cdots, x_\ell, 0) = 0$ since $p^*(y_i) = x_i$ and $p^*(e) = 0$ from the short exact sequence. The relation $f(x_1, \cdots, x_\ell, 0) = 0$ takes place in the polynomial ring $R[x_1, \cdots, x_\ell]$, so $f(y_1, \cdots, y_\ell, 0) = 0$ in $R[y_1, \cdots, y_\ell, e]$, hence $f(y_1, \cdots, y_\ell, e)$ must be divisible by e, say $f = ge$ for some polynomial g. The relation $f(y_1, \cdots, y_\ell, e) = 0$ in $H^*(B;R)$ then has the form $g(y_1, \cdots, y_\ell, e) \smile e = 0$. Since $\smile e$ is injective, this gives a polynomial relation $g(y_1, \cdots, y_\ell, e) = 0$ with g having lower degree than f. By induction we deduce that g must be the zero polynomial, hence also f. $\qquad\square$

Example 4D.12. Let us apply this to give another proof that $H^*(G_n(\mathbb{C}^\infty); \mathbb{Z})$ is a polynomial ring $\mathbb{Z}[c_1, \cdots, c_n]$ with $|c_i| = 2i$. We use two fiber bundles:

$$S^{2n-1} \to E \to G_n(\mathbb{C}^\infty) \qquad\qquad S^\infty \to E \to G_{n-1}(\mathbb{C}^\infty)$$

The total space E in both cases is the space of pairs (P, v) where P is an n-plane in $\mathbb{C}^\infty$ and v is a unit vector in P. In the first bundle the map $E \to G_n(\mathbb{C}^\infty)$ is $(P, v) \mapsto P$, with fiber S^{2n-1}, and for the second bundle the map $E \to G_{n-1}(\mathbb{C}^\infty)$ sends (P, v) to the $(n-1)$-plane in P orthogonal to v, with fiber S^∞ consisting of all the unit vectors in $\mathbb{C}^\infty$ orthogonal to a given $(n-1)$-plane. Local triviality for the two bundles is verified in the usual way. Since S^∞ is contractible, the map $E \to G_{n-1}(\mathbb{C}^\infty)$ induces isomorphisms on homotopy groups, hence also on cohomology. By induction on n we then have $H^*(E; \mathbb{Z}) \approx \mathbb{Z}[c_1, \cdots, c_{n-1}]$. The first bundle is orientable since $G_n(\mathbb{C}^\infty)$ is simply-connected, so the proposition gives $H^*(G_n(\mathbb{C}^\infty); \mathbb{Z}) \approx \mathbb{Z}[c_1, \cdots, c_n]$ for $c_n = e$.

The same argument works in the quaternionic case. For a version of this argument in the real case see §3.3 of [VBKT].

Before giving our next example, let us observe that the Gysin sequence with a fixed coefficient ring R is valid for any orientable fiber bundle $F \to E \xrightarrow{p} B$ whose fiber is a CW complex F with $H^*(F;R) \approx H^*(S^{n-1};R)$. Orientability is defined just as before in terms of induced maps $L_y^* : H^{n-1}(F;R) \to H^{n-1}(F;R)$. No changes are needed in the derivation of the Gysin sequence to get this more general case, if the associated 'disk' bundle is again taken to be the mapping cylinder $CF \to M_p \to B$.

Example 4D.13. We have computed the cohomology of $\tilde{G}_n(\mathbb{R}^\infty)$ with $\mathbb{Z}_2$ coefficients, finding it to be a polynomial ring on generators in dimensions 2 through n, and now we compute the cohomology with $\mathbb{Z}_p$ coefficients for p an odd prime. The answer will again be a polynomial algebra, but this time on even-dimensional generators, depending on the parity of n. Consider first the case that n is odd, say $n = 2k + 1$. There are two fiber bundles

$$V_2(\mathbb{R}^{2k+1}) \longrightarrow E \longrightarrow \widetilde{G}_{2k+1}(\mathbb{R}^\infty) \qquad\qquad V_2(\mathbb{R}^\infty) \longrightarrow E \longrightarrow \widetilde{G}_{2k-1}(\mathbb{R}^\infty)$$

where E is the space of triples (P, v_1, v_2) with P an oriented $(2k + 1)$-plane in $\mathbb{R}^\infty$ and v_1 and v_2 two orthogonal unit vectors in P. The projection map in the first bundle is $(P, v_1, v_2) \mapsto P$, and for the second bundle the projection sends (P, v_1, v_2) to the oriented $(2k - 1)$-plane in P orthogonal to v_1 and v_2, with the orientation specified by saying for example that v_1, v_2 followed by a positively oriented basis for the orthogonal $(2k - 1)$-plane is a positively oriented basis for P. Both bundles are orientable since their base spaces $\widetilde{G}_n(\mathbb{R}^\infty)$ are simply-connected, from the bundle $SO(n) \to V_n(\mathbb{R}^\infty) \to \widetilde{G}_n(\mathbb{R}^\infty)$.

The fiber $V_2(\mathbb{R}^\infty)$ of the second bundle is contractible, so E has the same cohomology as $\widetilde{G}_{2k-1}(\mathbb{R}^\infty)$. The fiber of the first bundle has the same $\mathbb{Z}_p$ cohomology as S^{4k-1} if p is odd, by the calculation at the end of §3.D. So if we assume inductively that $H^*(\widetilde{G}_{2k-1}(\mathbb{R}^\infty); \mathbb{Z}_p) \approx \mathbb{Z}_p[p_1, \cdots, p_{k-1}]$ with $|p_i| = 4i$, then Proposition 4D.11 above implies that $H^*(\widetilde{G}_{2k+1}(\mathbb{R}^\infty); \mathbb{Z}_p) \approx \mathbb{Z}_p[p_1, \cdots, p_k]$ where $p_k = e$ has dimension $4k$. The induction can start with $\widetilde{G}_1(\mathbb{R}^\infty)$ which is just S^∞ since an oriented line in $\mathbb{R}^\infty$ contains a unique unit vector in the positive direction.

To handle the case of $\widetilde{G}_n(\mathbb{R}^\infty)$ with $n = 2k$ even, we proceed just as in Example 4D.12, considering the bundles

$$S^{2k-1} \longrightarrow E \longrightarrow \widetilde{G}_{2k}(\mathbb{R}^\infty) \qquad\qquad S^\infty \longrightarrow E \longrightarrow \widetilde{G}_{2k-1}(\mathbb{R}^\infty)$$

By the case n odd we have $H^*(\widetilde{G}_{2k-1}(\mathbb{R}^\infty); \mathbb{Z}_p) \approx \mathbb{Z}_p[p_1, \cdots, p_{k-1}]$ with $|p_i| = 4i$, so the corollary implies that $H^*(\widetilde{G}_{2k}(\mathbb{R}^\infty); \mathbb{Z}_p)$ is a polynomial ring on these generators and also a generator in dimension $2k$.

Summarizing, for p an odd prime we have shown:

$$H^*(\widetilde{G}_{2k+1}(\mathbb{R}^\infty); \mathbb{Z}_p) \approx \mathbb{Z}_p[p_1, \cdots, p_k], \quad |p_i| = 4i$$
$$H^*(\widetilde{G}_{2k}(\mathbb{R}^\infty); \mathbb{Z}_p) \approx \mathbb{Z}_p[p_1, \cdots, p_{k-1}, e], \quad |p_i| = 4i, \ |e| = 2k$$

The same result holds also with $\mathbb{Q}$ coefficients. In fact, our proof applies for any coefficient ring in which 2 has a multiplicative inverse, since all that is needed is that $H^*(V_2(\mathbb{R}^{2k+1}); R) \approx H^*(S^{4k-1}; R)$. For a calculation of the cohomology of $\widetilde{G}_n(\mathbb{R}^\infty)$ with $\mathbb{Z}$ coefficients, see [VBKT]. It turns out that all torsion elements have order 2, and modulo this torsion the integral cohomology is again a polynomial ring on the generators p_i and e. Similar results hold also for the cohomology of the unoriented Grassmann manifold $G_n(\mathbb{R}^\infty)$, but with the generator e replaced by p_k when $n = 2k$.

Exercises

1. By Exercise 35 in §4.2 there is a bundle $S^2 \to \mathbb{CP}^3 \to S^4$. Let $S^2 \to E_k \to S^4$ be the pullback of this bundle via a degree k map $S^4 \to S^4$, $k > 1$. Use the Leray-Hirsch theorem to show that $H^*(E_k; \mathbb{Z})$ is additively isomorphic to $H^*(\mathbb{CP}^3; \mathbb{Z})$ but has a different cup product structure in which the square of a generator of $H^2(E_k; \mathbb{Z})$ is k times a generator of $H^4(E_k; \mathbb{Z})$.

2. Apply the Leray-Hirsch theorem to the bundle $S^1 \to S^\infty / \mathbb{Z}_p \to \mathbb{CP}^\infty$ to compute $H^*(K(\mathbb{Z}_p, 1); \mathbb{Z}_p)$ from $H^*(\mathbb{CP}^\infty; \mathbb{Z}_p)$.

3. Use the Leray-Hirsch theorem as in Corollary 4D.3 to compute $H^*(V_n(\mathbb{C}^k); \mathbb{Z}) \approx \Lambda_{\mathbb{Z}}[x_{2k-2n+1}, x_{2k-2n+3}, \cdots, x_{2k-1}]$ and similarly in the quaternionic case.

4. For the flag space $F_n(\mathbb{C}^n)$ show that $H^*(F_n(\mathbb{C}^n); \mathbb{Z}) \approx \mathbb{Z}[x_1, \cdots, x_n]/(\sigma_1, \cdots, \sigma_n)$ where σ_i is the ith elementary symmetric polynomial.

5. Use the Gysin sequence to show that for a fiber bundle $S^k \to S^m \xrightarrow{p} S^n$ we must have $k = n - 1$ and $m = 2n - 1$. Then use the Thom isomorphism to show that the Hopf invariant of p must be ± 1. [Hence $n = 1, 2, 4, 8$ by Adams' theorem.]

6. Show that if M is a manifold of dimension $2n$ for which there exists a fiber bundle $S^1 \to S^{2n+1} \to M$, then M is simply-connected and $H^*(M; \mathbb{Z}) \approx H^*(\mathbb{CP}^n; \mathbb{Z})$ as rings. Conversely, if M is simply-connected and $H^*(M; \mathbb{Z}) \approx H^*(\mathbb{CP}^n; \mathbb{Z})$ as rings, show there is a bundle $S^1 \to E \to M$ where $E \simeq S^{2n+1}$. [When $n > 1$ there are examples where M is not homeomorphic to $\mathbb{CP}^n$.]

7. Show that if a disk bundle $D^n \to E \to B$ has a Thom class with $\mathbb{Z}$ coefficients, then it is orientable.

8. If E is the product bundle $B \times D^n$ with B a CW complex, show that the Thom space $T(E)$ is the n-fold reduced suspension $\Sigma^n(B_+)$, where B_+ is the union of B with a disjoint basepoint, and that the Thom isomorphism specializes to the suspension isomorphism $\tilde{H}^i(B; R) \approx \tilde{H}^{n+i}(\Sigma^n B; R)$ given by the reduced cross product in §3.2.

9. Show that the inclusion $T^n \hookrightarrow U(n)$ of the n-torus of diagonal matrices is homotopic to the map $T^n \to U(1) \hookrightarrow U(n)$ sending an n-tuple of unit complex numbers $(z_1, \cdots, z_n)$ to the 1×1 matrix $(z_1 \cdots z_n)$. Do the same for the diagonal subgroup of $Sp(n)$. [Hint: Diagonal matrices in $U(n)$ are compositions of scalar multiplication in n lines in $\mathbb{C}^n$, and $\mathbb{CP}^{n-1}$ is connected.]

10. Fill in the details of the following argument to show that every $n \times n$ matrix A with entries in $\mathbb{H}$ has an eigenvalue in $\mathbb{H}$. (The usual argument over $\mathbb{C}$ involving roots of the characteristic polynomial does not work due to the lack of a good quaternionic determinant function.) For $t \in [0, 1]$ and $\lambda \in S^3 \subset \mathbb{H}$, consider the matrix $t\lambda I + (1 - t)A$. If A has no eigenvalues, this is invertible for all t. Thus the map $S^3 \to GL_n(\mathbb{H})$, $\lambda \mapsto \lambda I$, is nullhomotopic. But by the preceding problem and Exercise 10(b) in §3.C, this map represents n times a generator of $\pi_3 GL_n(\mathbb{H})$.

4.E The Brown Representability Theorem

In Theorem 4.58 in §4.3 we showed that Ω-spectra define cohomology theories, and now we will prove the converse statement that all cohomology theories on the CW category arise in this way from Ω-spectra.

Theorem 4E.1. *Every reduced cohomology theory on the category of basepointed CW complexes and basepoint-preserving maps has the form $h^n(X) = \langle X, K_n \rangle$ for some Ω-spectrum $\{K_n\}$.*

We will also see that the spaces K_n are unique up to homotopy equivalence.

This theorem gives another proof that ordinary cohomology is representable as maps into Eilenberg–MacLane spaces, since for the spaces K_n in an Ω-spectrum representing $\tilde{H}^*(-;G)$ we have $\pi_i(K_n) = \langle S^i, K_n \rangle = \tilde{H}^n(S^i;G)$, so K_n is a $K(G,n)$.

Before getting into the proof of the theorem let us observe that cofibration sequences, as constructed in §4.3, allow us to recast the definition of a reduced cohomology theory in a slightly more concise form: A reduced cohomology theory on the category $\mathcal{C}$ whose objects are CW complexes with a chosen basepoint 0-cell and whose morphisms are basepoint-preserving maps is a sequence of functors h^n, $n \in \mathbb{Z}$, from $\mathcal{C}$ to abelian groups, together with natural isomorphisms $h^n(X) \approx h^{n+1}(\Sigma X)$ for all X in $\mathcal{C}$, such that the following axioms hold for each h^n:

(i) If $f \simeq g : X \to Y$ in the basepointed sense, then $f^* = g^* : h^n(Y) \to h^n(X)$.

(ii) For each inclusion $A \hookrightarrow X$ in $\mathcal{C}$ the sequence $h^n(X/A) \to h^n(X) \to h^n(A)$ is exact.

(iii) For a wedge sum $X = \bigvee_\alpha X_\alpha$ with inclusions $i_\alpha : X_\alpha \hookrightarrow X$, the product map $\prod_\alpha i_\alpha^* : h^n(X) \to \prod_\alpha h^n(X_\alpha)$ is an isomorphism.

To see that these axioms suffice to define a cohomology theory, the main thing to note is that the cofibration sequence $A \to X \to X/A \to \Sigma A \to \cdots$ allows us to construct the long exact sequence of a pair, just as we did in the case of the functors $h^n(X) = \langle X, K_n \rangle$. In the converse direction, if we have natural long exact sequences of pairs, then by applying these to pairs of the form (CX, X) we get natural isomorphisms $h^n(X) \approx h^{n+1}(\Sigma X)$. Note that these natural isomorphisms coming from coboundary maps of pairs (CX, X) uniquely determine the coboundary maps for all pairs (X, A) via the diagram at the right, where the maps from $h^n(A)$ are coboundary maps of pairs and the diagram commutes by naturality of these coboundary maps. The isomorphism comes from a deforma-

$$
\begin{array}{ccc}
h^n(A) & \longrightarrow & h^{n+1}(X/A) \\
\downarrow & \nearrow & \uparrow \\
h^{n+1}(CA/A) & \xleftarrow{\approx} & h^{n+1}(CX/A)
\end{array}
$$

tion retraction of CX onto CA. It is easy to check that these processes for converting one definition of a cohomology theory into the other are inverses of each other.

Most of the work in representing cohomology theories by Ω-spectra will be in realizing a single functor h^n of a cohomology theory as $\langle -, K_n \rangle$ for some space K_n.

So let us consider what properties the functor $h(X) = \langle X, K \rangle$ has, where K is a fixed space with basepoint. First of all, it is a contravariant functor from the category of basepointed CW complexes to the category of pointed sets, that is, sets with a distinguished element, the homotopy class of the constant map in the present case. Morphisms in the category of pointed sets are maps preserving the distinguished element. We have already seen in §4.3 that $h(X)$ satisfies the three axioms (i)–(iii). A further property is the following **Mayer–Vietoris axiom**:

- Suppose the CW complex X is the union of subcomplexes A and B containing the basepoint. Then if $a \in h(A)$ and $b \in h(B)$ restrict to the same element of $h(A \cap B)$, there exists an element $x \in h(X)$ whose restrictions to A and B are the given elements a and b.

Here and in what follows we use the term 'restriction' to mean the map induced by inclusion. In the case that $h(X) = \langle X, K \rangle$, this axiom is an immediate consequence of the homotopy extension property. The functors h^n in any cohomology theory also satisfy this axiom since there are Mayer–Vietoris exact sequences in any cohomology theory, as we observed in §2.3 in the analogous setting of homology theories.

> **Theorem 4E.2.** *If h is a contravariant functor from the category of connected base-pointed CW complexes to the category of pointed sets, satisfying the homotopy axiom (i), the Mayer–Vietoris axiom, and the wedge axiom (iii), then there exists a connected CW complex K and an element $u \in h(K)$ such that the transformation $T_u : \langle X, K \rangle \to h(X)$, $T_u(f) = f^*(u)$, is a bijection for all X.*

Such a pair (K, u) is called **universal** for the functor h. It is automatic from the definition that the space K in a universal pair (K, u) is unique up to homotopy equivalence. For if (K', u') is also universal for h, then, using the notation $f : (K, u) \to (K', u')$ to mean $f : K \to K'$ with $f^*(u') = u$, universality implies that there are maps $f : (K, u) \to (K', u')$ and $g : (K', u') \to (K, u)$ that are unique up to homotopy. Likewise the compositions $gf : (K, u) \to (K, u)$ and $fg : (K', u') \to (K', u')$ are unique up to homotopy, hence are homotopic to the identity maps.

Before starting the proof of this theorem we make a few preliminary comments on the axioms.

(1) The wedge axiom implies that $h(point)$ is trivial. To see this, just use the fact that for any X we have $X \vee point = X$, so the map $h(X) \times h(point) \to h(X)$ induced by inclusion of the first summand is a bijection, but this map is the projection $(a, b) \mapsto a$, hence $h(point)$ must have only one element.

(2) Axioms (i), (iii), and the Mayer–Vietoris axiom imply axiom (ii). Namely, (ii) is equivalent to exactness of $h(A) \leftarrow h(X) \leftarrow h(X \cup CA)$, where CA is the reduced cone since we are in the basepointed category. The inclusion $\mathrm{Im} \subset \mathrm{Ker}$ holds since the composition $A \to X \cup CA$ is nullhomotopic, so the induced map factors through $h(point) = 0$.

To obtain the opposite inclusion $\text{Ker} \subset \text{Im}$, decompose $X \cup CA$ into two subspaces Y and Z by cutting along a copy of A halfway up the cone CA, so Y is a smaller copy of CA and Z is the reduced mapping cylinder of the inclusion $A \hookrightarrow X$. Given an element $x \in h(X)$, this extends to an element $z \in h(Z)$ since Z deformation retracts to X. If x restricts to the trivial element of $h(A)$, then z restricts to the trivial element of $h(Y \cap Z)$. The latter element extends to the trivial element of $h(Y)$, so the Mayer–Vietoris axiom implies there is an element of $h(X \cup CA)$ restricting to z in $h(Z)$ and hence to x in $h(X)$.

(3) If h satisfies axioms (i) and (iii) then $h(\Sigma Y)$ is a group and $T_u : \langle \Sigma Y, K \rangle \to h(\Sigma Y)$ is a homomorphism for all suspensions ΣY and all pairs (K, u). [See the Corrections.]

The proof of Theorem 4E.2 will use two lemmas. To state the first, consider pairs (K, u) with K a basepointed connected CW complex and $u \in h(K)$, where h satisfies the hypotheses of the theorem. Call such a pair (K, u) **n-universal** if the homomorphism $T_u : \pi_i(K) \to h(S^i)$, $T_u(f) = f^*(u)$, is an isomorphism for $i < n$ and surjective for $i = n$. Call (K, u) **π_*-universal** if it is n-universal for all n.

> **Lemma 4E.3.** *Given any pair (Z, z) with Z a connected CW complex and $z \in h(Z)$, there exists a π_*-universal pair (K, u) with Z a subcomplex of K and $u \,|\, Z = z$.*

Proof: We construct K from Z by an inductive process of attaching cells. To begin, let $K_1 = Z \vee_\alpha S_\alpha^1$ where α ranges over the elements of $h(S^1)$. By the wedge axiom there exists $u_1 \in h(K_1)$ with $u_1 \,|\, Z = z$ and $u_1 \,|\, S_\alpha^1 = \alpha$, so (K_1, u_1) is 1-universal.

For the inductive step, suppose we have already constructed (K_n, u_n) with u_n n-universal, $Z \subset K_n$, and $u_n \,|\, Z = z$. Represent each element α in the kernel of $T_{u_n} : \pi_n(K_n) \to h(S^n)$ by a map $f_\alpha : S^n \to K_n$. Let $f = \vee_\alpha f_\alpha : \vee_\alpha S_\alpha^n \to K_n$. The reduced mapping cylinder M_f deformation retracts to K_n, so we can regard u_n as an element of $h(M_f)$, and this element restricts to the trivial element of $h(\vee_\alpha S_\alpha^n)$ by the definition of f. The exactness property of h then implies that for the reduced mapping cone $C_f = M_f / \vee_\alpha S_\alpha^n$ there is an element $w \in h(C_f)$ restricting to u_n on K_n. Note that C_f is obtained from K_n by attaching cells e_α^{n+1} by the maps f_α. To finish the construction of K_{n+1}, set $K_{n+1} = C_f \vee_\beta S_\beta^{n+1}$ where β ranges over $h(S^{n+1})$. By the wedge axiom, there exists $u_{n+1} \in h(K_{n+1})$ restricting to w on C_f and β on S_β^{n+1}.

To see that (K_{n+1}, u_{n+1}) is $(n+1)$-universal, consider the commutative diagram displayed at the right. Since K_{n+1} is obtained from K_n by attaching $(n+1)$-cells, the upper map is an isomorphism for $i < n$ and a surjection for $i = n$. By induction the same is true for T_{u_n}, hence it is also true for $T_{u_{n+1}}$. The

$$
\begin{array}{ccc}
\pi_i(K_n) & \longrightarrow & \pi_i(K_{n+1}) \\
& \searrow{\scriptstyle T_{u_n}} \quad \swarrow{\scriptstyle T_{u_{n+1}}} & \\
& h(S^i) &
\end{array}
$$

kernel of $T_{u_{n+1}}$ is trivial for $i = n$ since an element of this kernel pulls back to $\text{Ker}\, T_{u_n} \subset \pi_n(K_n)$, by surjectivity of the upper map when $i = n$, and we attached cells to K_n by maps representing all elements of $\text{Ker}\, T_{u_n}$. Also, $T_{u_{n+1}}$ is surjective for $i = n + 1$ by construction.

Now let $K = \bigcup_n K_n$. We apply a mapping telescope argument as in the proofs of Lemma 2.34 and Theorem 3F.8 to show there is an element $u \in h(K)$ restricting to u_n on K_n, for all n. The mapping telescope of the inclusions $K_1 \hookrightarrow K_2 \hookrightarrow \cdots$ is the subcomplex $T = \bigcup_i K_i \times [i, i+1]$ of $K \times [1, \infty)$. We take '$\times$' to be the reduced product here, with *basepoint $\times$ interval* collapsed to a point. The natural projection $T \to K$ is a homotopy equivalence since $K \times [1, \infty)$ deformation retracts onto T, as we showed in the proof of Lemma 2.34. Let $A \subset T$ be the union of the subcomplexes $K_i \times [i, i+1]$ for i odd and let B be the corresponding union for i even. Thus $A \cup B = T$, $A \cap B = \bigvee_i K_i$, $A \simeq \bigvee_i K_{2i-1}$, and $B \simeq \bigvee_i K_{2i}$. By the wedge axiom there exist $a \in h(A)$ and $b \in h(B)$ restricting to u_i on each K_i. Then using the fact that $u_{i+1} | K_i = u_i$, the Mayer–Vietoris axiom implies that a and b are the restrictions of an element $t \in h(T)$. Under the isomorphism $h(T) \approx h(K)$, t corresponds to an element $u \in h(K)$ restricting to u_n on K_n for all n.

To verify that (K, u) is π_*-universal we use the commutative diagram at the right. For $n > i + 1$ the upper map is an isomorphism and T_{u_n} is surjective with trivial kernel, so the same is true of T_u. $\qquad\square$

$$\begin{array}{ccc} \pi_i(K_n) & \longrightarrow & \pi_i(K) \\ & \searrow T_{u_n} \quad T_u \swarrow & \\ & h(S^i) & \end{array}$$

Lemma 4E.4. *Let (K, u) be a π_*-universal pair and let (X, A) be a basepointed CW pair. Then for each $x \in h(X)$ and each map $f : A \to K$ with $f^*(u) = x | A$ there exists a map $g : X \to K$ extending f with $g^*(u) = x$.*

Schematically, this is saying that the diagonal arrow in the diagram at the right always exists, where the map i is inclusion.

$$\begin{array}{ccc} (A, a) & \xrightarrow{\ f\ } & (K, u) \\ {\scriptstyle i}\downarrow & \nearrow {\scriptstyle g} & \\ (X, x) & & \end{array}$$

Proof: Replacing K by the reduced mapping cylinder of f reduces us to the case that f is the inclusion of a subcomplex. Let Z be the union of X and K with the two copies of A identified. By the Mayer–Vietoris axiom, there exists $z \in h(Z)$ with $z | X = x$ and $z | K = u$. By the previous lemma, we can embed (Z, z) in a π_*-universal pair (K', u'). The inclusion $(K, u) \hookrightarrow (K', u')$ induces an isomorphism on homotopy groups since both u and u' are π_*-universal, so K' deformation retracts onto K. This deformation retraction induces a homotopy rel A of the inclusion $X \hookrightarrow K'$ to a map $g : X \to K$. The relation $g^*(u) = x$ holds since $u' | K = u$ and $u' | X = x$. $\qquad\square$

Proof of Theorem 4E.2: It suffices to show that a π_*-universal pair (K, u) is universal. Applying the preceding lemma with A a point shows that $T_u : \langle X, K \rangle \to h(X)$ is surjective. To show injectivity, suppose $T_u(f_0) = T_u(f_1)$, that is, $f_0^*(u) = f_1^*(u)$. We apply the preceding lemma with $(X \times I, X \times \partial I)$ playing the role of (X, A), using the maps f_0 and f_1 on $X \times \partial I$ and taking x to be $p^* f_0^*(u) = p^* f_1^*(u)$ where p is the projection $X \times I \to X$. Here $X \times I$ should be the reduced product, with *basepoint $\times I$* collapsed to a point. The lemma then gives a homotopy from f_0 to f_1. $\qquad\square$

Proof of Theorem 4E.1: Since suspension is an isomorphism in any reduced cohomology theory, and the suspension of any CW complex is connected, it suffices to restrict attention to connected CW complexes. Each functor h^n satisfies the homotopy, wedge, and Mayer–Vietoris axioms, as we noted earlier, so the preceding theorem gives CW complexes K_n with $h^n(X) = \langle X, K_n \rangle$. It remains to show that the natural isomorphisms $h^n(X) \approx h^{n+1}(\Sigma X)$ correspond to weak homotopy equivalences $K_n \to \Omega K_{n+1}$. The natural isomorphism $h^n(X) \approx h^{n+1}(\Sigma X)$ corresponds to a natural bijection $\langle X, K_n \rangle \approx \langle \Sigma X, K_{n+1} \rangle = \langle X, \Omega K_{n+1} \rangle$ which we call Φ. The naturality of this bijection gives, for any map $f : X \to K_n$, a commutative diagram as at the right. Let $\varepsilon_n = \Phi(\mathbb{1}) : K_n \to \Omega K_{n+1}$. Then using commutativity we have $\Phi(f) = \Phi f^*(\mathbb{1}) = f^* \Phi(\mathbb{1}) = f^*(\varepsilon_n) =$

$$
\begin{array}{ccc}
\langle K_n, K_n \rangle & \xrightarrow{\ f^*\ } & \langle X, K_n \rangle \\
\downarrow{\scriptstyle \Phi} & & \downarrow{\scriptstyle \Phi} \\
\langle K_n, \Omega K_{n+1} \rangle & \xrightarrow{\ f^*\ } & \langle X, \Omega K_{n+1} \rangle
\end{array}
$$

$\varepsilon_n f$, which says that the map $\Phi : \langle X, K_n \rangle \to \langle X, \Omega K_{n+1} \rangle$ is composition with ε_n. Since Φ is a bijection, if we take X to be S^i, we see that ε_n induces an isomorphism on π_i for all i, so ε_n is a weak homotopy equivalence and we have an Ω-spectrum.

There is one final thing to verify, that the bijection $h^n(X) = \langle X, K_n \rangle$ is a group isomorphism, where $\langle X, K_n \rangle$ has the group structure that comes from identifying it with $\langle X, \Omega K_{n+1} \rangle = \langle \Sigma X, K_{n+1} \rangle$. Via the natural isomorphism $h^n(X) \approx h^{n+1}(\Sigma X)$ this is equivalent to showing the bijection $h^{n+1}(\Sigma X) = \langle \Sigma X, K_{n+1} \rangle$ preserves group structure. For maps $f, g : \Sigma X \to K$, the relation $T_u(f + g) = T_u(f) + T_u(g)$ means $(f + g)^*(u) = f^*(u) + g^*(u)$, and this holds since $(f + g)^* = f^* + g^* : h(K) \to h(\Sigma X)$ by Lemma 4.60. $\qquad\square$

4.F Spectra and Homology Theories

We have seen in §4.3 and the preceding section how cohomology theories have a homotopy-theoretic interpretation in terms of Ω-spectra, and it is natural to look for a corresponding description of homology theories. In this case we do not already have a homotopy-theoretic description of ordinary homology to serve as a starting point. But there is another homology theory we have encountered which does have a very homotopy-theoretic flavor:

Proposition 4F.1. *Stable homotopy groups $\pi_n^s(X)$ define a reduced homology theory on the category of basepointed CW complexes and basepoint-preserving maps.*

Proof: In the preceding section we reformulated the axioms for a cohomology theory so that the exactness axiom asserts just the exactness of $h^n(X/A) \to h^n(X) \to h^n(A)$ for CW pairs (X, A). In order to derive long exact sequences, the reformulated axioms require also that natural suspension isomorphisms $h^n(X) \approx h^{n+1}(\Sigma X)$ be specified as part of the cohomology theory. The analogous reformulation of the axioms for a homology theory is valid as well, by the same argument, and we shall use this in what follows.

For stable homotopy groups, suspension isomorphisms $\pi_n^s(X) \approx \pi_{n+1}^s(\Sigma X)$ are automatic, so it remains to verify the three axioms. The homotopy axiom is apparent. The exactness of a sequence $\pi_n^s(A) \to \pi_n^s(X) \to \pi_n^s(X/A)$ follows from exactness of $\pi_n(A) \to \pi_n(X) \to \pi_n(X,A)$ together with the isomorphism $\pi_n(X,A) \approx \pi_n(X/A)$ which holds under connectivity assumptions that are achieved after sufficiently many suspensions. The wedge sum axiom $\pi_n^s(\vee_\alpha X_\alpha) \approx \bigoplus_\alpha \pi_n^s(X_\alpha)$ reduces to the case of finitely many summands by the usual compactness argument, and the case of finitely many summands reduces to the case of two summands by induction. Then we have isomorphisms $\pi_{n+i}(\Sigma^i X \vee \Sigma^i Y) \approx \pi_{n+i}(\Sigma^i X \times \Sigma^i Y) \approx \pi_{n+i}(\Sigma^i X) \oplus \pi_{n+i}(\Sigma^i Y)$, the first of these isomorphisms holding when $n + i < 2i - 1$, or $i > n + 1$, since $\Sigma^i X \vee \Sigma^i Y$ is the $(2i - 1)$-skeleton of $\Sigma^i X \times \Sigma^i Y$. Passing to the limit over increasing i, we get the desired isomorphism $\pi_n^s(X \vee Y) \approx \pi_n^s(X) \oplus \pi_n^s(Y)$. $\square$

A modest generalization of this homology theory can be obtained by defining $h_n(X) = \pi_n^s(X \wedge K)$ for a fixed complex K. Verifying the homology axioms reduces to the case of stable homotopy groups themselves by basic properties of smash product:

- $h_n(X) \approx h_{n+1}(\Sigma X)$ since $\Sigma(X \wedge K) = (\Sigma X) \wedge K$, both spaces being $S^1 \wedge X \wedge K$.
- The exactness axiom holds since $(X \wedge K)/(A \wedge K) = (X/A) \wedge K$, both spaces being quotients of $X \times K$ with $A \times K \cup X \times \{k_0\}$ collapsed to a point.
- The wedge axiom follows from distributivity: $(\vee_\alpha X_\alpha) \wedge K = \vee_\alpha (X_\alpha \wedge K)$.

The coefficients of this homology theory are $h_n(S^0) = \pi_n^s(S^0 \wedge K) = \pi_n^s(K)$. Suppose for example that K is an Eilenberg-MacLane space $K(G,n)$. Because $K(G,n)$ is $(n - 1)$-connected, its stable homotopy groups are the same as its unstable homotopy groups below dimension $2n$. Thus if we shift dimensions by defining $h_i(X) = \pi_{i+n}^s(X \wedge K(G,n))$ we obtain a homology theory whose coefficient groups below dimension n are the same as ordinary homology with coefficients in G. It follows as in Theorem 4.59 that this homology theory agrees with ordinary homology for CW complexes of dimension less than $n - 1$.

This dimension restriction could be removed if there were a 'stable Eilenberg–MacLane space' whose stable homotopy groups were zero except in one dimension. However, this is a lot to ask for, so instead one seeks to form a limit of the groups $\pi_{i+n}^s(X \wedge K(G,n))$ as n goes to infinity. The spaces $K(G,n)$ for varying n are related by weak homotopy equivalences $K(G,n) \to \Omega K(G,n+1)$. Since suspension plays such a large role in the current discussion, let us consider instead the corresponding map $\Sigma K(G,n) \to K(G,n + 1)$, or to write this more concisely, $\Sigma K_n \to K_{n+1}$ This induces a map $\pi_{i+n}^s(X \wedge K_n) = \pi_{i+n+1}^s(X \wedge \Sigma K_n) \to \pi_{i+n+1}^s(X \wedge K_{n+1})$. Via these maps, it then makes sense to consider the direct limit as n goes to infinity, the group $h_i(X) = \varinjlim \pi_{i+n}^s(X \wedge K_n)$. This gives a homology theory since direct limits preserve exact sequences so the exactness axiom holds, and direct limits preserve isomorphisms so the suspension isomorphism and the wedge axiom hold. The coefficient groups of this homology theory are the same as for ordinary homology with G coefficients since $h_i(S^0) = \varinjlim \pi_{i+n}^s(K_n)$ is zero unless $i = 0$, when it is G. Hence this homology theory coincides with ordinary homology by Theorem 4.59.

To place this result in its natural generality, define a **spectrum** to be a sequence of CW complexes K_n together with basepoint-preserving maps $\Sigma K_n \to K_{n+1}$. This generalizes the notion of an Ω-spectrum, where the maps $\Sigma K_n \to K_{n+1}$ come from weak homotopy equivalences $K_n \to \Omega K_{n+1}$. Another obvious family of examples is **suspension spectra**, where one starts with an arbitrary CW complex X and defines $K_n = \Sigma^n X$ with $\Sigma K_n \to K_{n+1}$ the identity map.

The homotopy groups of a spectrum K are defined to be $\pi_i(K) = \varinjlim \pi_{i+n}(K_n)$ where the direct limit is computed using the compositions

$$\pi_{i+n}(K_n) \xrightarrow{\ \Sigma\ } \pi_{i+n+1}(\Sigma K_n) \longrightarrow \pi_{i+n+1}(K_{n+1})$$

with the latter map induced by the given map $\Sigma K_n \to K_{n+1}$. Thus in the case of the suspension spectrum of a space X, the homotopy groups of the spectrum are the same as the stable homotopy groups of X. For a general spectrum K we could also describe $\pi_i(K)$ as $\varinjlim \pi_{i+n}^s(K_n)$ since the composition $\pi_{i+n}(K_n) \to \pi_{i+n+j}(K_{n+j})$ factors through $\pi_{i+n+j}(\Sigma^j K_n)$. So the homotopy groups of a spectrum are 'stable homotopy groups' essentially by definition.

Returning now to the context of homology theories, if we are given a spectrum K and a CW complex X, then we have a spectrum $X \wedge K$ with $(X \wedge K)_n = X \wedge K_n$, using the obvious maps $\Sigma(X \wedge K_n) = X \wedge \Sigma K_n \to X \wedge K_{n+1}$. The groups $\pi_i(X \wedge K)$ are the groups $\varinjlim \pi_{i+n}^s(X \wedge K_n)$ considered earlier in the case of an Eilenberg–MacLane spectrum, and the arguments given there show:

Proposition 4F.2. *For a spectrum K, the groups $h_i(X) = \pi_i(X \wedge K)$ form a reduced homology theory. When K is the Eilenberg–MacLane spectrum with $K_n = K(G, n)$, this homology theory is ordinary homology, so $\pi_i(X \wedge K) \approx \widetilde{H}_i(X; G)$.*   □

If one wanted to associate a cohomology theory to an arbitrary spectrum K, one's first inclination would be to set $h^i(X) = \varinjlim \langle \Sigma^n X, K_{n+i} \rangle$, the direct limit with respect to the compositions

$$\langle \Sigma^n X, K_{n+i} \rangle \xrightarrow{\ \Sigma\ } \langle \Sigma^{n+1} X, \Sigma K_{n+i} \rangle \longrightarrow \langle \Sigma^{n+1} X, K_{n+i+1} \rangle$$

For example, in the case of the sphere spectrum $S = \{S^n\}$ this definition yields the **stable cohomotopy groups** $\pi_s^i(X) = \varinjlim \langle \Sigma^n X, S^{n+i} \rangle$. Unfortunately the definition $h^i(X) = \varinjlim \langle \Sigma^n X, K_{n+i} \rangle$ runs into problems with the wedge sum axiom since the direct limit of a product need not equal the product of the direct limits. For finite wedge sums there is no difficulty, so we do have a cohomology theory for finite CW complexes. But for general CW complexes a different definition is needed. The simplest thing to do is to associate to each spectrum K an Ω-spectrum K' and let $h^n(X) = \langle X, K'_n \rangle$. We obtain K' from K by setting $K'_n = \varinjlim \Omega^i K_{n+i}$, the mapping telescope of the sequence $K_n \to \Omega K_{n+1} \to \Omega^2 K_{n+2} \to \cdots$. The Ω-spectrum structure is given by equivalences

$$K'_n = \varinjlim \Omega^i K_{n+i} \simeq \varinjlim \Omega^{i+1} K_{n+i+1} \xrightarrow{\ \kappa\ } \Omega \varinjlim \Omega^i K_{n+i+1} = \Omega K'_{n+1}$$

The first homotopy equivalence comes from deleting the first term of the sequence $K_n \to \Omega K_{n+1} \to \Omega^2 K_{n+2} \to \cdots$, which has negligible effect on the mapping telescope. The next map κ is a special case of the natural weak equivalence $\varinjlim \Omega Z_n \to \Omega \varinjlim Z_n$ that holds for any sequence $Z_1 \to Z_2 \to \cdots$. Strictly speaking, we should let K'_n be a CW approximation to the mapping telescope $\varinjlim \Omega^i K_{n+i}$ in order to obtain a spectrum consisting of CW complexes, in accordance with our definition of a spectrum.

In case one starts with a suspension spectrum $K_n = \Sigma^n K$ it is not necessary to take mapping telescopes since one can just set $K'_n = \bigcup_i \Omega^i \Sigma^{i+n} K = \bigcup_i \Omega^i \Sigma^i K_n$, the union with respect to the natural inclusions $\Omega^i \Sigma^i K_n \subset \Omega^{i+1} \Sigma^{i+1} K_n$. The union $\bigcup_i \Omega^i \Sigma^i X$ is usually abbreviated to $\Omega^\infty \Sigma^\infty X$. Another common notation for this union is QX. Thus $\pi_i(QX) - \pi_i^s(X)$, so Q is a functor converting stable homotopy groups into ordinary homotopy groups.

It follows routinely from the definitions that the homology theory defined by a spectrum is the same as the homology theory defined by the associated Ω-spectrum. One may ask whether every homology theory is defined by a spectrum, as we showed for cohomology. The answer is yes if one replaces the wedge axiom by a stronger **direct limit axiom**: $h_i(X) = \varinjlim h_i(X_\alpha)$, the direct limit over the finite subcomplexes X_α of X. The homology theory defined by a spectrum satisfies this axiom, and the converse is proved in [Adams 1971].

Spectra have become the preferred language for describing many stable phenomena in algebraic topology. The increased flexibility of spectra is not without its price, however, since a number of concepts that are elementary for spaces become quite a bit more subtle for spectra, such as the proper definition of a map between spectra, or the smash product of two spectra. For the reader who wants to learn more about this language a good starting point is [Adams 1974].

Exercises

1. Assuming the first two axioms for a homology theory on the CW category, show that the direct limit axiom implies the wedge sum axiom. Show that the converse also holds for countable CW complexes.

2. For CW complexes X and Y consider the suspension sequence

$$\langle X, Y \rangle \xrightarrow{\ \Sigma\ } \langle \Sigma X, \Sigma Y \rangle \xrightarrow{\ \Sigma\ } \langle \Sigma^2 X, \Sigma^2 Y \rangle \longrightarrow \cdots$$

Show that if X is a finite complex, these maps eventually become isomorphisms. [Use induction on the number of cells of X and the five-lemma.]

3. Show that for any sequence $Z_1 \to Z_2 \to \cdots$, the natural map $\varinjlim \Omega Z_n \to \Omega \varinjlim Z_n$ is a weak homotopy equivalence, where the direct limits mean mapping telescopes.

4.G Gluing Constructions

It is a common practice in algebraic topology to glue spaces together to form more complicated spaces. In this section we describe two general procedures for making such constructions. The first is fairly straightforward but also rather rigid, lacking some homotopy invariance properties an algebraic topologist would like to see. The second type of gluing construction avoids these drawbacks by systematic use of mapping cylinders. We have already seen many special cases of both types of constructions, and having a general framework covering all these special cases should provide some conceptual clarity.

A **diagram of spaces** consists of an oriented graph Γ with a space X_v for each vertex v of Γ and a map $f_e : X_v \to X_w$ for each edge e of Γ from a vertex v to a vertex w, the words 'from' and 'to' referring to the given orientation of e. Commutativity of the diagram is not assumed. Denoting such a diagram of spaces simply by X, we define a space $\sqcup X$ to be the quotient of the disjoint union of all the spaces X_v associated to vertices of Γ under the identifications $x \sim f_e(x)$ for all maps f_e associated to edges of Γ. To give a name to this construction, let us call $\sqcup X$ the *amalgamation* of the diagram X. Here are some examples:

- If the diagram of spaces has the simple form $X_0 \xleftarrow{f} A \hookrightarrow X_1$ then $\sqcup X$ is the space $X_0 \sqcup_f X_1$ obtained from X_0 by attaching X_1 along A via f.

- A sequence of inclusions $X_0 \hookrightarrow X_1 \hookrightarrow \cdots$ determines a diagram of spaces X for which $\sqcup X$ is $\bigcup_i X_i$ with the weak topology. This holds more generally when the spaces X_i are indexed by any directed set.

- From a cover $\mathcal{U} = \{X_i\}$ of a space X by subspaces X_i we can form a diagram of spaces $X_{\mathcal{U}}$ whose vertices are the nonempty finite intersections $X_{i_1} \cap \cdots \cap X_{i_n}$ with distinct indices i_j, and whose edges are the various inclusions obtained by omitting some of the subspaces in such an intersection, for example the inclusions $X_i \cap X_j \hookrightarrow X_i$. Then $\sqcup X_{\mathcal{U}}$ equals X as a set, though possibly with a different topology. If the cover is an open cover, or if X is a CW complex and the X_i's are subcomplexes, then the topology will be the original topology on X.

- An action of a group G on a space X determines a diagram of spaces X_G, with X itself as the only space and with maps the homeomorphisms $g : X \to X$, $g \in G$, given by the action. In this case $\sqcup X_G$ is the orbit space X/G.

- A Δ-complex X can be viewed as a diagram of spaces X_Δ where each simplex of X gives a vertex space X_v which is a simplex of the same dimension, and the edge maps are the inclusions of faces into the simplices that contain them. Then $\sqcup X_\Delta = X$.

It can very easily happen that for a diagram of spaces X the amalgamation $\sqcup X$ is rather useless because so much collapsing has occurred that little of the original diagram remains. For example, consider a diagram X of the form $X_0 \leftarrow X_0 \times X_1 \rightarrow X_1$ whose maps are the projections onto the two factors. In this case $\sqcup X$ is simply a point. To correct for problems like this, and to get a notion with nicer homotopy-theoretic properties, we introduce the homotopy version of $\sqcup X$, which we shall denote ΔX and call the *realization* of X. Here we again start with the disjoint union of all the vertex spaces X_v, but instead of passing to a quotient space of this disjoint union, we enlarge it by filling in a mapping cylinder M_f for each map f of the diagram, identifying the two ends of this cylinder with the appropriate X_v's. In the case of the projection diagram $X_0 \leftarrow X_0 \times X_1 \rightarrow X_1$, the union of the two mapping cylinders is the same as the quotient of $X_0 \times X_1 \times I$ with $X_0 \times X_1 \times \{0\}$ collapsed to X_0 and $X_0 \times X_1 \times \{1\}$ collapsed to X_1. Thus ΔX is the join $X_0 * X_1$ defined in Chapter 0.

We have seen a number of other special cases of the construction ΔX. For a diagram consisting of just one map $f : X_0 \rightarrow X_1$ one gets of course the mapping cylinder M_f itself. For a diagram $X_0 \xleftarrow{f} X_1 \xrightarrow{g} X_2$ the realization ΔX is a double mapping cylinder. In case X_2 is a point this is the mapping cone of f. When the diagram has just one space and one map from this space to itself, then ΔX is the mapping torus. For a diagram consisting of two maps $f, g : X_0 \rightarrow X_1$ the space ΔX was studied in Example 2.48. Mapping telescopes are the case of a sequence of maps $X_0 \rightarrow X_1 \rightarrow \cdots$. In §1.B we considered general diagrams in which the spaces are $K(G, 1)$'s.

There is a natural generalization of ΔX in which one starts with a Δ-complex Γ and a diagram of spaces associated to the 1-skeleton of Γ such that the maps corresponding to the edges of each n-simplex of Γ, $n > 1$, form a commutative diagram. We call this data a **complex of spaces**. If X is a complex of spaces, then for each n-simplex of Γ we have a sequence of maps $X_0 \xrightarrow{f_1} X_1 \xrightarrow{f_2} \cdots \xrightarrow{f_n} X_n$, and we define the *iterated mapping cylinder* $M(f_1, \cdots, f_n)$ to be the usual mapping cylinder for $n = 1$, and inductively for $n > 1$, the mapping cylinder of the composition $M(f_1, \cdots, f_{n-1}) \rightarrow X_{n-1} \xrightarrow{f_n} X_n$ where the first map is the canonical projection of a mapping cylinder onto its target end. There is a natural projection $M(f_1, \cdots, f_n) \rightarrow \Delta^n$, and over each face of Δ^n one has the iterated mapping cylinder for the maps associated to the edges in this face. For example when $n = 2$ one has the three mapping cylinders $M(f_1)$, $M(f_2)$, and $M(f_2 f_1)$ over the three edges of Δ^2. All these iterated mapping cylinders over the various simplices of Γ thus fit together to form a space ΔX with a canonical projection $\Delta X \rightarrow \Gamma$. We again call ΔX the realization of the complex of spaces X, and we call Γ the *base* of X or ΔX.

Some of our earlier examples of diagrams of spaces can be regarded in a natural way as complexes of spaces:

- For a cover $\mathcal{U} = \{X_i\}$ of a space X the diagram of spaces $X_{\mathcal{U}}$ whose vertices are the finite intersections of X_i's and whose edges are inclusions is a complex of spaces with n-simplices the n-fold inclusions. The base Γ for this complex of spaces is the barycentric subdivision of the nerve of the cover. Recall from the end of §3.3 that the nerve of a cover is the simplicial complex with n-simplices the nonempty $(n + 1)$-fold intersections of sets in the cover.

- The diagram of spaces X_G associated to an action of a group G on a space X is a complex of spaces, with n-simplices corresponding to the n-fold compositions $X \xrightarrow{g_1} X \xrightarrow{g_2} \cdots \xrightarrow{g_n} X$. The base Δ-complex Γ is the $K(G, 1)$ called BG in §1.B. This was the orbit space of a free action of G on a contractible Δ-complex EG. Checking through the definitions, one sees that the space ΔX_G in this case can be regarded as the quotient of $X \times EG$ under the diagonal action of G, $g(x, y) = (g(x), g(y))$. This is the space we called the Borel construction in §3.G, with the notation $X \times_G EG$.

By a map $f : X \to Y$ of complexes of spaces over the same base Γ we mean a collection of maps $f_v : X_v \to Y_v$ for all the vertices of Γ, with commutative squares over all edges of Γ. There is then an induced map $\Delta f : \Delta X \to \Delta Y$.

‖ **Proposition 4G.1.** *If all the maps f_v making up a map of complexes of spaces $f : X \to Y$ are homotopy equivalences, then so is the map $\Delta f : \Delta X \to \Delta Y$.*

Proof: The mapping cylinders $M(f_v)$ form a complex of spaces $M(f)$ over the same base Γ, and the space $\Delta M(f)$ is the mapping cylinder $M(\Delta f)$. This deformation retracts onto ΔY, so it will suffice to show that it also deformation retracts onto ΔX.

Let $M^n(\Delta f)$ be the part of $M(\Delta f)$ lying over the n-skeleton of Γ. We claim that $M^n(\Delta f) \cup \Delta X$ deformation retracts onto $M^{n-1}(\Delta f) \cup \Delta X$. It is enough to show this when $\Gamma = \Delta^n$. In this case f is a map from $X_0 \to \cdots \to X_n$ to $Y_0 \to \cdots \to Y_n$. By Corollary 0.20 it suffices to show that the inclusion $M^{n-1}(\Delta f) \cup \Delta X \hookrightarrow M(\Delta f)$ is a homotopy equivalence and the pair $(M(\Delta f), M^{n-1}(\Delta f) \cup \Delta X)$ satisfies the homotopy extension property. The latter assertion is evident from Example 0.15 since a mapping cylinder neighborhood is easily constructed for this pair. For the other condition, note that by induction on the dimension of Γ we may assume that $M^{n-1}(\Delta f)$ deformation retracts onto the part of ΔX over $\partial \Delta^n$. Also, the inclusion $\Delta X \hookrightarrow M(\Delta f)$ is a homotopy equivalence since it is equivalent to the map $X_n \to Y_n$, which is a homotopy equivalence by hypothesis. So Corollary 0.20 applies, and the claim that $M^n(\Delta f) \cup \Delta X$ deformation retracts onto $M^{n-1}(\Delta f) \cup \Delta X$ is proved.

Letting n vary, the infinite concatenation of these deformation retractions in the t-intervals $[1/2^{n+1}, 1/2^n]$ gives a deformation retraction of $M(\Delta f)$ onto ΔX.   □

There is a canonical map $\Delta X \to \sqcup X$ induced by retracting each mapping cylinder onto its target end. In some cases this is a homotopy equivalence, for example, for a diagram $X_0 \leftarrow A \hookrightarrow X_1$ where the pair (X_1, A) has the homotopy extension property. Another example is a sequence of inclusions $X_0 \hookrightarrow X_1 \hookrightarrow \cdots$ for which the pairs (X_n, X_{n-1}) satisfy the homotopy extension property, by the argument involving mapping telescopes in the proof of Lemma 2.34. However, without some conditions on the maps it need not be true that $\Delta X \to \sqcup X$ is a homotopy equivalence, as the earlier example of the projections $X_0 \leftarrow X_0 \times X_1 \to X_1$ shows. Even with inclusion maps one need not have $\Delta X \simeq \sqcup X$ if the base Γ is not contractible. A trivial example is the diagram consisting of the two spaces Δ^0 and Δ^1 and two maps $f_0, f_1 : \Delta^0 \to \Delta^1$ that happen to have the same image.

Thus one can expect the map $\Delta X \to \sqcup X$ to be a homotopy equivalence only in special circumstances. Here is one such situation:

Proposition 4G.2. *When $X_{\mathcal{U}}$ is the complex of spaces associated to an open cover $\mathcal{U} = \{X_i\}$ of a paracompact space X, the map $p : \Delta X_{\mathcal{U}} \to \sqcup X_{\mathcal{U}} = X$ is a homotopy equivalence.*

Proof: The realization $\Delta X_{\mathcal{U}}$ can also be described as the quotient space of the disjoint union of all the products $X_{i_0} \cap \cdots \cap X_{i_n} \times \Delta^n$, as the subscripts range over sets of $n + 1$ distinct indices and $n \geq 0$, with the identifications over the faces of Δ^n using inclusions $X_{i_0} \cap \cdots \cap X_{i_n} \hookrightarrow X_{i_0} \cap \cdots \cap \hat{X}_{i_j} \cap \cdots \cap X_{i_n}$. From this viewpoint, points of $\Delta X_{\mathcal{U}}$ in a given 'fiber' $p^{-1}(x)$ can be written as finite linear combinations $\sum_i t_i x_i$ where $\sum_i t_i = 1$ and x_i is x regarded as a point of X_i, for those X_i's that contain x.

Since X is paracompact there is a partition of unity subordinate to the cover $\mathcal{U}$. This is a family of maps $\varphi_\alpha : X \to [0, 1]$ satisfying three conditions: The support of each φ_α is contained in some $X_{i(\alpha)}$, only finitely many φ_α's are nonzero near each point of X, and $\sum_\alpha \varphi_\alpha = 1$. Define a section $s : X \to \Delta X_{\mathcal{U}}$ of p by setting $s(x) = \sum_\alpha \varphi_\alpha(x) x_{i(\alpha)}$. The figure shows the case $X = S^1$ with a cover by two arcs, the heavy line indicating the image of s. In the general case the section s embeds X as a retract of $\Delta X_{\mathcal{U}}$, and it is a deformation retract since points in fibers $p^{-1}(x)$ can move linearly along line segments to $s(x)$. $\square$

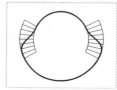

Corollary 4G.3. *If $\mathcal{U}$ is an open cover of a paracompact space X such that every nonempty intersection of finitely many sets in $\mathcal{U}$ is contractible, then X is homotopy equivalent to the nerve $N\mathcal{U}$.*

Proof: The proposition gives a homotopy equivalence $X \simeq \Delta X_{\mathcal{U}}$. Since the nonempty finite intersections of sets in $\mathcal{U}$ are contractible, the earlier proposition implies that the map $\Delta X_{\mathcal{U}} \to \Gamma$ induced by sending each intersection to a point is a homotopy equivalence. Since Γ is the barycentric subdivision of $N\mathcal{U}$, the result follows. $\square$

Let us conclude this section with a few comments about terminology. For some diagrams of spaces such as sequences $X_1 \to X_2 \to \cdots$ the amalgamation $\sqcup X$ can be regarded as the direct limit of the vertex spaces X_v with respect to the edge maps f_e. Following this cue, the space $\sqcup X$ is commonly called the direct limit for arbitrary diagrams, even finite ones. If one views $\sqcup X$ as a direct limit, then ΔX becomes a sort of homotopy direct limit. For reasons that are explained in the next section, direct limits are often called 'colimits.' This has given rise to the rather unfortunate name of 'hocolim' for ΔX, short for 'homotopy colimit.' In preference to this we have chosen the term 'realization,' both for its intrinsic merits and because ΔX is closely related to what is called the geometric realization of a simplicial space.

Exercises

1. Show that for a sequence of maps $X_0 \xrightarrow{f_1} X_1 \xrightarrow{f_2} \cdots$, the infinite iterated mapping cylinder $M(f_1, f_2, \cdots)$, which is the union of the finite iterated mapping cylinders $M(f_1, \cdots, f_n)$, deformation retracts onto the mapping telescope.

2. Show that if X is a complex of spaces in which all the maps are homeomorphisms, then the projection $\Delta X \to \Gamma$ is a fiber bundle.

3. What is the nerve of the cover of a simplicial complex by the open stars of its vertices? [See Lemma 2C.2.]

4. Show that Proposition 4G.2 and its corollary hold also for CW complexes and covers by families of subcomplexes. [CW complexes are paracompact; see [VBKT].]

4.H   Eckmann-Hilton Duality

There is a very nice duality principle in homotopy theory, called Eckmann–Hilton duality in its more refined and systematic aspects, but which in its most basic form involves the simple idea of reversing the direction of all arrows in a given construction. For example, if in the definition of a fibration as a map satisfying the homotopy lifting property we reverse the direction of all the arrows, we obtain the dual notion of a **cofibration**. This is a map $i : A \to B$ satisfying the following property: Given $\tilde{g}_0 : B \to X$ and a homotopy $g_t : A \to X$ such that $\tilde{g}_0 i = g_0$, there exists a homotopy $\tilde{g}_t : B \to X$ such that $\tilde{g}_t i = g_t$. In the special case that i is the inclusion of a subspace, this is the homotopy extension property, and the next proposition says that this is indeed the general case. So a cofibration is the same as an inclusion satisfying the homotopy extension property.

Proposition 4H.1. *If $i : A \to B$ is a cofibration, then i is injective, and in fact a homeomorphism onto its image.*

Proof: Consider the mapping cylinder M_i, the quotient of $A \times I \amalg B$ in which $(a, 1)$ is identified with $i(a)$. Let $g_t : A \to M_i$ be the homotopy mapping $a \in A$ to the image of $(a, 1 - t) \in A \times I$ in M_i, and let $\tilde{g}_0$ be the inclusion $B \hookrightarrow M_i$. The cofibration property gives $\tilde{g}_t : B \to M_i$ with $\tilde{g}_t i = g_t$. Restricting to a fixed $t > 0$, this implies i is injective since g_t is. Furthermore, since g_t is a homeomorphism onto its image $A \times \{1 - t\}$, the relation $\tilde{g}_t i = g_t$ implies that the map $g_t^{-1} \tilde{g}_t : i(A) \to A$ is a continuous inverse of $i : A \to i(A)$. $\qquad\qquad\square$

Many constructions for fibrations have analogs for cofibrations, and vice versa. For example, for an arbitrary map $f : A \to B$ the inclusion $A \hookrightarrow M_f$ is readily seen to be a cofibration, so the analog of the factorization $A \hookrightarrow E_f \to B$ of f into a homotopy equivalence followed by a fibration is the factorization $A \hookrightarrow M_f \to B$ into a cofibration followed by a homotopy equivalence. Even the definition of M_f is in some way dual to the definition of E_f, since E_f can be defined as a pullback and M_f can be defined as a dual pushout. In general, the **pushout** of maps $f : Z \to X$ and $g : Z \to Y$ is defined as the quotient of $X \amalg Y$ under the identifications $f(z) \sim g(z)$.

$$
\begin{array}{ccc}
E_f & \longrightarrow & B^I \\
\downarrow & & \downarrow \\
A & \longrightarrow & B
\end{array}
\qquad
\begin{array}{ccc}
A & \longrightarrow & B \\
\downarrow & & \downarrow \\
A \times I & \longrightarrow & M_f
\end{array}
$$

Thus the pushout is a quotient of $X \amalg Y$, while the pullback of maps $X \to Z$ and $Y \to Z$ is a subobject of $X \times Y$, so we see here two instances of duality: a duality between disjoint union and product, and a duality between subobjects and quotients. The first of these is easily explained, since a collection of maps $X_\alpha \to X$ is equivalent to a map $\amalg_\alpha X_\alpha \to X$, while a collection of maps $X \to X_\alpha$ is equivalent to a map $X \to \prod_\alpha X_\alpha$. The notation $\amalg$ for the 'coproduct' was chosen to indicate that it is dual to $\prod$. If we were dealing with basepointed spaces and maps, the coproduct would be wedge sum. In the category of abelian groups the coproduct is direct sum.

The duality between subobjects and quotient objects is clear for abelian groups, where subobjects are kernels and quotient objects cokernels. The strict topological analog of a kernel is a fiber of a fibration. Dually, the topological analog of a cokernel is the **cofiber** B/A of a cofibration $A \hookrightarrow B$. If we make an arbitrary map $f : A \to B$ into a cofibration $A \hookrightarrow M_f$, the cofiber is the mapping cone $C_f = M_f / (A \times \{0\})$.

In the diagram showing E_f and M_f as pullback and pushout, there also appears to be some sort of duality involving the terms $A \times I$ and B^I. This leads us to ask whether $X \times I$ and X^I are in some way dual. Indeed, if we ignore topology and just think set-theoretically, this is an instance of the familiar product-coproduct duality since the product of copies of X indexed by I is X^I, all functions $I \to X$, while the coproduct of copies of X indexed by I is $X \times I$, the disjoint union of the sets $X \times \{t\}$ for $t \in I$. Switching back from the category of sets to the topological category, we can view X^I as a 'continuous product' of copies of X and $X \times I$ as a 'continuous coproduct.'

On a less abstract level, the fact that maps $A \times I \to B$ are the same as maps $A \to B^I$ indicates a certain duality between $A \times I$ and B^I. This leads to a duality between suspension and loopspace, since ΣA is a quotient of $A \times I$ and ΩB is a subspace of B^I. This duality is expressed in the adjoint relation $\langle \Sigma X, Y \rangle = \langle X, \Omega Y \rangle$ from §4.3. Combining this duality between Σ and Ω with the duality between fibers and cofibers, we see a duality relationship between the fibration and cofibration sequences of §4.3:

$$\cdots \longrightarrow \Omega F \longrightarrow \Omega E \longrightarrow \Omega B \longrightarrow F \longrightarrow E \longrightarrow B$$

$$A \longrightarrow X \longrightarrow X/A \longrightarrow \Sigma A \longrightarrow \Sigma X \longrightarrow \Sigma(X/A) \longrightarrow \cdots$$

Pushout and pullback constructions can be generalized to arbitrary diagrams. In the case of pushouts, this was done in §4.G where we associated a space $\sqcup X$ to a diagram of spaces X. This was the quotient of the coproduct $\coprod_v X_v$, with v ranging over vertices of the diagram, under the identifications $x \sim f_e(x)$ for all maps f_e associated to edges e of the diagram. The dual construction $\sqcap X$ would be the subspace of the product $\prod_v X_v$ consisting of tuples (x_v) with $f_e(x_v) = x_w$ for all maps $f_e : X_v \to X_w$ in the diagram. Perhaps more useful in algebraic topology is the homotopy variant of this notion obtained by dualizing the definition of ΔX in the previous section. This is the space ∇X consisting of all choices of a point x_v in each X_v and a path y_e in the target space of each edge map $f_e : X_v \to X_w$, with $y_e(0) = f(x_v)$ and $y_e(1) = x_w$. The subspace with all paths constant is $\sqcap X$. In the case of a diagram $\cdots \to X_2 \to X_1$ such as a Postnikov tower this construction gives something slightly different from simply turning each successive map into a fibration via the usual pathspace construction, starting with $X_2 \to X_1$ and proceeding up the tower, as we did in §4.3. The latter construction is rather the dual of an iterated mapping cylinder, involving spaces of maps $\Delta^n \to X_v$ instead of simply pathspaces. One could use such mapping spaces to generalize the definition of ∇X from diagrams of spaces to complexes of spaces.

As special cases of the constructions $\sqcup X$ and $\sqcap X$ we have direct limits and inverse limits for diagrams $X_0 \to X_1 \to \cdots$ and $\cdots \to X_1 \to X_0$, respectively. Since inverse limit is related to product and direct limit to coproduct, it is common practice in some circles to use reverse logic and call inverse limit simply 'limit' and direct limit 'colimit.' The homotopy versions are then called 'holim' for ∇X and 'hocolim' for ΔX. This terminology is frequently used for more general diagrams as well.

Homotopy Groups with Coefficients

There is a somewhat deeper duality between homotopy groups and cohomology, which one can see in the fact that cohomology groups are homotopy classes of maps into a space with a single nonzero homotopy group, while homotopy groups are homotopy classes of maps from a space with a single nonzero cohomology group. This duality is in one respect incomplete, however, in that the cohomology statement holds for an arbitrary coefficient group, but we have not yet defined homotopy groups with coefficients. In view of the duality, one would be tempted to define $\pi_n(X; G)$ to be the set of basepoint-preserving homotopy classes of maps from the cohomology analog of a Moore space $M(G, n)$ to X. The cohomology analog of $M(G, n)$ would be a space

Y whose only nonzero cohomology group $\tilde{H}^i(Y;\mathbb{Z})$ is G for $i = n$. Unfortunately, such a space does not exist for arbitrary G, for example for $G = \mathbb{Q}$, since we showed in Proposition 3F.12 that if the cohomology groups of a space are all countable, then they are all finitely generated.

As a first approximation to $\pi_n(X;G)$ let us consider $\langle M(G,n), X \rangle$, the set of basepoint-preserving homotopy classes of maps $M(G,n) \to X$. To give this set a more suggestive name, let us call it $\mu_n(X;G)$. We should assume $n > 1$ to guarantee that the homotopy type of $M(G,n)$ is well-defined, as shown in Example 4.34. For $n > 1$, $\mu_n(X;G)$ is a group since we can choose $M(G,n)$ to be the suspension of an $M(G, n-1)$. And if $n > 2$ then $\mu_n(X;G)$ is abelian since we can choose $M(G,n)$ to be a double suspension.

There is something like a universal coefficient theorem for these groups $\mu_n(X;G)$:

Proposition 4H.2. *For $n > 1$ there are natural short exact sequences*

$$0 \longrightarrow \text{Ext}(G, \pi_{n+1}(X)) \longrightarrow \mu_n(X;G) \longrightarrow \text{Hom}(G, \pi_n(X)) \longrightarrow 0.$$

The similarity with the universal coefficient theorem for cohomology is apparent, but with a reversal of the variables in the Ext and Hom terms, reflecting the fact that $\mu_n(X;G)$ is covariant as a functor of X and contravariant as a functor of G, just like the Ext and Hom terms.

Proof: Let $0 \to R \xrightarrow{i} F \to G \to 0$ be a free resolution of G. The inclusion map i is realized by a map $M(R,n) \to M(F,n)$, where $M(R,n)$ and $M(F,n)$ are wedges of S^n's corresponding to bases for F and R. Converting this map into a cofibration via the mapping cylinder, the cofiber is an $M(G,n)$, as one sees from the long exact sequence of homology groups. As in §4.3, the cofibration sequence

$$M(R,n) \to M(F,n) \to M(G,n) \to M(R, n+1) \to M(F, n+1)$$

gives rise to the exact sequence across the top of the following diagram:

$$\mu_{n+1}(X;F) \longrightarrow \mu_{n+1}(X;R) \longrightarrow \mu_n(X;G) \longrightarrow \mu_n(X;F) \longrightarrow \mu_n(X;R)$$

$$\text{Hom}(F, \pi_{n+1}(X)) \xrightarrow{i^*} \text{Hom}(R, \pi_{n+1}(X)) \qquad \text{Hom}(F, \pi_n(X)) \xrightarrow{i^*} \text{Hom}(R, \pi_n(X))$$

The four outer terms of the exact sequence can be identified with the indicated Hom terms since mapping a wedge sum of S^n's into X amounts to choosing an element of $\pi_n(X)$ for each wedge summand. The kernel and cokernel of i^* are $\text{Hom}(G, -)$ and $\text{Ext}(G, -)$ by definition, and so we obtain the short exact sequence we are looking for. Naturality will be left for the reader to verify. $\square$

Unlike in the universal coefficient theorems for homology and cohomology, the short exact sequence in this proposition does not split in general. For an example, take $G = \mathbb{Z}_2$ and $X = M(\mathbb{Z}_2, n)$ for $n \geq 2$, where the identity map of $M(\mathbb{Z}_2, n)$

defines an element of $\mu_n(M(\mathbb{Z}_2, n); \mathbb{Z}_2) = \langle M(\mathbb{Z}_2, n), M(\mathbb{Z}_2, n) \rangle$ having order 4, as we show in Example 4L.7, whereas the two outer terms in the short exact sequence can only contain elements of order 2 since $G = \mathbb{Z}_2$. This example shows also that $\mu_n(X; \mathbb{Z}_m)$ need not be a module over $\mathbb{Z}_m$, as homology and cohomology groups with $\mathbb{Z}_m$ coefficients are.

The proposition implies that the first nonzero $\mu_i(S^n; \mathbb{Z}_m)$ is $\mu_{n-1}(S^n; \mathbb{Z}_m) = \mathbb{Z}_m$, from the Ext term. This result would look more reasonable if we changed notation to replace the subscript $n - 1$ by n. So let us make the definition

$$\pi_n(X; \mathbb{Z}_m) = \langle M(\mathbb{Z}_m, n - 1), X \rangle = \mu_{n-1}(X; \mathbb{Z}_m)$$

There are two good reasons to expect this to be the right definition. The first is formal: $M(\mathbb{Z}_m, n - 1)$ is a 'cohomology $M(\mathbb{Z}_m, n)$' since its only nontrivial cohomology group $\tilde{H}^i(M(\mathbb{Z}_m, n - 1); \mathbb{Z})$ is $\mathbb{Z}_m$ in dimension n. The second reason is more geometric: Elements of $\pi_n(X; \mathbb{Z}_m)$ should be homotopy classes of 'homotopy-theoretic cycles mod m,' meaning maps $D^n \to X$ whose boundary is not necessarily a constant map as would be the case for $\pi_n(X)$, but rather whose boundary is m times a cycle $S^{n-1} \to X$. This is precisely what a map $M(\mathbb{Z}_m, n - 1) \to X$ is, if we choose $M(\mathbb{Z}_m, n - 1)$ to be S^{n-1} with a cell e^n attached by a map of degree m.

Besides the calculation $\pi_n(S^n; \mathbb{Z}_m) \approx \mathbb{Z}_m$, the proposition also yields an isomorphism $\pi_n(M(\mathbb{Z}_m, n); \mathbb{Z}_m) \approx \mathrm{Ext}(\mathbb{Z}_m, \mathbb{Z}_m) = \mathbb{Z}_m$. Both these results are in fact special cases of a Hurewicz-type theorem relating $\pi_n(X; \mathbb{Z}_m)$ and $H_n(X; \mathbb{Z}_m)$, which is proved in [Neisendorfer 1980].

Along with $\mathbb{Z}$ and $\mathbb{Z}_m$, another extremely useful coefficient group for homology and cohomology is $\mathbb{Q}$. We pointed out above the difficulty that there is no cohomology analog of $M(\mathbb{Q}, n)$. The groups $\mu_n(X; \mathbb{Q})$ are also problematic. For example the proposition gives $\mu_{n-1}(S^n; \mathbb{Q}) \approx \mathrm{Ext}(\mathbb{Q}, \mathbb{Z})$, which is a somewhat complicated uncountable group as we showed in §3.F. However, there is an alternative approach that turns out to work rather well. One defines rational homotopy groups simply as $\pi_n(X) \otimes \mathbb{Q}$, analogous to the isomorphism $H_n(X; \mathbb{Q}) \approx H_n(X; \mathbb{Z}) \otimes \mathbb{Q}$ from the universal coefficient theorem for homology. See [SSAT] for more on this.

Homology Decompositions

Eckmann–Hilton duality can be extremely helpful as an organizational principle, reducing significantly what one has to remember, and providing valuable hints on how to proceed in various situations. To illustrate, let us consider what would happen if we dualized the notion of a Postnikov tower of principal fibrations, where a space is represented as an inverse limit of a sequence of fibers of maps to Eilenberg–MacLane spaces. In the dual representation, a space would be realized as a direct limit of a sequence of cofibers of maps from Moore spaces.

In more detail, suppose we are given a sequence of abelian groups G_n, $n \geq 1$, and we build a CW complex X with $H_n(X) \approx G_n$ for all n by constructing inductively an increasing sequence of subcomplexes $X_1 \subset X_2 \subset \cdots$ with $H_i(X_n) \approx G_i$ for $i \leq n$ and $H_i(X_n) = 0$ for $i > n$, where:

(1) X_1 is a Moore space $M(G_1, 1)$.

(2) X_{n+1} is the mapping cone of a cellular map $h_n : M(G_{n+1}, n) \rightarrow X_n$ such that the induced map $h_{n*} : H_n(M(G_{n+1}, n)) \rightarrow H_n(X_n)$ is trivial.

(3) $X = \bigcup_n X_n$.

One sees inductively that X_{n+1} has the desired homology groups by comparing the long exact sequences of the pairs (X_{n+1}, X_n) and (CM, M) where $M = M(G_{n+1}, n)$ and CM is the cone $M \times I / M \times \{0\}$:

$$
\begin{array}{ccccccccc}
0 & \longrightarrow & H_{n+1}(X_{n+1}) & \longrightarrow & H_{n+1}(X_{n+1}, X_n) & \overset{\partial}{\longrightarrow} & H_n(X_n) & \longrightarrow & H_n(X_{n+1}) & \longrightarrow & 0 \\
& & & & \Big\uparrow{\scriptstyle\approx} & & \Big\uparrow{\scriptstyle h_{n*}} & & & & \\
& & & & H_{n+1}(CM, M) & \underset{\approx}{\overset{\partial}{\longrightarrow}} & H_n(M) \approx G_{n+1} & & & &
\end{array}
$$

The assumption that h_{n*} is trivial means that the boundary map in the upper row is zero, hence $H_{n+1}(X_{n+1}) \approx G_{n+1}$. The other homology groups of X_{n+1} are the same as those of X_n since $H_i(X_{n+1}, X_n) \approx H_i(CM, M)$ for all i by excision, and $H_i(CM, M) \approx \widetilde{H}_{i-1}(M)$ since CM is contractible.

In case all the maps h_n are trivial, X is the wedge sum of the Moore spaces $M(G_n, n)$ since in this case the mapping cone construction in (2) produces a wedge sum with the suspension of $M(G_{n+1}, n)$, a Moore space $M(G_{n+1}, n+1)$.

For a space Y, a homotopy equivalence $f : X \rightarrow Y$ where X is constructed as in (1)–(3) is called a **homology decomposition** of Y.

║ **Theorem 4H.3.** *Every simply-connected CW complex has a homology decomposition.*

Proof: Given a simply-connected CW complex Y, let $G_n = H_n(Y)$. Suppose inductively that we have constructed X_n via maps h_i as in (2), together with a map $f : X_n \rightarrow Y$ inducing an isomorphism on H_i for $i \leq n$. The induction can start with X_1 a point since Y is simply-connected. To construct X_{n+1} we first replace Y by the mapping cylinder of $f : X_n \rightarrow Y$, converting f into an inclusion. By the Hurewicz theorem and the homology exact sequence of the pair (Y, X_n) we have $\pi_{n+1}(Y, X_n) \approx H_{n+1}(Y, X_n) \approx H_{n+1}(Y) = G_{n+1}$. We will use this isomorphism to construct a map $h_n : M(G_{n+1}, n) \rightarrow X_n$ and an extension $f : X_{n+1} \rightarrow Y$.

The standard construction of an $M(G_{n+1}, n)$ consists of a wedge of spheres S_α^n corresponding to generators g_α of G_{n+1}, with cells e_β^{n+1} attached according to certain linear combinations $r_\beta = \sum_\alpha n_{\alpha\beta} g_\alpha$ that are zero in G_{n+1}. Under the isomorphism $G_{n+1} \approx \pi_{n+1}(Y, X_n)$ each g_α corresponds to a basepoint-preserving map $f_\alpha : (CS^n, S^n) \rightarrow (Y, X_n)$ where CS^n is the cone on S^n. The restrictions of these f_α's to S^n define $h_n : \bigvee_\alpha S_\alpha^n \rightarrow X_n$, and the maps $f_\alpha : CS^n \rightarrow Y$ themselves give an extension of $f : X_n \rightarrow Y$ to the mapping cone of $h_n : \bigvee_\alpha S_\alpha^n \rightarrow X_n$. Each relation r_β gives a homotopy $F_\beta : (CS^n, S^n) \times I \rightarrow (Y, X_n)$ from $\sum_\alpha n_{\alpha\beta} f_\alpha$ to the constant map. We use

$F_\beta | S^n \times \{0\}$ to attach e_β^{n+1}, and then $F_\beta | S^n \times I$ gives h_n on e_β^{n+1} and F_β gives an extension of f over the cone on e_β^{n+1}.

    This construction assures that $f_* : H_{n+1}(X_{n+1}, X_n) \to H_{n+1}(Y, X_n)$ is an isomorphism, so from the five-lemma applied to the long exact sequences of these pairs we deduce that $f_* : H_i(X_{n+1}) \to H_i(Y)$ is an isomorphism for $i \le n+1$. This finishes the induction step. We may assume the maps f_α and F_β are cellular, so $X = \bigcup_n X_n$ is a CW complex with subcomplexes X_n. Since $f : X \to Y$ is a homology isomorphism between simply-connected CW complexes, it is a homotopy equivalence.   □

    As an example, suppose that Y is a simply-connected CW complex having all its homology groups free. Then the Moore spaces used in the construction of X can be taken to be wedges of spheres, and so X_n is obtained from X_{n-1} by attaching an n-cell for each $\mathbb{Z}$ summand of $H_n(Y)$. The attaching maps may be taken to be cellular, making X into a CW complex whose cellular chain complex has trivial boundary maps. Similarly, finite cyclic summands of $H_n(Y)$ can be realized by wedge summands of the form $S^{n-1} \cup e^n$ in $M(H_n(Y), n-1)$, contributing an n-cell and an $(n+1)$-cell to X. This is Proposition 4C.1, but the present result is stronger because it tells us that a finite cyclic summand of H_n can be realized in one step by attaching the cone on a Moore space $M(\mathbb{Z}_k, n-1)$, rather than in two steps of attaching an n-cell and then an $(n+1)$-cell.

Exercises

1. Show that if $A \hookrightarrow X$ is a cofibration of compact Hausdorff spaces, then for any space Y, the map $Y^X \to Y^A$ obtained by restriction of functions is a fibration. [If $A \hookrightarrow X$ is a cofibration, so is $A \times Y \hookrightarrow X \times Y$ for any space Y.]

2. Consider a pushout diagram as at the right, where $B \sqcup_f X$ is B with X attached along A via f. Show that if $A \hookrightarrow X$ is a cofibration, so is $B \hookrightarrow B \sqcup_f X$.

$$
\begin{array}{ccc}
A & \xrightarrow{\;f\;} & B \\
\downarrow & & \downarrow \\
X & \longrightarrow & B \sqcup_f X
\end{array}
$$

3. For fibrations $E_1 \to B$ and $E_2 \to B$, show that a fiber-preserving map $E_1 \to E_2$ that is a homotopy equivalence is in fact a fiber homotopy equivalence. [This is dual to Proposition 0.19.]

4. Define the dual of an iterated mapping cylinder precisely, in terms of maps from Δ^n, and use this to give a definition of ∇X, the dual of ΔX, for X a complex of spaces.

4.I   Stable Splittings of Spaces

    It sometimes happens that suspending a space has the effect of simplifying its homotopy type, as the suspension becomes homotopy equivalent to a wedge sum of smaller spaces. Much of the interest in such stable splittings comes from the fact that they provide a geometric explanation for algebraic splittings of homology and cohomology groups, as well as other algebraic invariants of spaces that are unaffected by suspension such as the cohomology operations studied in §4.L.

The simplest example of a stable splitting occurs for the torus $S^1 \times S^1$. Here the reduced suspension $\Sigma(S^1 \times S^1)$ is homotopy equivalent to $S^2 \vee S^2 \vee S^3$ since $\Sigma(S^1 \times S^1)$ is $S^2 \vee S^2$ with a 3-cell attached by the suspension of the attaching map of the 2-cell of the torus, but the latter attaching map is the commutator of the two inclusions $S^1 \hookrightarrow S^1 \vee S^1$, and the suspension of this commutator is trivial since it lies in the abelian group $\pi_2(S^2 \vee S^2)$.

By an easy geometric argument we will prove more generally:

Proposition 4I.1. *If X and Y are CW complexes, then $\Sigma(X \times Y) \simeq \Sigma X \vee \Sigma Y \vee \Sigma(X \wedge Y)$.*

For example, $\Sigma(S^m \times S^n) \simeq S^{m+1} \vee S^{n+1} \vee S^{m+n+1}$. In view of the cup product structure on $H^*(S^m \times S^n)$ there can be no such splitting of $S^m \times S^n$ before suspension.

Proof: Consider the join $X * Y$ defined in Chapter 0, consisting of all line segments joining points in X to points in Y. For our present purposes it is convenient to use the reduced version of the join, obtained by collapsing to a point the line segment joining the basepoints $x_0 \in X$ and $y_0 \in Y$. We will still denote this reduced join by $X * Y$. Consider the space obtained from $X * Y$ by attaching reduced cones CX and CY to the copies of X and Y at the two ends of $X * Y$. If we collapse each of these cones to a point, we get the reduced suspension $\Sigma(X \times Y)$.

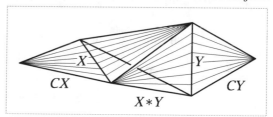

Since each cone is contractible, collapsing the cones gives a homotopy equivalence $X * Y \cup CX \cup CY \simeq \Sigma(X \times Y)$. Inside $X * Y$ there are also cones $x_0 * Y$ and $X * y_0$ intersecting in a point. Collapsing these cones converts $X * Y$ into $\Sigma(X \wedge Y)$ and $X * Y \cup CX \cup CY$ into $\Sigma(X \wedge Y) \vee \Sigma X \vee \Sigma Y$. $\quad\square$

This result can be applied inductively to obtain splittings for suspensions of products of more than two spaces, using the fact that reduced suspension is smash product with S^1, and smash product is associative and commutative. For example,
$$\Sigma(X \times Y \times Z) \simeq \Sigma X \vee \Sigma Y \vee \Sigma Z \vee \Sigma(X \wedge Y) \vee \Sigma(X \wedge Z) \vee \Sigma(Y \wedge Z) \vee \Sigma(X \wedge Y \wedge Z)$$

Our next example involves the reduced product $J(X)$ defined in §3.2. An interesting case is $J(S^n)$, which has a CW structure of the form $S^n \cup e^{2n} \cup e^{3n} \cup \cdots$. All the cells e^{in} for $i > 1$ are attached nontrivially since $H^*(J(S^n); \mathbb{Q})$ is a polynomial ring $\mathbb{Q}[x]$ for n even and a tensor product $\mathbb{Q}[x] \otimes \Lambda_\mathbb{Q}[y]$ for n odd. However, after we suspend to $\Sigma J(S^n)$, it is a rather surprising fact that all the attaching maps become trivial:

Proposition 4I.2. $\Sigma J(S^n) \simeq S^{n+1} \vee S^{2n+1} \vee S^{3n+1} \vee \cdots$. *More generally, if X is a connected CW complex then $\Sigma J(X) \simeq \bigvee_n \Sigma X^{\wedge n}$ where $X^{\wedge n}$ denotes the smash product of n copies of X.*

Proof: The space $J(X)$ is the union of an increasing sequence of subcomplexes $J_k(X)$ with $J_k(X)$ a quotient of the k-fold product $X^{\times k}$. The quotient $J_k(X)/J_{k-1}(X)$ is $X^{\wedge k}$. Thus we have maps

$$X^{\times k} \longrightarrow J_k(X) \longrightarrow X^{\wedge k} = J_k(X)/J_{k-1}(X)$$

By repeated application of the preceding proposition, $\Sigma X^{\wedge k}$ is a wedge summand of $\Sigma X^{\times k}$, up to homotopy equivalence. The proof shows moreover that there is a map $\Sigma X^{\wedge k} \to \Sigma X^{\times k}$ such that the composition $\Sigma X^{\wedge k} \to \Sigma X^{\times k} \to \Sigma X^{\wedge k}$ is homotopic to the identity. This composition factors as

$$\Sigma X^{\wedge k} \longrightarrow \Sigma X^{\times k} \longrightarrow \Sigma J_k(X) \longrightarrow \Sigma X^{\wedge k}$$

so we obtain a map $s_k : \Sigma X^{\wedge k} \to \Sigma J_k(X)$ such that $\Sigma X^{\wedge k} \xrightarrow{s_k} \Sigma J_k(X) \to \Sigma X^{\wedge k}$ is homotopic to the identity.

The map s_k induces a splitting of the long exact sequence of homology groups for the pair $(\Sigma J_k(X), \Sigma J_{k-1}(X))$. Hence the map $i \vee s_k : \Sigma J_{k-1}(X) \vee \Sigma X^{\wedge k} \to \Sigma J_k(X)$ induces an isomorphism on homology, where i denotes the inclusion map. It follows by induction that the map $\bigvee_{k=1}^n s_k : \bigvee_{k=1}^n \Sigma X^{\wedge k} \to \Sigma J_n(X)$ induces an isomorphism on homology for all finite n. This implies the corresponding statement for $n = \infty$ since $X^{\wedge n}$ is $(n-1)$-connected if X is connected. Thus we have a map $\bigvee_k \Sigma X^{\wedge k} \to \Sigma J(X)$ inducing an isomorphism on homology. By Whitehead's theorem this map is a homotopy equivalence since the spaces are simply-connected CW complexes. $\square$

For our final example the stable splitting will be constructed using the group structure on $\langle \Sigma X, Y \rangle$, the set of basepointed homotopy classes of maps $\Sigma X \to Y$.

Proposition 4I.3. *For any prime power p^n the suspension $\Sigma K(\mathbb{Z}_{p^n}, 1)$ is homotopy equivalent to a wedge sum $X_1 \vee \cdots \vee X_{p-1}$ where X_i is a CW complex having $\tilde{H}_*(X_i; \mathbb{Z})$ nonzero only in dimensions congruent to $2i \mod 2p - 2$.*

This result is best possible in a strong sense: No matter how many times any one of the spaces X_i is suspended, it never becomes homotopy equivalent to a nontrivial wedge sum. This will be shown in Example 4L.3 by studying cohomology operations in $H^*(K(\mathbb{Z}_{p^n}, 1); \mathbb{Z}_p)$. There is also a somewhat simpler K–theoretic explanation for this; see [VBKT].

Proof: Let $K = K(\mathbb{Z}_{p^n}, 1)$. The multiplicative group of nonzero elements in the field $\mathbb{Z}_p$ is cyclic, so let the integer r represent a generator. By Proposition 1B.9 there is a map $f : K \to K$ inducing multiplication by r on $\pi_1(K)$. We will need to know that f induces multiplication by r^i on $H_{2i-1}(K; \mathbb{Z}) \approx \mathbb{Z}_{p^n}$, and this can be seen as follows. Via the natural isomorphism $\pi_1(K) \approx H_1(K;\mathbb{Z})$ we know that f induces multiplication by r on $H_1(K;\mathbb{Z})$. Via the universal coefficient theorem, f also induces multiplication by r on $H^1(K;\mathbb{Z}_{p^n})$ and $H^2(K;\mathbb{Z}_{p^n})$. The cup product structure in $H^*(K;\mathbb{Z}_{p^n})$ computed in Examples 3.41 and 3E.2 then implies that f induces multiplication by r^i on $H^{2i-1}(K;\mathbb{Z}_{p^n})$, so the same is true for $H_{2i-1}(K;\mathbb{Z})$ by another application of the universal coefficient theorem.

For each integer $j \geq 0$ let $h_j : \Sigma K \to \Sigma K$ be the difference $\Sigma f - r^j \mathbb{1}$, so h_j induces multiplication by $r^i - r^j$ on $H_{2i}(\Sigma K;\mathbb{Z}) \approx \mathbb{Z}_{p^n}$. By the choice of r we know that p divides $r^i - r^j$ iff $i \equiv j \bmod p - 1$. This means that the map induced by h_j on $\tilde{H}_{2i}(\Sigma K;\mathbb{Z})$ has nontrivial kernel iff $i \equiv j \bmod p - 1$. Therefore the composition $m_i = h_1 \circ \cdots h_{i-1} \circ h_{i+1} \cdots h_{p-1}$ induces an isomorphism on $\tilde{H}_*(\Sigma K;\mathbb{Z})$ in dimensions congruent to $2i \bmod 2p - 2$ and has a nontrivial kernel in other dimensions where the homology group is nonzero. When there is a nontrivial kernel, some power of m_i will induce the zero map since we are dealing with homomorphisms $\mathbb{Z}_{p^n} \to \mathbb{Z}_{p^n}$.

Let X_i be the mapping telescope of the sequence $\Sigma K \to \Sigma K \to \cdots$ where each map is m_i. Since homology commutes with direct limits, the inclusion of the first factor $\Sigma K \hookrightarrow X_i$ induces an isomorphism on $\tilde{H}_*$ in dimensions congruent to $2i \bmod 2p - 2$, and $\tilde{H}_*(X_i;\mathbb{Z}) = 0$ in all other dimensions. The sum of these inclusions is a map $\Sigma K \to X_1 \vee \cdots \vee X_{p-1}$ inducing an isomorphism on all homology groups. Since these complexes are simply-connected, the result follows by Whitehead's theorem. $\square$

The construction of the spaces X_i as mapping telescopes produces rather large spaces, with infinitely many cells in each dimension. However, by Proposition 4C.1 each X_i is homotopy equivalent to a CW complex with the minimum configuration of cells consistent with its homology, namely, a 0-cell and a k-cell for each k congruent to $2i$ or $2i + 1 \bmod 2p - 2$.

Stable splittings of $K(G, 1)$'s for finite groups G have been much studied and are a complicated and subtle business. To take the simplest noncyclic example, Proposition 4I.1 implies that $\Sigma K(\mathbb{Z}_2 \times \mathbb{Z}_2, 1)$ splits as the wedge sum of two copies of $\Sigma K(\mathbb{Z}_2, 1)$ and $\Sigma(K(\mathbb{Z}_2, 1) \wedge K(\mathbb{Z}_2, 1))$, but the latter summand can be split further, according to a result in [Harris & Kuhn 1988] which says that for G the direct sum of k copies of $\mathbb{Z}_{p^n}$, $\Sigma K(G, 1)$ splits canonically as the wedge sum of pieces having exactly $p^k - 1$ distinct homotopy types. Some of these summands occur more than once, as we see in the case of $\mathbb{Z}_2 \times \mathbb{Z}_2$.

Exercises

1. If a connected CW complex X retracts onto a subcomplex A, show that $\Sigma X \simeq \Sigma A \vee \Sigma(X/A)$. [One approach: Show the map $\Sigma r + \Sigma q : \Sigma X \to \Sigma A \vee \Sigma(X/A)$ induces an isomorphism on homology, where $r : X \to A$ is the retraction and $q : X \to X/A$ is the quotient map.]

2. Using the Künneth formula, show that $\Sigma K(\mathbb{Z}_m \times \mathbb{Z}_n, 1) \simeq \Sigma K(\mathbb{Z}_m, 1) \vee \Sigma K(\mathbb{Z}_n, 1)$ if m and n are relatively prime. Thus to determine stable splittings of $K(\mathbb{Z}_n, 1)$ it suffices to do the case that n is a prime power, as in Proposition 4I.3.

3. Extending Proposition 4I.3, show that the $(2k + 1)$-skeleton of the suspension of a high-dimensional lens space with fundamental group of order p^n is homotopy equivalent to the wedge sum of the $(2k + 1)$-skeleta of the spaces X_i, if these X_i's are chosen to have the minimum number of cells in each dimension, as described in the remarks following the proof.

4.J   The Loopspace of a Suspension

Loopspaces appear at first glance to be hopelessly complicated objects, but if one is only interested in homotopy type, there are many cases when great simplifications are possible. One of the nicest of these cases is the loopspace of a sphere. We show in this section that ΩS^{n+1} has the weak homotopy type of the James reduced product $J(S^n)$ introduced in §3.2. More generally, we show that $\Omega \Sigma X$ has the weak homotopy type of $J(X)$ for every connected CW complex X. If one wants, one can strengthen 'weak homotopy type' to 'homotopy type' by quoting Milnor's theorem, mentioned in §4.3, that the loopspace of a CW complex has the homotopy type of a CW complex.

Part of the interest in $\Omega \Sigma X$ can be attributed to its close connection with the suspension homomorphism $\pi_i(X) \to \pi_{i+1}(\Sigma X)$. We will use the weak homotopy equivalence of $\Omega \Sigma X$ with $J(X)$ to give another proof that the suspension homomorphism is an isomorphism in dimensions up to approximately double the connectivity of X. In addition, we will obtain an exact sequence that measures the failure of the suspension map to be an isomorphism in dimensions between double and triple the connectivity of X. An easy application of this, together with results proved elsewhere in the book, will be to compute $\pi_{n+1}(S^n)$ and $\pi_{n+2}(S^n)$ for all n.

As a rough first approximation to ΩS^{n+1} there is a natural inclusion of S^n into ΩS^{n+1} obtained by regarding S^{n+1} as the reduced suspension ΣS^n, the quotient $(S^n \times I)/(S^n \times \partial I \cup \{e\} \times I)$ where e is the basepoint of S^n, then associating to each point $x \in S^n$ the loop $\lambda(x)$ in ΣS^n given by $t \mapsto (x, t)$. The figure shows what a few such loops look like. However, we cannot expect this inclusion $S^n \hookrightarrow \Omega S^{n+1}$ to be a homotopy equivalence since ΩS^{n+1} is an H–space but S^n is only an H–space when $n = 1, 3, 7$ by the theorem of 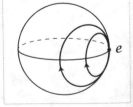 Adams discussed in §4.B. The simplest way to correct this deficiency in S^n would be to replace it by the free H–space that it generates, the reduced product $J(S^n)$. Recall from §3.2 that a point in $J(S^n)$ is a formal product $x_1 \cdots x_k$ of points $x_i \in S^n$, with the basepoint e acting as an identity element for the multiplication obtained by juxtaposition of formal products. We would like to define a map $\lambda : J(S^n) \to \Omega S^{n+1}$ by setting $\lambda(x_1 \cdots x_k) = \lambda(x_1) \cdots \lambda(x_k)$, the product of the loops $\lambda(x_i)$. The only difficulty is in the parametrization of this product, which needs to be adjusted so that λ is continuous. The problem is that when some x_i approaches the basepoint $e \in S^n$, one wants the loop $\lambda(x_i)$ to disappear gradually from the product $\lambda(x_1) \cdots \lambda(x_k)$, without disrupting the parametrization as simply deleting $\lambda(e)$ would do. This can be achieved by first making the time it takes to traverse each loop $\lambda(x_i)$ equal to the distance from x_i to the basepoint of S^n, then normalizing the resulting product of loops so that it takes unit time, giving a map $I \to \Sigma S^n$.

More generally, this same procedure defines a map $\lambda : J(X) \to \Omega \Sigma X$ for any connected CW complex X, where 'distance to the basepoint' is replaced by any map $d : X \to [0,1]$ with $d^{-1}(0) = e$, the basepoint of X.

Theorem 4J.1. *The map $\lambda : J(X) \to \Omega \Sigma X$ is a weak homotopy equivalence for every connected CW complex X.*

Proof: The main step will be to compute the homology of $\Omega \Sigma X$. After this is done, it will be easy to deduce that λ induces an isomorphism on homology using the calculation of the homology of $J(X)$ in Proposition 3C.8, and from this conclude that λ is a weak homotopy equivalence. It will turn out to be sufficient to consider homology with coefficients in a field F. We know that $H_*(J(X);F)$ is the tensor algebra $T\tilde{H}_*(X;F)$ by Proposition 3C.8, so we want to show that $H_*(\Omega \Sigma X;F)$ has this same structure, a result first proved in [Bott & Samelson 1953].

Let us write the reduced suspension $Y = \Sigma X$ as the union of two reduced cones $Y_+ = C_+X$ and $Y_- = C_-X$ intersecting in the equatorial $X \subset \Sigma X$. Consider the path fibration $p : PY \to Y$ with fiber ΩY. Let $P_+Y = p^{-1}(Y_+)$ and $P_-Y = p^{-1}(Y_-)$, so P_+Y consists of paths in Y starting at the basepoint and ending in Y_+, and similarly for P_-Y. Then $P_+Y \cap P_-Y$ is $p^{-1}(X)$, the paths from the basepoint to X. Since Y_+ and Y_- are deformation retracts of open neighborhoods U_+ and U_- in Y such that $U_+ \cap U_-$ deformation retracts onto $Y_+ \cap Y_- = X$, the homotopy lifting property implies that P_+Y, P_-Y, and $P_+Y \cap P_-Y$ are deformation retracts, in the weak sense, of open neighborhoods $p^{-1}(U_+)$, $p^{-1}(U_-)$, and $p^{-1}(U_+) \cap p^{-1}(U_-)$, respectively. Therefore there is a Mayer–Vietoris sequence in homology for the decomposition of PY as $P_+Y \cup P_-Y$. Taking reduced homology and using the fact that PY is contractible, this gives an isomorphism

(i)
$$\Phi : \tilde{H}_*(P_+Y \cap P_-Y;F) \xrightarrow{\approx} \tilde{H}_*(P_+Y;F) \oplus \tilde{H}_*(P_-Y;F)$$

The two coordinates of Φ are induced by the inclusions, with a minus sign in one case, but Φ will still be an isomorphism if this minus sign is eliminated, so we may assume this has been done.

We claim that the isomorphism Φ can be rewritten as an isomorphism

(ii)
$$\Theta : \tilde{H}_*(\Omega Y \times X; F) \xrightarrow{\approx} \tilde{H}_*(\Omega Y; F) \oplus \tilde{H}_*(\Omega Y; F)$$

To see this, we observe that the fibration $P_+Y \to Y_+$ is fiber-homotopically trivial. This is true since the cone Y_+ is contractible, but we shall need an explicit fiber homotopy equivalence $P_+Y \simeq \Omega Y \times Y_+$, and this is easily constructed as follows. Define $f_+ : P_+Y \to \Omega Y \times Y_+$ by $f_+(y) = (y \cdot y_y^+, y)$ where $y = y(1)$ and y_y^+ is the obvious path in Y_+ from $y = (x, t)$ to the basepoint along the segment $\{x\} \times I$. In the other direction, define $g_+ : \Omega Y \times Y_+ \to P_+Y$ by $g_+(y, y) = y \cdot \overline{y}_y^+$ where the bar denotes the inverse path. Then f_+g_+ and g_+f_+ are fiber-homotopic to the respective identity maps since $\overline{y}_y^+ \cdot y_y^+$ and $y_y^+ \cdot \overline{y}_y^+$ are homotopic to the constant paths.

In similar fashion the fibration $P_-Y \to Y_-$ is fiber-homotopically trivial via maps f_- and g_-. By restricting a fiber-homotopy trivialization of either P_+Y or P_-Y to $P_+Y \cap P_-Y$, we see that the fibration $P_+Y \cap P_-Y$ is fiber-homotopy equivalent to the product $\Omega Y \times X$. Let us do this using the fiber-homotopy trivialization of P_-Y. The groups in (i) can now be replaced by those in (ii). The map Φ has coordinates induced by inclusion, and it follows that the corresponding map Θ in (ii) has coordinates induced by the two maps $\Omega Y \times X \to \Omega Y$, $(y, x) \mapsto y \cdot \lambda(x)$ and $(y, x) \mapsto y$. Namely, the first coordinate of Θ is induced by $f_+g_- | \Omega Y \times X$ followed by projection to ΩY, and the second coordinate is the same but with f_-g_- in place of f_+g_-.

Writing the two coordinates of Θ as Θ_1 and Θ_2, the fact that Θ is an isomorphism means that the restriction of Θ_1 to the kernel of Θ_2 is an isomorphism. Via the Künneth formula we can write $\tilde{H}_*(\Omega Y \times X; F)$ as $(H_*(\Omega Y; F) \otimes \tilde{H}_*(X; F)) \oplus \tilde{H}_*(\Omega Y; F)$ where projection onto the latter summand is Θ_2. Hence Θ_1 gives an isomorphism from the first summand $H_*(\Omega Y; F) \otimes \tilde{H}_*(X; F)$ onto $\tilde{H}_*(\Omega Y; F)$. Since $\Theta_1(y, x) = (y \cdot \lambda(x))$, this means that the composed map

$$H_*(\Omega Y; F) \otimes \tilde{H}_*(X; F) \xrightarrow{\mathbb{1} \otimes \lambda_*} H_*(\Omega Y; F) \otimes \tilde{H}_*(\Omega Y; F) \longrightarrow \tilde{H}_*(\Omega Y; F)$$

with the second map Pontryagin product, is an isomorphism. Now to finish the calculation of $H_*(\Omega Y; F)$ as the tensor algebra $T\tilde{H}_*(X; F)$, we apply the following algebraic lemma, with $A = H_*(\Omega Y; F)$, $V = \tilde{H}_*(X; F)$, and $i = \lambda_*$.

Lemma 4J.2. *Let A be a graded algebra over a field F with $A_0 = F$ and let V be a graded vector space over F with $V_0 = 0$. Suppose we have a linear map $i : V \to A$ preserving grading, such that the multiplication map $\mu : A \otimes V \to \tilde{A}$, $\mu(a \otimes v) = a i(v)$, is an isomorphism. Then the canonical algebra homomorphism $i : TV \to A$ extending the previous i is an isomorphism.*

For example, if V is a 1-dimensional vector space over F, as happens in the case $X = S^n$, then this says that if the map $A \to \tilde{A}$ given by right-multiplication by an element $a = i(v)$ is an isomorphism, then A is the polynomial algebra $F[a]$. The general case can be viewed as the natural generalization of this to polynomials in any number of noncommuting variables.

Proof: Since μ is an isomorphism, each element $a \in A_n$ with $n > 0$ can be written uniquely in the form $\mu(\sum_j a_j \otimes v_j) = \sum_j a_j i(v_j)$ for $v_j \in V$ and $a_j \in A_{n(j)}$, with $n(j) < n$ since $V_0 = 0$. By induction on n, $a_j = i(\alpha_j)$ for a unique $\alpha_j \in (TV)_{n(j)}$. Thus $a = i(\sum_j \alpha_j \otimes v_j)$ so i is surjective. Since these representations are unique, i is also injective. The induction starts with the hypothesis that $A_0 = F$, the scalars in TV. $\square$

Returning now to the proof of the theorem, we observe that λ is an H–map: The two maps $J(X) \times J(X) \to \Omega\Sigma X$, $(x, y) \mapsto \lambda(xy)$ and $(x, y) \mapsto \lambda(x)\lambda(y)$, are homotopic since the loops $\lambda(xy)$ and $\lambda(x)\lambda(y)$ differ only in their parametrizations. Since λ is an H–map, the maps $X \hookrightarrow J(X) \xrightarrow{\lambda} \Omega\Sigma X$ induce the commutative diagram at the right. We have shown that the downward map on the right is an isomorphism, and the same is true of the one on the left by the calculation of $H_*(J(X); F)$ in Proposition 3C.8. Hence λ_* is an isomorphism. By Corollary 3A.7 this is also true for $\mathbb{Z}$ coefficients. When X is simply-connected, so are $J(X)$ and $\Omega\Sigma X$, so after taking a CW approximation to $\Omega\Sigma X$, Whitehead's theorem implies that λ is a weak homotopy equivalence. In the general case that X is only connected, we obtain the same conclusion from the generalization of Whitehead's theorem to abelian spaces, Proposition 4.74, since $J(X)$ and $\Omega\Sigma X$ are H–spaces, with trivial action of π_1 on all homotopy groups by Example 4A.3. $\square$

$$TH_*(X;F)$$
$$H_*(J(X);F) \xrightarrow{\lambda_*} H_*(\Omega\Sigma X;F)$$

Using the natural identification $\pi_i(\Omega\Sigma X) = \pi_{i+1}(\Sigma X)$, the inclusion $X \hookrightarrow \Omega\Sigma X$ induces the suspension map $\pi_i(X) \to \pi_{i+1}(\Sigma X)$. Since this inclusion factors through $J(X)$, we can identify the relative groups $\pi_i(\Omega\Sigma X, X)$ with $\pi_i(J(X), X)$. If X is n-connected then the pair $(J(X), X)$ is $(2n + 1)$-connected since we can replace X by a complex with n-skeleton a point, and then the $(2n + 1)$-skeleton of $J(X)$ is contained in X. Thus we have:

Corollary 4J.3. *The suspension map $\pi_i(X) \to \pi_{i+1}(\Sigma X)$ for an n-connected CW complex X is an isomorphism if $i \leq 2n$ and a surjection if $i = 2n + 1$.* $\square$

In the case of a sphere we can describe what happens in the first dimension when suspension is not an isomorphism, namely the suspension $\pi_{2n-1}(S^n) \to \pi_{2n}(S^{n+1})$ which the corollary guarantees only to be a surjection. The CW structure on $J(S^n)$ consists of a single cell in each dimension a multiple of n, so from exactness of $\pi_{2n}(J(S^n), S^n) \xrightarrow{\partial} \pi_{2n-1}(S^n) \xrightarrow{\Sigma} \pi_{2n}(S^{n+1})$ we see that the kernel of the suspension $\pi_{2n-1}(S^n) \to \pi_{2n}(S^{n+1})$ is generated by the attaching map of the $2n$-cell of $J(S^n)$. This attaching map is the Whitehead product $[\mathbb{1}, \mathbb{1}]$, as we noted in §4.2 when we defined Whitehead products following Example 4.52. When n is even, the Hopf invariant homomorphism $\pi_{2n-1}(S^n) \to \mathbb{Z}$ has the value ± 2 on $[\mathbb{1}, \mathbb{1}]$, as we saw in §4.B. If there is no map of Hopf invariant ± 1, it follows that $[\mathbb{1}, \mathbb{1}]$ generates a $\mathbb{Z}$ summand of $\pi_{2n-1}(S^n)$, and so the suspension homomorphism simply cancels this summand from $\pi_{2n-1}(S^n)$. By Adams' theorem, this is the situation for all even n except 2, 4, and 8.

When $n = 2$ we have $\pi_3(S^2) \approx \mathbb{Z}$ generated by the Hopf map η with Hopf invariant 1, so $2\eta = \pm[\mathbb{1}, \mathbb{1}]$, generating the kernel of the suspension, hence:

Corollary 4J.4. $\pi_{n+1}(S^n)$ is $\mathbb{Z}_2$ for $n \geq 3$, generated by the suspension or iterated suspension of the Hopf map.   □

The situation for $n = 4$ and 8 is more subtle. We do not have the tools available here to do the actual calculation, but if we consult the table near the beginning of §4.1 we see that the suspension $\pi_7(S^4) \to \pi_8(S^5)$ is a map $\mathbb{Z} \oplus \mathbb{Z}_{12} \to \mathbb{Z}_{24}$. By our preceding remarks we know this map is surjective with kernel generated by the single element $[\mathbb{1}, \mathbb{1}]$. Algebraically, what must be happening is that the coordinate of $[\mathbb{1}, \mathbb{1}]$ in the $\mathbb{Z}$ summand is twice a generator, while the coordinate in the $\mathbb{Z}_{12}$ summand is a generator. Thus a generator of the $\mathbb{Z}$ summand, which we may take to be the Hopf map $S^7 \to S^4$, suspends to a generator of the $\mathbb{Z}_{24}$. For $n = 8$ the situation is entirely similar, with the suspension $\pi_{15}(S^8) \to \pi_{16}(S^9)$ a homomorphism $\mathbb{Z} \oplus \mathbb{Z}_{120} \to \mathbb{Z}_{240}$.

We can also obtain some information about suspension somewhat beyond the edge of the stable dimension range. Since S^n is $(n-1)$-connected and $(J(S^n), S^n)$ is $(2n-1)$-connected, we have isomorphisms $\pi_i(J(S^n), S^n) \approx \pi_i(J(S^n)/S^n)$ for $i \leq 3n - 2$ by Proposition 4.28. The group $\pi_i(J(S^n)/S^n)$ is isomorphic to $\pi_i(S^{2n})$ in the same range $i \leq 3n - 2$ since $J(S^n)/S^n$ has S^{2n} as its $(3n-1)$-skeleton. Thus the terminal portion of the long exact sequence of the pair $(J(S^n), S^n)$ starting with the term $\pi_{3n-2}(S^n)$ can be written in the form

$$\pi_{3n-2}(S^n) \xrightarrow{\Sigma} \pi_{3n-1}(S^{n+1}) \to \pi_{3n-2}(S^{2n}) \to \pi_{3n-3}(S^n) \xrightarrow{\Sigma} \pi_{3n-2}(S^{n+1}) \to \cdots$$

This is known as the **EHP sequence** since its three maps were originally called E, H, and P. (The German word for 'suspension' begins with E, the H refers to a generalization of the Hopf invariant, and the P denotes a connection with Whitehead products; see [Whitehead 1978] for more details.) Note that the terms $\pi_i(S^{2n})$ in the EHP sequence are stable homotopy groups since $i \leq 3n - 2$. Thus we have the curious situation that stable homotopy groups are measuring the lack of stability of the groups $\pi_i(S^n)$ in the range $2n - 1 \leq i \leq 3n - 2$, the so-called *metastable* range.

Specializing to the first interesting case $n = 2$, the sequence becomes

$$\pi_4(S^2) \xrightarrow{\Sigma} \pi_5(S^3) \longrightarrow \pi_4(S^4) \longrightarrow \pi_3(S^2) \xrightarrow{\Sigma} \pi_4(S^3) \longrightarrow 0$$
$$\mathbb{Z}_2 \qquad\qquad\qquad\qquad \mathbb{Z} \qquad\qquad \mathbb{Z} \qquad\qquad \mathbb{Z}_2$$

From the Hopf bundle $S^1 \to S^3 \to S^2$ we have $\pi_4(S^2) \approx \pi_4(S^3) \approx \mathbb{Z}_2$, with $\pi_4(S^2)$ generated by the composition $\eta(\Sigma\eta)$ where η is the Hopf map $S^3 \to S^2$. From exactness of the latter part of the sequence we deduce that the map $\pi_4(S^4) \to \pi_3(S^2)$ is injective, and hence that the suspension $\pi_4(S^2) \to \pi_5(S^3)$ is surjective, so $\pi_5(S^3)$ is either $\mathbb{Z}_2$ or 0. From the general suspension theorem, the suspension $\pi_5(S^3) \to \pi_6(S^4)$ is surjective as well, and the latter group is in the stable range. We show in Proposition 4L.11 that the stable group π_2^s is nonzero, and so we conclude that $\pi_{n+2}(S^n) \approx \mathbb{Z}_2$ for all $n \geq 2$, generated by the composition $(\Sigma^{n-2}\eta)(\Sigma^{n-1}\eta)$.

We will see in [SSAT] that the EHP sequence extends all the way to the left to form an infinite exact sequence when n is odd, and when n is even a weaker statement holds: The sequence extends after factoring out all odd torsion.

Replacing S^n by any $(n-1)$-connected CW complex X, our derivation of the finite EHP sequence generalizes immediately to give an exact sequence

$$\pi_{3n-2}(X) \xrightarrow{\Sigma} \pi_{3n-1}(\Sigma X) \to \pi_{3n-2}(X \wedge X) \to \pi_{3n-3}(X) \xrightarrow{\Sigma} \pi_{3n-2}(\Sigma X) \to \cdots$$

using the fact that $J_2(X)/X = X \wedge X$.

The generalization of the results of this section to $\Omega^n \Sigma^n X$ turns out to be of some importance in homotopy theory. In case we do not get to this topic in [SSAT], the reader can begin to learn about it by looking at [Carlsson & Milgram 1995].

Exercise

1. Show that $\Omega\Sigma X$ for a nonconnected CW complex X reduces to the connected case by showing that each path-component of $\Omega\Sigma X$ is homotopy equivalent to $\Omega\Sigma(\bigvee_\alpha X_\alpha)$ where the X_α's are the components of X.

4.K The Dold–Thom Theorem

In the preceding section we studied the free monoid $J(X)$ generated by a space X, and in this section we take up its commutative analog, the free abelian monoid generated by X. This is the infinite symmetric product $SP(X)$ introduced briefly in §3.C. The main result will be a theorem of [Dold & Thom 1958] asserting that $\pi_* SP(X) \approx \widetilde{H}_*(X;\mathbb{Z})$ for all connected CW complexes X. In particular this yields the surprising fact that $SP(S^n)$ is a $K(\mathbb{Z},n)$, and more generally that the functor SP takes Moore spaces $M(G,n)$ to Eilenberg–MacLane spaces $K(G,n)$. This leads to the general result that for all connected CW complexes X, $SP(X)$ has the homotopy type of a product of Eilenberg–MacLane spaces. In other words, the k-invariants of $SP(X)$ are all trivial.

The main step in the proof of the Dold–Thom theorem will be to show that the homotopy groups $\pi_* SP(X)$ define a homology theory. An easy computation of the coefficient groups $\pi_* SP(S^n)$ will then show that this must be ordinary homology with $\mathbb{Z}$ coefficients. A new idea needed for the proof of the main step is the notion of a *quasifibration*, generalizing fibrations and fiber bundles. In order to establish a few basic facts about quasifibrations we first make a small detour to prove an essentially elementary fact about relative homotopy groups.

A Mayer–Vietoris Property of Homotopy Groups

In this subsection we will be concerned largely with relative homotopy groups, and it will be impossible to avoid the awkward fact that there is no really good way to define the relative π_0. What we will do as a compromise is to take $\pi_0(X, A, x_0)$ to be the quotient set $\pi_0(X, x_0)/\pi_0(A, x_0)$. This at least allows the long exact sequence of homotopy groups for (X, A) to end with the terms

$$\pi_0(A, x_0) \to \pi_0(X, x_0) \to \pi_0(X, A, x_0) \to 0$$

An exercise for §4.1 shows that the five-lemma can be applied to the map of long exact sequences induced by a map $(X, A) \to (Y, B)$, provided the basepoint is allowed to vary. However, the long exact sequence of a triple cannot be extended through the π_0 terms with this definition, so one must proceed with some caution.

The excision theorem for homology involves a space X with subspaces A and B such that X is the union of the interiors of A and B. In this situation we call $(X; A, B)$ an **excisive triad**. By a map $f : (X; A, B) \to (Y; C, D)$ we mean $f : X \to Y$ with $f(A) \subset C$ and $f(B) \subset D$.

Proposition 4K.1. *Let $f : (X; A, B) \to (Y; C, D)$ be a map of excisive triads. If the induced maps $\pi_i(A, A \cap B) \to \pi_i(C, C \cap D)$ and $\pi_i(B, A \cap B) \to \pi_i(D, C \cap D)$ are bijections for $i < n$ and surjections for $i = n$, for all choices of basepoints, then the same is true of the induced maps $\pi_i(X, A) \to \pi_i(Y, C)$. By symmetry the conclusion holds also for the maps $\pi_i(X, B) \to \pi_i(Y, D)$.*

The corresponding statement for homology is a trivial consequence of excision which says that $H_i(X, A) \approx H_i(B, A \cap B)$ and $H_i(Y, C) \approx H_i(D, C \cap D)$, so it is not necessary to assume anything about the map $H_i(A, A \cap B) \to H_i(C, C \cap D)$. With the failure of excision for homotopy groups, however, it is not surprising that the assumption on $\pi_i(A, A \cap B) \to \pi_i(C, C \cap D)$ cannot be dropped. An example is provided by the quotient map $f : D^2 \to S^2$ collapsing ∂D^2 to the north pole of S^2, with C and D the northern and southern hemispheres of S^2, and A and B their preimages under f.

Proof: First we will establish a general fact about relative homotopy groups. Consider an inclusion $(X, A) \hookrightarrow (Y, C)$. We will show the following three conditions are equivalent for each $n \geq 1$:

(i) For all choices of basepoints the map $\pi_i(X, A) \to \pi_i(Y, C)$ induced by inclusion is surjective for $i = n$ and has trivial kernel for $i = n - 1$.

(ii) Let ∂D^n be written as the union of hemispheres $\partial_+ D^n$ and $\partial_- D^n$ intersecting in S^{n-2}. Then every map

$$(D^n \times \{0\} \cup \partial_+ D^n \times I, \partial_- D^n \times \{0\} \cup S^{n-2} \times I) \longrightarrow (Y, C)$$

taking $(\partial_+ D^n \times \{1\}, S^{n-2} \times \{1\})$ to (X, A) extends to a map $(D^n \times I, \partial_- D^n \times I) \to (Y, C)$ taking $(D^n \times \{1\}, \partial_- D^n \times \{1\})$ to (X, A).

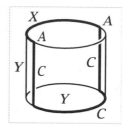

(iii) Condition (ii) with the added hypothesis that the restriction of the given map to $\partial_+ D^n \times I$ is independent of the I coordinate.

It is obvious that (ii) and (iii) are equivalent since the stronger hypothesis in (iii) can always be achieved by composing with a homotopy of $D^n \times I$ that shrinks $\partial_+ D^n \times I$ to $\partial_+ D^n \times \{1\}$.

To see that (iii) implies (i), let $f : (\partial_+ D^n \times \{1\}, S^{n-2} \times \{1\}) \to (X, A)$ represent an element of $\pi_{n-1}(X, A)$. If this is in the kernel of the map to $\pi_{n-1}(Y, C)$, then we get an extension of f over $D^n \times \{0\} \cup \partial_+ D^n \times I$, with the constant homotopy on $\partial_+ D^n \times I$ and $(D^n \times \{0\}, \partial_- D^n \times \{0\})$ mapping to (Y, C). Condition (iii) then gives an extension over $D^n \times I$, whose restriction to $D^n \times \{1\}$ shows that f is zero in $\pi_{n-1}(X, A)$, so the kernel of $\pi_{n-1}(X, A) \to \pi_{n-1}(Y, C)$ is trivial. To check surjectivity of the map $\pi_n(X, A) \to \pi_n(Y, C)$, represent an element of $\pi_n(Y, C)$ by a map $f : D^n \times \{0\} \to Y$ taking $\partial_- D^n \times \{0\}$ to C and $\partial_+ D^n \times \{0\}$ to a chosen basepoint. Extend f over $\partial_+ D^n \times I$ via the constant homotopy, then extend over $D^n \times I$ by applying (iii). The result is a homotopy of the given f to a map representing an element of the image of $\pi_n(X, A) \to \pi_n(Y, C)$.

Now we show that (i) implies (ii). Given a map f as in the hypothesis of (ii), the injectivity part of (i) gives an extension of f over $D^n \times \{1\}$. Choose a small disk $E^n \subset \partial_- D^n \times I$, shown shaded in the figure, intersecting $\partial_- D^n \times \{1\}$ in a hemisphere $\partial_+ E^n$ of its boundary. We may assume the extended f has a constant value $x_0 \in A$ on $\partial_+ E^n$. Viewing the extended f as representing an element of $\pi_n(Y, C, x_0)$, the surjectivity part

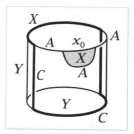

of (i) then gives an extension of f over $D^n \times I$ taking $(E^n, \partial_- E^n)$ to (X, A) and the rest of $\partial_- D^n \times I$ to C. The argument is finished by composing this extended f with a deformation of $D^n \times I$ pushing E^n into $D^n \times \{1\}$.

Having shown the equivalence of (i)–(iii), let us prove the proposition. We may reduce to the case that the given $f : (X; A, B) \to (Y; C, D)$ is an inclusion by using mapping cylinders. One's first guess would be to replace $(Y; C, D)$ by the triad of mapping cylinders $(M_f; M_{f|A}, M_{f|B})$, where we view $f|A$ as a map $A \to C$ and $f|B$ as a map $B \to D$. However, the triad $(M_f; M_{f|A}, M_{f|B})$ need not be excisive, for example if X

consists of two points A and B and Y is a single point. To remedy this problem, replace $M_{f|A}$ by its union with $f^{-1}(C) \times (1/2, 1)$ in M_f, and enlarge $M_{f|B}$ similarly.

Now we prove the proposition for an inclusion $(X; A, B) \hookrightarrow (Y; C, D)$. The case $n = 0$ is trivial from the definitions, so let us assume $n \geq 1$. In view of the equivalence of condition (i) with (ii) and (iii), it suffices to show that condition (ii) for the inclusions $(A, A \cap B) \hookrightarrow (C, C \cap D)$ and $(B, A \cap B) \hookrightarrow (D, C \cap D)$ implies (iii) for the inclusion $(X, A) \hookrightarrow (Y, C)$. Let a map $f : D^n \times \{0\} \cup \partial_+ D^n \times I \to Y$ as in the hypothesis of (iii) be given. The argument will involve subdivision of D^n into smaller disks, and for this it is more convenient to use the cube I^n instead of D^n, so let us identify I^n with D^n in such a way that $\partial_- D^n$ corresponds to the face $I^{n-1} \times \{1\}$, which we denote by $\partial_- I^n$, and $\partial_+ D^n$ corresponds to the remaining faces of I^n, which we denote by $\partial_+ I^n$. Thus we are given f on $I^n \times \{0\}$ taking $\partial_+ I^n \times \{0\}$ to X and $\partial_- I^n \times \{0\}$ to C, and on $\partial_+ I^n \times I$ we have the constant homotopy.

Since $(Y; C, D)$ is an excisive triad, we can subdivide each of the I factors of $I^n \times \{0\}$ into subintervals so that f takes each of the resulting n-dimensional subcubes of $I^n \times \{0\}$ into either C or D. The extension of f we construct will have the following key property:

($*$)   If K is a one of the subcubes of $I^n \times \{0\}$, or a lower-dimensional face of such a cube, then the extension of f takes $(K \times I, K \times \{1\})$ to (C, A) or (D, B) whenever f takes K to C or D, respectively.

Initially we have f defined on $\partial_+ I^n \times I$ with image in X, independent of the I coordinate, and we may assume the condition ($*$) holds here since we may assume that $A = X \cap C$ and $B = X \cap D$, these conditions holding for the mapping cylinder construction described above.

Consider the problem of extending f over $K \times I$ for K one of the subcubes. We may assume that f has already been extended to $\partial_+ K \times I$ so that ($*$) is satisfied, by induction on n and on the sequence of subintervals of the last coordinate of $I^n \times \{0\}$. To extend f over $K \times I$, let us first deal with the cases that the given f takes $(K, \partial_- K)$ to $(C, C \cap D)$ or $(D, C \cap D)$. Then by (ii) for the inclusion $(A, A \cap B) \hookrightarrow (C, C \cap D)$ or $(B, A \cap B) \hookrightarrow (D, C \cap D)$ we may extend f over $K \times I$ so that ($*$) is still satisfied. If neither of these two cases applies, then 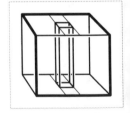 the given f takes $(K, \partial_- K)$ just to (C, C) or (D, D), and we can apply (ii) trivially to construct the desired extension of f over $K \times I$. $\square$

‖ **Corollary 4K.2.** *Given a map $f : X \to Y$ and open covers $\{U_i\}$ of X and $\{V_i\}$ of Y with $f(U_i) \subset V_i$ for all i, then if each restriction $f : U_i \to V_i$ and more generally each $f : U_{i_1} \cap \cdots \cap U_{i_n} \to V_{i_1} \cap \cdots \cap V_{i_n}$ is a weak homotopy equivalence, so is $f : X \to Y$.*

Proof: First let us do the case of covers by two sets. By the five-lemma, the hypotheses imply that $\pi_n(U_i, U_1 \cap U_2) \to \pi_n(V_i, V_1 \cap V_2)$ is bijective for $i = 1, 2$, $n \geq 0$, and all choices of basepoints. The preceding proposition then implies that the maps $\pi_n(X, U_1) \to \pi_n(Y, V_1)$ are isomorphisms. Hence by the five-lemma again, so are the maps $\pi_n(X) \to \pi_n(Y)$.

By induction, the case of finite covers by $k > 2$ sets reduces to the case of covers by two sets, by letting one of the two sets be the union of the first $k - 1$ of the given sets and the other be the kth set. The case of infinite covers reduces to the finite case since for surjectivity of $\pi_n(X) \to \pi_n(Y)$, a map $S^n \to Y$ has compact image covered by finitely many V_i's, and similarly for injectivity. $\square$

Quasifibrations

A map $p : E \to B$ with B path-connected is a **quasifibration** if the induced map $p_* : \pi_i(E, p^{-1}(b), x_0) \to \pi_i(B, b)$ is an isomorphism for all $b \in B$, $x_0 \in p^{-1}(b)$, and $i \geq 0$. We have shown in Theorem 4.41 that fiber bundles and fibrations have this property for $i > 0$, as a consequence of the homotopy lifting property, and the same reasoning applies for $i = 0$ since we assume B is path-connected.

For example, consider the natural projection $M_f \to I$ of the mapping cylinder of a map $f : X \to Y$. This projection will be a quasifibration iff f is a weak homotopy equivalence, since the latter condition is equivalent to having $\pi_i(M_f, p^{-1}(b)) = 0 = \pi_i(I, b)$ for all i and all $b \in I$. Note that if f is not surjective, there are paths in I that do not lift to paths in M_f with a prescribed starting point, so p will not be a fibration in such cases.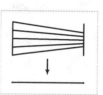

An alternative condition for a map $p : E \to B$ to be a quasifibration is that the inclusion of each fiber $p^{-1}(b)$ into the homotopy fiber F_b of p over b is a weak homotopy equivalence. Recall that F_b is the space of all pairs (x, y) with $x \in E$ and y a path in B from $p(x)$ to b. The actual fiber $p^{-1}(b)$ is included in F_b as the pairs with $x \in p^{-1}(b)$ and y the constant path at x. To see the equivalence of the two definitions, consider the commutative triangle at the right, where $F_b \to E_p \to B$ is the usual path-fibration construction applied to p. The right-hand map in the diagram is an isomorphism for all $i \geq 0$, and the

$$\pi_i(E, p^{-1}(b)) \longrightarrow \pi_i(E_p, F_b)$$
$$\searrow \qquad \swarrow$$
$$\pi_i(B, b)$$

upper map will be an isomorphism for all $i \geq 0$ iff the inclusion $p^{-1}(b) \hookrightarrow F_b$ is a weak equivalence since $E \simeq E_p$. Hence the two definitions are equivalent.

Recall from Proposition 4.61 that all fibers of a fibration over a path-connected base are homotopy equivalent. Since we are only considering quasifibrations over path-connected base spaces, this implies that all the fibers of a quasifibration have the same weak homotopy type. Quasifibrations over a base that is not path-connected are considered in the exercises, but we will not need this generality in what follows.

The following technical lemma gives various conditions for recognizing that a map is a quasifibration, which will be needed in the proof of the Dold–Thom theorem.

Lemma 4K.3. *A map $p : E \to B$ is a quasifibration if any one of the following conditions is satisfied:*

(a) *B can be decomposed as the union of open sets V_1 and V_2 such that each of the restrictions $p^{-1}(V_1) \to V_1$, $p^{-1}(V_2) \to V_2$, and $p^{-1}(V_1 \cap V_2) \to V_1 \cap V_2$ is a quasifibration.*

(b) *B is the union of an increasing sequence of subspaces $B_1 \subset B_2 \subset \cdots$ with the property that each compact set in B lies in some B_n, and such that each restriction $p^{-1}(B_n) \to B_n$ is a quasifibration.*

(c) *There is a deformation F_t of E into a subspace E', covering a deformation $\overline{F}_t$ of B into a subspace B', such that the restriction $E' \to B'$ is a quasifibration and $F_1 : p^{-1}(b) \to p^{-1}(\overline{F}_1(b))$ is a weak homotopy equivalence for each $b \in B$.*

By a 'deformation' in (c) we mean a deformation retraction in the weak sense as defined in the exercises for Chapter 0, where the homotopy is not required to be the identity on the subspace.

Proof: (a) To avoid some tedious details we will consider only the case that the fibers of p are path-connected, which will suffice for our present purposes, leaving the general case as an exercise for the reader. This hypothesis on fibers guarantees that all π_0's arising in the proof are trivial. In particular, by an exercise for §4.1 this allows us to terminate long exact sequences of homotopy groups of triples with zeros in the π_0 positions.

Let $U_1 = p^{-1}(V_1)$ and $U_2 = p^{-1}(V_2)$. The five-lemma for the long exact sequences of homotopy groups of the triples $(U_k, U_1 \cap U_2, p^{-1}(b))$ and $(V_k, V_1 \cap V_2, b)$ implies that the maps $\pi_i(U_k, U_1 \cap U_2) \to \pi_i(V_k, V_1 \cap V_2)$ are isomorphisms for $k = 1, 2$ and all i. Then Proposition 4K.1 implies that the maps $\pi_i(E, U_k) \to \pi_i(B, V_k)$ are isomorphisms for all choices of basepoints. The maps $\pi_i(U_k, p^{-1}(b)) \to \pi_i(V_k, b)$ are isomorphisms by hypothesis, so from the five-lemma we can then deduce that the maps $\pi_i(E, p^{-1}(b)) \to \pi_i(B, b)$ are isomorphisms for all $b \in V_k$, hence for all $b \in B$.

(b) Since each compact set in B lies in some B_n, each compact set in E lies in some subspace $E_n = p^{-1}(B_n)$, so $\pi_i(E, p^{-1}(b))$ is the direct limit $\varinjlim \pi_i(E_n, p^{-1}(b))$ just as $\pi_i(B, b) = \varinjlim \pi_i(B_n, b)$. It follows that the map $\pi_i(E, p^{-1}(b)) \to \pi_i(B, b)$ is an isomorphism since each of the maps $\pi_i(E_n, p^{-1}(b)) \to \pi_i(B_n, b)$ is an isomorphism by assumption. We can take the point b to be an arbitrary point in B and then discard any initial spaces B_n in the sequence that do not contain b, so we can assume b lies in B_n for all n.

(c) Consider the commutative diagram

$$
\begin{array}{ccc}
\pi_i(E, p^{-1}(b)) & \xrightarrow{\;F_{1*}\;} & \pi_i(E', p^{-1}(\overline{F}_1(b))) \\
\big\downarrow & & \big\downarrow \\
\pi_i(B, b) & \xrightarrow{\;\overline{F}_{1*}\;} & \pi_i(B', \overline{F}_1(b))
\end{array}
$$

where b is an arbitrary point in B. The upper map in the diagram is an isomorphism by the five-lemma since the hypotheses imply that F_1 induces isomorphisms $\pi_i(E) \to \pi_i(E')$ and $\pi_i(p^{-1}(b)) \to \pi_i(p^{-1}(\overline{F}_1(b)))$ for all i. The hypotheses also imply that the lower map and the right-hand map are isomorphisms. Hence the left-hand map is an isomorphism. $\qquad\square$

Symmetric Products

Let us recall the definition from §3.C. For a space X the n-fold symmetric product $SP_n(X)$ is the quotient space of the product of n copies of X obtained by factoring out the action of the symmetric group permuting the factors. A choice of basepoint $e \in X$ gives inclusions $SP_n(X) \hookrightarrow SP_{n+1}(X)$ induced by $(x_1, \cdots, x_n) \mapsto (x_1, \cdots, x_n, e)$, and $SP(X)$ is defined to be the union of this increasing sequence of spaces, with the direct limit topology. Note that SP_n is a homotopy functor: A map $f: X \to Y$ induces $f_*: SP_n(X) \to SP_n(Y)$, and $f \simeq g$ implies $f_* \simeq g_*$. Hence $X \simeq Y$ implies $SP_n(X) \simeq SP_n(Y)$. In similar fashion SP is a homotopy functor on the category of basepointed spaces and basepoint-preserving homotopy classes of maps. It follows that $X \simeq Y$ implies $SP(X) \simeq SP(Y)$ for connected CW complexes X and Y since in this case requiring maps and homotopies to preserve basepoints does not affect the relation of homotopy equivalence.

Example 4K.4. An interesting special case is when $X = S^2$ because in this case $SP(S^2)$ can be identified with $\mathbb{CP}^\infty$ in the following way. We first identify $\mathbb{CP}^n$ with the nonzero polynomials of degree at most n with coefficients in $\mathbb{C}$, modulo scalar multiplication, by letting $a_0 + \cdots + a_n z^n$ correspond to the line containing $(a_0, \cdots, a_n)$. The sphere S^2 we view as $\mathbb{C} \cup \{\infty\}$, and then we define $f: (S^2)^n \to \mathbb{CP}^n$ by setting $f(a_1, \cdots, a_n) = (z + a_1) \cdots (z + a_n)$ with factors $z + \infty$ omitted, so in particular $f(\infty, \cdots, \infty) = 1$. To check that f is continuous, suppose some a_i approaches ∞, say a_n, and all the other a_j's are finite. Then if we write

$$(z + a_1) \cdots (z + a_n) =$$
$$z^n + (a_1 + \cdots + a_n)z^{n-1} + \cdots + \sum_{i_1 < \cdots < i_k} a_{i_1} \cdots a_{i_k} z^{n-k} + \cdots + a_1 \cdots a_n$$

we see that, dividing through by a_n and letting a_n approach ∞, this polynomial approaches $z^{n-1} + (a_1 + \cdots + a_{n-1})z^{n-2} + \cdots + a_1 \cdots a_{n-1} = (z + a_1) \cdots (z + a_{n-1})$. The same argument would apply if several a_i's approach ∞ simultaneously.

The value $f(a_1, \cdots, a_n)$ is unchanged under permutation of the a_i's, so there is an induced map $SP_n(S^2) \to \mathbb{CP}^n$ which is a continuous bijection, hence a homeomorphism since both spaces are compact Hausdorff. Letting n go to ∞, we then get a homeomorphism $SP(S^2) \approx \mathbb{CP}^\infty$.

The same argument can be used to show that $SP_n(S^1) \simeq S^1$ for all n, including $n = \infty$. Namely, the argument shows that $SP_n(\mathbb{C} - \{0\})$ can be identified with the polynomials $z^n + a_{n-1} z^{n-1} + \cdots + a_0$ with $a_0 \neq 0$, or in other words, the n-tuples $(a_0, \cdots, a_{n-1}) \in \mathbb{C}^n$ with $a_0 \neq 0$, and this subspace of $\mathbb{C}^n$ deformation retracts onto a circle.

The symmetric products of higher-dimensional spheres are more complicated, though things are not so bad for the 2-fold symmetric product:

Example 4K.5. Let us show that $SP_2(S^n)$ is homeomorphic to the mapping cone of a map $S^n \mathbb{R}P^{n-1} \to S^n$ where $S^n \mathbb{R}P^{n-1}$ is the n-fold unreduced suspension of $\mathbb{R}P^{n-1}$. Hence $H_*(SP_2(S^n)) \approx H_*(S^n) \oplus \tilde{H}_*(S^{n+1} \mathbb{R}P^{n-1})$ from the long exact sequence of homology groups for the pair $(SP_2(S^n), S^n)$, since $SP_2(S^n)/S^n$ is $S^{n+1} \mathbb{R}P^{n-1}$ with no reduced homology below dimension $n + 2$.

If we view S^n as $D^n / \partial D^n$, then $SP_2(S^n)$ becomes a certain quotient of $D^n \times D^n$. Viewing $D^n \times D^n$ as the cone on its boundary $D^n \times \partial D^n \cup \partial D^n \times D^n$, the identifications that produce $SP_2(S^n)$ respect the various concentric copies of this boundary which fill up the interior of $D^n \times D^n$, so it suffices to analyze the identifications in all these copies of the boundary. The identifications on the boundary of $D^n \times D^n$ itself yield S^n. This is clear since the identification $(x, y) \sim (y, x)$ converts $D^n \times \partial D^n \cup \partial D^n \times D^n$ to $D^n \times \partial D^n$, and all points of ∂D^n are identified in S^n.

It remains to see that the identifications $(x, y) \sim (y, x)$ on each concentric copy of the boundary in the interior of $D^n \times D^n$ produce $S^n \mathbb{R}P^{n-1}$. Denote by Z the quotient of $D^n \times \partial D^n \cup \partial D^n \times D^n$ under these identifications. This is the same as the quotient of $D^n \times \partial D^n$ under the identifications $(x, y) \sim (y, x)$ for $(x, y) \in \partial D^n \times \partial D^n$.

Define $f : D^n \times \mathbb{R}P^{n-1} \to Z$ by $f(x, L) = (w, z)$ where x is equidistant from $z \in \partial D^n$ and $w \in D^n$ along the line through x parallel to L, as in the figure. If x is the midpoint of the segment zz' then $w = z'$ and there is no way to distinguish between w and z, but since f takes values in the quotient space Z, this is not a problem. If $x \in \partial D^n$ then

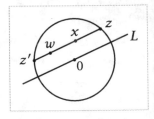

$w = z = x$, independent of L. If $x \in D^n - \partial D^n$ then $w \neq z$, and conversely, given $(w, z) \in D^n \times \partial D^n$ with $w \neq z$ there is a unique (x, L) with $f(x, L) = (w, z)$, namely x is the midpoint of the segment wz and L is the line parallel to this segment. In view of these remarks, we see that Z is the quotient space of $D^n \times \mathbb{R}P^{n-1}$ under the identifications $(x, L) \sim (x, L')$ if $x \in \partial D^n$. This quotient is precisely $S^n \mathbb{R}P^{n-1}$.

This example illustrates that passing from a CW structure on X to a CW structure on $SP_n(X)$ or $SP(X)$ is not at all straightforward. However, if X is a simplicial complex, there is a natural way of putting Δ-complex structures on $SP_n(X)$ and $SP(X)$, as follows. A simplicial complex structure on X gives a CW structure on the product of n copies of X, with cells n-fold products of simplices. Such a product has a canonical barycentric subdivision as a simplicial complex, with vertices the points whose coordinates are barycenters of simplices of X. By induction over skeleta, this just amounts to coning off a simplicial structure on the boundary of each product cell. This simplicial structure on the product of n copies of X is in fact a Δ-complex structure since the vertices of each of its simplices have a natural ordering given by the dimensions of the cells of which they are barycenters. The action of the symmetric group permuting coordinates respects this Δ-complex structure, taking simplices homeomorphically to simplices, preserving vertex-orderings, so there is an induced Δ-complex structure on the quotient $SP_n(X)$. With the basepoint of X chosen to be a vertex, $SP_n(X)$ is a subcomplex of $SP_{n+1}(X)$ so there is a natural Δ-complex structure on the infinite symmetric product $SP(X)$ as well.

As usual with products, the CW topology on $SP_n(X)$ and $SP(X)$ is in general different from the topology arising from the original definition in terms of product topologies, but one can check that the two topologies have the same compact sets, so the distinction will not matter for our present purposes. For definiteness, we will use the CW topology in what follows, which means restricting X to be a simplicial complex. Since every CW complex is homotopy equivalent to a simplicial complex by Theorem 2C.5, and SP_n and SP are homotopy functors, there is no essential loss of generality in restricting from CW complexes to simplicial complexes.

Here is the main result of this section, the Dold–Thom theorem:

Theorem 4K.6. *The functor $X \mapsto \pi_i SP(X)$ for $i \geq 1$ coincides with the functor $X \mapsto H_i(X; \mathbb{Z})$ on the category of basepointed connected CW complexes.*

In particular this says that $SP(S^n)$ is a $K(\mathbb{Z}, n)$, and more generally that for a Moore space $M(G, n)$, $SP(M(G, n))$ is a $K(G, n)$.

The fact that $SP(X)$ is a commutative, associative H–space with a strict identity element limits its weak homotopy type considerably:

Corollary 4K.7. *A path-connected, commutative, associative H–space with a strict identity element has the weak homotopy type of a product of Eilenberg–MacLane spaces.*

In particular, if X is a connected CW complex then $SP(X)$ is path-connected and has the weak homotopy type of $\prod_n K(H_n(X), n)$. Thus the functor SP essentially reduces to Eilenberg–MacLane spaces.

Proof: Let X be a path-connected, commutative, associative H–space with a strict identity element, and let $G_n = \pi_n(X)$. By Lemma 4.31 there is a map $M(G_n, n) \to X$ inducing an isomorphism on π_n when $n > 1$ and an isomorphism on H_1 when $n = 1$. We can take these maps to be basepoint-preserving, and then they combine to give a map $\bigvee_n M(G_n, n) \to X$. The very special H–space structure on X allows us to extend this to a homomorphism $f : SP(\bigvee_n M(G_n, n)) \to X$. In general, $SP(\bigvee_\alpha X_\alpha)$ can be identified with $\prod_\alpha SP(X_\alpha)$ where this is the 'weak' infinite product, the union of the finite products. This, together with the general fact that the map

$\pi_i(X) \to \pi_i SP(X) = H_i(X; \mathbb{Z})$ induced by the inclusion $X = SP_1(X) \hookrightarrow SP(X)$ is the Hurewicz homomorphism, as we will see at the end of the proof of the Dold–Thom theorem, implies that the map f induces an isomorphism on all homotopy groups. Thus we have a weak homotopy equivalence $\prod_n SP(M(G_n, n)) \to X$, and as we noted above, $SP(M(G_n, n))$ is a $K(G_n, n)$. Finally, since each factor $SP(M(G_n, n))$ has only one nontrivial homotopy group, the weak infinite product has the same weak homotopy type as the ordinary infinite product.   □

The main step in the proof of the theorem will be to show that for a simplicial pair (X, A) with both X and A connected, there is a long exact sequence

$$\cdots \to \pi_i SP(A) \to \pi_i SP(X) \to \pi_i SP(X/A) \to \pi_{i-1} SP(A) \to \cdots$$

This would follow if the maps $SP(A) \to SP(X) \to SP(X/A)$ formed a fiber bundle or fibration. There is some reason to think this might be true, because all the fibers of the projection $SP(X) \to SP(X/A)$ are homeomorphic to $SP(A)$. In fact, in terms of the H-space structure on $SP(X)$ as the free abelian monoid generated by X, the fibers are exactly the cosets of the submonoid $SP(A)$. The projection $SP(X) \to SP(X/A)$, however, fails to have the homotopy lifting property, even the special case of lifting paths. For if x_t, $t \in [0, 1)$, is a path in $X - A$ approaching a point $x_1 = a \in A$ other than the basepoint, then regarding x_t as a path in $SP(X/A)$, any lift to $SP(X)$ would have the form $x_t \alpha_t$, $\alpha_t \in SP(A)$, ending at $x_1 \alpha_1 = a\alpha_1$, a point of $SP(A)$ which is a multiple of a, so in particular there would be no lift ending at the basepoint of $SP(X)$.

What we will show is that the projection $SP(X) \to SP(X/A)$ has instead the weaker structure of a quasifibration, which is still good enough to deduce a long exact sequence of homotopy groups.

Proof of 4K.6: As we have said, the main step will be to associate a long exact sequence of homotopy groups to each simplicial pair (X, A) with X and A connected. This will be the long exact sequence of homotopy groups coming from the quasifibration $SP(A) \to SP(X) \to SP(X/A)$, so the major work will be in verifying the quasifibration property. Since SP is a homotopy functor, we are free to replace (X, A) by a homotopy equivalent pair, so let us replace (X, A) by (M, A) where M is the mapping cylinder of the inclusion $A \hookrightarrow X$. This new pair, which we still call (X, A), has some slight technical advantages, as we will see later in the proof.

To begin the proof that the projection $p : SP(X) \to SP(X/A)$ is a quasifibration, let $B_n = SP_n(X/A)$ and $E_n = p^{-1}(B_n)$. Thus E_n consists of those points in $SP(X)$ having at most n coordinates in $X - A$. By Lemma 4K.3(b) it suffices to show that $p : E_n \to B_n$ is a quasifibration. The proof of the latter fact will be by induction on n, starting with the trivial case $n = 0$ when B_0 is a point. The induction step will consist of showing that p is a quasifibration over a neighborhood of B_{n-1} and over

$B_n - B_{n-1}$, then applying Lemma 4K.3(a). We first tackle the problem of showing the quasifibration property over a neighborhood of B_{n-1}.

Let $f_t : X \to X$ be a homotopy of the identity map deformation retracting a neighborhood N of A onto A. Since we have replaced the original X by the mapping cylinder of the inclusion $A \hookrightarrow X$, we can take f_t simply to slide points along the segments $\{a\} \times I$ in the mapping cylinder, with $N = A \times [0, 1/2)$. Let $U \subset E_n$ consist of those points having at least one coordinate in N, or in other words, products with at least one factor in N. Thus U is a neighborhood of E_{n-1} in E_n and $p(U)$ is a neighborhood of B_{n-1} in B_n.

The homotopy f_t induces a homotopy $F_t : E_n \to E_n$ whose restriction to U is a deformation of U into E_{n-1}, where by 'deformation' we mean deformation retraction in the weak sense. Since f_t is the identity on A, F_t is the lift of a homotopy $\overline{F}_t : B_n \to B_n$ which restricts to a deformation of $\overline{U} = p(U)$ into B_{n-1}. We will deduce that the projection $U \to \overline{U}$ is a quasifibration by using Lemma 4K.3(c). To apply this to the case at hand we need to verify that $F_1 : p^{-1}(b) \to p^{-1}(\overline{F}_1(b))$ is a weak equivalence for all b. Each point $w \in p^{-1}(b)$ is a commuting product of points in X. Let $\widehat{w}$ be the subproduct whose factors are points in $X - A$, so we have $w = \widehat{w}v$ for v a product of points in A. Since f_1 is the identity on A and F_1 is a homomorphism, we have $F_1(w) = F_1(\widehat{w})v$, which can be written $\widehat{F_1(\widehat{w})}v'v$ with v' also a product of points in A. If we fix $\widehat{w}$ and let v vary over $SP(A)$, we get all points of $p^{-1}(b)$ exactly once, or in other words, we have $p^{-1}(b)$ expressed as the coset $\widehat{w}SP(A)$. The map F_1, $\widehat{w}v \mapsto \widehat{F_1(\widehat{w})}v'v$, takes this coset to the coset $\widehat{F_1(\widehat{w})}SP(A)$ by a map that would be a homeomorphism if the factor v' were not present. Since A is connected, there is a path v'_t from v' to the basepoint, and so by replacing v' with v'_t in the product $\widehat{F_1(\widehat{w})}v'v$ we obtain a homotopy from $F_1 : p^{-1}(b) \to p^{-1}(\overline{F}_1(b))$ to a homeomorphism, so this map is a homotopy equivalence, as desired.

It remains to see that p is a quasifibration over $B_n - B_{n-1}$ and over the intersection of this set with $\overline{U}$. The argument will be the same in both cases.

Identifying $B_n - B_{n-1}$ with $SP_n(X - A)$, the projection $p : E_n - E_{n-1} \to B_n - B_{n-1}$ is the same as the operator $w \mapsto \widehat{w}$. The inclusion $SP_n(X - A) \hookrightarrow E_n - E_{n-1}$ gives a section for $p : E_n - E_{n-1} \to B_n - B_{n-1}$, so $p_* : \pi_i(E_n - E_{n-1}, p^{-1}(b)) \to \pi_i(B_n - B_{n-1}, b)$ is surjective. To see that it is also injective, represent an element of its kernel by a map $g : (D^i, \partial D^i) \to (E_n - E_{n-1}, p^{-1}(b))$. A nullhomotopy of pg gives a homotopy of g changing only its coordinates in $X - A$. This homotopy is through maps $(D^i, \partial D^i) \to (E_n - E_{n-1}, p^{-1}(b))$, and ends with a map to $p^{-1}(b)$, so the kernel of p_* is trivial. Thus the projection $E_n - E_{n-1} \to B_n - B_{n-1}$ is a quasifibration, at least if $B_n - B_{n-1}$ is path-connected. But by replacing the original X with the mapping cylinder of the inclusion $A \hookrightarrow X$, we guarantee that $X - A$ is path-connected since it deformation retracts onto X. Hence the space $B_n - B_{n-1} = SP_n(X - A)$ is also path-connected.

This argument works equally well over any open subset of $B_n - B_{n-1}$ that is path-connected, in particular over $U \cap (B_n - B_{n-1})$, so via Lemma 4K.3(a) this finishes the proof that $SP(A) \to SP(X) \to SP(X/A)$ is a quasifibration.

Since the homotopy axiom is obvious, this gives us the first two of the three axioms needed for the groups $h_i(X) = \pi_i SP(X)$ to define a reduced homology theory. There remains only the wedge sum axiom, $h_i(\bigvee_\alpha X_\alpha) \approx \bigoplus_\alpha h_i(X_\alpha)$, but this is immediate from the evident fact that $SP(\bigvee_\alpha X_\alpha) = \prod_\alpha SP(X_\alpha)$, where this is the 'weak' product, the union of the products of finitely many factors.

The homology theory $h_*(X)$ is defined on the category of connected, basepointed simplicial complexes, with basepoint-preserving maps. The coefficients of this homology theory, the groups $h_i(S^n)$, are the same as for ordinary homology with $\mathbb{Z}$ coefficients since we know this is true for $n = 2$ by the homeomorphism $SP(S^2) \approx \mathbb{C}P^\infty$, and there are isomorphisms $h_i(X) \approx h_{i+1}(\Sigma X)$ in any reduced homology theory. If the homology theory $h_*(X)$ were defined on the category of all simplicial complexes, without basepoints, then Theorem 4.59 would give natural isomorphisms $h_i(X) \approx H_i(X; \mathbb{Z})$ for all X, and the proof would be complete. However, it is easy to achieve this by defining a new homology theory $h_i'(X) = h_{i+1}(\Sigma X)$, since the suspension of an arbitrary complex is connected and the suspension of an arbitrary map is basepoint-preserving, taking the basepoint to be one of the suspension points. Since $h_i'(X)$ is naturally isomorphic to $h_i(X)$ if X is connected, we are done. $\quad\square$

It is worth noting that the map $\pi_i(X) \to \pi_i SP(X) = H_i(X; \mathbb{Z})$ induced by the inclusion $X = SP_1(X) \hookrightarrow SP(X)$ is the Hurewicz homomorphism. For by definition of the Hurewicz homomorphism and naturality this reduces to the case $X = S^i$, where the map $SP_1(S^i) \hookrightarrow SP(S^i)$ induces on π_i a homomorphism $\mathbb{Z} \to \mathbb{Z}$, which one just has to check is an isomorphism, the Hurewicz homomorphism being determined only up to sign. The suspension isomorphism gives a further reduction to the case $i = 1$, where the inclusion $SP_1(S^1) \hookrightarrow SP(S^1)$ is a homotopy equivalence, hence induces an isomorphism on π_1.

Exercises

1. Show that Corollary 4K.2 remains valid when X and Y are CW complexes and the subspaces U_i and V_i are subcomplexes rather than open sets.

2. Show that a simplicial map $f : K \to L$ is a homotopy equivalence if $f^{-1}(x)$ is contractible for all $x \in L$. [Consider the cover of L by open stars of simplices and the cover of K by the preimages of these open stars.]

3. Show that $SP_n(I) = \Delta^n$.

4. Show that $SP_2(S^1)$ is a Möbius band, and that this is consistent with the description of $SP_2(S^n)$ as a mapping cone given in Example 4K.5.

5. A map $p : E \to B$ with B not necessarily path-connected is defined to be a quasifibration if the following equivalent conditions are satisfied:

(i) For all $b \in B$ and $x_0 \in p^{-1}(b)$, the map $p_* : \pi_i(E, p^{-1}(b), x_0) \to \pi_i(B, b)$ is an
isomorphism for $i > 0$ and $\pi_0(p^{-1}(b), x_0) \to \pi_0(E, x_0) \to \pi_0(B, b)$ is exact.

(ii) The inclusion of the fiber $p^{-1}(b)$ into the homotopy fiber F_b of p over b is a
weak homotopy equivalence for all $b \in B$.

(iii) The restriction of p over each path-component of B is a quasifibration according
to the definition in this section.

Show these three conditions are equivalent, and prove Lemma 4K.3 for quasifibrations
over non-pathconnected base spaces.

6. Let X be a complex of spaces over a Δ-complex Γ, as defined in §4.G. Show that
the natural projection $\Delta X \to \Gamma$ is a quasifibration if all the maps in X associated to
edges of Γ are weak homotopy equivalences.

4.L Steenrod Squares and Powers

The main objects of study in this section are certain homomorphisms called **Steenrod squares** and **Steenrod powers**:

$$Sq^i : H^n(X; \mathbb{Z}_2) \to H^{n+i}(X; \mathbb{Z}_2)$$
$$P^i : H^n(X; \mathbb{Z}_p) \to H^{n+2i(p-1)}(X; \mathbb{Z}_p) \quad \text{for odd primes } p$$

The terms 'squares' and 'powers' arise from the fact that Sq^i and P^i are related to
the maps $\alpha \mapsto \alpha^2$ and $\alpha \mapsto \alpha^p$ sending a cohomology class α to the 2-fold or p-fold
cup product with itself. Unlike cup products, however, the operations Sq^i and P^i are
stable, that is, invariant under suspension.

The operations Sq^i generate an algebra $\mathcal{A}_2$, called the Steenrod algebra, such that
$H^*(X; \mathbb{Z}_2)$ is a module over $\mathcal{A}_2$ for every space X, and maps between spaces induce
homomorphisms of $\mathcal{A}_2$-modules. Similarly, for odd primes p, $H^*(X; \mathbb{Z}_p)$ is a module
over a corresponding Steenrod algebra $\mathcal{A}_p$ generated by the P^i's and Bockstein homomorphisms. Like the ring structure given by cup product, these module structures
impose strong constraints on spaces and maps. For example, we will use them to
show that there do not exist spaces X with $H^*(X; \mathbb{Z})$ a polynomial ring $\mathbb{Z}[\alpha]$ unless
α has dimension 2 or 4, where there are the familiar examples of $\mathbb{C}P^\infty$ and $\mathbb{H}P^\infty$.

This rather lengthy section is divided into two main parts. The first part describes
the basic properties of Steenrod squares and powers and gives a number of examples
and applications. The second part is devoted to constructing the squares and powers
and showing they satisfy the basic properties listed in the first part. More extensive

applications will be given in [SSAT] after spectral sequences have been introduced. Most applications of Steenrod squares and powers do not depend on how these operations are actually constructed, but only on their basic properties. This is similar to the situation for ordinary homology and cohomology, where the axioms generally suffice for most applications. The construction of Steenrod squares and powers and the verification of their basic properties, or axioms, is rather interesting in its own way, but does involve a certain amount of work, particularly for the Steenrod powers, and this is why we delay the work until later in the section.

We begin with a few generalities. A **cohomology operation** is a transformation $\Theta = \Theta_X : H^m(X;G) \to H^n(X;H)$ defined for all spaces X, with a fixed choice of m, n, G, and H, and satisfying the naturality property that for all maps $f : X \to Y$ there is a commuting diagram as shown at the right. For example, with coefficients in a ring R the transformation $H^m(X;R) \to H^{mp}(X;R)$,

$$\begin{array}{ccc} H^m(Y;G) & \xrightarrow{\;\Theta_Y\;} & H^n(Y;H) \\ \downarrow{f^*} & & \downarrow{f^*} \\ H^m(X;G) & \xrightarrow{\;\Theta_X\;} & H^n(X;H) \end{array}$$

$\alpha \mapsto \alpha^p$, is a cohomology operation since $f^*(\alpha^p) = (f^*(\alpha))^p$. Taking $R = \mathbb{Z}$, this example shows that cohomology operations need not be homomorphisms. On the other hand, when $R = \mathbb{Z}_p$ with p prime, the operation $\alpha \mapsto \alpha^p$ is a homomorphism. Other examples of cohomology operations we have already encountered are the Bockstein homomorphisms defined in §3.E. As a more trivial example, a homomorphism $G \to H$ induces change-of-coefficient homomorphisms $H^m(X;G) \to H^m(X;H)$ which can be viewed as cohomology operations.

In spite of their rather general definition, cohomology operations can be described in somewhat more concrete terms:

Proposition 4L.1. *For fixed m, n, G, and H there is a bijection between the set of all cohomology operations $\Theta : H^m(X;G) \to H^n(X;H)$ and $H^n(K(G,m);H)$, defined by $\Theta \mapsto \Theta(\iota)$ where $\iota \in H^m(K(G,m);G)$ is a fundamental class.*

Proof: Via CW approximations to spaces, it suffices to restrict attention to CW complexes, so we can identify $H^m(X;G)$ with $\langle X, K(G,m) \rangle$ when $m > 0$ by Theorem 4.57, and with $[X, K(G,0)]$ when $m = 0$. If an element $\alpha \in H^m(X;G)$ corresponds to a map $\varphi : X \to K(G,m)$, so $\varphi^*(\iota) = \alpha$, then $\Theta(\alpha) = \Theta(\varphi^*(\iota)) = \varphi^*(\Theta(\iota))$ and Θ is uniquely determined by $\Theta(\iota)$. Thus $\Theta \mapsto \Theta(\iota)$ is injective. For surjectivity, given an element $\alpha \in H^n(K(G,m);H)$ corresponding to a map $\theta : K(G,m) \to K(H,n)$, then composing with θ defines a transformation $\langle X, K(G,m) \rangle \to \langle X, K(H,n) \rangle$, that is, $\Theta : H^m(X;G) \to H^n(X;H)$, with $\Theta(\iota) = \alpha$. The naturality property for Θ amounts to associativity of the compositions $X \xrightarrow{f} Y \xrightarrow{\varphi} K(G,m) \xrightarrow{\theta} K(H,n)$ and so Θ is a cohomology operation. $\qquad\square$

A consequence of the proposition is that cohomology operations that decrease dimension are all rather trivial since $K(G,m)$ is $(m-1)$-connected. Moreover, since

$H^m(K(G,m);H) \approx \mathrm{Hom}(G,H)$, it follows that the only cohomology operations that preserve dimension are given by coefficient homomorphisms.

The Steenrod squares $Sq^i:H^n(X;\mathbb{Z}_2) \to H^{n+i}(X;\mathbb{Z}_2)$, $i \geq 0$, will satisfy the following list of properties, beginning with naturality:

(1) $Sq^i(f^*(\alpha)) = f^*(Sq^i(\alpha))$ for $f:X \to Y$.

(2) $Sq^i(\alpha + \beta) = Sq^i(\alpha) + Sq^i(\beta)$.

(3) $Sq^i(\alpha \smile \beta) = \sum_j Sq^j(\alpha) \smile Sq^{i-j}(\beta)$ (the Cartan formula).

(4) $Sq^i(\sigma(\alpha)) = \sigma(Sq^i(\alpha))$ where $\sigma:H^n(X;\mathbb{Z}_2) \to H^{n+1}(\Sigma X;\mathbb{Z}_2)$ is the suspension isomorphism given by reduced cross product with a generator of $H^1(S^1;\mathbb{Z}_2)$.

(5) $Sq^i(\alpha) = \alpha^2$ if $i = |\alpha|$, and $Sq^i(\alpha) = 0$ if $i > |\alpha|$.

(6) $Sq^0 = \mathbb{1}$, the identity.

(7) Sq^1 is the $\mathbb{Z}_2$ Bockstein homomorphism β associated with the coefficient sequence $0 \to \mathbb{Z}_2 \to \mathbb{Z}_4 \to \mathbb{Z}_2 \to 0$.

The first part of (5) says that the Steenrod squares extend the squaring operation $\alpha \mapsto \alpha^2$, which has the nice feature of being a homomorphism with $\mathbb{Z}_2$ coefficients. Property (4) says that the Sq^i's are stable operations, invariant under suspension. The actual squaring operation $\alpha \mapsto \alpha^2$ does not have this property since in a suspension ΣX all cup products of positive-dimensional classes are zero, according to an exercise for §3.2.

The fact that Steenrod squares are stable operations extending the cup product square yields the following theorem, which implies that the stable homotopy groups of spheres π_1^s, π_3^s, and π_7^s are nontrivial:

Theorem 4L.2. *If $f:S^{2n-1} \to S^n$ has Hopf invariant 1, then $[f] \in \pi_{n-1}^s$ is nonzero, so the iterated suspensions $\Sigma^k f : S^{2n+k-1} \to S^{n+k}$ are all homotopically nontrivial.*

Proof: Associated to a map $f:S^\ell \to S^m$ is the mapping cone $C_f = S^m \cup_f e^{\ell+1}$ with the cell $e^{\ell+1}$ attached via f. Assuming f is basepoint-preserving, we have the relation $C_{\Sigma f} = \Sigma C_f$ where Σ denotes reduced suspension.

If $f:S^{2n-1} \to S^n$ has Hopf invariant 1, then by (5), $Sq^n:H^n(C_f;\mathbb{Z}_2) \to H^{2n}(C_f;\mathbb{Z}_2)$ is nontrivial. By (4) the same is true for $Sq^n:H^{n+k}(\Sigma^k C_f;\mathbb{Z}_2) \to H^{2n+k}(\Sigma^k C_f;\mathbb{Z}_2)$ for all k. If $\Sigma^k f$ were homotopically trivial we would have a retraction $r:\Sigma^k C_f \to S^{n+k}$. The diagram at the right would then commute by naturality of Sq^n, but since the group in the lower left corner of the diagram is zero, this gives a contradiction.   □

$$H^{n+k}(S^{n+k};\mathbb{Z}_2) \xrightarrow[\approx]{r^*} H^{n+k}(\Sigma^k C_f;\mathbb{Z}_2)$$
$$\downarrow Sq^n \qquad\qquad\qquad \downarrow Sq^n$$
$$H^{2n+k}(S^{n+k};\mathbb{Z}_2) \xrightarrow{r^*} H^{2n+k}(\Sigma^k C_f;\mathbb{Z}_2)$$

The Steenrod power operations $P^i:H^n(X;\mathbb{Z}_p) \to H^{n+2i(p-1)}(X;\mathbb{Z}_p)$ for p an odd prime will satisfy analogous properties:

(1) $P^i(f^*(\alpha)) = f^*(P^i(\alpha))$ for $f:X \to Y$.

(2) $P^i(\alpha + \beta) = P^i(\alpha) + P^i(\beta)$.

(3) $P^i(\alpha \smile \beta) = \sum_j P^j(\alpha) \smile P^{i-j}(\beta)$ (the Cartan formula).

(4) $P^i(\sigma(\alpha)) = \sigma(P^i(\alpha))$ where $\sigma : H^n(X; \mathbb{Z}_p) \to H^{n+1}(\Sigma X; \mathbb{Z}_p)$ is the suspension isomorphism given by reduced cross product with a generator of $H^1(S^1; \mathbb{Z}_p)$.

(5) $P^i(\alpha) = \alpha^p$ if $2i = |\alpha|$, and $P^i(\alpha) = 0$ if $2i > |\alpha|$.

(6) $P^0 = \mathbb{1}$, the identity.

The germinal property $P^i(\alpha) = \alpha^p$ in (5) can only be expected to hold for even-dimensional classes α since for odd-dimensional α the commutativity property of cup product implies that $\alpha^2 = 0$ with $\mathbb{Z}_p$ coefficients if p is odd, and then $\alpha^p = 0$ since $\alpha^2 = 0$. Note that the formula $P^i(\alpha) = \alpha^p$ for $|\alpha| = 2i$ implies that P^i raises dimension by $2i(p-1)$, explaining the appearance of this number.

The Bockstein homomorphism $\beta : H^n(X; \mathbb{Z}_p) \to H^{n+1}(X; \mathbb{Z}_p)$ is not included as one of the P^i's, but this is mainly a matter of notational convenience. As we shall see later when we discuss Adem relations, the operation Sq^{2i+1} is the same as the composition $Sq^1 Sq^{2i} = \beta Sq^{2i}$, so the Sq^{2i}'s can be regarded as the P^i's for $p = 2$.

One might ask if there are elements of π_*^s detectable by Steenrod powers in the same way that the Hopf maps are detected by Steenrod squares. The answer is yes for the operation P^1, as we show in Example 4L.6. It is a perhaps disappointing fact that no other squares or powers besides Sq^1, Sq^2, Sq^4, Sq^8, and P^1 detect elements of homotopy groups of spheres. (Sq^1 detects a map $S^n \to S^n$ of degree 2.) We will prove this for certain Sq^i's and P^i's later in this section. The general case for $p = 2$ is Adams' theorem on the Hopf invariant discussed in §4.B, while the case of odd p is proved in [Adams & Atiyah 1966]; see also [VBKT].

The Cartan formulas can be expressed in a more concise form by defining *total* Steenrod square and power operations by $Sq = Sq^0 + Sq^1 + \cdots$ and $P = P^0 + P^1 + \cdots$. These act on $H^*(X; \mathbb{Z}_p)$ since by property (5), only a finite number of Sq^i's or P^i's can be nonzero on a given cohomology class. The Cartan formulas then say that $Sq(\alpha \smile \beta) = Sq(\alpha) \smile Sq(\beta)$ and $P(\alpha \smile \beta) = P(\alpha) \smile P(\beta)$, so Sq and P are ring homomorphisms.

We can use Sq and P to compute the operations Sq^i and P^i for projective spaces and lens spaces via the following general formulas:

$$(*) \quad \begin{aligned} Sq^i(\alpha^n) &= \binom{n}{i} \alpha^{n+i} \text{ for } \alpha \in H^1(X; \mathbb{Z}_2) \\ P^i(\alpha^n) &= \binom{n}{i} \alpha^{n+i(p-1)} \text{ for } \alpha \in H^2(X; \mathbb{Z}_p) \end{aligned}$$

To derive the first formula, properties (5) and (6) give $Sq(\alpha) = \alpha + \alpha^2 = \alpha(1 + \alpha)$, so $Sq(\alpha^n) = Sq(\alpha)^n = \alpha^n(1 + \alpha)^n = \sum_i \binom{n}{i} \alpha^{n+i}$ and hence $Sq^i(\alpha^n) = \binom{n}{i} \alpha^{n+i}$. The second formula is obtained in similar fashion: $P(\alpha) = \alpha + \alpha^p = \alpha(1 + \alpha^{p-1})$ so $P(\alpha^n) = \alpha^n(1 + \alpha^{p-1})^n = \sum_i \binom{n}{i} \alpha^{n+i(p-1)}$.

In Lemma 3C.6 we described how binomial coefficients can be computed modulo a prime p:

$$\binom{m}{n} \equiv \prod_i \binom{m_i}{n_i} \bmod p, \qquad \text{where } m = \sum_i m_i p^i \text{ and } n = \sum_i n_i p^i \text{ are the}$$
$$p\text{-adic expansions of } m \text{ and } n.$$

When $p = 2$ for example, the extreme cases of a dyadic expansion consisting of a single 1 or all 1's give

$$Sq(\alpha^{2^k}) = \alpha^{2^k} + \alpha^{2^{k+1}}$$
$$Sq(\alpha^{2^k-1}) = \alpha^{2^k-1} + \alpha^{2^k} + \alpha^{2^{k+1}} + \cdots + \alpha^{2^{k+1}-2}$$

for $\alpha \in H^1(X;\mathbb{Z}_2)$. More generally, the coefficients of $Sq(\alpha^n)$ can be read off from the $(n+1)$st row of the mod 2 Pascal triangle, a portion of which is shown in the figure at the right, where dots denote zeros.

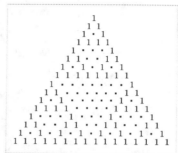

Example 4L.3: Stable Splittings. The formula $(*)$ tells us how to compute Steenrod squares for $\mathbb{R}P^\infty$, hence also for any suspension of $\mathbb{R}P^\infty$. The explicit formulas for $Sq(\alpha^{2^k})$ and $Sq(\alpha^{2^k-1})$ above show that all the powers of the generator $\alpha \in H^1(\mathbb{R}P^\infty;\mathbb{Z}_2)$ are tied together by Steenrod squares since the first formula connects α inductively to all the powers α^{2^k} and the second formula connects these powers to all the other powers. This shows that no suspension $\Sigma^k \mathbb{R}P^\infty$ has the homotopy type of a wedge sum $X \vee Y$ with both X and Y having nontrivial cohomology. In the case of $\mathbb{R}P^\infty$ itself we could have deduced this from the ring structure of $H^*(\mathbb{R}P^\infty;\mathbb{Z}_2) \approx \mathbb{Z}_2[\alpha]$, but cup products become trivial in a suspension.

The same reasoning shows that $\mathbb{C}P^\infty$ and $\mathbb{H}P^\infty$ have no nontrivial stable splittings. The $\mathbb{Z}_2$ cohomology in these cases is again $\mathbb{Z}_2[\alpha]$, though with α no longer 1-dimensional. However, we still have $Sq(\alpha) = \alpha + \alpha^2$ since these spaces have no nontrivial cohomology in the dimensions between α and α^2, so we have $Sq^{2i}(\alpha^n) = \binom{n}{i}\alpha^{n+i}$ for $\mathbb{C}P^\infty$ and $Sq^{4i}(\alpha^n) = \binom{n}{i}\alpha^{n+i}$ for $\mathbb{H}P^\infty$. Then the arguments from the real case carry over using the operations Sq^{2i} and Sq^{4i} in place of Sq^i.

Suppose we consider the same question for $K(\mathbb{Z}_3, 1)$ instead of $\mathbb{R}P^\infty$. Taking cohomology with $\mathbb{Z}_3$ coefficients, the Bockstein β is nonzero on odd-dimensional classes in $H^*(K(\mathbb{Z}_3,1);\mathbb{Z}_3)$, thus tying them to the even-dimensional classes, so we only need to see which even-dimensional classes are connected by P^i's. The even-dimensional part of $H^*(K(\mathbb{Z}_3,1);\mathbb{Z}_3)$ is a polynomial algebra $\mathbb{Z}_3[\alpha]$ with $|\alpha| = 2$, so we have $P^i(\alpha^n) = \binom{n}{i}\alpha^{n+i(p-1)} = \binom{n}{i}\alpha^{n+2i}$ by our earlier formula. Since P^i raises dimension by $4i$ when $p = 3$, there is no chance that all the even-dimensional cohomology will be connected by the P^i's. In fact, we showed in Proposition 4I.3 that $\Sigma K(\mathbb{Z}_3, 1) \simeq X_1 \vee X_2$ where X_1 has the cohomology of $\Sigma K(\mathbb{Z}_3, 1)$ in dimensions congruent to 2 and 3 mod 4, while X_2 has the remaining cohomology. Thus the best one could hope would be that all the odd powers of α are connected by P^i's and likewise all the even powers are connected, since this would imply that neither X_1 nor X_2 splits

nontrivially. This is indeed the case, as one sees by an exam-
ination of the coefficients in the formula $P^i(\alpha^n) = \binom{n}{i}\alpha^{n+2i}$.
In the Pascal triangle mod 3, shown at the right, $P(\alpha^n)$ is determined by the $(n+1)$st row. For example the sixth row
says that $P(\alpha^5) = \alpha^5 + 2\alpha^7 + \alpha^9 + \alpha^{11} + 2\alpha^{13} + \alpha^{15}$. A few
moments' thought shows that the rows that compute $P(\alpha^n)$

```
            1
           1 1
          1 2 1
         1 . . 1
        1 1 . 1 1
       1 2 1 1 2 1
      1 . . 2 . . 1
     1 1 . 2 2 . 1 1
    1 2 1 2 1 2 1 2 1
   1 . . . . . . . . 1
```

for $n = 3^k m - 1$ have all nonzero entries, and these rows together with the rows right
after them suffice to connect the powers of α in the desired way, so X_1 and X_2 have
no stable splittings. One can also see that $\Sigma^2 X_1$ and X_2 are not homotopy equivalent,
even stably, since the operations P^i act differently in the two spaces. For example P^2
is trivial on suspensions of α but not on suspensions of α^2.

The situation for $K(\mathbb{Z}_p, 1)$ for larger primes p is entirely similar, with $\Sigma K(\mathbb{Z}_p, 1)$
splitting as a wedge sum of $p - 1$ spaces. The same arguments work more generally
for $K(\mathbb{Z}_{p^i}, 1)$, though for $i > 1$ the usual Bockstein β is identically zero so one has
to use instead a Bockstein involving $\mathbb{Z}_{p^i}$ coefficients. We leave the details of these
arguments as exercises.

Example 4L.4: Maps of $\mathbb{HP}^\infty$. We can use the operations P^i together with a bit
of number theory to demonstrate an interesting distinction between $\mathbb{HP}^\infty$ and $\mathbb{CP}^\infty$,
namely, we will show that if a map $f : \mathbb{HP}^\infty \to \mathbb{HP}^\infty$ has $f^*(y) = dy$ for y a generator
of $H^4(\mathbb{HP}^\infty; \mathbb{Z})$, then the integer d, which we call the *degree* of f, must be a square.
By contrast, since $\mathbb{CP}^\infty$ is a $K(\mathbb{Z}, 2)$, there are maps $\mathbb{CP}^\infty \to \mathbb{CP}^\infty$ carrying a generator
$\alpha \in H^2(\mathbb{CP}^\infty; \mathbb{Z})$ onto any given multiple of itself. Explicitly, the map $z \mapsto z^d$, $z \in \mathbb{C}$,
induces a map f of $\mathbb{CP}^\infty$ with $f^*(\alpha) = d\alpha$, but commutativity of $\mathbb{C}$ is needed for this
construction so it does not extend to the quaternionic case.

We shall deduce the action of Steenrod powers on $H^*(\mathbb{HP}^\infty; \mathbb{Z}_p)$ from their action on $H^*(\mathbb{CP}^\infty; \mathbb{Z}_p)$, given by the earlier formula $(*)$ which says that $P^i(\alpha^n) =$
$\binom{n}{i}\alpha^{n+i(p-1)}$ for α a generator of $H^2(\mathbb{CP}^\infty; \mathbb{Z}_p)$. There is a natural quotient map
$\mathbb{CP}^\infty \to \mathbb{HP}^\infty$ arising from the definition of both spaces as quotients of S^∞. This map
takes the 4-cell of $\mathbb{CP}^\infty$ homeomorphically onto the 4-cell of $\mathbb{HP}^\infty$, so the induced
map on cohomology sends a generator $y \in H^4(\mathbb{HP}^\infty; \mathbb{Z}_p)$ to α^2, hence y^n to α^{2n}.
Thus the formula $P^i(\alpha^{2n}) = \binom{2n}{i}\alpha^{2n+i(p-1)}$ implies that $P^i(y^n) = \binom{2n}{i}y^{n+i(p-1)/2}$.
For example, $P^1(y) = 2y^{(p+1)/2}$.

Now let $f : \mathbb{HP}^\infty \to \mathbb{HP}^\infty$ be any map. Applying the formula $P^1(y) = 2y^{(p+1)/2}$ in
two ways, we get

$$P^1 f^*(y) = f^* P^1(y) = f^*(2y^{(p+1)/2}) = 2d^{(p+1)/2}y^{(p+1)/2}$$

$$\text{and} \quad P^1 f^*(y) = P^1(dy) = 2dy^{(p+1)/2}$$

Hence the degree d satisfies $d^{(p+1)/2} \equiv d \bmod p$ for all odd primes p. Thus either
$d \equiv 0 \bmod p$ or $d^{(p-1)/2} \equiv 1 \bmod p$. In both cases d is a square mod p since the

congruence $d^{(p-1)/2} \equiv 1 \bmod p$ is equivalent to d being a nonzero square mod p, the multiplicative group of nonzero elements of the field $\mathbb{Z}_p$ being cyclic of order $p - 1$.

The argument is completed by appealing to the number theory fact that an integer which is a square mod p for all sufficiently large primes p must be a square. This can be deduced from quadratic reciprocity and Dirichlet's theorem on primes in arithmetic progressions as follows. Suppose on the contrary that the result is false for the integer d. Consider primes p not dividing d. Since the product of two squares in $\mathbb{Z}_p$ is again a square, we may assume that d is a product of distinct primes $q_1, \cdots, q_n$, where one of these primes is allowed to be -1 if d is negative. In terms of the Legendre symbol $\left(\frac{d}{p}\right)$ which is defined to be $+1$ if d is a square mod p and -1 otherwise, we have

$$\left(\frac{d}{p}\right) = \left(\frac{q_1}{p}\right) \cdots \left(\frac{q_n}{p}\right)$$

The left side is $+1$ for all large p by hypothesis, so it will suffice to see that p can be chosen to give each term on the right an arbitrary preassigned value. The values of $\left(\frac{-1}{p}\right)$ and $\left(\frac{2}{p}\right)$ depend only on p mod 8, and the four combinations of values are realized by the four residues $1, 3, 5, 7$ mod 8. Having specified the value of p mod 8, the quadratic reciprocity law then says that for odd primes q, specifying $\left(\frac{q}{p}\right)$ is equivalent to specifying $\left(\frac{p}{q}\right)$. Thus we need only choose p in the appropriate residue classes mod 8 and mod q_i for each odd q_i. By the Chinese remainder theorem, this means specifying p modulo 8 times a product of odd primes. Dirichlet's theorem guarantees that in fact infinitely many primes p exist satisfying this congruence condition.

It is known that the integers realizable as degrees of maps $\mathbb{H}P^\infty \to \mathbb{H}P^\infty$ are exactly the odd squares and zero. The construction of maps of odd square degree will be given in [SSAT] using localization techniques, following [Sullivan 1974]. Ruling out nonzero even squares can be done using K–theory; see [Feder & Gitler 1978], which also treats maps $\mathbb{H}P^n \to \mathbb{H}P^n$.

The preceding calculations can also be used to show that every map $\mathbb{H}P^n \to \mathbb{H}P^n$ must have a fixed point if $n > 1$. For, taking $p = 3$, the element $P^1(y)$ lies in $H^8(\mathbb{H}P^n; \mathbb{Z}_3)$ which is nonzero if $n > 1$, so, when the earlier argument is specialized to the case $p = 3$, the congruence $d^{(p+1)/2} \equiv d \bmod p$ becomes $d^2 = d$ in $\mathbb{Z}_3$, which is satisfied only by 0 and 1 in $\mathbb{Z}_3$. In particular, d is not equal to -1. The Lefschetz number $\lambda(f) = 1 + d + \cdots + d^n = (d^{n+1} - 1)/(d - 1)$ is therefore nonzero since the only integer roots of unity are ± 1. The Lefschetz fixed point theorem then gives the result.

Example 4L.5: Vector Fields on Spheres. Let us now apply Steenrod squares to determine the maximum number of orthonormal tangent vector fields on a sphere in all cases except when the dimension of the sphere is congruent to -1 mod 16. The first step is to rephrase the question in terms of Stiefel manifolds. Recall from the end of §3.D and Example 4.53 the space $V_{n,k}$ of orthonormal k-frames in $\mathbb{R}^n$. Projection of a k-frame onto its first vector gives a map $p : V_{n,k} \to S^{n-1}$, and a section for this projection, that is, a map $f : S^{n-1} \to V_{n,k}$ such that $pf = \mathbb{1}$, is exactly a set of $k - 1$ orthonormal tangent vector fields $v_1, \cdots, v_{k-1}$ on S^{n-1} since f assigns to each $x \in S^{n-1}$ an orthonormal k-frame $(x, v_1(x), \cdots, v_{k-1}(x))$.

We described a cell structure on $V_{n,k}$ at the end of §3.D, and we claim that the $(n - 1)$-skeleton of this cell structure is $\mathbb{RP}^{n-1} / \mathbb{RP}^{n-k-1}$ if $2k - 1 \le n$. The cells of $V_{n,k}$ were products $e^{i_1} \times \cdots \times e^{i_m}$ with $n > i_1 > \cdots > i_m \ge n - k$, so the products with a single factor account for all of the $(2n - 2k)$-skeleton, hence they account for all of the $(n - 1)$-skeleton if $n - 1 \le 2n - 2k$, that is, if $2k - 1 \le n$. The cells that are products with a single factor are the homeomorphic images of cells of $\mathbb{RP}^{n-1}$ under a map $\mathbb{RP}^{n-1} \to SO(n) \to SO(n)/SO(n - k) = V_{n,k}$. This map collapses $\mathbb{RP}^{n-k-1}$ to a point, so we get the desired conclusion that $\mathbb{RP}^{n-1} / \mathbb{RP}^{n-k-1}$ is the $(n - 1)$-skeleton of $V_{n,k}$ if $2k - 1 \le n$.

Now suppose we have $f : S^{n-1} \to V_{n,k}$ with $pf = \mathbb{1}$. In particular, f^* is surjective on $H^{n-1}(-; \mathbb{Z}_2)$. If we deform f to a cellular map, with image in the $(n - 1)$-skeleton, then by the preceding paragraph this will give a map $g : S^{n-1} \to \mathbb{RP}^{n-1} / \mathbb{RP}^{n-k-1}$ if $2k - 1 \le n$, and this map will still induce a surjection on $H^{n-1}(-; \mathbb{Z}_2)$, hence an isomorphism. If the number k happens to be such that $\binom{n-k}{k-1} \equiv 1 \bmod 2$, then by the earlier formula $(*)$ the operation

$$Sq^{k-1} : H^{n-k}(\mathbb{RP}^{n-1}/\mathbb{RP}^{n-k-1}; \mathbb{Z}_2) \to H^{n-1}(\mathbb{RP}^{n-1}/\mathbb{RP}^{n-k-1}; \mathbb{Z}_2)$$

will be nonzero, contradicting the existence of the map g since obviously the operation $Sq^{k-1} : H^{n-k}(S^{n-1}; \mathbb{Z}_2) \to H^{n-1}(S^{n-1}; \mathbb{Z}_2)$ is zero.

In order to guarantee that $\binom{n-k}{k-1} \equiv 1 \bmod 2$, write $n = 2^r(2s + 1)$ and choose $k = 2^r + 1$. Assume for the moment that $s \ge 1$. Then $\binom{n-k}{k-1} = \binom{2^{r+1}s-1}{2^r}$, and in view of the rule for computing binomial coefficients in $\mathbb{Z}_2$, this is nonzero since the dyadic expansion of $2^{r+1}s - 1$ ends with a string of 1's including a 1 in the single digit where the expansion of 2^r is nonzero. Note that the earlier condition $2k - 1 \le n$ is satisfied since it becomes $2^{r+1} + 1 \le 2^{r+1}s + 2^r$ and we assume $s \ge 1$.

Summarizing, we have shown that for $n = 2^r(2s + 1)$, the sphere S^{n-1} cannot have 2^r orthonormal tangent vector fields if $s \ge 1$. This is also trivially true for $s = 0$ since S^{n-1} cannot have n orthonormal tangent vector fields.

It is easy to see that this result is best possible when $r \le 3$ by explicitly constructing $2^r - 1$ orthonormal tangent vector fields on S^{n-1} when $n = 2^r m$. When $r = 1$, view S^{n-1} as the unit sphere in $\mathbb{C}^m$, and then $x \mapsto ix$ defines a tangent vector field since the unit complex numbers 1 and i are orthogonal and multiplication by a unit complex number is an isometry of $\mathbb{C}$, so x and ix are orthogonal in each coordinate of $\mathbb{C}^m$, hence are orthogonal. When $r = 2$ the same construction works with $\mathbb{H}$ in place of $\mathbb{C}$, using the maps $x \mapsto ix$, $x \mapsto jx$, and $x \mapsto kx$ to define three orthonormal tangent vector fields on the unit sphere in $\mathbb{H}^m$. When $r = 3$ we can follow the same procedure with the octonions, constructing seven orthonormal tangent vector fields to the unit sphere in $\mathbb{O}^m$ via an orthonormal basis $1, i, j, k, \cdots$ for $\mathbb{O}$.

The upper bound of $2^r - 1$ for the number of orthonormal vector fields on S^{n-1} is not best possible in the remaining case $n \equiv 0$ mod 16. The optimal upper bound is obtained instead using K–theory; see [VBKT] or [Husemoller 1966]. The construction of the requisite number of vector fields is again algebraic, this time using Clifford algebras.

Example 4L.6: A Map of mod p Hopf Invariant One. Let us describe a construction for a map $f : S^{2p} \to S^3$ such that in the mapping cone $C_f = S^3 \cup_f e^{2p+1}$, the first Steenrod power $P^1 : H^3(C_f; \mathbb{Z}_p) \to H^{2p+1}(C_f; \mathbb{Z}_p)$ is nonzero, hence f is nonzero in π^s_{2p-3}. The construction starts with the fact that a generator of $H^2(K(\mathbb{Z}_p, 1); \mathbb{Z}_p)$ has nontrivial p^{th} power, so the operation $P^1 : H^2(K(\mathbb{Z}_p, 1); \mathbb{Z}_p) \to H^{2p}(K(\mathbb{Z}_p, 1); \mathbb{Z}_p)$ is nontrivial by property (5). This remains true after we suspend to $\Sigma K(\mathbb{Z}_p, 1)$, and we showed in Proposition 4I.3 that $\Sigma K(\mathbb{Z}_p, 1)$ has the homotopy type of a wedge sum of CW complexes X_i, $1 \le i \le p - 1$, with $\widetilde{H}_*(X_i; \mathbb{Z})$ consisting only of a $\mathbb{Z}_p$ in each dimension congruent to $2i$ mod $2p - 2$. We are interested here in the space $X = X_1$, which has nontrivial $\mathbb{Z}_p$ cohomology in dimensions $2, 3, 2p, 2p + 1, \cdots$. Since X is, up to homotopy, a wedge summand of $\Sigma K(\mathbb{Z}_p, 1)$, the operation $P^1 : H^3(X; \mathbb{Z}_p) \to H^{2p+1}(X; \mathbb{Z}_p)$ is nontrivial. Since X is simply-connected, the construction in §4.C shows that we may take X to have $(2p + 1)$-skeleton of the form $S^2 \cup e^3 \cup e^{2p} \cup e^{2p+1}$. In fact, using the notion of homology decomposition in §4.H, we can take this skeleton to be the reduced mapping cone C_g of a map of Moore spaces $g : M(\mathbb{Z}_p, 2p - 1) \to M(\mathbb{Z}_p, 2)$. It follows that the quotient C_g/S^2 is the reduced mapping cone of the composition $h : M(\mathbb{Z}_p, 2p - 1) \xrightarrow{g} M(\mathbb{Z}_p, 2) \to M(\mathbb{Z}_p, 2)/S^2 = S^3$. The restriction $h \mid S^{2p-1}$ represents an element of $\pi_{2p-1}(S^3)$ that is either trivial or has order p since this restriction extends over the $2p$-cell of $M(\mathbb{Z}_p, 2p - 1)$ which is attached by a map $S^{2p-1} \to S^{2p-1}$ of degree p. In fact, $h \mid S^{2p-1}$ is nullhomotopic since, as we will see in [SSAT] using the Serre spectral sequence, $\pi_i(S^3)$ contains no elements of order p if $i \le 2p - 1$. This implies that the space $C_h = C_g/S^2$ is homotopy equivalent to a CW complex Y obtained from $S^3 \vee S^{2p}$ by attaching a cell e^{2p+1}. The quotient Y/S^{2p} then has the form $S^3 \cup e^{2p+1}$, so it is the mapping cone of a map $f : S^{2p} \to S^3$. By construction there is a map $C_g \to C_f$ inducing an isomorphism on $\mathbb{Z}_p$ cohomology in dimensions 3 and $2p + 1$, so the operation P^1 is nontrivial in $H^*(C_f; \mathbb{Z}_p)$ since this was true for C_g, the $(2p + 1)$-skeleton of X.

Example 4L.7: Moore Spaces. Let us use the operation Sq^2 to show that for $n \ge 2$, the identity map of $M(\mathbb{Z}_2, n)$ has order 4 in the group of basepoint-preserving homotopy classes of maps $M(\mathbb{Z}_2, n) \to M(\mathbb{Z}_2, n)$, with addition defined via the suspension structure on $M(\mathbb{Z}_2, n) = \Sigma M(\mathbb{Z}_2, n - 1)$. According to Proposition 4H.2, this group is the middle term of a short exact sequence, the remaining terms of which contain only elements of order 2. Hence if the identity map of $M(\mathbb{Z}_2, n)$ has order 4, this short exact sequence cannot split.

In view of the short exact sequence just referred to, it will suffice to show that twice the identity map of $M(\mathbb{Z}_2, n)$ is not nullhomotopic. If twice the identity were nullhomotopic, then the mapping cone C of this map would have the homotopy type of $M(\mathbb{Z}_2, n) \vee \Sigma M(\mathbb{Z}_2, n)$. This would force $Sq^2 : H^n(C; \mathbb{Z}_2) \to H^{n+2}(C; \mathbb{Z}_2)$ to be trivial since the source and target groups would come from different wedge summands. However, we will now show that this Sq^2 operation is nontrivial. Twice the identity map of $M(\mathbb{Z}_2, n)$ can be regarded as the smash product of the degree 2 map $S^1 \to S^1$, $z \mapsto z^2$, with the identity map of $M(\mathbb{Z}_2, n-1)$. If we smash the cofibration sequence $S^1 \to S^1 \to \mathbb{R}P^2$ for this degree 2 map with $M(\mathbb{Z}_2, n-1)$ we get the cofiber sequence $M(\mathbb{Z}_2, n) \to M(\mathbb{Z}_2, n) \to C$, in view of the identity $(X/A) \wedge Y = (X \wedge Y)/(A \wedge Y)$. This means we can view C as $\mathbb{R}P^2 \wedge M(\mathbb{Z}_2, n-1)$. The Cartan formula translated to cross products gives $Sq^2(\alpha \times \beta) = Sq^0 \alpha \times Sq^2 \beta + Sq^1 \alpha \times Sq^1 \beta + Sq^2 \alpha \times Sq^0 \beta$. This holds for smash products as well as ordinary products, by naturality. Taking α to be a generator of $H^1(\mathbb{R}P^2; \mathbb{Z}_2)$ and β a generator of $H^{n-1}(M(\mathbb{Z}_2, n-1); \mathbb{Z}_2)$, we have $Sq^2 \alpha = 0 = Sq^2 \beta$, but $Sq^1 \alpha$ and $Sq^1 \beta$ are nonzero since Sq^1 is the Bockstein. By the Künneth formula, $Sq^1 \alpha \times Sq^1 \beta$ then generates $H^{n+2}(\mathbb{R}P^2 \wedge M(\mathbb{Z}_2, n-1); \mathbb{Z}_2)$ and we are done.

Adem Relations and the Steenrod Algebra

When Steenrod squares or powers are composed, the compositions satisfy certain relations, unfortunately rather complicated, known as **Adem relations**:

$$Sq^a Sq^b = \sum_j \binom{b-j-1}{a-2j} Sq^{a+b-j} Sq^j \qquad \text{if } a < 2b$$

$$P^a P^b = \sum_j (-1)^{a+j} \binom{(p-1)(b-j)-1}{a-pj} P^{a+b-j} P^j \qquad \text{if } a < pb$$

$$P^a \beta P^b = \sum_j (-1)^{a+j} \binom{(p-1)(b-j)}{a-pj} \beta P^{a+b-j} P^j$$

$$- \sum_j (-1)^{a+j} \binom{(p-1)(b-j)-1}{a-pj-1} P^{a+b-j} \beta P^j \qquad \text{if } a \le pb$$

By convention, the binomial coefficient $\binom{m}{n}$ is taken to be zero if m or n is negative or if $m < n$. Also $\binom{m}{0} = 1$ for $m \ge 0$.

For example, taking $a = 1$ in the Adem relation for the Steenrod squares we have $Sq^1 Sq^b = (b-1)Sq^{b+1}$, so $Sq^1 Sq^{2i} = Sq^{2i+1}$ and $Sq^1 Sq^{2i+1} = 0$. The relations $Sq^1 Sq^{2i} = Sq^{2i+1}$ and $Sq^1 = \beta$ explain the earlier comment that Sq^{2i} is the analog of P^i for $p = 2$.

The **Steenrod algebra** $\mathcal{A}_2$ is defined to be the algebra over $\mathbb{Z}_2$ that is the quotient of the algebra of polynomials in the noncommuting variables $Sq^1, Sq^2, \cdots$ by the two-sided ideal generated by the Adem relations, that is, by the polynomials given by the differences between the left and right sides of the Adem relations. In similar fashion, $\mathcal{A}_p$ for odd p is defined to be the algebra over $\mathbb{Z}_p$ formed by polynomials in the noncommuting variables $\beta, P^1, P^2, \cdots$ modulo the Adem relations and the relation

$\beta^2 = 0$. Thus for every space X, $H^*(X;\mathbb{Z}_p)$ is a module over $\mathcal{A}_p$, for all primes p. The Steenrod algebra is a graded algebra, the elements of degree k being those that map $H^n(X;\mathbb{Z}_p)$ to $H^{n+k}(X;\mathbb{Z}_p)$ for all n.

The next proposition implies that $\mathcal{A}_2$ is generated as an algebra by the elements Sq^{2^k}, while $\mathcal{A}_p$ for p odd is generated by β and the elements P^{p^k}.

Proposition 4L.8. *There is a relation $Sq^i = \sum_{0<j<i} a_j Sq^{i-j} Sq^j$ with coefficients $a_j \in \mathbb{Z}_2$ whenever i is not a power of 2. Similarly, if i is not a power of p there is a relation $P^i = \sum_{0<j<i} a_j P^{i-j} P^j$ with $a_j \in \mathbb{Z}_p$.*

Proof: The argument is the same for $p = 2$ and p odd, so we describe the latter case. The idea is to write i as the sum $a + b$ of integers $a > 0$ and $b > 0$ with $a < pb$, such that the coefficient of the $j = 0$ term in the Adem relation for $P^a P^b$ is nonzero. Then one can solve this relation for $P^{a+b} = P^i$.

Let the p-adic representation of i be $i = i_0 + i_1 p + \cdots + i_k p^k$ with $i_k \neq 0$. Let $b = p^k$ and $a = i - p^k$, so $b > 0$ and $a > 0$ if i is not a power of p. The claim is that $\binom{(p-1)b-1}{a}$ is nonzero in $\mathbb{Z}_p$. The p-adic expansion of $(p-1)b - 1 = (p^{k+1} - 1) - p^k$ is $(p-1) + (p-1)p + \cdots + (p-2)p^k$, and the p-adic expansion of a is $i_0 + i_1 p + \cdots + (i_k - 1)p^k$. Hence $\binom{(p-1)b-1}{a} \equiv \binom{p-1}{i_0} \cdots \binom{p-2}{i_k-1}$ and in each factor of the latter product the numerator is nonzero in $\mathbb{Z}_p$ so the product is nonzero in $\mathbb{Z}_p$. When $p = 2$ the last factor is omitted, and the product is still nonzero in $\mathbb{Z}_2$. $\square$

This proposition says that most of the Sq^i's and P^i's are decomposable, where an element a of a graded algebra such as $\mathcal{A}_p$ is **decomposable** if it can be expressed in the form $\sum_i a_i b_i$ with each a_i and b_i having lower degree than a. The operation Sq^{2^k} is indecomposable since for α a generator of $H^1(\mathbb{R}P^\infty;\mathbb{Z}_2)$ we saw that $Sq^{2^k}(\alpha^{2^k}) = \alpha^{2^{k+1}}$ but $Sq^i(\alpha^{2^k}) = 0$ for $0 < i < 2^k$. Similarly P^{p^k} is indecomposable since if $\alpha \in H^2(\mathbb{C}P^\infty;\mathbb{Z}_p)$ is a generator then $P^{p^k}(\alpha^{p^k}) = \alpha^{p^{k+1}}$ but $P^i(\alpha^{p^k}) = 0$ for $0 < i < p^k$ and also $\beta(\alpha^{p^k}) = 0$.

Here is an application of the preceding proposition:

Theorem 4L.9. *Suppose $H^*(X;\mathbb{Z}_p)$ is the polynomial algebra $\mathbb{Z}_p[\alpha]$ on a generator α of dimension n, possibly truncated by the relation $\alpha^m = 0$ for $m > p$. Then if $p = 2$, n must be a power of 2, and if p is an odd prime, n must be of the form $p^k \ell$ where ℓ is an even divisor of $2(p - 1)$.*

As we mentioned in §3.2, there is a stronger theorem that n must be 1, 2, 4, or 8 when $p = 2$, and n must be an even divisor of $2(p - 1)$ when p is an odd prime. We also gave examples showing the necessity of the hypothesis $m > p$ in the case of a truncated polynomial algebra.

Proof: In the case $p = 2$, $Sq^n(\alpha) = \alpha^2 \neq 0$. If n is not a power of 2 then Sq^n decomposes into compositions $Sq^{n-j}Sq^j$ with $0 < j < n$. Such compositions must be zero since they pass through the group $H^{n+j}(X;\mathbb{Z}_2)$ which is zero for $0 < j < n$.

For odd p, the fact that α^2 is nonzero implies that n is even, say $n = 2k$. Then $P^k(\alpha) = \alpha^p \neq 0$. Since P^k can be expressed in terms of P^{p^i}'s, some P^{p^i} must be nonzero in $H^*(X;\mathbb{Z}_p)$. This implies that $2p^i(p-1)$, the amount by which P^{p^i} raises dimension, must be a multiple of n since $H^*(X;\mathbb{Z}_p)$ is concentrated in dimensions that are multiples of n. Since n divides $2p^i(p-1)$, it must be a power of p times a divisor of $2(p-1)$, and this divisor must be even since n is even and p is odd. $\square$

Corollary 4L.10. *If $H^*(X;\mathbb{Z})$ is a polynomial algebra $\mathbb{Z}[\alpha]$, possibly truncated by $\alpha^m = 0$ with $m > 3$, then $|\alpha| = 2$ or 4.*

Proof: Passing from $\mathbb{Z}$ to $\mathbb{Z}_2$ coefficients, the theorem implies that $|\alpha|$ is a power of 2, and taking $\mathbb{Z}_3$ coefficients we see that $|\alpha|$ is a power of 3 times a divisor of $2(3-1) = 4$. $\square$

In particular, the octonionic projective plane $\mathbb{O}P^2$, constructed in Example 4.47 by attaching a 16-cell to S^8 via the Hopf map $S^{15} \to S^8$, does not generalize to an octonionic projective n-space $\mathbb{O}P^n$ with $n \geq 3$.

In a similar vein, decomposability implies that if an element of π_*^s is detected by a Sq^i or P^i then i must be a power of 2 for Sq^i and a power of p for P^i. For if Sq^i is decomposable, then the map $Sq^i : H^n(C_f;\mathbb{Z}_2) \to H^{n+i}(C_f;\mathbb{Z}_2)$ must be trivial since it is a sum of compositions that pass through trivial cohomology groups, and similarly for P^i.

Interestingly enough, the Adem relations can also be used in a positive way to detect elements of π_*^s, as the proof of the following result will show.

Proposition 4L.11. *If $\eta \in \pi_1^s$ is represented by the Hopf map $S^3 \to S^2$, then η^2 is nonzero in π_2^s. Similarly, the other two Hopf maps represent elements $\nu \in \pi_3^s$ and $\sigma \in \pi_7^s$ whose squares are nontrivial in π_6^s and π_{14}^s.*

Proof: Let $\eta : S^{n+1} \to S^n$ be a suspension of the Hopf map, with mapping cone C_η obtained from S^n by attaching a cell e^{n+2} via η. If we assume the composition $(\Sigma\eta)\eta$ is nullhomotopic, then we can define a map $f : S^{n+3} \to C_\eta$ in the following way. Decompose S^{n+3} as the union of two cones CS^{n+2}. On one of these cones let f be a nullhomotopy of $(\Sigma\eta)\eta$. On the other cone let f be the composition $CS^{n+2} \to CS^{n+1} \to C_\eta$ where the first map is obtained by coning $\Sigma\eta$ and the second map is a characteristic map for the cell e^{n+2}.

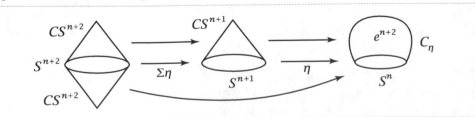

We use the map f to attach a cell e^{n+4} to C_n, forming a space X. This has C_n as its $(n+2)$-skeleton, so $Sq^2 : H^n(X; \mathbb{Z}_2) \to H^{n+2}(X; \mathbb{Z}_2)$ is an isomorphism. The map $Sq^2 : H^{n+2}(X; \mathbb{Z}_2) \to H^{n+4}(X; \mathbb{Z}_2)$ is also an isomorphism since the quotient map $X \to X/S^n$ induces an isomorphism on cohomology groups above dimension n and X/S^n is homotopy equivalent to the mapping cone of $\Sigma^2 \eta$. Thus the composition $Sq^2 Sq^2 : H^n(X; \mathbb{Z}_2) \to H^{n+4}(X; \mathbb{Z}_2)$ is an isomorphism. But this is impossible in view of the Adem relation $Sq^2 Sq^2 = Sq^3 Sq^1$, since Sq^1 is trivial on $H^n(X; \mathbb{Z}_2)$.

The same argument shows that ν^2 and σ^2 are nontrivial using the relations $Sq^4 Sq^4 = Sq^7 Sq^1 + Sq^6 Sq^2$ and $Sq^8 Sq^8 = Sq^{15} Sq^1 + Sq^{14} Sq^2 + Sq^{12} Sq^4$. $\square$

This line of reasoning does not work for odd primes and the element $\alpha \in \pi^s_{2p-3}$ detected by P^1 since the Adem relation for $P^1 P^1$ is $P^1 P^1 = 2P^2$, which is not helpful. And in fact $\alpha^2 = 0$ by the commutativity property of the product in π^s_*.

When dealing with $\mathcal{A}_2$ it is often convenient to abbreviate notation by writing a monomial $Sq^{i_1} Sq^{i_2} \cdots$ as Sq^I where I is the finite sequence of nonnegative integers $i_1, i_2, \cdots$. Call Sq^I **admissible** if no Adem relation can be applied to it, that is, if $i_j \geq 2i_{j+1}$ for all j. The Adem relations imply that every monomial Sq^I can be written as a sum of admissible monomials. For if Sq^I is not admissible, it contains a pair $Sq^a Sq^b$ to which an Adem relation can be applied, yielding a sum of terms Sq^J for which $J > I$ with respect to the lexicographic ordering on finite sequences of integers. These Sq^J's have the same degree $i_1 + \cdots + i_k$ as Sq^I, and since the number of monomials Sq^I of a fixed degree is finite, successive applications of the Adem relations eventually reduce any Sq^I to a sum of admissible monomials.

For odd p, elements of $\mathcal{A}_p$ are linear combinations of monomials $\beta^{\varepsilon_1} P^{i_1} \beta^{\varepsilon_2} P^{i_2} \cdots$ with each $\varepsilon_j = 0$ or 1. Such a monomial is **admissible** if $i_j \geq \varepsilon_{j+1} + p i_{j+1}$ for all j, which again means that no Adem relation can be applied to the monomial. As with $\mathcal{A}_2$, the Adem relations suffice to reduce every monomial to a linear combination of admissible monomials, by the same argument as before but now using the lexicographic ordering on tuples $(\varepsilon_1 + p i_1, \varepsilon_2 + p i_2, \cdots)$.

Define the **excess** of the admissible monomial Sq^I to be $\sum_j (i_j - 2i_{j+1})$, the amount by which Sq^I exceeds being admissible. For odd p one might expect the excess of an admissible monomial $\beta^{\varepsilon_1} P^{i_1} \beta^{\varepsilon_2} P^{i_2} \cdots$ to be defined as $\sum_j (i_j - p i_{j+1} - \varepsilon_{j+1})$, but instead it is defined to be $\sum_j (2i_j - 2p i_{j+1} - \varepsilon_{j+1})$, for reasons which will become clear below.

As we explained at the beginning of this section, cohomology operations correspond to elements in the cohomology of Eilenberg–MacLane spaces. Here is a rather important theorem which will be proved in [SSAT] since the proof makes heavy use of spectral sequences:

Theorem. *For each prime p, $H^*(K(\mathbb{Z}_p, n); \mathbb{Z}_p)$ is the free commutative algebra on the generators $\Theta(\iota_n)$ where $\iota_n \in H^n(K(\mathbb{Z}_p, n); \mathbb{Z}_p)$ is a generator and Θ ranges over all admissible monomials of excess less than n.*

Here 'free commutative algebra' means 'polynomial algebra' when $p = 2$ and 'polynomial algebra on even-dimensional generators tensor exterior algebra on odd-dimensional generators' when p is odd. We will say something about the rationale behind the 'excess less than n' condition in a moment.

Specializing the theorem to the first two cases $n = 1, 2$, we have the following cohomology algebras:

$$K(\mathbb{Z}_2, 1): \quad \mathbb{Z}_2[\iota]$$
$$K(\mathbb{Z}_p, 1): \quad \Lambda_{\mathbb{Z}_p}[\iota] \otimes \mathbb{Z}_p[\beta\iota]$$
$$K(\mathbb{Z}_2, 2): \quad \mathbb{Z}_2[\iota, Sq^1\iota, Sq^2Sq^1\iota, Sq^4Sq^2Sq^1\iota, \cdots]$$
$$K(\mathbb{Z}_p, 2): \quad \mathbb{Z}_p[\iota, \beta P^1\beta\iota, \beta P^p P^1\beta\iota, \beta P^{p^2} P^p P^1\beta\iota, \cdots]$$
$$\otimes \Lambda_{\mathbb{Z}_p}[\beta\iota, P^1\beta\iota, P^p P^1\beta\iota, P^{p^2} P^p P^1\beta\iota, \cdots]$$

The theorem implies that the admissible monomials in $\mathcal{A}_p$ are linearly independent, hence form a basis for $\mathcal{A}_p$ as a vector space over $\mathbb{Z}_p$. For if some linear combination of admissible monomials were zero, then it would be zero when applied to the class ι_n, but if we choose n larger than the excess of each monomial in the linear combination, this would contradict the freeness of the algebra $H^*(K(\mathbb{Z}_p, n); \mathbb{Z}_p)$. Even though the multiplicative structure of the Steenrod algebra is rather complicated, the Adem relations provide a way of performing calculations algorithmically by systematically reducing all products to sums of admissible monomials. A proof of the linear independence of admissible monomials using more elementary techniques can be found in [Steenrod & Epstein 1962].

Another consequence of the theorem is that all cohomology operations with $\mathbb{Z}_p$ coefficients are polynomials in the Sq^i's when $p = 2$ and polynomials in the P^i's and β when p is odd, in view of Proposition 4L.1. We can also conclude that $\mathcal{A}_p$ consists precisely of all the $\mathbb{Z}_p$ cohomology operations that are stable, commuting with suspension. For consider the map $\Sigma K(\mathbb{Z}_p, n) \to K(\mathbb{Z}_p, n + 1)$ that pulls ι_{n+1} back to the suspension of ι_n. This map induces an isomorphism on homotopy groups π_i for $i \le 2n$ and a surjection for $i = 2n + 1$ by Corollary 4.24, hence the same is true for homology and cohomology. Letting n go to infinity, the limit $\varprojlim \tilde{H}^*(K(\mathbb{Z}_p, n); \mathbb{Z}_p)$ then exists in a strong sense. On the one hand, this limit is exactly the stable operations by Proposition 4L.1 and the definition of a stable operation. On the other hand, the preceding theorem implies that this limit is $\mathcal{A}_p$ since it says that all elements of $H^*(K(\mathbb{Z}_p, n); \mathbb{Z}_p)$ below dimension $2n$ are uniquely expressible as sums of admissible monomials applied to ι_n.

Now let us explain why the condition 'excess less than n' in the theorem is natural. For a monomial $Sq^I = Sq^{i_1}Sq^{i_2} \cdots$ the definition of the excess $e(I)$ can be rewritten as an equation $i_1 = e(I) + i_2 + i_3 + \cdots$. Thus if $e(I) > n$, we have $i_1 > |Sq^{i_2} Sq^{i_3} \cdots (\iota_n)|$, hence $Sq^I(\iota_n) = 0$. And if $e(I) = n$ then $Sq^I(\iota_n) = (Sq^{i_2} Sq^{i_3} \cdots (\iota_n))^2$ and either $Sq^{i_2} Sq^{i_3} \cdots$ has excess less than n or it has excess equal to n and we can repeat the process to write $Sq^{i_2} Sq^{i_3} \cdots (\iota_n) = (Sq^{i_3} \cdots (\iota_n))^2$, and so on, until we obtain an equation $Sq^I(\iota_n) = (Sq^J(\iota_n))^{2^k}$ with $e(J) < n$, so that $Sq^I(\iota_n)$ is already in the algebra generated by the elements $Sq^J(\iota_n)$ with $e(J) < n$. The situation for odd p is similar. For an admissible monomial $P^I = \beta^{\varepsilon_1} P^{i_1} \beta^{\varepsilon_2} P^{i_2} \cdots$ the definition of excess gives $2i_1 = e(I) + \varepsilon_2 + 2(p-1)i_2 + \cdots$, so if $e(I) > n$ we must have $P^I(\iota_n) = 0$, and if $e(I) = n$ then either $P^I(\iota_n)$ is a power $(P^J(\iota_n))^{p^k}$ with $e(J) < n$, or, if P^I begins with β, then $P^I(\iota_n) = \beta((P^J(\iota_n))^{p^k}) = 0$ by the formula $\beta(x^m) = mx^{m-1}\beta(x)$, which is valid when $|x|$ is even, as we may assume is the case here, otherwise $(P^J(\iota_n))^{p^k} = 0$ by commutativity of cup product.

There is another set of relations among Steenrod squares equivalent to the Adem relations and somewhat easier to remember:

$$\sum_j \binom{k}{j} Sq^{2n-k+j-1} Sq^{n-j} = 0$$

When $k = 0$ this is simply the relation $Sq^{2n-1} Sq^n = 0$, and the cases $k > 0$ are obtained from this via Pascal's triangle. For example, from $Sq^7 Sq^4 = 0$ we obtain the following table of relations:

$$
\begin{array}{llllll}
Sq^7 Sq^4 & & & & & = 0 \\
Sq^6 Sq^4 + Sq^7 Sq^3 & & & & & = 0 \\
Sq^5 Sq^4 & + & Sq^7 Sq^2 & & & = 0 \\
Sq^4 Sq^4 + Sq^5 Sq^3 & + & Sq^6 Sq^2 & + & Sq^7 Sq^1 & = 0 \\
Sq^3 Sq^4 & & + & & Sq^7 Sq^0 & = 0 \\
Sq^2 Sq^4 + Sq^3 Sq^3 & & + & & Sq^6 Sq^0 & = 0 \\
Sq^1 Sq^4 & + & Sq^3 Sq^2 & + & Sq^5 Sq^0 & = 0 \\
Sq^0 Sq^4 + Sq^1 Sq^3 + Sq^2 Sq^2 + Sq^3 Sq^1 + Sq^4 Sq^0 & & & & & = 0
\end{array}
$$

These relations are not in simplest possible form. For example, $Sq^5 Sq^3 = 0$ in the fourth row and $Sq^3 Sq^2 = 0$ in the seventh row, instances of $Sq^{2n-1} Sq^n = 0$. For Steenrod powers there are similar relations $\sum_j \binom{k}{j} P^{pn-k+j-1} P^{n-j} = 0$ derived from the basic relation $P^{pn-1} P^n = 0$. We leave it to the interested reader to show that these relations follow from the Adem relations.

Constructing the Squares and Powers

Now we turn to the construction of the Steenrod squares and powers, and the proof of their basic properties including the Adem relations. As will be seen, this all hinges on the fact that cohomology is maps into Eilenberg-MacLane spaces. The case $p = 2$ is in some ways simpler than the case p odd, so in the first part of the development we will specialize p to 2 whenever there is a significant difference between the two cases.

Before giving the construction in detail, let us describe the idea in the case $p = 2$. The cup product square α^2 of an element $\alpha \in H^n(X; \mathbb{Z}_2)$ can be viewed as a composition $X \to X \times X \to K(\mathbb{Z}_2, 2n)$, with the first map the diagonal map and the second map representing the cross product $\alpha \times \alpha$. Since we have $\mathbb{Z}_2$ coefficients, cup product and cross product are strictly commutative, so if $T : X \times X \to X \times X$ is the map $T(x_1, x_2) = (x_2, x_1)$ transposing the two factors, then $T^*(\alpha \times \alpha) = \alpha \times \alpha$. Thinking of $\alpha \times \alpha$ as a map $X \times X \to K(\mathbb{Z}_2, 2n)$, this says there is a homotopy f_t from $\alpha \times \alpha$ to $(\alpha \times \alpha)T$. If we follow the homotopy f_t by the homotopy $f_t T$, we obtain a homotopy from $\alpha \times \alpha$ to $(\alpha \times \alpha)T$ and then to $(\alpha \times \alpha)T^2 = \alpha \times \alpha$, in other words a loop of maps $X \times X \to K(\mathbb{Z}_2, 2n)$. We can view this loop as a map $S^1 \times X \times X \to K(\mathbb{Z}_2, 2n)$. As we will see, if the homotopy f_t is chosen appropriately, the loop of maps will be null-homotopic, extending to a map $D^2 \times X \times X \to K(\mathbb{Z}_2, 2n)$. Regarding D^2 as the upper hemisphere of S^2, this gives half of a map $S^2 \times X \times X \to K(\mathbb{Z}_2, 2n)$, and once again we obtain the other half by composition with T. This process can in fact be repeated infinitely often to yield a map $S^\infty \times X \times X \to K(\mathbb{Z}_2, 2n)$ with the property that each pair of points (s, x_1, x_2) and $(-s, x_2, x_1)$ is sent to the same point in $K(\mathbb{Z}_2, 2n)$. This means that when we compose with the diagonal map $S^\infty \times X \to S^\infty \times X \times X$, $(s, x) \mapsto (s, x, x)$, there is an induced quotient map $\mathbb{R}P^\infty \times X \to K(\mathbb{Z}_2, 2n)$ extending $\alpha^2 : X \to K(\mathbb{Z}_2, 2n)$. This extended map represents a class in $H^{2n}(\mathbb{R}P^\infty \times X; \mathbb{Z}_2)$. By the Künneth formula and the fact that $H^*(\mathbb{R}P^\infty; \mathbb{Z}_2)$ is the polynomial ring $\mathbb{Z}_2[\omega]$, this cohomology class in $H^{2n}(\mathbb{R}P^\infty \times X; \mathbb{Z}_2)$ can be written in the form $\sum_i \omega^{n-i} \times a_i$ with $a_i \in H^{n+i}(X; \mathbb{Z}_2)$. Then we define $Sq^i(\alpha) = a_i$.

The construction of the map $S^\infty \times X \times X \to K(\mathbb{Z}_2, 2n)$ will proceed cell by cell, so it will be convenient to eliminate any unnecessary cells. This is done by replacing $X \times X$ by the smash product $X \wedge X$ and factoring out a cross-sectional slice S^∞ in $S^\infty \times X \wedge X$. A further simplification will be to use naturality to reduce to the case $X = K(\mathbb{Z}_2, n)$.

Now we begin the actual construction. For a space X with basepoint x_0, let $X^{\wedge p}$ denote the smash product $X \wedge \cdots \wedge X$ of p copies of X. There is a map $T : X^{\wedge p} \to X^{\wedge p}$, $T(x_1, \cdots, x_p) = (x_2, \cdots, x_p, x_1)$, permuting the factors cyclically. Note that when $p = 2$ this is just the transposition $(x_1, x_2) \mapsto (x_2, x_1)$. The map T generates an action of $\mathbb{Z}_p$ on $X^{\wedge p}$. There is also the standard action of $\mathbb{Z}_p$ on S^∞ viewed as the union of the unit spheres S^{2n-1} in $\mathbb{C}^n$, a generator of $\mathbb{Z}_p$ rotating each $\mathbb{C}$ factor through an angle $2\pi/p$, with quotient space an infinite-dimensional lens space L^∞, or $\mathbb{R}P^\infty$ when $p = 2$. On the product $S^\infty \times X^{\wedge p}$ there is then the diagonal action $g(s, x) = (g(s), g(x))$ for $g \in \mathbb{Z}_p$. Let ΓX denote the orbit space $(S^\infty \times X^{\wedge p})/\mathbb{Z}_p$ of this diagonal action. This is the same as the Borel construction $S^\infty \times_{\mathbb{Z}_p} X^{\wedge p}$ described in §3.G. The projection $S^\infty \times X^{\wedge p} \to S^\infty$ induces a projection $\pi : \Gamma X \to L^\infty$ with $\pi^{-1}(z) = X^{\wedge p}$ for all $z \in L^\infty$ since the action of $\mathbb{Z}_p$ on S^∞ is free. This projection $\Gamma X \to L^\infty$ is in fact a fiber bundle, though we shall not need this fact and so we leave the proof as an exercise. The $\mathbb{Z}_p$ action on $X^{\wedge p}$ fixes the basepoint $x_0 \in X^{\wedge p}$, so the inclusion $S^\infty \times \{x_0\} \hookrightarrow S^\infty \times X^{\wedge p}$

induces an inclusion $L^\infty \hookrightarrow \Gamma X$. The composition $L^\infty \hookrightarrow \Gamma X \to L^\infty$ is the identity, so in fiber bundle terminology this subspace $L^\infty \subset \Gamma X$ is a section of the bundle. Let ΛX denote the quotient $\Gamma X / L^\infty$ obtained by collapsing the section L^∞ to a point. Note that the fibers $X^{\wedge p}$ in ΓX are still embedded in the quotient ΛX since each fiber meets the section L^∞ in a single point.

If we replace S^∞ by S^1 in these definitions, we get subspaces $\Gamma^1 X \subset \Gamma X$ and $\Lambda^1 X \subset \Lambda X$. All these spaces have natural CW structures if X is a CW complex having x_0 as a 0-cell. To see this, let L^∞ be given its standard CW structure with one cell in each dimension. This lifts to a CW structure on S^∞ with p cells in each dimension, and then T freely permutes the product cells of $S^\infty \times X^{\wedge p}$ so there is an induced quotient CW structure on ΓX. The section $L^\infty \subset \Gamma X$ is a subcomplex, so the quotient ΛX inherits a CW structure from ΓX. In particular, note that if the n-skeleton of X is S^n with its usual CW structure, then the pn-skeleton of ΛX is S^{pn} with its usual CW structure.

We remark also that Γ, Γ^1, Λ, and Λ^1 are functors: A map $f : (X, x_0) \to (Y, y_0)$ induces maps $\Gamma f : \Gamma X \to \Gamma Y$, etc., in the evident way.

For brevity we write $H^*(-; \mathbb{Z}_p)$ simply as $H^*(-)$. For $n > 0$ let K_n denote a CW complex $K(\mathbb{Z}_p, n)$ with $(n-1)$-skeleton a point and n-skeleton S^n. Let $\iota \in H^n(K_n)$ be the canonical fundamental class described in the discussion following Theorem 4.57. It will be notationally convenient to regard an element $\alpha \in H^n(X)$ also as a map $\alpha : X \to K_n$ such that $\alpha^*(\iota) = \alpha$. Here we are assuming X is a CW complex.

From §3.2 we have a reduced p-fold cross product $\tilde{H}^*(X)^{\otimes p} \to \tilde{H}^*(X^{\wedge p})$ where $\tilde{H}^*(X)^{\otimes p}$ denotes the p-fold tensor product of $\tilde{H}^*(X)$ with itself. This cross product map $\tilde{H}^*(X)^{\otimes p} \to \tilde{H}^*(X^{\wedge p})$ is an isomorphism since we are using $\mathbb{Z}_p$ coefficients. With this isomorphism in mind, we will use the notation $\alpha_1 \otimes \cdots \otimes \alpha_p$ rather than $\alpha_1 \times \cdots \times \alpha_p$ for p-fold cross products in $\tilde{H}^*(X^{\wedge p})$. In particular, for each element $\alpha \in H^n(X)$, $n > 0$, we have its p-fold cross product $\alpha^{\otimes p} \in \tilde{H}^{pn}(X^{\wedge p})$. Our first task will be to construct an element $\lambda(\alpha) \in H^{pn}(\Lambda X)$ restricting to $\alpha^{\otimes p}$ in each fiber $X^{\wedge p} \subset \Lambda X$. By naturality it will suffice to construct $\lambda(\iota) \in H^{pn}(\Lambda K_n)$.

The key point in the construction of $\lambda(\iota)$ is the fact that $T^*(\iota^{\otimes p}) = \iota^{\otimes p}$. In terms of maps $K_n^{\wedge p} \to K_{pn}$, this says the composition $\iota^{\otimes p} T$ is homotopic to $\iota^{\otimes p}$, preserving basepoints. Such a homotopy can be constructed as follows. The pn-skeleton of $K_n^{\wedge p}$ is $(S^n)^{\wedge p} = S^{pn}$, with T permuting the factors cyclically. Thinking of S^n as $(S^1)^{\wedge n}$, the permutation T is a product of $(p-1)n^2$ transpositions of adjacent factors, so T has degree $(-1)^{(p-1)n^2}$ on S^{pn}. If p is odd, this degree is $+1$, so the restriction of T to this skeleton is homotopic to the identity, hence $\iota^{\otimes p} T$ is homotopic to $\iota^{\otimes p}$ on this skeleton. This conclusion also holds when $p = 2$, signs being irrelevant in this case since we are dealing with maps $S^{2n} \to K_{2n}$ and $\pi_{2n}(K_{2n}) = \mathbb{Z}_2$. Having a homotopy $\iota^{\otimes p} T \simeq \iota^{\otimes p}$ on the pn-skeleton, there are no obstructions to extending the homotopy over all higher-dimensional cells $e^i \times (0, 1)$ since $\pi_i(K_{pn}) = 0$ for $i > pn$.

The homotopy $\iota^{\otimes p} T \simeq \iota^{\otimes p} : K_n^{\wedge p} \to K_{pn}$ defines a map $\Gamma^1 K_n \to K_{pn}$ since $\Gamma^1 X$ is the quotient of $I \times X^{\wedge p}$ under the identifications $(0, x) \sim (1, T(x))$. The homotopy is basepoint-preserving, so the map $\Gamma^1 K_n \to K_{pn}$ passes down to a quotient map $\lambda_1 : \Lambda^1 K_n \to K_{pn}$. Since K_n is obtained from S^n by attaching cells of dimension greater than n, ΛK_n is obtained from $\Lambda^1 K_n$ by attaching cells of dimension greater than $pn + 1$. There are then no obstructions to extending λ_1 to a map $\lambda : \Lambda K_n \to K_{pn}$ since $\pi_i(K_{pn}) = 0$ for $i > pn$.

The map λ gives the desired element $\lambda(\iota) \in H^{pn}(\Lambda K_n)$ since the restriction of λ to each fiber $K_n^{\wedge p}$ is homotopic to $\iota^{\otimes p}$. Note that this property determines λ uniquely up to homotopy since the restriction map $H^{pn}(\Lambda K_n) \to H^{pn}(K_n^{\wedge p})$ is injective, the pn-skeleton of ΛK_n being contained in $K_n^{\wedge p}$. We shall have occasion to use this argument again in the proof, so we refer to it as 'the uniqueness argument.'

For any $\alpha \in H^n(X)$ let $\lambda(\alpha)$ be the composition $\Lambda X \xrightarrow{\Lambda \alpha} \Lambda K_n \xrightarrow{\lambda} K_{pn}$. This restricts to $\alpha^{\otimes p}$ in each fiber $X^{\wedge p}$ since $\Lambda \alpha$ restricts to $\alpha^{\otimes p}$ in each fiber.

Now we are ready to define some cohomology operations. There is an inclusion $L^\infty \times X \hookrightarrow \Gamma X$ as the quotient of the diagonal embedding $S^\infty \times X \hookrightarrow S^\infty \times X^{\wedge p}$, $(s, x) \mapsto (s, x, \cdots, x)$. Composing with the quotient map $\Gamma X \to \Lambda X$,' we get a map $\nabla : L^\infty \times X \to \Lambda X$ inducing $\nabla^* : H^*(\Lambda X) \to H^*(L^\infty \times X) \approx H^*(L^\infty) \otimes H^*(X)$. For each $\alpha \in H^n(X)$ the element $\nabla^*(\lambda(\alpha)) \in H^{pn}(L^\infty \times X)$ may be written in the form

$$\nabla^*(\lambda(\alpha)) = \sum_i \omega_{(p-1)n-i} \otimes \theta_i(\alpha)$$

where ω_j is a generator of $H^j(L^\infty)$ and $\theta_i(\alpha) \in H^{n+i}(X)$. Thus θ_i increases dimension by i. When $p = 2$ there is no ambiguity about ω_j. For odd p we choose ω_1 to be the class dual to the 1-cell of L^∞ in its standard cell structure, then we take ω_2 to be the Bockstein $\beta \omega_1$ and we set $\omega_{2j} = \omega_2^j$ and $\omega_{2j+1} = \omega_1 \omega_2^j$.

It is clear that θ_i is a cohomology operation since $\theta_i(\alpha) = \alpha^*(\theta_i(\iota))$. Note that $\theta_i = 0$ for $i < 0$ since $H^{n+i}(K_n) = 0$ for $i < 0$ except for $i = -n$, and in this special case $\theta_i = 0$ since $\nabla : L^\infty \times X \to \Lambda X$ sends $L^\infty \times \{x_0\}$ to a point.

For $p = 2$ we set $Sq^i(\alpha) = \theta_i(\alpha)$. For odd p we will show that $\theta_i = 0$ unless $i = 2k(p - 1)$ or $2k(p - 1) + 1$. The operation P^k will be defined to be a certain constant times $\theta_{2k(p-1)}$, and $\theta_{2k(p-1)+1}$ will be a constant times βP^k, for β the mod p Bockstein.

|| **Theorem 4L.12.** *The operations Sq^i satisfy the properties* (1)–(7).

Proof: We have already observed that the θ_i's are cohomology operations, so property (1) holds. The basic property that $\lambda(\alpha)$ restricts to $\alpha^{\otimes p}$ in each fiber implies that $\theta_{(p-1)n}(\alpha) = \alpha^p$ since $\omega_0 = 1$. This gives the first half of property (5) for Sq^i. The second half follows from the fact that $\theta_i = 0$ for $i > (p - 1)n$ since the factor $\omega_{(p-1)n-i}$ vanishes in this case.

Next we turn to the Cartan formula. For any prime p we will show that $\lambda(\alpha \smile \beta) = (-1)^{p(p-1)mn/2} \lambda(\alpha) \smile \lambda(\beta)$ for $m = |\alpha|$ and $n = |\beta|$. This implies (3) when $p = 2$

since if we let $\omega = \omega_1$, hence $\omega_j = \omega^j$, then

$$\sum_i Sq^i(\alpha \smile \beta) \otimes \omega^{n+m-i} = \nabla^*(\lambda(\alpha \smile \beta)) = \nabla^*(\lambda(\alpha) \smile \lambda(\beta))$$

$$= \nabla^*(\lambda(\alpha)) \smile \nabla^*(\lambda(\beta))$$

$$= \sum_j Sq^j(\alpha) \otimes \omega^{n-j} \smile \sum_k Sq^k(\beta) \otimes \omega^{m-k}$$

$$= \sum_i \left(\sum_{j+k=i} Sq^j(\alpha) \smile Sq^k(\beta) \right) \otimes \omega^{n+m-i}$$

To show that $\lambda(\alpha \smile \beta) = (-1)^{p(p-1)mn/2}\lambda(\alpha) \smile \lambda(\beta)$ we use the following diagram:

$$\begin{array}{ccccccc}
\Lambda X & \xrightarrow{\Lambda(\Delta)} & \Lambda(X \wedge X) & \xrightarrow{\Lambda(\alpha \wedge \beta)} & \Lambda(K_m \wedge K_n) & \xrightarrow{\Lambda(\iota_m \otimes \iota_n)} & \Lambda K_{m+n} \\
& \searrow^{\Delta} & \downarrow & & \downarrow & \searrow^{\lambda(\iota_m \otimes \iota_n)} & \downarrow^{\lambda} \\
& & \Lambda X \wedge \Lambda X & \xrightarrow{\Lambda \alpha \wedge \Lambda \beta} & \Lambda K_m \wedge \Lambda K_n & \xrightarrow{\lambda(\iota_m) \otimes \lambda(\iota_n)} & K_{pm+pn}
\end{array}$$

Here Δ is a generic symbol for diagonal maps $x \mapsto (x,x)$. These relate cross product to cup product via $\Delta^*(\varphi \otimes \psi) = \varphi \smile \psi$. The two unlabeled vertical maps are induced by $(s, x_1, y_1, \cdots, x_p, y_p) \mapsto (s, x_1, \cdots, x_p, s, y_1, \cdots, y_p)$. The composition $\Lambda X \to K_{pm+pn}$ going across the top of the diagram is $\lambda(\alpha \smile \beta)$ since the composition $\Lambda X \to \Lambda K_{m+n}$ is $\Lambda(\alpha \smile \beta)$. The composition $\Lambda X \wedge \Lambda X \to K_{pm+pn}$ is $\lambda(\alpha) \otimes \lambda(\beta)$ so the composition $\Lambda X \to K_{pm+pn}$ across the bottom of the diagram is $\lambda(\alpha) \smile \lambda(\beta)$. The triangle on the left, the square, and the upper triangle on the right obviously commute from the definitions. It remains to see that the third triangle commutes up to the sign $(-1)^{p(p-1)mn/2}$. Since $(K_m \wedge K_n)^{\wedge p}$ includes the $(pm+pn)$-skeleton of $\Lambda(K_m \wedge K_n)$, restriction to this fiber is injective on H^{pm+pn}. On this fiber the two routes around the triangle give $(\iota_m \smile \iota_n)^{\otimes p}$ and $\iota_m^{\otimes p} \otimes \iota_n^{\otimes p}$. These differ by a permutation that is the product of $(p-1) + (p-2) + \cdots + 1 = p(p-1)/2$ transpositions of adjacent factors. Since ι_m and ι_n have dimensions m and n, this permutation introduces a sign $(-1)^{p(p-1)mn/2}$ by the commutativity property of cup product. This finishes the verification of the Cartan formula when $p = 2$.

Before proceeding further we need to make an explicit calculation to show that Sq^0 is the identity on $H^1(S^1)$. Viewing S^1 as the one-point compactification of $\mathbb{R}$, with the point at infinity as the basepoint, the 2-sphere $S^1 \wedge S^1$ becomes the one-point compactification of $\mathbb{R}^2$. The map $T: S^1 \wedge S^1 \to S^1 \wedge S^1$ then corresponds to reflecting $\mathbb{R}^2$ across the line $x = y$, so after a rotation of coordinates this becomes reflection of S^2 across the equator. Hence $\Gamma^1 S^1$ is obtained from the shell $I \times S^2$ by identifying its inner and outer boundary spheres via a reflection across the equator. The diagonal $\mathbb{R}P^1 \times S^1 \subset \Gamma^1 S^1$ is a torus, obtained from the equatorial annulus $I \times S^1 \subset I \times S^2$ by identifying the two ends via the

identity map since the equator is fixed by the reflection. This $\mathbb{R}P^1 \times S^1$ represents the same element of $H_2(\Gamma^1 S^1; \mathbb{Z}_2)$ as the fiber sphere $S^1 \wedge S^1$ since the upper half of the shell is a 3-cell whose mod 2 boundary in $\Gamma^1 S^1$ is the union of these two surfaces.

For a generator $\alpha \in H^1(S^1)$, consider the element $\nabla^*(\lambda(\alpha))$ in $H^2(\mathbb{R}\mathrm{P}^\infty \times S^1) \approx$ $\mathrm{Hom}(H_2(\mathbb{R}\mathrm{P}^\infty \times S^1; \mathbb{Z}_2), \mathbb{Z}_2)$. A basis for $H_2(\mathbb{R}\mathrm{P}^\infty \times S^1; \mathbb{Z}_2)$ is represented by $\mathbb{R}\mathrm{P}^2 \times \{x_0\}$ and $\mathbb{R}\mathrm{P}^1 \times S^1$. A cocycle representing $\nabla^*(\lambda(\alpha))$ takes the value 0 on $\mathbb{R}\mathrm{P}^2 \times \{x_0\}$ since $\mathbb{R}\mathrm{P}^\infty \times \{x_0\}$ collapses to a point in ΛS^1 and $\lambda(\alpha)$ lies in $H^2(\Lambda S^1)$. On $\mathbb{R}\mathrm{P}^1 \times S^1$, $\nabla^*(\lambda(\alpha))$ takes the value 1 since when $\lambda(\alpha)$ is pulled back to ΓS^1 it takes the same value on the homologous cycles $\mathbb{R}\mathrm{P}^1 \times S^1$ and $S^1 \wedge S^1$, namely 1 by the defining property of $\lambda(\alpha)$ since $\alpha \otimes \alpha \in H^2(S^1 \wedge S^1)$ is a generator. Thus $\nabla^*(\lambda(\alpha)) = \omega_1 \otimes \alpha$ and hence $Sq^0(\alpha) = \alpha$ by the definition of Sq^0.

We use this calculation to prove that Sq^i commutes with the suspension σ, where σ is defined by $\sigma(\alpha) = \varepsilon \otimes \alpha \in H^*(S^1 \wedge X)$ for ε a generator of $H^1(S^1)$ and $\alpha \in H^*(X)$. We have just seen that $Sq^0(\varepsilon) = \varepsilon$. By (5), $Sq^1(\varepsilon) = \varepsilon^2 = 0$ and $Sq^i(\varepsilon) = 0$ for $i > 1$. The Cartan formula then gives $Sq^i(\sigma(\alpha)) = Sq^i(\varepsilon \otimes \alpha) = \sum_j Sq^j(\varepsilon) \otimes Sq^{i-j}(\alpha) = \varepsilon \otimes Sq^i(\alpha) = \sigma(Sq^i(\alpha))$.

From this it follows that Sq^0 is the identity on $H^n(S^n)$ for all $n > 0$. Since S^n is the n-skeleton of K_n, this implies that Sq^0 is the identity on the fundamental class ι_n, hence Sq^0 is the identity on all positive-dimensional classes.

Property (7) is proved similarly: Sq^1 coincides with the Bockstein β on the generator $\omega \in H^1(\mathbb{R}\mathrm{P}^2)$ since both equal ω^2. Hence $Sq^1 = \beta$ on the iterated suspensions of ω, and the n-fold suspension of $\mathbb{R}\mathrm{P}^2$ is the $(n+2)$-skeleton of K_{n+1}.

Finally we have the additivity property (2). This holds in fact for any cohomology operation that commutes with suspension. For such operations, it suffices to prove additivity in spaces that are suspensions. Consider a composition

$$\Sigma X \xrightarrow{c} \Sigma X \vee \Sigma X \xrightarrow{\alpha \vee \beta} K_n \xrightarrow{\theta} K_m$$

where c is the map that collapses an equatorial copy of X in ΣX to a point. The composition of the first two maps is $\alpha + \beta$, as in Lemma 4.60. Composing with the third map then gives $\theta(\alpha + \beta)$. On the other hand, if we first compose the second and third maps we get $\theta(\alpha) \vee \theta(\beta)$, and then composing with the first map gives $\theta(\alpha) + \theta(\beta)$. The two ways of composing are equal, so $\theta(\alpha + \beta) = \theta(\alpha) + \theta(\beta)$. $\square$

‖ **Theorem 4L.13.** *The Adem relations hold for Steenrod squares.*

Proof: The idea is to imitate the construction of ΛX using $\mathbb{Z}_p \times \mathbb{Z}_p$ in place of $\mathbb{Z}_p$. The Adem relations will come from the symmetry of $\mathbb{Z}_p \times \mathbb{Z}_p$ interchanging the factors.

The group $\mathbb{Z}_p \times \mathbb{Z}_p$ acts on $S^\infty \times S^\infty$ via $(g, h)(s, t) = (g(s), h(t))$, with quotient $L^\infty \times L^\infty$. There is also an action of $\mathbb{Z}_p \times \mathbb{Z}_p$ on $X^{\wedge p^2}$, obtained by writing points of $X^{\wedge p^2}$ as p^2-tuples (x_{ij}) with subscripts i and j varying from 1 to p, and then letting the first $\mathbb{Z}_p$ act on the first subscript and the second $\mathbb{Z}_p$ act on the second. Factoring out the diagonal action of $\mathbb{Z}_p \times \mathbb{Z}_p$ on $S^\infty \times S^\infty \times X^{\wedge p^2}$ gives a quotient space $\Gamma_2 X$. This projects to $L^\infty \times L^\infty$ with a section, and collapsing the section gives $\Lambda_2 X$. The fibers of the projection $\Lambda_2 X \to L^\infty \times L^\infty$ are $X^{\wedge p^2}$ since the action of $\mathbb{Z}_p \times \mathbb{Z}_p$ on $S^\infty \times S^\infty$ is free. We could also obtain $\Lambda_2 X$ from the product $S^\infty \times S^\infty \times X^{p^2}$ by first collapsing the subspace of points having at least one X coordinate equal to the basepoint x_0 and then factoring out the $\mathbb{Z}_p \times \mathbb{Z}_p$ action.

It will be useful to compare $\Lambda_2 X$ with $\Lambda(\Lambda X)$. The latter space is the quotient of $S^\infty \times (S^\infty \times X^p)^p$ in which one first identifies all points having at least one X coordinate equal to x_0 and then one factors out by an action of the wreath product $\mathbb{Z}_p \wr \mathbb{Z}_p$, the group of order p^{p+1} defined by a split exact sequence $0 \to \mathbb{Z}_p^p \to \mathbb{Z}_p \wr \mathbb{Z}_p \to \mathbb{Z}_p \to 0$ with conjugation by the quotient group $\mathbb{Z}_p$ given by cyclic permutations of the p $\mathbb{Z}_p$ factors of $\mathbb{Z}_p^p$. In the coordinates $(s, t_1, x_{11}, \cdots, x_{1p}, \cdots, t_p, x_{p1}, \cdots, x_{pp})$ the ith factor $\mathbb{Z}_p$ of $\mathbb{Z}_p^p$ acts in the block $(t_i, x_{i1}, \cdots, x_{ip})$, and the quotient $\mathbb{Z}_p$ acts by cyclic permutation of the index i and by rotation in the s coordinate. There is a natural map $\Lambda_2 X \to \Lambda(\Lambda X)$ induced by $(s, t, x_{11}, \cdots, x_{1p}, \cdots, x_{p1}, \cdots, x_{pp}) \mapsto$ $(s, t, x_{11}, \cdots, x_{1p}, \cdots, t, x_{p1}, \cdots, x_{pp})$. In $\Lambda_2 X$ one is factoring out by the action of $\mathbb{Z}_p \times \mathbb{Z}_p$. This is the subgroup of $\mathbb{Z}_p \wr \mathbb{Z}_p$ obtained by restricting the action of the quotient $\mathbb{Z}_p$ on $\mathbb{Z}_p^p$ to the diagonal subgroup $\mathbb{Z}_p \subset \mathbb{Z}_p^p$, where this action becomes trivial so that one has the direct product $\mathbb{Z}_p \times \mathbb{Z}_p$.

Since it suffices to prove that the Adem relations hold on the class $\iota \in H^n(K_n)$, we take $X = K_n$. There is a map $\lambda_2 : \Lambda_2 K_n \to K_{p^2 n}$ restricting to $\iota^{\otimes p^2}$ in each fiber. This is constructed by the same method used to construct λ. One starts with a map representing $\iota^{\otimes p^2}$ in a fiber, then extends this over the part of $\Lambda_2 K_n$ projecting to the 1-skeleton of $L^\infty \times L^\infty$, and finally one extends inductively over higher-dimensional cells of $\Lambda_2 K_n$ using the fact that $K_{p^2 n}$ is an Eilenberg–MacLane space. The map λ_2 fits into the diagram at the right, where ∇_2 is induced by the map $(s, t, x) \mapsto (s, t, x, \cdots, x)$ and the unlabeled map is the one defined above. It is clear that the square commutes.

$$
\begin{array}{ccccc}
L^\infty \times L^\infty \times K_n & \xrightarrow{\ \nabla_2\ } & \Lambda_2 K_n & \xrightarrow{\ \lambda_2\ } & K_{p^2 n} \\
\downarrow{\scriptstyle \mathbb{1} \times \nabla} & & \downarrow & \nearrow{\scriptstyle \lambda(\lambda)} & \\
L^\infty \times \Lambda K_n & \xrightarrow{\ \nabla\ } & \Lambda(\Lambda K_n) & &
\end{array}
$$

Commutativity of the triangle up to homotopy follows from the fact that λ_2 is uniquely determined, up to homotopy, by its restrictions to fibers.

The element $\nabla_2^* \lambda_2^*(\iota)$ may be written in the form $\sum_{r,s} \omega_r \otimes \omega_s \otimes \varphi_{rs}$, and we claim that the elements φ_{rs} satisfy the symmetry relation $\varphi_{rs} = (-1)^{rs + p(p-1)n/2} \varphi_{sr}$. To verify this we use the commutative diagram at the right where the map τ on the left switches the two L^∞ factors and the τ on

$$
\begin{array}{ccc}
L^\infty \times L^\infty \times K_n & \xrightarrow{\ \nabla_2\ } & \Lambda_2 K_n \\
\downarrow{\scriptstyle \tau} & & \downarrow{\scriptstyle \tau} \\
L^\infty \times L^\infty \times K_n & \xrightarrow{\ \nabla_2\ } & \Lambda_2 K_n
\end{array}
$$

the right is induced by switching the two S^∞ factors of $S^\infty \times S^\infty \times K_n^{\wedge p^2}$ and permuting the K_n factors of the smash product by interchanging the two subscripts in p^2-tuples (x_{ij}). This permutation is a product of $p(p-1)/2$ transpositions, one for each pair (i, j) with $1 \le i < j \le p$, so in a fiber the class $\iota^{\otimes p^2}$ is sent to $(-1)^{p(p-1)n/2} \iota^{\otimes p^2}$. By the uniqueness property of λ_2 this means that $\tau^* \lambda_2^*(\iota) = (-1)^{p(p-1)n/2} \lambda_2^*(\iota)$. Commutativity of the square then gives

$$
(-1)^{p(p-1)n/2} \nabla_2^* \lambda_2^*(\iota) = \nabla_2^* \tau^* \lambda_2^*(\iota) = \tau^* \nabla_2^* \lambda_2^*(\iota) = \sum_{r,s} (-1)^{rs} \omega_s \otimes \omega_r \otimes \varphi_{rs}
$$

where the last equality follows from the commutativity property of cross products. The symmetry relation $\varphi_{rs} = (-1)^{rs+p(p-1)n/2}\varphi_{sr}$ follows by interchanging the indices r and s in the last summation.

If we compute $\nabla_2^* \lambda_2^*(\iota)$ using the lower route across the earlier diagram containing the map λ_2, we obtain

$$\nabla_2^* \lambda_2^*(\iota) = \sum_i \omega_{(p-1)pn-i} \otimes \theta_i \left(\sum_j \omega_{(p-1)n-j} \otimes \theta_j(\iota) \right)$$

$$= \sum_{i,j} \omega_{(p-1)pn-i} \otimes \theta_i (\omega_{(p-1)n-j} \otimes \theta_j(\iota))$$

Now we specialize to $p = 2$, so $\theta_i = Sq^i$ for all i. The Cartan formula converts the last summation above into $\sum_{i,j,k} \omega^{2n-i} \otimes Sq^k(\omega^{n-j}) \otimes Sq^{i-k} Sq^j(\iota)$. Plugging in the value for $Sq^k(\omega^{n-j})$ computed in the discussion preceding Example 4L.3, we obtain $\sum_{i,j,k} \binom{n-j}{k} \omega^{2n-i} \otimes \omega^{n-j+k} \otimes Sq^{i-k} Sq^j(\iota)$. To write this summation more symmetrically with respect to the two ω terms, let $n - j + k = 2n - \ell$. Then we get

$$\sum_{i,j,\ell} \binom{n-j}{n+j-\ell} \omega^{2n-i} \otimes \omega^{2n-\ell} \otimes Sq^{i+\ell-n-j} Sq^j(\iota)$$

In view of the symmetry property of φ_{rs}, which becomes $\varphi_{rs} = \varphi_{sr}$ for $p = 2$, switching i and ℓ in this formula leaves it unchanged. Hence we get the relation

$$(*) \qquad \sum_j \binom{n-j}{n+j-\ell} Sq^{i+\ell-n-j} Sq^j(\iota) = \sum_j \binom{n-j}{n+j-i} Sq^{i+\ell-n-j} Sq^j(\iota)$$

This holds for all n, i, and ℓ, and the idea is to choose these numbers so that the left side of this equation has only one nonzero term. Given integers r and s, let $n = 2^r - 1 + s$ and $\ell = n + s$, so that $\binom{n-j}{n+j-\ell} = \binom{2^r-1-(j-s)}{j-s}$. If r is sufficiently large, this will be 0 unless $j = s$. This is because the dyadic expansion of $2^r - 1$ consists entirely of 1's, so the expansion of $2^r - 1 - (j - s)$ will have 0's in the positions where the expansion of $j - s$ has 1's, hence these positions contribute factors of $\binom{0}{1} = 0$ to $\binom{2^r-1-(j-s)}{j-s}$. Thus with n and ℓ chosen as above, the relation $(*)$ becomes

$$Sq^i Sq^s(\iota) = \sum_j \binom{2^r-1+s-j}{2^r-1+s+j-i} Sq^{i+s-j} Sq^j(\iota) = \sum_j \binom{2^r+s-j-1}{i-2j} Sq^{i+s-j} Sq^j(\iota)$$

where the latter equality comes from the general relation $\binom{x}{y} = \binom{x}{x-y}$.

The final step is to show that $\binom{2^r+s-j-1}{i-2j} = \binom{s-j-1}{i-2j}$ if $i < 2s$. Both of these binomial coefficients are zero if $i < 2j$. If $i \geq 2j$ then we have $2j \leq i < 2s$, so $j < s$, hence $s - j - 1 \geq 0$. The term 2^r then makes no difference in $\binom{2^r+s-j-1}{i-2j}$ if r is large since this 2^r contributes only a single 1 to the dyadic expansion of $2^r + s - j - 1$, far to the left of all the nonzero entries in the dyadic expansions of $s - j - 1$ and $i - 2j$.

This gives the Adem relations for the classes ι of dimension $n = 2^r - 1 + s$ with r large. This implies the relations hold for all classes of these dimensions, by naturality. Since we can suspend repeatedly to make any class have dimension of this form, the Adem relations must hold for all cohomology classes. $\qquad\square$

Steenrod Powers

Our remaining task is to verify the axioms and Adem relations for the Steenrod powers for an odd prime p. Unfortunately this is quite a bit more complicated than the $p = 2$ case, largely because one has to be very careful in computing the many coefficients in $\mathbb{Z}_p$ that arise. Even for the innocent-looking axiom $P^0 = 1\!\!1$ it will take three pages to calculate the normalization constants needed to make the axiom hold. One could wish that the whole process was a lot cleaner.

Lemma 4L.14. $\theta_i = 0$ *unless* $i = 2k(p-1)$ *or* $2k(p-1) + 1$.

Proof: The group of automorphisms of $\mathbb{Z}_p$ is the multiplicative group $\mathbb{Z}_p^*$ of nonzero elements of $\mathbb{Z}_p$. Since p is prime, $\mathbb{Z}_p$ is a field and $\mathbb{Z}_p^*$ is cyclic of order $p - 1$. Let r be a generator of $\mathbb{Z}_p^*$. Define a map $\varphi : S^\infty \times X^{\wedge p} \to S^\infty \times X^{\wedge p}$ permuting the factors X_j of $X^{\wedge p}$ by $\varphi(s, X_j) = (s^r, X_{rj})$ where subscripts are taken mod p and s^r means raise each coordinate of s, regarded as a unit vector in $\mathbb{C}^\infty$, to the rth power and renormalize the resulting vector to have unit length. Then if γ is a generator of the $\mathbb{Z}_p$ action on $S^\infty \times X^{\wedge p}$, we have $\varphi(\gamma(s, X_j)) = \varphi(e^{2\pi i/p}s, X_{j-1}) = (e^{2r\pi i/p}s^r, X_{rj-r}) = \gamma^r(\varphi(s, X_j))$. This says that φ takes orbits to orbits, so φ induces maps $\varphi : \Gamma X \to \Gamma X$ and $\varphi : \Lambda X \to \Lambda X$. Restricting to the first coordinate, there is also an induced map $\varphi : L^\infty \to L^\infty$. Taking $X = K_n$, these maps fit into the diagram at the right. The square obviously commutes. The triangle commutes up to homotopy and a sign of $(-1)^n$ since it suffices to verify this on the pn-skeleton $(S^n)^{\wedge p}$, and here the map φ is an odd permutation of the S^n factors since it is a cyclic permutation of order $p-1$, which is even, and a transposition of two S^n factors has degree 1 if n is even and degree -1 if n is odd.

Suppose first that n is even. Then commutativity of the diagram means that $\sum_i \omega_{(p-1)n-i} \otimes \theta_i(\iota)$ is invariant under $\varphi^* \otimes 1\!\!1$, hence $\varphi^*(\omega_{(p-1)n-i}) = \omega_{(p-1)n-i}$ if $\theta_i(\iota)$ is nonzero. The map φ induces multiplication by r in $\pi_1(L^\infty)$, hence also in $H_1(L^\infty)$ and $H^1(L^\infty; \mathbb{Z}_p)$, sending ω_1 to $r\omega_1$. Since ω_2 was chosen to be the Bockstein of ω_1, it is also multiplied by r. We chose r to have order $p - 1$ in $\mathbb{Z}_p^*$, so $\varphi^*(\omega_\ell) = \omega_\ell$ only when the total number of ω_1 and ω_2 factors in ω_ℓ is a multiple of $p - 1$. For $\omega_\ell = \omega_2^k$ this means $\ell = (p-1)n - i = 2k(p-1)$, while for $\omega_\ell = \omega_1\omega_2^{k-1}$ it means ℓ is 1 less than this, $2k(p-1) - 1$. Solving these equations for i gives $i = (n - 2k)(p - 1)$ or $i = (n - 2k)(p - 1) + 1$. Since n is even this says that i is congruent to 0 or 1 mod $2(p - 1)$, which is what the lemma asserts.

When n is odd the condition $\varphi^*(\omega_\ell) = \omega_\ell$ becomes $\varphi^*(\omega_\ell) = -\omega_\ell$. In the cyclic group $\mathbb{Z}_p^*$ the element -1 is the only element of order 2, and this element is $(p - 1)/2$ times a generator, so the total number of ω_1 and ω_2 factors in ω_ℓ must be $(2k + 1)(p - 1)/2$ for some integer k. This implies that $\ell = (p - 1)n - i =$

$2(2k+1)(p-1)/2$ or 1 less than this, hence $i = (n-2k-1)(p-1)$ or 1 greater than this. As n is odd, this again says that i is congruent to 0 or 1 mod $2(p-1)$.   □

Since $\theta_0 : H^n(X) \to H^n(X)$ is a cohomology operation that preserves dimension, it must be defined by a coefficient homomorphism $\mathbb{Z}_p \to \mathbb{Z}_p$, multiplication by some $a_n \in \mathbb{Z}_p$. We claim that these a_n's satisfy

$$a_{m+n} = (-1)^{p(p-1)mn/2} a_m a_n \quad \text{and} \quad a_n = (-1)^{p(p-1)n(n-1)/4} a_1^n$$

To see this, recall the formula $\lambda(\alpha \smile \beta) = (-1)^{p(p-1)mn/2} \lambda(\alpha) \lambda(\beta)$ for $|\alpha| = m$ and $|\beta| = n$. From the definition of the θ_i's it then follows that $\theta_0(\alpha \smile \beta) = (-1)^{p(p-1)mn/2} \theta_0(\alpha) \theta_0(\beta)$, which gives the first part of the claim. The second part follows from this by induction on n.

‖ Lemma 4L.15. $a_1 = \pm m!$ for $m = (p-1)/2$, so $p = 2m+1$.

Proof: It suffices to compute $\theta_0(\alpha)$ where α is any nonzero 1-dimensional class, so the simplest thing is to choose α to be a generator of $H^1(S^1)$, say a generator coming from a generator of $H^1(S^1; \mathbb{Z})$. This determines α up to a sign. Since $H^i(S^1) = 0$ for $i > 1$, we have $\theta_i(\alpha) = 0$ for $i > 0$, so the defining formula for $\theta_0(\alpha)$ has the form $\nabla^*(\lambda(\alpha)) = \omega_{p-1} \otimes \theta_0(\alpha) = a_1 \omega_{p-1} \otimes \alpha$ in $H^p(L^\infty \times S^1)$. To compute a_1 there is no harm in replacing L^∞ by a finite-dimensional lens space, say L^p, the p-skeleton of L^∞. Thus we may restrict the bundle $\Lambda S^1 \to L^\infty$ to a bundle $\Lambda^p S^1 \to L^p$ with the same fibers $(S^1)^{\wedge p} = S^p$. We regard S^1 as the one-point compactification of $\mathbb{R}$ with basepoint the added point at infinity, and then $(S^1)^{\wedge p}$ becomes the one-point compactification of $\mathbb{R}^p$ with $\mathbb{Z}_p$ acting by permuting the coordinates of $\mathbb{R}^p$ cyclically, preserving the origin and the point at infinity. This action defines the bundle $\Gamma^p S^1 \to L^p$ with fibers S^p, containing a zero section and a section at infinity, and $\Lambda^p S^1$ is obtained by collapsing the section at infinity. We can also describe $\Lambda^p S^1$ as the one-point compactification of the complement of the section at infinity in $\Gamma^p S^1$, since the base space L^p is compact. The complement of the section at infinity is a bundle $E \to L^p$ with fibers $\mathbb{R}^p$. In general, the one-point compactification of a fiber bundle E over a compact base space with fibers $\mathbb{R}^n$ is called the **Thom space** $T(E)$ of the bundle, and a class in $H^n(T(E))$ that restricts to a generator of H^n of the one-point compactification of each fiber $\mathbb{R}^n$ is called a **Thom class**. In our situation, $\lambda(\alpha)$ is such a Thom class.

Our first task is to construct subbundles $E_0, E_1, \cdots, E_m$ of E, where E_0 has fiber $\mathbb{R}$ and the other E_j's have fiber $\mathbb{R}^2$, so $p = 2m+1$. The bundle E comes from the linear transformation $T : \mathbb{R}^p \to \mathbb{R}^p$ permuting the coordinates cyclically. We claim there is a decomposition $\mathbb{R}^p = V_0 \oplus V_1 \oplus \cdots \oplus V_m$ with V_0 1-dimensional and the other V_j's 2-dimensional, such that $T(V_j) = V_j$ for all j, with $T|V_0$ the identity and $T|V_j$ a rotation by the angle $2\pi j/p$ for $j > 0$. Thus T defines an action of $\mathbb{Z}_p$ on V_j and we can define E_j just as E was defined, as the quotient $(S^p \times V_j)/\mathbb{Z}_p$ with respect to the diagonal action.

An easy way to get the decomposition $\mathbb{R}^p = V_0 \oplus V_1 \oplus \cdots \oplus V_m$ is to regard $\mathbb{R}^p$ as a module over the principal ideal domain $\mathbb{R}[t]$ by setting $tv = T(v)$ for $v \in \mathbb{R}^p$. Then $\mathbb{R}^p$ is isomorphic as a module to the module $\mathbb{R}[t]/(t^p - 1)$ since T permutes coordinates cyclically; this amounts to identifying the standard basis vectors $v_1, \cdots, v_p$ in $\mathbb{R}^p$ with $1, t, \cdots, t^{p-1}$. The polynomial $t^p - 1$ factors over $\mathbb{C}$ into the linear factors $t - e^{2\pi i j/p}$ for $j = 0, \cdots, p - 1$. Combining complex conjugate factors, this gives a factorization over $\mathbb{R}$, $t^p - 1 = (t - 1)\prod_{1 \le j \le m}(t^2 - 2(\cos\varphi_j)t + 1)$, where $\varphi_j = 2\pi j/p$. These are distinct monic irreducible factors, so the module $\mathbb{R}[t]/(t^p - 1)$ splits as $\mathbb{R}[t]/(t - 1) \oplus_{1 \le j \le m} \mathbb{R}[t]/(t^2 - 2(\cos\varphi_j)t + 1)$ by the basic structure theory of modules over a principal ideal domain. This translates into a decomposition $\mathbb{R}^p = V_0 \oplus V_1 \oplus \cdots \oplus V_m$ with $T(V_j) \subset V_j$. Here V_0 corresponds to $\mathbb{R}[t]/(t - 1) \approx \mathbb{R}$ with t acting as the identity, and V_j for $j > 0$ corresponds to $\mathbb{R}[t]/(t^2 - 2(\cos\varphi_j)t + 1)$. The latter module is isomorphic to $\mathbb{R}^2$ with t acting as rotation by the angle φ_j since the characteristic polynomial of this rotation is readily computed to be $t^2 - 2(\cos\varphi_j)t + 1$, hence this rotation satisfies $t^2 - 2(\cos\varphi_j)t + 1 = 0$ so there is a module homomorphism $\mathbb{R}[t]/(t^2 - 2(\cos\varphi_j)t + 1) \to \mathbb{R}^2$ which is obviously an isomorphism.

From the decomposition $\mathbb{R}^p = V_0 \oplus V_1 \oplus \cdots \oplus V_m$ and the action of T on each factor we can see that the only vectors fixed by T are those in the line V_0. The vectors $(x, \cdots, x)$ are fixed by T, so V_0 must be this diagonal line.

Next we compute Thom classes for the bundles E_j. This is easy for E_0 which is the product $L^p \times \mathbb{R}$, so the projection $E_0 \to \mathbb{R}$ one-point compactifies to a map $T(E_0) \to S^1$ and we can pull back the chosen generator $\alpha \in H^1(S^1)$ to a Thom class for E_0. The other E_j's have 2-dimensional fibers, which we now view as $\mathbb{C}$ rather than $\mathbb{R}^2$. Just as E_j is the quotient of $S^p \times \mathbb{C}$ via the identifications $(v, z) \sim (e^{2\pi i/p}v, e^{2\pi ij/p}z)$, we can define a bundle $\overline{E}_j \to \mathbb{C}\mathrm{P}^m$ with fiber $\mathbb{C}$ by the identifications $(v, z) \sim (\lambda v, \lambda^j z)$ for $\lambda \in S^1 \subset \mathbb{C}$. We then have the left half of the commutative diagram shown at the right, where the quotient map $\widetilde{q}$ restricts to a homeomorphism on each fiber. The maps $\widetilde{f}$ and f are induced by the map $S^p \times \mathbb{C} \to S^p \times \mathbb{C}$ sending

$$\begin{array}{ccccc} E_j & \xrightarrow{\widetilde{q}} & \overline{E}_j & \xrightarrow{\widetilde{f}} & \overline{E}_1 \\ \downarrow & & \downarrow & & \downarrow \\ L^p & \xrightarrow{q} & \mathbb{C}\mathrm{P}^m & \xrightarrow{f} & \mathbb{C}\mathrm{P}^m \end{array}$$

(v, z) to (v^j, z) where v^j means raise each coordinate of v to the jth power and then rescale to get a vector of unit length. The map $\widetilde{f}$ is well-defined since equivalent pairs $(v, z) \sim (\lambda v, \lambda^j z)$ in $\overline{E}_j$ are carried to pairs (v^j, z) and $(\lambda^j v^j, \lambda^j z)$ that are equivalent in $\overline{E}_1$.

Since both $\widetilde{q}$ and $\widetilde{f}$ restrict to homeomorphisms in each fiber, they extend to maps of Thom spaces that pull a Thom class for $\overline{E}_1$ back to Thom classes for $\overline{E}_j$ and E_j. To construct a Thom class for $\overline{E}_1$, observe that the Thom space $T(\overline{E}_1)$ is homeomorphic to $\mathbb{C}\mathrm{P}^{m+1}$, namely, view the sphere $S^p = S^{2m+1}$ as the unit sphere in $\mathbb{C}^{m+1}$, and then the inclusion $S^p \times \mathbb{C} \hookrightarrow \mathbb{C}^{m+1} \times \mathbb{C} = \mathbb{C}^{m+2}$ induces a map $g: \overline{E}_1 \to \mathbb{C}\mathrm{P}^{m+1}$ since the equivalence relation defining $\overline{E}_1$ is $(v, z) \sim (\lambda v, \lambda z)$ for $\lambda \in S^1$. It is evident that g is a homeomorphism onto the complement of the point $[0, \cdots, 0, 1]$ in $\mathbb{CP}^{m+1}$, so sending the point at infinity in $T(\overline{E}_1)$ to $[0, \cdots, 0, 1]$ gives an extension of g to a homeomorphism $T(\overline{E}_1) \approx \mathbb{CP}^{m+1}$. Under this homeomorphism the one-point compactifications of the fibers of $\overline{E}_1$ correspond to the 2-spheres S_v^2 consisting of $[0, \cdots, 0, 1]$ and the points $[v, z] \in \mathbb{CP}^{m+1}$ with fixed $v \in S^p$ and varying $z \in \mathbb{C}$. Each S_v^2 is a $\mathbb{CP}^1$ in $\mathbb{CP}^{m+1}$ equivalent to the standard $\mathbb{CP}^1$ under a homeomorphism of $\mathbb{CP}^{m+1}$ coming from a linear isomorphism of $\mathbb{C}^{m+2}$, so a generator y of $H^2(\mathbb{CP}^{m+1})$ is a Thom class, restricting to a generator of $H^2(S_v^2)$ for each v. We choose y to be the $\mathbb{Z}_p$ reduction of a generator of $H^2(\mathbb{CP}^{m+1}; \mathbb{Z})$, so y is determined up to a sign.

A slightly different view of Thom classes will be useful. For the bundle $E \to L^p$, for example, we have isomorphisms

$$\widetilde{H}^*(T(E)) \approx H^*(T(E), \infty) \quad \text{where } \infty \text{ is the compactification point}$$

$$\approx H^*(T(E), T(E) - L^p) \quad \text{where } L^p \text{ is embedded in } T(E) \text{ as the zero}$$
$$\text{section, so } T(E) - L^p \text{ deformation retracts onto } \infty$$

$$\approx H^*(E, E - L^p) \quad \text{by excision}$$

Thus we can view a Thom class as lying in $H^*(E, E - L^p)$, and similarly for the bundles E_j.

We have projections $\pi_j : E \to E_j$ via the projections $V_0 \oplus V_1 \oplus \cdots \oplus V_m \to V_j$ in fibers. If $\tau_j \in H^*(E_j, E_j - L^p)$ denotes the Thom class constructed above, then we have the pullback $\pi_j^*(\tau_j) \in H^*(E, E - \pi_j^{-1}(L^p))$, and the cup product $\prod_j \pi_j^*(\tau_j)$ in $H^*(E, E - L^p)$ is a Thom class for E, as one sees by applying the calculation at the end of Example 3.11 in each fiber. Under the isomorphism $H^*(E, E - L^p) \approx \widetilde{H}^*(T(E))$, the class $\prod_j \pi_j^*(\tau_j)$ corresponds to $\pm \lambda(\alpha)$ since both classes restrict to $\pm \alpha^{\otimes p}$ in each fiber $S^p \subset T(E)$ and $\lambda(\alpha)$ is uniquely determined by its restriction to fibers.

Now we can finish the proof of the lemma. The class $\nabla^*(\lambda(\alpha))$ is obtained by restricting $\lambda(\alpha) \in H^p(T(E))$ to the diagonal $T(E_0)$, then pulling back to $L^p \times S^1$ via the quotient map $L^p \times S^1 \to T(E_0)$ which collapses the section at infinity to a point. Restricting $\prod_j \pi_j^*(\tau_j)$ to $H^p(E_0, E_0 - L^p) \approx H^p(T(E_0))$ gives $\tau_0 \smile e_1 \smile \cdots \smile e_m$ where $e_j \in H^2(E_0)$ is the image of τ_j under $H^2(E_j, E_j - L^p) \to H^2(E_j) \approx H^2(L^p) \approx H^2(E_0)$, these last two isomorphisms coming from including L^p in E_j and E_0 as the zero section, to which they deformation retract. To compute e_j, we use the diagram

$$
\begin{array}{ccccc}
H^2(E_j, E_j - L^p) & \xleftarrow{\widetilde{q}^*} & H^2(\overline{E}_j, \overline{E}_j - \mathbb{CP}^m) & \xleftarrow{\widetilde{f}^*} & H^2(\overline{E}_1, \overline{E}_1 - \mathbb{CP}^m) \\
\downarrow & & \downarrow & & \downarrow \\
H^2(E_j) & \xleftarrow{\widetilde{q}^*} & H^2(\overline{E}_j) & \xleftarrow{\widetilde{f}^*} & H^2(\overline{E}_1) \\
\downarrow{\approx} & & \downarrow{\approx} & & \downarrow{\approx} \\
H^2(L^p) & \xleftarrow{q^*} & H^2(\mathbb{CP}^m) & \xleftarrow{f^*} & H^2(\mathbb{CP}^m)
\end{array}
$$

The Thom class for $\overline{E}_1$ lies in the upper right group. Following this class across the top of the diagram and then down to the lower left corner gives the element e_j. To compute e_j we take the alternate route through the lower right corner of the diagram. The image of the Thom class for $\overline{E}_1$ in the lower right $H^2(\mathbb{C}\mathrm{P}^m)$ is the generator y since $T(\overline{E}_1) = \mathbb{C}\mathrm{P}^{m+1}$. The map f^* is multiplication by j since f has degree j on $\mathbb{C}\mathrm{P}^1 \subset \mathbb{C}\mathrm{P}^m$. And $q^*(y) = \pm\omega_2$ since q restricts to a homeomorphism on the 2-cell of L^p in the CW structure defined in Example 2.43. Thus $e_j = \pm j\omega_2$, and so $\tau_0 \smile e_1 \smile \cdots \smile e_m = \pm m!\tau_0 \smile \omega_2^m = \pm m!\tau_0 \smile \omega_{p-1}$. Since τ_0 was the pullback of α via the projection $T(E_0) \to S^1$, when we pull τ_0 back to $L^p \times S^1$ via ∇ we get $1 \otimes \alpha$, so $\tau_0 \smile e_1 \smile \cdots \smile e_m$ pulls back to $\pm m!\omega_{p-1} \otimes \alpha$. Hence $a_1 = \pm m!$. $\square$

The lemma implies in particular that a_n is not zero in $\mathbb{Z}_p$, so a_n has a multiplicative inverse a_n^{-1} in $\mathbb{Z}_p$. We then define

$$P^i(\alpha) = (-1)^i a_n^{-1}\theta_{2i(p-1)}(\alpha) \qquad \text{for } \alpha \in H^n(X)$$

The factor a_n^{-1} guarantees that P^0 is the identity. The factor $(-1)^i$ is inserted in order to make $P^i(\alpha) = \alpha^p$ if $|\alpha| = 2i$, as we show next. We know that $\theta_{2i(p-1)}(\alpha) = \alpha^p$, so what must be shown is that $(-1)^i a_{2i}^{-1} = 1$, or equivalently, $a_{2i} = (-1)^i$.

To do this we need a number theory fact: $((p-1)/2)!^2 \equiv (-1)^{(p+1)/2} \bmod p$. To derive this, note first that the product of all the elements $\pm 1, \pm 2, \cdots, \pm(p-1)/2$ of $\mathbb{Z}_p^*$ is $((p-1)/2)!^2 (-1)^{(p-1)/2}$. On the other hand, this group is cyclic of even order, so the product of all its elements is the unique element of order 2, which is -1, since all the other nontrivial elements cancel their inverses in this product. Thus $((p-1)/2)!^2(-1)^{(p-1)/2} \equiv -1$ and hence $((p-1)/2)!^2 \equiv (-1)^{(p+1)/2} \bmod p$.

Using the formulas $a_n = (-1)^{p(p-1)n(n-1)/4}a_1^n$ and $a_1 = \pm((p-1)/2)!$ we then have

$$\begin{aligned}
a_{2i} &= (-1)^{p(p-1)2i(2i-1)/4}((p-1)/2)!^{2i} \\
&= (-1)^{p[(p-1)/2]i(2i-1)}(-1)^{i(p+1)/2} \\
&= (-1)^{i(p-1)/2}(-1)^{i(p+1)/2} \quad \text{since } p \text{ and } 2i-1 \text{ are odd} \\
&= (-1)^{ip} = (-1)^i \quad \text{since } p \text{ is odd.}
\end{aligned}$$

Theorem 4L.16. *The operations P^i satisfy the properties* (1)–(6) *and the Adem relations.*

Proof: Naturality and the fact that $P^i(\alpha) = 0$ if $2i > |\alpha|$ are inherited from the θ_i's. Property (6) and the other half of (5) have just been shown above. For the Cartan formula we have, for $\alpha \in H^m$ and $\beta \in H^n$, $\lambda(\alpha \smile \beta) = (-1)^{p(p-1)mn/2}\lambda(\alpha)\lambda(\beta)$ and hence

$$\sum_i \omega_{(p-1)(m+n)-i} \otimes \theta_i(\alpha \smile \beta) =$$
$$(-1)^{p(p-1)mn/2}\left(\sum_j \omega_{(p-1)m-j} \otimes \theta_j(\alpha)\right)\left(\sum_k \omega_{(p-1)n-k} \otimes \theta_k(\beta)\right)$$

Recall that $\omega_{2r} = \omega_2^r$ and $\omega_{2r+1} = \omega_1\omega_2^r$, with $\omega_1^2 = 0$. Therefore terms with i even on the left side of the equation can only come from terms with j and k even on the right side. This leads to the second equality in the following sequence:

$$P^i(\alpha \smile \beta) = (-1)^i a_{m+n}^{-1} \theta_{2i(p-1)}(\alpha \smile \beta)$$

$$= (-1)^i a_{m+n}^{-1}(-1)^{p(p-1)mn/2} \sum_j \theta_{2(i-j)(p-1)}(\alpha)\theta_{2j(p-1)}(\beta)$$

$$= \sum_j (-1)^{i-j} a_m^{-1} \theta_{2(i-j)(p-1)}(\alpha)(-1)^j a_n^{-1} \theta_{2j(p-1)}(\beta)$$

$$= \sum_j P^{i-j}(\alpha)P^j(\beta)$$

Property (4), the invariance of P^i under suspension, follows from the Cartan formula just as in the case $p = 2$, using the fact that P^0 is the only P^i that can be nonzero on 1-dimensional classes, by (5). The additivity property follows just as before.

It remains to prove the Adem relations for Steenrod powers. We will need a Bockstein calculation:

‖ **Lemma 4L.17.** $\beta\theta_{2k} = -\theta_{2k+1}$.

Proof: Let us first reduce the problem to showing that $\beta\nabla^*(\lambda(\iota)) = 0$. If we compute $\beta\nabla^*(\lambda(\iota))$ using the product formula for β, we get

$$\beta\Big(\sum_i \omega_{(p-1)n-i} \otimes \theta_i(\iota)\Big) = \sum_i \Big(\beta\omega_{(p-1)n-i} \otimes \theta_i(\iota) + (-1)^i \omega_{(p-1)n-i} \otimes \beta\theta_i(\iota)\Big)$$

Since $\beta\omega_{2j-1} = \omega_{2j}$ and $\beta\omega_{2j} = 0$, the terms with $i = 2k$ and $i = 2k + 1$ give $\sum_k \omega_{(p-1)n-2k} \otimes \beta\theta_{2k}(\iota)$ and $\sum_k \omega_{(p-1)n-2k} \otimes \theta_{2k+1}(\iota) - \sum_k \omega_{(p-1)n-2k-1} \otimes \beta\theta_{2k+1}(\iota)$, respectively. Thus the coefficient of $\omega_{(p-1)n-2k}$ in $\beta\nabla^*(\lambda(\iota))$ is $\beta\theta_{2k}(\iota)+\theta_{2k+1}(\iota)$, so if we assume that $\beta\nabla^*(\lambda(\iota)) = 0$, this coefficient must vanish since we are in the tensor product $H^*(L^\infty) \otimes H^*(K_n)$. So we get $\beta\theta_{2k}(\iota) = -\theta_{2k+1}(\iota)$ and hence $\beta\theta_{2k}(\alpha) = -\theta_{2k+1}(\alpha)$ for all α. Note that $\beta\theta_{2k+1} = 0$ from the coefficient of $\omega_{(p-1)n-2k-1}$. This also follows from the formula $\beta\theta_{2k} = -\theta_{2k+1}$ since $\beta^2 = 0$.

In order to show that $\beta\nabla^*(\lambda(\iota)) = 0$ we first compute $\beta\lambda(\iota)$. We may assume K_n has a single n-cell and a single $(n + 1)$-cell, attached by a map of degree p. Let φ and ψ be the cellular cochains assigning the value 1 to the n-cell and the $(n+1)$-cell, respectively, so $\delta\varphi = p\psi$. In $K_n^{\wedge p}$ we then have

$$(*) \qquad \delta(\varphi^{\otimes p}) = \sum_i (-1)^{in} \varphi^{\otimes i} \otimes \delta\varphi \otimes \varphi^{\otimes(p-i-1)} = p \sum_i (-1)^{in} \varphi^{\otimes i} \otimes \psi \otimes \varphi^{\otimes(p-i-1)}$$

where the tensor notation means cellular cross product, so for example $\varphi^{\otimes p}$ is the cellular cochain dual to the np-cell $e^n \times \cdots \times e^n$ of $K_n^{\wedge p}$. The formula $(*)$ holds also in ΛK_n since the latter space has only one $(np + 1)$-cell not in $K_n^{\wedge p}$, with cellular boundary zero. Namely, this cell is the product of the 1-cell of L^∞ and the np-cell of $K_n^{\wedge p}$ with one end of this product attached to the np-cell by the identity map and the other end by the cyclic permutation T, which has degree $+1$ since p is odd, so these two terms in the boundary of this cell cancel, and there are no other terms since the rest of the attachment of this cell is at the basepoint.

Bockstein homomorphisms can be computed using cellular cochain complexes, so the formula $(*)$ says that $\sum_i (-1)^{in} \varphi^{\otimes i} \otimes \psi \otimes \varphi^{\otimes(p-i-1)}$ represents $\beta\lambda(\iota)$. Via the quotient map $\Gamma K_n \to \Lambda K_n$, the class $\lambda(\iota)$ pulls back to a class $y(\iota)$ with $\beta y(\iota)$ also represented by $\sum_i (-1)^{in} \varphi^{\otimes i} \otimes \psi \otimes \varphi^{\otimes(p-i-1)}$. To see what happens when we pull $\beta y(\iota)$ back to $\beta \nabla^*(\lambda(\iota))$ via the inclusion $L^\infty \times K_n \hookrightarrow \Gamma K_n$, consider the commutative diagram at the right. In the left-hand square the maps π^* are induced by the covering space projections $\pi : S^\infty \times K_n^{\wedge p} \to \Gamma K_n$

$$
\begin{array}{ccccc}
H^*(\Gamma K_n) & \xrightarrow{\pi^*} & H^*(S^\infty \times K_n^{\wedge p}) & \xrightarrow{\tau} & H^*(\Gamma K_n) \\
\downarrow & & \downarrow & & \downarrow \\
H^*(L^\infty \times K_n) & \xrightarrow{\pi^*} & H^*(S^\infty \times K_n) & \xrightarrow{\tau} & H^*(L^\infty \times K_n)
\end{array}
$$

and $\pi : S^\infty \times K_n \to L^\infty \times K_n$ arising from the free $\mathbb{Z}_p$ actions. The vertical maps are induced by the diagonal inclusion $S^\infty \times K \hookrightarrow S^\infty \times K_n^{\wedge p}$. The maps τ are the transfer homomorphisms defined in §3.G. Recall the definition: If $\pi : \tilde{X} \to X$ is a p-sheeted covering space, a chain map $C_*(X) \to C_*(\tilde{X})$ is defined by sending a singular simplex $\sigma : \Delta^k \to X$ to the sum of its p lifts to $\tilde{X}$, and τ is the induced map on cohomology. The key property of τ is that $\tau \pi^* : H^*(X) \to H^*(X)$ is multiplication by p, for any choice of coefficient group, since when we project the p lifts of $\sigma : \Delta^k \to X$ back to X we get $p\sigma$. When X is a CW complex and $\tilde{X}$ is given the lifted CW structure, then τ can also be defined in cellular cohomology by the same procedure.

Let us compute the value of the upper τ in the diagram on $1 \otimes \psi \otimes \varphi^{\otimes(p-1)}$ where '1' is the cellular cocycle assigning the value 1 to each 0-cell of S^∞. By the definition of τ we have $\tau(1 \otimes \psi \otimes \varphi^{\otimes(p-1)}) = \sum_i T^i(\psi \otimes \varphi^{\otimes(p-1)})$ where $T : K_n^{\wedge p} \to K_n^{\wedge p}$ permutes the factors cyclically. It does not matter whether T moves coordinates one unit leftwards or one unit rightwards since we are summing over all the powers of T, so let us say T moves coordinates rightward. Then $T(\psi \otimes \varphi^{\otimes(p-1)}) = \varphi \otimes \psi \otimes \varphi^{\otimes(p-2)}$, with the last φ moved into the first position. This move is achieved by transposing this φ with each of the preceding $p-2$ φ's and with ψ. Transposing two φ's introduces a sign $(-1)^{n^2}$, and transposing φ with ψ introduces a sign $(-1)^{n(n+1)} = +1$, by the commutativity property of cross product. Thus the total sign introduced by T is $(-1)^{n^2(p-2)}$, which equals $(-1)^n$ since p is odd. Each successive iterate of T also introduces a sign of $(-1)^n$, so T^i introduces a sign $(-1)^{in}$ for $0 \le i \le p-1$. Thus

$$
\tau(1 \otimes \psi \otimes \varphi^{\otimes(p-1)}) = \sum_i T^i(\psi \otimes \varphi^{\otimes(p-1)}) = \sum_i (-1)^{in} \varphi^{\otimes i} \otimes \psi \otimes \varphi^{\otimes(p-i-1)}
$$

As observed earlier, this last cocycle represents the class $\beta y(\iota)$.

Since $\beta y(\iota)$ is in the image of the upper τ in the diagram, the image of $\beta y(\iota)$ in $H^*(L^\infty \times K_n)$, which is $\nabla^*(\beta\lambda(\iota))$, is in the image of the lower τ since the right-hand square commutes. The map π^* in the lower row is obviously onto since S^∞ is contractible, so $\nabla^*(\beta\lambda(\iota))$ is in the image of the composition $\tau\pi^*$ across the bottom of the diagram. But this composition is multiplication by p, which is zero for $\mathbb{Z}_p$ coefficients, so $\beta\nabla^*(\lambda(\iota)) = \nabla^*(\beta\lambda(\iota)) = 0$. $\square$

The derivation of the Adem relations now follows the pattern for the case $p = 2$. We had the formula $\nabla_2^* \lambda_2^*(\iota) = \sum_{i,j} \omega_{(p-1)pn-i} \otimes \theta_i(\omega_{(p-1)n-j} \otimes \theta_j(\iota))$. Since we are letting $p = 2m + 1$, this can be rewritten as $\sum_{i,j} \omega_{2mpn-i} \otimes \theta_i(\omega_{2mn-j} \otimes \theta_j(\iota))$. The only nonzero θ_i's are $\theta_{2i(p-1)} = (-1)^i a_n P^i$ and $\theta_{2i(p-1)+1} = -\beta \theta_{2i(p-1)}$ so we have

$$\sum_{i,j} \omega_{2mpn-i} \otimes \theta_i(\omega_{2mn-j} \otimes \theta_j(\iota)) =$$
$$\sum_{i,j}(-1)^{i+j} a_{2mn} a_n \omega_{2m(pn-2i)} \otimes P^i(\omega_{2m(n-2j)} \otimes P^j(\iota))$$
$$- \sum_{i,j}(-1)^{i+j} a_{2mn} a_n \omega_{2m(pn-2i)} \otimes P^i(\omega_{2m(n-2j)-1} \otimes \beta P^j(\iota))$$
$$- \sum_{i,j}(-1)^{i+j} a_{2mn} a_n \omega_{2m(pn-2i)-1} \otimes \beta P^i(\omega_{2m(n-2j)} \otimes P^j(\iota))$$
$$+ \sum_{i,j}(-1)^{i+j} a_{2mn} a_n \omega_{2m(pn-2i)-1} \otimes \beta P^i(\omega_{2m(n-2j)-1} \otimes \beta P^j(\iota))$$

Since m and n will be fixed throughout the discussion, we may factor out the nonzero constant $a_{2mn} a_n$. Then applying the Cartan formula to expand the P^i terms, using also the formulas $P^k(\omega_{2r}) = \binom{r}{k}\omega_{2r+2k(p-1)}$ and $P^k(\omega_{2r+1}) = \binom{r}{k}\omega_{2r+2k(p-1)+1}$ derived earlier in the section, we obtain

$$\sum_{i,j,k}(-1)^{i+j}\binom{m(n-2j)}{k}\omega_{2m(pn-2i)} \otimes \omega_{2m(n-2j+2k)} \otimes P^{i-k}P^j(\iota)$$
$$- \sum_{i,j,k}(-1)^{i+j}\binom{m(n-2j)-1}{k}\omega_{2m(pn-2i)} \otimes \omega_{2m(n-2j+2k)-1} \otimes P^{i-k}\beta P^j(\iota)$$
$$- \sum_{i,j,k}(-1)^{i+j}\binom{m(n-2j)}{k}\omega_{2m(pn-2i)-1} \otimes \omega_{2m(n-2j+2k)} \otimes \beta P^{i-k}P^j(\iota)$$
$$+ \sum_{i,j,k}(-1)^{i+j}\binom{m(n-2j)-1}{k}\omega_{2m(pn-2i)-1} \otimes \omega_{2m(n-2j+2k)} \otimes P^{i-k}\beta P^j(\iota)$$
$$- \sum_{i,j,k}(-1)^{i+j}\binom{m(n-2j)-1}{k}\omega_{2m(pn-2i)-1} \otimes \omega_{2m(n-2j+2k)-1} \otimes \beta P^{i-k}\beta P^j(\iota)$$

Letting $\ell = mn+j-k$, so that $n-2j+2k = pn-2\ell$, the first of these five summations becomes

$$\sum_{i,j,\ell}(-1)^{i+j}\binom{m(n-2j)}{mn+j-\ell}\omega_{2m(pn-2i)} \otimes \omega_{2m(pn-2\ell)} \otimes P^{i+\ell-mn-j}P^j(\iota)$$

and similarly for the other four summations.

Now we bring in the symmetry property $\varphi_{rs} = (-1)^{rs+mnp}\varphi_{sr}$, where, as before, $\nabla_2^* \lambda_2^*(\iota) = \sum_{r,s} \omega_r \otimes \omega_s \otimes \varphi_{rs}$. Of the five summations, only the first has both ω terms with even subscripts, namely $r = 2m(pn - 2i)$ and $s = 2m(pn - 2\ell)$, so the coefficient of $\omega_r \otimes \omega_s$ in this summation must be symmetric with respect to switching i and ℓ, up to a sign which will be $+$ if we choose n to be even, as we will do. This gives the relation

(1) $\sum_j(-1)^{i+j}\binom{m(n-2j)}{mn+j-\ell}P^{i+\ell-mn-j}P^j(\iota) = \sum_j(-1)^{\ell+j}\binom{m(n-2j)}{mn+j-i}P^{i+\ell-mn-j}P^j(\iota)$

Similarly, the second, third, and fourth summations involve ω's with subscripts of opposite parity, yielding the relation

(2) $\sum_j(-1)^{i+j}\binom{m(n-2j)-1}{mn+j-\ell}P^{i+\ell-mn-j}\beta P^j(\iota) =$
$$\sum_j(-1)^{\ell+j}\binom{m(n-2j)}{mn+j-i}\beta P^{i+\ell-mn-j}P^j(\iota) - \sum_j(-1)^{\ell+j}\binom{m(n-2j)-1}{mn+j-i}P^{i+\ell-mn-j}\beta P^j(\iota)$$

The relations (1) and (2) will yield the two Adem relations, so we will not need to consider the relation arising from the fifth summation.

To get the first Adem relation from (1) we choose n and ℓ so that the left side of (1) has only one term, namely we take $n = 2(1 + p + \cdots + p^{r-1}) + 2s$ and $\ell = mn + s$ for given integers r and s. Then

$$\binom{m(n-2j)}{mn+j-\ell} = \binom{p^r-1-(p-1)(j-s)}{j-s}$$

and if r is large, this binomial coefficient is 1 if $j = s$ and 0 otherwise since if the rightmost nonzero digit in the p-adic expansion of the 'denominator' $j - s$ is x, then the corresponding digit of the 'numerator' $(p - 1)[(1 + p + \cdots + p^{r-1}) - (j - s)]$ is obtained by reducing $(p - 1)(1 - x)$ mod p, giving $x - 1$, and $\binom{x-1}{x} = 0$. Then (1) becomes

$$(-1)^{i+s} P^i P^s(\iota) = \sum_j (-1)^{\ell+j} \binom{m(n-2j)}{mn+j-i} P^{i+s-j} P^j(\iota)$$

$$\text{or} \quad P^i P^s(\iota) = \sum_j (-1)^{i+j} \binom{m(n-2j)}{mn+j-i} P^{i+s-j} P^j(\iota) \qquad \text{since } \ell \equiv s \bmod 2$$

$$= \sum_j (-1)^{i+j} \binom{m(n-2j)}{i-pj} P^{i+s-j} P^j(\iota) \qquad \text{since } \binom{x}{y} = \binom{x}{x-y}$$

$$= \sum_j (-1)^{i+j} \binom{p^r+(p-1)(s-j)-1}{i-pj} P^{i+s-j} P^j(\iota)$$

If r is large and $i < ps$, the term p^r in the binomial coefficient can be omitted since we may assume $i \geq pj$, hence $j < s$, so $-1 + (p - 1)(s - j) \geq 0$ and the p^r has no effect on the binomial coefficient if r is large. This shows the first Adem relation holds for the class ι, and the general case follows as in the case $p = 2$.

To get the second Adem relation we choose $n = 2p^r + 2s$ and $\ell = mn + s$. Reasoning as before, the left side of (2) then reduces to $(-1)^{i+s} P^i \beta P^s(\iota)$ and (2) becomes

$$P^i \beta P^s(\iota) = \sum_j (-1)^{i+j} \binom{(p-1)(p^r+s-j)}{i-pj} \beta P^{i+s-j} P^j(\iota)$$
$$- \sum_j (-1)^{i+j} \binom{(p-1)(p^r+s-j)-1}{i-pj-1} P^{i+s-j} \beta P^j(\iota)$$

This time the term p^r can be omitted if r is large and $i \leq ps$. $\qquad \square$

Exercises

1. Determine all cohomology operations $H^1(X; \mathbb{Z}) \to H^n(X; \mathbb{Z})$, $H^2(X; \mathbb{Z}) \to H^n(X; \mathbb{Z})$, and $H^1(X; \mathbb{Z}_p) \to H^n(X; \mathbb{Z}_p)$ for p prime.

2. Use cohomology operations to show that the spaces $(S^1 \times \mathbb{CP}^\infty)/(S^1 \times \{x_0\})$ and $S^3 \times \mathbb{CP}^\infty$ are not homotopy equivalent.

3. Since there is a fiber bundle $S^2 \to \mathbb{CP}^5 \to \mathbb{HP}^2$ by Exercise 35 in §4.2, one might ask whether there is an analogous bundle $S^4 \to \mathbb{HP}^5 \to \mathbb{OP}^2$. Use Steenrod powers for the prime 3 to show that such a bundle cannot exist. [The Gysin sequence can be used to determine the map on cohomology induced by the bundle projection $\mathbb{HP}^5 \to \mathbb{OP}^2$.]

4. Show there is no fiber bundle $S^7 \to S^{23} \to \mathbb{OP}^2$. [Compute the cohomology ring of the mapping cone of the projection $S^{23} \to \mathbb{OP}^2$ via Poincaré duality or the Thom isomorphism.]

5. Show that the subalgebra of $\mathcal{A}_2$ generated by Sq^i for $i \leq 2$ has dimension 8 as a vector space over $\mathbb{Z}_2$, with multiplicative structure encoded in the following diagram, where diagonal lines indicate left-multiplication by Sq^1 and horizontal lines indicate left-multiplication by Sq^2.

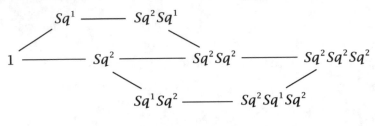

Appendix

Topology of Cell Complexes

Here we collect a number of basic topological facts about CW complexes for convenient reference. A few related facts about manifolds are also proved.

Let us first recall from Chapter 0 that a CW complex is a space X constructed in the following way:

(1) Start with a discrete set X^0, the 0-cells of X.

(2) Inductively, form the n-skeleton X^n from X^{n-1} by attaching n-cells e_α^n via maps $\varphi_\alpha : S^{n-1} \to X^{n-1}$. This means that X^n is the quotient space of $X^{n-1} \coprod_\alpha D_\alpha^n$ under the identifications $x \sim \varphi_\alpha(x)$ for $x \in \partial D_\alpha^n$. The cell e_α^n is the homeomorphic image of $D_\alpha^n - \partial D_\alpha^n$ under the quotient map.

(3) $X = \bigcup_n X^n$ with the weak topology: A set $A \subset X$ is open (or closed) iff $A \cap X^n$ is open (or closed) in X^n for each n.

Note that condition (3) is superfluous when X is finite-dimensional, so that $X = X^n$ for some n. For if A is open in $X = X^n$, the definition of the quotient topology on X^n implies that $A \cap X^{n-1}$ is open in X^{n-1}, and then by the same reasoning $A \cap X^{n-2}$ is open in X^{n-2}, and similarly for all the skeleta X^{n-i}.

Each cell e_α^n has its **characteristic map** Φ_α, which is by definition the composition $D_\alpha^n \hookrightarrow X^{n-1} \coprod_\alpha D_\alpha^n \to X^n \hookrightarrow X$. This is continuous since it is a composition of continuous maps, the inclusion $X^n \hookrightarrow X$ being continuous by (3). The restriction of Φ_α to the interior of D_α^n is a homeomorphism onto e_α^n.

An alternative way to describe the topology on X is to say that a set $A \subset X$ is open (or closed) iff $\Phi_\alpha^{-1}(A)$ is open (or closed) in D_α^n for each characteristic map Φ_α. In one direction this follows from continuity of the Φ_α's, and in the other direction, suppose $\Phi_\alpha^{-1}(A)$ is open in D_α^n for each Φ_α, and suppose by induction on n that $A \cap X^{n-1}$ is open in X^{n-1}. Then since $\Phi_\alpha^{-1}(A)$ is open in D_α^n for all α, $A \cap X^n$ is open in X^n by the definition of the quotient topology on X^n. Hence by (3), A is open in X.

A consequence of this characterization of the topology on X is that X is a quotient space of $\coprod_{n,\alpha} D_\alpha^n$.

A **subcomplex** of a CW complex X is a subspace $A \subset X$ which is a union of cells of X, such that the closure of each cell in A is contained in A. Thus for each cell in A, the image of its attaching map is contained in A, so A is itself a CW complex. Its CW complex topology is the same as the topology induced from X, as one sees by noting inductively that the two topologies agree on $A^n = A \cap X^n$. It is easy to see by induction over skeleta that a subcomplex is a closed subspace. Conversely, a subcomplex could be defined as a closed subspace which is a union of cells.

A finite CW complex, that is, one with only finitely many cells, is compact since attaching a single cell preserves compactness. A sort of converse to this is:

Proposition A.1. *A compact subspace of a CW complex is contained in a finite subcomplex.*

Proof: First we show that a compact set C in a CW complex X can meet only finitely many cells of X. Suppose on the contrary that there is an infinite sequence of points $x_i \in C$ all lying in distinct cells. Then the set $S = \{x_1, x_2, \cdots\}$ is closed in X. Namely, assuming $S \cap X^{n-1}$ is closed in X^{n-1} by induction on n, then for each cell e_α^n of X, $\varphi_\alpha^{-1}(S)$ is closed in ∂D_α^n, and $\Phi_\alpha^{-1}(S)$ consists of at most one more point in D_α^n, so $\Phi_\alpha^{-1}(S)$ is closed in D_α^n. Therefore $S \cap X^n$ is closed in X^n for each n, hence S is closed in X. The same argument shows that any subset of S is closed, so S has the discrete topology. But it is compact, being a closed subset of the compact set C. Therefore S must be finite, a contradiction.

Since C is contained in a finite union of cells, it suffices to show that a finite union of cells is contained in a finite subcomplex of X. A finite union of finite subcomplexes is again a finite subcomplex, so this reduces to showing that a single cell e_α^n is contained in a finite subcomplex. The image of the attaching map φ_α for e_α^n is compact, hence by induction on dimension this image is contained in a finite subcomplex $A \subset X^{n-1}$. So e_α^n is contained in the finite subcomplex $A \cup e_\alpha^n$. $\square$

Now we can explain the mysterious letters 'CW,' which refer to the following two properties satisfied by CW complexes:

(1) Closure-finiteness: The closure of each cell meets only finitely many other cells. This follows from the preceding proposition since the closure of a cell is compact, being the image of a characteristic map.

(2) Weak topology: A set is closed iff it meets the closure of each cell in a closed set. For if a set meets the closure of each cell in a closed set, it pulls back to a closed set under each characteristic map, hence is closed by an earlier remark.

In J. H. C. Whitehead's original definition of CW complexes these two properties played a more central role. The following proposition contains essentially this definition.

Proposition A.2. *Given a Hausdorff space X and a family of maps $\Phi_\alpha : D_\alpha^n \to X$, then these maps are the characteristic maps of a CW complex structure on X iff:*

(i) *Each Φ_α restricts to a homeomorphism from $\operatorname{int} D_\alpha^n$ onto its image, a cell $e_\alpha^n \subset X$, and these cells are all disjoint and their union is X.*

(ii) *For each cell e_α^n, $\Phi_\alpha(\partial D_\alpha^n)$ is contained in the union of a finite number of cells of dimension less than n.*

(iii) *A subset of X is closed iff it meets the closure of each cell of X in a closed set.*

Condition (iii) can be restated as saying that a set $C \subset X$ is closed iff $\Phi_\alpha^{-1}(C)$ is closed in D_α^n for all α, since a map from a compact space onto a Hausdorff space is a quotient map. In particular, if there are only finitely many cells then (iii) is automatic since in this case the projection $\coprod_\alpha D_\alpha^n \to X$ is a map from a compact space onto a Hausdorff space, hence is a quotient map.

For an example where all the conditions except the finiteness hypothesis in (ii) are satisfied, take X to be D^2 with its interior as a 2-cell and each point of ∂D^2 as a 0-cell. The identity map of D^2 serves as the Φ_α for the 2-cell. Condition (iii) is satisfied since it is a nontrivial condition only for the 2-cell.

Proof: We have already taken care of the 'only if' implication. For the converse, suppose inductively that X^{n-1}, the union of all cells of dimension less than n, is a CW complex with the appropriate Φ_α's as characteristic maps. The induction can start with $X^{-1} = \varnothing$. Let $f : X^{n-1} \coprod_\alpha D_\alpha^n \to X^n$ be given by the inclusion on X^{n-1} and the maps Φ_α for all the n-cells of X. This is a continuous surjection, and if we can show it is a quotient map, then X^n will be obtained from X^{n-1} by attaching the n-cells e_α^n. Thus if $C \subset X^n$ is such that $f^{-1}(C)$ is closed, we need to show that $C \cap \overline{e}_\beta^m$ is closed for all cells e_β^m of X, the bar denoting closure.

There are three cases. If $m < n$ then $f^{-1}(C)$ closed implies $C \cap X^{n-1}$ closed, hence $C \cap \overline{e}_\beta^m$ is closed since $\overline{e}_\beta^m \subset X^{n-1}$. If $m = n$ then e_β^m is one of the cells e_α^n, so $f^{-1}(C)$ closed implies $f^{-1}(C) \cap D_\alpha^n$ is closed, hence compact, hence its image $C \cap \overline{e}_\alpha^n$ under f is compact and therefore closed. Finally there is the case $m > n$. Then $C \subset X^n$ implies $C \cap \overline{e}_\beta^m \subset \Phi_\beta(\partial D_\beta^m)$. The latter space is contained in a finite union of $\overline{e}_\gamma^\ell$'s with $\ell < m$. By induction on m, each $C \cap \overline{e}_\gamma^\ell$ is closed. Hence the intersection of C with the union of the finite collection of $\overline{e}_\gamma^\ell$'s is closed. Intersecting this closed set with $\overline{e}_\beta^m$, we conclude that $C \cap \overline{e}_\beta^m$ is closed.

It remains only to check that X has the weak topology with respect to the X^n's, that is, a set in X is closed iff it intersects each X^n in a closed set. The preceding argument with $C = X^n$ shows that X^n is closed, so a closed set intersects each X^n in a closed set. Conversely, if a set C intersects X^n in a closed set, then C intersects each $\overline{e}_\alpha^n$ in a closed set, so C is closed in X by (iii). $\qquad\square$

Next we describe a convenient way of constructing open neighborhoods $N_\varepsilon(A)$ of subsets A of a CW complex X, where ε is a function assigning a number $\varepsilon_\alpha > 0$ to each cell e_α^n of X. The construction is inductive over the skeleta X^n, so suppose we have already constructed $N_\varepsilon^n(A)$, a neighborhood of $A \cap X^n$ in X^n, starting the process with $N_\varepsilon^0(A) = A \cap X^0$. Then we define $N_\varepsilon^{n+1}(A)$ by specifying its preimage under the characteristic map $\Phi_\alpha : D^{n+1} \to X$ of each cell e_α^{n+1}, namely, $\Phi_\alpha^{-1}(N_\varepsilon^{n+1}(A))$ is the union of two parts: an open ε_α-neighborhood of $\Phi_\alpha^{-1}(A) - \partial D^{n+1}$ in $D^{n+1} - \partial D^{n+1}$, and a product $(1 - \varepsilon_\alpha, 1] \times \Phi_\alpha^{-1}(N_\varepsilon^n(A))$ with respect to 'spherical' coordinates (r, θ) in D^{n+1}, where $r \in [0, 1]$ is the radial coordinate and θ lies in $\partial D^{n+1} = S^n$. Then we define $N_\varepsilon(A) = \bigcup_n N_\varepsilon^n(A)$. This is an open set in X since it pulls back to an open set under each characteristic map.

Proposition A.3. *CW complexes are normal, and in particular, Hausdorff.*

Proof: Points are closed in a CW complex X since they pull back to closed sets under all characteristic maps Φ_α. For disjoint closed sets A and B in X, we show that $N_\varepsilon(A)$ and $N_\varepsilon(B)$ are disjoint for small enough ε_α's. In the inductive process for building these open sets, assume $N_\varepsilon^n(A)$ and $N_\varepsilon^n(B)$ have been chosen to be disjoint. For a characteristic map $\Phi_\alpha : D^{n+1} \to X$, observe that $\Phi_\alpha^{-1}(N_\varepsilon^n(A))$ and $\Phi_\alpha^{-1}(B)$ are a positive distance apart, since otherwise by compactness we would have a sequence in $\Phi_\alpha^{-1}(B)$ converging to a point of $\Phi_\alpha^{-1}(B)$ in ∂D^{n+1} of distance zero from $\Phi_\alpha^{-1}(N_\varepsilon^n(A))$, but this is impossible since $\Phi_\alpha^{-1}(N_\varepsilon^n(B))$ is a neighborhood of $\Phi_\alpha^{-1}(B) \cap \partial D^{n+1}$ in ∂D^{n+1} disjoint from $\Phi_\alpha^{-1}(N_\varepsilon^n(A))$. Similarly, $\Phi_\alpha^{-1}(N_\varepsilon^n(B))$ and $\Phi_\alpha^{-1}(A)$ are a positive distance apart. Also, $\Phi_\alpha^{-1}(A)$ and $\Phi_\alpha^{-1}(B)$ are a positive distance apart. So a small enough ε_α will make $\Phi_\alpha^{-1}(N_\varepsilon^{n+1}(A))$ disjoint from $\Phi_\alpha^{-1}(N_\varepsilon^{n+1}(B))$ in D^{n+1}. $\square$

Proposition A.4. *Each point in a CW complex has arbitrarily small contractible open neighborhoods, so CW complexes are locally contractible.*

Proof: Given a point x in a CW complex X and a neighborhood U of x in X, we can choose the ε_α's small enough so that $N_\varepsilon(x) \subset U$ by requiring that the closure of $N_\varepsilon^n(x)$ be contained in U for each n. It remains to see that $N_\varepsilon(x)$ is contractible. If $x \in X^m - X^{m-1}$ and $n > m$ we can construct a deformation retraction of $N_\varepsilon^n(x)$ onto $N_\varepsilon^{n-1}(x)$ by sliding outward along radial segments in cells e_β^n, the images under the characteristic maps Φ_β of radial segments in D^n. A deformation retraction of $N_\varepsilon(x)$ onto $N_\varepsilon^m(x)$ is then obtained by performing the deformation retraction of $N_\varepsilon^n(x)$ onto $N_\varepsilon^{n-1}(x)$ during the t-interval $[1/2^n, 1/2^{n-1}]$, points of $N_\varepsilon^n(x) - N_\varepsilon^{n-1}(x)$ being stationary outside this t-interval. Finally, $N_\varepsilon^m(x)$ is an open ball about x, and so deformation retracts onto x. $\square$

In particular, CW complexes are locally path-connected. So a CW complex is path-connected iff it is connected.

Proposition A.5. *For a subcomplex A of a CW complex X, the open neighborhood $N_\varepsilon(A)$ deformation retracts onto A if $\varepsilon_\alpha < 1$ for all α.*

Proof: In each cell of $X - A$, $N_\varepsilon(A)$ is a product neighborhood of the boundary of this cell, so a deformation retraction of $N_\varepsilon(A)$ onto A can be constructed just as in the previous proof. $\qquad\qquad\square$

Note that for subcomplexes A and B of X, we have $N_\varepsilon(A) \cap N_\varepsilon(B) = N_\varepsilon(A \cap B)$. This implies for example that the van Kampen theorem and Mayer-Vietoris sequences hold for decompositions $X = A \cup B$ into subcomplexes A and B as well as into open sets A and B.

A map $f : X \to Y$ with domain a CW complex is continuous iff its restrictions to the closures $\overline{e}_\alpha^n$ of all cells e_α^n are continuous, and it is useful to know that the same is true for homotopies $f_t : X \to Y$. With this objective in mind, let us introduce a little terminology. A topological space X is said to be **generated** by a collection of subspaces X_α if $X = \bigcup_\alpha X_\alpha$ and a set $A \subset X$ is closed iff $A \cap X_\alpha$ is closed in X_α for each α. Equivalently, we could say 'open' instead of 'closed' here, but 'closed' is more convenient for our present purposes. As noted earlier, though not in these words, a CW complex X is generated by the closures $\overline{e}_\alpha^n$ of its cells e_α^n. Since every finite subcomplex of X is a finite union of closures $\overline{e}_\alpha^n$, X is also generated by its finite subcomplexes. It follows that X is also generated by its compact subspaces, or more briefly, X is **compactly generated**.

Proposition A.15 later in the Appendix asserts that if X is a compactly generated Hausdorff space and Z is locally compact, then $X \times Z$, with the product topology, is compactly generated. In particular, $X \times I$ is compactly generated if X is a CW complex. Since every compact set in $X \times I$ is contained in the product of a compact subspace of X with I, hence in the product of a finite subcomplex of X with I, such product subspaces also generate $X \times I$. Since such a product subspace is a finite union of products $\overline{e}_\alpha^n \times I$, it is also true that $X \times I$ is generated by its subspaces $\overline{e}_\alpha^n \times I$. This implies that a homotopy $F : X \times I \to Y$ is continuous iff its restrictions to the subspaces $\overline{e}_\alpha^n \times I$ are continuous, which is the statement we were seeking.

Products of CW Complexes

There are some unexpected point-set-topological subtleties that arise with products of CW complexes. As we shall show, the product of two CW complexes does have a natural CW structure, but its topology is in general finer, with more open sets, than the product topology. However, the distinctions between the two topologies are rather small, and indeed nonexistent in most cases of interest, so there is no real problem for algebraic topology.

Given a space X and a collection of subspaces X_α whose union is X, these subspaces generate a possibly finer topology on X by defining a set $A \subset X$ to be open iff $A \cap X_\alpha$ is open in X_α for all α. The axioms for a topology are easily verified for this definition. In case $\{X_\alpha\}$ is the collection of compact subsets of X, we write X_c for this new compactly generated topology. It is easy to see that X and X_c have the same compact subsets, and the two induced topologies on these compact subsets coincide. If X is compact, or even locally compact, then $X = X_c$, that is, X is compactly generated.

Theorem A.6. *For CW complexes X and Y with characteristic maps Φ_α and Ψ_β, the product maps $\Phi_\alpha \times \Psi_\beta$ are the characteristic maps for a CW complex structure on $(X \times Y)_c$. If either X or Y is compact or more generally locally compact, then $(X \times Y)_c = X \times Y$. Also, $(X \times Y)_c = X \times Y$ if both X and Y have countably many cells.*

Proof: For the first statement it suffices to check that the three conditions in Proposition A.2 are satisfied when we take the space 'X' there to be $(X \times Y)_c$. The first two conditions are obvious. For the third, which says that $(X \times Y)_c$ is generated by the products $\overline{e}_\alpha^m \times \overline{e}_\beta^n$, observe that every compact set in $X \times Y$ is contained in the product of its projections onto X and Y, and these projections are compact and hence contained in finite subcomplexes of X and Y, so the original compact set is contained in a finite union of products $\overline{e}_\alpha^m \times \overline{e}_\beta^n$. Hence the products $\overline{e}_\alpha^m \times \overline{e}_\beta^n$ generate $(X \times Y)_c$.

The second assertion of the theorem is a special case of Proposition A.15, having nothing to do with CW complexes, which says that a product $X \times Y$ is compactly generated if X is compactly generated Hausdorff and Y is locally compact.

For the last statement of the theorem, suppose X and Y each have at most countably many cells. For an open set $W \subset (X \times Y)_c$ and a point $(a, b) \in W$ we need to find a product $U \times V \subset W$ with U an open neighborhood of a in X and V an open neighborhood of b in Y. Choose finite subcomplexes $X_1 \subset X_2 \subset \cdots$ of X with $X = \bigcup_i X_i$, and similarly for Y. We may assume $a \in X_1$ and $b \in Y_1$. Since the two topologies agree on $X_1 \times Y_1$, there is a compact product neighborhood $K_1 \times L_1 \subset W$ of (a, b) in $X_1 \times Y_1$. Assuming inductively that $K_i \times L_i \subset W$ has been constructed in $X_i \times Y_i$, we would like to construct $K_{i+1} \times L_{i+1} \subset W$ as a compact neighborhood of $K_i \times L_i$ in $X_{i+1} \times Y_{i+1}$. To do this, we first choose for each $x \in K_i$ compact neighborhoods K_x of x in X_{i+1} and L_x of L_i in Y_{i+1} such that $K_x \times L_x \subset W$, using the compactness of L_i. By compactness of K_i, a finite number of the K_x's cover K_i. Let K_{i+1} be the union of these K_x's and let L_{i+1} be the intersection of the corresponding L_x's. This defines the desired $K_{i+1} \times L_{i+1}$. Let U_i be the interior of K_i in X_i, so $U_i \subset U_{i+1}$ for each i. The union $U = \bigcup_i U_i$ is then open in X since it intersects each X_i in a union of open sets and the X_i's generate X. In the same way the L_i's yield an open set V in Y. Thus we have a product of open sets $U \times V \subset W$ containing (a, b). $\qquad\square$

We will describe now an example from [Dowker 1952] where the product topology on $X \times Y$ differs from the CW topology. Both X and Y will be graphs consisting of infinitely many edges emanating from a single vertex, with uncountably many edges for X and countably many for Y.

Let $X = \bigvee_s I_s$ where I_s is a copy of the interval $[0, 1]$ and the index s ranges over all infinite sequences $s = (s_1, s_2, \cdots)$ of positive integers. The wedge sum is formed at the 0 endpoint of I_s. Similarly we let $Y = \bigvee_j I_j$ but with j varying just over positive integers. Let p_{sj} be the point $(1/s_j, 1/s_j) \in I_s \times I_j \subset X \times Y$ and let P be the union of all these points p_{sj}. Thus P consists of a single point in each 2-cell of $X \times Y$, so P is closed in the CW topology on $X \times Y$. We will show it is not closed in the product topology by showing that (x_0, y_0) lies in its closure, where x_0 is the common endpoint of the intervals I_s and y_0 is the common endpoint of the intervals I_j.

A basic open set containing (x_0, y_0) in the product topology has the form $U \times V$ where $U = \bigvee_s [0, a_s)$ and $V = \bigvee_j [0, b_j)$. It suffices to show that P has nonempty intersection with $U \times V$. Choose a sequence $t = (t_1, t_2, \cdots)$ with $t_j > j$ and $t_j > 1/b_j$ for all j, and choose an integer $k > 1/a_t$. Then $t_k > k > 1/a_t$ hence $1/t_k < a_t$. We also have $1/t_k < b_k$. So $(1/t_k, 1/t_k)$ is a point of P that lies in $[0, a_t) \times [0, b_k)$ and hence in $U \times V$.

Euclidean Neighborhood Retracts

At certain places in this book it is desirable to know that a given compact space is a retract of a finite simplicial complex, or equivalently (as we shall see) a retract of a neighborhood in some Euclidean space. For example, this condition occurs in the Lefschetz fixed point theorem, and it was used in the proof of Alexander duality. So let us study this situation in more detail.

Theorem A.7. *A compact subspace K of $\mathbb{R}^n$ is a retract of some neighborhood iff K is locally contractible in the weak sense that for each $x \in K$ and each neighborhood U of x in K there exists a neighborhood $V \subset U$ of x such that the inclusion $V \hookrightarrow U$ is nullhomotopic.*

Note that if K is a retract of some neighborhood, then it is a retract of every smaller neighborhood, just by restriction of the retraction. So it does not matter if we require the neighborhoods to be open. Similarly it does not matter if the neighborhoods U and V in the statement of the theorem are required to be open.

Proof: Let us do the harder half first, constructing a retraction of a neighborhood of K onto K under the local contractibility assumption. The first step is to put a CW structure on the open set $X = \mathbb{R}^n - K$, with the size of the cells approaching zero near K. Consider the subdivision of $\mathbb{R}^n$ into unit cubes of dimension n with vertices at the points with integer coordinates. Call this collection of cubes C_0. For an integer $k > 0$, we can subdivide the cubes of C_0 by taking n-dimensional cubes of edgelength $1/2^k$ with vertices having coordinates of the form $i/2^k$ for $i \in \mathbb{Z}$. Denote this collection of cubes by C_k. Let $A_0 \subset C_0$ be the set of cubes disjoint from K, and inductively, let $A_k \subset C_k$ be the set of cubes disjoint from K and not contained in cubes of A_j for $j < k$. The open set X is then the union of all the cubes in the combined collection $A = \bigcup_k A_k$. Note that the collection A is locally finite: Each point of X has a neighborhood meeting only finitely many cubes in A, since the point has a positive distance from the closed set K.

If two cubes of A intersect, their intersection is an i-dimensional face of one of them for some $i < n$. Likewise, when two faces of cubes of A intersect, their intersection is a face of one of them. This implies that the open faces of cubes of A that are minimal with respect to inclusion among such faces form the cells of a CW structure on X, since the boundary of such a face is a union of such faces. The vertices of this CW structure are thus the vertices of all the cubes of A, and the n-cells are the interiors of the cubes of A.

Next we define inductively a subcomplex Z of this CW structure on X and a map $r : Z \to K$. The 0-cells of Z are exactly the 0-cells of X, and we let r send each 0-cell to the closest point of K, or if this is not unique, any one of the closest points of K. Assume inductively that Z^k and $r : Z^k \to K$ have been defined. For a cell e^{k+1} of X with boundary in Z^k, if the restriction of r to this boundary extends over e^{k+1} then we include e^{k+1} in Z^{k+1} and we let r on e^{k+1} be such an extension that is not too large, say an extension for which the diameter of its image $r(e^{k+1})$ is less than twice the infimum of the diameters for all possible extensions. This defines Z^{k+1} and $r : Z^{k+1} \to K$. At the end of the induction we set $Z = Z^n$.

It remains to verify that by letting r equal the identity on K we obtain a continuous retraction $Z \cup K \to K$, and that $Z \cup K$ contains a neighborhood of K. Given a point $x \in K$, let U be a ball in the metric space K centered at x. Since K is locally contractible, we can choose a finite sequence of balls in K centered at x, of the form $U = U_n \supset V_n \supset U_{n-1} \supset V_{n-1} \supset \cdots \supset U_0 \supset V_0$, each ball having radius equal to some small fraction of the radius of the preceding one, and with V_i contractible in U_i. Let $B \subset \mathbb{R}^n$ be a ball centered at x with radius less than half the radius of V_0, and let Y be the subcomplex of X formed by the cells whose closures are contained in B. Thus $Y \cup K$ contains a neighborhood of x in $\mathbb{R}^n$. By the choice of B and the definition of r on 0-cells we have $r(Y^0) \subset V_0$. Since V_0 is contractible in U_0, r is defined on the 1-cells of Y. Also, $r(Y^1) \subset V_1$ by the definition of r on 1-cells and the fact that U_0 is much smaller than V_1. Similarly, by induction we have r defined on Y^i with $r(Y^i) \subset V_i$ for all i. In particular, r maps Y to U. Since U could be arbitrarily small, this shows that extending r by the identity map on K gives a continuous map $r : Z \cup K \to K$. And since $Y \subset Z$, we see that $Z \cup K$ contains a neighborhood of K by the earlier observation that $Y \cup K$ contains a neighborhood of x. Thus $r : Z \cup K \to K$ retracts a neighborhood of K onto K.

Now for the converse. Since open sets in $\mathbb{R}^n$ are locally contractible, it suffices to show that a retract of a locally contractible space is locally contractible. Let $r : X \to A$

be a retraction and let $U \subset A$ be a neighborhood of a given point $x \in A$. If X is locally contractible, then inside the open set $r^{-1}(U)$ there is a neighborhood V of x that is contractible in $r^{-1}(U)$, say by a homotopy $f_t : V \to r^{-1}(U)$. Then $V \cap A$ is contractible in U via the restriction of the composition $r f_t$. $\square$

A space X is called a **Euclidean neighborhood retract** or **ENR** if for some n there exists an embedding $i : X \hookrightarrow \mathbb{R}^n$ such that $i(X)$ is a retract of some neighborhood in $\mathbb{R}^n$. The preceding theorem implies that the existence of the retraction is independent of the choice of embedding, at least when X is compact.

Corollary A.8. *A compact space is an ENR iff it can be embedded as a retract of a finite simplicial complex. Hence the homology groups and the fundamental group of a compact ENR are finitely generated.*

Proof: A finite simplicial complex K with n vertices is a subcomplex of a simplex Δ^{n-1}, and hence embeds in $\mathbb{R}^n$. The preceding theorem then implies that K is a retract of some neighborhood in $\mathbb{R}^n$, so any retract of K is also a retract of such a neighborhood, via the composition of the two retractions. Conversely, let K be a compact space that is a retract of some open neighborhood U in $\mathbb{R}^n$. Since K is compact it is bounded, lying in some large simplex $\Delta^n \subset \mathbb{R}^n$. Subdivide Δ^n, say by repeated barycentric subdivision, so that all simplices of the subdivision have diameter less than the distance from K to the complement of U. Then the union of all the simplices in this subdivision that intersect K is a finite simplicial complex that retracts onto K via the restriction of the retraction $U \to K$. $\square$

Corollary A.9. *Every compact manifold, with or without boundary, is an ENR.*

Proof: Manifolds are locally contractible, so it suffices to show that a compact manifold M can be embedded in $\mathbb{R}^k$ for some k. If M is not closed, it embeds in the closed manifold obtained from two copies of M by identifying their boundaries. So it suffices to consider the case that M is closed. By compactness there exist finitely many closed balls $B_i^n \subset M$ whose interiors cover M, where n is the dimension of M. Let $f_i : M \to S^n$ be the quotient map collapsing the complement of the interior of B_i^n to a point. These f_i's are the components of a map $f : M \to (S^n)^m$ which is injective since if x and y are distinct points of M with x in the interior of B_i^n, say, then $f_i(x) \neq f_i(y)$. Composing f with an embedding $(S^n)^m \hookrightarrow \mathbb{R}^k$, for example the product of the standard embeddings $S^n \hookrightarrow \mathbb{R}^{n+1}$, we obtain a continuous injection $M \hookrightarrow \mathbb{R}^k$, and this is a homeomorphism onto its image since M is compact. $\square$

Corollary A.10. *Every finite CW complex is an ENR.*

Proof: Since CW complexes are locally contractible, it suffices to show that a finite CW complex can be embedded in some $\mathbb{R}^n$. This is proved by induction on the number of cells. Suppose the CW complex X is obtained from a subcomplex A by attaching a cell e^k via a map $f : S^{k-1} \to A$, and suppose that we have an embedding $A \hookrightarrow \mathbb{R}^m$. Then we can embed X in $\mathbb{R}^k \times \mathbb{R}^m \times \mathbb{R}$ as the union of $D^k \times \{0\} \times \{0\}$, $\{0\} \times A \times \{1\}$, and all line segments joining points $(x, 0, 0)$ and $(0, f(x), 1)$ for $x \in S^{k-1}$. $\square$

Spaces Dominated by CW Complexes

We have been considering spaces which are retracts of finite simplicial complexes, and now we show that such spaces have the homotopy type of CW complexes. In fact, we can just as easily prove something a little more general than this. A space Y is said to be **dominated** by a space X if there are maps $Y \xrightarrow{\ i\ } X \xrightarrow{\ r\ } Y$ with $ri \simeq \mathbb{1}$. This makes the notion of a retract into something that depends only on the homotopy types of the spaces involved.

> $\mathbf{P}$**roposition A.11.** *A space dominated by a CW complex is homotopy equivalent to a CW complex.*

Proof: Recall from §3.F that the mapping telescope $T(f_1, f_2, \cdots)$ of a sequence of maps $X_1 \xrightarrow{f_1} X_2 \xrightarrow{f_2} X_3 \longrightarrow \cdots$ is the quotient space of $\coprod_i (X_i \times [i, i+1])$ obtained by identifying $(x, i+1) \in X_i \times [i, i+1]$ with $(f(x), i+1) \in X_{i+1} \times [i+1, i+2]$. We shall need the following elementary facts:

(1) $T(f_1, f_2, \cdots) \simeq T(g_1, g_2, \cdots)$ if $f_i \simeq g_i$ for each i.

(2) $T(f_1, f_2, \cdots) \simeq T(f_2, f_3, \cdots)$.

(3) $T(f_1, f_2, \cdots) \simeq T(f_2 f_1, f_4 f_3, \cdots)$.

The second of these is obvious. To prove the other two we will use Proposition 0.18, whose proof applies not just to CW pairs but to any pair (X_1, A) for which there is a deformation retraction of $X_1 \times I$ onto $X_1 \times \{0\} \cup A \times I$. To prove (1) we regard $T(f_1, f_2, \cdots)$ as being obtained from $\coprod_i (X_i \times \{i\})$ by attaching $\coprod_i (X_i \times [i, i+1])$. Then we can obtain $T(g_1, g_2, \cdots)$ by varying the attaching map by homotopy. To prove (3) we view $T(f_1, f_2, \cdots)$ as obtained from the disjoint union of the mapping cylinders $M(f_{2i})$ by attaching $\coprod_i (X_{2i-1} \times [2i-1, 2i])$. By sliding the attachment of $X_{2i-1} \times [2i-1, 2i]$ to $X_{2i} \subset M(f_{2i})$ down the latter mapping cylinder to X_{2i+1} we convert $M(f_{2i-1}) \cup M(f_{2i})$ into $M(f_{2i} f_{2i-1}) \cup M(f_{2i})$. This last space deformation retracts onto $M(f_{2i} f_{2i-1})$. Doing this for all i gives the homotopy equivalence in (3).

Now to prove the proposition, suppose that the space Y is dominated by the CW complex X via maps $Y \xrightarrow{\ i\ } X \xrightarrow{\ r\ } Y$ with $ri \simeq \mathbb{1}$. By (2) and (3) we have $T(ir, ir, \cdots) \simeq T(r, i, r, i, \cdots) \simeq T(i, r, i, r, \cdots) \simeq T(ri, ri, \cdots)$. Since $ri \simeq \mathbb{1}$, $T(ri, ri, \cdots)$ is homotopy equivalent to the telescope of the identity maps $Y \to Y \to Y \to \cdots$, which is $Y \times [0, \infty) \simeq Y$. On the other hand, the map ir is homotopic to a cellular map $f : X \to X$, so $T(ir, ir, \cdots) \simeq T(f, f, \cdots)$, which is a CW complex. $\square$

One might ask whether a space dominated by a finite CW complex is homotopy equivalent to a finite CW complex. In the simply-connected case this follows from Proposition 4C.1 since such a space has finitely generated homology groups. But there are counterexamples in the general case; see [Wall 1965].

In view of Corollary A.10 the preceding proposition implies:

Corollary A.12. *A compact manifold is homotopy equivalent to a CW complex.*   □

One could ask more refined questions. For example, do all compact manifolds have CW complex structures, or even simplicial complex structures? Answers here are considerably harder to come by. Restricting attention to closed manifolds for simplicity, the present status of these questions is the following. For manifolds of dimensions less than 4, simplicial complex structures always exist. In dimension 4 there are closed manifolds that do not have simplicial complex structures, while the existence of CW structures is an open question. In dimensions greater than 4, CW structures always exist, but whether simplicial structures always exist is unknown, though it is known that there are n-manifolds not having simplicial structures locally isomorphic to any linear simplicial subdivision of $\mathbb{R}^n$, for all $n \geq 4$. For more on these questions, see [Kirby & Siebenmann 1977] and [Freedman & Quinn 1990].

Exercises

1. Show that a covering space of a CW complex is also a CW complex, with cells projecting homeomorphically onto cells.

2. Let X be a CW complex and x_0 any point of X. Construct a new CW complex structure on X having x_0 as a 0-cell, and having each of the original cells a union of the new cells. The latter condition is expressed by saying the new CW structure is a **subdivision** of the old one.

3. Show that a CW complex is path-connected iff its 1-skeleton is path-connected.

4. Show that a CW complex is locally compact iff each point has a neighborhood that meets only finitely many cells.

5. For a space X, show that the identity map $X_c \to X$ induces an isomorphism on π_1, where X_c denotes X with the compactly generated topology.

The Compact-Open Topology

By definition, the compact-open topology on the space X^Y of maps $f : Y \to X$ has a subbasis consisting of the sets $M(K, U)$ of mappings taking a compact set $K \subset Y$ to an open set $U \subset X$. Thus a basis for X^Y consists of sets of maps taking a finite number of compact sets $K_i \subset Y$ to open sets $U_i \subset X$. If Y is compact, which is the only case we consider in this book, convergence to $f \in X^Y$ means, loosely speaking, that finer and finer compact covers $\{K_i\}$ of Y are taken to smaller and smaller open covers $\{U_i\}$ of $f(Y)$. One of the main cases of interest in homotopy theory is when

$Y = I$, so X^I is the space of paths in X. In this case one can check that a system of basic neighborhoods of a path $f : I \to X$ consists of the open sets $\bigcap_i M(K_i, U_i)$ where the K_i's are a partition of I into nonoverlapping closed intervals and U_i is an open neighborhood of $f(K_i)$.

The compact-open topology is the same as the topology of uniform convergence in many cases:

Proposition A.13. *If X is a metric space and Y is compact, then the compact-open topology on X^Y is the same as the metric topology defined by the metric $d(f, g) = \sup_{y \in Y} d(f(y), g(y))$.*

Proof: First we show that every open ε-ball $B_\varepsilon(f)$ about $f \in X^Y$ contains a neighborhood of f in the compact-open topology. Since $f(Y)$ is compact, it is covered by finitely many balls $B_{\varepsilon/3}(f(y_i))$. Let $K_i \subset Y$ be the closure of $f^{-1}(B_{\varepsilon/3}(f(y_i)))$, so K_i is compact, $Y = \bigcup_i K_i$, and $f(K_i) \subset B_{\varepsilon/2}(f(y_i)) = U_i$, hence $f \in \bigcap_i M(K_i, U_i)$. To show that $\bigcap_i M(K_i, U_i) \subset B_\varepsilon(f)$, suppose that $g \in \bigcap_i M(K_i, U_i)$. For any $y \in Y$, say $y \in K_i$, we have $d(g(y), f(y_i)) < \varepsilon/2$ since $g(K_i) \subset U_i$. Likewise we have $d(f(y), f(y_i)) < \varepsilon/2$, so $d(f(y), g(y)) \leq d(f(y), f(y_i)) + d(g(y), f(y_i)) < \varepsilon$. Since y was arbitrary, this shows $g \in B_\varepsilon(f)$.

Conversely, we show that for each open set $M(K, U)$ and each $f \in M(K, U)$ there is a ball $B_\varepsilon(f) \subset M(K, U)$. Since $f(K)$ is compact, it has a distance $\varepsilon > 0$ from the complement of U. Then $d(f, g) < \varepsilon/2$ implies $g(K) \subset U$ since $g(K)$ is contained in an $\varepsilon/2$-neighborhood of $f(K)$. So $B_{\varepsilon/2}(f) \subset M(K, U)$. $\square$

The next proposition contains the essential properties of the compact-open topology from the viewpoint of algebraic topology.

Proposition A.14. *If Y is locally compact, then:*
(a) *The evaluation map $e : X^Y \times Y \to X$, $e(f, y) = f(y)$, is continuous.*
(b) *A map $f : Y \times Z \to X$ is continuous iff the map $\hat{f} : Z \to X^Y$, $\hat{f}(z)(y) = f(y, z)$, is continuous.*

Different definitions of local compactness are common, but the definition we are using is that Y is **locally compact** if for each point $y \in Y$ and each neighborhood U of y there is a compact neighborhood V of y contained in U.

In particular, part (b) of the proposition provides the point-set topology justifying the adjoint relation $\langle \Sigma X, Y \rangle = \langle X, \Omega Y \rangle$ in §4.3, since it implies that a map $\Sigma X \to Y$ is continuous iff the associated map $X \to \Omega Y$ is continuous, and similarly for homotopies of such maps. Namely, think of a basepoint-preserving map $\Sigma X \to Y$ as a map $f : I \times X \to Y$ taking $\partial I \times X \cup \{x_0\} \times I$ to the basepoint of Y, so the associated map $\hat{f} : X \to Y^I$ has image in the subspace $\Omega Y \subset Y^I$. A homotopy $f_t : \Sigma X \to Y$ gives a map $F : I \times X \times I \to Y$ taking $\partial I \times X \times I \cup I \times \{x_0\} \times I$ to the basepoint, with $\hat{F}$ a map $X \times I \to \Omega Y \subset Y^I$ defining a basepoint-preserving homotopy $\hat{f}_t$.

Proof: (a) For $(f, y) \in X^Y \times Y$ let $U \subset X$ be an open neighborhood of $f(y)$. Since Y is locally compact, continuity of f implies there is a compact neighborhood $K \subset Y$ of y such that $f(K) \subset U$. Then $M(K, U) \times K$ is a neighborhood of (f, y) in $X^Y \times Y$ taken to U by e, so e is continuous at (f, y).

(b) Suppose $f : Y \times Z \to X$ is continuous. To show continuity of $\hat{f}$ it suffices to show that for a subbasic set $M(K, U) \subset X^Y$, the set $\hat{f}^{-1}(M(K, U)) = \{ z \in Z \mid f(K, z) \subset U \}$ is open in Z. Let $z \in \hat{f}^{-1}(M(K, U))$. Since $f^{-1}(U)$ is an open neighborhood of the compact set $K \times \{z\}$, there exist open sets $V \subset Y$ and $W \subset Z$ whose product $V \times W$ satisfies $K \times \{z\} \subset V \times W \subset f^{-1}(U)$. So W is a neighborhood of z in $\hat{f}^{-1}(M(K, U))$. (The hypothesis that Y is locally compact is not needed here.)

For the converse half of (b) note that f is the composition $Y \times Z \to Y \times X^Y \to X$ of $\mathbb{1} \times \hat{f}$ and the evaluation map, so part (a) gives the result. $\square$

Proposition A.15. *If X is a compactly generated Hausdorff space and Y is locally compact, then the product topology on $X \times Y$ is compactly generated.*

Proof: First a preliminary observation: A function $f : X \times Y \to Z$ is continuous iff its restrictions $f : C \times Y \to Z$ are continuous for all compact $C \subset X$. For, using (b) of the previous proposition, the first statement is equivalent to $\hat{f} : X \to Z^Y$ being continuous and the second statement is equivalent to $\hat{f} : C \to Z^Y$ being continuous for all compact $C \subset X$. Since X is compactly generated, the latter two statements are equivalent.

To prove the proposition we just need to show the identity map $X \times Y \to (X \times Y)_c$ is continuous. By the previous paragraph, this is equivalent to continuity of the inclusion maps $C \times Y \to (X \times Y)_c$ for all compact $C \subset X$. Since Y is locally compact, it is compactly generated, and C is compact Hausdorff hence locally compact, so the same reasoning shows that continuity of $C \times Y \to (X \times Y)_c$ is equivalent to continuity of $C \times C' \to (X \times Y)_c$ for all compact $C' \subset Y$. But on the compact set $C \times C'$, the two topologies on $X \times Y$ agree, so we are done. (This proof is from [Dugundji 1966].) $\square$

Proposition A.16. *The map $X^{Y \times Z} \to (X^Y)^Z$, $f \mapsto \hat{f}$, is a homeomorphism if Y is locally compact Hausdorff and Z is Hausdorff.*

Proof: First we show that a subbasis for $X^{Y \times Z}$ is formed by the sets $M(A \times B, U)$ as A and B range over compact sets in Y and Z respectively and U ranges over open sets in X. Given a compact $K \subset Y \times Z$ and $f \in M(K, U)$, let K_Y and K_Z be the projections of K onto Y and Z. Then $K_Y \times K_Z$ is compact Hausdorff, hence normal, so for each point $k \in K$ there are compact sets $A_k \subset Y$ and $B_k \subset Z$ such that $A_k \times B_k$ is a compact neighborhood of k in $f^{-1}(U) \cap (K_Y \times K_Z)$. By compactness of K a finite number of the products $A_k \times B_k$ cover K. Discarding the others, we then have $f \in \bigcap_k M(A_k \times B_k, U) \subset M(K, U)$, which shows that the sets $M(A \times B, U)$ form a subbasis.

Under the bijection $X^{Y \times Z} \to (X^Y)^Z$ these sets $M(A \times B, U)$ correspond to the sets $M(B, M(A, U))$, so it will suffice to show the latter sets form a subbasis for $(X^Y)^Z$. We show more generally that X^Y has as a subbasis the sets $M(K, V)$ as V ranges over a subbasis for X and K ranges over compact sets in Y, assuming that Y is Hausdorff.

Given $f \in M(K, U)$, write U as a union of basic sets U_α with each U_α an intersection of finitely many sets $V_{\alpha, j}$ of the given subbasis. The cover of K by the open sets $f^{-1}(U_\alpha)$ has a finite subcover, say by the open sets $f^{-1}(U_i)$. Since K is compact Hausdorff, hence normal, we can write K as a union of compact subsets K_i with $K_i \subset f^{-1}(U_i)$. Then f lies in $M(K_i, U_i) = M(K_i, \bigcap_j V_{ij}) = \bigcap_j M(K_i, V_{ij})$ for each i. Hence f lies in $\bigcap_{i,j} M(K_i, V_{ij}) = \bigcap_i M(K_i, U_i) \subset M(K, U)$. Since $\bigcap_{i,j} M(K_i, V_{ij})$ is a finite intersection, this shows that the sets $M(K, V)$ form a subbasis for X^Y. $\square$

Proposition A.17. *If $f : X \to Y$ is a quotient map then so is $f \times \mathbb{1} : X \times Z \to Y \times Z$ whenever Z is locally compact.*

This can be applied when $Z = I$ to show that a homotopy defined on a quotient space is continuous.

Proof: Consider the diagram at the right, where W is $Y \times Z$ with the quotient topology from $X \times Z$, with g the quotient map and h the identity. Every open set in $Y \times Z$ is open in W since $f \times \mathbb{1}$ is continuous, so it will suffice to show that h is continuous.

$$X \times Z \xrightarrow{f \times \mathbb{1}} Y \times Z$$
$$g \searrow \quad \swarrow h$$
$$W$$

Since g is continuous, so is the associated map $\hat{g} : X \to W^Z$, by Proposition A.14. This implies that $\hat{h} : Y \to W^Z$ is continuous since f is a quotient map. Applying Proposition A.14 again, we conclude that h is continuous. $\square$

Bibliography

Books

J. F. ADAMS, *Algebraic Topology: a Student's Guide*, Cambridge Univ. Press, 1972.

J. F. ADAMS, *Stable Homotopy and Generalised Homology*, Univ. of Chicago Press, 1974.

J. F. ADAMS, *Infinite Loop Spaces*, Ann. of Math. Studies 90, 1978.

A. ADEM and R. J. MILGRAM, *Cohomology of Finite Groups*, Springer-Verlag, 1994.

M. AGUILAR, S. GITLER, and C. PRIETO, *Algebraic Topology from a Homotopical Viewpoint*, Springer-Verlag, 2002.

M. AIGNER and G. ZIEGLER, *Proofs from THE BOOK*, Springer-Verlag, 1999.

P. ALEXANDROFF and H. HOPF, *Topologie*, Chelsea, 1972 (reprint of original 1935 edition).

M. A. ARMSTRONG, *Basic Topology*, Springer-Verlag, 1983.

M. F. ATIYAH, *K-Theory*, W. A. Benjamin, 1967.

H. J. BAUES, *Homotopy Type and Homology*, Oxford Univ. Press, 1996.

D. J. BENSON, *Representations and Cohomology, Volume II: Cohomology of Groups and Modules*, Cambridge Univ. Press, 1992.

D. J. BENSON, *Polynomial Invariants of Finite Groups*, Cambridge Univ. Press, 1993.

R. BOTT and L. TU, *Differential Forms in Algebraic Topology*, Springer-Verlag GTM 82, 1982.

G. BREDON, *Topology and Geometry*, Springer-Verlag GTM 139, 1993.

K. BROWN, *Cohomology of Groups*, Springer-Verlag GTM 87, 1982.

R. BROWN, *The Lefschetz Fixed Point Theorem*, Scott Foresman, 1971.

M. COHEN, *A Course in Simple-Homotopy Theory*, Springer-Verlag GTM 10, 1973.

T. TOM DIECK, *Algebraic Topology*, E.M.S., 2008.

J. DIEUDONNÉ, *A History of Algebraic and Differential Topology 1900-1960*, Birkhäuser, 1989.

A. DOLD, *Lectures on Algebraic Topology*, Springer-Verlag, 1980.

J. DUGUNDJI, *Topology*, Allyn & Bacon, 1966.

H.-D. EBBINGHAUS et al., *Numbers*, Springer-Verlag GTM 123, 1991.

S. EILENBERG and N. STEENROD, *Foundations of Algebraic Topology*, Princeton Univ. Press, 1952.

Y. FÉLIX, S. HALPERIN, and J.-C. THOMAS, *Rational Homotopy Theory*, Springer-Verlag GTM 205, 2001.

R. FENN, *Techniques of Geometric Topology*, Cambridge Univ. Press, 1983.

A. T. FOMENKO and D. B. FUKS, *A Course in Homotopic Topology*, Izd. Nauka, 1989. (In Russian; an English translation of an earlier version was published by Akadémiai Kiadó, Budapest, 1986.)

M. FREEDMAN and F. QUINN, *Topology of 4-Manifolds*, Princeton Univ. Press, 1990.

W. FULTON, *Algebraic Topology: A First Course*, Springer-Verlag GTM 153, 1995.

B. GRAY, *Homotopy Theory*, Academic Press, 1975.

M. GREENBERG and J. HARPER, *Algebraic Topology: A First Course*, Addison-Wesley, 1981.

P. GRIFFITHS and J. MORGAN, *Rational Homotopy Theory and Differential Forms*, Birkhäuser, 1981.

P. J. HILTON, *An Introduction to Homotopy Theory*, Cambridge Univ. Press, 1953.

P. J. HILTON, *Homotopy Theory and Duality*, Gordon and Breach, 1966.

P. J. HILTON and S. WYLIE, *Homology Theory*, Cambridge Univ. Press, 1967.

P. J. HILTON and U. STAMMBACH, *A Course in Homological Algebra*, Springer-Verlag GTM 4, 1970.

P. J. HILTON, G. MISLIN, and J. ROITBERG, *Localization of Nilpotent Groups and Spaces*, North-Holland, 1975.

D. HUSEMOLLER, *Fibre Bundles*, McGraw-Hill, 1966 (later editions by Springer-Verlag).

I. M. JAMES, ed., *Handbook of Algebraic Topology*, North-Holland, 1995.

I. M. JAMES, ed., *History of Topology*, North-Holland, 1999.

K. JÄNICH, *Topology*, Springer-Verlag, 1984.

R. M. KANE, *The Homology of Hopf Spaces*, North-Holland, 1988.

R. KIRBY and L. SIEBENMANN, *Foundational Essays on Topological Manifolds, Smoothings, and Triangulations*, Ann. of Math. Studies 88, 1977.

S. KOCHMAN, *Stable Homotopy Groups of Spheres*, Springer Lecture Notes 1423, 1990.

S. KOCHMAN, *Bordism, Stable Homotopy and Adams Spectral Sequences*, Fields Institute Monographs 7, A.M.S., 1996.

D. KÖNIG, *Theory of Finite and Infinite Graphs*, Birkhäuser, 1990.

A. LUNDELL and S. WEINGRAM, *The Topology of CW Complexes*, Van Nostrand Reinhold, 1969.

S. MACLANE, *Homology*, Springer-Verlag, 1963.

S. MACLANE, *Categories for the Working Mathematician*, Springer-Verlag GTM 5, 1971.

I. MADSEN and R. MILGRAM, *The Classifying Spaces for Surgery and Cobordism of Manifolds*, Ann. of Math. Studies 92, 1979.

W. MASSEY, *Algebraic Topology: An Introduction*, Harcourt, Brace & World, 1967 (reprinted by Springer-Verlag).

W. MASSEY, *A Basic Course in Algebraic Topology*, Springer-Verlag GTM 127, 1993.

J. MATOUŠEK, *Using the Borsuk-Ulam Theorem*, Springer-Verlag, 2003.

C. R. F. MAUNDER, *Algebraic Topology*, Cambridge Univ. Press, 1980 (reprinted by Dover Publications).

J. P. MAY, *Simplicial Objects in Algebraic Topology*, Van Nostrand, 1967 (reprinted by Univ. Chicago Press).

J. P. MAY, *A Concise Course in Algebraic Topology*, Univ. Chicago Press, 1999.

J. MILNOR, *Topology from the Differentiable Viewpoint*, Univ. Press of Virginia, 1965.

J. MILNOR and J. STASHEFF, *Characteristic Classes*, Ann. of Math. Studies 76, 1974.

R. MOSHER and M. TANGORA, *Cohomology Operations and Applications in Homotopy Theory*, Harper and Row, 1968.

V. V. PRASOLOV, *Elements of Homology Theory*, A.M.S., 2007.

D. RAVENEL, *Complex Cobordism and Stable Homotopy Groups of Spheres*, Academic Press, 1986.

D. RAVENEL, *Nilpotence and Periodicity in Stable Homotopy Theory*, Ann. of Math. Studies 128, 1992.

E. REES and J. D. S. JONES, eds., *Homotopy Theory: Proceeding of the Durham Symposium 1985*, Cambridge Univ. Press, 1987.

D. ROLFSEN, *Knots and Links*, Publish or Perish, 1976.

H. SEIFERT and W. THRELFALL, *Lehrbuch der Topologie*, Teubner, 1934.

P. SELICK, *Introduction to Homotopy Theory*, Fields Institute Monographs 9, A.M.S., 1997.

J.-P. SERRE, *A Course in Arithmetic*, Springer-Verlag GTM 7, 1973.

E. SPANIER, *Algebraic Topology*, McGraw-Hill, 1966 (reprinted by Springer-Verlag).

N. STEENROD, *The Topology of Fibre Bundles*, Princeton Univ. Press, 1951.

N. STEENROD and D. EPSTEIN, *Cohomology Operations*, Ann. of Math. Studies 50, 1962.

R. STONG, *Notes on Cobordism Theory*, Princeton Univ. Press, 1968.

D. SULLIVAN, *Geometric Topology*, xeroxed notes from MIT, 1970.

R. SWITZER, *Algebraic Topology*, Springer-Verlag, 1975.

H. TODA, *Composition Methods in Homotopy Groups of Spheres*, Ann. of Math. Studies 49, 1962.

K. VARADARAJAN, *The Finiteness Obstruction of C. T. C. Wall*, Wiley, 1989.

C. WEIBEL, *An Introduction to Homological Algebra*, Cambridge Univ. Press, 1994.

G. WHITEHEAD, *Elements of Homotopy Theory*, Springer-Verlag GTM 62, 1978.

J. WOLF, *Spaces of Constant Curvature*, Publish or Perish, 1984.

Papers

J. F. ADAMS, On the non-existence of elements of Hopf invariant one, *Ann. of Math.* 72 (1960), 20–104.

J. F. ADAMS, Vector fields on spheres, *Ann. of Math.* 75 (1962), 603–632.

J. F. ADAMS, On the groups $J(X)$ IV, *Topology* 5 (1966), 21-71.

J. F. ADAMS and M. F. ATIYAH, K-theory and the Hopf invariant, *Quart. J. Math.* 17 (1966), 31-38.

J. F. ADAMS, A variant of E. H. Brown's representability theorem, *Topology* 10 (1971), 185-198.

J. F. ADAMS and C. WILKERSON, Finite H-spaces and algebras over the Steenrod algebra, *Ann. of Math.* 111 (1980), 95-143.

K. ANDERSEN and J. GRODAL, The Steenrod problem of realizing polynomial cohomology rings, *Journal of Topology* 1 (2008), 747-760.

M. BARRATT and J. MILNOR, An example of anomalous singular homology, *Proc. A.M.S.* 13 (1962), 293-297.

M. BESTVINA and N. BRADY, Morse theory and finiteness properties of groups, *Invent. Math.* 129 (1997), 445-470.

A. BOREL, Sur la cohomologie des espaces fibrés principaux, *Ann. of Math.* 57 (1953), 115-207.

R. BOTT and H. SAMELSON, On the Pontryagin product in spaces of paths, *Comment. Math. Helv.* 27 (1953), 320-337.

R. BOTT and J. MILNOR, On the parallelizability of spheres, *Bull. A.M.S.* 64 (1958), 87-89.

R. BOTT, The stable homotopy of the classical groups, *Ann. of Math.* 70 (1959), 313-337.

E. BROWN and A. COPELAND, An homology analog of Postnikov systems, *Michigan Math. J.* 6 (1959), 313-330.

E. BROWN, Cohomology theories, *Ann. of Math.* 75 (1962), 467-484.

M. BROWN, A proof of the generalized Schoenflies theorem, *Bull. A.M.S.* 66 (1960), 74-76.

J. CANNON and G. CONNER, The combinatorial structure of the Hawaiian earring group, *Top. and its Appl.* 106 (2000), 225-271.

G. CARLSSON and R. J. MILGRAM, Stable homotopy and iterated loopspaces, pp. 505-583 in *Handbook of Algebraic Topology*, ed. I. M. James, Elsevier 1995.

J. DAVIS and R. J. MILGRAM, A survey of the spherical space form problem, *Math. Reports* 2 (1985), 223-283, Harwood Acad. Pub.

A. DOLD and R. THOM, Quasifaserungen und unendliche symmetrischen Produkte, *Ann. of Math.* 67 (1958), 239-281.

C. DOWKER, Topology of metric complexes, *Am. J. Math.* 74 (1952), 555-577.

W. DWYER, H. MILLER, and C. WILKERSON, Homotopical uniqueness of classifying spaces, *Topology* 31 (1992), 29-45.

W. DWYER and C. WILKERSON, A new finite loop space at the prime two, *Journal A.M.S.* 6 (1993), 37-63.

E. DYER and A. VASQUEZ, Some small aspherical spaces, *J. Austral. Math. Soc.* 16 (1973), 332-352.

S. EILENBERG, Singular homology theory, *Ann. of Math.* 45 (1944), 407-447.

S. EILENBERG and J. A. ZILBER, Semi-simplicial complexes and singular homology, *Ann. of Math.* 51 (1950), 499-513.

R. EDWARDS, A contractible, nowhere locally connected compactum, *Abstracts A.M.S.* 20 (1999), 494.

S. FEDER and S. GITLER, Mappings of quaternionic projective spaces, *Bol. Soc. Mat. Mexicana* 18 (1978), 33-37.

J. GUBELADZE, The isomorphism problem for commutative monoid rings, *J. Pure Appl. Alg.* 129 (1998), 35-65.

J. HARRIS and N. KUHN, Stable decompositions of classifying spaces of finite abelian p-groups, *Math. Proc. Camb. Phil. Soc.* 103 (1988), 427-449.

P. HOFFMAN and G. PORTER, Cohomology realizations of $\mathbb{Q}[x]$, *Ox. Q.* 24 (1973), 251-255.

R. HOLZSAGER, Stable splitting of $K(G,1)$, *Proc. A.M.S.* 31 (1972), 305-306.

H. HOPF, Über die Abbildungen der dreidimensionalen Sphäre auf die Kugelfläche, *Math. Ann.* 104 (1931), 637-665.

I. M. JAMES, Reduced product spaces, *Ann. of Math.* 62 (1955), 170-197.

M. KERVAIRE, Non-parallelizability of the n-sphere for $n > 7$, *Proc. N.A.S.* 44 (1958), 280-283.

S. KOCHMAN and M. MAHOWALD, On the computation of stable stems, *Contemp. Math.* 181 (1995), 299-316.

I. MADSEN, C. B. THOMAS, and C. T. C. WALL, The topological spherical space form problem, II: existence of free actions, *Topology* 15 (1976), 375-382.

J. P. MAY, A general approach to Steenrod operations, *Springer Lecture Notes* 168 (1970), 153-231.

J. P. MAY, Weak equivalences and quasifibrations, *Springer Lecture Notes* 1425 (1990), 91-101.

C. MILLER, The topology of rotation groups, *Ann. of Math.* 57 (1953), 95-110.

J. MILNOR, Construction of universal bundles I, II, *Ann. of Math.* 63 (1956), 272-284, 430-436.

J. MILNOR, Groups which act on S^n without fixed points, *Am. J. Math* 79 (1957), 623-630.

J. MILNOR, Some consequences of a theorem of Bott, *Ann. of Math.* 68 (1958), 444-449.

J. MILNOR, On spaces having the homotopy type of a CW complex, *Trans. A.M.S.* 90 (1959), 272-280.

J. MILNOR, On axiomatic homology theory, *Pac. J. Math.* 12 (1962), 337-341.

J. MILNOR and J. MOORE, On the structure of Hopf algebras, *Ann. of Math.* 81 (1965), 211-264.

J. MILNOR, On the Steenrod homology theory, pp. 79-96 in *Novikov Conjectures, Index Theorems, and Rigidity*, ed. S. Ferry, A. Ranicki, and J. Rosenberg, Cambridge Univ. Press, 1995.

J. NEISENDORFER, Primary homotopy theory, *Mem. A.M.S.* 232 (1980).

D. NOTBOHM, Spaces with polynomial mod-p cohomology, *Math. Proc. Camb. Phil. Soc.*, (1999), 277-292.

D. QUILLEN, Rational homotopy theory, *Ann. of Math.* 90 (1969), 205–295.

D. RECTOR, Loop structures on the homotopy type of S^3, *Springer Lecture Notes* 249 (1971), 99–105.

C. ROBINSON, Moore-Postnikov systems for non-simple fibrations, *Ill. J. Math.* 16 (1972), 234–242.

P. SCOTT and T. WALL, Topological methods in group theory, *London Math. Soc. Lecture Notes* 36 (1979), 137–203.

J.-P. SERRE, Homologie singulière des espaces fibrés, *Ann. of Math.* 54 (1951), 425–505.

J.-P. SERRE, Groupes d'homotopie et classes de groupes abéliens, *Ann. of Math.* 58 (1953), 258–294.

A. SIERADSKI, Algebraic topology for two dimensional complexes, pp. 51–96 in *Two-Dimensional Homotopy and Combinatorial Group Theory*, ed. C. Hog-Angeloni, W. Metzler, and A. Sieradski., Cambridge Univ. Press, 1993.

S. SHELAH, Can the fundamental group of a space be the rationals?, *Proc. A.M.S.* 103 (1988), 627–632.

J. STALLINGS, A finitely presented group whose 3-dimensional integral homology is not finitely generated, *Am. J. Math.* 85 (1963), 541–543.

A. STRØM, Note on cofibrations II, *Math. Scand.* 22 (1968), 130–142.

D. SULLIVAN, Genetics of homotopy theory and the Adams conjecture, *Ann. of Math.* 100 (1974), 1–79.

H. TODA, Note on the cohomology ring of certain spaces, *Proc. A.M.S.* 14 (1963), 89–95.

E. VAN KAMPEN, On the connection between the fundamental groups of some related spaces, *Am. J. Math.* 55 (1933), 261–267.

C. T. C. WALL, Finiteness conditions for CW complexes, *Ann. of Math.* 81 (1965), 56–69.

G. WHITEHEAD, Generalized homology theories, *Trans. A.M.S.* 102 (1962), 227–283.

J. H. C. WHITEHEAD, Combinatorial homotopy II, *Bull. A.M.S.* 55 (1949), 453–496.

Index

Printed in the United States
By Bookmasters